FRACTURE MECHANICS

Fundamentals and Applications

T.L. Anderson, Ph.D.

Department of Mechanical Engineering
Texas A&M University
College Station, Texas

CRC Press
Boca Raton Ann Arbor Boston

Library of Congress Cataloging-in-Publication Data

Catalog record is available from the Library of Congress.

This book represents information obtained from authentic and highly regarded sources. Reprinted material is quoted with permission, and sources are indicated. A wide variety of references are listed. Every reasonable effort has been made to give reliable data and information, but the author and the publisher cannot assume responsibility for the validity of all materials or for the consequences of their use.

Direct all inquiries to CRC Press, Inc., 2000 Corporate Blvd., N.W., Boca Raton, Florida, 33431.

International Standard Book Number 0-8493-4277-5

Printed in the United States

PREFACE

The field of fracture mechanics was virtually nonexistent prior to World War II, but has since matured into an established discipline. Most universities with an engineering program offer at least one fracture mechanics course on the graduate level, and an increasing number of undergraduates have been exposed to this subject. Applications of fracture mechanics in industry are relatively common, as knowledge that was once confined to a few specialists is becoming more widespread.

While there are a number of books on fracture mechanics, most are geared to a specific audience. Some treatments of this subject emphasize material testing, while others concentrate on detailed mathematical derivations. A few books address the microscopic aspects of fracture, but most consider only continuum models. Many books are restricted to a particular material system, such as metals or polymers. Current offerings include advanced, highly specialized books, as well as introductory texts. While the former are valuable to researchers in this field, they are unsuitable for students with no prior background. On the other hand, introductory treatments of the subject are sometimes simplistic and misleading.

This book provides a comprehensive treatment of fracture mechanics that should appeal to a relatively wide audience. Theoretical background and practical applications are both covered in detail. This book is suitable as a graduate text, as well as a reference for engineers and researchers. Selected portions of this book would also be appropriate for an undergraduate course in fracture mechanics.

The subject matter is organized in a unique fashion. The book is intended to be readable without being superficial. The fundamental concepts are first described qualitatively, with a minimum of higher level mathematics. This enables a student with a reasonable grasp of undergraduate calculus to gain physical insight into the subject. For the more advanced reader, appendices at the end of certain chapters give the detailed mathematical background.

In outlining the basic principles and applications of fracture mechanics, I have attempted to integrate materials science and solid mechanics to a much greater extent than previous texts. Although continuum theory has proved to be a very powerful tool in fracture mechanics, one cannot ignore microstructural aspects. Continuum theory can predict the stresses and strains near a crack tip, but it is the material's microstructure that determines the critical conditions for fracture.

The first chapter introduces the subject of fracture mechanics and provides an overview; this chapter includes a review of dimensional analysis, which proves to be a useful tool in later chapters. Chapters 2 and 3 describe the fundamental concepts of linear elastic and elastic-plastic fracture mechanics, respectively. One of the most important and most often misunderstood concepts in fracture mechanics is the single parameter assumption, which enables the prediction of structural behavior from small scale laboratory tests. When a single parameter uniquely describes the crack tip conditions, fracture toughness, which is a critical value of this parameter, is independent of specimen size. When the single parameter assumption breaks down, fracture toughness becomes size dependent, and a small scale fracture toughness test may not be indicative of structural behavior. Chapters 2 and 3 describe the basis of the single parameter assumption in detail, and outline the requirements for its validity. Chapter 3 includes the results of recent research that quantifies the size dependence of fracture toughness for cleavage in metals. The main bodies of Chapters 2 and 3 are written in such a way as to be accessible to the beginning student. Appendices 2 and 3, which follow Chapters 2 and 3, respectively, give mathematical derivations of several important relationships in linear elastic and elastic-plastic fracture mechanics. Most of the material in these appendices requires a graduate-level background in solid mechanics.

Chapter 4 introduces dynamic and time-dependent fracture mechanics. The section on dynamic fracture includes a brief discussion of rapid loading of a stationary crack, as well as rapid crack propagation and arrest. The C^*, $C(t)$, and C_t parameters for characterizing creep crack growth are introduced, together with analogous quantities that characterize fracture in viscoelastic materials.

Chapter 5 outlines micromechanisms of fracture in metals and alloys, while Chapter 6 describes fracture mechanisms in polymers, ceramics and composites. These chapters emphasize the importance of microstructure and material properties on the fracture behavior. An appendix at the end of Chapter 5 illustrates that the micromechanism of fracture can have a profound effect on the validity of the single parameter assumption.

The applications portion of this book begins with Chapter 7, which gives practical advice on fracture toughness testing in metals. This chapter describes standard test methods, such as K_{IC}, J_{IC}, and CTOD, as well as recent research results. Chapter 7 includes a section on weldment testing, which has yet to be standardized in the U.S. Chapter 8 describes fracture testing of nonmetallic materials. Most of these test methods are still experimental in nature, since this is a relatively new field. Currently, a number of researchers are characterizing fracture behavior of plastics with test methods that were originally developed for metals; Chapter 8 discusses the validity of such tests for polymers, and suggests improvements in current methodology. Chapter 9 outlines the available methods for applying fracture mechanics to structures, including linear elastic approaches, the EPRI J estimation scheme, the R-6 method, and the British Standards PD 6493 approach. A brief description of probabilistic fracture mechanics is also included, as well as a discussion of the shortcomings of existing analyses. Chapter 10 describes the fracture mechanics approach to fatigue crack propagation, and discusses some of the critical issues in this area, including crack closure and the behavior of short cracks. Chapter 11 outlines some of the most recent developments in computational fracture mechanics. Procedures for determining stress intensity and the J integral in structure are described, with particular emphasis on the energy domain integral approach.

Chapter 12 provides reference material that is usually found in fracture mechanics handbooks. This material includes stress intensity factors for common configurations, as well as limit load, elastic compliance, and fully plastic J solutions. Chapter 13 contains a series of practice problems that correspond to material in Chapters 1 to 11.

If this book is used as a college text, it is unlikely that all of the material can be covered in a single semester. Thus the instructor should select the portions of the book that suit the needs and background of the students. The first three chapters, excluding appendices, should form the foundation of any course. In addition, I strongly recommend the inclusion of at least one of the materials chapters (5 or 6), regardless of whether or not materials science is the students' major field of study. A course that is oriented toward applications could include Chapters 7 to 10, in addition to the earlier chapters. A graduate level course in a solid mechanics curriculum might include Appendices 2 and 3, Chapter 4, Appendix 4, and Chapter 11.

Desk-top publishing enthusiasts may be interested to know that this book was produced on a Macintosh personal computer. The text was written in Microsoft Word 4.0 and the graphics were produced with Canvas 2.1 and KaleidaGraph 2.0. The final camera-ready copy was printed on a 300 dpi LaserWriter II NT. Since I am responsible for virtually all of the key strokes and mouse movements that went into this book, I have no one to blame but myself for any mistakes that may have occurred.

I do, however, have many people to thank. I am grateful to Joel Claypool, Russ Hall and Sandy Perlman at CRC Press for their support and advice. A number of colleagues and friends reviewed portions of the draft manuscript and/or provided photographs and homework problems, including W.L. Bradley, M. Cayard, R Chona, M.G. Dawes, R.H. Dodds Jr., A.G. Evans, S.J. Garwood, J.P. Gudas, E.G. Guynn, A.L. Highsmith, R.E. Jones Jr., J. Keeney-Walker, Y.W. Kwon, E.J. Lavernia, A. Letton, R.C. McClung, D.L. McDowell, J.G. Merkle, M.T. Miglin, D.M. Parks, P.T. Purtscher, R.A. Schapery, and C.F. Shih. I apologize to anyone whose name I have inadvertently omitted from this list. I received valuable assistance from Twyla Ray and Amy Cummings, who performed some of the tedious clean-up work on the manuscript. Mr. Sun Yongqi produced a number of SEM fractographs especially for this book. I would like to express my appreciation to Walter Bradley, the Head of the Mechanical Engineering Department at Texas A&M University, for providing an environment conducive to the preparation of this book. Finally I wish to express my gratitude and apologies to my wife Sarah and my daughter Molly for enduring this past year, when much of my time and energy that should have been devoted to them was instead focused on this book.

T.L. Anderson

To

Sarah and Molly

CONTENTS

PART I: INTRODUCTION

1. HISTORY AND OVERVIEW

Fracture is a problem that society has faced for as long as there have been man-made structures. The problem may actually be worse today than in previous centuries, because more can go wrong in our complex technological society. Major airline crashes, for instance, would not be possible without modern aerospace technology.

Fortunately, advances in the field of fracture mechanics have helped to offset some of the potential dangers posed by increasing technological complexity. Our understanding of how materials fail and our ability to prevent such failures has increased considerably since World War II. Much remains to be learned, however, and existing knowledge of fracture mechanics is not always applied when appropriate.

While catastrophic failures provide income for attorneys and consulting engineers, such events are detrimental to the economy as a whole. An economic study [1] estimated the cost of fracture in the United States in 1978 at $119 billion (in 1982 dollars), about 4% of the gross national product. Furthermore, this study estimated that the annual cost could be reduced by $35 billion if current technology were applied, and that further fracture mechanics research could reduce this figure by an additional $28 billion.

1.1 WHY STRUCTURES FAIL

The cause of most structural failures generally falls into one of the following categories:

(1) Negligence during design, construction or operation of the structure.

(2) Application of a new design or material, which produces an unexpected (and undesirable) result.

In the first instance, existing procedures are sufficient to avoid failure, but are not followed by one or more of the parties involved, due to human error, ignorance, or willful misconduct. Poor workmanship, inappropriate or substandard materials, errors in stress analysis, and operator error are examples of where the appropriate technology and experience are available, but not applied.

The second type of failure is much more difficult to prevent. When an "improved" design is introduced, there are invariably factors that the designer does not anticipate. New materials can offer tremendous advantages, but also potential problems. Consequently, a new design or material should be placed into service only after extensive testing and analysis. Such an approach will reduce the frequency of failures, but not eliminate them entirely; there may be important factors that are overlooked during testing and analysis.

One of the most famous Type 2 failures is the brittle fracture of the World War II Liberty ships (see Section 1.2.2). These ships, which were the first to have an all-welded hull, could be fabricated much faster and cheaper than earlier riveted designs, but a significant number of these vessels sustained serious fractures as a result of the design change. Today, virtually all steel ships are welded, but sufficient knowledge was gained from the Liberty ship failures to avoid similar problems in present structures.

Knowledge must be applied in order to be useful, however. Figure 1.1 shows an example of a Type 1 failure, where poor workmanship in a seemingly inconsequential structural detail caused a more recent fracture in a welded ship. In 1979, the Kurdistan oil tanker broke completely in two while sailing in the north Atlantic [2]. The combination of warm oil in the tanker with cold water in contact with the outer hull produced substantial thermal stresses. The fracture initiated from a bilge keel that was improperly welded. The weld failed to penetrate the structural detail, resulting in a severe stress concentration. Although the hull steel had adequate toughness to prevent fracture initiation, it failed to stop the propagating crack.

Polymers, which are becoming more common in structural applications, provide a number of advantages over metals, but also have the potential for causing Type 2 failures. For example, polyethylene (PE) is currently the material of choice in natural gas transportation systems in the United States. One advantage of PE piping is that maintenance can be performed on a small branch of the line without shutting down the entire system; a local area is shut down by applying a clamping tool

(a) Fractured vessel in dry dock.

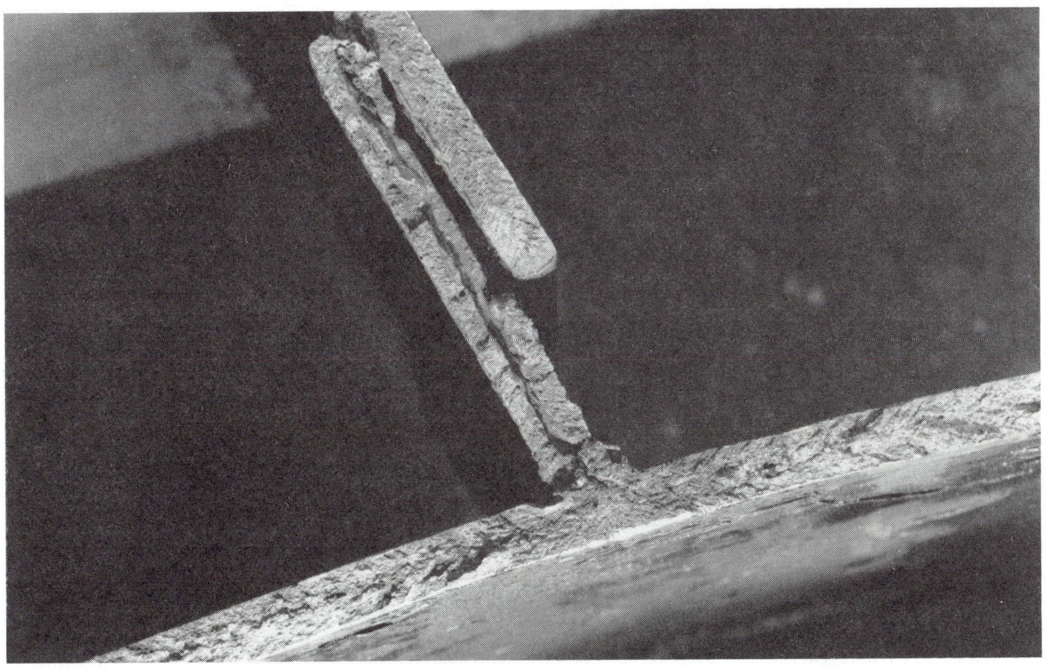

(b) Bilge keel from which the fracture initiated.

FIGURE 1.1 The Kurdistan oil tanker, which sustained a brittle fracture while sailing in the north Atlantic in 1979. Photographs provided by S.J. Garwood to the PE pipe and stopping the flow of gas. The practice of pinch clamping has undoubtedly saved vast sums of money, but has also led to an unexpected problem.

In 1983 a section of 4 in diameter PE pipe developed a major leak. The gas collected beneath a residence where it ignited, resulting in severe damage to the house. Maintenance records and a visual inspection of the pipe indicated that it had been pinch clamped 6 years earlier in the region where the leak developed. A failure investigation [3] concluded that the pinch clamping operation was responsible for the failure. Microscopic examination of the pipe revealed that a small flaw apparently initiated on the inner surface of the pipe and grew through the wall. Figure 1.2 shows a low magnification photograph of the fracture surface. Laboratory tests simulated the pinch clamping operation on sections of PE pipe; small thumbnail-shaped flaws (Fig. 1.3) formed on the inner wall of the pipes, as a result of the severe strains that were applied. Fracture mechanics tests and analyses [3, 4] indicated that stresses in the pressurized pipe were sufficient to cause the observed time-dependent crack growth; i.e., growth from a small thumbnail flaw to a through-thickness crack over a period of 6 years.

The introduction of flaws in PE pipe by pinch clamping represents a Type 2 failure. The pinch clamping process was presumably tested thoroughly before it was applied in service, but no one anticipated that the procedure would introduce damage in the material that could lead to failure after several years in service. Although specific data are not available, pinch clamping has undoubtedly led to a significant number of gas leaks. The practice of pinch clamping is still widespread in the natural gas industry, but many companies and some states now require that a sleeve be fitted to the affected region in order to relieve the stresses locally. In addition, newer grades of PE pipe material have lower density and are less susceptible to damage by pinch clamping.

Some catastrophic events include elements both of Types 1 and 2 failures. On January 28, 1986, the Challenger Space Shuttle exploded because an O-ring seal in one of the main boosters did not respond well to cold weather. The Shuttle represents relatively new technology, where service experience is limited (Type 2), but engineers from the booster manufacturer suspected a potential problem with the O-ring seals and recommended that the launch be delayed (Type 1). Unfortunately, these engineers had little or no data to support their po-

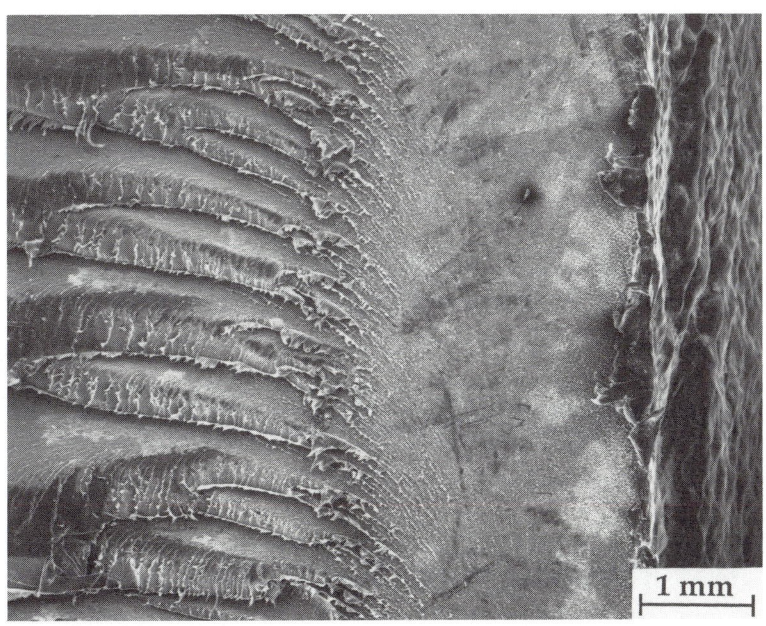

FIGURE 1.2 Fracture surface of a PE pipe that sustained time-dependent crack growth as a result of pinch clamping [3]. (Photograph provided by R.E. Jones Jr.)

FIGURE 1.3 Thumbnail crack produced in a PE pipe after pinch clamping for 72 hours. (Photograph provided by R.E. Jones Jr.)

sition and were unable to convince their managers or NASA officials. The tragic results of the decision to launch are well known.

Over the past few decades, the field of fracture mechanics has undoubtedly prevented a substantial number of structural failures. We will never know how many lives have been saved or how much property damage has been avoided by applying this technology, because it is impossible to quantify disasters that *don't* happen. When applied correctly, fracture mechanics not only helps to prevent Type 1 failures but also reduces the frequency of failures of the second type, because designers can rely on rational analysis rather than trial and error.

1.2 HISTORICAL PERSPECTIVE

Designing structures to avoid fracture is not a new idea. The fact that many structures commissioned by the Pharaohs of ancient Egypt and the Caesars of Rome are still standing is a testimony to the ability of early architects and engineers. In Europe, numerous buildings and bridges constructed during the Renaissance Period are still used for their intended purpose.

The ancient structures that are still standing today obviously represent successful designs. There were undoubtedly many more unsuccessful designs that endured a much shorter life span. Since mankind's knowledge of mechanics was limited prior to the time of Isaac Newton, workable designs were probably achieved largely by trial and error. The Romans supposedly tested each new bridge by requiring the design engineer to stand underneath while chariots drove over the it. Such a practice would not only provide an incentive for developing good designs, but would also result in a Darwinian natural selection, where the worst engineers are "removed" from the profession.

The durability of ancient structures is particularly amazing when one considers that the choice of building materials prior to the Industrial Revolution was rather limited. Metals could not be produced in sufficient quantity to be formed into load-bearing members for buildings and bridges. The primary construction materials prior to the 19th century were timber, brick, and mortar; only the latter two materials were usually practical for large structures such as cathedrals, because trees of sufficient size for support beams were rare.

Brick and mortar are relatively brittle and are unreliable for carrying tensile loads. Consequently, pre-Industrial Revolution structures were usually designed to be loaded in compression. Figure 1.4 schematically illustrates a Roman bridge design. The arch shape causes compressive rather than tensile stresses to be transmitted through the structure.

The arch is the predominate shape in pre-Industrial Revolution architecture. Windows and roof spans were arched in order to maintain compressive loading. For example, Fig. 1.5 shows two windows and a portion of the ceiling in Kings College Chapel in Cambridge, England. Although these shapes are aesthetically pleasing, their primary purpose is more pragmatic.

Compressively loaded structures are obviously stable, since some have lasted for many centuries. The pyramids in Egypt are probably the most extreme example of a stable design.

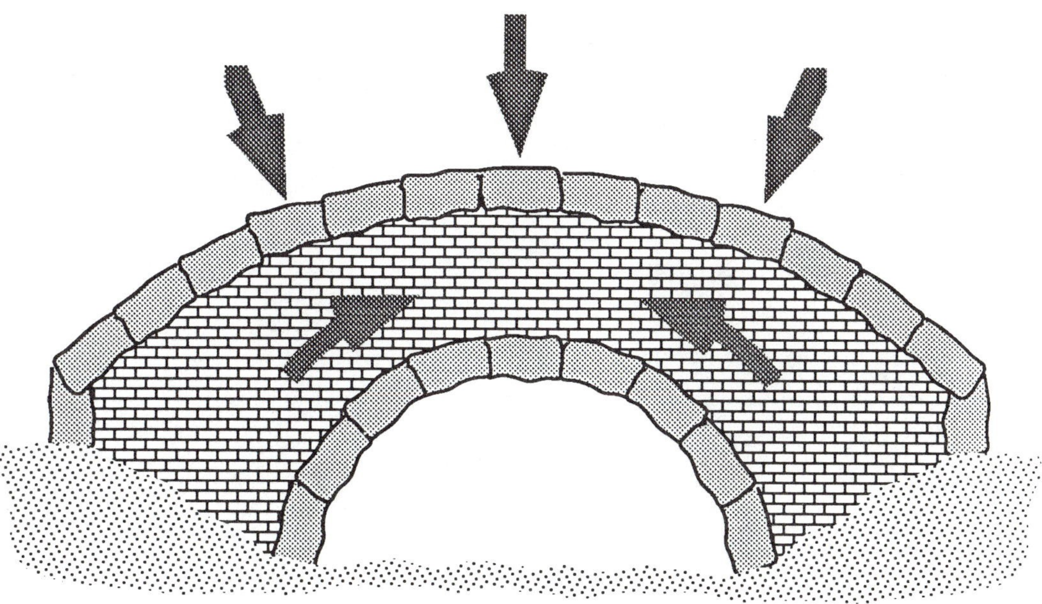

FIGURE 1.4 Schematic Roman bridge design. The arch shape of the bridge causes loads to be transmitted through the structure as compressive stresses.

FIGURE 1.5 Kings College Chapel in Cambridge, England. This structure was completed in 1515.

FIGURE 1.6 The Tower Bridge in London, completed in 1894. Note the modern beam design, made possible by the availability of steel support beams.

With the Industrial Revolution came mass production of iron and steel. (Or, conversely, one might argue that mass production of iron and steel fueled the Industrial Revolution.) The availability of relatively ductile construction materials removed the earlier restrictions on design. It was finally feasible to build structures that carried tensile stresses. Note the difference between the design of the Tower Bridge in London (Fig. 1.6) and the earlier bridge design (Fig. 1.4).

The change from brick and mortar structures loaded in compression to steel structures in tension brought problems, however. Occasionally, a steel structure would fail unexpectedly at stresses well below the anticipated tensile strength. One of the most famous of these failures was the rupture of a molasses tank in Boston in January 1919 [5]. Over 2 million gallons of molasses were spilled, resulting in 12 deaths, 40 injuries, massive property damage, and several drowned horses.

The cause of failures as the molasses tank was largely a mystery at the time. In the first edition of his elasticity text published in 1892, Love [6] remarked that "the conditions of rupture are but vaguely understood." Designers typically applied safety factors of 10 or more (based on the tensile strength) in an effort to avoid these seemingly random failures.

1.2.1 Early Fracture Research

Experiments performed by Leonardo da Vinci several centuries earlier provided some clues as to the root cause of fracture. He measured the strength of iron wires and found that the strength varied inversely with wire length. These results implied that flaws in the material controlled the strength; a longer wire corresponded to a larger sample volume and a higher probability of sampling a region containing a flaw. These results were only qualitative, however.

A quantitative connection between fracture stress and flaw size came from the work of Griffith, which was published in 1920 [7]. He applied a stress analysis of an elliptical hole (performed by Inglis [8] seven years earlier) to the unstable propagation of a crack. Griffith invoked the first law of thermodynamics to formulate a fracture theory based on a simple energy balance. According to this theory, a flaw becomes unstable, and thus fracture occurs, when the strain energy change that results from an increment of crack growth is sufficient to overcome the surface energy of the material (See Section 2.3). Griffith's model correctly predicted the relationship between strength and flaw size in glass specimens. Subsequent efforts to apply the Griffith model to metals were unsuccessful. Since this model assumes that the work of fracture comes exclusively from the surface energy of the material, the Griffith approach only applies to ideally brittle solids. A modification to Griffith's model that made it applicable to metals did not come until 1948.

1.2.2 The Liberty Ships

The mechanics of fracture progressed from being a scientific curiosity to an engineering discipline, primarily because of what happened to the Liberty ships during World War II [9].

In the early days of World War II, the United States was supplying ships and planes to Great Britain under the Lend-Lease Act. Britain's greatest need at the time was for cargo ships to carry supplies. The German Navy was sinking cargo ships at three times the rate at which they could be replaced with existing ship-building procedures.

Under the guidance of Henry Kaiser, a famous construction engineer whose previous projects included the Hoover Dam, the United States developed a revolutionary procedure for fabricating ships quickly. These new vessels, which became known as the Liberty ships, had an all-welded hull, as opposed to the riveted construction of traditional ship designs.

The Liberty ship program was a resounding success, until one day in 1943, when one of the vessels broke completely in two while sailing between Siberia and Alaska. Subsequent fractures occurred in other Liberty ships. Of the roughly 2700 liberty ships build during World War II, approximately 400 sustained fractures, of which 90 were considered serious. In 20 ships the failure was essentially total, and about half of these broke completely in two.

Investigations revealed that the Liberty ship failures were caused by a combination of three factors:

- The welds, which were produced by a semi-skilled work force, contained crack-like flaws.

- Most of the fractures initiated on the deck at square hatch corners, where there was a local stress concentration.

- The steel from which the Liberty ships were made had poor toughness, as measured by Charpy impact tests.

The steel in question had always been adequate for riveted ships because fracture could not propagate across panels that were joined by rivets. A welded structure, however, is essentially a single piece of metal; propagating cracks in the Liberty ships encountered no significant barriers, and were sometimes able to traverse the entire hull.

Once the causes of failure were identified, the remaining Liberty ships were retro-fitted with rounded reinforcements at the hatch corners. In addition, high toughness steel crack arrester plates were riveted to the deck at strategic locations. These corrections prevented further serious fractures.

In the longer term, structural steels were developed with vastly improved toughness, and weld quality control standards were developed. Also, a group of researchers at the Naval Research Laboratory in Washington D.C. studied the fracture problem in detail. The field we now know as fracture mechanics was born in this lab during the decade following the War.

1.2.3 Post-War Fracture Mechanics Research[1]

The fracture mechanics research group at the Naval Research Laboratory was led by Dr. G.R. Irwin. After studying the early work of Inglis, Griffith, and others, Irwin concluded that the basic tools needed to analyzed fracture were already available. Irwin's first major contribution was to extend the Griffith approach to metals by including the energy dissipated by local plastic flow [10]. Orowan independently proposed a similar modification to the Griffith theory [11]. During this

---

[1] For an excellent summary of early fracture mechanics research, refer to *Fracture Mechanics Retrospective: Early Classic Papers (1913-1965)*, John M. Barsom, ed., American Society of Testing and Materials (RPS 1), Philadelphia, 1987. This volume contains reprints of 17 classic papers, as well as a complete bibliography of fracture mechanics papers published up to 1965.

same period, Mott [12] extended the Griffith theory to a rapidly propagating crack.

In 1956, Irwin [13] developed the energy release rate concept, which is related to the Griffith theory but is in a form that is more useful for solving engineering problems. Shortly afterward, several of Irwin's colleagues brought to his attention a paper by Westergaard [14] that was published in 1938. Westergaard had developed a semi-inverse technique for analyzing stresses and displacements ahead of a sharp crack. Irwin [15] used the Westergaard approach to show that the stresses and displacements near the crack tip could be described by a single constant that was related to the energy release rate. This crack tip characterizing parameter later became known as the stress intensity factor. During this same period of time, Williams [16] applied a somewhat different technique to derive crack tip solutions that were essentially identical to Irwin's results.

A number of successful early applications of fracture mechanics bolstered the standing of this new field in the engineering community. In 1956, Wells [17] used fracture mechanics to show that the fuselage failures in several Comet jet aircraft resulted from fatigue cracks reaching a critical size. These cracks initiated at windows and were caused by insufficient reinforcement locally, combined with square corners which produced a severe stress concentration. (Recall the unfortunate hatch design in the Liberty ships.) A second early application of fracture mechanics occurred at General Electric in 1957. Winne and Wundt [18] applied Irwin's energy release rate approach to the failure of large rotors from steam turbines. They were able to predict the bursting behavior of large disks extracted from rotor forgings, and applied this knowledge to the prevention of fracture in actual rotors.

It seems that all great ideas encounter stiff opposition initially, and fracture mechanics is no exception. Although the U.S. military and the electric power generating industry were very supportive of the early work in this field, such was not the case in all provinces of government and industry. Several government agencies openly discouraged research in this area.

In 1960, Paris and his co-workers [19] failed to find a receptive audience for their ideas on applying fracture mechanics principles to fatigue crack growth. Although Paris et al. provided convincing experimental and theoretical arguments for their approach, it seems that design engineers were not yet ready to abandon their S-N curves in favor of a more rational approach to fatigue design. The resistance to this work was so intense that Paris and his colleagues were unable to find a peer-reviewed technical journal that was willing to publish their manuscript. They finally opted to publish their work in a University of Washington periodical entitled *The Trend in Engineering*.

1.2.4 Fracture Mechanics from 1960 to 1980

The Second World War obviously separates two distinct eras in the history of fracture mechanics. There is, however, some disagreement as to how the period between the end of the War and the present should be divided. One possible historical boundary occurs around 1960, when the fundamentals of linear elastic fracture mechanics were fairly well established, and researchers turned their attention to crack tip plasticity.

Linear elastic fracture mechanics (LEFM) ceases to be valid when significant plastic deformation precedes failure. During a relatively short time period (1960-61) several researchers developed analyses to correct for yielding at the crack tip, including Irwin [20], Dugdale [21], Barenblatt [22], and Wells [23]. The Irwin plastic zone correction [20] was a relatively simple extension of LEFM, while Dugdale [21] and Barenblatt [22] each developed somewhat more elaborate models based on a narrow strip of yielded material at the crack tip.

Wells [23] proposed the displacement of the crack faces as an alternative fracture criterion when significant plasticity precedes failure. Previously, Wells had worked with Irwin while on sabbatical at the Naval Research Laboratory. When Wells returned to his post at the British Welding Research Association, he attempted to apply LEFM to low- and medium-strength structural steels. These materials were too ductile for LEFM to apply, but Wells noticed that the crack faces moved apart with plastic deformation. This observation led to the development of the parameter now known as the crack tip opening displacement (CTOD).

In 1968, Rice [24] developed another parameter to characterize nonlinear material behavior ahead of a crack. By idealizing plastic deformation as nonlinear elastic, Rice was able to generalize the energy release rate to nonlinear materials. He showed that this nonlinear energy release rate can be expressed as a line integral, which he called the J integral, evaluated along an arbitrary contour around the crack. At the time his work was being published, Rice discovered that Eshelby [25] had previously published several so-called conservation integrals, one of which was equivalent to Rice's J integral. Eshelby, however, did not apply his integrals to crack problems.

That same year, Hutchinson [26] and Rice and Rosengren [27] related the J integral to crack tip stress fields in nonlinear materials. These analyses showed that J can be viewed as a nonlinear stress intensity parameter as well as an energy release rate.

Rice's work might have been relegated to obscurity had it not been for the active research effort by nuclear power industry in the United States in the early 1970s. Because of legitimate concerns for safety, as well as political and public relations considerations, the nuclear power industry endeavored to apply state-of-the-art technology, including fracture mechanics, to the design and construction of nuclear power plants. The difficulty with applying fracture mechanics in this instance was that most nuclear pressure vessel steels were too tough to be characterized with LEFM without resorting to enormous laboratory specimens. In 1971, Begley and Landes [28], who were research engineers at Westinghouse, came across Rice's article and decided, despite skepticism from their co-workers, to characterize fracture toughness of these steels with the J integral. Their experiments were very successful and led to the publication of a standard procedure for J testing of metals ten years later [29].

Material toughness characterization is only one aspect of fracture mechanics. In order to apply fracture mechanics concepts to design, one must have a mathematical relationship between toughness, stress and flaw size. Although these relationships were well established for linear elastic problems, a fracture design analysis based on the J integral was not available until Shih and Hutchinson [30] provided the theoretical framework for such an approach in 1976. A few years later, the Electric Power Research Institute (EPRI) published a fracture design handbook [31] based on the Shih and Hutchinson methodology.

In the United Kingdom, Well's CTOD parameter was applied extensively to fracture analysis of welded structures, beginning in the late 1960s. While fracture research in the U.S. was driven primarily by the nuclear power industry during the 1970s, fracture research in the

UK was motivated largely by the development of oil resources in the North Sea. In 1971, Burdekin and Dawes [32] applied several ideas proposed by Wells [33] several years earlier and developed the CTOD design curve, a semiempirical fracture mechanics methodology for welded steel structures. The nuclear power industry in the UK developed their own fracture design analysis [34], based on the strip yield model of Dugdale [21] and Barenblatt [22].

Shih [35] demonstrated a relationship between the J integral and CTOD, implying that both parameters are equally valid for characterizing fracture. The J-based material testing and structural design approaches developed in the U.S. and the British CTOD methodology have begun to merge in recent years, with positive aspects of each approach combined to yield improved analyses. Both parameters are currently applied throughout the world to a range of materials.

Much of the theoretical foundation of dynamic fracture mechanics was developed in the period between 1960 and 1980. Significant contributions were made by a number of researchers, as discussed in Chapter 4.

1.2.5 Recent Trends in Fracture Research

It is difficult to discuss fracture mechanics research performed since 1980 in a historical context. Identifying major breakthroughs usually requires the passage of time; what seems important today may be obsolete later, while a major discovery may be overlooked when it is first published. It is possible, however, to identify a few trends in recent work.

The field of fracture mechanics has matured in recent years. Current research tends to result in incremental advances rather than major gains.

More sophisticated models for material behavior are being incorporated into fracture mechanics analyses. While plasticity was the important concern in 1960, more recent work has gone a step further, incorporating time-dependent nonlinear material behavior such as viscoplasticity and viscoelasticity. The former is motivated by the need for tough, creep-resistant high temperature materials, while the latter reflects the increasing proportion of plastics in structural applications.

Fracture mechanics has also been used (and sometimes abused) in the characterization of composite materials.

Another trend in recent research is the development of microstuctural models for fracture and models to relate local and global fracture behavior of materials. A related topic is the efforts to characterize and predict geometry dependence of fracture toughness. Such approaches are necessary when traditional, so-called single-parameter fracture mechanics break down.

1.3 THE FRACTURE MECHANICS APPROACH TO DESIGN

Figure 1.7 contrasts the fracture mechanics approach with the traditional approach to structural design and material selection. In the latter case, the anticipated design stress is compared to the flow properties of candidate materials; a material is assumed to be adequate if its strength is greater than the expected applied stress. Such an approach may attempt to guard against brittle fracture by imposing a safety factor on stress, combined with minimum tensile elongation requirements on the material. The fracture mechanics approach (Fig. 1.7(b)) has three important variables, rather than two as in Fig. 1.7(a). The additional structural variable is flaw size, and fracture toughness replaces strength as the relevant material property. Fracture mechanics quantifies the critical combinations of these three variables.

There are two alternative approaches to fracture analysis: the energy criterion and the stress intensity approach. These two approaches are equivalent in certain circumstances. Both are discussed briefly below.

1.3.1 The Energy Criterion

The energy approach states that crack extension (i.e. fracture) occurs when the energy available for crack growth is sufficient to overcome the resistance of the material. The material resistance may include the surface energy, plastic work, or other type of energy dissipation associated with a propagating crack.

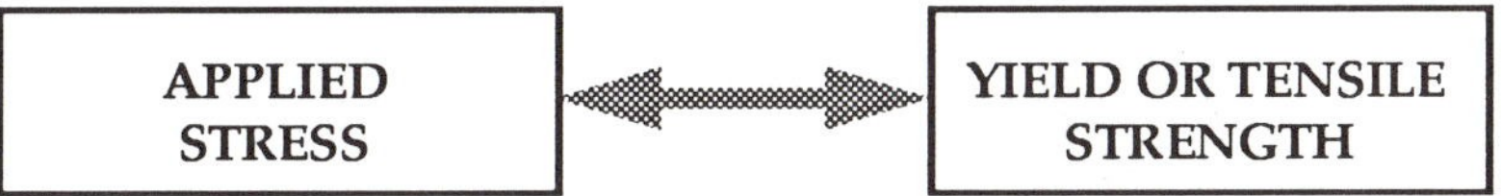

(a) The strength of materials approach.

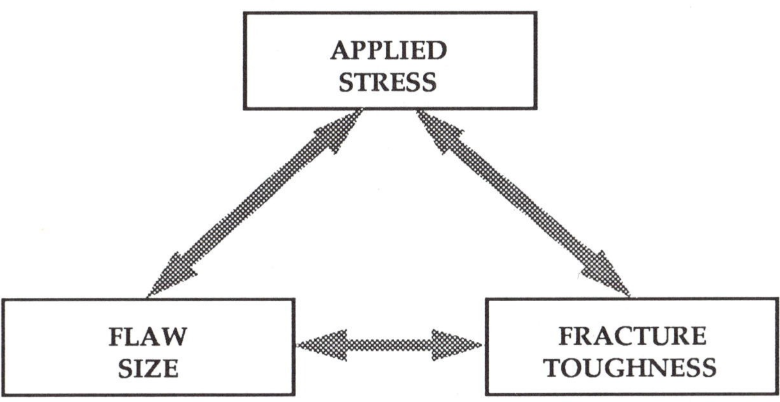

(b) The fracture mechanics approach

FIGURE 1.7 Comparison of the fracture mechanics approach to design with the traditional strength of materials approach.

Griffith [7] was the first to propose the energy criterion for fracture, but Irwin [13] is primarily responsible for developing the present version of this approach: the energy release rate, $\mathcal{G}$, which is defined as the rate of change in potential energy with crack area for a linear elastic material. At the moment of fracture, $\mathcal{G} = \mathcal{G}_c$, the critical energy release rate, which is a measure of fracture toughness.

For a crack of length 2a in an infinite plate subject to a remote tensile stress (Fig. 1.8), the energy release rate is given by

$$\mathcal{G} = \frac{\pi \sigma^2 a}{E} \tag{1.1}$$

where E is Youngs modulus, σ is the remotely applied stress, and a is the half crack length. At fracture, $\mathcal{G} = \mathcal{G}_c$, and Eq. (1.1) describes the critical combinations of stress and crack size for failure:

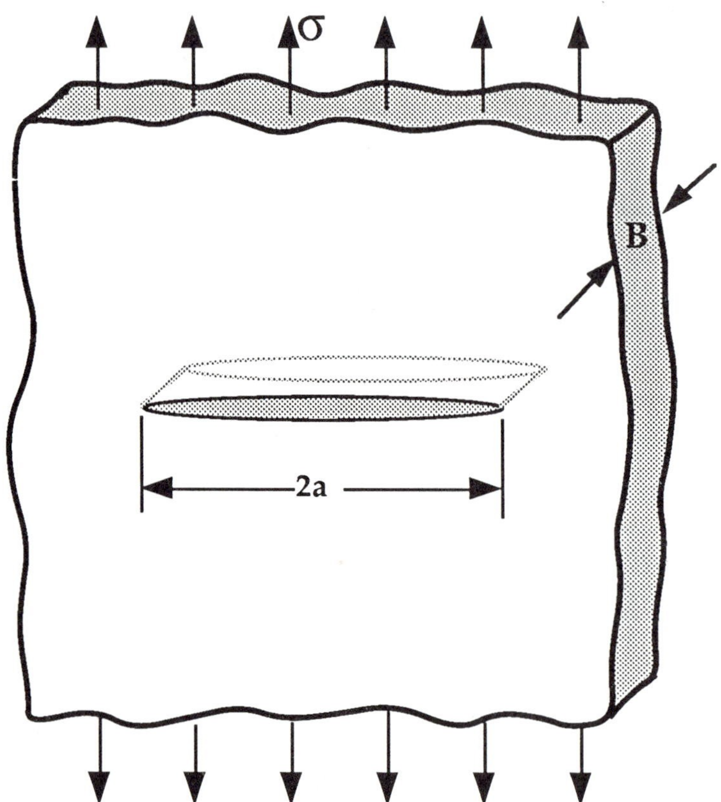

FIGURE 1.8 Through-thickness crack in an infinite plate subject to a remote tensile stress. In practical terms, "infinite" means that the width of the plate is >> 2a.

$$\mathcal{G}_c = \frac{\pi \sigma_f^2 a_c}{E} \tag{1.2}$$

Note that for a constant $\mathcal{G}_c$ value, failure stress, σ_f, varies with $1/\sqrt{a}$. The energy release rate, $\mathcal{G}$, is the driving force for fracture, while $\mathcal{G}_c$ is the material's resistance to fracture. To draw an analogy to the strength of materials approach of Fig. 1.7(a), the applied stress can be viewed as the driving force for plastic deformation, while the yield strength is a measure of the material's resistance to deformation.

The tensile stress analogy is also useful for illustrating the concept of similitude. A yield strength value measured with a laboratory specimen should be applicable to a large structure; yield strength does not depend on specimen size, provided the material is reasonably homogeneous. One of the fundamental assumptions of fracture mechanics is that fracture toughness ($\mathcal{G}_c$ in this case) is independent of the size and geometry of the cracked body; a fracture toughness measurement on a laboratory specimen should be applicable to a structure. As long as this assumption is valid, all configuration effects are taken into account by the driving force, $\mathcal{G}$. The similitude assumption is valid as long as the material behavior is predominantly linear elastic.

1.3.2 The Stress Intensity Approach

Figure 1.9 schematically shows an element near the tip of a crack in an elastic material, together with the in-plane stresses on this element. Note that each stress component is proportional to a single constant, K_I. If this constant is known, the entire stress distribution at the crack tip can be computed with the equations in Fig. 1.9. This constant, which is called the stress intensity factor, completely characterizes the crack tip conditions in a linear elastic material. (The meaning of the subscript on K is explained in Chapter 2.) If one assumes that the material fails locally at some critical combination of stress and strain, then it follows that fracture must occur at a critical stress intensity, K_{IC}. Thus K_{IC} is an alternate measure of fracture toughness.

For the plate illustrated in Fig. 1.8, the stress intensity factor is given by

$$K_I = \sigma\sqrt{\pi a} \tag{1.3}$$

Failure occurs when $K_I = K_{IC}$. In this case, K_I is the driving force for fracture and K_{IC} is a measure of material resistance. As with $\mathcal{G}_c$, the property of similitude should apply to K_{IC}. That is, K_{IC} is assumed to be a size-independent material property.

Comparing Eqs. (1.1) and (1.3) results in a relationship between K_I and $\mathcal{G}$:

$$\mathcal{G} = \frac{K_I^2}{E} \tag{1.4}$$

This same relationship obviously holds for $\mathcal{G}_c$ and K_{IC}. Thus the energy and stress intensity approaches to fracture mechanics are essentially equivalent for linear elastic materials.

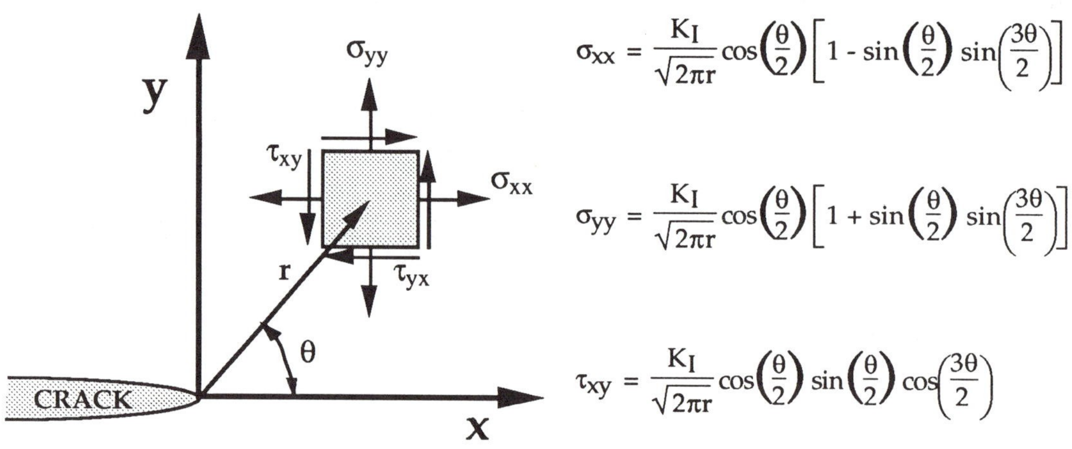

FIGURE 1.9 Stresses near the tip of a crack in an elastic material.

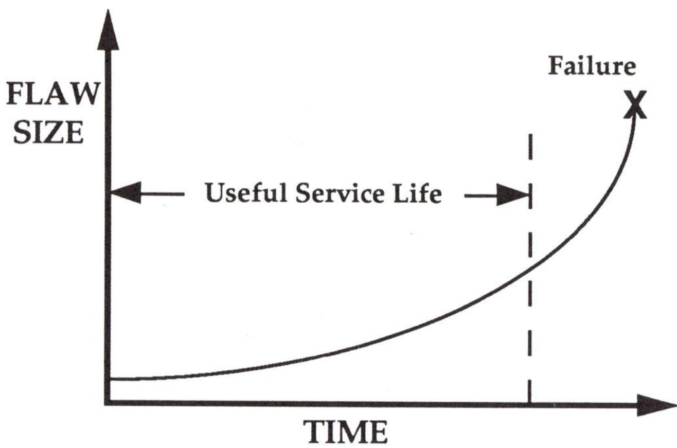

FIGURE 1.10 The damage tolerance approach to design.

1.3.3 Damage Tolerance

Damage tolerance, as its name suggests, entails allowing subcritical flaws to remain in a structure. Repairing flawed material or scrapping a flawed structure is expensive and is often unnecessary. Fracture mechanics provides a rational basis for establishing flaw tolerance limits.

Consider a flaw in a structure that grows with time (e.g. a fatigue crack or a stress corrosion crack) as illustrated schematically in Fig. 1.10. If the fracture toughness of the material is known, fracture mechanics relationships provide an estimate of the critical flaw size required to cause failure in the structure. Normally, an allowable flaw size would be defined by dividing the critical size by a safety factor. The structure would then be allowed to operate until the flaw grew to the maximum allowable size.

Examples of time-dependent crack growth include fatigue, environmental-assisted cracking, creep crack growth, and viscoelastic crack growth.

1.4 EFFECT OF MATERIAL PROPERTIES ON FRACTURE

Figure 1.11 shows a simplified family tree for the field of fracture mechanics. Most early work was only applicable to linear elastic materials under quasistatic conditions, while subsequent advances in fracture research incorporated other types of material behavior. Elastic-plastic fracture mechanics considers plastic deformation under quasistatic conditions, while dynamic, viscoelastic, and viscoplastic fracture mechanics include time as a variable. A dashed line is drawn between linear elastic and dynamic fracture mechanics because some early research considered dynamic linear elastic behavior. The chapters that describe the various types of fracture behavior are shown in Fig. 1.11. Elastic-plastic, viscoelastic, and viscoplastic fracture behavior are sometimes included in the more general heading of *nonlinear fracture mechanics*. The branch of fracture mechanics one should apply to a particular problem obviously depends on material behavior.

Consider a cracked plate (Fig. 1.8) that is loaded to failure. Figure 1.12 is a schematic plot of failure stress versus fracture toughness (K_{IC}). For low toughness materials, brittle fracture is the governing failure mechanism, and critical stress varies linearly with K_{IC}, as predicted by Eq. (1.3). At very high toughness values, LEFM is no longer valid, and failure is governed by the flow properties of the material. At intermediate toughness levels, there is a transition between brittle fracture under linear elastic conditions and ductile overload. Nonlinear fracture mechanics bridges the gap between LEFM and collapse. If toughness is low, LEFM is applicable to the problem, but if toughness is sufficiently high, fracture mechanics ceases to be relevant to the problem because failure stress is insensitive to toughness; a simple limit load analysis is all that is required to predict failure stress in a material with very high fracture toughness.

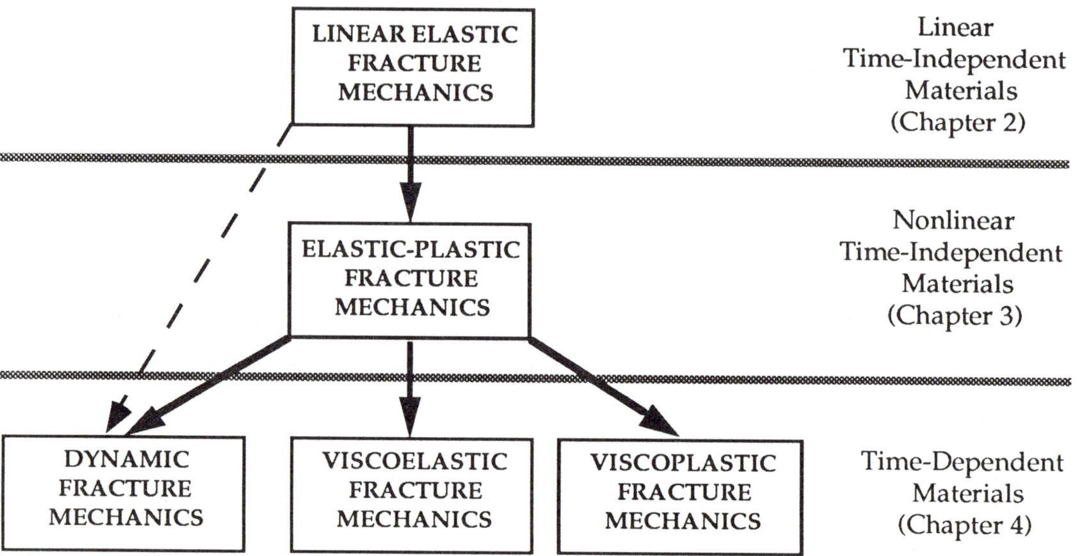

FIGURE 1.11 Simplified family tree of fracture mechanics.

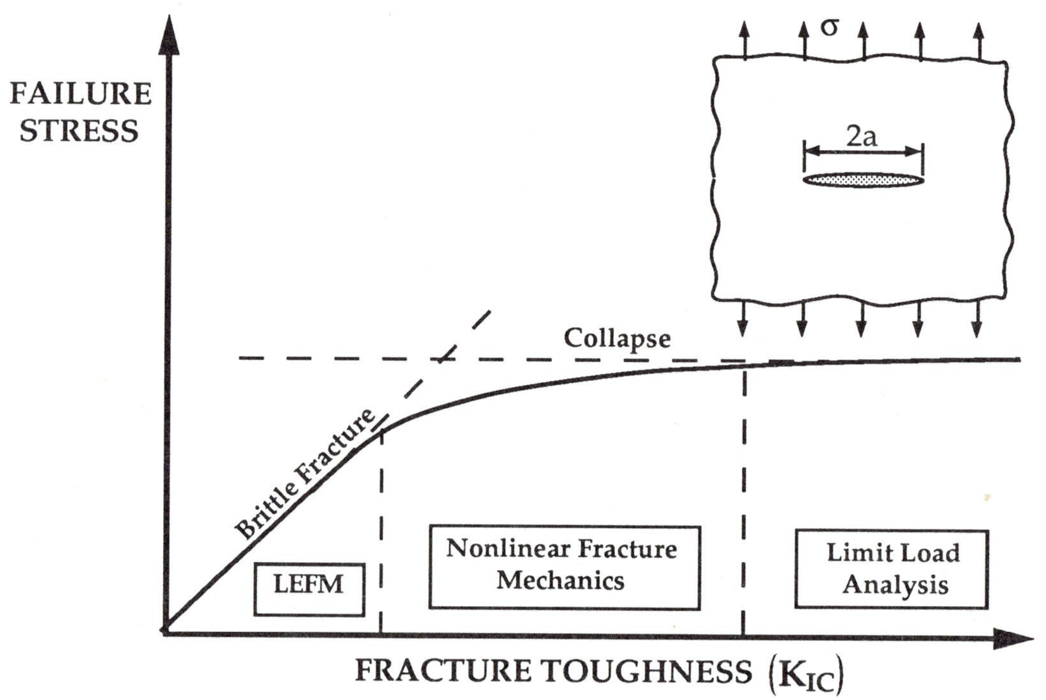

FIGURE 1.12 Effect of fracture toughness on the governing failure mechanism.

Table 1.1 lists various materials, together with the typical fracture regime for each material.

TABLE 1.1
Typical fracture behavior of selected materials. Temperature is ambient unless otherwise specified.

Material	Typical Fracture Behavior
High strength steel	Linear elastic
Low- and medium-strength steel	Elastic-plastic/Fully plastic
Austenitic stainless steel	Fully plastic
Precipitation-hardened aluminum	Linear elastic
Metals at high temperatures	Viscoplastic
Metals at high strain rates	Dynamic-viscoplastic
Polymers (below T_g)*	Linear elastic/Viscoelastic
Polymers (above T_g)*	Viscoelastic
Monolithic ceramics	Linear elastic
Ceramic composites	Linear elastic
Ceramics at high temperatures	Viscoplastic

*T_g - Glass transition temperature.

1.5 A BRIEF REVIEW OF DIMENSIONAL ANALYSIS

At first glance, a section on dimensional analysis may seem out of place in the introductory chapter of a book on fracture mechanics. However, dimensional analysis is an important tool for developing mathematical models of physical phenomena, and it can help us understand existing models. Many difficult concepts in fracture mechanics become relatively transparent when one considers the relevant dimensions of the problem. For example, dimensional analysis gives us a clue as to when a particular model, such as linear elastic fracture mechanics, is no longer valid.

Let us review the fundamental theorem of dimensional analysis and then look at a few simple applications to fracture mechanics.

1.5.1 The Buckinham Π-Theorem

The first step in building a mathematical model of a physical phenomenon is to identify all of the parameters that may influence the phenomenon. Assume that a problem, or at least an idealized version of it, can be described by the following set of scalar quantities: $\{u, W_1, W_2, \ldots, W_n\}$. The dimensions of all quantities in this set is denoted by $\{[u], [W_1], [W_2], \ldots, [W_n]\}$. Now suppose that we wish to express the first variable, u, as a function of the remaining parameters:

$$u = f(W_1, W_2, \ldots, W_n) \tag{1.5}$$

Thus the process of modeling the problem is reduced to finding a mathematical relationship that represents f as best as possible. We might accomplish this by performing a set of experiments in which we measure u while varying each W_i independently. The number of experiments can be greatly reduced, and the modeling processes simplified, through dimensional analysis. The first step is to identify all of the *fundamental dimensional units* (fdu's) in the problem: $\{L_1, L_2, \ldots L_m\}$. For example, a typical mechanics problem may have $\{L_1$ = length, L_2 = mass, L_3 = time$\}$. We can express the dimensions of each quantity in our problem as the product of powers of the fdu's; i.e. for any quantity X, we have

$$[X] = L_1^{a_1} L_2^{a_2} \ldots L_m^{a_m} \tag{1.6}$$

The quantity X is dimensionless if $[X] = 1$.

In the set of W's, we can identify m *primary quantities* that contain all of the fdu's in the problem. The remaining variables are secondary quantities, and their dimensions can be expressed in terms of the primary quantities:

$$[W_{m+j}] = [W_1]^{a_{m+j}(1)} \ldots [W_m]^{a_{m+j}(m)} \quad (j = 1, 2, \ldots, n\text{-}m) \tag{1.7}$$

Thus we can define a set of new quantities, π_i, that are dimensionless:

$$\pi_i = \frac{W_{m+j}}{W_1{}^{a_{m+j}(1)} \ldots W_m{}^{a_{m+j}(m)}} \tag{1.8}$$

Similarly, the dimensions of u can be expressed in terms of the dimensions of the primary quantities:

$$[u] = [W_1]^{a_1} \ldots [W_m]^{a_m} \tag{1.9}$$

and we can form the following dimensionless quantity:

$$\pi = \frac{u}{W_1{}^{a_1} \ldots W_m{}^{a_m}} \tag{1.10}$$

According to the Buckingham Π-theorem, π depends only on the other dimensionless groups:

$$\pi = F(\pi_1, \pi_2, \ldots, \pi_{n\text{-}m}) \tag{1.11}$$

This new function, F, is independent of the system of measurement units. Note that the number of quantities in F has been reduced from the old function by m, the number of fdu's. Thus dimensional analyses has reduced the degrees of freedom in our model, and we need only vary n-m quantities in our experiments or computer simulations.

The Buckingham Π-theorem gives guidance on how to scale a problem to different sizes or to other systems of measurement units. Each dimensionless group, (π_i) must be scaled in order to obtain equivalent conditions at two different scales. Suppose, for example, that we want to perform wind tunnel tests on a model of a new airplane design. Dimensional analysis tells us that we should reduce all length dimensions in the same proportion; thus we would build a "scale" model of the airplane. The length dimensions of the plane are not the only important quantities in the problem, however. In order to model the aerodynamic behavior accurately, we would need to scale the wind velocity and the viscosity of the air in accordance with the reduced size of the airplane model. (Modifying the viscosity of the air is not practical in most cases. In real wind tunnel tests, the size of the model is usually close enough to full scale that the errors introduced by not scaling viscosity are minor.)

1.5.2 Dimensional Analysis in Fracture Mechanics

Dimensional analysis proves to be a very useful tool in fracture mechanics. Later chapters describe how dimensional arguments play a key role in developing mathematical descriptions for important phenomena. For now, let us explore a few simple examples.

Consider a series of cracked plates under a remote tensile stress, σ^∞, as illustrated in Fig. 1.13. Assume that each is a two-dimensional problem; that is, the thickness dimension does not enter into the problem. The first case, Fig. 1.13(a), is an edge crack of length a in an elastic, semi-infinite plate. In this case infinite means that the plate width is much larger than the crack size. Suppose that we wish to know how one of the stress components, σ_{ij}, varies with position. We will adopt a polar coordinate system with the origin at the crack tip, as illustrated in Fig. 1.9. A generalized functional relationship can be written as

$$\sigma_{ij} = f_1(\sigma^\infty, E, \nu, \sigma_{kl}, \varepsilon_{kl}, a, r, \theta) \tag{1.12}$$

where ν is Poisson's ratio, σ_{kl} represents the other stress components, and ε_{kl} represents all nonzero components of the strain tensor. We can eliminate σ_{kl} and ε_{kl} from f_1 by noting that for a linear elastic problem, strain is uniquely defined by stress through Hooke's law and the stress components at a point increase in proportion to one another. Let σ^∞ and a be the primary quantities. Invoking the Buckingham Π-theorem gives

$$\frac{\sigma_{ij}}{\sigma^\infty} = F_1\left(\frac{E}{\sigma^\infty}, \frac{r}{a}, \nu, \theta\right) \tag{1.13}$$

When the plate width is finite (Fig. 1.13(b)), an additional dimension is required to describe the problem:

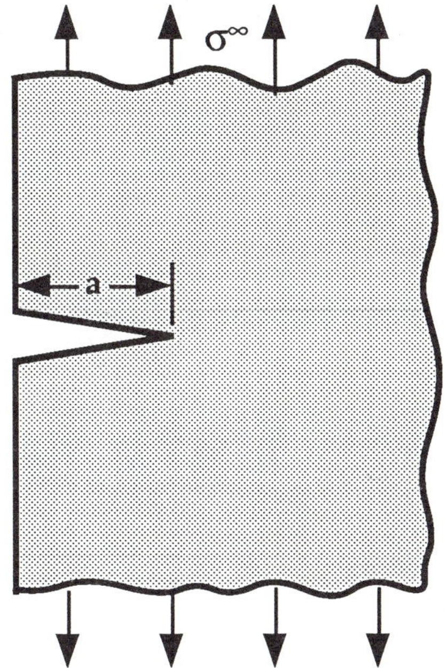

(a) Edge crack in a wide elastic plate.

(b) Edge crack in a finite width elastic plate.

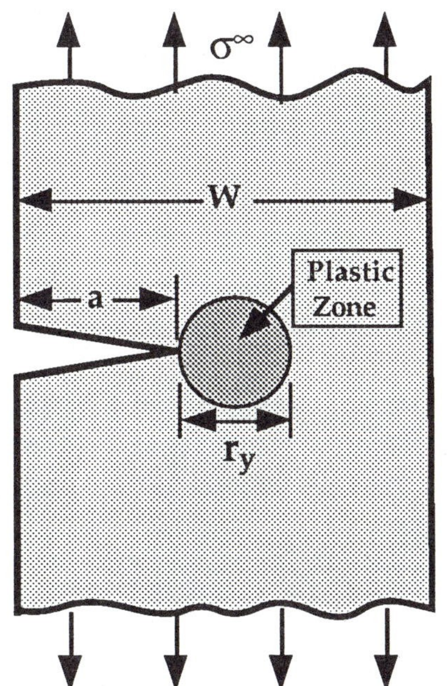

(c) Edge crack with a plastic zone at the crack tip.

FIGURE 1.13 Edge cracked plates subject to a remote tensile stress.

$$\frac{\sigma_{ij}}{\sigma^{\infty}} = F_2\left(\frac{E}{\sigma^{\infty}}, \frac{r}{a}, \frac{W}{a}, \nu, \theta\right) \tag{1.14}$$

Thus, one might expect Eq. (1.13) to give erroneous results when the crack extends across a significant fraction of the plate width. Consider a large plate and a small plate made of the same material (same E and ν), with the same a/W ratio, loaded to the same remote stress. The local stress at an angle θ from the crack plane in each plate would depend only on the r/a ratio, as long as both plates remained elastic.

When a plastic zone forms ahead of the crack tip (Fig. 1.13(c)), the problem is complicated further. If we assume that the material does not strain harden, the yield strength is sufficient to define the flow properties. The stress field is given by

$$\frac{\sigma_{ij}}{\sigma^{\infty}} = F_3\left(\frac{E}{\sigma^{\infty}}, \frac{\sigma_{YS}}{\sigma^{\infty}}, \frac{r}{a}, \frac{W}{a}, \frac{r_y}{a}, \nu, \theta\right) \tag{1.15}$$

The first two functions, F_1 and F_2, correspond to linear elastic fracture mechanics (LEFM), while F_3 is an elastic-plastic relationship. Thus, dimensional analysis tells us that LEFM is only valid when $r_y << a$ and $\sigma^{\infty} << \sigma_{YS}$. In Chapter 2, the same conclusion is reached through a somewhat more complicated argument.

REFERENCES

1. Duga, J.J., Fisher, W.H., Buxbaum, R.W., Rosenfield, A.R., Burh, A.R., Honton, E.J., and McMillan, S.C., "The Economic Effects of Fracture in the United States." NBS Special Publication 647-2, United States Department of Commerce, Washington, DC, March 1983.

2. Garwood, S.J., Private Communication, 1990.

3. Jones, R.E. and Bradley, W.L., "Failure Analysis of a Polyethylene Natural Gas Pipeline." *Forensic Engineering*, Vol. 1, 1987, pp. 47-59.

4. Jones, R.E. and Bradley, W.L., "Fracture Toughness Testing of Polyethylene Pipe Materials." ASTM STP 995, Vol. 1, 1989, American Society for Testing and Materials, Philadelphia, pp. 447-456.

5. Shank, M.E., "A Critical Review of Brittle Failure in Carbon Plate Steel Structures Other than Ships." Ship Structure Committee Report SSC-65,

National Academy of Science-National Research Council, Washington, DC, December, 1953.

6. Love A.E.H, *A Treatise on The Mathematical Theory of Elasticity.* Dover Publications, New York, 1944.

7. Griffith, A.A. "The Phenomena of Rupture and Flow in Solids." *Philosophical Transactions,* Series A, Vol. 221, 1920, pp. 163-198.

8. Inglis, C.E., "Stresses in a Plate Due to the Presence of Cracks and Sharp Corners." *Transactions of the Institute of Naval Architects,* Vol. 55, 1913, pp. 219-241.

9. Bannerman, D.B. and Young, R.T., "Some Improvements Resulting from Studies of Welded Ship Failures." *Welding Journal,* Vol. 25, 1946.

10. Irwin, G.R., "Fracture Dynamics." *Fracturing of Metals,* American Society for Metals, Cleveland, 1948, pp. 147-166.

11. Orowan, E., "Fracture and Strength of Solids." *Reports on Progress in Physics,* Vol. XII, 1948, p. 185.

12. Mott, N.F., "Fracture of Metals: Theoretical Considerations." *Engineering,* Vol. 165, 1948, pp. 16-18.

13. Irwin, G.R., "Onset of Fast Crack Propagation in High Strength Steel and Aluminum Alloys." *Sagamore Research Conference Proceedings,* Vol. 2, 1956, pp. 289-305.

14. Westergaard, H.M., "Bearing Pressures and Cracks." *Journal of Applied Mechanics,* Vol. 6, 1939, pp. 49-53.

15. Irwin, G.R., "Analysis of Stresses and Strains near the End of a Crack Traversing a Plate." *Journal of Applied Mechanics,* Vol. 24, 1957, pp. 361-364.

16. Williams, M.L., "On the Stress Distribution at the Base of a Stationary Crack." *Journal of Applied Mechanics,* Vol. 24, 1957, pp. 109-114.

17. Wells, A.A., "The Condition of Fast Fracture in Aluminum Alloys with Particular Reference to Comet Failures." British Welding Research Association Report, April 1955.

18. Winne, D.H. and Wundt, B.M., "Application of the Griffith-Irwin Theory of Crack Propagation to the Bursting Behavior of Disks, Including Analytical and Experimental Studies." *Transactions of the American Society of Mechanical Engineers,* Vol. 80, 1958, pp. 1643-1655.

19. Paris, P.C., Gomez, M.P., and Anderson, W.P., "A Rational Analytic Theory of Fatigue." *The Trend in Engineering,* Vol. 13, 1961, pp. 9-14.

20. Irwin, G.R., "Plastic Zone Near a Crack and Fracture Toughness." *Sagamore Research Conference Proceedings,* Vol. 4, 1961.

21. Dugdale, D.S., "Yielding in Steel Sheets Containing Slits." *Journal of the Mechanics and Physics of Solids*, Vol 8, pp. 100-104.

22. Barenblatt, G.I., "The Mathematical Theory of Equilibrium Cracks in Brittle Fracture." *Advances in Applied Mechanics*, Vol VII, Academic Press, 1962, pp. 55-129.

23. Wells, A.A., "Unstable Crack Propagation in Metals: Cleavage and Fast Fracture." *Proceedings of the Crack Propagation Symposium*, Vol 1, Paper 84, Cranfield, UK, 1961.

24. Rice, J.R. "A Path Independent Integral and the Approximate Analysis of Strain Concentration by Notches and Cracks." *Journal of Applied Mechanics*, Vol. 35, 1968, pp. 379-386.

25. Eshelby, J.D., "The Continuum Theory of Lattice Defects." *Solid State Physics*, Vol. 3, 1956.

26. Hutchinson, J.W., "Singular Behavior at the End of a Tensile Crack Tip in a Hardening Material." *Journal of the Mechanics and Physics of Solids*, Vol. 16, 1968, pp. 13-31.

27. Rice, J.R. and Rosengren, G.F., "Plane Strain Deformation near a Crack Tip in a Power-Law Hardening Material." *Journal of the Mechanics and Physics of Solids*, Vol. 16, 1968, pp. 1-12.

28. Begley, J. A. and Landes, J.D., "The J-Integral as a Fracture Criterion." ASTM STP 514, American Society for Testing and Materials, Philadelphia, 1972, pp. 1-20.

29. E 813-81, "Standard Test Method for J_{Ic}, a Measure of Fracture Toughness." American Society for Testing and Materials, Philadelphia, 1981.

30. Shih, C.F. and Hutchinson, J.W., "Fully Plastic Solutions and Large-Scale Yielding Estimates for Plane Stress Crack Problems." *Journal of Engineering Materials and Technology*, Vol. 98, 1976, pp. 289-295.

31. Kumar, V., German, M.D., and Shih, C.F.,"An Engineering Approach for Elastic-Plastic Fracture Analysis." EPRI Report NP-1931, Electric Power Research Institute, Palo Alto, CA, 1981.

32. Burdekin, F.M. and Dawes, M.G., "Practical Use of Linear Elastic and Yielding Fracture Mechanics with Particular Reference to Pressure Vessels." *Proceedings of the Institute of Mechanical Engineers Conference*, London, May 1971, pp. 28-37.

33. Wells, A.A., "Application of Fracture Mechanics at and Beyond General Yielding." *British Welding Journal*, Vol 10, 1963, pp. 563-570.

34. Harrison, R.P., Loosemore, K., Milne, I, and Dowling, A.R., "Assessment of the Integrity of Structures Containing Defects." Central Electricity Generating Board Report R/H/R6-Rev 2, April 1980.

35. Shih, C.F. "Relationship between the J-Integral and the Crack Opening Displacement for Stationary and Extending Cracks." *Journal of the Mechanics and Physics of Solids,* Vol 29, 1981, pp. 305-326.

PART II: FUNDAMENTAL CONCEPTS

2. LINEAR ELASTIC FRACTURE MECHANICS

The concepts of fracture mechanics that were derived prior to 1960 are applicable only to materials that obey Hooke's law. Although corrections for small scale plasticity were proposed as early as 1948, these analyses are restricted to structures whose global behavior is linear elastic.

Since 1960, fracture mechanics theories have been developed to account for various types of nonlinear material behavior (i.e. plasticity, viscoplasticity, and viscoelasticity) as well as dynamic effects. All of these more recent results, however, are extensions of linear elastic fracture mechanics (LEFM). Thus a solid background in the fundamentals of LEFM is essential to an understanding of more advanced concepts in fracture mechanics.

This chapter describes both the energy and stress intensity approaches to linear fracture mechanics. The early work of Inglis and Griffith is summarized, followed by an introduction to the energy release rate and stress intensity parameters. The appendix at the end of this chapter includes mathematical derivations of several important results in LEFM. We begin with a brief discussion of fracture on the atomic level.

2.1 AN ATOMIC VIEW OF FRACTURE

A material fractures when sufficient stress and work are applied on the atomic level to break the bonds that hold atoms together. The bond strength is supplied by the attractive forces between atoms.

Figure 2.1 shows schematic plots of the potential energy and force versus separation distance between atoms. The equilibrium spacing occurs where the potential energy is at a minimum. A tensile force is required to increase the separation distance from the equilibrium value; this force must exceed the cohesive force to sever the bond completely. The bond energy is given by

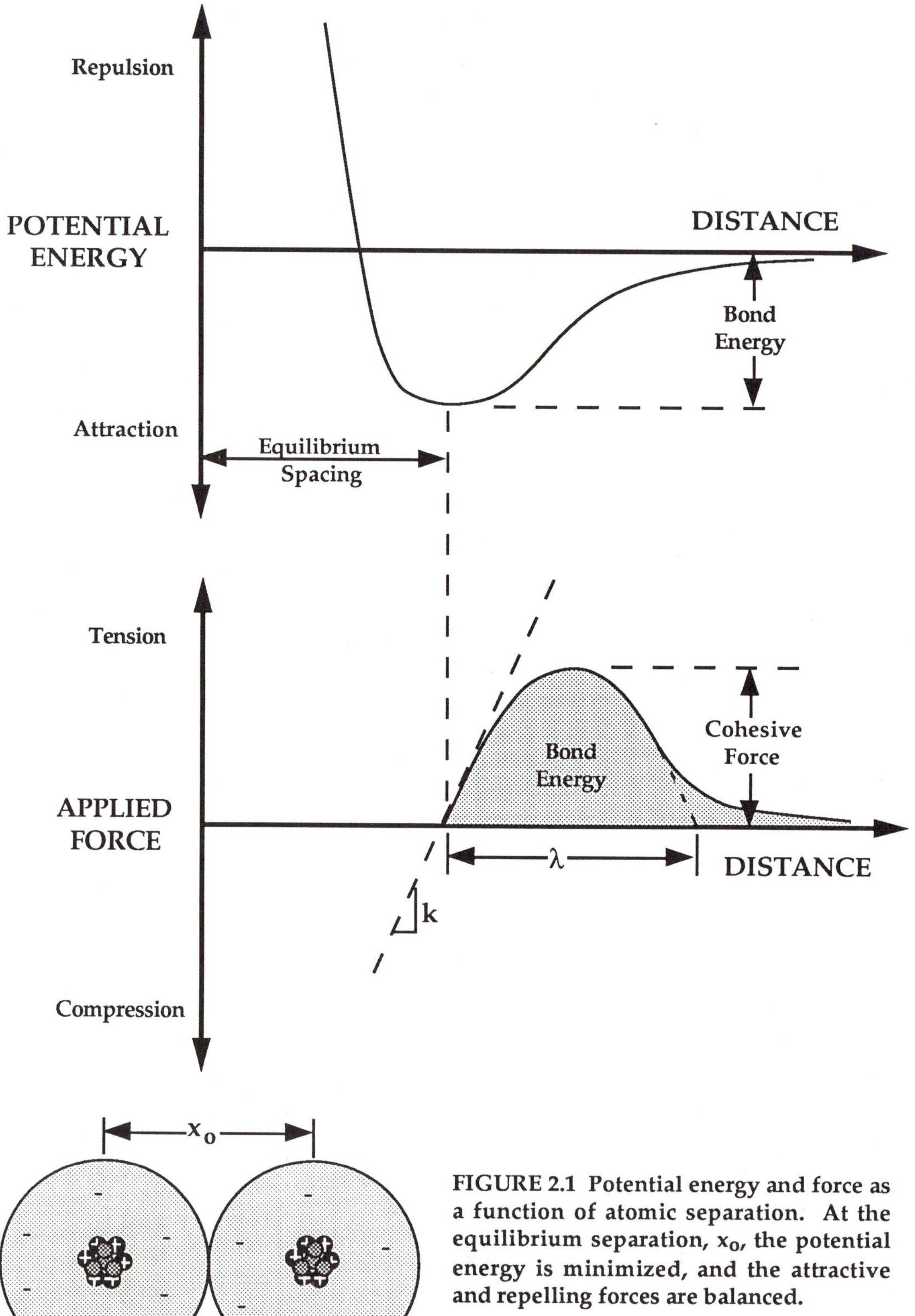

FIGURE 2.1 Potential energy and force as a function of atomic separation. At the equilibrium separation, x_o, the potential energy is minimized, and the attractive and repelling forces are balanced.

$$E_b = \int_{x_o}^{\infty} P\,dx \tag{2.1}$$

where x_o is the equilibrium spacing and P is the applied force.

It is possible to estimate the cohesive strength at the atomic level by idealizing the interatomic force-displacement relationship as one half the period of a sine wave:

$$P = P_c \sin\left(\frac{\pi x}{\lambda}\right) \tag{2.2}$$

where the distance λ is defined in Fig. 2.1. For the sake of simplicity, the origin is defined at x_o. For small displacements, the force-displacement relationship is linear:

$$P = P_c\left(\frac{\pi x}{\lambda}\right)$$

and the bond stiffness (i.e., the spring constant) is given by

$$k = \frac{P_c \pi}{\lambda} \tag{2.3}$$

Multiplying both sides of this equation by the number of bonds per unit area and the gage length, x_o, converts k to Young's modulus, E, and P_c to the cohesive stress, σ_c. Solving for σ_c gives

$$\sigma_c = \frac{E\lambda}{\pi x_o} \tag{2.4}$$

or

$$\sigma_c \approx \frac{E}{\pi} \tag{2.5}$$

if λ is assumed to be approximately equal to the atomic spacing.

The surface energy can be estimated as follows:

$$\gamma_s = \frac{1}{2}\int_0^{\lambda} \sigma_c \sin\left(\frac{\pi x}{\lambda}\right)dx = \sigma_c\frac{\lambda}{\pi} \tag{2.6}$$

The surface energy per unit area, γ_s, is equal to one half the fracture energy because two surfaces are created when a material fractures. Substituting Eq. (2.4) into Eq. (2.6) and solving for σ_c gives

$$\sigma_c = \sqrt{\frac{E\gamma_s}{x_o}} \tag{2.7}$$

2.2 STRESS CONCENTRATION EFFECT OF FLAWS

The derivation in the previous section showed that the theoretical cohesive strength of a material is approximately E/π, but experimental fracture strengths are typically three or four orders of magnitude below this value. As discussed in Chapter 1, experiments by Leonardo da Vinci, Griffith, and others indicated that the discrepancy between the actual strengths of brittle materials and theoretical estimates was due to flaws in these materials. Fracture cannot occur unless the stress at the atomic level exceeds the cohesive strength of the material. Thus the flaws must lower the global strength by magnifying the stress locally.

The first quantitative evidence for the stress concentration effect of flaws was provided by Inglis [1], who analyzed elliptical holes in flat plates. His analyses included an elliptical hole 2a long by 2b wide with an applied stress perpendicular to the major axis of the ellipse (see Fig. 2.2). He assumed that the hole is not influenced by the plate boundary; i.e., the plate width >> 2a and the plate height >> 2b. The stress at the tip of the major axis (Point A) is given by

$$\sigma_A = \sigma\left(1 + \frac{2a}{b}\right) \tag{2.8}$$

The ratio σ_A/σ is defined as the stress concentration factor, k_t. When a = b, the hole is circular and k_t = 3.0, a well-known result that can be found in most strength of materials text books.

As the major axis, a, increases relative to b, the elliptical hole begins to take on the appearance of a sharp crack. For this case, Inglis found it more convenient to express Eq. (2.8) in terms of the radius of curvature, ρ:

$$\sigma_A = \sigma\left(1 + 2\sqrt{\frac{a}{\rho}}\right) \tag{2.9}$$

where

$$\rho = \frac{b^2}{a} \tag{2.10}$$

When a >> b, Eq. (2.9) becomes

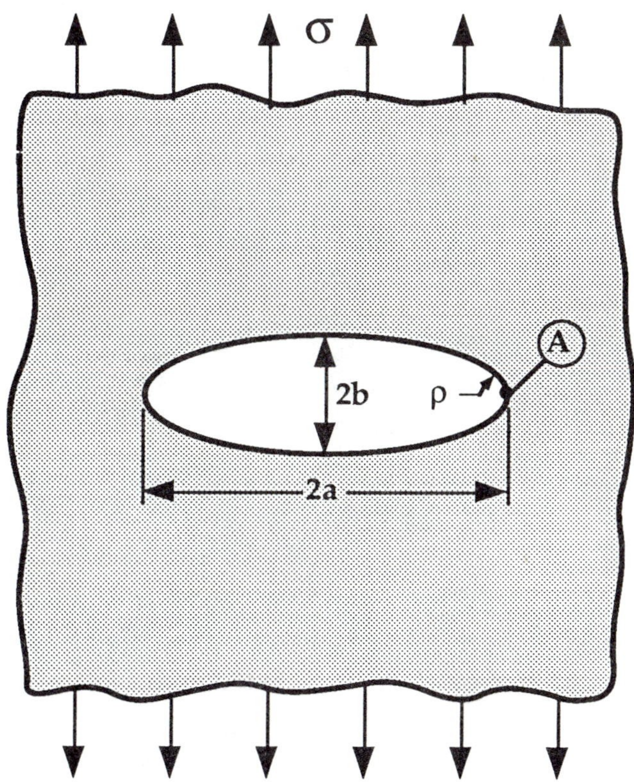

FIGURE 2.2 Elliptical hole in a flat plate.

$$\sigma_A = 2\sigma\sqrt{\frac{a}{\rho}} \tag{2.11}$$

Inglis showed that Eq. (2.11) gives a good approximation for the stress concentration due to a notch that is not elliptical except at the tip.

Equation (2.11) predicts an infinite stress at the tip of an infinitely sharp crack, where $\rho = 0$. This result caused concern when it was first discovered, because no material is capable of withstanding infinite stress. A material that contains a sharp crack theoretically should fail upon the application of an infinitesimal load. The paradox of a sharp crack motivated Griffith [2] to develop a fracture theory based on energy rather than local stress (Section 2.3).

An infinitely sharp crack in a continuum is a mathematical abstraction that is not relevant to real materials, which are made of atoms. Metals, for instance, deform plastically, which causes an initially sharp crack to blunt. In the absence of plastic deformation, the minimum radius a crack tip can have is on the order of the atomic radius. By substituting $\rho = x_o$ into Eq. [2.11] we obtain an estimate of the local stress concentration at the tip of an atomically sharp crack:

$$\sigma_A = 2\sigma\sqrt{\frac{a}{x_o}} \tag{2.12}$$

If it is assumed that fracture occurs when $\sigma_A = \sigma_c$, Eq. (2.12) can be set equal to Eq. (2.7), resulting in the following expression for the remote stress at failure:

$$\sigma_f = \left(\frac{E\,\gamma_s}{4\,a}\right)^{1/2} \tag{2.13}$$

Equation (2.13) must be viewed as a rough estimate of failure stress, because the continuum assumption upon which the Inglis analysis is based is not valid at the atomic level. However, Gehlen and Kanninen [3] obtained similar results from a numerical simulation of a crack in a two-dimensional lattice, where discrete "atoms" were connected by nonlinear springs:

$$\sigma_f = \alpha \left(\frac{E \gamma_s}{a} \right)^{1/2} \tag{2.14}$$

where α is a constant, on the order of unity, which depends slightly on the assumed atomic force-displacement law (Eq. (2.2)).

2.3 THE GRIFFITH ENERGY BALANCE

According to the First Law of thermodynamics, when a system goes from a nonequilibrium state to equilibrium, there will be a net decrease in energy. In 1920 Griffith [2] applied this idea to the formation of a crack:

> It may be supposed, for the present purpose, that the crack is formed by the sudden annihilation of the tractions acting on its surface. At the instant following this operation, the strains, and therefore the potential energy under consideration, have their original values; but in general, the new state is not one of equilibrium. If it is not a state of equilibrium, then, by the theorem of minimum potential energy, the potential energy is reduced by the attainment of equilibrium; if it is a state of equilibrium the energy does not change.

A crack can form (or an existing crack can grow) only if such a process causes the total energy to decrease or remain constant. Thus the critical conditions for fracture can be defined as the point where crack growth occurs under equilibrium conditions, with no net change in total energy.

Consider a plate subjected to a constant stress, σ, which contains a crack 2a long (Fig. 2.3). Assume that the plate width >> 2a and that plane stress conditions prevail. (Note that the plates in Figs. 2.2 and 2.3 are identical when a >> b). In order for this crack to increase in size, sufficient potential energy must be available in the plate to overcome the surface energy of the material. The Griffith energy balance for an incremental increase in the crack area, $d\mathcal{A}$, under equilibrium conditions can be expressed in the following way:

$$\frac{dE}{d\mathcal{A}} = \frac{d\Pi}{d\mathcal{A}} + \frac{dW_s}{d\mathcal{A}} = 0 \tag{2.15}$$

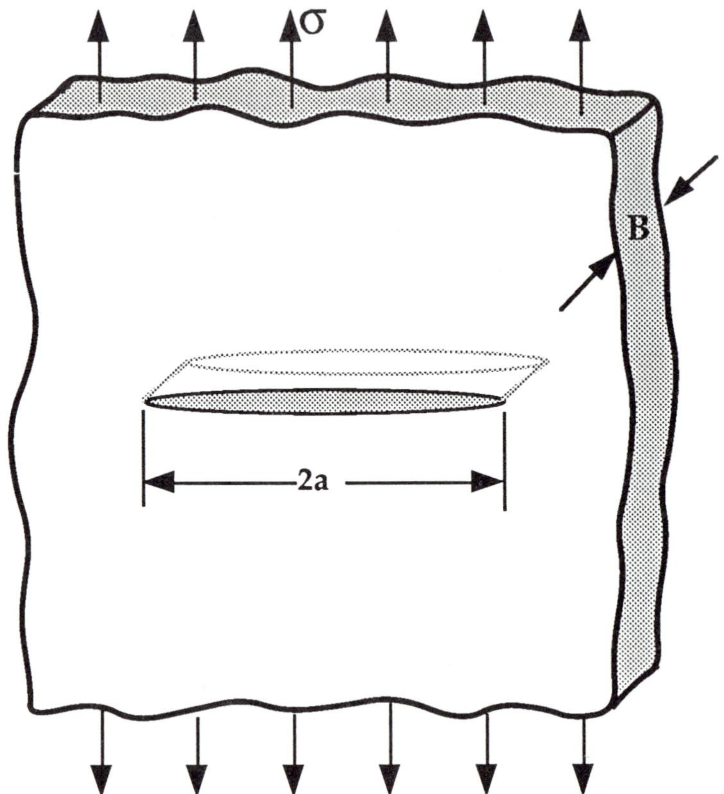

FIGURE 2.3 A through-thickness crack in an infinitely wide plate subjected to a remote tensile stress.

or

$$-\frac{d\Pi}{d\mathcal{A}} = \frac{dW_S}{d\mathcal{A}}$$

where E is the total energy, Π is the potential energy supplied by the internal strain energy and external forces, and W_S is the work required to create new surfaces. For the cracked plate illustrated in Fig. 2.3, Griffith used the stress analysis of Inglis [1] to show that

$$\Pi = \Pi_o - \frac{\pi \sigma^2 a^2 B}{E} \tag{2.16}$$

where Π_o is the potential energy of an uncracked plate and B is the plate thickness. Since the formation of a crack requires the creation of two surfaces, W_S is given by

$$W_S = 4\, a\, B\, \gamma_s \tag{2.17}$$

where γ_s is the surface energy of the material. Thus:

$$-\frac{d\Pi}{d\mathcal{A}} = \frac{\pi\, \sigma^2\, a}{E} \tag{2.18a}$$

and

$$\frac{dW_S}{d\mathcal{A}} = 2\, \gamma_s \tag{2.18b}$$

Equating (2.18a) and (2.18b) and solving for fracture stress gives

$$\sigma_f = \left(\frac{2\, E\, \gamma_s}{\pi\, a}\right)^{1/2} \tag{2.19}$$

It is important to note the distinction between *crack area* and *surface area*. The crack area is defined as the projected area of the crack (2aB in the present example), but since a crack includes two matching surfaces, the surface area = $2\mathcal{A}$.

The Griffith approach can be applied to other crack shapes. For example, the fracture stress for a penny-shaped flaw embedded in the material (Fig. 2.4) is given by

$$\sigma_f = \left(\frac{\pi\, E\, \gamma_s}{2\,(1 - \nu^2)\, a}\right)^{1/2} \tag{2.20}$$

where a is the crack radius and ν is Poisson's ratio.

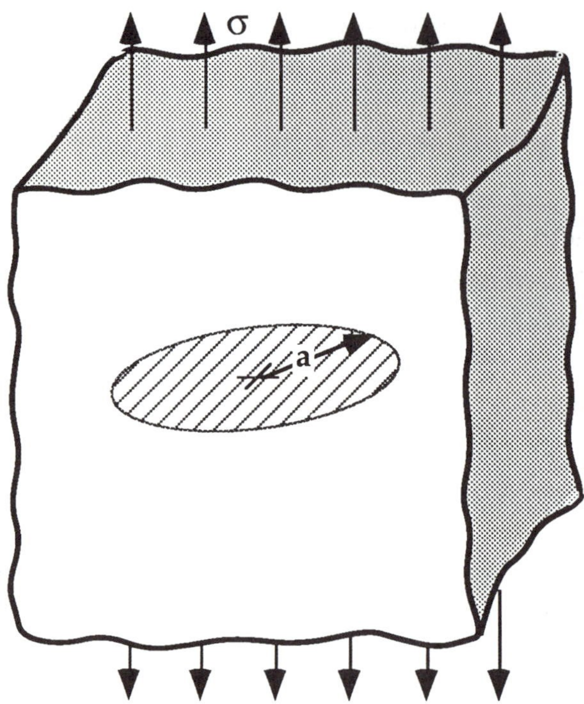

FIGURE 2.4 A penny-shaped (circular) crack embedded in a solid subjected to a remote tensile stress.

2.3.1 Comparison with the Critical Stress Criterion

The Griffith model is based on a global energy balance: for fracture to occur, the energy stored in the structure must be sufficient to overcome the surface energy of the material. Since fracture involves breaking bonds, the stress on the atomic level must be ≥ the cohesive stress. This local stress intensification can be provided by flaws in the material, as discussed in Section 2.2.

The similarity between Eqs. (2.13), (2.14), and (2.19) is obvious. Predictions of the global fracture stress from the Griffith approach and the local stress criterion differ by less than 40 percent. Thus these two approaches are consistent with one another, at least in the case of a sharp crack in an ideally brittle solid.

An apparent contradiction emerges when the crack tip radius is significantly greater than the atomic spacing. The change in stored energy with crack formation (Eq. (2.16)) is insensitive to the notch radius as long as $a >> b$; thus the Griffith model implies that the fracture stress is insensitive to ρ. According to the Inglis stress analysis, however, in order for σ_c to be attained at the tip of the notch, σ_f must vary with $1/\sqrt{\rho}$.

Consider a crack with $\rho = 5 \times 10^{-6}$ m. Such a crack would appear sharp under a light microscope, but ρ would be four orders of magnitude larger than the atomic spacing in a typical crystalline solid. Thus the local stress approach would predict a global fracture strength 100 times larger than the Griffith equation. Actual material behavior is somewhere between these extremes; fracture stress does depend on notch root radius, but not to the extent implied by the Inglis stress analysis.

The apparent discrepancy between the critical stress criterion and the energy criterion based on thermodynamics can be resolved by viewing fracture as a nucleation and growth process. When the global stress and crack size satisfy the Griffith energy criterion, there is sufficient thermodynamic driving force to grow the crack; but fracture must first be nucleated. This situation is analogous to the solidification of liquids. Water, for example, is in equilibrium with ice at 0°C, but the liquid-solid reaction requires ice crystals to be nucleated, usually on the surface of another solid (e.g. your car windshield on a January morning). When nucleation is suppressed, liquid water can be supercooled (at least momentarily) to as much as 30°C below the equilibrium freezing point.

Nucleation of fracture can come from a number of sources. For example, microscopic surface roughness at the tip of the flaw could produce sufficient local stress concentration to nucleate failure. Another possibility, illustrated in Fig. 2.5, involves a sharp microcrack near the tip of a macroscopic flaw with a finite notch radius. The macroscopic crack magnifies the stress in the vicinity of the microcrack, which propagates when it satisfies the Griffith equation. The microcrack links with the large flaw, which then propagates if the Griffith criterion is satisfied globally. This type of mechanism controls cleavage fracture in ferritic steels, as discussed in Chapter 5.

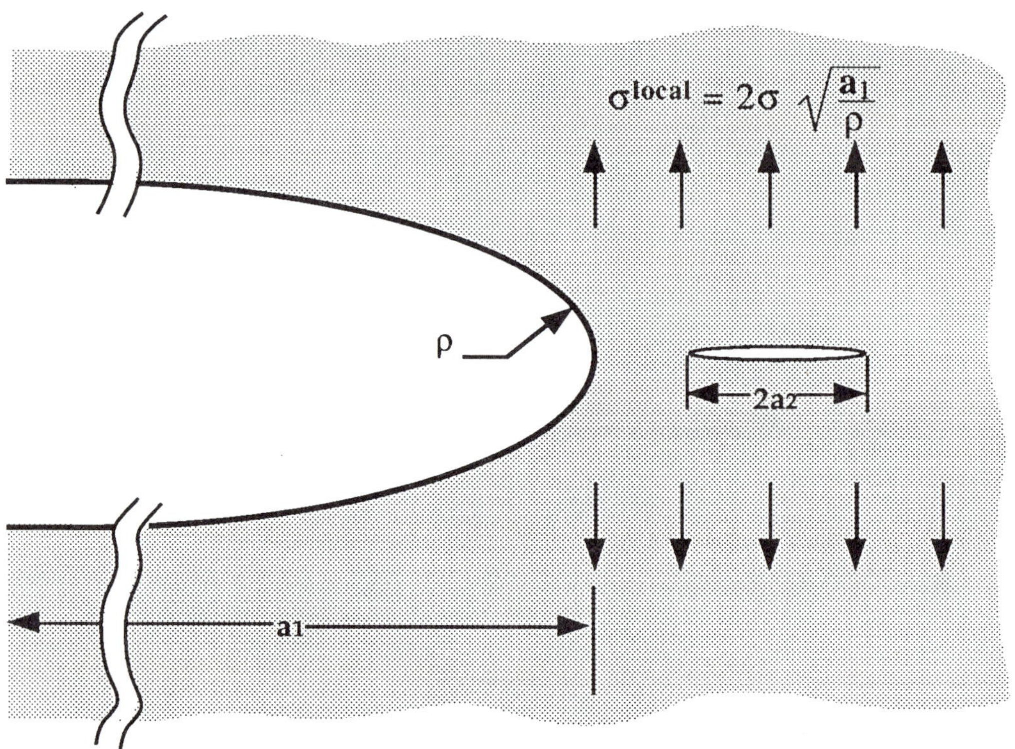

FIGURE 2.5 A sharp microcrack at the tip of a macroscopic crack.

EXAMPLE 2.1

A flat plate made from a brittle material contains a macroscopic through-thickness crack with half length a_1 and notch tip radius ρ. A sharp penny-shaped microcrack with radius a_2 is located near the tip of the larger flaw, as illustrated in Fig. 2.5. Estimate the minimum size of the microcrack to cause failure in the plate when the Griffith equation is satisfied by the global stress and a_1.

Solution: The nominal stress at failure is obtained by substituting a_1 into Eq. (2.19). The stress in the vicinity of the microcrack can be estimated from Eq. (2.11), which is set equal to the Griffith criterion for the penny-shaped microcrack (Eq. 2.20):

EXAMPLE 2.1 (cont.)

$$2\left(\frac{2\ E\ \gamma_s}{\pi\ a_1}\right)^{1/2}\sqrt{\frac{a_1}{\rho}} = \left(\frac{\pi\ E\ \gamma_s}{2(1-\nu^2)\ a_2}\right)^{1/2}$$

Solving for a_2 gives

$$a_2 = \frac{\pi^2\ \rho}{16(1-\nu^2)}$$

For $\nu = 0.3$, $a_2 = 0.68\ \rho$. Thus the nucleating microcrack must be approximately the size of the macroscopic crack tip radius.

This derivation contains a number of simplifying assumptions. The notch tip stress computed from Eq. (2.11) is assumed to act uniformly ahead of the notch, in the region of the microcrack; the actual stress would decay away from the notch tip. Also, this derivation neglects free boundary effects from the tip of the macroscopic notch.

2.3.2 Modified Griffith Equation

Equation (2.19) is valid only for ideally brittle solids. Griffith obtained good agreement between Eq. (2.19) and experimental fracture strengths of glass, but the Griffith equation severely underestimates the fracture strength of metals.

Irwin [4] and Orowan [5] independently modified the Griffith expression to account for materials that are capable of plastic flow. The revised expression is given by

$$\sigma_f = \left(\frac{2\ E\ (\gamma_s + \gamma_p)}{\pi\ a}\right)^{1/2} \tag{2.21}$$

where γ_p is the plastic work per unit area of surface created, and is typically much larger than γ_s.

In an ideally brittle solid, a crack can be formed merely by breaking atomic bonds; γ_s reflects the total energy of broken bonds in a unit area.

When a crack propagates through a metal, however, dislocation motion occurs in the vicinity of the crack tip, resulting in additional energy dissipation.

Although, Irwin and Orowan originally derived Eq. (2.21) for metals, it is possible to generalize the Griffith model to account for any type of energy dissipation:

$$\sigma_f = \left(\frac{2\,E\,w_f}{\pi\,a} \right)^{1/2} \tag{2.22}$$

where w_f is the fracture energy, which could include plastic, viscoelastic, or viscoplastic effects, depending on the material. The fracture energy can also be influenced by crack meandering and branching, which increase the the surface area. Figure 2.6 illustrates various types of material behavior and the corresponding fracture energy.

A word of caution is necessary when applying Eq. (2.22) to materials that exhibit nonlinear deformation. The Griffith model, in particular Eq. (2.16), applies only to linear elastic material behavior. Thus the global behavior of the structure must be elastic. Any nonlinear effects, such as plasticity, must be confined to a small region near the crack tip. In addition, Eq. (2.22) assumes that w_f is constant; in many ductile materials, the fracture energy increases with crack growth, as discussed in Section 2.5.

2.4 THE ENERGY RELEASE RATE

In 1956, Irwin [6] proposed an energy approach for fracture that is essentially equivalent to the Griffith model, except that Irwin's approach is in a form that is more convenient for solving engineering problems. Irwin defined an *energy release rate*, $\mathcal{G}$, which is a measure of the energy available for an increment of crack extension:

$$\mathcal{G} = -\frac{d\Pi}{d\mathcal{A}} \tag{2.23}$$

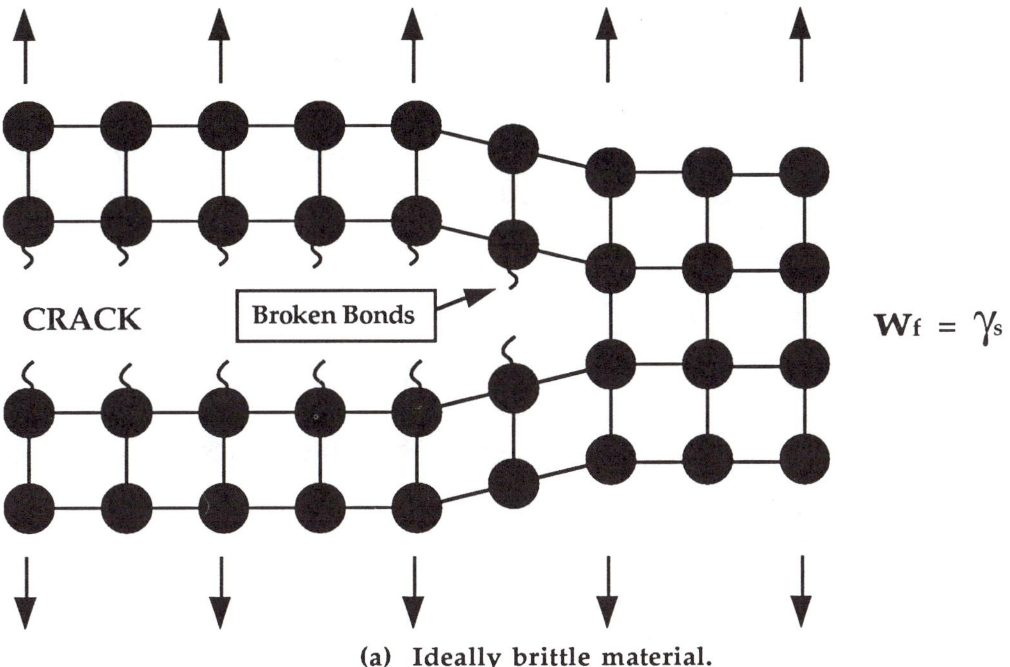

(a) Ideally brittle material.

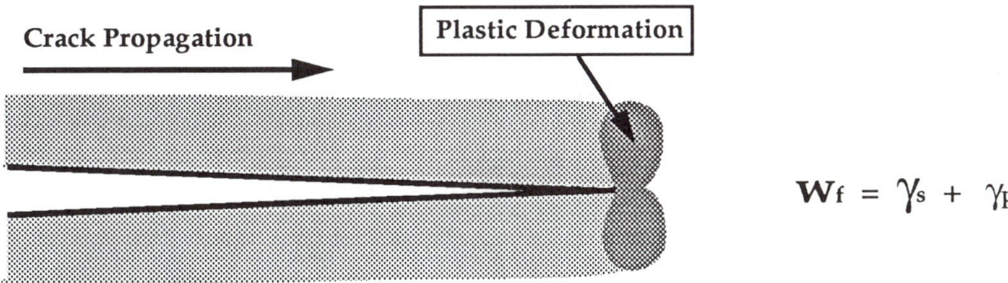

(b) Quasi-brittle elastic-plastic material.

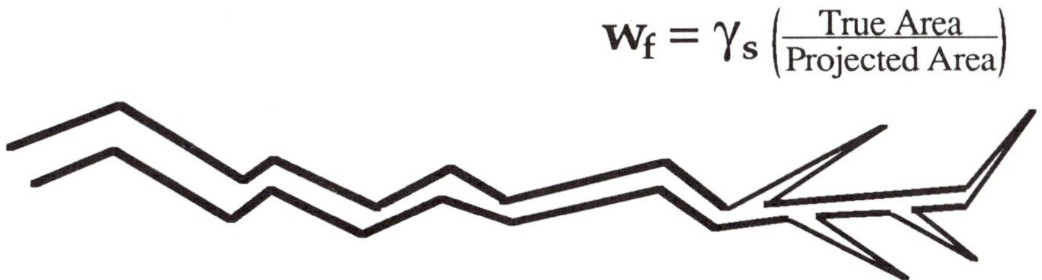

(c) Brittle material with crack meandering and branching.

FIGURE 2.6 Crack propagation in various types of materials, with the corresponding fracture energy.

The term *rate,* as it is used in this context, does not refer to a derivative with respect to time; $\mathcal{G}$ is the rate of change in potential energy with crack area. Since $\mathcal{G}$ is obtained from the derivative of a potential, it is also called the *crack extension force* or the *crack driving force.* According to Eq. (2.18a), the energy release rate for a wide plate in plane stress with a crack of length 2a (Fig. 2.3) is given by

$$\mathcal{G} = \frac{\pi \sigma^2 a}{E} \tag{2.24}$$

Referring to the previous section, crack extension occurs when $\mathcal{G}$ reaches a critical value; i.e.,

$$\mathcal{G}_c = \frac{dW_s}{d\mathcal{A}} = 2\,w_f \tag{2.25}$$

where $\mathcal{G}_c$ is a measure of the *fracture toughness* of the material.

The potential energy of an elastic body, Π, is defined as follows:

$$\Pi = U - F \tag{2.26}$$

where U is the strain energy stored in the body and F is the work done by external forces.

Consider a cracked plate that is dead loaded, as illustrated in Fig. 2.7. Since the load is fixed at P, the structure is said to be *load controlled.* For this case,

$$F = P\Delta$$

and

$$U = \int_0^\Delta P\,d\Delta = \frac{P\Delta}{2}$$

Therefore,

$$\Pi = -U$$

and

$$\mathcal{G} = \frac{1}{B}\left(\frac{dU}{da}\right)_P = \frac{P}{2B}\left(\frac{d\Delta}{da}\right)_P \tag{2.27}$$

When displacement is fixed (Fig. 2.8), the plate is *displacement controlled;* F = 0 and Π = U. Thus

$$\mathcal{G} = -\frac{1}{B}\left(\frac{dU}{da}\right)_{\Delta} = -\frac{\Delta}{2B}\left(\frac{dP}{da}\right)_{\Delta} \tag{2.28}$$

It is convenient at this point to introduce the compliance, which is the inverse of the plate stiffness:

$$C = \frac{\Delta}{P} \tag{2.29}$$

By substituting Eq. (2.29) into Eqs. (2.27) and (2.28) it can easily be shown that

$$\mathcal{G} = \frac{P^2}{2B}\frac{dC}{da} \tag{2.30}$$

for both load control and displacement control. Therefore, the energy release rate, as defined in Eq. (2.23), is the same for load control and displacement control. Also,

$$\left(\frac{dU}{da}\right)_P = -\left(\frac{dU}{da}\right)_{\Delta} \tag{2.31}$$

Equation (2.31) is demonstrated graphically in Figs. 2.7b and 2.8b. In load control, a crack extension da results in a net *increase* in strain energy because of the contribution of the external force P:

$$(dU)_P = Pd\Delta - \frac{Pd\Delta}{2} = \frac{Pd\Delta}{2}$$

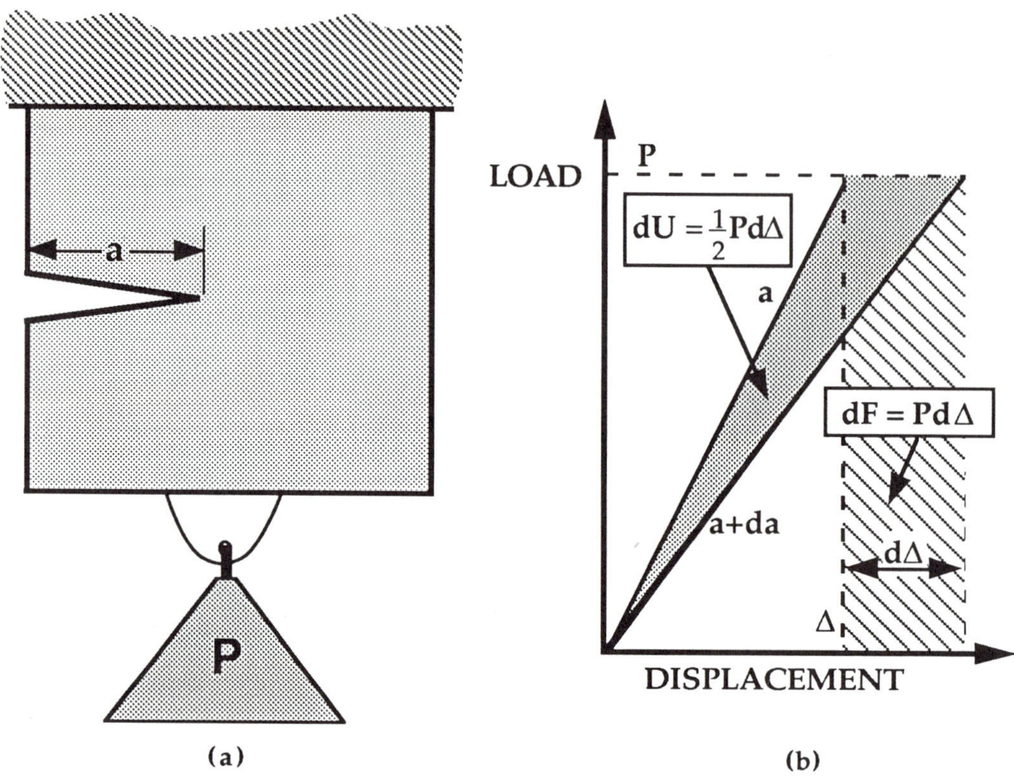

FIGURE 2.7 Cracked plate at a fixed load, P.

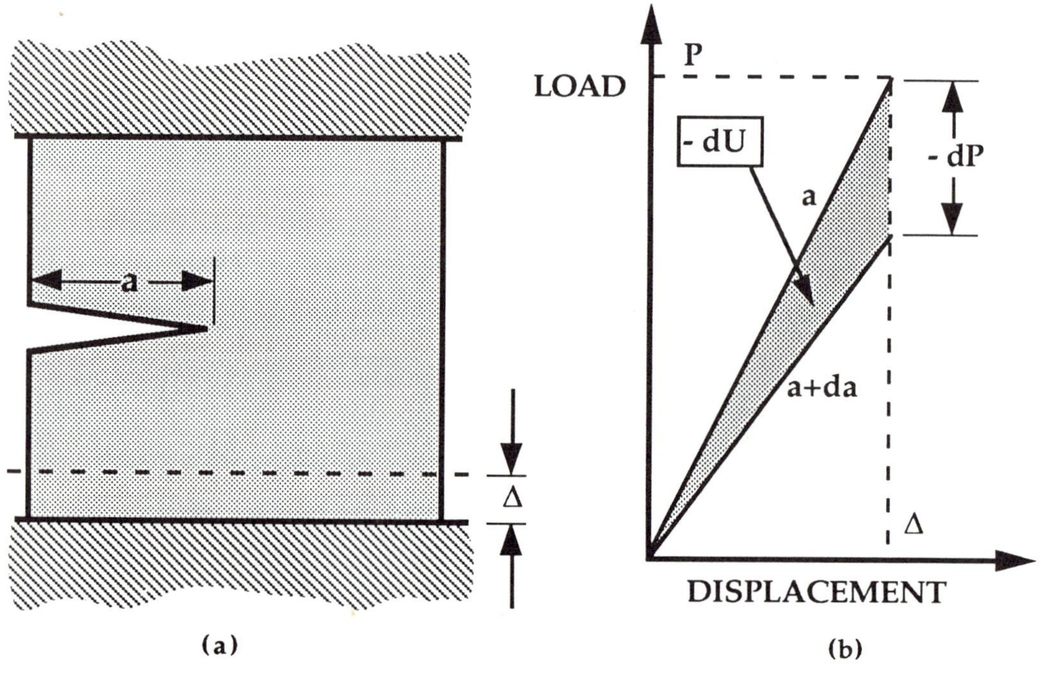

FIGURE 2.8 Cracked plate at a fixed displacement, Δ.

When displacement is fixed, dF = 0 and the strain energy *decreases*:

$$(dU)_\Delta = \frac{\Delta dP}{2}$$

where dP is negative. As can be seen in Figs 2.7b and 2.8b, the absolute values of these energies differ by the amount $dPd\Delta/2$, which is negligible. Thus

$$(dU)_P = -(dU)_\Delta$$

for an increment of crack growth at a given P and Δ.

EXAMPLE 2.2

Determine the energy release rate for a double cantilever beam (DCB) specimen (Fig. 2.9)

Solution: From beam theory,

$$\frac{\Delta}{2} = \frac{P\,a^3}{3\,E\,I} \quad \text{where} \quad I = \frac{B\,h^3}{12}$$

The elastic compliance is given by

$$C = \frac{\Delta}{P} = \frac{2\,a^3}{3\,E\,I}$$

Substituting C into Eq. 2.30 gives

$$\mathcal{G} = \frac{P^2\,a^2}{B\,E\,I} = \frac{12\,P^2\,a^2}{B^2\,h^3\,E}$$

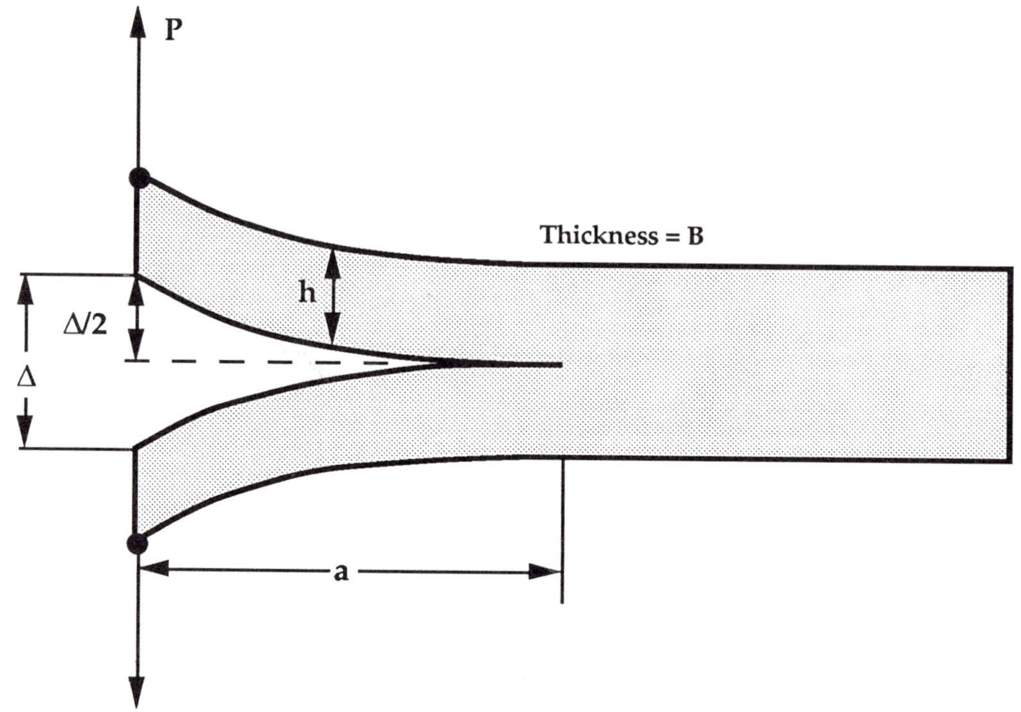

FIGURE 2.9 Double cantilever beam (DCB) specimen.

2.5 INSTABILITY AND THE R CURVE

Crack extension occurs when $\mathcal{G} = 2w_f$; but crack growth may be stable or unstable, depending on how $\mathcal{G}$ and w_f vary with crack size. To illustrate stable and unstable behavior, it is convenient to replace $2w_f$ with R, the material resistance to crack extension. A plot of R versus crack extension is called a *resistance curve* or *R curve*. The corresponding plot of G versus crack extension is the *driving force curve.*

Consider a wide plate with a through crack of initial length $2a_o$ (Fig. 2.3). At a fixed remote stress, σ, the energy release rate varies linearly with crack size (Eq. (2.24). Figure 2.10 shows schematic driving force/R curves for two types of material behavior.

The first case, Fig. 2.10a, shows a flat R curve, where the material resistance is constant with crack growth. When the stress = σ_1, the crack is stable. Fracture occurs when the stress reaches σ_2; the crack propagation is unstable because the driving force increases with crack growth, but the material resistance remains constant.

Figure 2.10b illustrates a material with a rising R curve. The crack grows a small amount when the stress reaches σ_2, but cannot grow further unless the stress increases. The driving force, when stress is fixed at σ_2, increases at a slower rate than R. Stable crack growth continues as the stress increases to σ_3. Finally, when the stress reaches σ_4, the driving force curve is tangent to the R curve. The plate is unstable with further crack growth because the rate of change in driving force exceeds the slope of the R curve.

The conditions for *stable* crack growth can be expressed as follows:

$$\mathcal{G} = R \tag{2.32a}$$

and

$$\frac{d\mathcal{G}}{da} \leq \frac{dR}{da} \tag{2.32b}$$

Unstable crack growth occurs when

$$\frac{d\mathcal{G}}{da} > \frac{dR}{da} \tag{2.33}$$

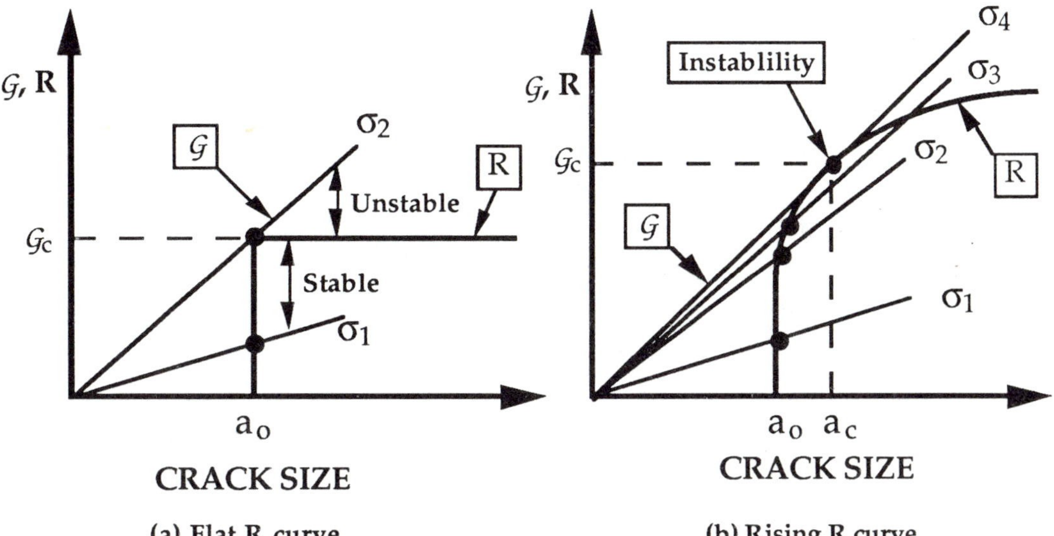

(a) Flat R curve **(b) Rising R curve**

FIGURE 2.10 Schematic driving force/R curve diagrams.

When the resistance curve is flat, as in Fig 2.10a, one can define a critical value of energy release rate, $\mathcal{G}_C$, unambiguously. A material with a rising R curve, however, cannot be uniquely characterized with a single toughness value. According to Eq. (2.33) a flawed structure fails when the driving force curve is tangent with the R curve, but this point of tangency depends on the shape of the driving force curve, which depends on configuration of the structure. The driving force curve for the through crack configuration is linear, but $\mathcal{G}$ in the DCB specimen (Example 2.2) varies with a^2; these two configurations would have different $\mathcal{G}_C$ values for a given R curve.

Materials with rising R curves can be characterized by the value of $\mathcal{G}$ at initiation of crack growth. Although the initiation toughness is usually not sensitive to structural geometry, there are other problems with this measurement. It is virtually impossible to determine the precise moment of crack initiation in most materials; an engineering definition of initiation, analogous to the 0.2 percent offset yield strength in tensile tests, is usually required. Another limitation of initiation toughness is that it characterizes only the onset of crack growth; it provides no information on the shape of the R curve.

2.5.1 Reasons for the R Curve Shape

Some materials exhibit a rising R curve, while the R curve for other materials is flat. The shape of the R curve depends on material behavior and, to a lesser extent, on the configuration of the cracked structure.

The R curve for an ideally brittle material is flat because the surface energy is an invariant material property. When nonlinear material behavior accompanies fracture, however, the R curve can take on a variety of shapes. For example, ductile fracture in metals usually results in a rising R curve; a plastic zone at the tip of the crack increases in size as the crack grows. The driving force must increase in such materials to maintain crack growth. If the cracked body is infinite (i.e. if the plastic zone is small compared to relevant dimensions of the body) the plastic zone size and R eventually reach steady-state values, and the R curve becomes flat with further growth (see Section 3.5.2).

Some materials can display a falling R curve. When a metal fails by cleavage, for example, the material resistance is provided by the surface energy and local plastic dissipation, as illustrated in Fig. 2.6b. The

R curve would be relatively flat if the crack growth were stable. However, cleavage propagation is normally unstable; the material near the tip of the growing crack is subject to very high strain rates, which suppress plastic deformation. Thus the resistance of a rapidly growing cleavage crack is less than the initial resistance at the onset of fracture.

The size and geometry of the cracked structure can exert some influence on the shape of the R curve. A crack in a thin sheet tends to produce a steeper R curve than a crack in a thick plate because the thin sheet is loaded predominantly in plane stress, while material near the tip of the crack in the thick plate may be in plane strain. The R curve can also be affected if the growing crack approaches a free boundary in the structure; Thus a wide plate may exhibit a somewhat different crack growth resistance behavior than a narrow plate of the same material.

Ideally, the R curve, as well as other measures of fracture toughness, should only be a property of the material and not depend on the size or shape of the cracked body. Much of fracture mechanics is predicated on the assumption that fracture toughness is a material property. Configurational effects can occur, however; a practitioner of fracture mechanics should be aware of these effects and their potential influence on the accuracy of an analysis. This issue is explored in detail in Sections 2.10, 3.5, and 3.6.

2.5.2 Load Control Versus Displacement Control

According to Eqs. (2.32) and (2.33), the stability of crack growth depends on the rate of change in $\mathcal{G}$, i.e., the second derivative of potential energy. Although the driving force ($\mathcal{G}$) is the same for both load control and displacement control, the *rate of change* of the driving force curve depends on how the structure is loaded.

Displacement control tends to be more stable than load control. With some configurations, the driving force actually decreases with crack growth in displacement control. A typical example is illustrated in Fig. 2.11.

Referring to Fig. 2.11, consider a cracked structure subjected to a load P_3 and a displacement Δ_3. If the structure is load controlled, it is at the point of instability, where the driving force curve is tangent to the R curve. In displacement control, however the structure is stable because the driving force decreases with crack growth; the displacement must be increased for further crack growth.

When an R curve is determined experimentally, the specimen is usually tested in displacement control (or as near to pure displacement control as is possible in the test machine). Since most of the common test specimen geometries exhibit falling driving force curves in displacement control, it is possible to obtain a significant amount of stable crack growth. If an instability occurs during the test, the R curve cannot be defined beyond the point of ultimate failure.

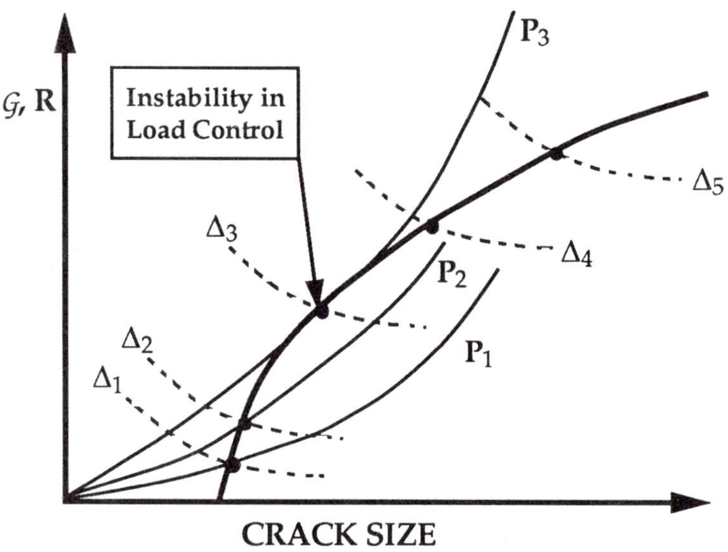

FIGURE 2.11 Schematic driving force/R curve diagram which compares load control and displacement control.

EXAMPLE 2.3

Evaluate the relative stability of a DCB specimen (Fig. 2.9) in load control and displacement control.

Solution: From the result derived in Example 2.2, the slope of the driving force curve in load control is given by

EXAMPLE 2.3 (cont.)

$$\left(\frac{d\mathcal{G}}{da}\right)_P = \frac{2\,P^2\,a}{B\,E\,I} = \frac{2\,\mathcal{G}}{a}$$

In order to evaluate displacement control, it is necessary to express $\mathcal{G}$ in terms of Δ and a. From beam theory, load is related to displacement as follows:

$$P = \frac{3\,\Delta\,E\,I}{2\,a^3}$$

substituting the above equation into expression for energy release rate gives

$$\mathcal{G} = \frac{9\,\Delta^2\,E\,I}{4\,B\,a^4}$$

Thus

$$\left(\frac{d\mathcal{G}}{da}\right)_\Delta = -\frac{9\,\Delta^2\,E\,I}{B\,a^5} = -\frac{4\,\mathcal{G}}{a}$$

Therefore, the driving force increases with crack growth in load control and decreases in displacement control. For a flat R curve, crack growth in load control is always unstable, while displacement control is always stable.

2.5.3 Structures with Finite Compliance

Most real structures are subject to conditions between pure load control and pure displacement control. This intermediate situation can be schematically represented by a spring in series with the flawed structure (Fig. 2.12). The structure is fixed at a constant remote displacement, Δ_T; the spring represents the system compliance, C_m. Pure displacement control corresponds to an infinitely stiff spring, where $C_m = 0$. Load control (dead loading) implies an infinitely soft spring; i.e., $C_m = \infty$.

When the system compliance is finite, the point of fracture instability obviously lies somewhere between the extremes of pure load control and pure displacement control. However, determining the precise point of instability requires a rather complex analysis.

At the moment of instability, the following conditions are satisfied:

$$\mathcal{G} = R \tag{2.34a}$$

and

$$\left(\frac{d\mathcal{G}}{da}\right)_{\Delta_T} = \frac{dR}{da} \tag{2.34b}$$

The left side of Eq. (2.34b) is given by [7]:

$$\left(\frac{d\mathcal{G}}{da}\right)_{\Delta_T} = \left(\frac{\partial\mathcal{G}}{\partial a}\right)_P - \left(\frac{\partial\mathcal{G}}{\partial P}\right)_a \left(\frac{\partial\Delta}{\partial a}\right)_P \left[C_m + \left(\frac{\partial\Delta}{\partial P}\right)_a\right]^{-1} \tag{2.35}$$

Equation (2.35) is derived in Appendix 2.2.

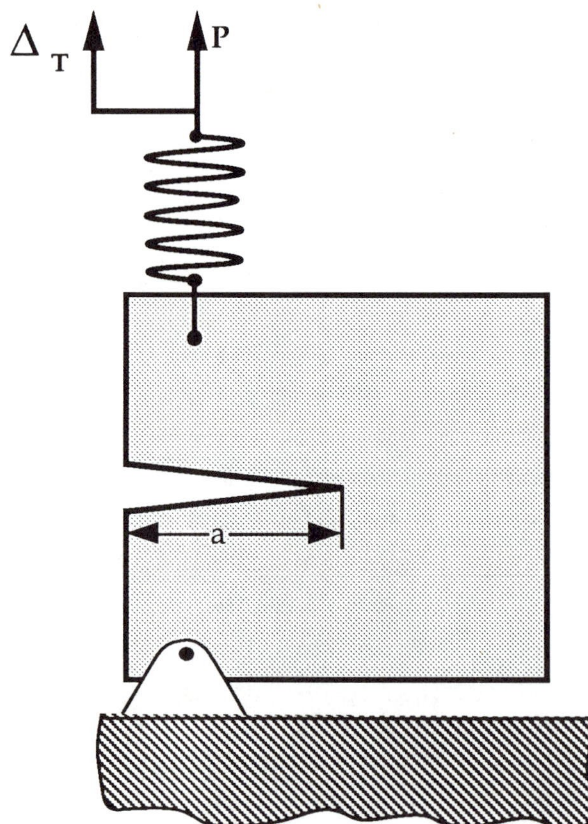

FIGURE 2.12 A cracked structure with finite compliance, represented schematically by a spring in series.

2.6 STRESS ANALYSIS OF CRACKS

For certain cracked configurations subjected to external forces, it is possible to derive closed-form expressions for the stresses in the body, assuming isotropic linear elastic material behavior. Westergaard [8], Irwin [9], Sneddon [10] and Williams [11] were among the first to publish such solutions. If we define a polar coordinate axis with the origin at the crack tip (Fig. 2.13), it can be shown that the stress field in any linear elastic cracked body is given by

$$\sigma_{ij} = \left(\frac{k}{\sqrt{r}}\right) f_{ij}(\theta) + \text{other terms} \tag{2.36}$$

where σ_{ij} is the stress tensor, r and θ are as defined in Fig. 2.13, k is a constant, and f_{ij} is a dimensionless function of θ. The higher order terms depend on geometry, but the solution for any given configuration contains a leading term that is proportional to $1/\sqrt{r}$. As $r \to 0$, the leading term approaches infinity, but the other terms remain finite or approach zero. Thus stress near the crack tip varies with $1/\sqrt{r}$, regardless of the configuration of the cracked body. It can also be shown that displacement near the crack tip varies with $\sqrt{r}$. Equation (2.36) describes a stress *singularity*, since stress is asymptotic to $r = 0$. The basis of this relationship is explored in more detail in Appendix 2.3.

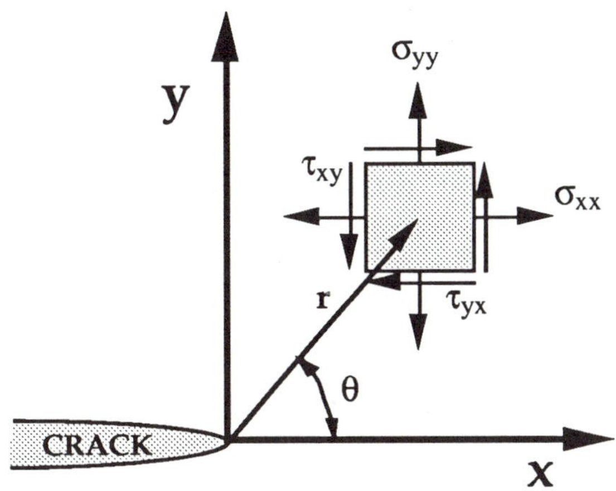

FIGURE 2.13 Definition of the coordinate axis ahead of a crack tip. The z direction is normal to the page.

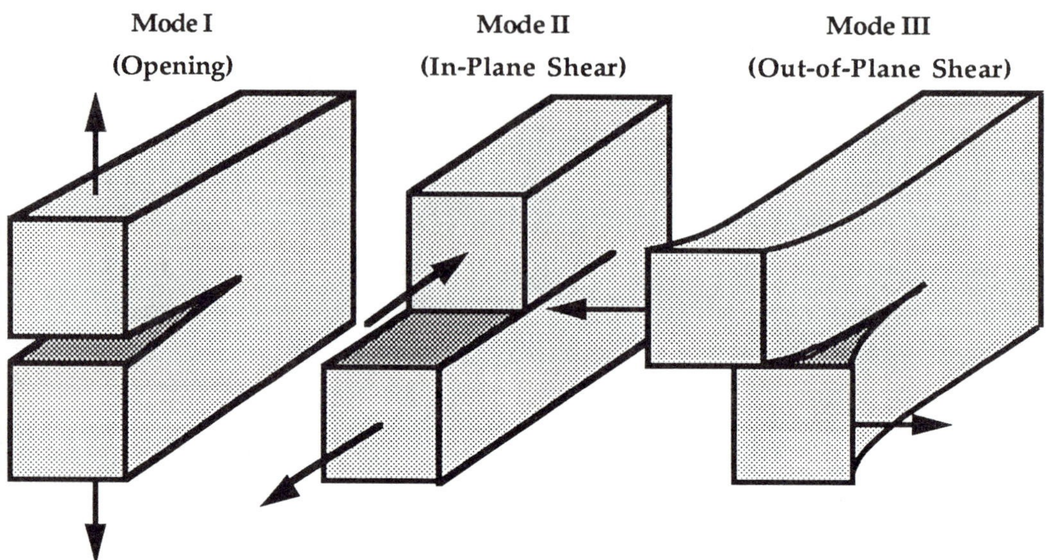

FIGURE 2.14 The three modes of loading that can be applied to a crack.

There are three types of loading that a crack can experience, as illustrated in Fig. 2.14. Mode I loading, where the principal load is applied normal to the crack plane, tends to open the crack. Mode II corresponds to in-plane shear loading and tends to slide one crack face with respect to the other. Mode III refers to out-of-plane shear. A cracked body can be loaded in any one of these modes, or a combination of two or three modes.

2.6.1 The Stress Intensity Factor

Each mode of loading produces the $1/\sqrt{r}$ singularity at the crack tip, but the proportionality constant, k, and f_{ij} depend on mode. It is convenient at this point to replace k by the *stress intensity factor*, K, where $K = k\sqrt{2\pi}$. The stress intensity factor is usually given a subscript to denote the mode of loading; i.e., K_I, K_{II}, or K_{III}. Thus the stress fields ahead of a crack tip in an isotropic linear elastic material can be written as

$$\lim_{r \to 0} \sigma_{ij}^{(I)} = \frac{K_I}{\sqrt{2\pi r}} f_{ij}^{(I)}(\theta) \tag{2.37a}$$

$$\lim_{r \to 0} \sigma_{ij}^{(II)} = \frac{K_{II}}{\sqrt{2\pi r}} f_{ij}^{(II)}(\theta) \tag{2.37b}$$

$$\lim_{r \to 0} \sigma_{ij}^{(III)} = \frac{K_{III}}{\sqrt{2\pi r}} f_{ij}^{(III)}(\theta) \tag{2.37c}$$

for Modes I, II, and III, respectively. In a mixed-mode problem (i.e., when more than one loading mode is present), the individual contributions to a given stress component are additive:

$$\sigma_{ij}^{(total)} = \sigma_{ij}^{(I)} + \sigma_{ij}^{(II)} + \sigma_{ij}^{(III)} \tag{2.38}$$

Equation (2.38) stems from the principle of linear superposition.

Detailed expressions for the singular stress fields for Modes I and II are given in Table 2.1. Displacement relationships for Modes I and II are listed in Table 2.2. Table 2.3 lists the nonzero stress and displacement components for Mode III.

Consider the Mode I singular field on the crack plane, where $\theta = 0$. According to Table 2.1, the stresses in the x and y direction are equal:

$$\sigma_{xx} = \sigma_{yy} = \frac{K_I}{\sqrt{2\pi r}} \tag{2.39}$$

When $\theta = 0$, the shear stress is zero, which means that the crack plane is a principal plane for pure Mode I loading. Figure 2.15 is a schematic plot of σ_{yy}, the stress normal to the crack plane, versus distance from the crack tip. Equation (2.39) is only valid near the crack tip, where the $1/\sqrt{r}$ singularity dominates the stress field. Stresses far from the crack tip are governed by the remote boundary conditions. For example, if the cracked structure is subjected to a uniform remote tensile stress, σ_{yy} approaches a constant value, σ^{∞}. We can define a *singularity dominated zone* as the region where the equations in Tables 2.1 to 2.3 describe the crack tip fields.

TABLE 2.1

Stress fields ahead of a crack tip for Mode I and Mode II in a linear elastic, isotropic material.

	Mode I	Mode II
σ_{xx}	$\frac{K_I}{\sqrt{2\pi r}}\cos\left(\frac{\theta}{2}\right)\left[1 - \sin\left(\frac{\theta}{2}\right)\sin\left(\frac{3\theta}{2}\right)\right]$	$-\frac{K_{II}}{\sqrt{2\pi r}}\sin\left(\frac{\theta}{2}\right)\left[2 + \cos\left(\frac{\theta}{2}\right)\cos\left(\frac{3\theta}{2}\right)\right]$
σ_{yy}	$\frac{K_I}{\sqrt{2\pi r}}\cos\left(\frac{\theta}{2}\right)\left[1 + \sin\left(\frac{\theta}{2}\right)\sin\left(\frac{3\theta}{2}\right)\right]$	$\frac{K_{II}}{\sqrt{2\pi r}}\sin\left(\frac{\theta}{2}\right)\cos\left(\frac{\theta}{2}\right)\cos\left(\frac{3\theta}{2}\right)$
τ_{xy}	$\frac{K_I}{\sqrt{2\pi r}}\cos\left(\frac{\theta}{2}\right)\sin\left(\frac{\theta}{2}\right)\cos\left(\frac{3\theta}{2}\right)$	$\frac{K_{II}}{\sqrt{2\pi r}}\cos\left(\frac{\theta}{2}\right)\left[1 - \sin\left(\frac{\theta}{2}\right)\sin\left(\frac{3\theta}{2}\right)\right]$
σ_{zz}	0 (Plane Stress) $\nu(\sigma_{xx} + \sigma_{yy})$ (Plane Strain)	0 (Plane Stress) $\nu(\sigma_{xx} + \sigma_{yy})$ (Plane Strain)
τ_{xz}, τ_{yz}	0	0

ν is Poisson's ratio.

TABLE 2.2

Crack tip displacement fields for Mode I and Mode II (linear elastic, isotropic material).

	Mode I	Mode II
u_x	$\frac{K_I}{2\mu}\sqrt{\frac{r}{2\pi}}\cos\left(\frac{\theta}{2}\right)\left[\kappa - 1 + 2\sin^2\left(\frac{\theta}{2}\right)\right]$	$\frac{K_{II}}{2\mu}\sqrt{\frac{r}{2\pi}}\sin\left(\frac{\theta}{2}\right)\left[\kappa + 1 + 2\cos^2\left(\frac{\theta}{2}\right)\right]$
u_y	$\frac{K_I}{2\mu}\sqrt{\frac{r}{2\pi}}\sin\left(\frac{\theta}{2}\right)\left[\kappa + 1 - 2\cos^2\left(\frac{\theta}{2}\right)\right]$	$-\frac{K_{II}}{2\mu}\sqrt{\frac{r}{2\pi}}\cos\left(\frac{\theta}{2}\right)\left[\kappa - 1 - 2\sin^2\left(\frac{\theta}{2}\right)\right]$

μ is the shear modulus

$\kappa = 3 - 4\nu$ (plane strain)

$\kappa = (3 - \nu)/(1 + \nu)$ (plane stress)

$\tau_{xz} = -\dfrac{K_{III}}{\sqrt{2\pi r}} \sin\left(\dfrac{\theta}{2}\right)$
$\tau_{yz} = \dfrac{K_{III}}{\sqrt{2\pi r}} \cos\left(\dfrac{\theta}{2}\right)$
$u_z = \dfrac{K_{III}}{\mu} \sqrt{\dfrac{r}{2\pi}} \sin\left(\dfrac{\theta}{2}\right)$

TABLE 2.3 Non-zero stress and displacement components in Mode III (linear elastic, isotropic material).

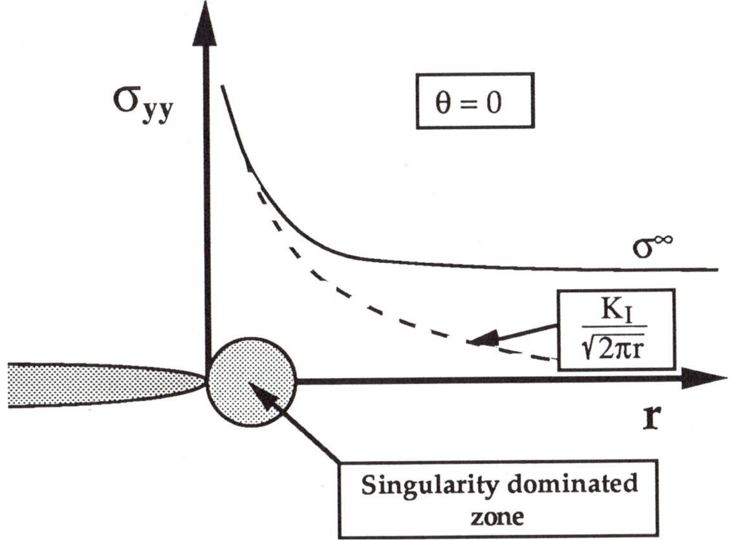

FIGURE 2.15 Stress normal to the crack plane in Mode I.

The stress intensity factor defines the amplitude of the crack tip singularity. That is, stresses near the crack tip increase in proportion to K. Moreover, the stress intensity factor completely defines the crack tip conditions; if K is known, it is possible to solve for all components of stress, strain, and displacement as a function of r and θ. This single parameter description of crack tip conditions turns out to be one of the most important concepts in fracture mechanics.

2.6.2 Relationship between K and Global Behavior

In order for the stress intensity factor to be useful, one must be able to determine K from remote loads and the geometry. Closed-form solutions for K have been derived for a number of simple configurations. For more complex situations the stress intensity factor can be estimated by experiment or numerical analysis (see Chapter 11).

One configuration for which a closed-form solution exists is a through crack in an infinite plate subjected to a remote tensile stress (Fig. 2.3). Since the remote stress, σ, is perpendicular to the crack plane, the loading is pure Mode I. Linear elastic bodies must undergo proportional stressing; i.e., all stress components at all locations increase in proportion to the remotely applied forces. Thus the crack tip stresses must be proportional to the remote stress, and $K_I \propto \sigma$. According to Eq. (2.37), stress intensity has units of stress•length$^{1/2}$. Since the only relevant length scale in Fig. 2.3 is the crack size, the relationship between K_I and the global conditions must have the following form:

$$K_I \propto \sigma \sqrt{a} \tag{2.40}$$

The actual solution, which is derived in Appendix 2.3, is given by

$$K_I = \sigma \sqrt{\pi a} \tag{2.41}$$

Thus the amplitude of the crack tip singularity for this configuration is proportional to the remote stress and the square root of crack size. The stress intensity factor for Mode II loading of the plate in Fig. 2.3 can be obtained by replacing σ in Eq. (2.41) by the remotely applied shear stress (see Fig 2.18 and Eq. (2.43) below).

A related solution is that for a semi-infinite plate with an edge crack (Fig. 2.16). Note that this configuration can be obtained by slicing the plate in Fig. 2.3 through the middle of the crack. The stress intensity factor for the edge crack is given by

$$K_I = 1.12\,\sigma \sqrt{\pi a} \tag{2.42}$$

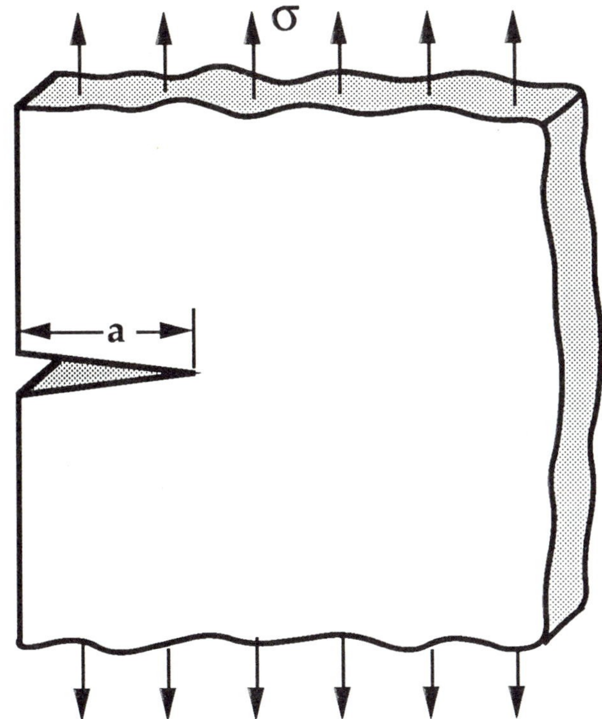

FIGURE 2.16 Edge crack in a semi-infinite plate subject to a remote tensile stress.

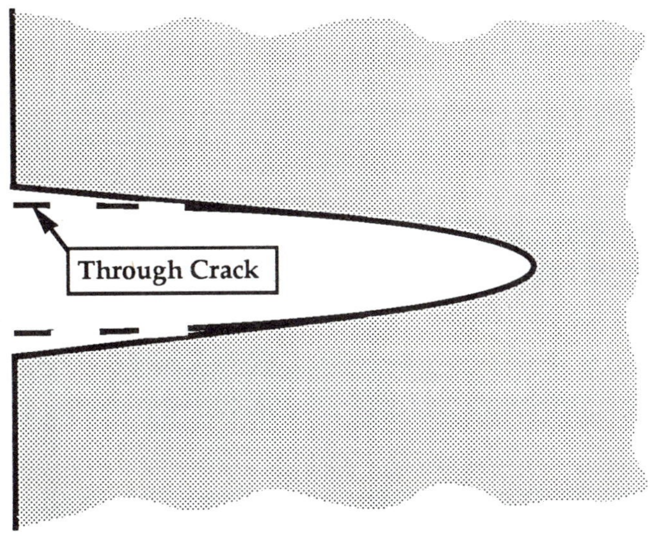

FIGURE 2.17 Comparison of crack opening displacements for an edge crack and through crack. The edge crack opens wider at a given stress, resulting in a stress intensity that is 12% higher.

which is similar to Eq. (2.41). The 12% increase in K_I for the edge crack is caused by different boundary conditions at the free edge. As Fig. 2.17 illustrates, the edge crack opens more because it is less restrained than the through crack, which forms an elliptical shape when loaded.

Consider a through crack in an infinite plate where the normal to the crack plane is oriented at an angle β with the stress axis (Fig. 2.18a). If $\beta \neq 0$, the crack experiences combined Mode I and Mode II loading; $K_{III} = 0$ as long as the stress axis and the crack normal both lie in the plane of the plate. If we redefine the coordinate axis to coincide with the crack orientation (Fig. 2.18b), we see that the applied stress can be resolved into normal and shear components. The stress normal to the crack plane, $\sigma_{y'y'}$ produces pure Mode I loading, while $\tau_{y'x'}$ applies Mode II loading to the crack. The stress intensity factors for the plate in Fig. 2.18 can be inferred by relating $\sigma_{y'y'}$ and $\tau_{y'x'}$ to σ and β through Mohr's circle:

$$K_I = \sigma_{y'y'} \sqrt{\pi a}$$

$$= \sigma \cos^2(\beta) \sqrt{\pi a} \tag{2.43a}$$

and

$$K_{II} = \tau_{y'x'} \sqrt{\pi a}$$

$$= \sigma \sin(\beta) \cos(\beta) \sqrt{\pi a} \tag{2.43b}$$

Note that Eq. (2.43) reduces to the pure Mode I solution when $\beta = 0$. The maximum K_{II} occurs at $\beta = 45°$, where the shear stress is also at a maximum.

The penny-shaped crack in an infinite medium (Fig. 2.4) is another configuration for which a closed-form K_I solution exists [11]:

$$K_I = \frac{2}{\pi} \sigma \sqrt{\pi a} \tag{2.44}$$

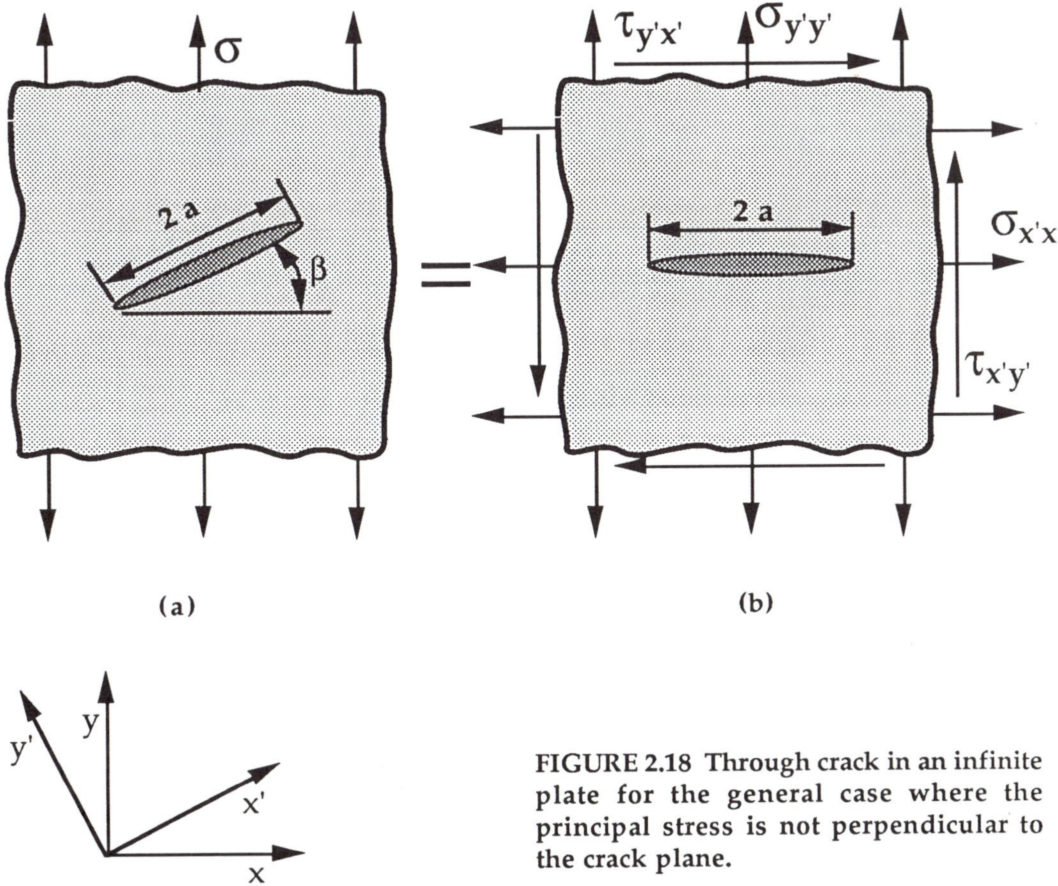

FIGURE 2.18 Through crack in an infinite plate for the general case where the principal stress is not perpendicular to the crack plane.

where a is crack radius. Note that Eq. (2.44) has the same form as the previous relationships for a through crack, except that the crack radius is the characteristic length in the above equation. The stress intensity factor for a semicircular surface crack (in an infinite half-space) with radius a can be obtained by multiplying Eq. (2.44) by 1.12, the surface correction factor. The more general case of an elliptical or semi-elliptical flaw is illustrated in Fig. 2.19. In this instance, two length dimensions are needed to characterize the crack size: 2c and 2a, the major and minor axes of the ellipse, respectively (see Fig. 2.19). Also, when $a \neq c$, the stress intensity factor varies along the crack front, with the maximum K_I at $\phi = 90°$. The flaw shape parameter, ψ, is obtained from an elliptic integral, as discussed in Appendix 2.4 Figure 2.19 gives an approximate solution for ψ.

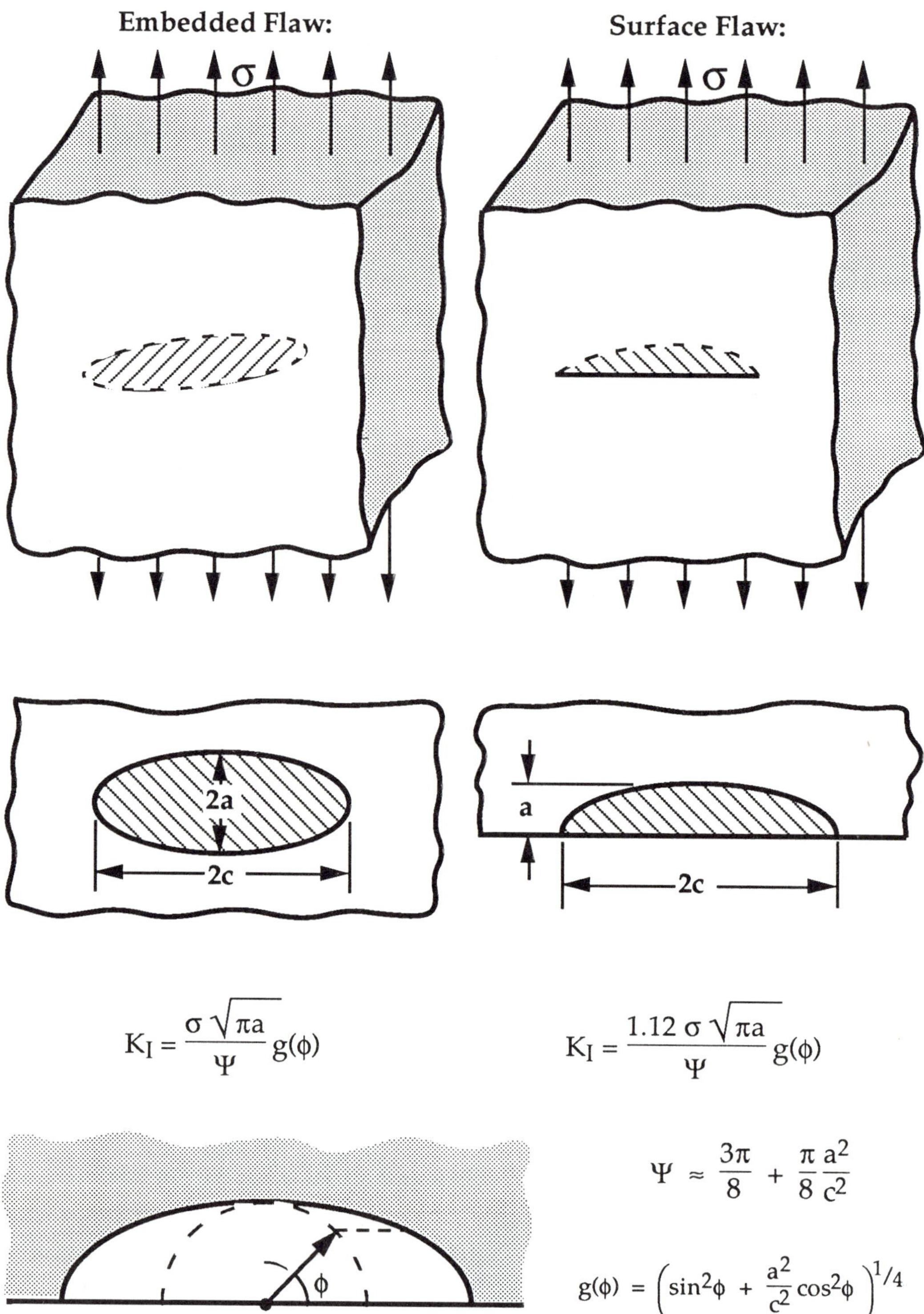

FIGURE 2.19 Mode I stress intensity factors for elliptical and semi-elliptical cracks. These solutions are only valid as long as the crack is small compared to the plate dimensions.

2.6.3 Effect of Finite Size

Most configurations for which there is a closed-form K solution consist of a crack with a simple shape (e.g. a rectangle or ellipse) in an infinite plate. Stated another way, the crack dimensions are small compared to the size of the plate; the crack tip conditions are not influenced by external boundaries. As the crack size increases, or as the plate dimensions decrease, the outer boundaries begin to exert an influence on the crack tip. In such cases, a closed-form stress intensity solution is usually not possible.

Consider a cracked plate subjected to a remote tensile stress. Figure 2.20 schematically illustrates the effect of finite width on the crack tip stress distribution, which is represented by lines of force; the local stress is proportional to the spacing between lines of force. Since a tensile stress cannot be transmitted through a crack, the lines of force are diverted around the crack, resulting in a local stress concentration. In the infinite plate, the line of force at a distance W from the crack center line has force components in the x and y directions. If the plate width is restricted to 2W, the x force must be zero on the free edge; this boundary condition causes the lines of force to be compressed, which results in a higher stress intensification at the crack tip.

One technique to approximate the finite width boundary condition is to assume a periodic array of collinear cracks in an infinite plate (Fig. 2.21). The Mode I stress intensity factor for this situation is given by

$$K_I = \sigma\sqrt{\pi a}\left[\frac{2W}{\pi a}\tan\left(\frac{\pi a}{2W}\right)\right]^{1/2} \tag{2.45}$$

The stress intensity approaches the infinite plate value as a/W approaches zero; K_I is asymptotic to a/W = 1.

More accurate solutions for a through crack in a finite plate have been obtained from finite element analysis; solutions of this type are usually fit to a polynomial expression. One such solution [12] is given by

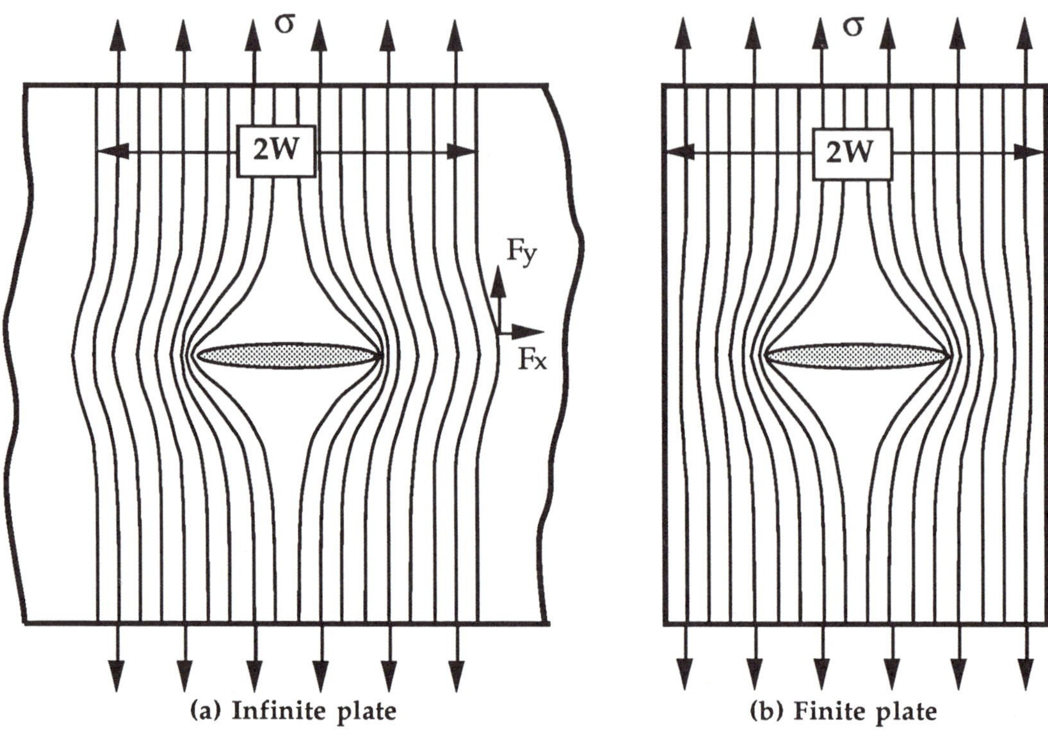

FIGURE 2.20 Stress concentration effects due to a through crack in finite and infinite width plates.

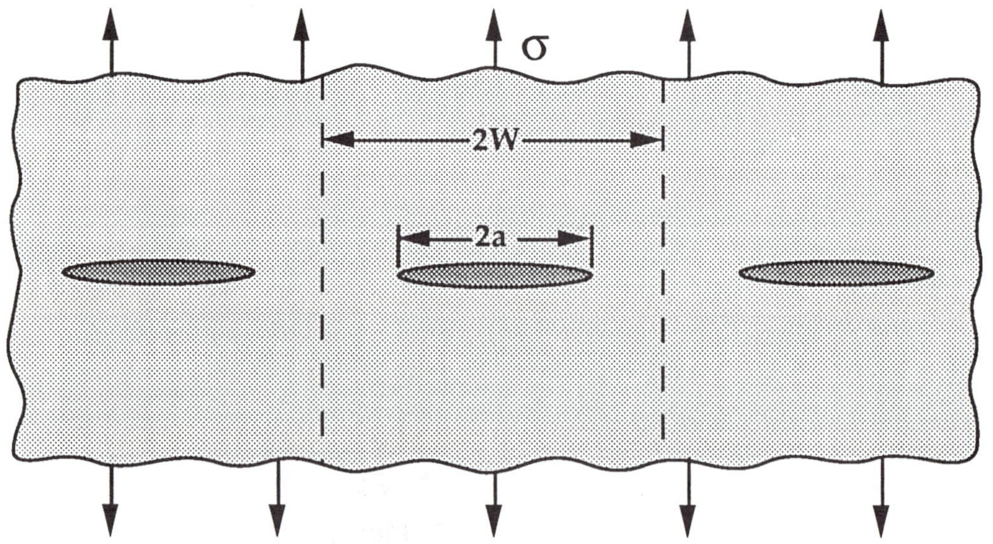

FIGURE 2.21 Collinear cracks in an infinite plate subject to remote tension.

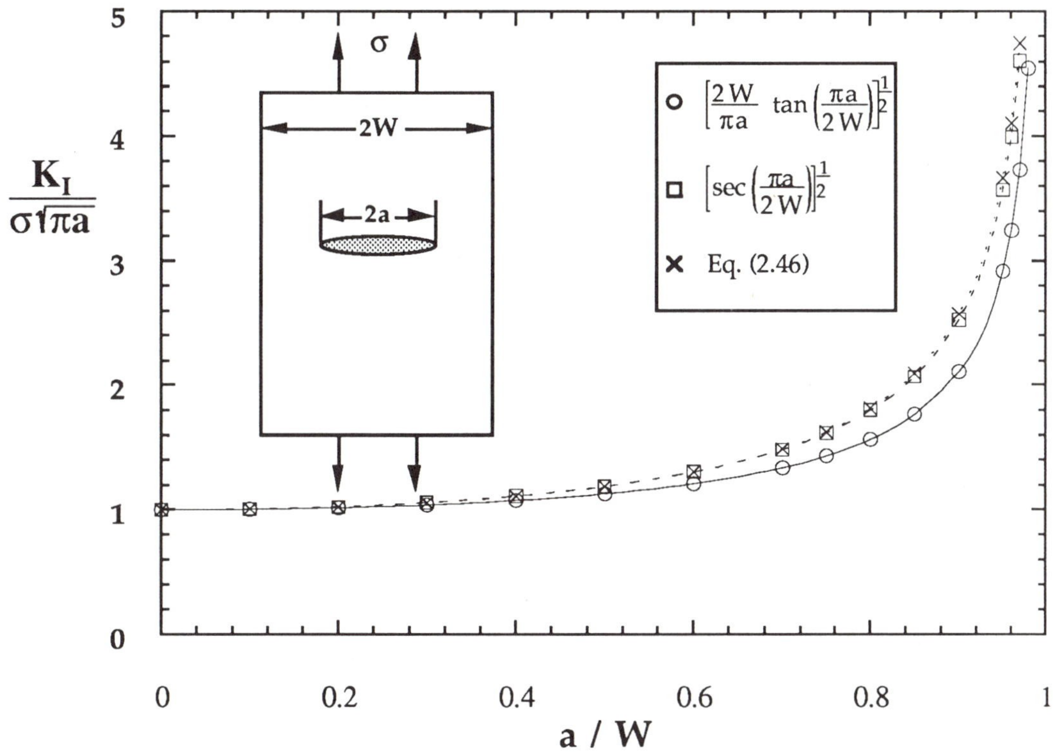

FIGURE 2.22 Comparison of finite width corrections for a center cracked plate in tension.

$$K_I = \sigma\sqrt{\pi a}\left[\sec\left(\frac{\pi a}{2W}\right)\right]^{1/2}\left[1 - 0.025\left(\frac{a}{W}\right)^2 + 0.06\left(\frac{a}{W}\right)^4\right] \quad (2.46)$$

Figure 2.22 compares the finite width corrections in Eqs. (2.45) and (2.46). The secant term (without the polynomial term) in Eq. (2.46) is also plotted. Equation (2.45) agrees with the finite element solution to within 7% for $a/W < 0.6$. The secant correction is much closer to the finite element solution; the error is less than 2% for $a/W < 0.9$. Thus the polynomial term in Eq. [2.46] contributes little and can be neglected in most cases.

Table 2.4 lists stress intensity solutions for several common configurations. Chapter 12 contains a more extensive collection of K solutions. Several handbooks devoted solely to stress intensity solutions have also been published [12-14].

TABLE 2.4
K_I solutions for common test specimens [12].

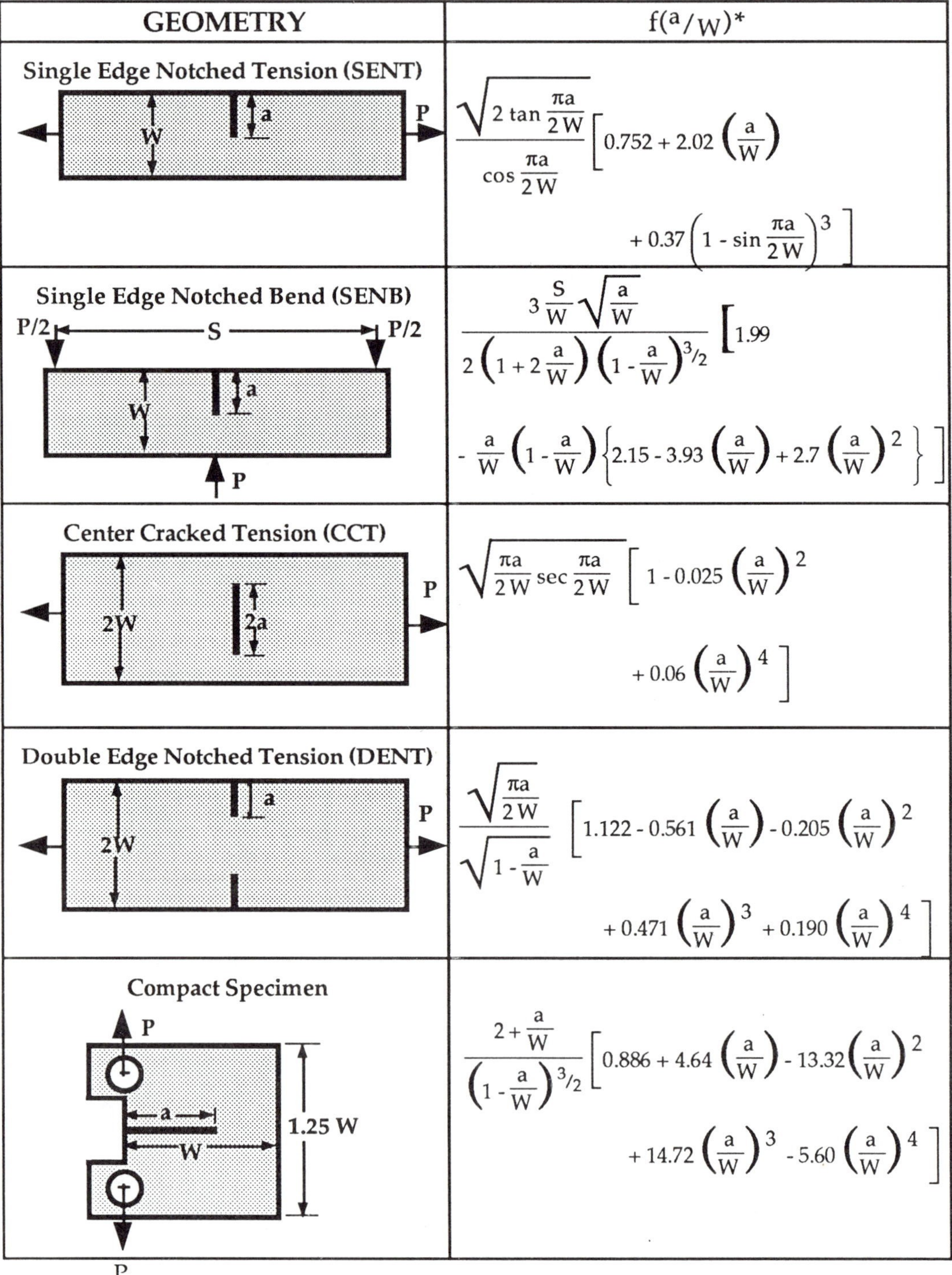

GEOMETRY	$f(a/W)$*
Single Edge Notched Tension (SENT)	$\dfrac{\sqrt{2\tan\frac{\pi a}{2W}}}{\cos\frac{\pi a}{2W}}\left[0.752+2.02\left(\frac{a}{W}\right)+0.37\left(1-\sin\frac{\pi a}{2W}\right)^3\right]$
Single Edge Notched Bend (SENB)	$\dfrac{3\frac{S}{W}\sqrt{\frac{a}{W}}}{2\left(1+2\frac{a}{W}\right)\left(1-\frac{a}{W}\right)^{3/2}}\left[1.99-\frac{a}{W}\left(1-\frac{a}{W}\right)\left\{2.15-3.93\left(\frac{a}{W}\right)+2.7\left(\frac{a}{W}\right)^2\right\}\right]$
Center Cracked Tension (CCT)	$\sqrt{\frac{\pi a}{2W}\sec\frac{\pi a}{2W}}\left[1-0.025\left(\frac{a}{W}\right)^2+0.06\left(\frac{a}{W}\right)^4\right]$
Double Edge Notched Tension (DENT)	$\dfrac{\sqrt{\frac{\pi a}{2W}}}{\sqrt{1-\frac{a}{W}}}\left[1.122-0.561\left(\frac{a}{W}\right)-0.205\left(\frac{a}{W}\right)^2+0.471\left(\frac{a}{W}\right)^3+0.190\left(\frac{a}{W}\right)^4\right]$
Compact Specimen	$\dfrac{2+\frac{a}{W}}{\left(1-\frac{a}{W}\right)^{3/2}}\left[0.886+4.64\left(\frac{a}{W}\right)-13.32\left(\frac{a}{W}\right)^2+14.72\left(\frac{a}{W}\right)^3-5.60\left(\frac{a}{W}\right)^4\right]$

*$K_I = \dfrac{P}{B\sqrt{W}} f(a/W)$ where B is the specimen thickness.

EXAMPLE 2.4

Show that the K_I solution for the single edge notched tensile panel reduces to Eq. (2.42) when $a << W$.

Solution: All of the K_I expressions in Table 2.4 are of the form:

$$K_I = \frac{P}{B\sqrt{W}} f\left(\frac{a}{W}\right)$$

where P is the applied force, B is plate thickness, and f(a/W) is a dimensionless function of a/W. The above equation can be expressed in the form of Eq. (2.47):

$$\frac{P}{B\sqrt{W}} f\left(\frac{a}{W}\right) = \frac{P}{BW} f\left(\frac{a}{W}\right)\sqrt{\frac{W}{\pi a}}\sqrt{\pi a} = C\sigma\sqrt{\pi a}$$

where

$$C = f\left(\frac{a}{W}\right)\sqrt{\frac{W}{\pi a}}$$

In the limit of a small flaw, the geometry correction factor in Table 2.4 becomes

$$\lim_{a/W \to 0} f\left(\frac{a}{W}\right) = \sqrt{\frac{\pi a}{W}}\,[0.752 + 0.37]$$

Thus

$$\lim_{a/W \to 0} (C) = 1.12$$

Although stress intensity solutions are given in a variety of forms, K can always be related to the through crack (Fig. 2.4) through the appropriate correction factor:

$$K_{(I,\ II\ or\ III)} = C\,\sigma\sqrt{\pi a} \tag{2.47}$$

where σ is a characteristic stress, a is a characteristic crack dimension, and C is a dimensionless constant that depends on geometry and mode of loading.

2.6.4 Principle of Superposition

For linear elastic materials, individual components of stress, strain, and displacement are additive. For example, two normal stresses in the x direction imposed by different external forces can be added to obtain the total σ_{xx}, but a normal stress cannot be summed with a shear stress. Similarly, stress intensity factors are additive as long as the mode of loading is consistent. That is,

$$K_I^{(total)} = K_I^{(A)} + K_I^{(B)} + K_I^{(C)} + \ldots$$

but

$$K_{(total)} \neq K_I + K_{II} + K_{III}$$

In many instances, the principle of superposition allows stress intensity solutions for complex configurations to be built from simple cases for which the solutions are well established. Consider, for example, an edge cracked panel (Table 2.4) subject to combined membrane (axial) loading, P_m, and three-point bending, P_b. Since both types of loading produce pure Mode I conditions, the K_I values can be added:

$$K_I^{(total)} = K_I^{(membrane)} + K_I^{(bending)}$$

$$= \frac{1}{B\sqrt{W}}\left[P_m f_m\left(\frac{a}{W}\right) + P_b f_b\left(\frac{a}{W}\right)\right] \tag{2.48}$$

where f_m and f_b are the geometry correction factors for membrane and bending loading, respectively, listed in Table 2.4.

EXAMPLE 2.5

Determine the stress intensity factor for a semicircular surface crack subjected to an internal pressure, p (Fig. 2.23(a)).

Solution: The principle of superposition enables us to construct the solution from known cases. One relevant case is the semicircular surface flaw under uniform remote tension, p (Fig. 2.23(b)). If we impose a uniform compressive stress, -p, on the crack surface (Fig. 2.23(c)), $K_I = 0$ because the crack faces close and the plate behaves as if the crack were not present. The loading configuration of interest is obtained by subtracting the stresses in Fig. 2.23(c) from those of Fig. 2.23(b):

$$K_I^{(a)} = K_I^{(b)} - K_I^{(c)}$$

$$= 1.12 \frac{2}{\pi} p \sqrt{\pi a} - 0 = \frac{2.24}{\pi} p \sqrt{\pi a}$$

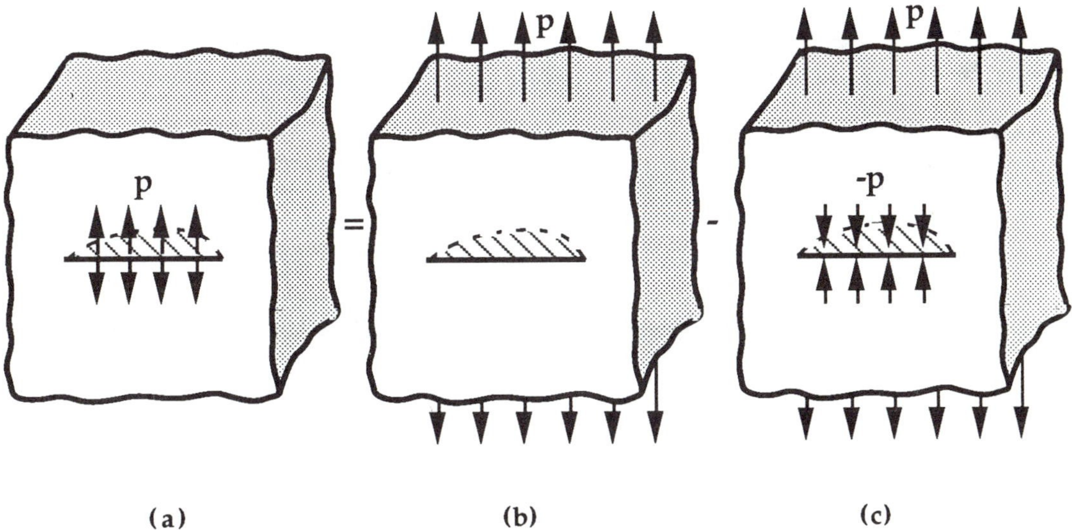

FIGURE 2.23 Determination of K_I for a semicircular surface crack under internal pressure, p, by means of the principle of superposition.

The principle of superposition leads to the highly useful concept of influence coefficients. When a cracked structure is subject to a nonuniform global stress field, (Fig. 2.24) the stress intensity factor can be inferred with the aid of influence coefficients if the stress field can be fit to a polynomial expression. For example if the remote stress normal to the crack plane can be fit to a cubic polynomial:

$$\sigma_{yy} = A_o + A_1x + A_2x^2 + A_3x^3$$

the Mode I stress intensity factor is equal to the sum of the contributions of each term in the polynomial, which depends on the magnitude of the constants A_o, A_1, A_2, and A_3. Separate K_I solutions must be obtained for power law stress distributions; i.e.,

$$\sigma_{yy}^{\infty} = \lambda x^n \quad (n = 0, 1, 2, 3)$$

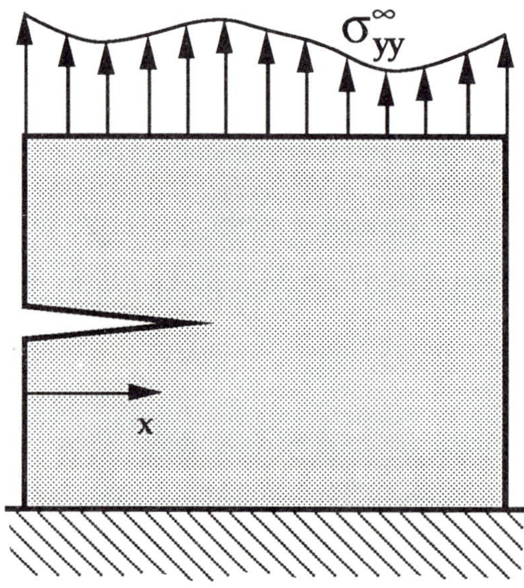

FIGURE 2.24 Cracked structure subject to a nonuniform remote stress distribution.

Thus by determining four separate K_I solutions (or more if a higher degree polynomial is desired) for a given geometry, it is possible to generate an infinite number of solutions by fitting the stress field to a polynomial and determining the relative influence each term has on the stress intensity. The combined membrane and bending example described earlier in this section represents a special case where the stress varies linearly through the cross section and $A_2 = A_3 = 0$.

Influence coefficients are described in more detail in Chapter 9.

2.7 RELATIONSHIP BETWEEN K AND $\mathcal{G}$

Two parameters that describe the behavior of cracks have been introduced so far: the energy release rate and the stress intensity factor. The former parameter quantifies the net change in potential energy that accompanies an increment of crack extension; the latter quantity characterizes the stresses, strains, and displacements near the crack tip. The energy release rate describes global behavior, while K is a local parameter. For linear elastic materials, K and $\mathcal{G}$ are uniquely related.

For a through crack in an infinite plate subject to a uniform tensile stress (Fig. 2.3), $\mathcal{G}$ and K_I are given by Eqs. (2.24) and (2.41), respectively. Combining these two equations leads to the following relationship between $\mathcal{G}$ and K_I for plane stress:

$$\mathcal{G} = \frac{K_I^2}{E} \tag{2.49}$$

For plane strain conditions, E must be replaced by $E/(1 - \nu^2)$. To avoid writing separate expressions for plane stress and plane strain, the following notation will be adopted throughout this book:

$$E' = E \text{ for plane stress} \tag{2.50a}$$

and

$$E' = \frac{E}{(1 - \nu^2)} \text{ for plane strain} \tag{2.50b}$$

Thus the $\mathcal{G}$ - K_I relationship for both plane stress and plane strain becomes

$$\mathcal{G} = \frac{K_I^2}{E'} \tag{2.51}$$

Since Eqs. (2.24) and (2.41) apply only to a through crack in an infinite plate, we have yet to prove that Eq. (2.51) is a general relationship that applies to all configurations. Irwin [9] performed a crack closure analysis that provides such a proof.

Consider a crack of initial length a + Δa subject to Mode I loading, as illustrated in Fig. 2.25(a). It is convenient in this case to place the origin a distance Δa behind the crack tip. Assume that the plate has unit thickness. Let us now apply a compressive stress field to the crack faces between x = 0 and x = Δa of sufficient magnitude to close the crack in this region. The work required to close the crack at the tip is related to the energy release rate:

$$\mathcal{G} = \lim_{\Delta a \to 0} \left(\frac{\Delta U}{\Delta a} \right)_{\text{fixed load}} \tag{2.52}$$

where ΔU is the work of crack closure, which is equal to the sum of contributions to work from x = 0 to x = Δa:

$$\Delta U = \int_{x=0}^{x=\Delta a} dU(x) \tag{2.53}$$

and the incremental work at x is equal to the area under the force-displacement curve:

$$dU(x) = 2 \bullet \frac{1}{2} F_y(x)\, u_y(x) = \sigma_{yy}(x)\, u_y(x)\, dx \tag{2.54}$$

The factor of 2 on work is required because both crack faces are displaced an absolute distance $u_y(x)$. The crack opening displacement, u_y, for Mode I is obtained from Table 2.2 by setting θ = π:

$$u_y = \frac{(\kappa + 1)\, K_I(a+\Delta a)}{\mu} \sqrt{\frac{\Delta a - x}{2\pi}} \tag{2.55}$$

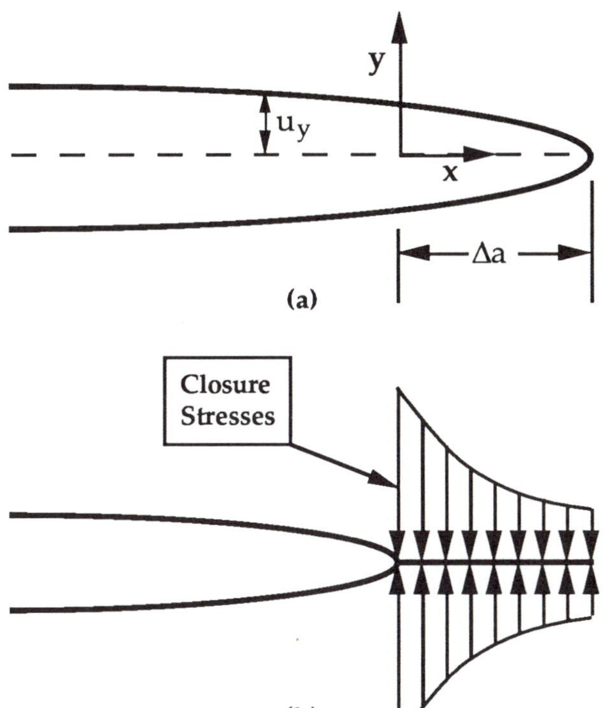

FIGURE 2.25 Application of closure stresses which shorten a crack by Δa.

where $K_I(a+\Delta a)$ denotes the stress intensity factor at the original crack tip. The normal stress required to close the crack is related to K_I for the shortened crack:

$$\sigma_{yy} = \frac{K_I(a)}{\sqrt{2\pi x}} \tag{2.56}$$

Combining Eqs. (2.52) to (2.56) gives

$$\mathcal{G} = \lim_{\Delta a \to 0} \frac{(\kappa + 1)\, K_I(a)\, K_I(a+\Delta a)}{4\pi\, \mu\, \Delta a} \int_0^{\Delta a} \sqrt{\frac{\Delta a - x}{x}}\, dx$$

$$= \frac{(\kappa + 1)\, K_I^2}{8\, \mu} = \frac{K_I^2}{E'} \tag{2.57}$$

Thus Eq. (2.51) is a general relationship for Mode I. The above analysis can be repeated for other modes of loading; the relevant closure stress and displacement for Mode II is τ_{yx} and u_x, and the corresponding quantities for Mode III are τ_{yz} and u_z. When all three modes of loading are present, the energy release rate is given by

$$\mathcal{G} = \frac{K_I^2}{E'} + \frac{K_{II}^2}{E'} + \frac{K_{III}^2}{2\mu} \tag{2.58}$$

Contributions to $\mathcal{G}$ from the three modes are additive because energy release rate, like energy, is a scalar quantity.

2.8 CRACK TIP PLASTICITY

Linear elastic stress analysis of sharp cracks predicts infinite stresses at the crack tip. In real materials, however, stresses at the crack tip are finite because the crack tip radius must be finite (Section 2.2). Inelastic material deformation, such as plasticity in metals and crazing in polymers, leads to further relaxation of crack tip stresses.

The elastic stress analysis becomes increasingly inaccurate as the inelastic region at the crack tip grows. Simple corrections to linear elastic fracture mechanics (LEFM) are available when moderate crack tip yielding occurs. For more extensive yielding, one must apply alternative crack tip parameters that take nonlinear material behavior into account (see Chapter 3).

The size of the crack tip yielding zone is estimated below by two methods: the Irwin approach, where the elastic stress analysis is used to estimate the elastic-plastic boundary, and the strip yield model. Both approaches lead to simple corrections for crack tip yielding. The term *plastic zone* usually applies to metals, but will be adopted here to describe inelastic crack tip behavior in a more general sense. Differences in the yielding behavior between metals and polymers are discussed briefly in the next section and in more detail in Chapter 6.

2.8.1 The Irwin Approach

On the crack plane ($\theta = 0$) the normal stress, σ_{yy}, in a linear elastic material is given by Eq. (2.39). As a first approximation, we can assume that the boundary between elastic and plastic behavior occurs when the stresses given by Eq. (2.39) satisfy a yield criterion. For plane stress conditions, yielding occurs when $\sigma_{yy} = \sigma_{YS}$, the uniaxial yield strength of the material. Substituting yield strength into the left side of Eq. (2.39) and solving for r gives a first order estimate of plastic zone size:

$$r_y = \frac{1}{2\pi}\left(\frac{K_I}{\sigma_{YS}}\right)^2 \tag{2.59}$$

If we neglect strain hardening, the stress distribution for $r \leq r_y$ can be represented by a horizontal line at $\sigma_{yy} = \sigma_{YS}$, as illustrated in Fig. 2.26; the stress singularity is truncated by yielding at the crack tip.

The simple analysis in the preceding paragraph is not strictly correct because it was based on an elastic crack tip solution. When yielding occurs, stresses must redistribute in order to satisfy equilibrium. The cross-hatched region in Fig. 2.26 represents forces that would be present in an elastic material but cannot be carried in the elastic-plastic material because the stress cannot exceed yield. The plastic zone must increase in size in order to accommodate these forces. A simple force balance leads to a second order estimate of the plastic zone size, r_p:

$$\sigma_{YS}\, r_p = \int_0^{r_y} \sigma_{yy}\, dr = \int_0^{r_y} \frac{K_I}{\sqrt{2\pi r}}\, dr \tag{2.60}$$

Integrating and solving for r_p gives

$$r_p = \frac{1}{\pi}\left(\frac{K_I}{\sigma_{YS}}\right)^2 \tag{2.61}$$

which is twice as large as r_y, the first order estimate.

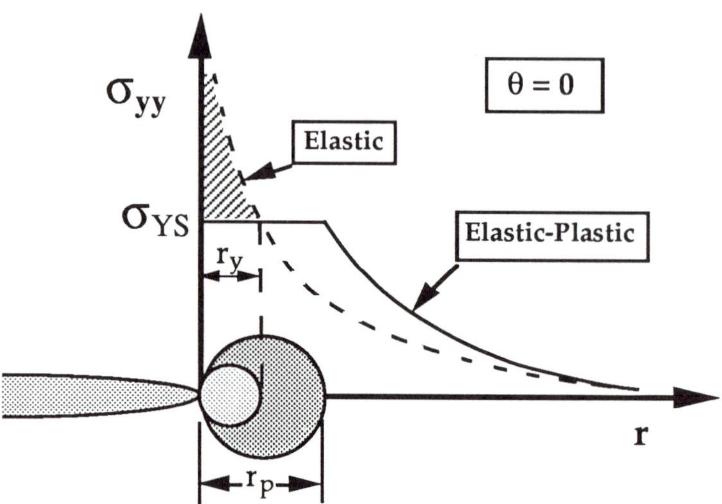

FIGURE 2.26 First-order and second-order estimates of plastic zone size (r_y and r_p, respectively). The cross-hatched area represents load that must be resdistributed, resulting in a larger plastic zone.

Referring to Fig. 2.26, note that the redistributed stress in the elastic region is higher than predicted by Eq. (2.39), implying a higher *effective* stress intensity factor. Irwin [15] accounted for this increase in K by defining an effective crack length that is slightly longer than the actual crack size. He found that a good approximation of K_{eff} can be obtained by placing the tip of the effective crack in the center of the plastic zone, as illustrated in Fig. 2.27. Thus the effective crack length is defined as the sum of the actual crack size and a plastic zone correction:

$$a_{eff} = a + r_y \tag{2.62}$$

where r_y for plane stress is given by Eq. (2.59). In plane strain, yielding is suppressed by the triaxial stress state, and the plastic zone correction is smaller by a factor of three:

$$r_y = \frac{1}{6\pi}\left(\frac{K_I}{\sigma_{YS}}\right)^2 \tag{2.63}$$

The effective stress intensity is obtained by inserting a_{eff} into the K expression for the geometry of interest:

$$K_{eff} = C(a_{eff})\,\sigma\sqrt{\pi a_{eff}} \tag{2.64}$$

Since the effective crack size is taken into account in the geometry correction factor, C, an iterative solution is usually required to solve for K_{eff}. That is, K is first determined in the absence of a plasticity correction; a first order estimate of a_{eff} is then obtained from Eq. (2.59) or (2.63), which in turn used to estimate K_{eff}. A new a_{eff} is computed from the K_{eff} estimate, and the process is repeated until successive K_{eff} estimates converge. Typically, no more than three or four iterations are required for reasonable convergence.

In certain cases, this iterative procedure is unnecessary because a closed-form solution is possible. For example, the effective Mode I stress intensity factor for a through crack in an infinite plate in plane stress is given by

$$K_{eff} = \frac{\sigma\sqrt{\pi a}}{\sqrt{1 - \frac{1}{2}\left(\frac{\sigma}{\sigma_{YS}}\right)^2}} \tag{2.65}$$

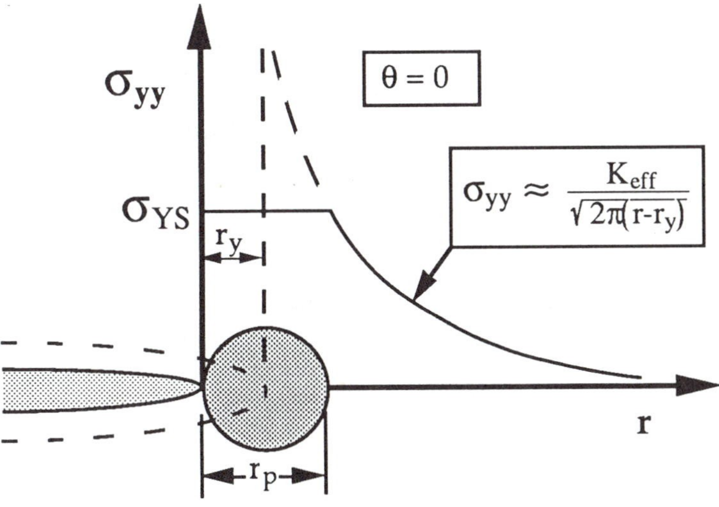

FIGURE 2.27 The Irwin plastic zone correction. The increase in the effective stress intensity is taken into account by assuming the crack is longer by r_y.

The elliptical and semi-elliptical flaws (Fig. 2.20) also have a closed form plastic zone correction, provided the flaw is small compared to plate dimensions. In the case of the embedded elliptical flaw, K_{eff} is given by

$$K_{eff} = \sigma\sqrt{\frac{\pi a}{Q}}\left(\sin^2\phi + \frac{a^2}{c^2}\cos^2\phi\right)^{1/4} \tag{2.66}$$

where Q is the flaw shape parameter, defined as

$$Q = \Psi^2 - 0.212\left(\frac{\sigma}{\sigma_{YS}}\right)^2 \tag{2.67}$$

Equation (2.66) must be multiplied by 1.12 for a semi-elliptical surface flaw.

2.8.2 The Strip Yield Model

The strip yield model, which is illustrated in Fig 2.28, was first proposed by Dugdale [16] and Barenblatt [17]. They assumed a long, slender plastic zone at the crack tip in a nonhardening material in plane stress. These early analyses considered only a through crack in an infinite plate. The strip yield plastic zone is modeled by assuming a crack of length $2a + 2\rho$, where ρ is the length of the plastic zone, with a closure stress equal to σ_{YS} applied at each crack tip (Fig. 2.28(b)).

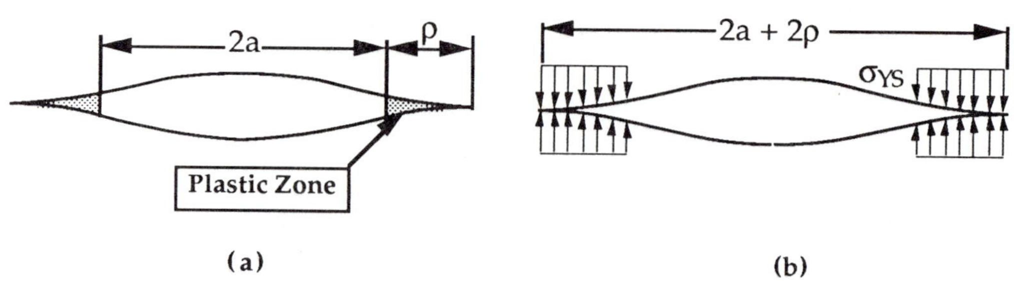

FIGURE 2.28 The strip yield model. The plastic zone is modeled by yield magnitude compressive stresses at each crack tip (b).

This model approximates elastic-plastic behavior superimposing two elastic solutions: a through crack under remote tension and a through crack with closure stresses at the tip. Thus the strip yield model is a classical application of the principle of superposition.

Since the stresses are finite in the strip yield zone, there cannot be a stress singularity at the crack tip. Therefore, the leading term in the crack tip field that varies with $1/\sqrt{r}$ (Eq. 2.36) must be zero. The plastic zone length, ρ, must be chosen such that the stress intensity factors from the remote tension and closure stress cancel one another.

The stress intensity due to the closure stress can be estimated by considering a normal force P applied to the crack at a distance x from the center line of the crack (Fig. 2.29). The stress intensities for the two crack tips are given by

$$K_{I(+a)} = \frac{P}{\sqrt{\pi a}}\sqrt{\frac{a+x}{a-x}}$$

$$K_{I(-a)} = \frac{P}{\sqrt{\pi a}}\sqrt{\frac{a-x}{a+x}} \tag{2.68}$$

assuming the plate is of unit thickness. The closure force at a point within the strip yield zone is equal to

$$P = -\sigma_{YS}\, dx \tag{2.69}$$

Thus the total stress intensity at each crack tip resulting from the closure stresses is obtained by replacing a with a + ρ in Eq. (2.68) and summing the contribution from both crack tips:

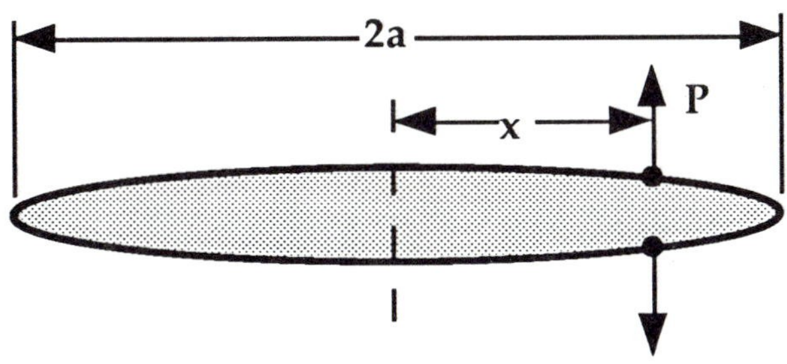

FIGURE 2.29 Crack opening force applied at a distance x from the center line.

$$K_{closure} = -\frac{\sigma_{YS}}{\sqrt{\pi(a+\rho)}} \int_a^{a+\rho} \left\{ \sqrt{\frac{a+\rho+x}{a+\rho-x}} + \sqrt{\frac{a+\rho-x}{a+\rho+x}} \right\} dx$$

$$= -2\,\sigma_{YS}\sqrt{\frac{a+\rho}{\pi}} \int_a^{a+\rho} \frac{dx}{\sqrt{(a+\rho)^2 - x^2}} \tag{2.70}$$

Solving this integral gives

$$K_{closure} = -2\,\sigma_{YS}\sqrt{\frac{a+\rho}{\pi}}\cos^{-1}\left(\frac{a}{a+\rho}\right) \tag{2.71}$$

The stress intensity from the remote tensile stress, $K_\sigma = \sigma\sqrt{\pi(a+\rho)}$, must balance with $K_{closure}$. Therefore,

$$\frac{a}{a+\rho} = \cos\left(\frac{\pi\,\sigma}{2\,\sigma_{YS}}\right) \tag{2.72}$$

Note that ρ approaches infinity as $\sigma \rightarrow \sigma_{YS}$. Let us explore the strip yield model further by performing a Taylor series expansion on Eq. (2.72):

$$\frac{a}{a+\rho} = 1 - \frac{1}{2!}\left(\frac{\pi\,\sigma}{2\,\sigma_{YS}}\right)^2 + \frac{1}{4!}\left(\frac{\pi\,\sigma}{2\,\sigma_{YS}}\right)^4 - \frac{1}{6!}\left(\frac{\pi\,\sigma}{2\,\sigma_{YS}}\right)^6 + \ldots \tag{2.73}$$

Neglecting all but the first two terms and solving for the plastic zone size gives

$$\rho = \frac{\pi^2\,\sigma^2\,a}{8\,\sigma_{YS}^2} = \frac{\pi}{8}\left(\frac{K_I}{\sigma_{YS}}\right)^2 \tag{2.74}$$

for $\sigma << \sigma_{YS}$. Note the similarity between Eqs. (2.74) and (2.61); since $1/\pi = 0.318$ and $\pi/8 = 0.392$, the Irwin and strip yield approaches predict similar plastic zone sizes.

One way to estimate the effective stress intensity with the strip yield model is to set a_{eff} equal to $a + \rho$:

$$K_{eff} = \sigma \sqrt{\pi a \sec\left(\frac{\pi \sigma}{2 \sigma_{YS}}\right)} \tag{2.75}$$

However, Eq. (2.75) tends to overestimate K_{eff}; the actual a_{eff} is somewhat less than $a + \rho$ because the strip yield zone is loaded to σ_{YS}. Burdekin and Stone [18] obtained a more realistic estimate of K_{eff} for the strip yield model:

$$K_{eff} = \sigma_{YS} \sqrt{\pi a} \left[\frac{8}{\pi^2} \ln \sec\left(\frac{\pi \sigma}{2 \sigma_{YS}}\right)\right]^{1/2} \tag{2.76}$$

2.8.3 Comparison of Plastic Zone Corrections

Figure 2.30 shows a comparison between a pure LEFM analysis (Eq. 2.41), the Irwin correction for plane stress (2.65), and the strip yield correction on stress intensity (Eq. (2.76)). The effective stress intensity, nondimensionalized by $\sigma_{YS} \sqrt{\pi a}$, is plotted against the normalized stress. The LEFM analysis predicts a linear relationship between K and stress. Both the Irwin and strip yield corrections deviate from LEFM theory at stresses greater than 0.5 σ_{YS}. The two plasticity corrections agree with each other up to approximately 0.85 σ_{YS}. According to the strip yield model, K_{eff} is infinite at yield; the strip yield zone extends completely across the plate, which has reached its maximum load capacity.

The plastic zone shape predicted by the strip yield model bears little resemblance to actual plastic zones in metals (see below), but many polymers produce crack tip craze zones which look very much like Fig. 2.28. Thus although Dugdale originally proposed the strip yield model to account for yielding in thin steel sheets, this model is better suited to polymers (see Chapter 6).

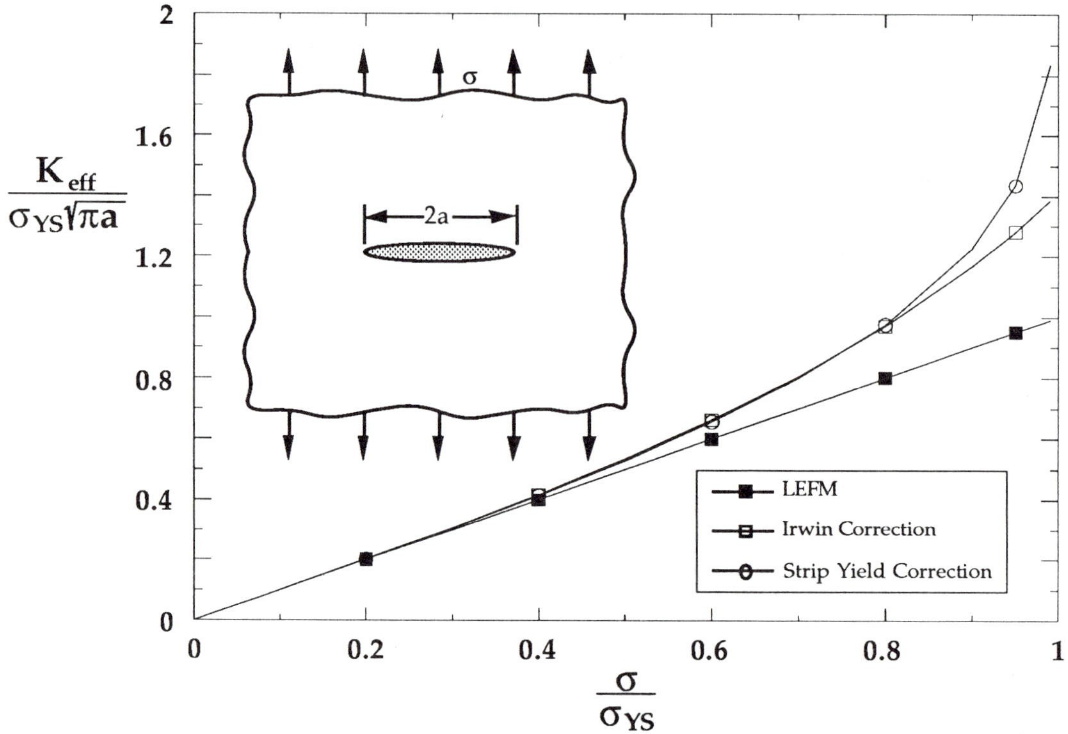

FIGURE 2.30 Comparison of plastic zone corrections for a through crack in plane strain.

2.8.4 Plastic Zone Shape

The estimates of plastic zone size that have been presented so far consider only the crack plane ($\theta = 0$). It is possible to estimate the extent of plasticity at all angles by applying an appropriate yield criterion to the equations in Tables 2.1 and 2.3. Consider the von Mises equation:

$$\sigma_e = \frac{1}{\sqrt{2}}\left[(\sigma_1 - \sigma_2)^2 + (\sigma_1 - \sigma_3)^2 + (\sigma_2 - \sigma_3)^2\right]^{1/2} \tag{2.77}$$

where σ_e is the effective stress, and σ_1, σ_2, and σ_3 are the three principal normal stresses. According to the von Mises criterion, yielding occurs when $\sigma_e = \sigma_{YS}$, the uniaxial yield strength. For plane stress or plane strain conditions, the principal stresses can be computed from the two-dimensional Mohr's circle relationship:

$$\sigma_1, \sigma_2 = \frac{\sigma_{xx} + \sigma_{yy}}{2} \pm \left[\left(\frac{\sigma_{xx} - \sigma_{yy}}{2}\right)^2 + \tau_{xy}^2\right]^{1/2} \tag{2.78}$$

For plane stress, $\sigma_3 = 0$, and $\sigma_3 = \nu(\sigma_1 + \sigma_2)$ for plane strain. Substituting the Mode I stress fields into Eq. (2.78) gives

$$\sigma_1 = \frac{K_I}{\sqrt{2\pi r}} \cos\left(\frac{\theta}{2}\right)\left[1 + \sin\left(\frac{\theta}{2}\right)\right] \tag{2.79a}$$

$$\sigma_2 = \frac{K_I}{\sqrt{2\pi r}} \cos\left(\frac{\theta}{2}\right)\left[1 - \sin\left(\frac{\theta}{2}\right)\right] \tag{2.79b}$$

$$\sigma_3 = 0 \quad \text{(plane stress)}$$

$$= \frac{2\nu K_I}{\sqrt{2\pi r}} \cos\left(\frac{\theta}{2}\right) \quad \text{(plane strain)} \tag{2.79c}$$

By substituting Eq. (2.79) into Eq. (2.77), setting $\sigma_e = \sigma_{YS}$, and solving for r, we obtain estimates of the Mode I plastic zone radius as a function of θ:

$$r_y(\theta) = \frac{1}{4\pi}\left(\frac{K_I}{\sigma_{YS}}\right)^2\left[1 + \cos\theta + \frac{3}{2}\sin^2\theta\right] \tag{2.80a}$$

for plane stress, and

$$r_y(\theta) = \frac{1}{4\pi}\left(\frac{K_I}{\sigma_{YS}}\right)^2\left[(1 - 2\nu)^2(1 + \cos\theta) + \frac{3}{2}\sin^2\theta\right] \tag{2.80b}$$

for plane strain. Equations (2.80a) and (2.80b), which are plotted in Fig. 2.31(a), define the approximate boundary between elastic and plastic behavior. The corresponding equations for Modes II and III are plotted in Figs 2.31(b) and 2.31(c), respectively.

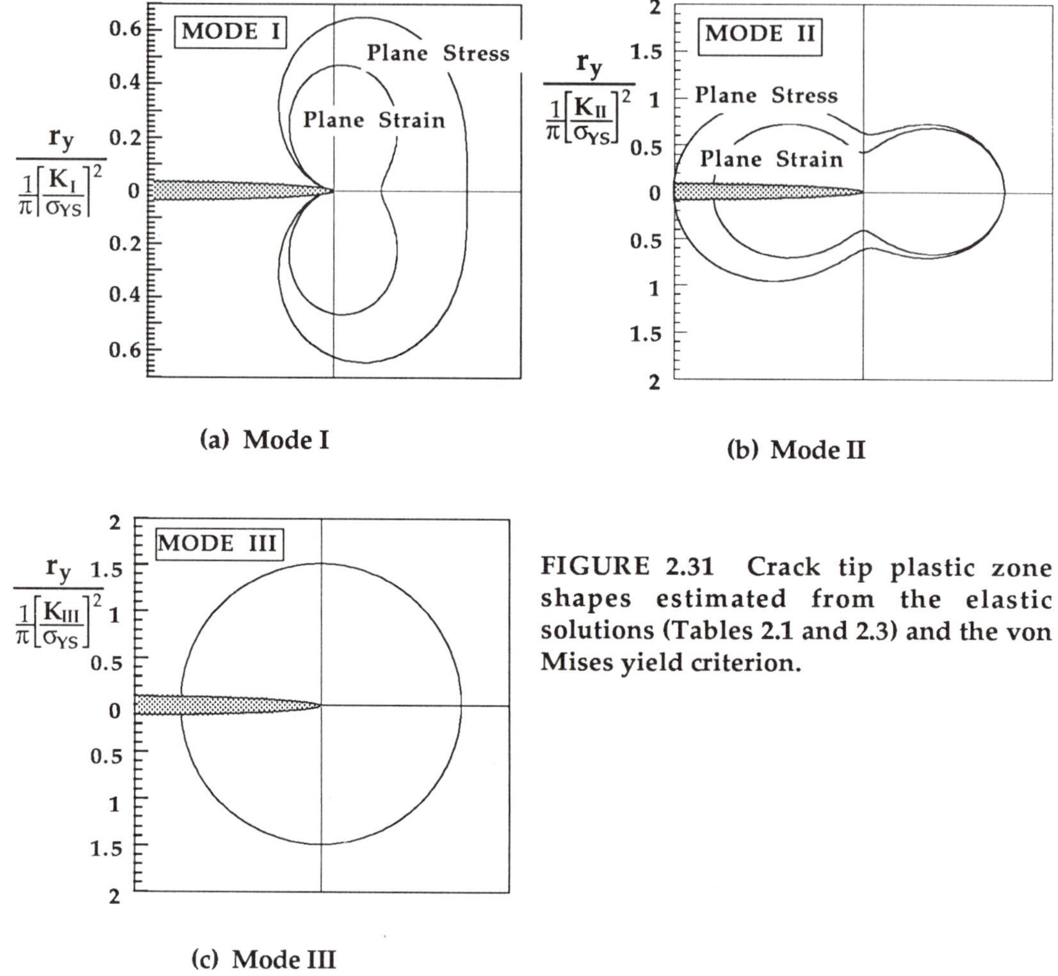

FIGURE 2.31 Crack tip plastic zone shapes estimated from the elastic solutions (Tables 2.1 and 2.3) and the von Mises yield criterion.

Note the significant difference in the size and shape of the Mode I plastic zones for plane stress and plane strain. The latter condition suppresses yielding, resulting in a smaller plastic zone for a given K_I value.

Equations (2.80a) and (2.80b) are not strictly correct because they are based on a purely elastic analysis. Recall Fig. 2.26, which schematically illustrates how crack tip plasticity causes stress redistribution, which is not taken into account in Fig. 2.31. The Irwin plasticity correction, which accounts for stress redistribution by means of an effective crack length, is also simplistic and not totally correct.

Figure 2.32 compares the plane strain plastic zone shape predicted from Eq. (2.80b) with a detailed elastic-plastic crack tip stress solution obtained from finite element analysis. The latter, which was published by Dodds, et al. [19], assumed a material with the following uniaxial stress-strain relationship:

$$\frac{\varepsilon}{\varepsilon_o} = \frac{\sigma}{\sigma_o} + \alpha\left(\frac{\sigma}{\sigma_o}\right)^n \tag{2.81}$$

where ε_o, σ_o, α, and n are material constants. We will examine the above relationship in more detail in Chapter 3; for now it is sufficient to note that the exponent, n, characterizes the strain hardening rate of a material. Dodds, et al. analyzed materials with n = 5, 10, and 50, which corresponds to high, medium, and low strain hardening, respectively. Figure 2.32 shows contours of constant σ_e for n = 50. The definition of the elastic-plastic boundary is somewhat arbitrary, since materials that can be described by Eq. (2.81) do not have a definite yield point. When the plastic zone boundary is defined at $\sigma_e = \sigma_{YS}$ (the 0.2% offset yield strength), the plane strain plastic zone is considerably smaller than predicted by Eq. (2.80b). Defining the boundary at a slightly lower effective stress results in a much larger plastic zone. Given the difficulties of defining the plastic zone unambiguously with a detailed analysis, the estimates of plastic zone size and shape from the elastic analysis (Fig. 2.31) appear to be reasonable.

Figure 2.33 illustrates the effect of strain hardening on the plastic zone. A high strain hardening rate results in a smaller plastic zone because the material inside of the plastic zone is capable of carrying higher stresses, and less stress redistribution is necessary.

2.9 PLANE STRESS VERSUS PLANE STRAIN

Most of the classical solutions in fracture mechanics reduce the problem to two dimensions. That is, at least one of the principal stresses or strains is assumed to equal zero (plane stress and plane strain, respectively).

In general, the conditions ahead of a crack are neither plane stress nor plane strain, but are three-dimensional. There are, however, limiting cases where a two-dimensional assumption is valid, or at least provides a good approximation.

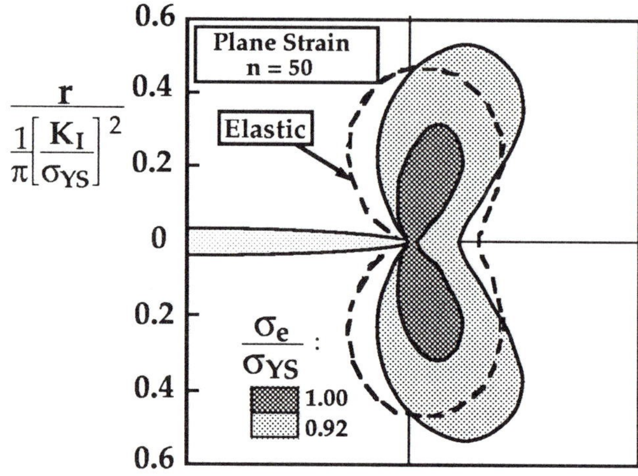

FIGURE 2.32 Contours of constant effective stress in Mode I, obtained from finite element analysis [19]. The elastic-plastic boundary estimated from Eq. (2.80a) is shown for comparison.

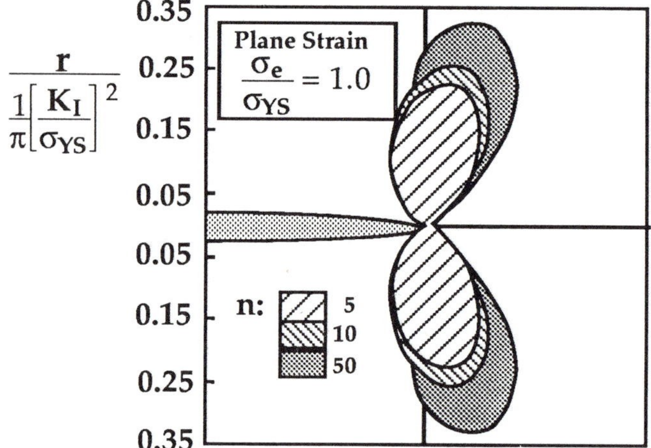

FIGURE 2.33 Effect of strain hardening on the Mode I plastic zone; n = 5 corresponds to a high strain hardening material, while n = 50 corresponds to very low hardening (cf. Eq. (2.81)).

Consider a cracked plate with thickness B subject to in-plane loading, as illustrated in Fig. 2.34. For the moment, assume that the plastic zone is small; the effect of crack tip plasticity is considered later. If there were no crack, the plate would obviously be in a state of plane stress. Thus, regions of the plate that are sufficiently far from the crack tip must also be loaded in plane stress. Material near the crack tip is loaded to higher stresses than the surrounding material. Because of the large stress normal to the crack plane, the crack tip material tries to contract in the x and z directions, but is prevented from doing so by the surrounding material (Fig. 2.34 (b)). This constraint causes a triaxial state of stress near the crack tip. For $r \ll B$, plane strain conditions exist in the interior of the plate. Material on the plate surface is in a state of plane stress, however, because there are no stresses normal to the free surface.

Figure 2.35 schematically illustrates the through-thickness variation of stress and strain in the z direction for $r \ll B$. At the plate surface, $\sigma_{zz} = 0$ and ε_{zz} is at its maximum (absolute) value. At the mid-plane (z=0), plane strain conditions exist and $\sigma_{zz} = \nu(\sigma_{xx} + \sigma_{yy})$ (assuming $r \gg r_y$). There is a region near the plate surface where the stress state is neither plane stress nor plane strain.

Figure 2.36 is a plot of σ_{zz} as a function of z/B and r/B. These results were obtained from a three-dimensional elastic-plastic finite element analysis performed by Narasimhan and Rosakis [20]. Note the transition from plane strain (at mid thickness) to plane stress as r increases relative to thickness.

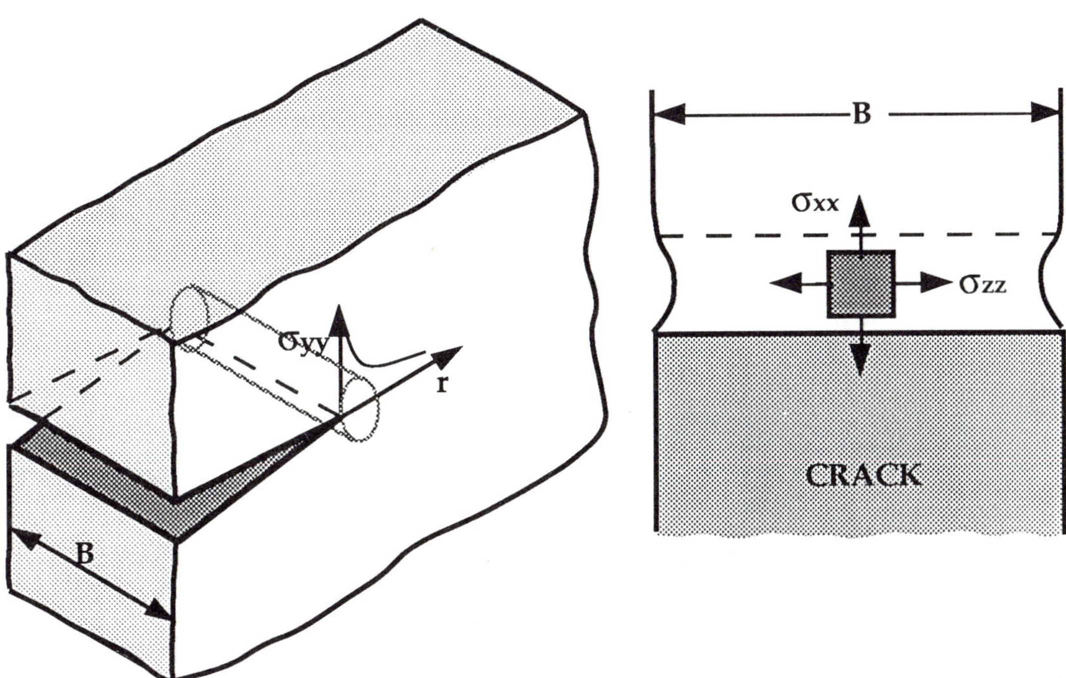

FIGURE 2.34 Three-dimensional deformation at the tip of a crack. The high normal stress at the crack tip causes material near the surface to contract, but material in the interior is constrained, resulting in a triaxial stress state.

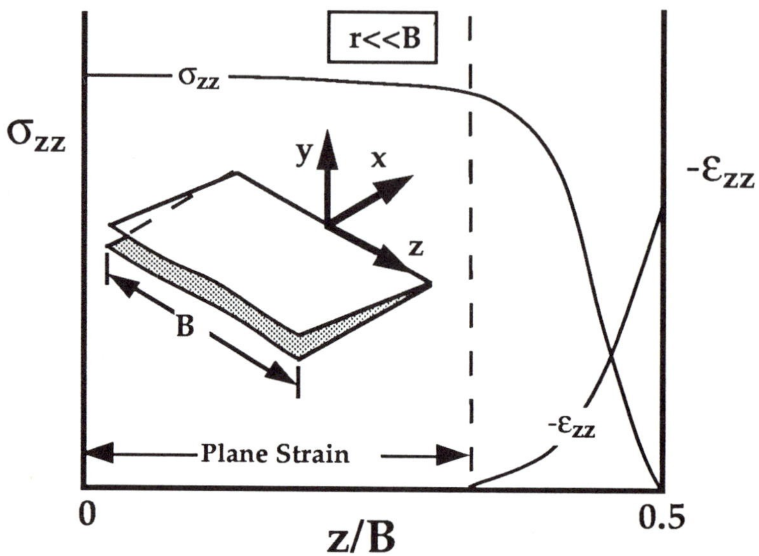

FIGURE 2.35 Schematic variation of transverse stress and strain through the thickness at a point near the crack tip.

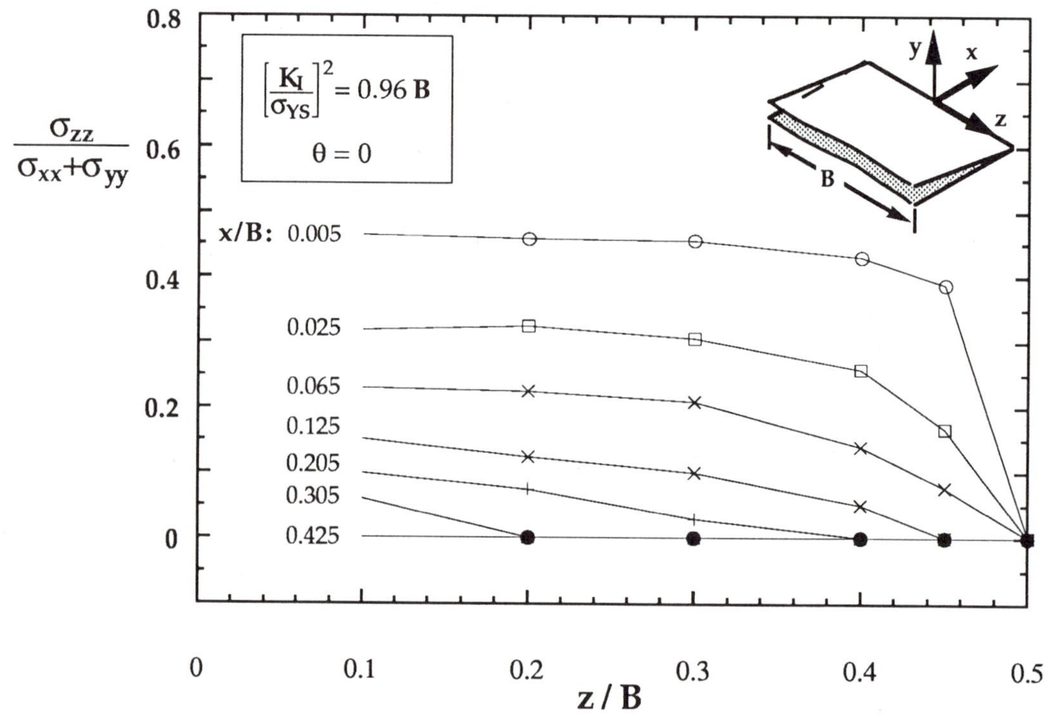

FIGURE 2.36 Transverse stress through the thickness as a function of distance from the crack tip. [20].

The stress state in the plastic zone depends on its size relative to the plate thickness. Plane strain conditions exist if the plastic zone is small compared to the thickness, but the stress state is predominantly plane stress if the plastic zone is of the same order as the thickness. Figure 2.37 shows Mode I plastic zones at mid-thickness computed from a three-dimensional elastic-plastic finite element analysis performed by Nakamura and Parks [21]. The elastic-plastic boundary is defined at $\sigma_e = \sigma_{YS}$ in this case. As $(K_I/\sigma_{YS})^2$ increases relative to thickness, the plastic zone grows, as one might expect. It is interesting, however, to note the change in shape of the elastic-plastic boundary: at low K_I values, the plastic zone has a typical plane strain shape, but takes on a plane stress shape as K_I increases (cf. Fig. 2.31(a)). If the stress state remained constant, the plastic zone size would increase in proportion to $(K_I/\sigma_{YS})^2$ and would retain a constant shape; the plastic zone actually increases at a faster rate because the stress state changes from plane strain to plane stress as K_I increases.

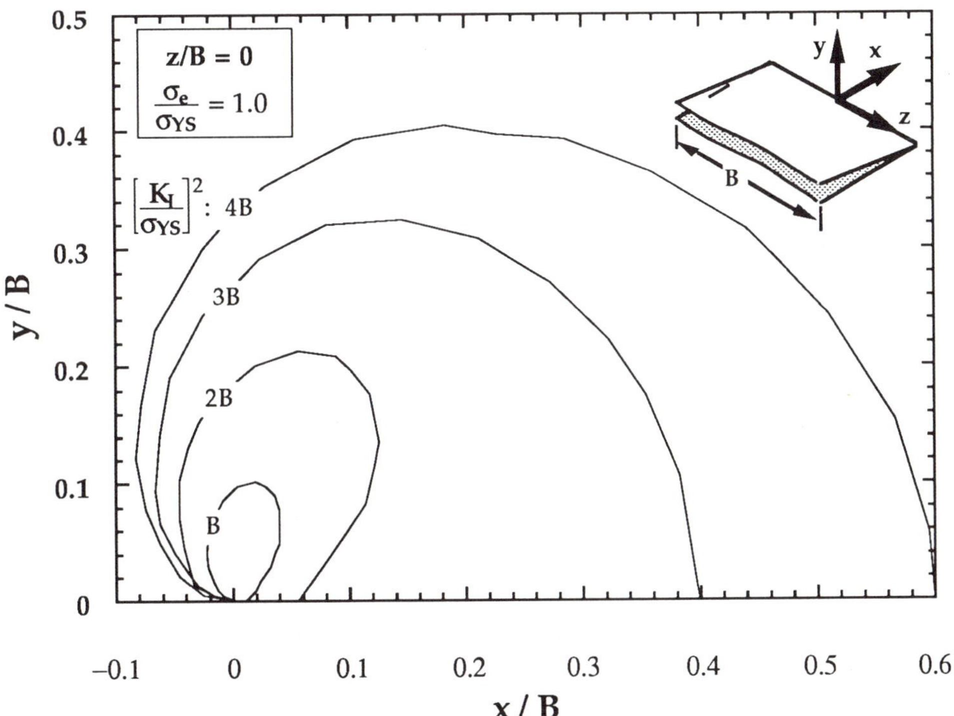

Figure 2.37 Effect of K_I, relative to thickness, of the plastic zone size and shape [21].

2.10 K AS A FAILURE CRITERION

Section 2.6.1 introduced the concept of the singularity dominated zone and alluded to single-parameter characterization of crack tip conditions. The stresses near the crack tip in a linear elastic material vary as $1/\sqrt{r}$; the stress intensity factor defines the amplitude of the singularity. Given the equations in Tables 2.1 to 2.3, one can completely define the stresses, strains, and displacements in the singularity dominated zone if the stress intensity factor is known. If we assume a material fails locally at some combination of stresses and strains, then crack extension must occur at a critical K value. This K_c value, which is a measure of *fracture toughness*, is a material constant that is independent of the size and geometry of the cracked body. Since energy release rate is uniquely related to stress intensity (Section 2.7), $\mathcal{G}$ also provides a single-parameter description of crack tip conditions, and $\mathcal{G}_c$ is an alternative measure of toughness.

The forgoing discussion does not consider plasticity or other types of nonlinear material behavior at the crack tip. Recall that the $1/\sqrt{r}$ singularity applies only to linear elastic materials. The equations in Tables 2.1 to 2.3 do not describe the stress distribution inside the plastic zone. As discussed in Chapters 5 and 6, the microscopic events that lead to fracture in various materials generally occur well within the plastic zone (or damage zone, to use a more generic term). Thus even if the plastic zone is very small, fracture may not nucleate in the singularity dominated zone. This fact raises an important question: is stress intensity a useful failure criterion in materials that exhibit inelastic deformation at the crack tip?

Under certain conditions, K still uniquely characterizes crack tip conditions when a plastic zone is present. In such cases, K_c is a geometry-independent material constant, as discussed below.

Consider a test specimen and structure loaded to the same K_I level, as illustrated in Fig. 2.38. Assume that the plastic zone is small compared to all length dimensions in the structure and test specimen. Let us construct a free-body diagram with a small region removed from the crack tip of each material. If this region is sufficiently small to be within the singularity dominated zone, the stresses and displacements at the boundary are defined by the relationships in Tables 2.1 and 2.2. The disk-shaped region in Fig. 2.38 can be viewed as an independent problem. Imposition of the $1/\sqrt{r}$ singularity at the boundary results in a plastic zone at the crack tip. The size of the plastic zone and the stress distribution within the disc-shaped region are a function only of the boundary conditions and material properties. Therefore, even though we do not know the actual stress distribution in the plastic zone, we can argue that it is uniquely characterized by the boundary conditions; i.e., K_I characterizes crack tip conditions even though the $1/\sqrt{r}$ singularity does not apply to the plastic zone. Since the structure and test specimen in Fig. 2.38 are loaded to the same K_I value, the crack tip conditions must be identical in the two configurations. Furthermore, as load is increased, both configurations will fail at the same critical stress intensity, provided the plastic zone remains small in each case. Similarly, if both structures are loaded in fatigue at the same ΔK, the crack growth rates will be similar as long as the cyclic plastic zone is embedded within the singularity dominated zone in each case (see Chapter 10).

Figure 2.39 schematically illustrates the stress distributions in the structure and test specimen from the previous figure. In the singularity dominated zone, a log-log plot of the stress distribution is linear with a slope of $-1/2$. Inside of the plastic zone, the stresses are lower than predicted by the elastic solution, but are identical for the two configurations. Outside of the singularity dominated zone, higher order terms become significant (Eq. 2.36) and the stress fields are different for the structure and test specimen; K does not uniquely characterize the magnitude of the higher order terms.

2.10.1 Effect of Loading Mode

A brief word of caution is necessary with respect to the mode of loading. Although the critical stress intensity factor for a given mode is a material constant (when crack tip plasticity is limited), K_c generally varies with loading mode. That is,

$$K_{IC} \neq K_{IIC} \neq K_{IIIC}$$

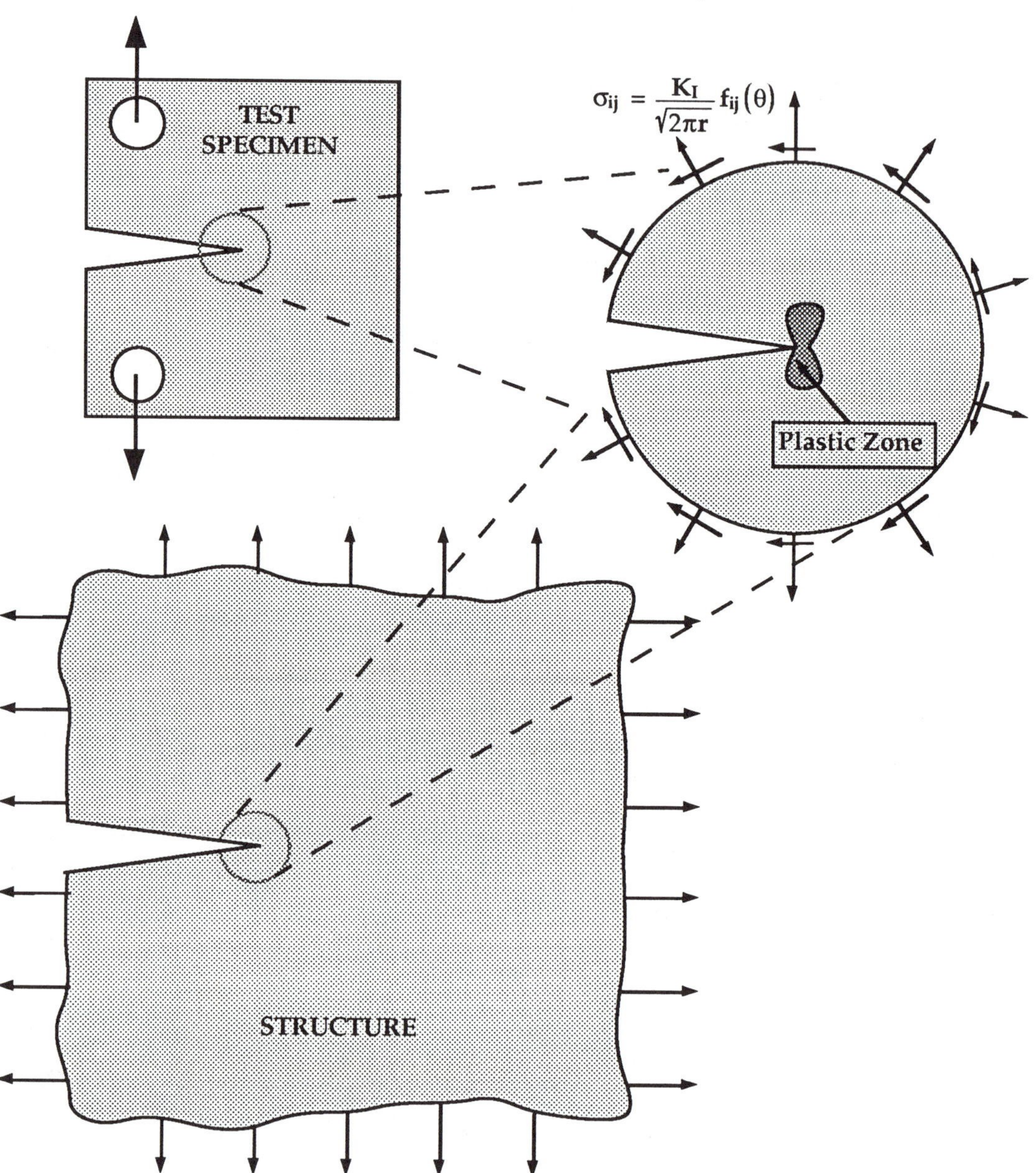

FIGURE 2.38 Schematic test specimen and structure loaded to the same stress intensity. The crack tip conditions should be identical in both configurations as long as the plastic zone is small compared to all relevant dimensions. Thus both will fail at the same critical K value.

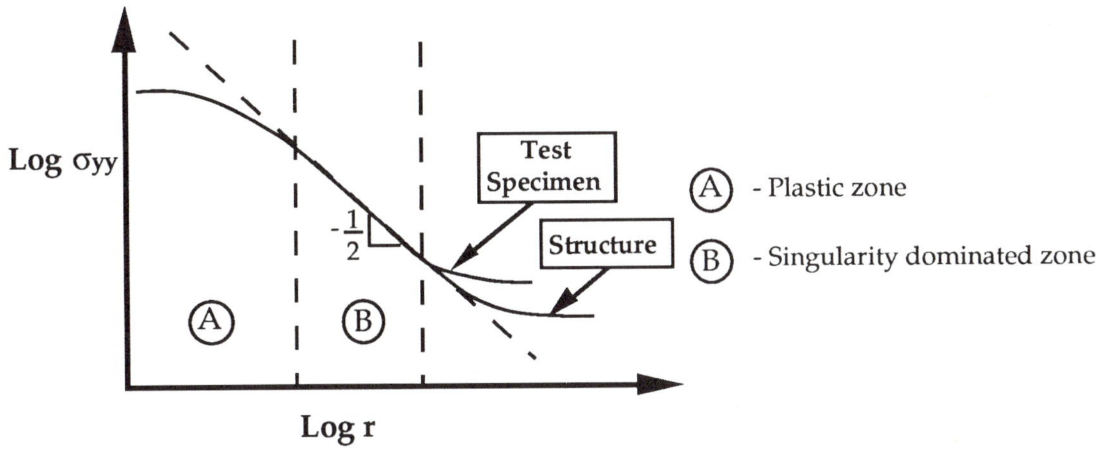

FIGURE 2.39 **Crack tip stress fields for the specimen and structure in Fig. 2.38.**

Most materials are more susceptible to fracture by normal tensile stresses than by shear stresses. Consequently, Mode I loading has the most practical importance. Mode II and Mode III loading usually do not lead to fracture. Stated another way, K_{IIC} and K_{IIIC} are generally much greater than K_{IC}.

The vast majority of practical applications of fracture mechanics consider only the Mode I component of loading. (The reader may have noticed that most of the examples in this chapter are pure Mode I problems.) Other modes of loading become important when they are applied to a weak interface in the material. For example, Mode II fiber/matrix debonding and Mode II delamination can occur in composite materials (see Chapter 6).

Mixed-mode fracture is a complex problem that is the subject of ongoing research. Most of the remainder of this book focuses on Mode I fracture, but other modes are considered where appropriate

2.10.2 Effect of Specimen Dimensions

The critical stress intensity factor is only a material constant when certain conditions are met. Otherwise, K_c values can be geometry dependent.

As stated in Section 2.9, the plastic zone must be small compared to the specimen thickness in order to achieve plane strain conditions at the crack tip. If the thickness is too small (or, equivalently, if the plastic zone is too large) the constraint at the crack tip relaxes. A lower degree of stress triaxiality usually results in higher toughness. Figure 2.40 illustrates the effect of thickness on the critical Mode I stress intensity factor. Small thickness corresponds to plane stress fracture. Fracture toughness decreases with thickness until a plateau is reached; further increases in thickness have little or no effect on toughness (see Appendix 5.2 for an exception to this rule). The critical K_I value at the plateau is defined as K_{IC}, the *plane strain fracture toughness.* (Critical K_I values corresponding to less than plane strain constraint are not called K_{IC} values. These are sometimes designated as K_c values, but this convention is avoided here because it can lead to confusion when other modes of loading are present.)

The through-thickness constraint can influence the shape of the R curve, particularly for ductile materials. Section 2.5.1 alluded to this effect. The R curve for a material in plane stress is often much steeper than the plane strain R curve for the same material. Some materials have a relatively flat plane strain R curve, resulting in toughness that is single valued, while the plane stress R curve rises with crack growth. (Refer to Fig. 2.10 for an illustration of flat and rising R curves.)

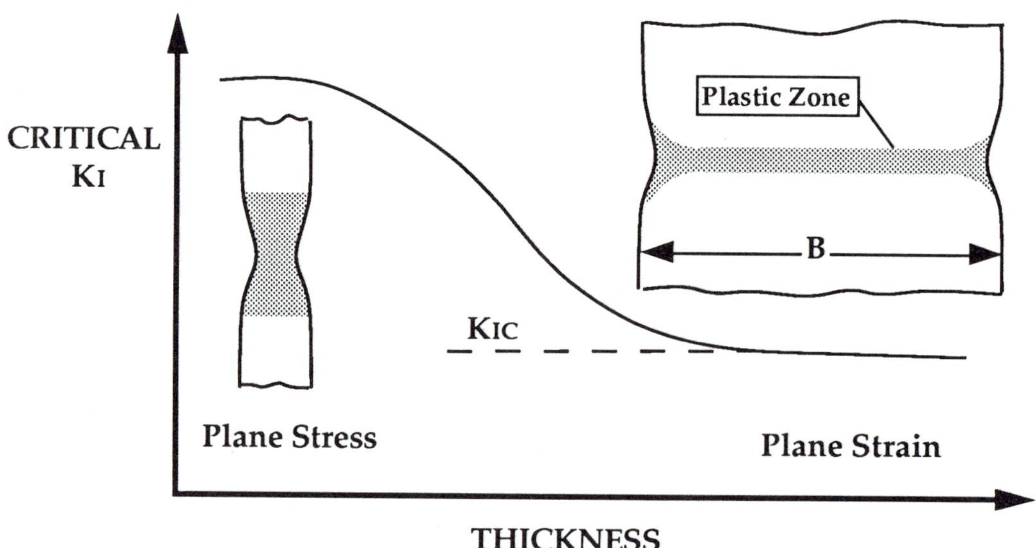

FIGURE 2.40 Effect of specimen thickness on Mode I fracture toughness.

The in-plane dimensions of a specimen or structure are as important as the thickness. In order for the stress intensity factor to have any meaning, there must be a singularity dominated zone near the crack tip. When the plastic zone becomes too large, the singularity dominated zone is destroyed, and K no longer characterizes crack tip conditions. Thus the plastic zone must be embedded within the singularity dominated zone. In general, the singularity zone is small relative to in-plane length scales in the structure (see Example 2.6). Therefore, the plastic zone must also be small compared to the relevant dimensions.

EXAMPLE 2.6

Estimate the relative size of the singularity dominated zone ahead of a through crack in an infinite plate subject to remote tension (Fig. 2.3), assuming the normal stress at $\theta = 0$ can be approximated as follows

$$\sigma_{yy} \approx \frac{K_I}{\sqrt{2\pi r}} + \sigma$$

where σ is the remotely applied tensile stress. Also, estimate the value of K_I where the plane strain plastic zone engulfs the singularity dominated zone.

Solution: Substituting Eq. (2.41) into the above relationship gives

$$\sigma_{yy} \approx \sigma\left(\sqrt{\frac{a}{2r}} + 1\right)$$

Thus r must be << a in order for the singular term to dominate. If we arbitrarily define the singularity dominated zone as the region where the first term is at least 5 times larger than the second term, the size of this zone is given by

$$r_s \approx \frac{a}{50}$$

EXAMPLE 2.6 (cont.)

In a plate with finite width, r_s would be small compared to all in-plane dimensions.

By setting the plane strain plastic zone correction (Eq. 2.63) equal to a/50, we obtain an estimate of the K_I value at which the singularity zone is engulfed by crack tip plasticity:

$$a = \frac{50}{6\pi}\left(\frac{K_I^*}{\sigma_{YS}}\right)^2 = 2.65\left(\frac{K_I^*}{\sigma_{YS}}\right)^2$$

or

$$K_I^* = 0.35\,\sigma_{YS}\sqrt{\pi a}$$

Therefore, when the nominal stress exceeds approximately 35% of yield in this case, the accuracy of K_I as a crack tip characterizing parameter is suspect.

2.10.3 Limits to the Validity of LEFM

According to the American Society for Testing and Materials (ASTM) standard for K_{IC} testing [22], the following specimen size requirements must be met to obtain a valid K_{IC} result in metals:

$$a,\ B,\ (W\text{-}a) \geq 2.5\left(\frac{K_I}{\sigma_{YS}}\right)^2 \tag{2.82}$$

Note the similarity between the crack length requirement and the result derived in Example 2.6. Thus Eq. (2.82) implies that r_y must be ≤ ~1/50 times specimen dimensions in order to obtain a size-independent critical K_I value. Equation (2.82) was based on experimental observations of the size dependence of fracture toughness in steel and aluminum [23]. The thickness requirement ensures nearly plane strain conditions, while the requirement on in-plane dimensions ensures that the nominal behavior is linear elastic and that K_I characterizes crack tip conditions.

Equation (2.82) gives the requirements for *plane strain, linear elastic* fracture. A valid K_{IC} result is a material property that does not depend on the size or geometry of the cracked body. While plane strain conditions are necessary to measure a valid K_{IC}, the lack of plane strain does not necessarily invalidate LEFM. As long as the in-plane dimensions are sufficiently large to confine the plastic zone to the singularity dominated zone, the stress intensity factor is a valid crack tip characterizing parameter. A fracture toughness value obtained from a laboratory specimen in plane stress or mixed conditions is applicable to a structure made of the same material, as long as the specimen and structure are the same thickness and the in-plane dimensions of both are large compared to the plastic zone. An example application of non-plane strain LEFM is fracture toughness testing of thin aluminum sheet used in aerospace structures.

Plasticity corrections such as those described in Section 2.8 can extend LEFM beyond its normal validity limits. One must remember, however, that the Irwin and strip yield corrections are only rough approximations of elastic-plastic behavior. When nonlinear material behavior becomes significant, one should discard stress intensity and adopt a crack tip parameter that takes the material behavior into account. Two such parameters, the crack tip opening displacement (CTOD) and the J integral, are the subject of Chapter 3.

REFERENCES

1. Inglis, C.E., "Stresses in a Plate Due to the Presence of Cracks and Sharp Corners." *Transactions of the Institute of Naval Architects*, Vol. 55, 1913, pp. 219-241.

2. Griffith, A.A. "The Phenomena of Rupture and Flow in Solids." *Philosophical Transactions*, Series A, Vol. 221, 1920, pp. 163-198.

3. Gehlen, P.C. and Kanninen, M.F., "An Atomic Model for Cleavage Crack Propagation in α iron." *Inelastic Behavior of Solids*, McGraw-Hill, New York, 1970, pp. 587-603.

4. Irwin, G.R., "Fracture Dynamics." *Fracturing of Metals*, American Society for Metals, Cleveland, 1948, pp. 147-166.

5. Orowan, E., "Fracture and Strength of Solids." *Reports on Progress in Physics*, Vol. XII, 1948, p. 185.

6. Irwin, G.R., "Onset of Fast Crack Propagation in High Strength Steel and Aluminum Alloys." *Sagamore Research Conference Proceedings*, Vol. 2, 1956, pp. 289-305.

7. Hutchinson, J.W. and Paris, P.C., "Stability Analysis of J-Controlled Crack Growth." ASTM STP 668, American Society for Testing and Materials, Philadelphia, 1979, pp. 37-64.

8. Westergaard, H.M., "Bearing Pressures and Cracks." *Journal of Applied Mechanics*, Vol. 6, 1939, pp. 49-53.

9. Irwin, G.R., "Analysis of Stresses and Strains near the End of a Crack Traversing a Plate." *Journal of Applied Mechanics*, Vol. 24, 1957, pp. 361-364.

10. Sneddon, I.N., "The Distribution of Stress in the Neighbourhood of a Crack in an Elastic Solid." *Proceedings, Royal Society of London*, Vol. A-187, 1946, pp. 229-260.

11. Williams, M.L., "On the Stress Distribution at the Base of a Stationary Crack." *Journal of Applied Mechanics*, Vol. 24, 1957, pp. 109-114.

12. Tada, H., Paris, P.C., and Irwin, G.R. *The Stress Analysis of Cracks Handbook.* (2nd Ed.) Paris Productions, Inc., St. Louis, 1985.

13. Murakami, Y. *Stress Intensity Factors Handbook.* Pergamon Press, New York, 1987.

14. Rooke, D.P. and Cartwright, D.J., *Compendium of Stress Intensity Factors.* Her Majesty's Stationary Office, London, 1976.

15. Irwin, G.R., "Plastic Zone Near a Crack and Fracture Toughness." *Sagamore Research Conference Proceedings*, Vol. 4, 1961.

16. Dugdale, D.S., "Yielding in Steel Sheets Containing Slits." *Journal of the Mechanics and Physics of Solids*, Vol 8, pp. 100-104.

17. Barenblatt, G.I., "The Mathematical Theory of Equilibrium Cracks in Brittle Fracture." *Advances in Applied Mechanics*, Vol VII, Academic Press, 1962, pp. 55-129.

18. Burdekin, F.M. and Stone, D.E.W., "The Crack Opening Displacement Approach to Fracture Mechanics in Yielding Materials." *Journal of Strain Analysis*, Vol. 1, 1966, pp. 145-153.

19. Dodds, R.H. Jr., Anderson T.L. and Kirk, M.T., "A Framework to Correlate a/W Effects on Elastic-Plastic Fracture Toughness (J_c)." to be published in *International Journal of Fracture.*

20. Narasimhan, R. and Rosakis A.J., "Three Dimensional Effects Near a Crack Tip in a Ductile Three Point Bend Specimen - Part I: A Numerical Investigation."

California Institute of Technology, Division of Engineering and Applied Science, Report SM 88-6, Pasadena, CA, January 1988.

21. Nakamura, T and Parks, D.M. "Conditions of J-Dominance in Three-Dimensional Thin Cracked Plates." *Analytical, Numerical, and Experimental Aspects of Three-Dimensional Fracture Processes*, ASME AMD-91, American Society of Mechanical Engineers, New York, 1988, pp. 227-238.

22. E 399-83, "Standard Test Method for Plane-Strain Fracture Toughness of Metallic Materials." American Society for Testing and Materials, Philadelphia, 1983.

23. Brown W.F. Jr. and Srawley, J.E., *Plane Strain Crack Toughness Testing of High Strength Metallic Materials.* ASTM STP 410, American Society for Testing and Materials, Philadelphia, PA, 1966.

24. Williams, M.L., "Stress Singularities Resulting from Various Boundary Conditions in Angular Corners of Plates in Extension." *Journal of Applied Mechanics*, Vol. 74, 1952, pp. 526-528.

25. Irwin, G.R., discussion of Ref. 29, 1958.

26. Sih, G.C., "On the Westergaard Method of Crack Analysis." *International Journal of Fracture Mechanics*, Vol. 2, 1966, pp. 628-631.

27. Eftis, J. and Liebowitz, H., "On the Modified Westergaard Equations for Certain Plane Crack Problems." *International Journal of Fracture Mechanics*, Vol. 8, p. 383.

28. Sanford, R.J., "A Critical Re-Examination of the Westergaard Method for Solving Opening Mode Crack Problems." *Mechanics Research Communications*, Vol. 6, 1979, pp. 289-294.

29. Wells, A.A. and Post, D., "The Dynamic Stress Distribution Surrounding a Running Crack--A Photoelastic Analysis." *Proceedings of the Society for Experimental Stress Analysis*, Vol. 16, 1958, pp. 69-92.

30. Muskhelishvili, N.I., *Some Basic Problems in the Theory of Elasticity.* Noordhoff, Ltd., Netherlands, 1953.

31. Green, A.E. and Sneddon, I.N., "The Distribution of Stress in the Neighbourhood of a Flat Elliptical Crack in an Elastic Solid." *Proceedings, Cambridge Philosophical Society, Vol. 46, 1950, pp. 159-163.*

APPENDIX 2: MATHEMATICAL FOUNDATIONS OF LINEAR ELASTIC FRACTURE MECHANICS

(Selected Results)

A2.1 PLANE ELASTICITY

This section catalogs the governing equations from which linear fracture mechanics is derived. The reader is encouraged to review the basis of these relationships by consulting one of the many textbooks on elasticity theory.[1]

The equations that follow are simplifications of more general relationships in elasticity and are subject to the following restrictions:

- Two-dimensional stress state (plane stress or plane strain).
- Isotropic material.
- Quasistatic, isothermal deformation.
- Body forces are absent from the problem. (In problems where body forces are present, a solution can first be obtained in the absence of body forces, and then modified by superimposing the body forces.)

Imposing these restrictions simplifies crack problems considerably, and permits closed-form solutions in many cases.

The governing equations of plane elasticity are given below for rectangular Cartesian coordinates. Section A2.1.2 lists the same relationships in terms of polar coordinates.

---

[1]Appendix 2 is intended only for more advanced readers, who have at least taken one graduate-level course in the theory of elasticity.

A2.1.1 Cartesian Coordinates

Strain-displacement relationships:

$$\varepsilon_{xx} = \frac{\partial u_x}{\partial x} \qquad \varepsilon_{yy} = \frac{\partial u_y}{\partial y} \qquad \varepsilon_{xy} = \frac{1}{2}\left(\frac{\partial u_x}{\partial y} + \frac{\partial u_y}{\partial x}\right) \qquad \text{(A2.1)}$$

where x and y are the horizontal and vertical coordinates, respectively, ε_{xx}, ε_{yy}, etc. are the strain components, and u_x and u_y are the displacement components.

Stress-strain relationships:

1. Plane strain.

$$\sigma_{xx} = \frac{E}{(1+\nu)(1-2\nu)}\left[(1-\nu)\,\varepsilon_{xx} + \nu\,\varepsilon_{yy}\right]$$

$$\sigma_{yy} = \frac{E}{(1+\nu)(1-2\nu)}\left[(1-\nu)\,\varepsilon_{yy} + \nu\,\varepsilon_{xx}\right]$$

$$\tau_{xy} = 2\,\mu\,\varepsilon_{xy} = \frac{E}{1+\nu}\,\varepsilon_{xy} \qquad \text{(A2.2)}$$

$$\sigma_{zz} = \nu\,(\sigma_{xx} + \sigma_{yy})$$

$$\varepsilon_{zz} = \varepsilon_{xz} = \varepsilon_{yz} = \tau_{xz} = \tau_{yz} = 0$$

where σ and τ are the normal and shear stress components, respectively, E is Young's modulus, μ is the shear modulus, and ν is Poisson's ratio.

2. Plane stress.

$$\sigma_{xx} = \frac{E}{1-\nu^2}\left[\varepsilon_{xx} + \nu\,\varepsilon_{yy}\right]$$

$$\sigma_{yy} = \frac{E}{1 - \nu^2} [\varepsilon_{yy} + \nu \varepsilon_{xx}]$$

$$\tau_{xy} = 2 \mu \varepsilon_{xy} = \frac{E}{1 + \nu} \varepsilon_{xy} \tag{A2.3}$$

$$\varepsilon_{zz} = \frac{-\nu}{1 + \nu} (\varepsilon_{xx} + \varepsilon_{yy})$$

$$\sigma_{zz} = \varepsilon_{xz} = \varepsilon_{yz} = \tau_{xz} = \tau_{yz} = 0$$

Equilibrium equations:

$$\frac{\partial \sigma_{xx}}{\partial x} + \frac{\partial \tau_{xy}}{\partial y} = 0 \qquad \frac{\partial \sigma_{yy}}{\partial y} + \frac{\partial \tau_{xy}}{\partial x} = 0 \tag{A2.4}$$

Compatibility equation:

$$\nabla^2 (\sigma_{xx} + \sigma_{yy}) = 0 \tag{A2.5}$$

where

$$\nabla^2 = \frac{\partial^2}{\partial x^2} + \frac{\partial^2}{\partial y^2}$$

Airy stress function:

For a two-dimensional continuous elastic medium, there exists a function $\Phi(x,y)$, from which the stresses can be derived:

$$\sigma_{xx} = \frac{\partial^2 \Phi}{\partial y^2} \qquad \sigma_{yy} = \frac{\partial^2 \Phi}{\partial x^2} \qquad \tau_{xy} = \frac{-\partial^2 \Phi}{\partial x \partial y} \tag{A2.6}$$

where Φ is the Airy stress function. The equilibrium and compatibility equations are automatically satisfied if Φ has the following property:

$$\frac{\partial^4 \Phi}{\partial x^4} + 2 \frac{\partial^4 \Phi}{\partial x \partial y} + \frac{\partial^4 \Phi}{\partial y^4} = 0$$

or

$$\nabla^2\nabla^2\Phi = 0 \tag{A2.7}$$

A2.1.2 Polar Coordinates

Strain-displacement relationships:

$$\varepsilon_{rr} = \frac{\partial u_r}{\partial r} \qquad \varepsilon_{\theta\theta} = \frac{u_r}{r} + \frac{1}{r}\frac{\partial u_\theta}{\partial \theta} \qquad \varepsilon_{r\theta} = \frac{1}{2}\left(\frac{1}{r}\frac{\partial u_r}{\partial \theta} + \frac{\partial u_\theta}{\partial r} - \frac{u_\theta}{r}\right) \tag{A2.8}$$

where u_r and u_θ are the radial and tangential displacement components, respectively.

Stress-strain relationships:

The stress-strain relationships in polar coordinates can be obtained by substituting r and θ for x and y in Eqs. (A2.2) and (A2.3). For example, the radial stress is given by

$$\sigma_{rr} = \frac{E}{(1+\nu)(1-2\nu)}\left[(1-\nu)\,\varepsilon_{rr} + \nu\,\varepsilon_{\theta\theta}\right] \tag{A2.9a}$$

for plane strain, and

$$\sigma_{rr} = \frac{E}{1-\nu^2}\left[\varepsilon_{rr} + \nu\,\varepsilon_{\theta\theta}\right] \tag{A2.9b}$$

for plane stress.

Equilibrium equations:

$$\begin{aligned} &\frac{\partial \sigma_{rr}}{\partial r} + \frac{1}{r}\frac{\partial \tau_{r\theta}}{\partial \theta} + \frac{\sigma_{rr} - \sigma_{\theta\theta}}{r} = 0 \\ &\frac{1}{r}\frac{\partial \sigma_{\theta\theta}}{\partial \theta} + \frac{\partial \tau_{r\theta}}{\partial r} + \frac{2\,\tau_{r\theta}}{r} = 0 \end{aligned} \tag{A2.10}$$

Compatibility equation:

$$\nabla^2(\sigma_{rr} + \sigma_{\theta\theta}) = 0 \tag{A2.11}$$

where

$$\nabla^2 = \frac{\partial^2}{\partial r^2} + \frac{1}{r}\frac{\partial^2}{\partial r} + \frac{1}{r^2}\frac{\partial^2}{\partial \theta^2}$$

Airy stress function:

$$\nabla^2\nabla^2\Phi = 0 \tag{A2.12}$$

where $\Phi = \Phi(r; \theta)$ and

$$\sigma_{rr} = \frac{1}{r^2}\frac{\partial^2\Phi}{\partial\theta^2} + \frac{1}{r}\frac{\partial\Phi}{\partial r} \qquad \sigma_{\theta\theta} = \frac{\partial^2\Phi}{\partial r^2} \qquad \tau_{r\theta} = -\frac{1}{r}\frac{\partial^2\Phi}{\partial r\partial\theta} + \frac{1}{r^2}\frac{\partial\Phi}{\partial\theta} \tag{A2.13}$$

A2.2 CRACK GROWTH INSTABILITY ANALYSIS

Figure 2.12 schematically illustrates the general case of a cracked structure with finite system compliance, C_M. The structure is held at a fixed remote displacement, Δ_T, given by

$$\Delta_T = \Delta + C_M P \tag{A2.14}$$

where Δ is the local load line displacement and P is the applied load. Differentiating Eq. (A2.14) gives

$$d\Delta_T = \left(\frac{\partial\Delta}{\partial a}\right)_P da + \left(\frac{\partial\Delta}{\partial P}\right)_a dP + C_M\, dP = 0 \tag{A2.15}$$

assuming Δ depends only on load and crack length. We can make this same assumption about the energy release rate:

$$d\mathcal{G} = \left(\frac{\partial\mathcal{G}}{\partial a}\right)_P da + \left(\frac{\partial\mathcal{G}}{\partial P}\right)_a dP \tag{A2.16}$$

Dividing both sides of Eq. (A2.16) by da and fixing Δ_T yields

$$\left(\frac{d\mathcal{G}}{da}\right)_{\Delta_T} = \left(\frac{\partial\mathcal{G}}{\partial a}\right)_P + \left(\frac{\partial\mathcal{G}}{\partial P}\right)_a \left(\frac{dP}{da}\right)_{\Delta_T} \tag{A2.17}$$

which, upon substitution of Eq. (A2.15), leads to

$$\left(\frac{d\mathcal{G}}{da}\right)_{\Delta_T} = \left(\frac{\partial\mathcal{G}}{\partial a}\right)_P - \left(\frac{\partial\mathcal{G}}{\partial P}\right)_a \left(\frac{\partial\Delta}{\partial a}\right)_P \left[C_m + \left(\frac{\partial\Delta}{\partial P}\right)_a\right]^{-1} \tag{A2.18}$$

A virtually identical expression for the J integral (Eq. 3.52) can be derived by assuming J depends only on P and a, and expanding dJ into its partial derivatives.

Under dead-loading conditions, $C_M = \infty$, and all but the first term in Eq. (A2.18) vanish. Conversely, $C_M = 0$ corresponds to an infinitely stiff system, and Eq. (A2.18) reduces to the pure displacement control case.

A2.3 CRACK TIP STRESS ANALYSIS

A variety of techniques are available for analyzing stresses in cracked bodies. This section focuses on two early approaches developed by Williams [11,24] and Westergaard [8]. These two analyses are complimentary; the Williams approach considers the local crack tip fields under generalized in-plane loading, while Westergaard provided a means for connecting the local fields to global boundary conditions in certain configurations.

Space limitations preclude listing every minute step in each derivation. Moreover, stress strain and displacement distributions are not derived for all modes of loading. The derivations that follow serve as illustrative examples. The reader who is interested in further details should consult the original references.

A2.3.1 Generalized In-Plane Loading

Williams [11,24] was the first to demonstrate the universal nature of the $1/\sqrt{r}$ singularity for elastic crack problems, although Inglis [1],

Westergaard [8] and Sneddon [10] had earlier obtained this result in specific configurations. Williams actually began by considering stresses at the corner of a plate with various boundary conditions and included angles; a crack is a special case where the included angle of the plate corner is 2π and the surfaces are traction free (Fig. A2.1).

For the configuration shown in Fig. A2.1(b), Williams postulated the following stress function:

$$\Phi = r^{\lambda+1}\left[c_1 \sin(\lambda+1)\theta^* + c_2 \cos(\lambda+1)\theta^* + c_3 \sin(\lambda-1)\theta^* + c_4 \cos(\lambda-1)\theta^*\right]$$

$$= r^{\lambda+1} F(\theta^*; \lambda) \tag{A2.19}$$

where c_1, c_2, c_3, and c_4 are constants, and θ^* is defined in Fig. A2.1(b). Invoking Eq. (A2.13) gives the following expressions for the stresses:

$$\sigma_{rr} = r^{\lambda-1}\left[F''(\theta^*) + (\lambda+1)F(\theta^*)\right]$$

$$\sigma_{\theta\theta} = r^{\lambda-1}\left[\lambda(\lambda+1)F(\theta^*)\right] \tag{A2.20}$$

$$\tau_{r\theta} = r^{\lambda-1}\left[-\lambda F'(\theta^*)\right]$$

where the primes denote derivatives with respect to θ^*. Williams also showed that Eq. (A2.19) implies that the displacements vary with r^{λ}. In order for displacements to be finite in all regions of the body, λ must be > 0. If the crack faces are traction free, $\sigma_{\theta\theta}(0) = \sigma_{\theta\theta}(2\pi) = \tau_{r\theta}(0) = \tau_{r\theta}(2\pi) = 0$, which implies the following boundary conditions:

$$F(0) = F(2\pi) = F'(0) = F'(2\pi) = 0 \tag{A2.21}$$

Assuming the constants in Eq. (A2.19) are nonzero in the most general case, the boundary conditions can only be satisfied when $\sin(2\pi\lambda) = 0$. Thus

$$\lambda = {}^{n}/_{2}, \quad \text{where } n = 1, 2, 3, \ldots$$

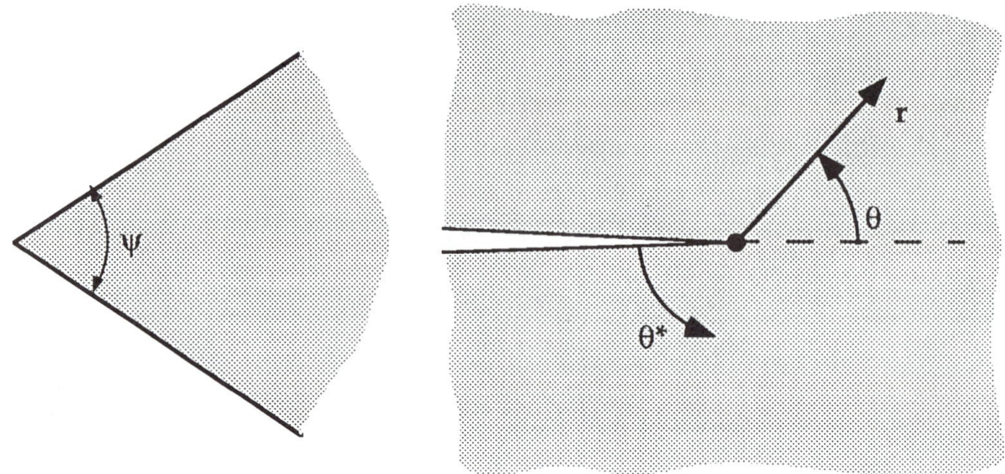

(a) Plate corner with included angle ψ. (b) Special case of a sharp crack.

FIGURE A2.1 Plate corner configuration analyzed by Williams [24]. A crack is formed when $\psi = 2\pi$.

There are an infinite number of λ values that satisfy the boundary conditions; the most general solution to a crack problem, therefore, is a polynomial of the form

$$\Phi = \sum_{n=1}^{N} \left(r^{n/2+1} \, F(\theta^*; n/2) \right) \tag{A2.22}$$

and the stresses are given by

$$\sigma_{ij} = \frac{\Gamma_{ij}(\theta^*; -1/2)}{\sqrt{r}} + \sum_{m=0}^{M} \left(r^{m/2} \, \Gamma_{ij}(\theta^*; m) \right) \tag{A2.23}$$

where Γ is a function that depends on F and its derivatives. The order of the stress function polynomial, N, must be sufficient to model the stresses in all regions of the body. When $r \to 0$, the first term in Eq. (A2.23) approaches infinity, while the higher order terms remain finite (when $m = 0$) or approach zero (for $m > 0$). Thus the higher order terms are negligible close to the crack tip, and stress exhibits an $1/\sqrt{r}$

singularity. Note that this result was obtained without assuming a specific configuration; thus it can be concluded that the inverse square-root singularity is universal for cracks in isotropic elastic media.

Further evaluation of Eqs. (A2.19) and (A2.20) with the appropriate boundary conditions reveals the precise nature of the function Γ. Recall that Eq. (A2.19) contains four, as yet unspecified, constants; by applying Eq. (A2.21), it is possible to eliminate two of these constants, resulting in

$$\Phi(r;\theta) = r^{n/2+1}\left\{c_3\left[\sin\left(\frac{n}{2}-1\right)\theta^* - \frac{n-2}{n+2}\sin\left(\frac{n}{2}+1\right)\theta^*\right] + c_4\left[\cos\left(\frac{n}{2}-1\right)\theta^* - \cos\left(\frac{n}{2}+1\right)\theta^*\right]\right\} \tag{A2.24}$$

for a given value of n. For crack problems it is more convenient to express the stress function in terms of, θ, the angle from the symmetry plane (Fig. A2.1). Substituting $\theta = \theta^* - \pi$ into Eq. (A2.24) yields, after some algebra, the following stress function for the first few values of n:

$$\Phi(r;\theta) = r^{3/2}\left[s_1\left(-\cos\frac{\theta}{2} - \frac{1}{3}\cos\frac{3\theta}{2}\right) + t_1\left(-\sin\frac{\theta}{2} - \sin\frac{3\theta}{2}\right)\right] + s_2 r^2\left[1-\cos 2\theta\right] + 0\left(r^{5/2}\right) + \ldots \tag{A2.25}$$

where s_i and t_i are constants to be defined. The stresses are given by

$$\sigma_{rr} = \frac{1}{4\sqrt{r}}\left\{s_1\left[-5\cos\frac{\theta}{2} + \cos\frac{3\theta}{2}\right] + t_1\left[-5\sin\frac{\theta}{2} + 3\sin\frac{3\theta}{2}\right]\right\} + 4s_2\cos^2\theta + 0\left(r^{1/2}\right) + \ldots$$

$$\sigma_{\theta\theta} = \frac{1}{4\sqrt{r}}\left\{s_1\left[-3\cos\frac{\theta}{2} - \cos\frac{3\theta}{2}\right] + t_1\left[-3\sin\frac{\theta}{2} - 3\sin\frac{3\theta}{2}\right]\right\} \tag{A2.26}$$

$$+ 4s_2 \sin^2\theta + 0\left(r^{1/2}\right) + \ldots$$

$$\tau_{r\theta} = \frac{1}{4\sqrt{r}}\left\{s_1\left[-\sin\frac{\theta}{2} - \sin\frac{3\theta}{2}\right] + t_1\left[\cos\frac{\theta}{2} + 3\cos\frac{3\theta}{2}\right]\right\}$$

$$- 2s_2 \sin 2\theta + 0\left(r^{1/2}\right) + \ldots$$

Note that the constants s_i in the stress function (Eq. (A2.25)) are multiplied by cosine terms while the t_i are multiplied by sine terms. Thus the stress function contains symmetric and antisymmetric components, with respect to $\theta = 0$. When the loading is symmetric about $\theta = 0$, $t_i = 0$, while $s_i = 0$ for the special case of pure antisymmetric loading. Examples of symmetric loading include pure bending and pure tension; in both cases the principal stress is normal to the crack plane. Therefore, symmetric loading corresponds to Mode I (Fig. 2.14); antisymmetric loading is produced by in-plane shear on the crack faces and corresponds to Mode II.

It is convenient in most cases to treat the symmetric and antisymmetric stresses separately. The constants s_1 and t_1 can be replaced by the Mode I and Mode II stress intensity factors, respectively:

$$s_1 = -\frac{K_I}{\sqrt{2\pi}} \qquad t_1 = \frac{K_{II}}{\sqrt{2\pi}} \tag{A2.27}$$

The crack tip stress fields for symmetric (Mode I) loading (assuming the higher order terms are negligible) are given by

$$\sigma_{rr} = \frac{K_I}{\sqrt{2\pi r}}\left[\frac{5}{4}\cos\left(\frac{\theta}{2}\right) - \frac{1}{4}\cos\left(\frac{3\theta}{2}\right)\right]$$

$$\sigma_{\theta\theta} = \frac{K_I}{\sqrt{2\pi r}}\left[\frac{3}{4}\cos\left(\frac{\theta}{2}\right) + \frac{1}{4}\cos\left(\frac{3\theta}{2}\right)\right] \tag{A2.28}$$

$$\tau_{r\theta} = \frac{K_I}{\sqrt{2\pi r}}\left[\frac{1}{4}\sin\left(\frac{\theta}{2}\right) + \frac{1}{4}\sin\left(\frac{3\theta}{2}\right)\right]$$

The singular stress fields for Mode II are given by

$$\sigma_{rr} = \frac{K_{II}}{\sqrt{2\pi r}}\left[-\frac{5}{4}\sin\left(\frac{\theta}{2}\right) + \frac{3}{4}\sin\left(\frac{3\theta}{2}\right)\right]$$

$$\sigma_{\theta\theta} = \frac{K_{II}}{\sqrt{2\pi r}}\left[-\frac{3}{4}\sin\left(\frac{\theta}{2}\right) - \frac{3}{4}\sin\left(\frac{3\theta}{2}\right)\right] \qquad \text{(A2.29)}$$

$$\tau_{r\theta} = \frac{K_{II}}{\sqrt{2\pi r}}\left[\frac{1}{4}\cos\left(\frac{\theta}{2}\right) + \frac{3}{4}\cos\left(\frac{3\theta}{2}\right)\right]$$

The relationships in Table 2.1 can be obtained by converting Eqs (A2.28) and (A2.29) to Cartesian coordinates.

The stress intensity factor defines the amplitude of the crack tip singularity; all stress and strain components at points near the crack tip increase in proportion to K, provided the crack is stationary. The precise definition of the stress intensity factor is arbitrary, however; the constants s_1 and t_1 would serve equally well for characterizing the singularity. The accepted definition of stress intensity stems from the early work of Irwin [9], who quantified the amplitude of the Mode I singularity with $\sqrt{\mathcal{G} E}$, where $\mathcal{G}$ is the energy release rate. It turns out that the $\sqrt{\pi}$ in the denominators of Eqs. (A2.28) and (A2.29) is superfluous (see Eqs. (A2.34) to (A2.36), below), but convention established over the last 35 years precludes redefining K in a more convenient form.

Williams also derived relationships for radial and tangential displacements near the crack tip. We will postpone evaluation of displacements until the next section, however, because the Westergaard approach for deriving displacements is somewhat more compact.

A2.3.2 The Westergaard Stress Function

Westergaard showed that a limited class of problems could be solved by introducing a complex stress function Z(z), where $z = x + iy$ and $i = \sqrt{-1}$. The Westergaard stress function is related to Airy stress function as follows:

$$\Phi = \text{Re}\bar{\bar{Z}} + y\,\text{Im}\bar{Z} \qquad \text{(A2.30)}$$

where Re and Im denote real and imaginary parts of the function, respectively, and the bars over Z represent integrations with respect to z; i.e.,

$$\bar{Z} = \frac{d\bar{\bar{Z}}}{dz} \quad \text{and} \quad Z = \frac{d\bar{Z}}{dz}$$

Applying Eq. (A2.6) gives

$$\sigma_{xx} = \text{Re}Z - y\,\text{Im}Z'$$

$$\sigma_{yy} = \text{Re}Z + y\,\text{Im}Z' \qquad \text{(A2.31)}$$

$$\tau_{xy} = -y\,\text{Re}Z'$$

Note that the imaginary part of the stresses vanishes when $y = 0$. In addition, the shear stress vanishes when $y = 0$, implying that the crack plane is a principal plane. Thus the stresses are symmetric about $\theta = 0$ and Eq. (A2.31) implies Mode I loading.

The Westergaard stress function, in its original form, is suitable for solving a limited range of Mode I crack problems. Subsequent modifications [25-28] generalized the Westergaard approach to be applicable to a wider range of cracked configurations.

Consider a through crack in an infinite plate subject to biaxial remote tension (Fig. A2.2). If the origin is defined at the center of the crack, the Westergaard stress function is given by

$$Z(z) = \frac{\sigma z}{\sqrt{z^2 - a^2}} \tag{A2.32}$$

where σ is the remote stress and a is the half crack length, as defined in Fig. A2.2. Consider the crack plane where y = 0. For $-a < x < a$, Z is pure imaginary, while Z is real for $|x| > |a|$. The normal stresses on the crack plane are given by

$$\sigma_{xx} = \sigma_{yy} = \text{Re}Z = \frac{\sigma x}{\sqrt{x^2 - a^2}} \tag{A2.33}$$

Let us now consider the horizontal distance from each crack tip, $x^* = x - a$; Eq. (A2.33) becomes

$$\sigma_{xx} = \sigma_{yy} = \frac{\sigma\sqrt{a}}{\sqrt{2x^*}} \tag{A2.34}$$

for $x^* << a$. Thus the Westergaard approach leads to the expected inverse square-root singularity. One advantage of this analysis is that it relates the local stresses to the global stress and crack size. From Eq. (A2.28), the stresses on the crack plane (θ = 0) are given by

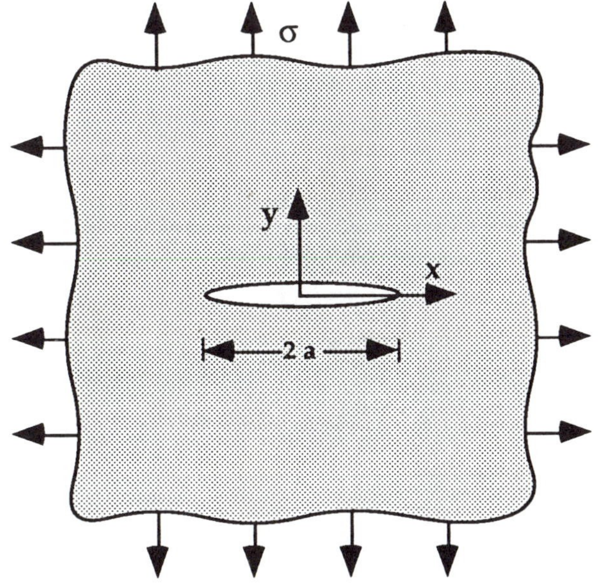

FIGURE A2.2 Through-thickness crack in an infinite plate loaded in biaxial tension.

$$\sigma_{rr} = \sigma_{\theta\theta} = \sigma_{xx} = \sigma_{yy} = \frac{K_I}{\sqrt{2\pi x^*}} \tag{A2.35}$$

Comparing Eqs. (A2.34) and (A2.35) gives

$$K_I = \sigma\sqrt{\pi a} \tag{A3.36}$$

for the configuration in Fig. A2.2. Note that $\sqrt{\pi}$ appears in Eq. (A3.36) because K was originally defined in terms of the energy release rate; a more appropriate definition of stress intensity might be

$$\sigma_{yy}(\theta = 0) = \frac{K_I^*}{\sqrt{2x^*}}, \quad \text{where } K_I^* = \sigma\sqrt{a} \text{ for the plate in Fig. A2.2.}$$

Substituting Eq. (A3.36) into Eq. (A2.32) results in an expression of the Westergaard stress function in terms of K_I:

$$Z(z^*) = \frac{K_I}{\sqrt{2\pi z^*}} \tag{A3.37}$$

where $z^* = z - a$. It is possible to solve for the singular stresses at other angles by making the following substitution in Eq. (A2.37):

$$z^* = r\,e^{i\theta}$$

where

$$r^2 = (x - a)^2 + y^2 \quad \text{and} \quad \theta = \tan^{-1}\left(\frac{y}{x-a}\right)$$

which leads to

$$\sigma_{xx} = \frac{K_I}{\sqrt{2\pi r}}\cos\left(\frac{\theta}{2}\right)\left[1 - \sin\left(\frac{\theta}{2}\right)\sin\left(\frac{3\theta}{2}\right)\right]$$

$$\sigma_{yy} = \frac{K_I}{\sqrt{2\pi r}}\cos\left(\frac{\theta}{2}\right)\left[1 + \sin\left(\frac{\theta}{2}\right)\sin\left(\frac{3\theta}{2}\right)\right] \tag{A2.38}$$

$$\tau_{xy} = \frac{K_I}{\sqrt{2\pi r}} \cos\left(\frac{\theta}{2}\right) \sin\left(\frac{\theta}{2}\right) \cos\left(\frac{3\theta}{2}\right)$$

assuming r << a. Equation (A2.38) is equivalent to Eq. (A2.28), except that the latter is expressed in terms of polar coordinates.

Westergaard published the following stress function for an array of collinear cracks in a plate in biaxial tension (Fig 2.21):

$$Z(z) = \frac{\sigma}{\left\{1 - \left[\frac{\sin\left(\frac{\pi a}{2W}\right)}{\sin\left(\frac{\pi z}{2W}\right)}\right]^2\right\}^{1/2}} \tag{A2.39}$$

where a is the half crack length and 2W is the spacing between the crack centers. The stress intensity for this case is given in Eq. (2.45); early investigators used this solution to approximate the behavior of center cracked tensile pane with finite width.

Irwin [9] published stress functions for several additional configurations, including a pair of crack opening forces located a distance X from the crack center (Fig. 2.30):

$$Z(z) = \frac{P\,a}{\pi\,(z - X)\,z} \sqrt{\frac{1 - (X/a)^2}{1 - (a/z)^2}} \tag{A2.40}$$

where P is the applied force. When there are matching forces at ± X, the appropriate stress function can be obtained by superposition:

$$Z(z) = \frac{2\,P\,a}{\pi\,(z^2 - X^2)} \sqrt{\frac{1 - (X/a)^2}{1 - (a/z)^2}} \tag{A2.41}$$

In each case, the stress function can be expressed in the form of Eq. (A2.37) and the near tip stresses are given by Eq. (A2.38). This is not surprising, since all of the above cases are pure Mode I and the

Williams analysis showed that the inverse square root singularity is universal.

The in-plane displacements are related to the Westergaard stress function as follows:

$$u_x = \frac{1-2\nu}{2\mu} \mathrm{Re}\bar{Z} - y\,\mathrm{Im}Z$$

$$u_y = \frac{1-\nu}{\mu} \mathrm{Im}\bar{Z} - y\,\mathrm{Re}Z \tag{A2.42}$$

For the plate in Fig. A2.2, the crack opening displacement is given by

$$2u_y = 2\frac{1-\nu}{\mu}\mathrm{Im}\bar{Z} = \frac{2(1-\nu^2)}{E}\mathrm{Im}\bar{Z} = \frac{4(1-\nu^2)\sigma}{E}\sqrt{a^2-x^2} \tag{A2.43}$$

Eq. (A2.43) predicts that a through crack forms an elliptical opening profile when subjected to tensile loading.

The near-tip displacements can be obtained by inserting Eq. (A2.37) into Eq. (A2.42):

$$u_x = \frac{K_I}{2\mu}\sqrt{\frac{r}{2\pi}}\cos\left(\frac{\theta}{2}\right)\left[\kappa - 1 + 2\sin^2\left(\frac{\theta}{2}\right)\right]$$

$$u_y = \frac{K_I}{2\mu}\sqrt{\frac{r}{2\pi}}\sin\left(\frac{\theta}{2}\right)\left[\kappa + 1 - 2\cos^2\left(\frac{\theta}{2}\right)\right] \tag{A2.44}$$

for $r \ll a$, where $\kappa = 3 - 4\nu$. Equation (A2.44) is also valid for plane stress if

$$\kappa = \frac{3-\nu}{1+\nu} \tag{A2.45}$$

Although the original Westergaard approach correctly describes the singular Mode I stresses in certain configurations, it is not sufficiently general to apply to all Mode I problems; this shortcoming has prompted various modifications to the Westergaard stress function. Irwin [25] noted that photoelastic fringe patterns observed by Wells and Post [29] on center cracked panels did not match the shear strain contours predicted by the Westergaard solution. Irwin achieved good agreement between theory and experiment by subtracting a uniform horizontal stress:

$$\sigma_{xx} = \text{Re}Z - y\text{Im}Z' - \sigma_{oxx} \tag{A2.46}$$

where σ_{oxx} depends on the remote stress. The other two stress components remain the same as in Eq. (A2.31). Subsequent analyses have revealed that when a center cracked panel is loaded in uniaxial tension, a transverse compressive stress develops in the plate. Thus Irwin's modification to the Westergaard solution has a physical basis in the case of a center cracked panel[2]. Equation (A2.46) has been used to interpret photoelastic fringe patterns in a variety of configurations.

Sih [26] provided a theoretical basis for the Irwin modification. A stress function for Mode I must lead to zero shear stress on the crack plane. Sih showed that the Westergaard function was more restrictive than it needed to be, and was thus unable to account for all situations. Sih generalized the Westergaard approach by applying a complex potential formulation for the Airy stress function [30]. He imposed the condition $\tau_{xy} = 0$ at $y = 0$, and showed that the stresses could be expressed in terms of a new function $\phi(z)$:

$$\begin{aligned}
\sigma_{xx} &= 2\,\text{Re}\phi'(z) - 2\,y\,\text{Im}\phi''(z) - A \\
\sigma_{yy} &= 2\,\text{Re}\phi'(z) + 2\,y\,\text{Im}\phi''(z) + A \\
\tau_{xy} &= 2\,y\,\text{Re}\phi''(z)
\end{aligned} \tag{A2.47}$$

---

[2]Recall that the stress functions in Eqs. (2.32) and (2.39) are strictly valid only for biaxial loading. Although this restriction was not imposed in Westergaard's original work, a transverse tensile stress is necessary in order to cancel with $-\sigma_{oxx}$. However, the transverse stresses, whether compressive or tensile, do not affect the singular term; thus the stress intensity factor is the same for uniaxial and biaxial tensile loading and is given by Eq. (A2.36).

where A is a real constant. Equation (A2.47) is equivalent to the Irwin modification of the Westergaard approach if [27]

$$2\,\phi'(z) = Z(z) - A \tag{A2.48}$$

Substituting Eq. (A2.48) into Eq. (A2.47) gives

$$\begin{aligned}\sigma_{xx} &= \mathrm{Re}Z - y\,\mathrm{Im}Z' - 2A \\ \sigma_{yy} &= \mathrm{Re}Z + y\,\mathrm{Im}Z' \\ \tau_{xy} &= -y\,\mathrm{Re}Z'\end{aligned} \tag{A2.49}$$

Comparing Eq. (A2.49) with Eqs. (A2.31) and (A2.46), it is obvious that the Sih and Irwin modifications are equivalent and $2A = \sigma_{oxx}$.

Sanford [28] showed that the Irwin-Sih approach is still too restrictive, and he proposed replacing A with a complex function $\eta(z)$:

$$2\,\phi'(z) = Z(z) - \eta(z) \tag{A2.50}$$

The modified stresses are given by

$$\begin{aligned}\sigma_{xx} &= \mathrm{Re}Z - y\,\mathrm{Im}Z' + y\mathrm{Im}\eta' - 2\,\mathrm{Re}\eta \\ \sigma_{yy} &= \mathrm{Re}Z + y\,\mathrm{Im}Z' + y\mathrm{Im}\eta' \\ \tau_{xy} &= -y\,\mathrm{Re}Z' + y\,\mathrm{Re}\eta' + \mathrm{Im}\eta\end{aligned} \tag{A2.51}$$

Equation (A2.51) represents the most general form of Westergaard-type stress functions. When $\eta(z)$ = a real constant for all z, Eq. (A2.51) reduces to the Irwin-Sih approach, while Eq. (A2.51) reduces to the original Westergaard solution when $\eta(z) = 0$ for all z.

The function η can be represented as a polynomial of the form

$$\eta(z) = \sum_{m=0}^{M} \alpha_m\, z^{m/2} \tag{A2.52}$$

Combining Eqs. (A2.37), (A2.50), and (A2.52) and defining the origin at the crack tip gives

$$2\,\phi' = \frac{K_I}{\sqrt{2\pi z}} - \sum_{m=0}^{M} \alpha_m\, z^{m/2} \tag{A2.53}$$

which is consistent with the Williams [11,24] asymptotic expansion.

A2.4 ELLIPTICAL INTEGRAL OF THE SECOND KIND

The solution of stresses in the vicinity of elliptical and semielliptical cracks in elastic solids [10,31] involves an elliptic integral of the second kind:

$$\Psi = \int_0^{\pi/2} \sqrt{1 - \frac{c^2 - a^2}{c^2}\sin^2\phi}\;\; d\phi \tag{A2.54}$$

where 2c and 2a are the major and minor axes of the elliptical flaw, respectively. Series expansion of Eq. (A2.54) gives

$$\Psi = \frac{\pi}{2}\left[1 - \frac{1}{4}\frac{c^2 - a^2}{c^2} - \frac{3}{64}\left(\frac{c^2 - a^2}{c^2}\right)^2 - \ldots\right] \tag{A2.55}$$

The first two terms of Eq. (A2.55) are large compared to higher order terms. Thus Ψ can be approximated by

$$\Psi \approx \frac{3\,\pi}{8} + \frac{\pi\, a^2}{8\, c^2} \tag{A2.56}$$

3. ELASTIC-PLASTIC FRACTURE MECHANICS

Linear elastic fracture mechanics (LEFM) is only valid as long as nonlinear material deformation is confined to a small region surrounding the crack tip. In many materials, it is virtually impossible to characterize the fracture behavior with LEFM, and an alternative fracture mechanics model is required.

Elastic-plastic fracture mechanics applies to materials that exhibit time-independent, nonlinear behavior (i.e., plastic deformation). Two elastic-plastic parameters are introduced in this chapter: the crack tip opening displacement (CTOD) and the J contour integral. Both parameters describe crack tip conditions in elastic-plastic materials, and each can be used as a fracture criterion. Critical values of CTOD or J give nearly size-independent measures of fracture toughness, even for relatively large amounts of crack tip plasticity. There are limits to the applicability of J and CTOD (Sections 3.5 & 3.6), but these limits are much less restrictive than the validity requirements of LEFM.

3.1 CRACK TIP OPENING DISPLACEMENT

When Wells [1] attempted to measure K_{IC} values in a number of structural steels, he found that these materials were too tough to be characterized by LEFM. This discovery brought both good news and bad news: high toughness is obviously desirable to designers and fabricators, but Wells' experiments indicated that existing fracture mechanics theory was not applicable to an important class of materials. While examining fractured test specimens, Wells noticed that the crack faces had moved apart prior to fracture; plastic deformation blunted an initially sharp crack, as illustrated in Fig. 3.1. The degree of crack blunting increased in proportion to the toughness of the material. This observation led Wells to propose the opening at the crack tip as a measure of fracture toughness. Today, this parameter is known as the crack tip opening displacement (CTOD).

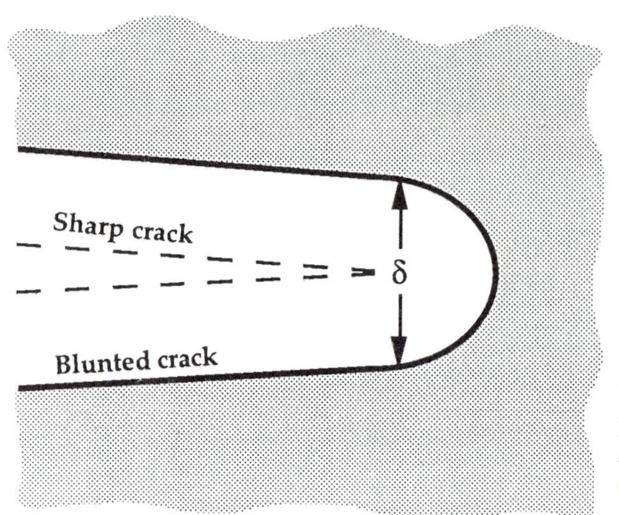

FIGURE 3.1 Crack tip opening displacement (CTOD). An initially sharp crack blunts with plastic deformation, resulting in a finite displacement (δ) at the crack tip.

In his original paper, Wells [1] performed an approximate analysis that related CTOD to the stress intensity factor in the limit of small scale yielding. Consider a crack with a small plastic zone, as illustrated in Fig. 3.2. Irwin [2] showed that crack tip plasticity makes the crack behave as if it were slightly longer. Thus, we can estimate CTOD by solving for the displacement at the physical crack tip, assuming an effective crack length of $a+r_y$. From Table 2.2, the displacement r_y behind the effective crack tip is given by

$$u_y = \frac{\kappa + 1}{2\mu} K_I \sqrt{\frac{r_y}{2\pi}} \tag{3.1}$$

and the Irwin plastic zone correction for plane stress is

$$r_y = \frac{1}{2\pi}\left(\frac{K_I}{\sigma_{YS}}\right)^2 \tag{3.2}$$

Substituting Eq. (3.1) into Eq. (3.2) gives

$$\delta = 2\,u_y = \frac{4}{\pi}\frac{K_I^2}{\sigma_{YS}\,E} \tag{3.3}$$

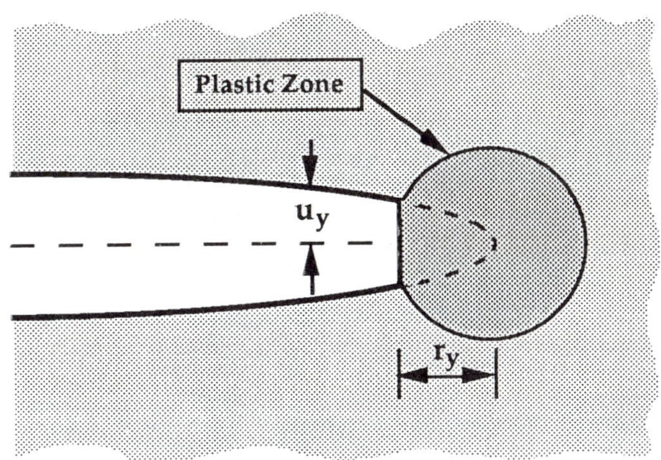

FIGURE 3.2 Estimation of CTOD from the displacement of the effective crack in the Irwin plastic zone correction.

where δ is the CTOD. Alternatively, CTOD can be related to the energy release rate by applying Eq. (2.51):

$$\delta = \frac{4}{\pi} \frac{\mathcal{G}}{\sigma_{YS}} \tag{3.4}$$

Thus in the limit of small scale yielding, CTOD is related to G and K_I. Wells postulated that CTOD is an appropriate crack tip characterizing parameter when LEFM is no longer valid. This assumption was shown to be correct several years later when a unique relationship between CTOD and the J integral was established (Section 3.3).

The strip yield model provides an alternate means for analyzing CTOD [3]. Recall Section 2.8.2, where the plastic zone was modelled by yield magnitude closure stresses. The size of the strip yield zone was defined by the requirement of finite stresses at the crack tip. The CTOD can be defined as the crack opening displacement at the end of the strip yield zone, as Fig. 3.3 illustrates. According to this definition, CTOD in a through crack in an infinite plate subject to a remote tensile stress (Fig. 2.3) is given by [3]

$$\delta = \frac{8 \sigma_{YS} a}{\pi E} \ln \sec\left(\frac{\pi}{2} \frac{\sigma}{\sigma_{YS}}\right) \tag{3.5}$$

Equation (3.5) is derived in Appendix 3.1. Series expansion of the ln sec term gives

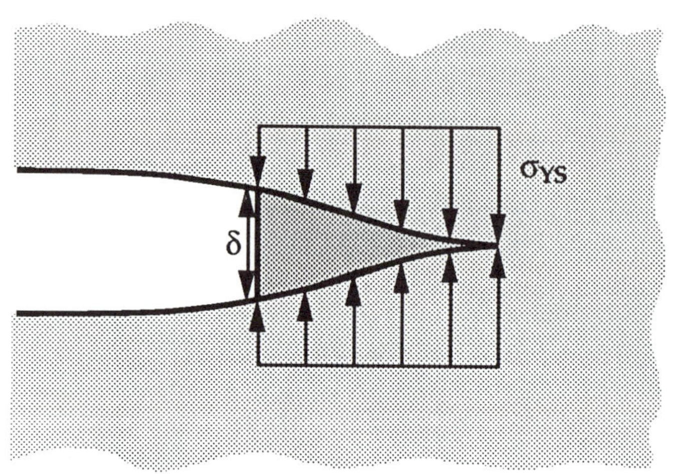

FIGURE 3.3 Estimation of CTOD from the strip yield model [3].

$$\delta = \frac{8\,\sigma_{YS}\,a}{\pi E}\left[\frac{1}{2}\left(\frac{\pi}{2}\frac{\sigma}{\sigma_{YS}}\right)^2 + \frac{1}{12}\left(\frac{\pi}{2}\frac{\sigma}{\sigma_{YS}}\right)^4 + \ldots\right]$$

$$= \frac{K_I^2}{\sigma_{YS} E}\left[1 + \frac{1}{6}\left(\frac{\pi}{2}\frac{\sigma}{\sigma_{YS}}\right)^2 + \ldots\right] \quad (3.6)$$

For $\sigma << \sigma_{YS}$, the higher order terms are negligible and CTOD is given by

$$\delta = \frac{K_I^2}{\sigma_{YS} E} = \frac{\mathcal{G}}{\sigma_{YS}} \quad (3.7)$$

which differs slightly from Eq. (3.3).

The strip yield model assumes plane stress conditions and a non-hardening material. The actual relationship between CTOD and K_I and $\mathcal{G}$ depends on stress state and strain hardening. The more general form of this relationship can be expressed as follows:

$$\delta = \frac{K_I^2}{m\,\sigma_{YS} E'} = \frac{\mathcal{G}}{m\,\sigma_{YS}} \quad (3.8)$$

where m is a dimensionless constant that is approximately 1.0 for plane stress and 2.0 for plane strain.

There are a number of alternative definitions of CTOD. The two most common definitions, which are illustrated in Fig. 3.4, are the displacement at the original crack tip and the 90° intercept. The latter definition was suggested by Rice [4] and is commonly used to infer CTOD in finite element measurements. Note that these two definitions are equivalent if the crack blunts in a semicircle.

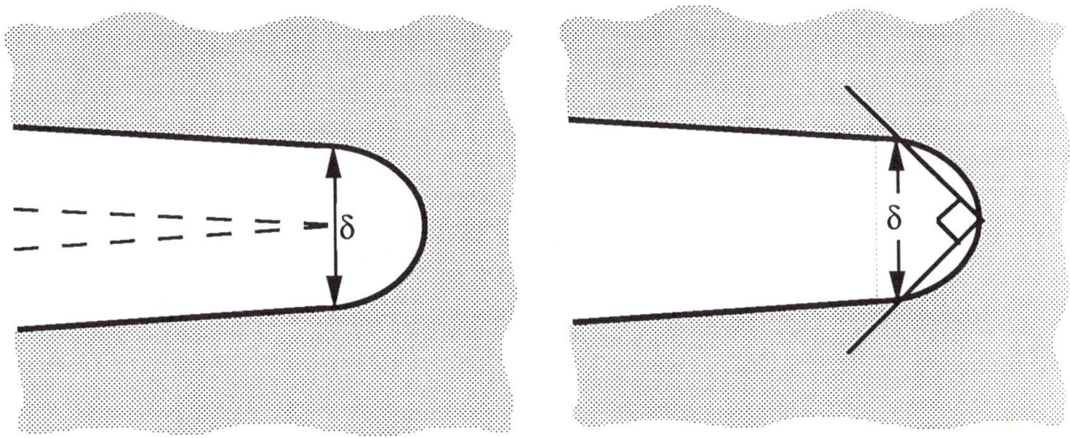

(a) Displacement at the original crack tip.

(b) Displacement at the intersection of a 90° vertex with the crack flanks.

FIGURE 3.4 Alternative definitions of CTOD.

Most laboratory measurements of CTOD have been made on edge-cracked specimens loaded in three-point bending (see Table 2.4). Early experiments utilized a flat paddle-shaped gage that was inserted into the crack; as the crack opened, the paddle gage rotated, and an electronic signal was sent to an x-y plotter. This method was inaccurate, however, because it was difficult to reach the crack tip with the paddle gage. Today, the displacement, V, at the crack mouth is measured, and the CTOD is inferred by assuming the specimen halves are rigid and rotate about a hinge point, as illustrated in Fig. 3.5. Referring to this figure, we can estimate CTOD from a similar triangles construction:

$$\frac{\delta}{r(W-a)} = \frac{V}{r(W-a)+a}$$

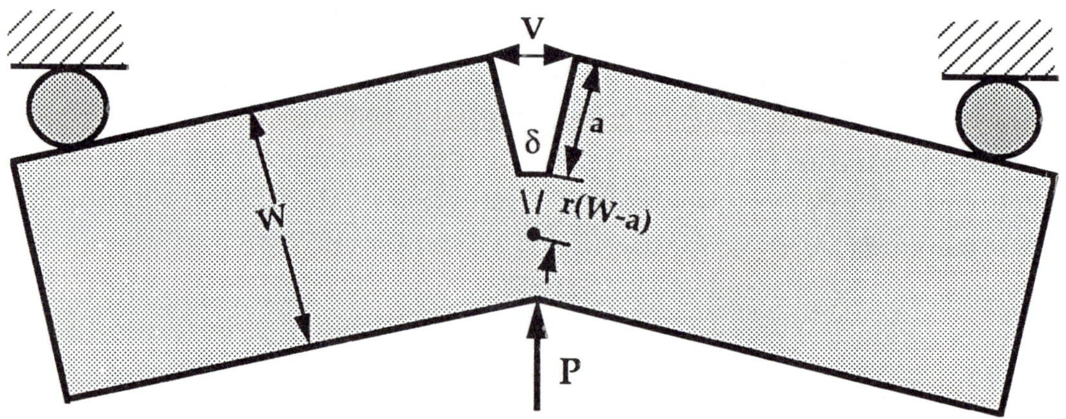

FIGURE 3.5 The hinge model for estimating CTOD from three-point bend specimens.

Therefore,

$$\delta = \frac{r(W - a)\,V}{r(W - a) + a} \tag{3.9}$$

where r is the rotational factor, a dimensionless constant between 0 and 1.

The hinge model is inaccurate when displacements are primarily elastic. Consequently, standard methods for CTOD testing [5,6] typically adopt a modified hinge model, in which displacements are separated into elastic and plastic components; the hinge assumption is applied only to plastic displacements. Figure 3.6 illustrates a typical load (P) versus displacement (V) curve from a CTOD test. The shape of the load-displacement curve is similar to a stress-strain curve: it is initially linear but deviates from linearity with plastic deformation. At a given point on the curve, the displacement is separated into elastic and plastic components by constructing a line parallel to the elastic loading line. The dashed line represents the path of unloading for this specimen, assuming the crack does not grow during the test. The CTOD in this specimen is estimated by

$$\delta = \delta_{el} + \delta_p = \frac{K_I^2}{m\,\sigma_{YS}\,E'} + \frac{r_p(W - a)\,V_p}{r_p(W - a) + a} \tag{3.10}$$

The subscripts "el" and "p" denote elastic and plastic components, respectively. The elastic stress intensity factor is computed by inserting the load and specimen dimensions into the appropriate expression in Table 2.4. The plastic rotational factor, r_p, is approximately 0.44 for typical materials and test specimens. Note that Equation (3.10) reduces to the small scale yielding result (Eq. 3.8) for linear elastic conditions, but the hinge model dominates when $V \approx V_p$.

Further details of CTOD testing are given in Chapter 7. Chapter 9 outlines how CTOD is used in design.

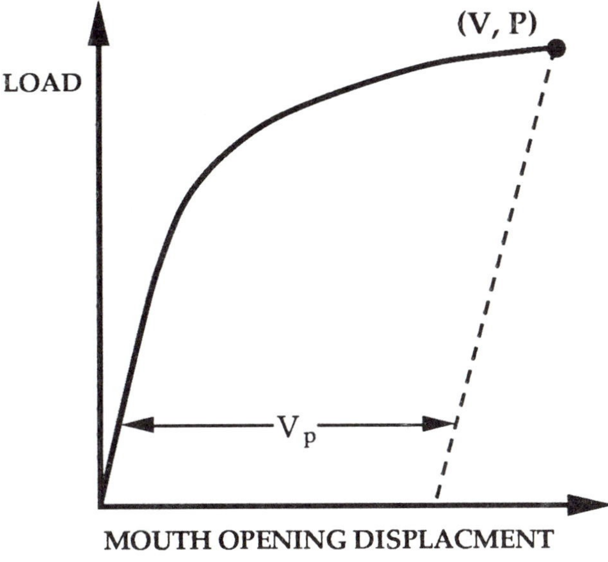

FIGURE 3.6 Determination of the plastic component of the crack mouth opening displacement.

3.2 THE J CONTOUR INTEGRAL

The J contour integral has enjoyed great success as a fracture characterizing parameter for nonlinear materials. By idealizing elastic-plastic deformation as nonlinear elastic, Rice [4] provided the basis for extending fracture mechanics methodology well beyond the validity limits of LEFM.

Figure 3.7 illustrates the uniaxial stress-strain behavior of elastic-plastic and nonlinear elastic materials. The loading behavior for the two materials is identical, but the material responses differ when each is unloaded. The elastic-plastic material follows a linear unloading path with the slope equal to Young's modulus, while the nonlinear elastic material unloads along the same path as it was loaded. There is a unique relationship between stress and strain in an elastic material, but a given strain in an elastic-plastic material can correspond to more than one stress value if the material is unloaded or cyclically loaded. Consequently, it is much easier to analyze an elastic material than a material that exhibits irreversible plasticity.

As long as the stresses in both materials in Fig. 3.7 increase monotonically, the mechanical response of the two materials is identical. When the problem is generalized to three dimensions, it does not necessarily follow that the loading behavior of the nonlinear elastic and elastic-plastic materials is identical, but there are many instances where this is a good assumption (see Appendix 3.6). Thus an analysis that assumes nonlinear elastic behavior may be valid for an elastic-plastic material, provided no unloading occurs. The *deformation theory of plasticity,* which relates total strains to stresses in a material, is equivalent to nonlinear elasticity.

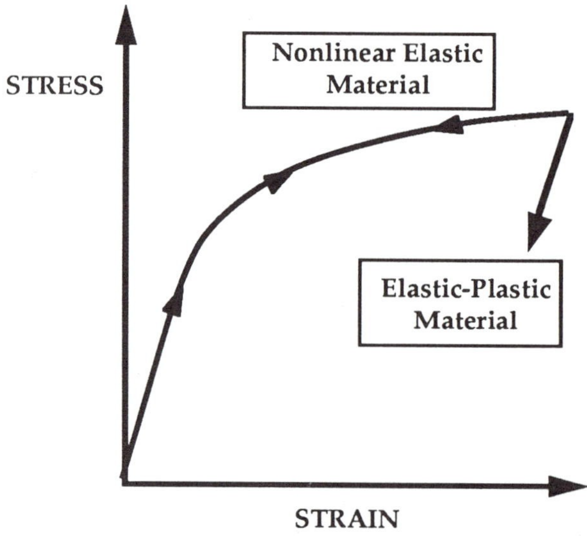

FIGURE 3.7 Schematic comparison of the stress-strain behavior of elastic-plastic and nonlinear elastic materials.

Rice [4] applied deformation plasticity (i.e., nonlinear elasticity) to the analysis of a crack in a nonlinear material. He showed that the nonlinear energy release rate, J, could be written as a path-independent line integral. Hutchinson [7] and Rice and Rosengren [8] also showed that J uniquely characterizes crack tip stresses and strains in nonlinear materials. Thus the J integral can be viewed as both an energy parameter and a stress intensity parameter.

3.2.1 Nonlinear Energy Release Rate

Rice [4] presented a path-independent contour integral for analysis of cracks. He then showed that the value of this integral, which he called J, is equal to the energy release rate in a nonlinear elastic body that contains a crack. In this section, however, the energy release rate interpretation is discussed first because it is closely related to concepts introduced in Chapter 2. The J contour integral is outlined in Section 3.2.2. Appendix 3.2 gives a mathematical proof, similar to what Rice [4] presented, that shows that this line integral is equivalent to the energy release rate in nonlinear elastic materials.

Equation (2.23) defines the energy release rate for linear materials. The same definition holds for nonlinear elastic materials, except that $\mathcal{G}$ is replaced by J:

$$J = -\frac{d\Pi}{d\mathcal{A}} \tag{3.11}$$

where Π is the potential energy and $\mathcal{A}$ is crack area. The potential energy is given by

$$\Pi = U - F \tag{3.12}$$

where U is the strain energy stored in the body and F is the work done by external forces. Consider a cracked plate which exhibits a nonlinear load-displacement curve, as illustrated in Fig. 3.8. If the plate has unit thickness, $\mathcal{A}$ = a.[1] For load control,

---

[1] It is important to remember that the energy release rate is defined in terms of crack area, not crack length. Failure to recognize this can lead to errors and confusion when

$$\Pi = U - P\Delta = -U^*$$

where U* is the complimentary strain energy, defined as

$$U^* = \int_0^P \Delta \, dP \tag{3.13}$$

Thus if the plate in Fig. 3.8 is in load control, J is given by

$$J = \left(\frac{dU^*}{da}\right)_P \tag{3.14}$$

If the crack advances at a fixed displacement, F = 0, and J is given by

$$J = -\left(\frac{dU}{da}\right)_\Delta \tag{3.15}$$

According to Fig. 3.8, dU* for load control differs from -dU for displacement control by the amount $\frac{1}{2}\,dPd\Delta$, which is vanishingly small compared to dU. Therefore, J for load control is equal to J for displacement control. Recall that we obtained this same result for $\mathcal{G}$ in Section 2.4.

By invoking the definitions for U and U*, we can express J in terms of load and displacement:

$$J = \left(\frac{\partial}{\partial a}\int_0^P \Delta \, dP\right)_P$$

$$= \int_0^P \left(\frac{\partial \Delta}{\partial a}\right)_P dP \tag{3.16}$$

---

computing $\mathcal{G}$ or J for configurations other than edge cracks; examples include a through crack, where $d\mathcal{A} = 2da$ (assuming unit thickness), and a penny-shaped crack, where $d\mathcal{A} = 2\pi a da$.

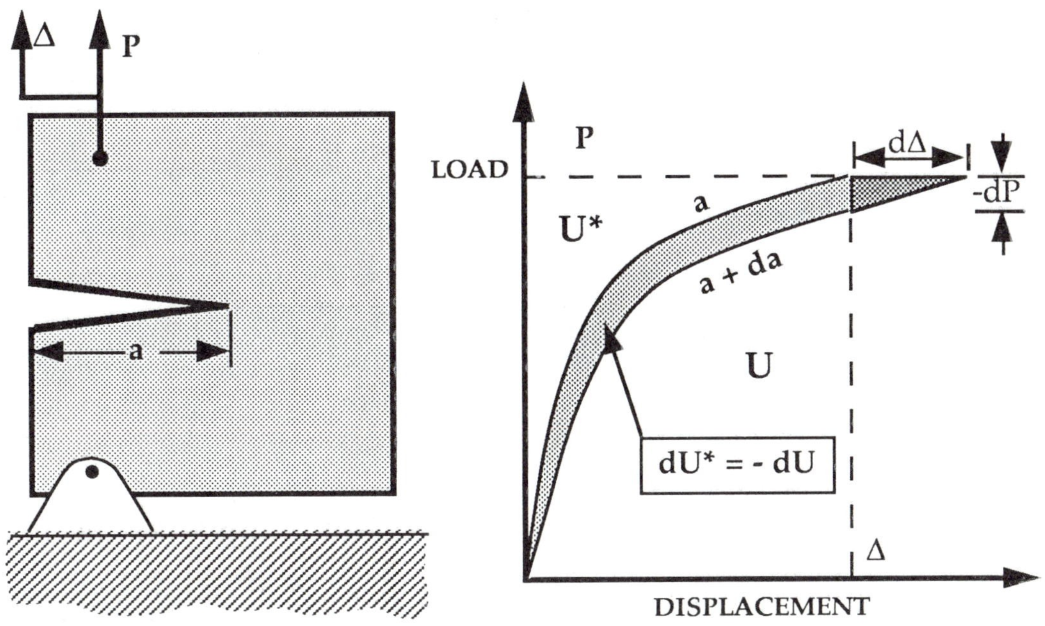

FIGURE 3.8 Nonlinear energy release rate.

or

$$J = -\left(\frac{\partial}{\partial a}\int_0^\Delta P\, d\Delta\right)_\Delta$$

$$= -\int_0^\Delta \left(\frac{\partial P}{\partial a}\right)_\Delta d\Delta \tag{3.17}$$

Integrating Eq. (3.17) by parts leads to a rigorous proof of what we have already inferred from Fig. 3.8. That is, Eqs. (3.16) and (3.17) are equal, and J is the same for fixed load and fixed grip conditions.

Thus, J is a more general version of the energy release rate. For the special case of a linear elastic material, $J = \mathcal{G}$. Also,

$$J = \frac{K_I^2}{E'} \tag{3.18}$$

for linear elastic Mode I loading. (For mixed mode loading. refer to Eq. (2.58).)

A word of caution is necessary when applying J to elastic-plastic materials. The energy release rate is normally defined as the potential energy that is *released* from a structure when the crack grows in an elastic material. However, much of the strain energy absorbed by an elastic-plastic material is not recovered when the crack grows or the specimen is unloaded; a growing crack in an elastic-plastic material leaves a plastic wake (Fig. 2.6(b)). Thus the energy release rate concept has a somewhat different interpretation for elastic-plastic materials. Rather than defining the energy released from the body when the crack grows, Eq. (3.15) relates J to the difference in energy absorbed by specimens with neighboring crack sizes. This distinction is important only when the crack grows (Section 3.5.3). See Appendix 4.2 and Chapter 11 for further discussion of the energy release rate concept.

The energy release rate definition of J is useful for elastic-plastic materials when applied in an appropriate manner. For example, Section 3.2.5 describes how Eqs. (3.15) to (3.17) can be exploited to measure J experimentally.

3.2.2 J as a Path-Independent Line Integral

Consider an arbitrary counter-clockwise path (Γ) around the tip of a crack, as illustrated in Fig. 3.9. The J integral is given by:

$$J = \int_{\Gamma} \left(w\, dy - T_i \frac{\partial u_i}{\partial x} ds \right) \tag{3.19}$$

where w is the strain energy density, T_i are components of the traction vector, u_i are the displacement vector components, and ds is a length increment along the contour Γ. The strain energy density is defined as

$$w = \int_0^{\varepsilon_{ij}} \sigma_{ij}\, d\varepsilon_{ij} \tag{3.20}$$

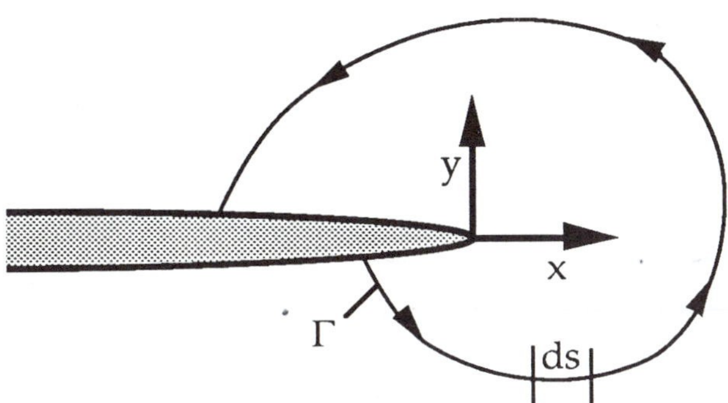

FIGURE 3.9 Arbitrary contour around the tip of a crack.

where σ_{ij} and ε_{ij} are the stress and strain tensors, respectively. The traction is a stress vector normal to the contour. That is, if we were to construct a free body diagram on the material inside of the contour, T_i would define the normal stresses acting at the boundaries. The components of the traction vector are given by

$$T_i = \sigma_{ij} n_j \tag{3.21}$$

where n_j are the components of the unit vector normal to Γ.

Rice [4] showed that the value of the J integral is independent of the path of integration around the crack. Thus J is called a *path-independent* integral. Appendix 3.2 demonstrates this path independence, and shows that Eq. (3.19) is equal to the energy release rate.

3.2.3 J as a Stress Intensity Parameter

Hutchinson [7] and Rice and Rosengren [8] independently showed that J characterizes crack tip conditions in a nonlinear elastic material. They each assumed a power law relationship between plastic strain and stress. If elastic strains are included, this relationship for uniaxial deformation is given by

$$\frac{\varepsilon}{\varepsilon_o} = \frac{\sigma}{\sigma_o} + \alpha\left(\frac{\sigma}{\sigma_o}\right)^n \tag{3.22}$$

where σ_o is a reference value stress that is usually equal to the yield strength, $\varepsilon_o = \sigma_o/E$, α is a dimensionless constant, and n is the strain hardening exponent[2]. Eq. (3.22) is known as the Ramberg-Osgood equation, and is widely used for curve-fitting stress-strain data. Hutchinson, Rice, and Rosengren showed that in order to remain path independent, stress•strain must vary as 1/r near the crack tip. At distances very close to the crack tip, well within the plastic zone, elastic strains are small in comparison to the total strain, and the stress-strain behavior reduces to a simple power law. These two conditions imply the following variation of stress and strain ahead of the crack tip:

$$\sigma_{ij} = k_1 \left(\frac{J}{r}\right)^{\frac{1}{n+1}} \tag{3.23a}$$

$$\varepsilon_{ij} = k_2 \left(\frac{J}{r}\right)^{\frac{n}{n+1}} \tag{3.23b}$$

where k_1 and k_2 are proportionality constants, which are defined more precisely below. For a linear elastic material, n=1, and Eq. (3.23) predicts a $1/\sqrt{r}$ singularity, which is consistent with LEFM theory.

The actual stress and strain distributions are obtained by applying the appropriate boundary conditions (see Appendix 3.4):

$$\sigma_{ij} = \sigma_o \left(\frac{E\,J}{\alpha\,\sigma_o^2\,I_n\,r}\right)^{\frac{1}{n+1}} \tilde{\sigma}_{ij}(n,\theta) \tag{3.24a}$$

and

$$\varepsilon_{ij} = \frac{\alpha\sigma_o}{E}\left(\frac{E\,J}{\alpha\,\sigma_o^2\,I_n\,r}\right)^{\frac{n}{n+1}} \tilde{\varepsilon}_{ij}(n,\theta) \tag{3.24b}$$

where I_n is an integration constant that depends on n, and $\tilde{\sigma}_{ij}$ and $\tilde{\varepsilon}_{ij}$ are dimensionless functions of n and θ. These parameters also depend on the stress state (i.e. plane stress or plane strain). Equations (3.24a) and (3.24b) are called the HRR singularity, named after Hutchinson, Rice, and Rosengren [7,8]. Figure 3.10 is a plot of I_n versus n for plane

---

[2]Although Eq. (3.22) contains four material constants, there are only two fitting parameters. The choice of σ_o, which is arbitrary, defines ε_o; a linear regression is then performed on a log-log plot of stress versus plastic strain to determine α and n.

stress and plane strain. Figures 3.11 shows the angular variation of $\tilde{\sigma}_{ij}(n,\theta)$ [7]. The stress components in Fig. 3.11 are defined in terms of polar coordinates rather than x and y.

The J integral defines the amplitude of the HRR singularity, just as the stress intensity factor characterizes the amplitude of the linear elastic singularity. Thus J completely describes the conditions within the plastic zone. A structure in small-scale yielding has two singularity-dominated zones: one in the elastic region, where stress varies as $1/\sqrt{r}$, and one in the plastic zone where stress varies as $r^{-1/(n+1)}$. The latter often persists long after the linear elastic singularity zone has been destroyed by crack tip plasticity.

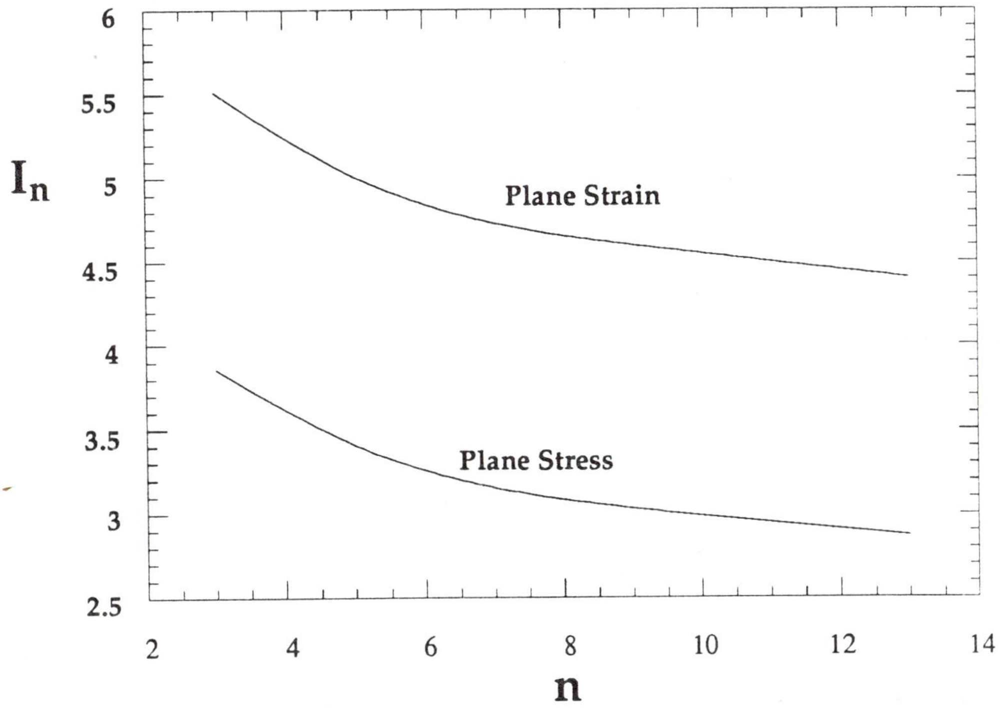

FIGURE 3.10 Effect of the strain hardening exponent on the HRR integration constant.

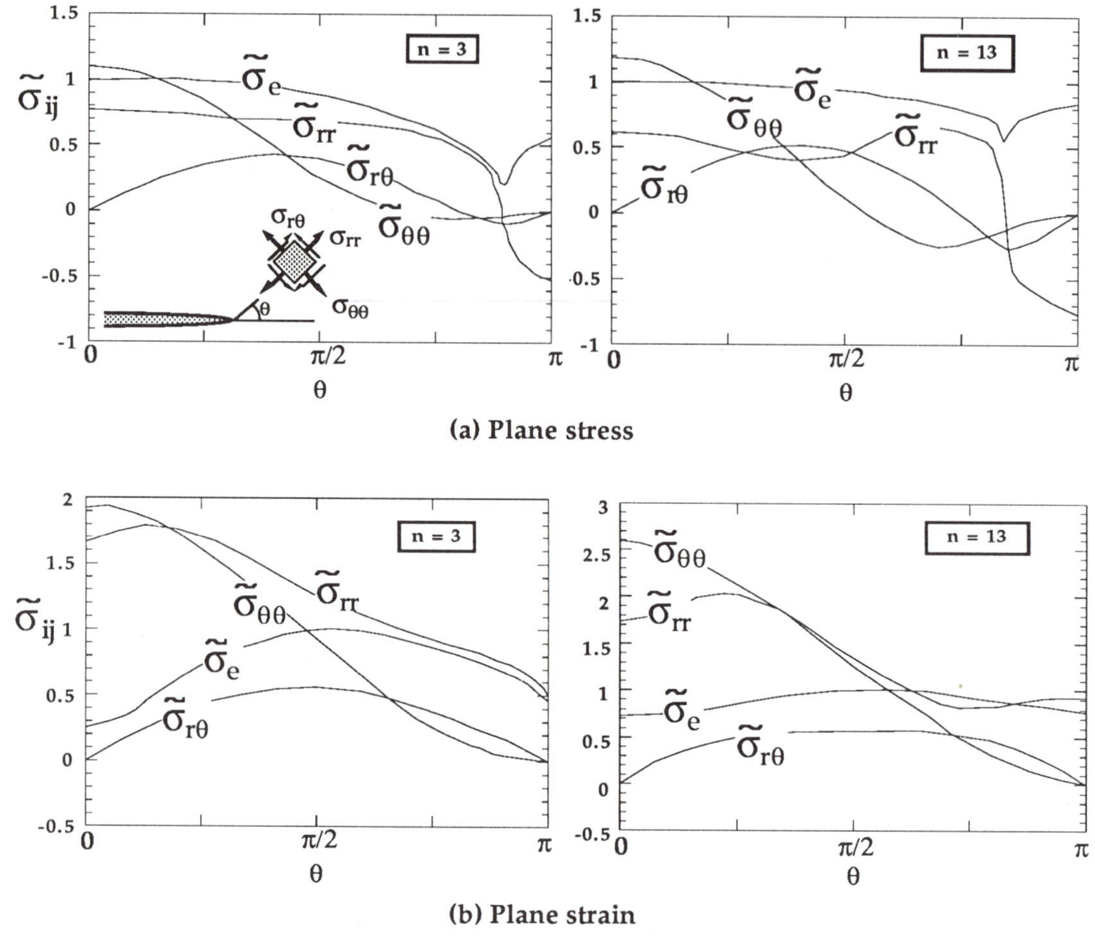

FIGURE 3.11 Angular variation of dimensionless stress for n = 3 and n = 13 [7].

3.2.4 The Large Strain Zone

The HRR singularity contains the same apparent anomaly as the LEFM singularity; namely, both predict infinite stresses as $r \to 0$. The singular field does not persist all the way to the crack tip, however. The large strains at the crack tip cause the crack to blunt, which reduces the stress triaxiality locally. The blunted crack tip is a free surface; thus σ_{xx} must vanish at $r = 0$.

The analysis that leads to the HRR singularity does not consider the effect of the blunted crack tip on the stress fields, nor does it take account of the large strains that are present near the crack tip. This analysis is based on small strain theory, which is the multi-axial equivalent of engineering strain in a tensile test. Small strain theory breaks down when strains are greater than ~ 0.10 (10%).

McMeeking and Parks [9] performed crack tip finite element analyses that incorporated large strain theory and finite geometry changes. Some of their results are shown in Fig. 3.12, which is a plot of stress normal to the crack plane versus distance. The HRR singularity (Eq. (3.24a)) is also shown on this plot. Note that both axes are nondimensionalized in such a way that both curves are invariant, as long as the plastic zone is small compared to specimen dimensions.

The solid curve in Fig. 3.12 reaches a peak when the ratio $x\sigma_o/J$ is approximately unity, and decreases as $x \rightarrow 0$. This distance corresponds approximately to twice the CTOD. The HRR singularity is invalid within this region, where the stresses are influenced by large strains and crack blunting.

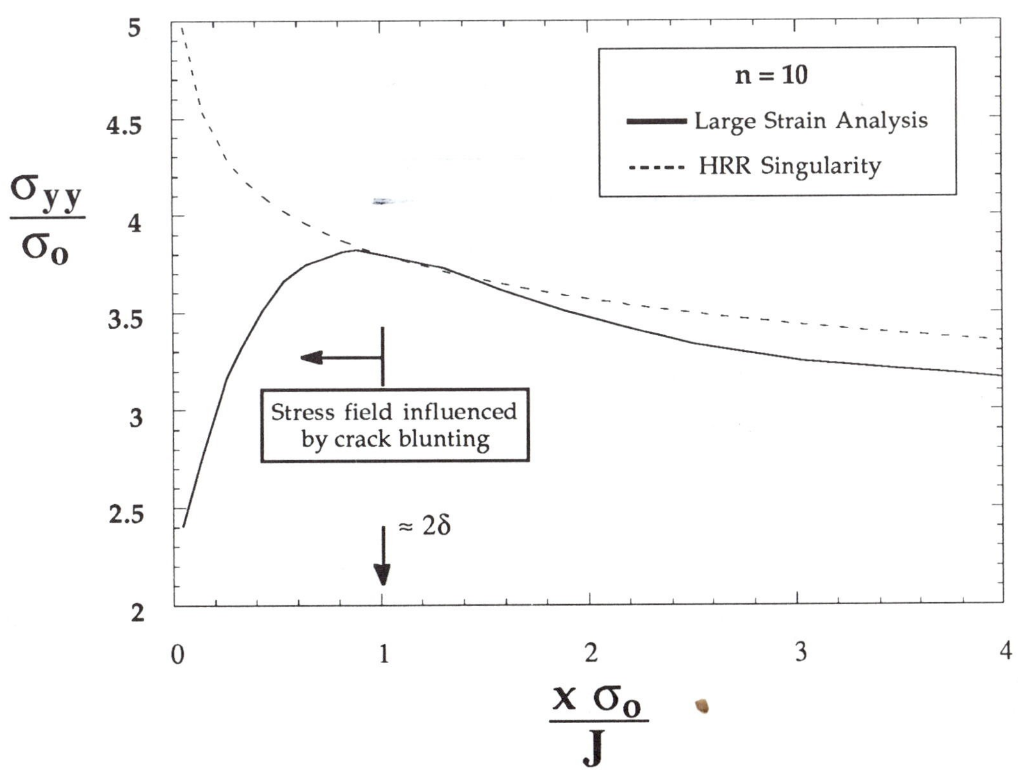

FIGURE 3.12 Large-strain crack tip finite element results of McMeeking and Parks [9]. Blunting causes the stresses to deviate from the HRR solution close to the crack tip

The break-down of the HRR solution at the crack tip leads to a similar question to one that was posed in Section 2.10: is the J integral a useful fracture criterion when a blunting zone forms at the crack tip? The answer is also similar to the argument offered in Section 2.10. That is, as long as there is a region *surrounding* the crack tip that can be described by Eq. (3.24), the J integral uniquely characterizes crack tip conditions, and a critical value of J is a size-independent measure of fracture toughness. The question of J controlled fracture is explored further in Section 3.5.

3.2.5 Laboratory Measurement of J

When the material behavior is linear elastic, calculation of the J integral in a test specimen or structure is relatively straightforward because $J = \mathcal{G}$, and $\mathcal{G}$ is uniquely related to the stress intensity factor. The latter quantity can be computed from the load and crack size, assuming a K solution for that particular geometry is available. Table 2.4 and Chapter 12 give several examples of stress intensity solutions.

Computing the J integral is somewhat more difficult when the material is nonlinear. The principle of superposition no longer applies, and J is not proportional to the applied load. Thus a simple relationship between J, load, and crack length is usually not available.

One option for determining J is to apply the line integral definition Eq. (3.19) to the configuration of interest. Read [10] has measured the J integral in test panels by attaching an array of strain gages in a contour around the crack tip. Since J is path independent and the choice of contour is arbitrary, he selected a contour in such a way as to simplify the calculation of J as much as possible. This method can also be applied to finite element analysis; i.e. stresses, strains and displacements can be determined along a contour and J can then calculated according to Eq. (3.19). The contour method for determining J is impractical in most cases, however. The instrumentation required for experimental measurements of the contour integral is highly cumbersome, and the contour method is also not very attractive in numerical analysis (see Chapter 11). A much better method for determining J numerically is outlined in Chapter 11. More practical experimental approaches are developed below and are explored further in Chapter 7.

Landes and Begley [11,12], who were among the first to measure J experimentally, invoked the energy release rate definition of J (Eq. 3.11). Figure 3.13 schematically illustrates their approach. They obtained a series of test specimens of the same size, geometry, and material and introduced cracks of various lengths[3]. They deformed each specimen and plotted load versus displacement (Fig. 3.13 (a)). The area under a given curve is equal to U, the energy absorbed by the specimen. Landes and Begley plotted U versus crack length at various fixed displacements (Fig. 3.13 (b)). For an edge cracked specimen of thickness B, the J integral is given by

$$J = -\frac{1}{B}\left(\frac{\partial U}{\partial a}\right)_{\Delta} \tag{3.25}$$

Thus J can be computed by determining the slope of the tangent to the curves in Fig. 3.13 (b). Applying Eq. (3.25) leads to Fig. 3.13 (c), a plot of J versus displacement at various crack lengths. The latter is a calibration curve, which only applies to the material, specimen size, specimen geometry, and temperature for which it was obtained. The Landes and Begley approach has obvious disadvantages, since multiple specimens must be tested and analyzed to determine J in a particular set of circumstances.

Rice, et. al. [13] showed that it was possible, in certain cases, to determine J directly from the load displacement curve of a single specimen. Their derivations of J relationships for several specimen configurations demonstrate the usefulness of dimensional analysis[4].

Consider a double edge notched tension panel of unit thickness (Fig. 3.14). Cracks of length a on opposite sides of the panel are separated by a ligament of length 2 b. For this configuration, $d\mathcal{A} = 2\,da = -2\,db$ (see Footnote 1); Eq. (3.16) must be modified accordingly:

$$J = \frac{1}{2}\int_0^P \left(\frac{\partial \Delta}{\partial a}\right)_P dP = -\frac{1}{2}\int_0^P \left(\frac{\partial \Delta}{\partial b}\right)_P dP \tag{3.26}$$

---

[3]See Chapter 7 for a description of fatigue precracking procedures for test specimens.
[4]See Section 1.5 for a review of the fundamentals of dimensional analysis.

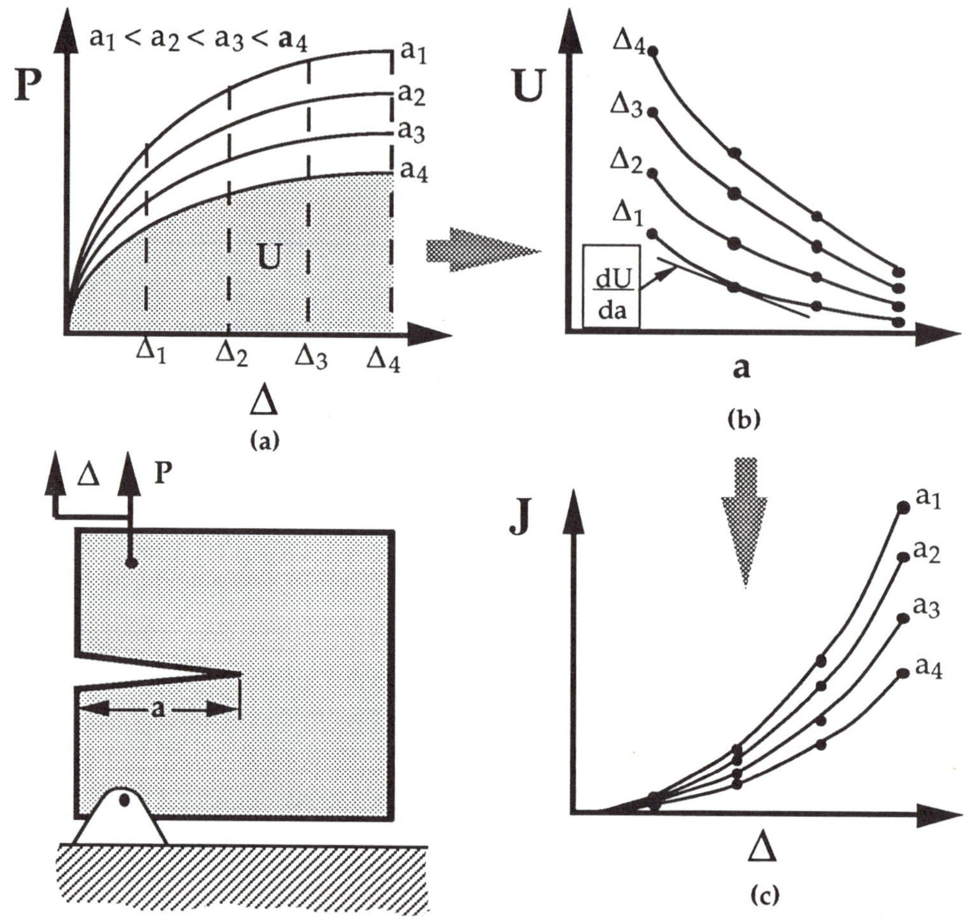

FIGURE 3.13 Schematic of early experimental measurements of J, performed by Landes and Begley [11,12]

In order to compute J from the above expression, it is necessary to determine the relationship between load, displacement, and panel dimensions. Assuming an isotropic material that obeys a Ramberg-Osgood stress-strain law (Eq. 3.22) dimensional analysis gives the following functional relationship for displacement:

$$\Delta = b\ \Phi\left(\frac{P}{\sigma_o b}; \frac{a}{b}; \frac{\sigma_o}{E}; \nu; \alpha; n\right) \tag{3.27}$$

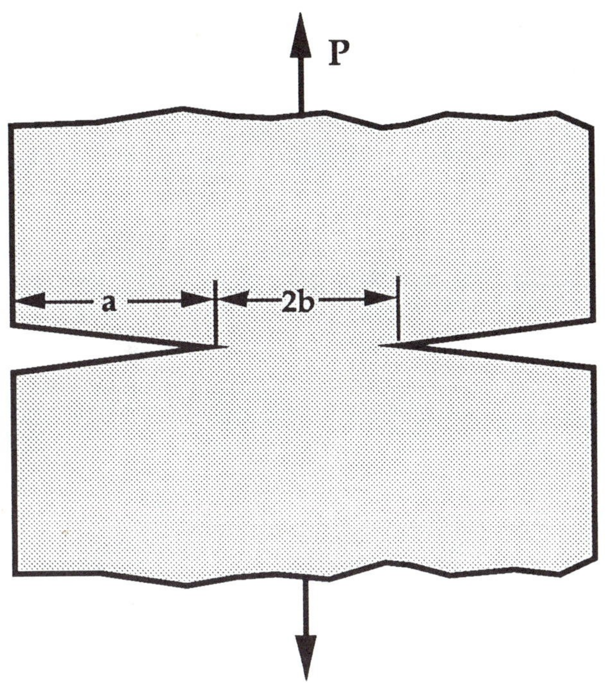

FIGURE 3.14 Double edge notched tension (DENT) panel.

where Φ is a dimensionless function. For fixed material properties, we need only consider load and specimen dimensions. For reasons described below, we can simplify the functional relationship for displacement by separating Δ into elastic and plastic components:

$$\Delta = \Delta_{el} + \Delta_p \tag{3.28}$$

Substituting Eq. (3.28) into Eq. (3.26) leads to a relationship for elastic and plastic components of J:

$$J = -\frac{1}{2}\int_0^P \left[\left(\frac{\partial \Delta_{el}}{\partial b}\right)_P + \left(\frac{\partial \Delta_p}{\partial b}\right)_P\right] dP$$

$$= \frac{K_I^2}{E'} - \frac{1}{2}\int_0^P \left(\frac{\partial \Delta_p}{\partial b}\right)_P dP \tag{3.29}$$

where E' = E for plane stress and E' = E/(1 - ν^2) for plane strain, as defined in Chapter 2. Thus we need only be concerned about plastic displacements because a solution for the elastic component of J is already available (Table 2.4). If plastic deformation is confined to the ligament between the crack tips (Fig. 3.14 (b)), we can assume that b is the only length dimension that influences Δ_p. This is a reasonable assumption, provided the panel is deeply notched so that the average stress in the ligament is substantially higher than the remote stress in the gross cross section. We can define a new function for Δ_p:

$$\Delta_p = b\,H\left(\frac{P}{b}\right) \tag{3.30}$$

Note that the net-section yielding assumption has eliminated the dependence on the a/b ratio. Taking a partial derivative with respect to the ligament length gives

$$\left(\frac{\partial \Delta_p}{\partial b}\right)_P = H\left(\frac{P}{b}\right) - H'\left(\frac{P}{b}\right)\frac{P}{b} \tag{3.31}$$

where H' denotes the first derivative of the function H. We can solve for H' by taking a partial derivative of Eq. (3.30) with respect to crack size:

$$\left(\frac{\partial \Delta_p}{\partial P}\right)_b = H'\left(\frac{P}{b}\right)$$

Therefore,

$$\left(\frac{\partial \Delta_p}{\partial b}\right)_P = \frac{1}{b}\left[\Delta_p - P\left(\frac{\partial \Delta_p}{\partial P}\right)_b\right] \tag{3.32}$$

Substituting Eq. (3.32) into Eq. (3.29) and integrating by parts gives

$$J = \frac{K_I^2}{E'} + \frac{1}{2b}\left[2\int_0^{\Delta_p} P \, d\Delta_p - P\,\Delta_p\right] \qquad (3.32)$$

Recall that we assumed a unit thickness at the beginning of this derivation. In general, the plastic term must be divided by the plate thickness; the term in square brackets, which depends on the load displacement curve, is normalized by the net cross-sectional area of the panel. The J integral has units of energy/area.

Another example from the Rice, et al. article [13] is an edge cracked plate in bending (Fig. 3.15). In this case they chose to separate displacements along somewhat different lines from the previous problem. If the plate is subject to a bending moment M, it would displace by an angle Ω_{nc} if no crack were present, and an additional amount, Ω_c, when the the plate is cracked. Thus the total angular displacement can be written as

$$\Omega = \Omega_{nc} + \Omega_c \qquad (3.33)$$

If the crack is deep, $\Omega_c >> \Omega_{nc}$. The energy absorbed by the plate is given by

$$U = \int_0^{\Omega} M \, d\Omega \qquad (3.34)$$

When we differentiate U with respect to crack area in order to determine J, only Ω_c contributes to the energy release rate because Ω_{nc} is not a function of crack size, by definition. By analogy with Eq. (3.16), J for the cracked plate in bending can be written as

$$J = \int_0^M \left(\frac{\partial \Omega_c}{\partial a}\right)_M dM = -\int_0^M \left(\frac{\partial \Omega_c}{\partial b}\right)_M dM \qquad (3.35)$$

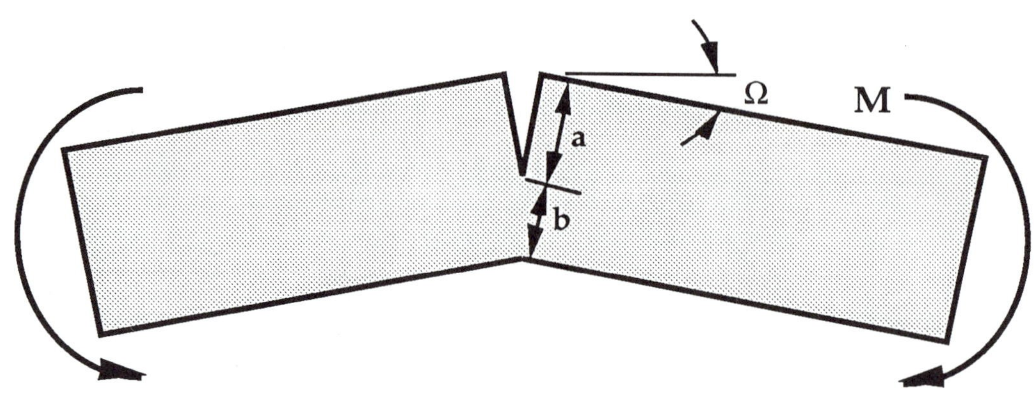

FIGURE 3.15 Edge cracked plate in pure bending.

If material properties are fixed, dimensional analysis leads to

$$\Omega_c = G\left(\frac{M}{b^2}\right) \tag{3.36}$$

assuming the ligament length is the only relevant length dimension, which is reasonable if the crack is deep. When Eq. (3.36) is differentiated with respect to b and inserted into Eq. (3.35), the resulting expression for J is as follows:

$$J = \frac{2}{b}\int_0^{\Omega_c} M \, d\Omega_c \tag{3.37}$$

The decision to separate Ω into "crack" and "no-crack" components was somewhat arbitrary. The angular displacement could have been divided into elastic and plastic components as in the previous example. If the crack is relatively deep, Ω_{nc} should be entirely elastic, while Ω_c may contain both elastic and plastic contributions. Therefore, Eq. (3.37) can be written as

$$J = \frac{2}{b}\left[\int_0^{\Omega_{c(el)}} M \, d\Omega_{c(el)} + \int_0^{\Omega_p} M \, d\Omega_p\right]$$

or

$$J = \frac{K_I^2}{E'} + \frac{2}{b}\int_0^{\Omega_p} M \, d\Omega_p \tag{3.38}$$

Conversely, the prior analysis on the double edged cracked plate in tension could have been written in terms of Δ_c and Δ_{nc}. Recall, however, that the dimensional analysis was simplified in each case (Eqs. (3.30) and (3.36)) by assuming a negligible dependence on a/b. This turns out to be a reasonable assumption for plastic displacements in deeply-notched DENT panels, but less so for elastic displacements. Thus while elastic and plastic displacements due the the crack can be combined to compute J in bending (Eq. (3.37)), it is not advisable to do so for tensile loading. The relative accuracy and the limitations of Eqs. (3.32) and (3.37) are evaluated in Section 9.5.

In general, the J integral for a variety of configurations can be written in the following form:

$$J = \frac{\eta \, U_c}{B \, b} \tag{3.39}$$

where η is a dimensionless constant. Note that Eq. (3.39) contains the actual thickness, while the above derivations assumed unit thickness for convenience. Equation (3.39) expresses J as the energy absorbed, divided by the cross-sectional area, times a dimensionless constant. For a deeply cracked plate in pure bending, $\eta = 2$. Equation (3.39) can be separated into elastic and plastic components:

$$J = \frac{\eta_{el} \, U_{c(el)}}{B \, b} + \frac{\eta_p \, U_p}{B \, b}$$

$$= \frac{K_I^2}{E'} + \frac{\eta_p \, U_p}{B \, b} \tag{3.40}$$

EXAMPLE 3.1

Determine the plastic η factor for the DENT configuration, assuming the load-plastic displacement curve follows a power law:

$$P = C\,\Delta_p^N$$

Solution: The plastic energy absorbed by the specimen is given by

$$U_p = C\int_0^{\Delta p} \Delta_p^N \, d\Delta_p = \frac{C\,\Delta_p^{N+1}}{N+1} = \frac{P\,\Delta p}{N+1}$$

Comparing Eqs. (3.32) and (3.40) and solving for η_p gives

$$\eta_p = \frac{P\,\Delta p\left(\frac{2}{N+1} - 1\right)}{\frac{P\,\Delta p}{N+1}} = \frac{1 - N}{N + 1}$$

For a nonhardening material, $N = 0$ and $\eta_p = 1$.

3.3 RELATIONSHIPS BETWEEN J AND CTOD

For linear elastic conditions, the relationship between CTOD and $\mathcal{G}$ is given by Eq. (3.8). Since $J = \mathcal{G}$ for linear elastic material behavior, these equations also describe the relationship between CTOD and J in the limit of small scale yielding. That is,

$$J = m\,\sigma_{YS}\,\delta \tag{3.41}$$

where m is a dimensionless constant that depends on stress state and material properties. It can be shown that Eq. (3.41) applies well beyond the validity limits of LEFM.

Consider, for example, a strip yield zone ahead of a crack tip, as illustrated in Fig. 3.16. Recall (from Chapter 2) that the strip yield zone is modeled by surface tractions along the crack face. Let us define a contour, Γ, along the boundary of this zone. If the damage zone is long and slender, i.e., if $\rho >> \delta$, the first term in the J contour integral (Eq. 3.19) vanishes because $dy = 0$. Since the only surface tractions within ρ are in the y direction, $n_y = 1$ and $n_x = n_z = 0$. Thus the J integral is given by

$$J = \int_{\Gamma} \sigma_{yy} \frac{\partial u_y}{\partial x} ds \tag{3.42}$$

Let us define a new coordinate system with the origin at the tip of the strip yield zone: $X = \rho - x$. For a fixed δ, σ_{yy} and u_y depend only on X, provided ρ is small compared to the in-plane dimensions of the cracked body. The J integral becomes

$$J = 2 \int_0^{\rho} \sigma_{yy}(X) \frac{du_y(X)}{dX} dX$$

$$= \int_0^{\delta} \sigma_{yy}(\delta)\; d\delta \tag{3.43}$$

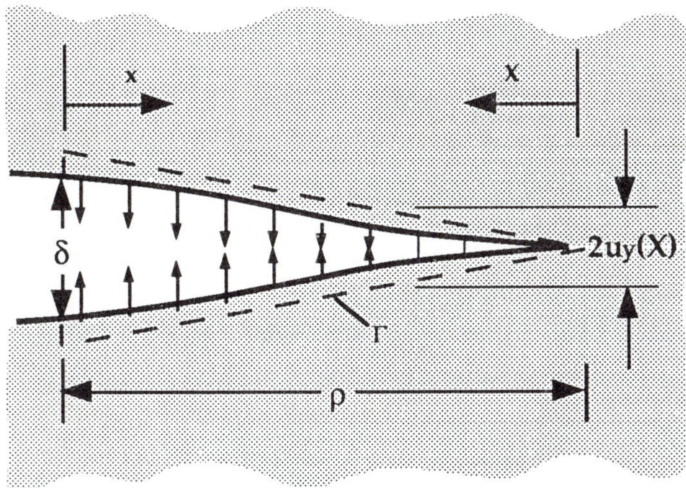

FIGURE 3.16 Contour along the boundary of the strip yield zone ahead of a crack tip.

where $\delta = 2\, u_y\, (X = \rho)$. Since the strip yield model assumes $\sigma_{yy} = \sigma_{YS}$ within the plastic zone, the J-CTOD relationship is given by

$$J = \sigma_{YS}\, \delta \tag{3.44}$$

Note the similarity between Eqs. (3.44) and (3.7). The latter was derived from the strip yield model by neglecting the higher order terms in a series expansion; no such assumption was necessary to derive Eq. (3.44). Thus the strip yield model, which assumes plane stress conditions and a nonhardening material, predicts that m = 1 for both linear elastic and elastic-plastic conditions.

Shih [14] provided further evidence that a unique J-CTOD relationship applies well beyond the validity limits of LEFM. He evaluated the displacements at the crack tip implied by the HRR solution and related the displacement at the crack tip to J and flow properties. According to the HRR solution, the displacements near the crack tip are as follows:

$$u_i = \frac{\alpha\sigma_o}{E}\left(\frac{E\,J}{\alpha\,\sigma_o^2\, I_n\, r}\right)^{\frac{n}{n+1}} r\; \tilde{u}_i\,(\theta, n) \tag{3.45}$$

where $\tilde{u}_i$ is a dimensionless function of θ and n, analogous to $\tilde{\sigma}_{ij}$ and $\tilde{\varepsilon}_{ij}$ (Eq. 3.24). Shih [14] invoked the 90° intercept definition of CTOD, as illustrated in Fig. 3.4(b). This 90° intercept construction is examined further in Fig. 3.17. The CTOD is obtained by evaluating u_x and u_y at r = r* and $\theta = \pi$:

$$\frac{\delta}{2} = u_y(r^*;\, \pi) = r^* - u_x(r^*;\, \pi) \tag{3.46}$$

Substituting Eq. (3.46) into Eq. (3.45) and solving for r* gives

$$r^* = \left(\frac{\alpha\sigma_o}{E}\right)^{\frac{1}{n}} \{\tilde{u}_x(\pi;\, n) + \tilde{u}_y(\pi;\, n)\}^{\frac{n+1}{n}} \frac{J}{\sigma_o\, I_n} \tag{3.47}$$

Setting $\delta = 2\, u_y(r^*;\, \pi)$ leads to

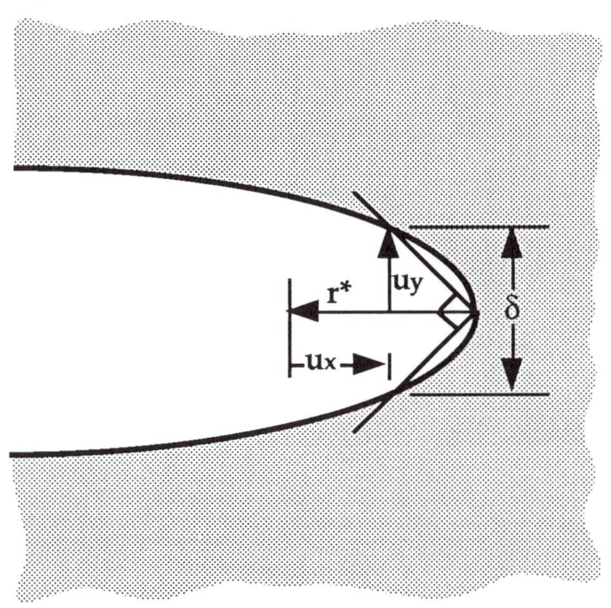

FIGURE 3.17 Estimation of CTOD from a 90° intercept construction and HRR displacements .

$$\delta = \frac{d_n J}{\sigma_o} \tag{3.48}$$

where d_n is a dimensionless constant, given by

$$d_n = \frac{2\,\tilde{u}_y(\pi; n)\left[\frac{\alpha\sigma_o}{E}\{\tilde{u}_x(\pi; n) + \tilde{u}_y(\pi; n)\}\right]^{\frac{1}{n}}}{I_n} \tag{3.49}$$

Figure 3.18 shows plots of d_n for $\alpha = 1.0$, which exhibits a strong dependence on the strain hardening exponent and a mild dependence on $\alpha\sigma_o/E$. A comparison Eqs. (3.41) and (3.48) indicates that $d_n = 1/m$, assuming $\sigma_o = \sigma_{YS}$ (see Footnote 2). According to Fig. 3.18(a), $d_n = 1.0$ for a nonhardening material ($n = \infty$) in plane stress, which agrees with the strip yield model (Eq. (3.44)).

The Shih analysis shows that there is a unique relationship between J and CTOD for a given material. Thus these two quantities are equally valid crack tip characterizing parameters for elastic-plastic materials. The fracture toughness of a material can be quantified either by a critical value of J or CTOD.

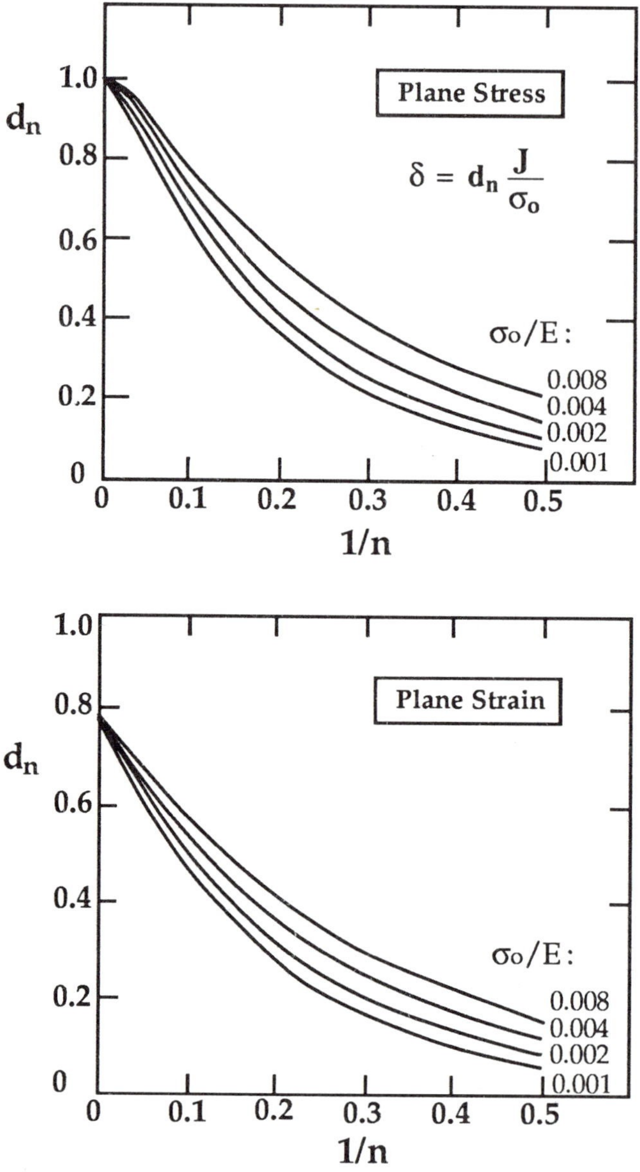

FIGURE 3.18 Predicted J-CTOD relationships for plane stress and plane strain, assuming $\alpha = 1$ [14]. For $\alpha \neq 1$, the above values should be multiplied by $\alpha^{1/n}$.

The above analysis contains an apparent inconsistency. Equation (3.48) is based on the HRR singularity, which does not account for large geometry changes at the crack tip. Figure 3.12 indicates that the stresses predicted by the HRR theory are inaccurate for $r < 2\delta$, but the Shih analysis uses the HRR solution to evaluate displacements well within the large strain region. Crack tip finite element analyses [14], however, are in general agreement with Eq. (3.48). Thus the displacement fields predicted from the HRR theory are reasonably accurate, despite the large plastic strains at the crack tip.

3.4 CRACK GROWTH RESISTANCE CURVES

Many materials with high toughness do not fail catastrophically at a particular value of J or CTOD. Rather, these materials display a rising R curve, where J and CTOD increase with crack growth.

Figure 3.19 schematically illustrates a typical J resistance curve for a ductile material. In the initial stages of deformation, the R curve is nearly vertical; there is a small amount of apparent crack growth due to blunting. As J increases, the material at the crack tip fails locally and the crack advances further. Because the R curve is rising, the initial crack growth is usually stable, but an instability can be encountered later, as discussed below.

One measure of fracture toughness, J_{IC}, is defined near the initiation of stable crack growth. The precise point at which crack growth begins is usually ill-defined. Consequently, the definition of J_{IC} is somewhat arbitrary, much like a 0.2% offset yield strength. The corresponding CTOD near the initiation of stable crack growth is denoted δ_i by U.S. and British testing standards. Chapter 7 describes experimental measurements of J_{IC} and δ_i in more detail

While initiation toughness provides some information about the fracture behavior of a ductile material, the entire R curve gives a more complete description. The slope of the R curve at a given amount of crack extension is indicative of the relative stability of the crack growth; a material with a steep R curve is less likely to experience unstable crack propagation. For J resistance curves, the slope is usually quantified by a dimensionless *tearing modulus:*

$$T_R = \frac{E}{\sigma_o^2} \frac{dJ_R}{da} \tag{3.49}$$

where the subscript R indicates a value of J on the resistance curve.

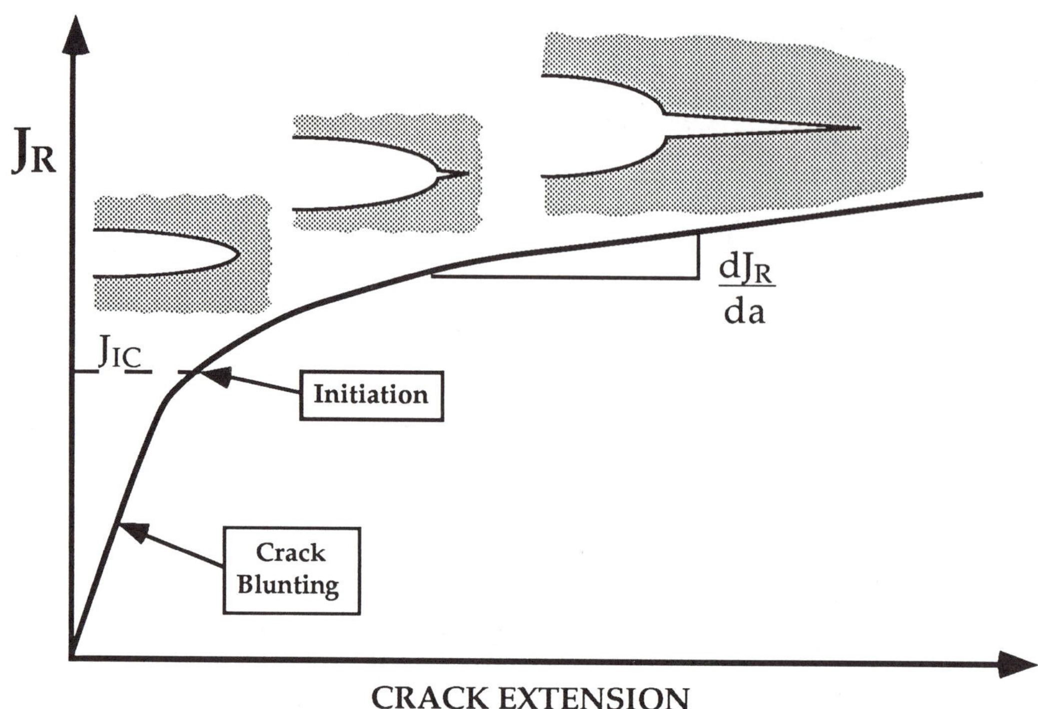

FIGURE 3.19 Schematic J resistance curve for a ductile material.

3.4.1 Stable and Unstable Crack Growth

The conditions that govern the stability of crack growth in elastic-plastic materials are virtually identical to the elastic case presented in Section 2.5. Instability occurs when the driving force curve is tangent to the R curve. As Fig. 3.20 indicates, load control is usually less stable than displacement control. The conditions in most structures are somewhere between the extremes of load control and displacement control. The intermediate case can be represented by a spring in series with the structure, where remote displacement is fixed (Fig. 2.12). Since the R curve slope has been represented by a dimensionless tearing modulus (Eq. 3.49), it is convenient to express the driving force in terms of an *applied tearing modulus:*

$$T_{app} = \frac{E}{\sigma_o^2}\left(\frac{dJ}{da}\right)_{\Delta_T} \tag{3.50}$$

where Δ_T is the total remote displacement defined as

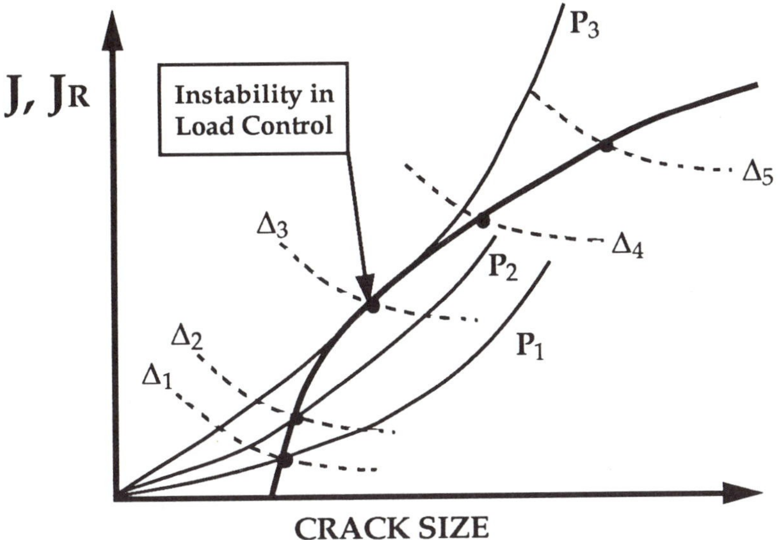

FIGURE 3.20 Schematic J driving force/R curve diagram which compares load control and displacement control.

$$\Delta_T = \Delta + C_m P \tag{3.51}$$

and C_m is the system compliance. The slope of the driving force curve for a fixed Δ_T is identical to the linear elastic case (Eq. 2.35), except that $\mathcal{G}$ is replaced by J:

$$\left(\frac{dJ}{da}\right)_{\Delta_T} = \left(\frac{\partial J}{\partial a}\right)_P - \left(\frac{\partial J}{\partial P}\right)_a \left(\frac{\partial \Delta}{\partial a}\right)_P \left[C_m + \left(\frac{\partial \Delta}{\partial P}\right)_a\right]^{-1} \tag{3.52}$$

For load control, $C_m = \infty$, and the second term in Eq. (3.52) vanishes:

$$\left(\frac{dJ}{da}\right)_{\Delta_T} = \left(\frac{\partial J}{\partial a}\right)_P$$

For displacement control, $C_m = 0$, and $\Delta_T = \Delta$. Equation (3.52) is derived in Appendix 2.1 for the linear elastic case.

The conditions during stable crack growth can be expressed as follows:

$$J = J_R \tag{3.53a}$$

and

$$T_{app} \leq T_R \tag{3.53b}$$

Unstable crack propagation occurs when

$$T_{app} > T_R \tag{3.54}$$

Chapter 9 gives practical guidance on assessing structural stability with Eqs. (3.50) to (3.55). A simple example is presented below.

EXAMPLE 3.2

Derive an expression for the applied tearing modulus in the double cantilever beam (DCB) specimen with a spring in series (Fig. 3.21), assuming linear elastic conditions.

Solution: From Example 2.1, we have the following relationships:

$$J = \mathcal{G} = \frac{P^2 a^2}{B E I} \quad \text{and} \quad \Delta = \frac{2 P a^3}{3 E I}$$

Therefore, the relevant partial derivatives are given by

$$\left(\frac{\partial J}{\partial a}\right)_P = \frac{2 P^2 a}{B E I} \qquad \left(\frac{\partial J}{\partial P}\right)_a = \frac{2 P a^2}{B E I}$$

$$\left(\frac{\partial \Delta}{\partial a}\right)_P = \frac{2 P a^2}{E I} \qquad \left(\frac{\partial \Delta}{\partial P}\right)_a = \frac{2 a^3}{3 E I}$$

Substituting the above relationships into Eqs. (3.50) and (3.52) gives

$$T_{app} = \frac{2 P^2 a}{\sigma_o^2 B I}\left\{1 - \frac{4 a^3}{E I}\left[C_M + \frac{2 a^3}{3 E I}\right]\right\}^{-1}$$

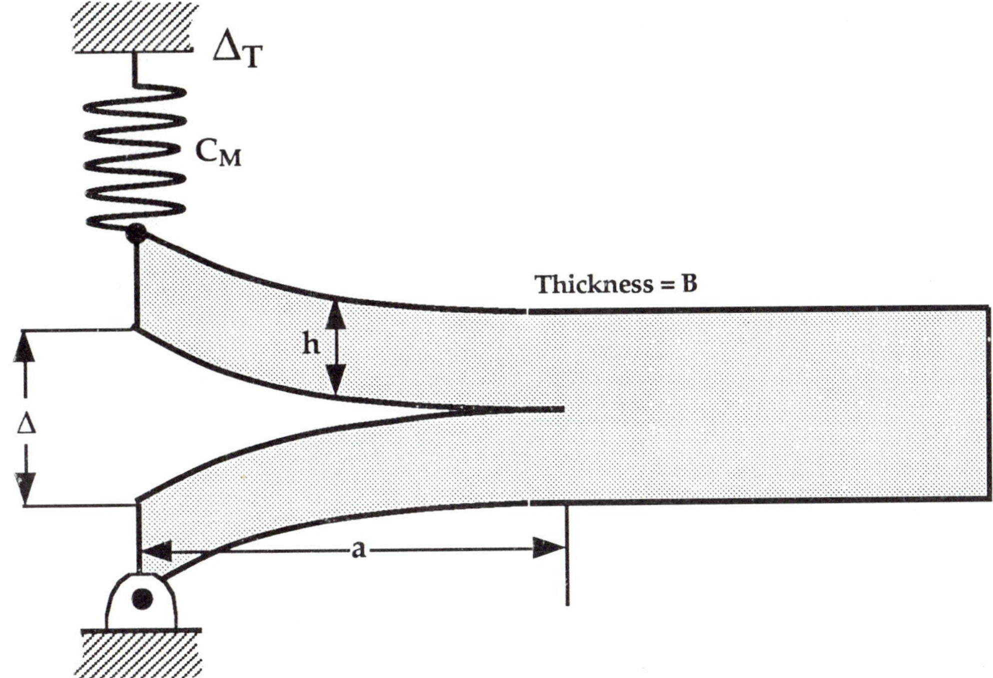

FIGURE 3.21 Double cantilever beam specimen with a spring in series.

As discussed in Section 2.5, the point of instability in a material with a rising R curve depends on the size and geometry of the cracked structure; a critical value of J at instability is not a material property if J increases with crack growth. It is usually assumed that the R curve, including the J_{IC} value, is a material property, independent of configuration. This is a reasonable assumption, within certain limitations.

3.4.2 Computing J for a Growing Crack

The geometry dependence of a J resistance curve is influenced by the way in which J is calculated. The equations derived in Section 3.2.5 are based on the pseudo energy release rate definition of J and are only valid for a stationary crack. There are various ways to compute J for a growing crack, including the deformation J, the far-field J, and the modified J, which are described below.

Figure 3.22 illustrates the load-displacement behavior in a specimen with a growing crack. Recall that the J integral is based on a deformation plasticity (or nonlinear elastic) assumption for material behavior. Consider the point on the load-displacement curve that is labeled in Fig. 3.22. The crack has grown to a length a_1 from an initial length a_o. The cross-hatched area represents energy that would be released if the material were elastic. In an elastic-plastic material, only the elastic portion of this energy is released; the remainder is dissipated in a plastic wake that forms behind the growing crack (see Figs 2.6(b) and 3.25).

In an elastic material, all quantities, including strain energy, are independent of the loading history. The energy absorbed during crack growth in an elastic-plastic material, however, exhibits a history dependence. The dashed curve in Fig. 3.22 represents the load-displacement behavior when the crack size is fixed at a_1. The area under this curve is the strain energy in an elastic material; this energy depends only on the current load and crack length:

$$U_D = U_D(P; a) = \left(\int_0^{\Delta} P \, d\Delta \right)_{a=a_1} \tag{3.55}$$

where the subscript D refers to deformation theory. Thus the J integral for a nonlinear elastic body with a growing crack is given by

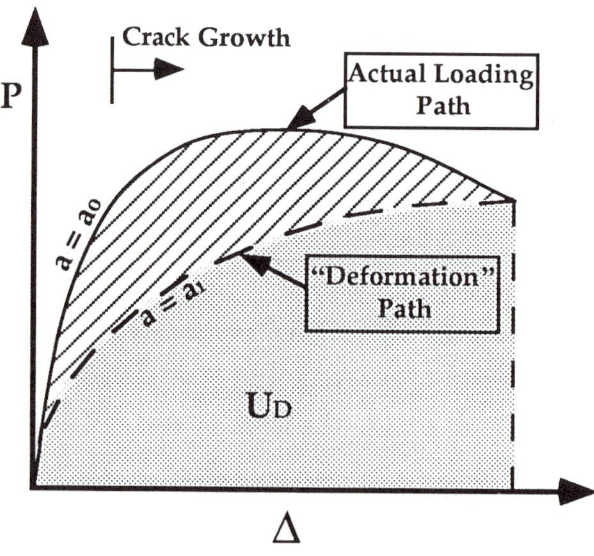

FIGURE 3.22 Schematic load-displacement curve for a specimen with a crack that grows to a_1 from an initial length a_o. U_D represents the strain energy in a nonlinear elastic material.

$$J_D = -\frac{1}{B}\left(\frac{\partial U_D}{\partial a}\right)_\Delta$$

$$= \frac{\eta\, U_D}{B\, b} \tag{3.56a}$$

or

$$J_D = \frac{K_I^2}{E'} + \frac{\eta_p\, U_{D(p)}}{B\, b} \tag{3.56b}$$

where b is the *current* ligament length. When the J integral for an elastic-plastic material is defined by Eq. (3.56), the history dependence is removed and the energy release rate interpretation of J is restored. The *deformation J* is usually computed from Eq. (3.56b) because no correction is required on the elastic term as long as K_I is determined from the current load and crack length. The calculation of $U_{D(p)}$ is usually performed incrementally, since the deformation theory load-displacement curve (Fig. 3.22 and Eq. (3.55)) depends on crack size. Specific procedures for computing the deformation J are outlined in Chapter 7.

One can determine a far-field J from the contour integral definition of Eq. (3.19), which may differ from J_D. For a deeply cracked bend specimen, Rice, et. al. [15] showed that the far-field J contour integral in a rigid, perfectly plastic material is given by

$$J_f = 0.73\, \sigma_o \int_0^\Omega b\, d\Omega \tag{3.57}$$

where the variation in b during the loading history is taken into account. Deformation theory leads to the following relationship for J in this specimen:

$$J_D = 0.73\ \sigma_o\, b\, \Omega \tag{3.58}$$

The two expressions are obviously identical when the crack is stationary.

There is no guarantee that either the deformation J_D or J_f will uniquely characterize crack tip conditions for a growing crack. Without this single parameter characterization, the J-R curve becomes geometry dependent.

Based on an earlier analysis by Rice, et al. [15], Ernst [16] proposed a modification to the deformation theory J in an attempt to remove the geometry dependence of resistance curves. He began by considering the deformation J, which depends only on the current load and crack size:

$$J_D = J_D(P, a) \tag{3.59}$$

An incremental change in J_D, dJ_D, is related to partial derivatives with respect to P and a:

$$dJ_D = \left(\frac{\partial J}{\partial P}\right)_a dP + \left(\frac{\partial J}{\partial a}\right)_P da \tag{3.60}$$

Separating J into elastic and plastic components gives

$$dJ_D = \left(\frac{\partial \mathcal{G}}{\partial P}\right)_a dP + \left(\frac{\partial \mathcal{G}}{\partial a}\right)_P da + \left(\frac{\partial J_p}{\partial P}\right)_a dP + \left(\frac{\partial J_p}{\partial a}\right)_P da \tag{3.61}$$

Ernst postulated that the last term in Eq. (3.61) was responsible for the geometry dependence in J-R curves. He proposed a modified J, where this final term is removed:

$$dJ_M = dJ_D - \left(\frac{\partial J_p}{\partial a}\right)_P da \tag{3.62}$$

or

$$J_M = J_D - \int_{a_o}^{a_1} \left(\frac{\partial J_p}{\partial a}\right)_P da \tag{3.63}$$

The modified J is history dependent; an energy release rate interpretation is no longer appropriate. The J_M correction has the effect of raising the R curve; i.e., $J_M \geq J_D$ at a given amount of crack extension. This effect is greatest in small specimens and for large amounts of crack growth.

Although J_M-R curves exhibit less geometry dependence in some cases [16], this effect is not universal [17]. In the late 1980s, there was a controversy as to whether deformation J_D or J_M is more appropriate for characterizing crack growth. The far-field J is usually not considered because it is too difficult to measure experimentally. The current U.S. standard method for J-R curve testing [18] utilizes the deformation J definition (see Chapter 7).

The J_D-J_M controversy has little practical importance. For small amounts of crack growth or when J_p is small, the difference between J_M and the deformation J is negligible. In order for the second term in Eq. (3.63) to be significant, the following conditions must be met in typical test specimens:

$$J_p >> G \tag{3.64a}$$

and

$$\frac{a - a_o}{b_o} > \sim 0.2 \tag{3.64b}$$

where b_o is the initial ligament length. These conditions are met most readily in small specimens or tough materials. In such cases, a single parameter description of crack tip conditions may be impossible; *neither* parameter will yield a geometry-independent R curve. Thus when the difference between J_D and J_M is significant, a J-based approach is likely to be invalid. The issue of J validity and size dependence is explored in detail in Sections 3.5 and 3.6.

3.5 J-CONTROLLED FRACTURE

The term *J-controlled fracture* corresponds to situations where J completely characterizes crack tip conditions. In such cases, there is a unique relationship between J and CTOD (Section 3.3); thus J-controlled fracture implies CTOD-controlled fracture, and vice versa. Just as there are limits to LEFM, fracture mechanics analyses based on J and CTOD become suspect when there is excessive plasticity or significant crack growth. In such cases, fracture toughness and the J-CTOD relationship depend on the size and geometry of the structure or test specimen.

The required conditions for J-controlled fracture are discussed below. Both fracture initiation from a stationary crack and stable crack growth are considered.

3.5.1 Stationary Cracks

Figure 3.23 schematically illustrates the effect of plasticity on the crack tip stresses; log (σ_{yy}) is plotted against normalized distance from the crack tip. The characteristic length scale L corresponds to the size of the structure; for example, L could represent the uncracked ligament length. Figure 3.23(a) shows the small scale yielding case, where both K and J characterize crack tip conditions. At a short distance from the crack tip, relative to L, the stress is proportional to $1/\sqrt{r}$; this area is called the *K-dominated region*. Assuming monotonic, quasistatic loading, a J-dominated region occurs in the plastic zone, where the elastic singularity no longer applies. Well inside of the plastic zone, the HRR solution is approximately valid and the stresses vary as $r^{-1/n+1}$. The finite strain region occurs within approximately 2δ from the crack tip, where large deformation invalidates the HRR theory. In small scale yielding, K uniquely characterizes crack tip conditions, despite the fact that the $1/\sqrt{r}$ singularity does not exist all the way to the crack tip. Similarly, J uniquely characterizes crack tip conditions even though the deformation plasticity and small strain assumptions are invalid within the finite strain region.

Figure 3.23(b) illustrates elastic-plastic conditions, where J is still approximately valid, but there is no longer a K field. As the plastic zone increases in size (relative to L), the K dominated zone disappears, but the J dominated zone persists. Thus although K has no meaning in this case, the J integral is still an appropriate fracture criterion. Since J dominance implies CTOD dominance, the latter parameter can also be applied in the elastic-plastic regime.

With large scale yielding (Fig. 3.23(c)), the size of the finite strain zone becomes significant relative to L, and there is no longer a region uniquely characterized by J. Single parameter fracture mechanics is invalid in large scale yielding, and critical J values exhibit a size and geometry dependence.

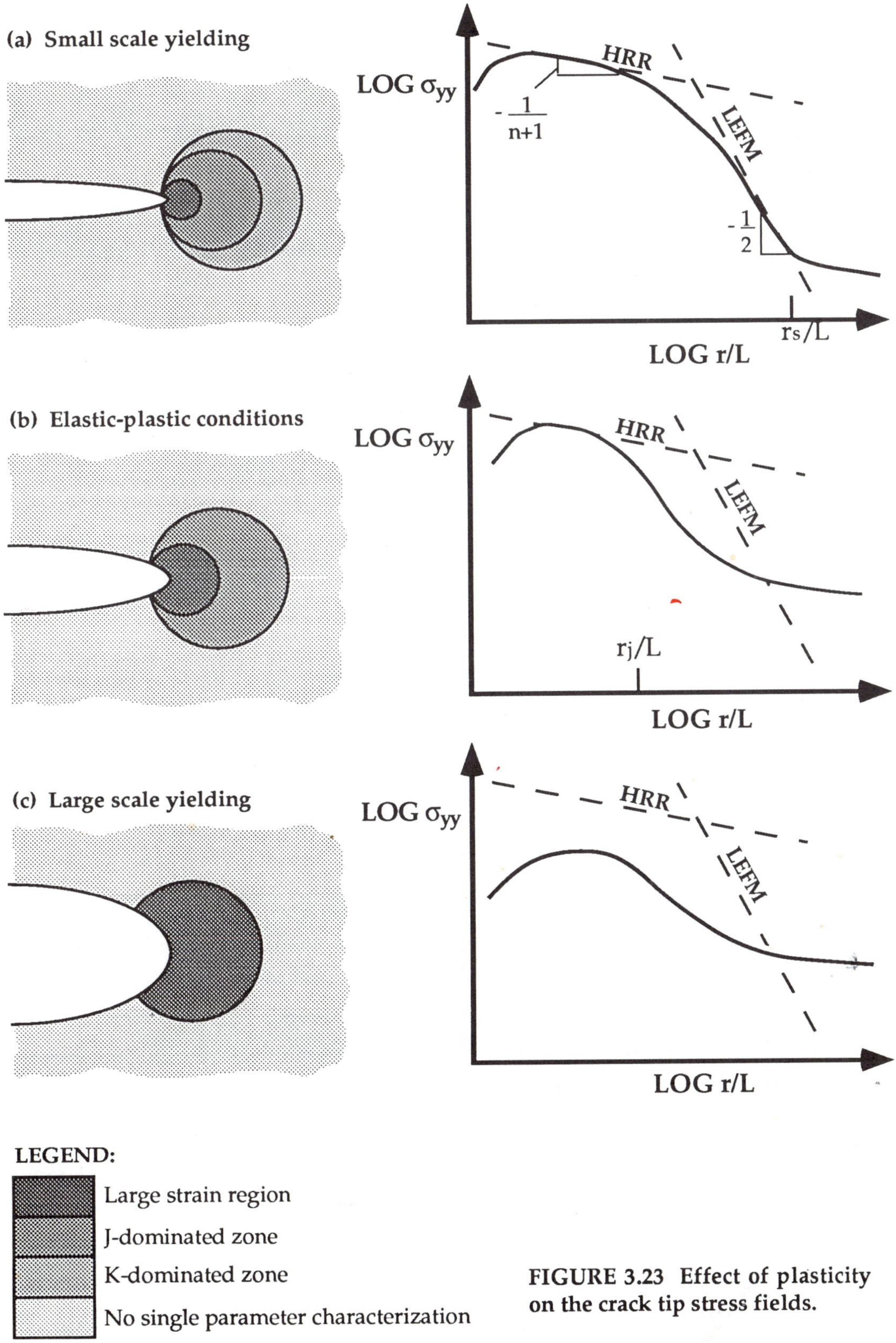

(a) Small scale yielding
LOG σyy
HRR
LEFM
- 1/(n+1)
- 1/2
rs/L
LOG r/L
(b) Elastic-plastic conditions
LOG σyy
HRR
LEFM
rj/L
LOG r/L
(c) Large scale yielding
LOG σyy
HRR
LEFM
LOG r/L
LEGEND:
Large strain region
J-dominated zone
K-dominated zone
No single parameter characterization

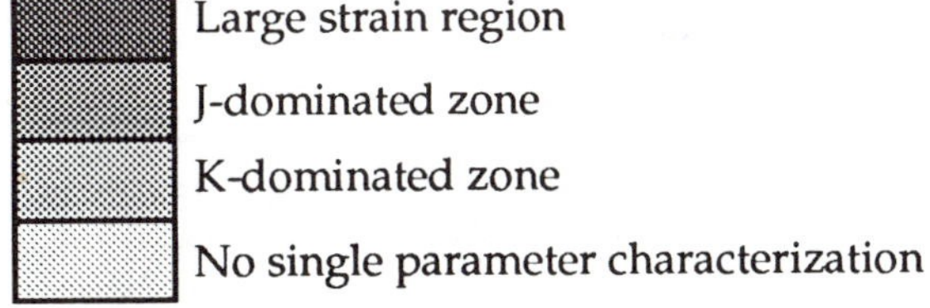

FIGURE 3.23 Effect of plasticity on the crack tip stress fields.

Recall Fig. 2.39, in which a free-body diagram was constructed from a disk-shaped region removed from the crack tip of a structure loaded in small scale yielding. Since the stresses on the boundary of this disk exhibit a $1/\sqrt{r}$ singularity, K_I uniquely defines the stresses and strains within the disk. For a given material[5], dimensional analysis leads to the following functional relationship for the stress distribution within this region:

$$\frac{\sigma_{ij}}{\sigma_o} = F_{ij}\left(\frac{K_I^2}{\sigma_o^2 r}, \theta\right) \quad (\text{for } 0 \le r \le r_s(\theta)) \tag{3.65}$$

where r_s is the radius of the elastic singularity dominated zone, which may depend on θ. Note that the $1/\sqrt{r}$ singularity is a special case of F, which exhibits a different dependence on r within the plastic zone. Invoking the relationship between J and K_I for small scale yielding (Eq. 3.18) gives

$$\frac{\sigma_{ij}}{\sigma_o} = F_{ij}\left(\frac{E' J}{\sigma_o^2 r}, \theta\right) \quad (\text{for } 0 \le r \le r_J(\theta)) \tag{3.66}$$

where r_J is the radius of the J dominated zone. The HRR singularity (Eq. (3.24a) is a special case of Eq. (3.66), but stress exhibits a $r^{-1/n+1}$ dependence only over a limited range of r.

For small scale yielding, $r_s = r_J$, but r_s vanishes when the plastic zone engulfs the elastic singularity dominated zone. The J dominated zone persists longer than the elastic singularity zone, as illustrated in Fig. 3.23.

It is important to emphasize that J dominance at the crack tip does not require the existence of an HRR singularity. In fact, J dominance requires only that Eq. (3.66) is valid in the *process zone* near the crack tip, where the microscopic events that lead to fracture occur. The HRR singularity is merely one possible solution to the more general requirement that J uniquely define crack tip stresses and strains. The flow properties of most materials do not conform to the idealization of

[5]A complete statement of the functional relationship of σ_{ij} should include all material flow properties (e.g. α and n for a Ramberg-Osgood material). These quantities were omitted from Eqs. (3.65) and (3.66) for the sake of clarity, since material properties are assumed to be fixed in this problem.

a Ramberg-Osgood power law, upon which the HRR analysis is based. Even in a Ramberg-Osgood material, the HRR singularity is valid over a limited range; large strain effects invalidate the HRR singularity close to the crack tip, and the computed stress lies below the HRR solution at greater distances. The latter effect can be understood by considering the analytical technique employed by Hutchinson [7], who represented the stress solution as an infinite series and showed that the leading term in the series was proportional to $r^{-1/n+1}$ (see Appendix 3.4). This singular term dominates as $r \rightarrow 0$; higher order terms are significant for moderate values of r. When the computed stress field deviates from HRR, it still scales with $J/(\sigma_o r)$, as required by Eq. (3.66). Thus J dominance does not necessarily imply agreement with the HRR fields.

Equations (3.65) and (3.66) gradually become invalid as specimen boundaries interact with the crack tip. We can apply dimensional arguments to infer when a single parameter description of crack tip conditions is suspect. As discussed in Chapter 2, the LEFM solution breaks down when the plastic zone size is a significant fraction of in-plane dimensions. Moreover, the crack tip conditions evolve from plane strain to plane stress as the plastic zone size grows to a significant fraction of the thickness. The J integral becomes invalid as a crack tip characterizing parameter when the large strain region reaches a finite size relative to in-plane dimensions. Section 3.6 provides quantitative information on size effects.

3.5.2 J-Controlled Crack Growth

According to the dimensional argument in the previous section, J controlled conditions exist at the tip of a stationary crack (loaded monotonically and quasistatically), provided the large strain region is small compared to in-plane dimensions of the cracked body. Stable crack growth, however, introduces another length dimension; i.e., the change in crack length from its original value. Thus J may not characterize crack tip conditions when the crack growth is significant compared to in-plane dimensions. Prior crack growth should not have any adverse effects in a purely elastic material, because the local crack tip fields depend only on current conditions. Prior history does influence the stresses and strains in an elastic-plastic material, however. Therefore, we might expect J integral theory to break down when there is a combination of significant plasticity and crack growth. This heuristic argument based on dimensional analysis agrees with experiment and with more complex analyses.

Figure 3.24 illustrates crack growth under J-controlled conditions. Material behind the growing crack tip has unloaded elastically. Recall Fig. 3.7, which compares the unloading behavior of nonlinear elastic and elastic-plastic materials; the material in the unloading region of Fig. 3.24 obviously violates the assumptions of deformation plasticity. The material directly in front of the crack also violates the single parameter assumption because the loading is highly nonproportional; i.e., the various stress components increase at different rates and some components actually decrease. In order for the crack growth to be J controlled, the elastic unloading and nonproportional plastic loading regions must be embedded within a zone of J dominance. When the crack grows out of the zone of J dominance, the measured R curve is no longer uniquely characterized by J.

In small scale yielding, there is always a zone of J dominance because the crack tip conditions are defined by the elastic stress intensity, which depends only on current values of load and crack size. The crack never grows out of the J-dominated zone as long as all specimen boundaries are remote from the the crack tip.

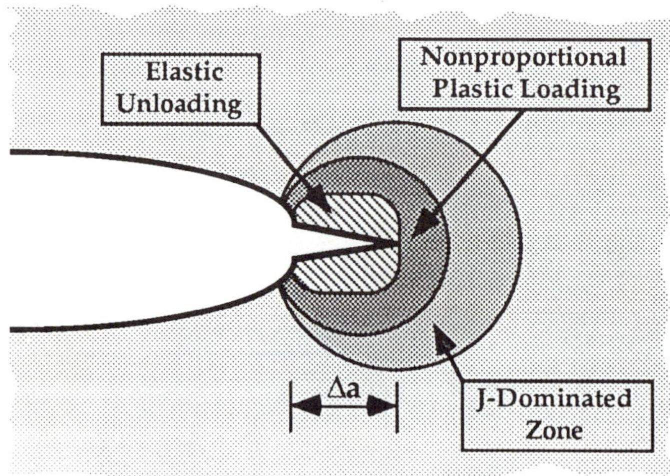

FIGURE 3.24 J-controlled crack growth.

Figure 3.25 illustrates three distinct stages of crack growth resistance behavior in small scale yielding. During the initial stage the crack is essentially stationary; the finite slope of the R curve is caused by blunting. The crack tip fields for Stage 1 are given by

$$\frac{\sigma_{ij}}{\sigma_o} = F_{ij}^{(1)}\left(\frac{E'J}{\sigma_o^2 r}, \theta\right) \tag{3.67}$$

which is a restatement of Eq. (3.66). The crack begins to grow in Stage 2. The crack tip stresses and strains are probably influenced by the original blunt crack tip during the early stages of crack growth. Dimensional analysis implies the following relationship:

$$\frac{\sigma_{ij}}{\sigma_o} = F_{ij}^{(2)}\left(\frac{E'J}{\sigma_o^2 r}, \theta, \frac{\Delta a}{\delta_i}\right) \tag{3.68}$$

(1) Crack blunting

(2) Fracture initiation

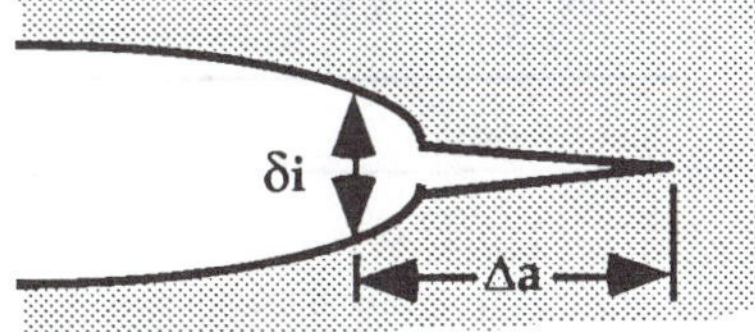

(3) Steady state crack growth

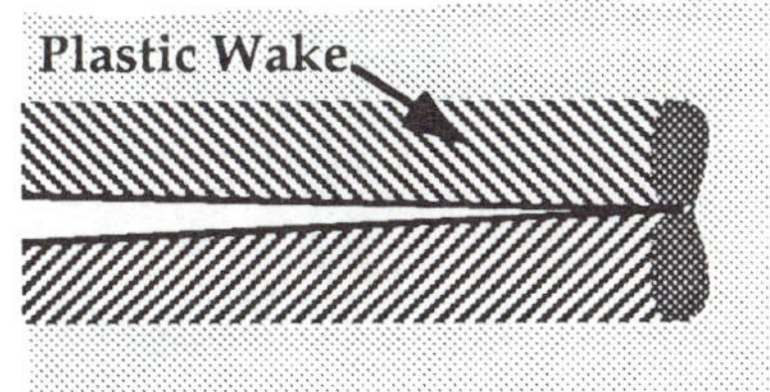

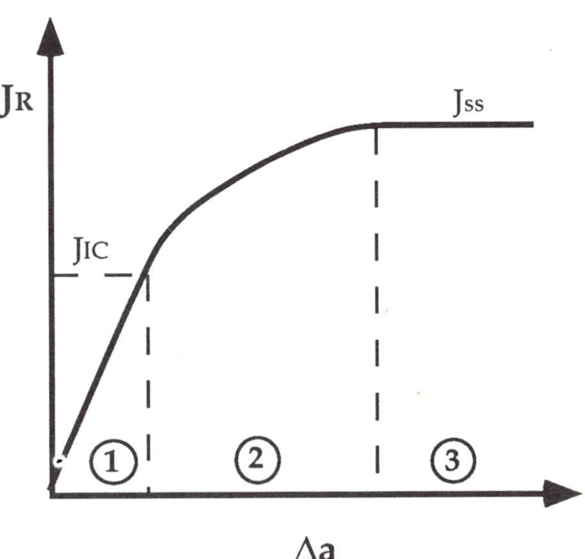

FIGURE 3.25 Three stages of crack growth in an infinite body.

where δ_i is the CTOD at initiation of stable tearing. When the crack grows well beyond the initial blunted tip, a steady-state condition is reached, where the local stresses and strain are independent of the extent of crack growth:

$$\frac{\sigma_{ij}}{\sigma_o} = F_{ij}^{(3)}\left(\frac{E' J}{\sigma_o^2 r}, \theta\right) \tag{3.69}$$

Although Eqs. (3.67) and (3.69) would predict identical conditions in the elastic singularity zone, material in the plastic zone at the tip of a growing crack is likely to experience a different loading history from material in the plastic zone of a blunting stationary crack; thus $F^{(1)} \neq F^{(3)}$ as $r \rightarrow 0$. During steady-state crack growth, a plastic zone of constant size sweeps through the material, leaving a plastic wake, as illustrated in Fig. 3.25. The R curve is flat; J does not increase with crack extension, provided the material properties do not vary with position. Appendix 3.5.2 presents a formal mathematical argument for a flat R curve during steady-state growth; a heuristic explanation is given below.

If Eq. (3.69) applies, J uniquely describes crack tip conditions, independent of crack extension. If the material fails at some critical combination of stresses and strains, then it follows that local failure at the crack tip must occur at a critical J value, as in the stationary crack case. This critical J value must remain constant with crack growth. A rising or falling R curve would imply that the local material properties varied with position.

The second stage in Fig. 3.25 corresponds to the transition between blunting of a stationary crack and crack growth under steady state conditions. A rising R curve is possible in Stage 2. For small scale yielding conditions the R curve depends only on crack extension:

$$J_R = J_R(\Delta a) \tag{3.70}$$

That is, the J-R curve is a material property.

The steady-state limit is usually not observed in laboratory tests on ductile materials. In typical test specimens, the ligament is fully plastic during crack growth, thereby violating the small scale yielding assumption. Moreover, the crack approaches a finite boundary while still in

Stage 2 growth. Enormous specimens would be required to observe steady state crack growth in tough materials.

Because flat J-R curves for ductile materials are rarely observed experimentally, some engineers do not believe that a steady-state limit exists. Appendix 3.5, however, offers (what should be) a convincing mathematical argument for such a limit. In addition, Rice, et al. [15] demonstrated that a steady-state limit occurs in small scale yielding when the crack advances by a critical crack tip opening angle (CTOA) criterion. Weatherby [19] performed a finite element simulation of a growing crack in an infinite body that showed that the R curve becomes flat and the plastic zone reaches a constant size once the crack grows well beyond the initial process zone.

3.6 EFFECT OF SPECIMEN DIMENSIONS

Under small scale yielding conditions, a single parameter (e.g. K, J or CTOD) characterizes crack tip conditions and can be used as a geometry-independent fracture criterion. Single parameter fracture mechanics breaks down in the presence of excessive plasticity, and fracture toughness depends on the size and geometry of the test specimen.

McClintock [20] applied slip line theory to estimate the stresses in a variety of configurations under plane strain, fully plastic conditions. Figure 3.26 summarizes some of these results. For small scale yielding, (Fig. 3.26(a)) the maximum stress at the crack tip is approximately $3\sigma_o$ in a nonhardening material. A deeply notched DENT specimen, illustrated in Fig. 3.26(b), maintains a high level of triaxiality under fully plastic conditions, such that the crack tip conditions are similar to the small scale yielding case. An edge cracked plate in bending (Fig. 3.26(c)) exhibits slightly less stress elevation, with the maximum principal stress approximately $2.5\sigma_o$. A panel in pure tension (Fig. 3.26(d)) is incapable of maintaining significant triaxiality under fully plastic conditions.

The results in Fig. 3.26 indicate that, for a nonhardening material under fully yielded conditions, the stresses near the crack tip are not unique, but depend on geometry. Traditional fracture mechanics approaches recognize that the stress and strain fields remote from the crack tip may depend on geometry, but it is assumed that the near-tip fields have a similar form in all configurations that can be scaled by a single parameter. The single parameter assumption is obviously not valid for nonhardening materials under fully plastic conditions, because the near tip fields depend on the configuration. Fracture toughness, whether quantified by J, K, or CTOD, must also depend on the configuration.

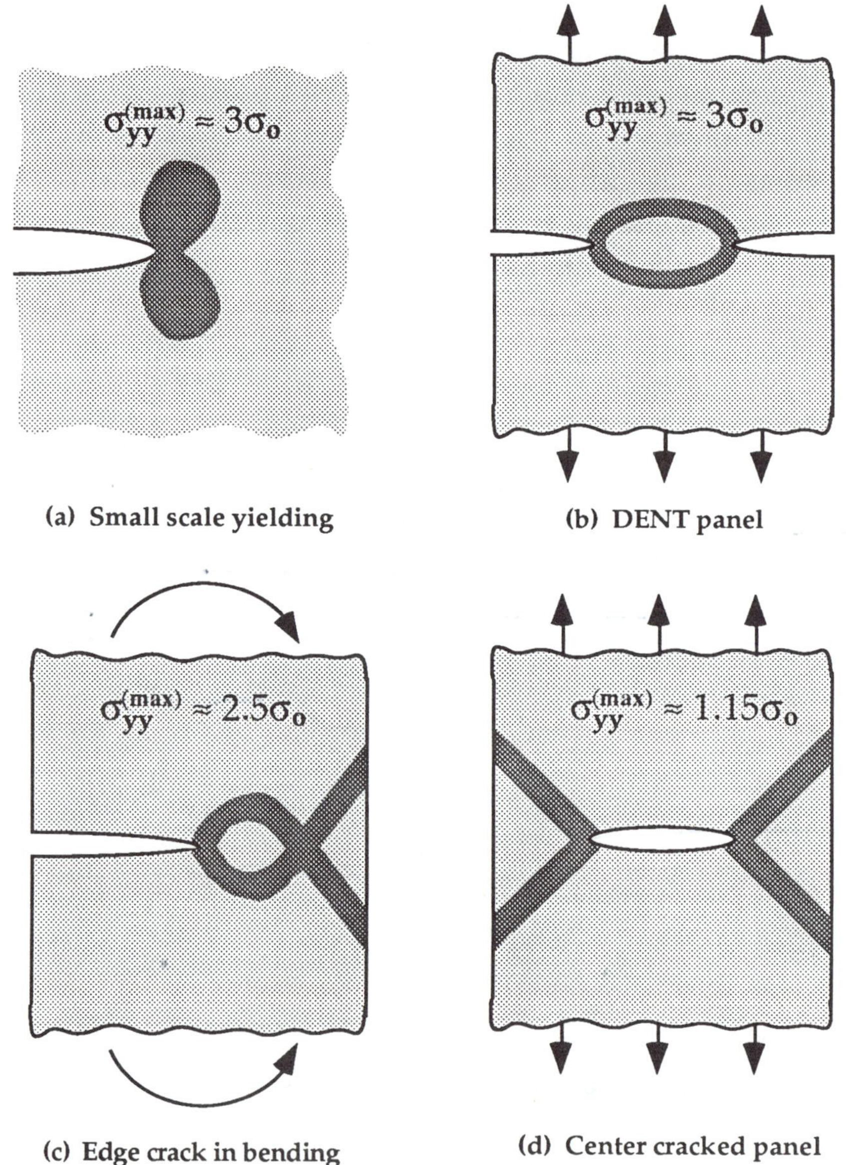

FIGURE 3.26 Comparison of the plastic deformation pattern in small scale yielding (a) with slip patterns under fully plastic conditions in three configurations. The estimated local stresses are based on the slip line analyses of McClintock [20], and only apply to nonhardening materials.

The prospects for applying fracture mechanics in the presence of large scale yielding are not quite as bleak as the McClintock analysis indicates. The configurational effects on the near-tip fields are much less severe when the material exhibits strain hardening. Moreover, single parameter fracture mechanics may be approximately valid in the presence of significant plasticity, provided the specimen maintains a relatively high level of triaxiality. Both the DENT specimen and the edge cracked plate in bending satisfy this requirement. Most laboratory measurements of fracture toughness are performed with bend-type specimens, such as the compact and three-point bend geometries, because these specimens present the fewest experimental difficulties

The loss of a single parameter description of the crack tip environment eventually occurs in all cases, but the effect is much less gradual in highly constrained specimens and high hardening materials. For the general case of large scale yielding (LSY) and stable crack growth, the near tip stress fields can be represented as follows:

$$\frac{\sigma_{ij}}{\sigma_o} = F_{ij}^{(LSY)}\left(\frac{E' J}{\sigma_o^2 r}, \theta, \frac{b\,\sigma_o}{J}, \frac{a}{b}, \frac{B}{b}, \frac{\Delta a}{b}, n, \mathcal{L}\right) \tag{3.71}$$

where $\mathcal{L}$ represents the loading conditions (e.g. bending or tension) and n is the strain hardening exponent (Eq. (3.22)). Equation (3.71) assumes a rectangular configuration, with crack length a, thickness B, and ligament length b; other geometries, such as a semielliptical surface flaw, can be represented by a slightly different function.

The first two terms in Eq. (3.71) are identical to Eq. (3.66), while the third term represents the extent of yielding. Since J/σ_o is proportional to CTOD (Eq. 3.48) the third term could be replaced by b/δ. Recall from the previous section that the finite strain zone, whose length is approximately 2δ, must be small compared to in-plane dimensions in order for Eq. (3.66) to be valid; thus the ratio $b\sigma_o/J$ should be large in order to maintain J-controlled conditions at the crack tip.

Equation (3.71) is obviously more complex than the single parameter conditions implied by Eq. (3.66). Although it is preferable to avoid complications whenever possible, the reader must bear in mind that single parameter fracture mechanics is a special case of a more general situation. In many instances, the single parameter approach provides a good approximation of actual material behavior, but one should always be aware of the potential for errors.

The sections that follow provide quantitative information on the sensitivity of specimen dimensions to the single parameter assumption.

3.6.1 In-Plane Dimensions

Much of the discussion that follows relies on computational results published by Dodds, et al. [21] and Anderson and Dodds [22]. They performed very detailed plane strain elastic-plastic finite element analyses on single edge notched bend (SENB) specimens. The meshes were sufficiently refined to infer the stress and strain distributions near the crack tip, although information at distances less than 2δ from the crack tip was disregarded because the solutions did not account for large strain effects. Solutions were generated for various strain hardening exponents and crack depths. They also generated stress solutions for the small scale yielding limit, where the in-plane dimensions are infinite relative to the plastic zone.

Figure 3.27 shows the effect of plasticity on the stress normal to the crack plane for n = 10 and a/W = 0.5. The ratio of σ_{yy} for the SENB specimen to the corresponding value for small scale yielding is plotted as a function of distance from the crack tip and applied load; P_o is a reference load, defined by

$$P_o = \frac{1.456\, \sigma_o\, b^2}{S} \tag{3.72}$$

for a specimen with unit thickness, where S is the loading span for three-point bending. Equation (3.72) is the estimated limit load for a nonhardening SENB specimen in plane strain, obtained from a slip line solution due to Green and Hundy [23]. For a hardening material, P_o corresponds approximately to the load at which the net cross section yields and produces a deformation pattern similar to Fig. 3.26(c).

Note that when the load is less than P_o, the computed stress ratios in Fig. 3.27 are close to 1.0, indicating approximate agreement with the small scale yielding solution. As load increases, the stress relaxes from the small scale yielding limit, but the loss in crack tip constraint is gradual. The curves in Fig. 3.27 are relatively flat; the ratio of σ_{yy} to the small scale yielding value is insensitive to distance from the crack tip within the range $2\delta \leq r \leq 10\,\delta$.

Figure 3.28(a) shows contours of maximum principal stress, where $\sigma_1 = 3\sigma_o$, as a function of J relative to the ligament length in the SENB specimen. Since r is normalized by J/σ_o, all contours should coincide in the limit of small scale yielding, according to Eq. (3.66). Because of relaxation in triaxiality, the normalized contours shrink as J increases.

Note that the contours in Fig. 3.28(a) all have a similar shape. For a given contour, it is possible to multiply r by a constant, κ, such that it coincides with the normalized contour for small scale yielding; κ should be close to 1.0 for low J values and increase with J. We can define an effective small scale yielding J as follows:

$$\kappa = \frac{J}{J_{ssy}} \qquad (3.73)$$

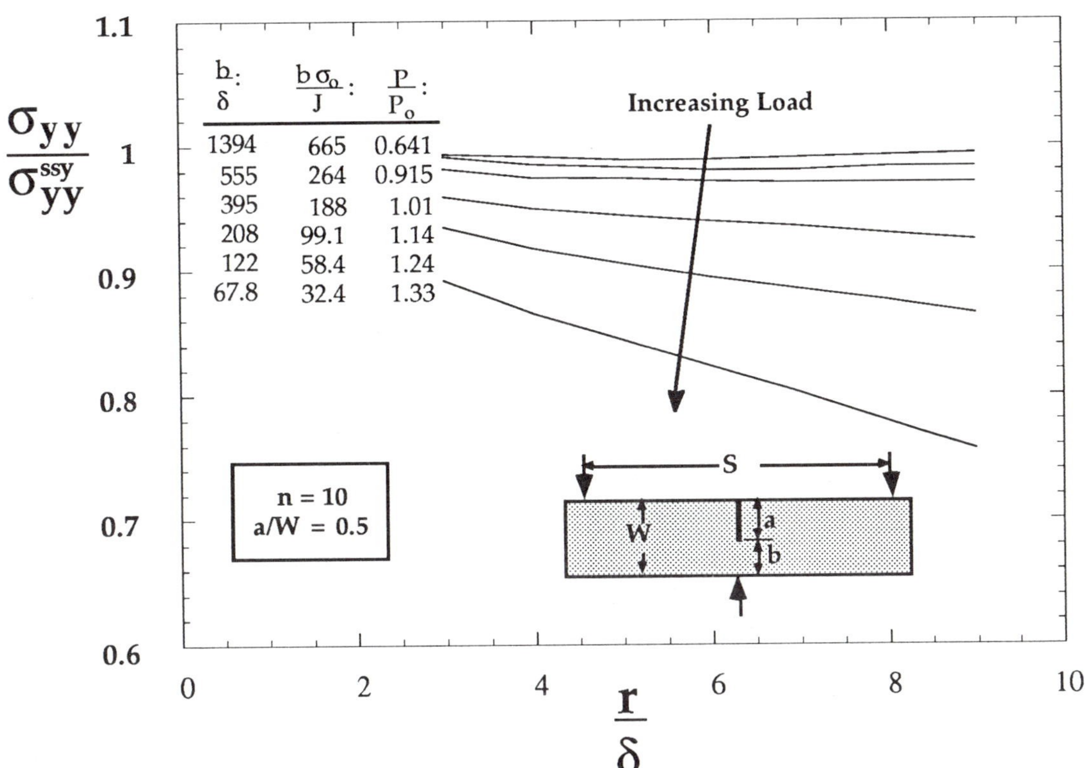

FIGURE 3.27 Stress normal to the crack plane, normalized by the small scale yielding value [22].

Figure 3.28(b) is a plot of the principal stress contours normalized by J_{ssy}/σ_o, where J_{ssy} was chosen in such a way that the contours would coincide at the various load levels.

The small scale yielding J, as defined in this case, can be interpreted as follows. The principal stress distribution for small scale yielding can be obtained by inserting the applied J into Eq. (3.66), assuming the form of this function is known. For large scale yielding, Eq. (3.66) will overestimate the principal stresses ahead of the crack tip, because the triaxiality is less than in the small scale yielding case. However, we can define an effective J (J_{ssy}), such that when it is inserted into Eq. (3.66), it gives the correct distribution of σ_1 for large scale yielding. Thus J_{ssy} uniquely characterizes the σ_1 distribution ahead of the crack tip. In the limit of small scale yielding, $J_{ssy} = J$, by definition.

One fracture mechanism that is controlled by the principal stress distribution is cleavage in metals, which is discussed in more detail in Chapter 5. Therefore, J_{ssy} can be viewed as the effective driving force for cleavage, and the J/J_{ssy} ratio characterizes the elevation in cleavage fracture toughness due to constraint loss. It is important to realize, however, that J_{ssy} characterizes only the maximum principal stress ahead of the crack tip; different parameters would be required for the other stress and strain components because the loading is nonproportional in large scale yielding. Thus other fracture mechanisms that are not controlled exclusively by the maximum principal stress would exhibit a different size dependence than cleavage.

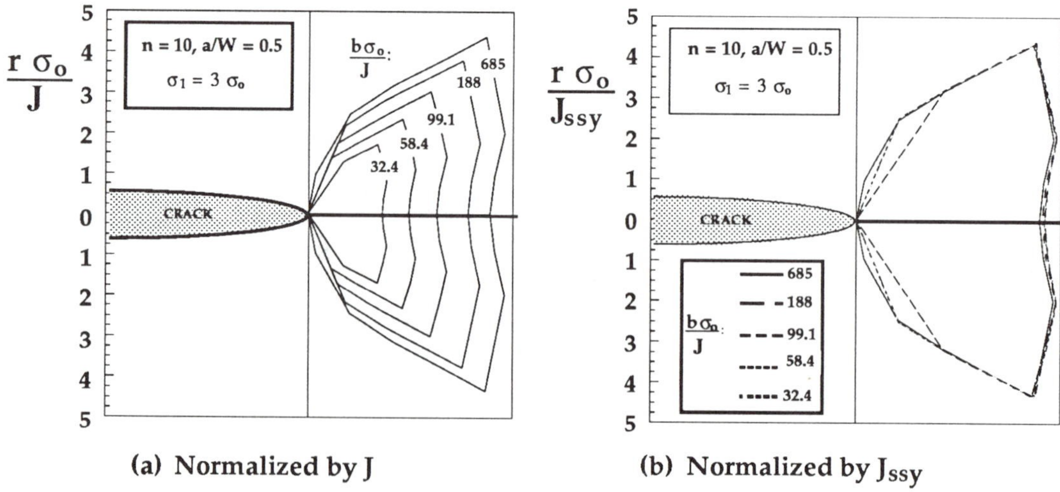

FIGURE 3.28 Principal stress contours ahead of the crack tip in an SENB specimen [22].

Figure 3.29 is a plot of J/J_{ssy} versus ligament length, relative to J/σ_o; n = 10 in this case, and solutions for three crack depths are shown. This graph indicates that cleavage fracture toughness is sensitive to specimen size in large scale yielding; the effect is particularly pronounced in specimens with shallow cracks. For a/W = 0.5, the ratio $a\sigma_o/J$ must be greater than ~200 in order to achieve cleavage under J-controlled conditions. There are two ways to ensure that this ratio is large in a laboratory test: (1) by testing a large specimen, or (2) by testing a material with a low cleavage toughness. A very large specimen or a very brittle material would be required to achieve J-controlled cleavage fracture when $a/W \leq 0.15$.

The results from Fig. 3.29 are plotted in Fig. 3.30 in a slightly different matter. The effective driving force for cleavage fracture, J_{ssy}, is plotted against J, the apparent driving force; both parameters are normalized by ligament length and σ_o. At low J values, small scale yielding conditions exist and $J = J_{ssy}$. At higher J values, the curves diverge from the small scale yielding line, and J_{ssy} increases at a slower rate than J. Eventually, the curves become relatively flat, implying that the driving force saturates at a constant value; further increases in J have little effect on J_{ssy}.

Figure 3.31(a) illustrates the effect of hardening exponent on the J/J_{ssy} ratio. The corresponding plot for CTOD is shown in Fig. 3.31(b). When n = 50, the material is virtually nonhardening, while n = 5 corresponds to a high hardening material. According to Fig. 3.31, low hardening materials exhibit the most size dependence in large scale yielding.

Specimen size requirements for obtaining J-controlled fracture are usually given in the following form:

$$B, b \geq M \frac{J}{\sigma_o} \tag{3.74}$$

where M is a dimensionless constant. According to Figs. 3.29 and 3.31(a), M should be approximately 200 for cleavage in deeply notched specimens, at least in the case of the ligament dimension. The size requirement is much less strict for initiation of ductile fracture in metals; M = 25 for J_{IC} testing [24]. The size requirements for cleavage are approximately eight times as strict as for ductile tearing because cleavage is more sensitive to deviations from the small scale yielding limit.

When toughness data do not satisfy the conditions necessary for J-controlled fracture, it may be possible to correct these data for the loss in crack tip constraint. The analyses presented above provide the means for correcting cleavage fracture toughness data for geometry dependence. Figure 3.32 is a plot of critical CTOD values for cleavage in an A 36 steel plate at two temperatures. These experimental data were obtained by Sorem [26]. The solid diamonds represent the actual experimental data, while the crosses denote δ_{ssy} values. For the uncorrected data, there is an obvious geometry dependence between the deeply notched specimens (a/W = 0.5) and the shallow notched specimens (a/W = 0.15). This geometry dependence is removed by computing δ_{ssy} values for each configuration.

Figure 3.26 shows that a fully yielded panel in pure tension has a low level of triaxiality, in the limit of a nonhardening material. Although this effect is not as pronounced in a material with finite strain hardening, it is still significant. Figure 3.33 is a plot of elastic-plastic finite element results published by Shih and German [25] The crack tip stress fields for an SENB specimen and a center cracked tension (CCT) panel are compared with the HRR solution. The stresses at the crack tip in the SENB specimen are reasonably close to the HRR limit, but the stresses in the tensile configuration are well below the HRR curve.

The numerical results in Fig. 3.33 imply that a material should have a higher apparent toughness when the loading is predominantly tensile and the specimen is fully yielded. Figure 3.34 compares critical CTOD values for cleavage in SENB specimens with corresponding toughness values for tensile panels [27]; the latter includes both CCT specimens and panels with semielliptical surface cracks. Although the scatter bands overlap slightly, it is obvious that the critical CTOD values are significantly higher on average in the tensile panels. These results are expected, in light of the finite element results in Fig. 3.33.

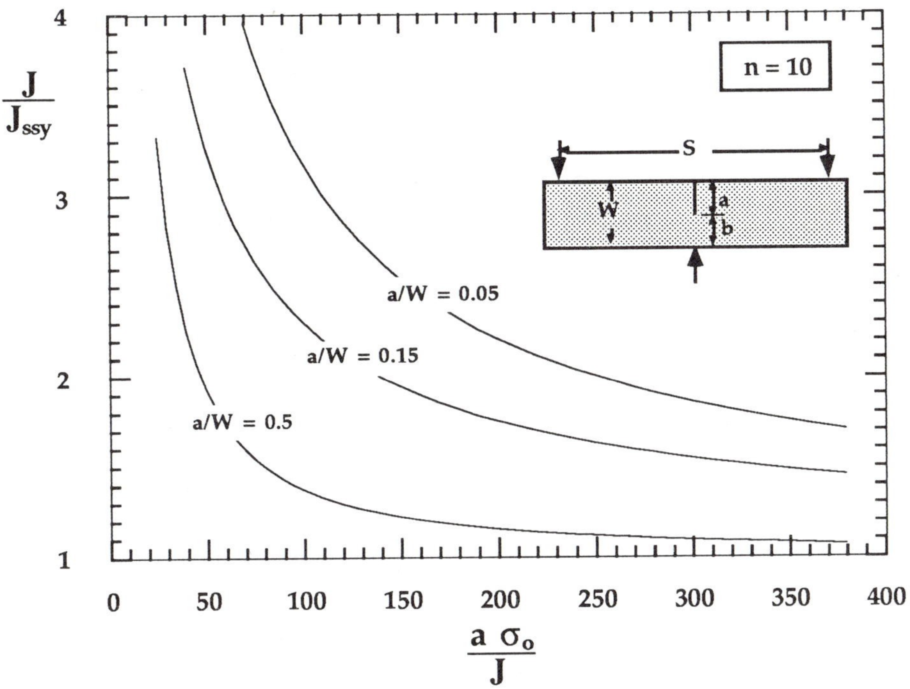

FIGURE 3.29 Effect of specimen size and a/W on J/J_{ssy} for SENB specimens [22].

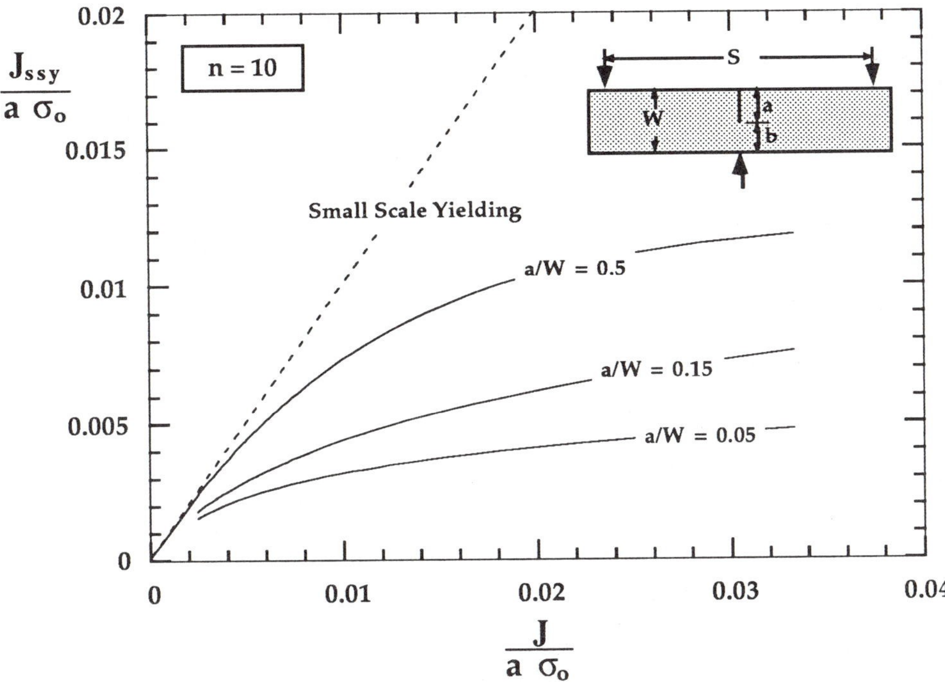

FIGURE 3.30 Effective driving force (J_{ssy}) versus the apparent driving force (J) for stress-controlled fracture [22].

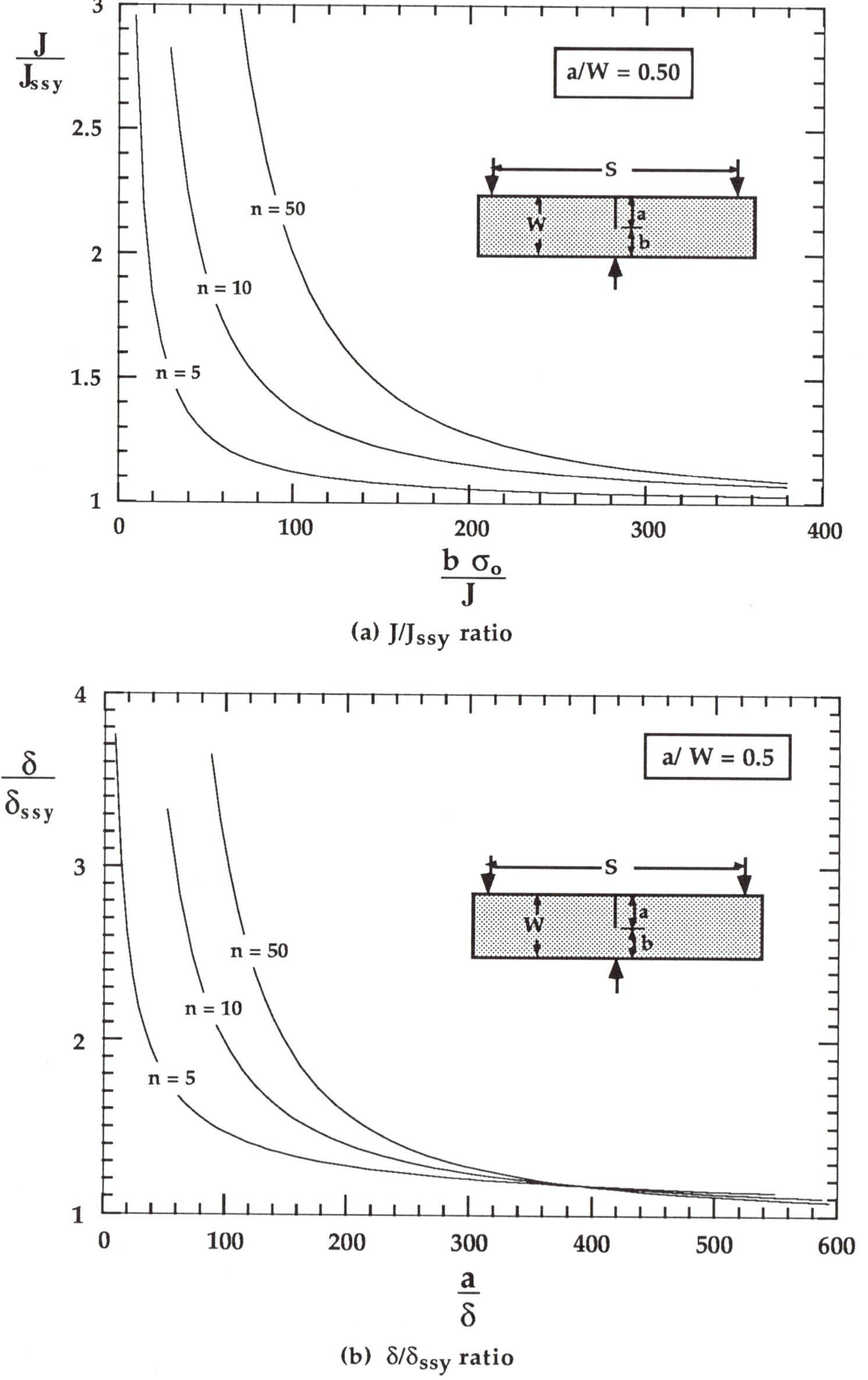

(a) J/J_{ssy} ratio

(b) δ/δ_{ssy} ratio

FIGURE 3.31 Effect of specimen size and strain hardening on the fracture toughness, relative to the small scale yielding value.

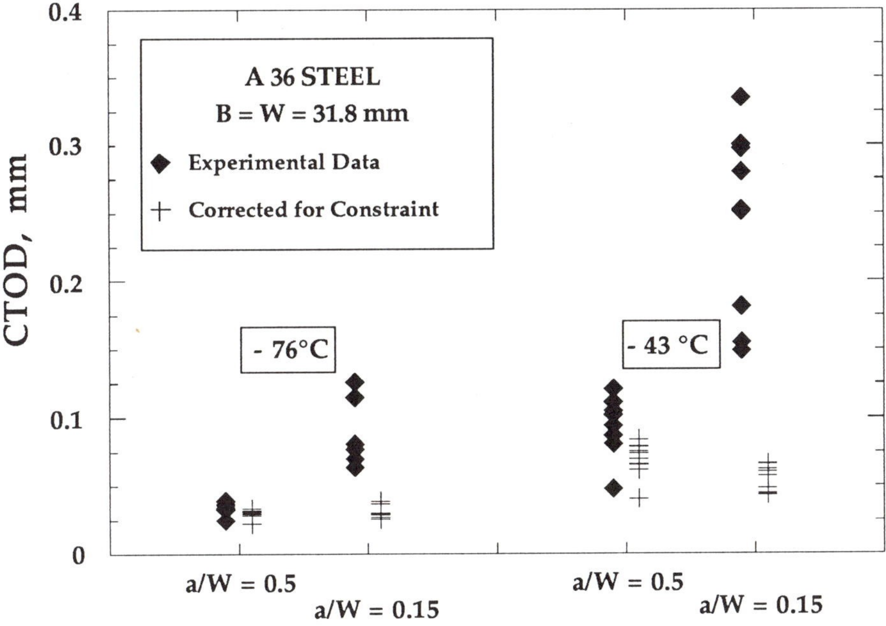

FIGURE 3.32 Fracture toughness data for a mild steel, corrected for constraint loss [22].

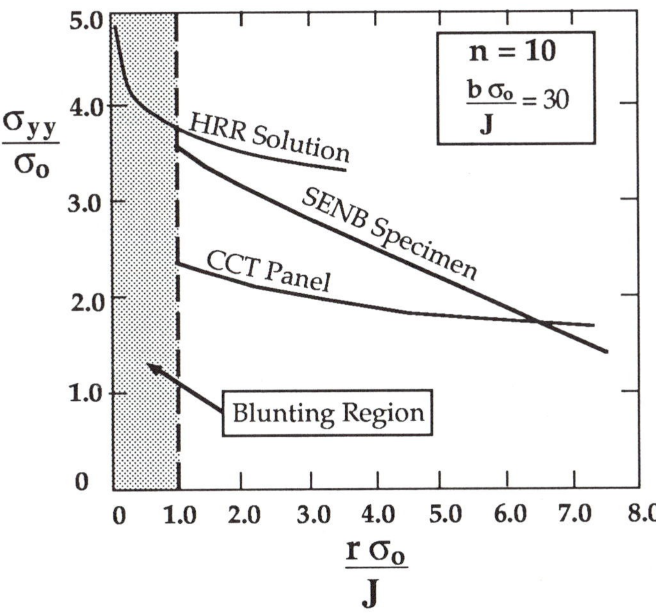

FIGURE 3.33 Comparison of computed crack tip stress fields for bending (SENB specimen) and tension (CCT panel) [25]. The ligaments have completely yielded in both cases.

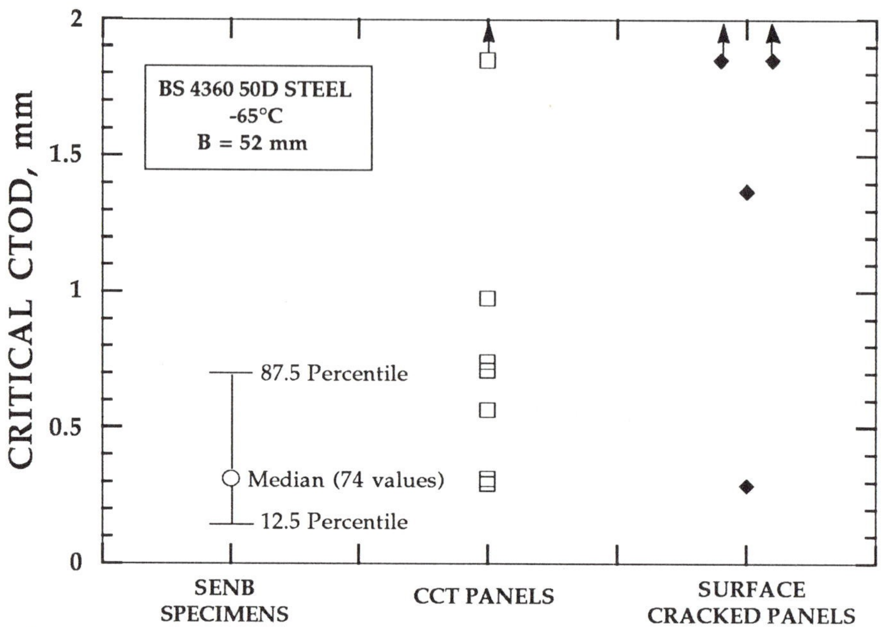

FIGURE 3.34 Critical CTOD values in bending and tensile loading for a low-alloy structural steel [27].

3.6.2 Three-Dimensional Effects

The previous section was limited to plane strain conditions; only the effect of in-plane dimensions (e.g. crack length, ligament length) was considered. The dimensions in the z direction along the crack front can also play a major role, however.

Section 2.9 showed that the stress state near the crack tip can vary from plane stress at the surface to plane strain at the midplane. The stress state remote from the crack tip is plane stress throughout (see Fig. 2.37) When the plastic zone size exceeds roughly half the thickness, the stress state is predominantly plane stress in the plastic zone (Fig. 2.38), except very close to the crack tip.

Figure 3.35 is a plot of stress normal to the crack plane versus position in thickness an SENB specimen, obtained from finite element results published by Narasimhan and Rosakis [28]. Curves for three J values are shown; in each case, r is approximately four times the CTOD. The stress is normalized by its value at the midplane. Note that for the two lowest J values, there is a region in the center of the specimen where σ_{yy} does not vary with z; the stress state is essentially two-dimensional in this region. Such a stress distribution does not necessarily imply plane strain, but the stresses at the mid plane in this three-dimensional analysis agreed with a two-dimensional plane strain analysis on the same configuration and flow behavior. Thus the mid plane of the three-dimensional specimen is approximately in plane strain at $r \sim 4\delta$, at least at the two lowest J values. The size of the plane strain zone decreases as J increases.

Recall the generalized size requirement of Eq. (3.74), where it was concluded that M = 200 for the ligament dimension was sufficient to guarantee J controlled cleavage fracture. This conclusion was based on a plane strain finite element analysis. Thus it is essential that plane strain conditions be maintained through a substantial portion of the thickness in order to apply the in-plane results from the previous section. From Fig. 3.35, when $B\sigma_o/J \geq 200$, approximately 80% of the thickness is in plane strain. Thus M = 200 appears to be appropriate for both the ligament and thickness dimensions in the case of cleavage fracture in deeply notched bend specimens.

One complicating feature of three-dimensional fracture mechanics is that the driving force is not constant through the thickness. Figure 3.36 shows the through-thickness variation of J in the bend specimen analyzed by Narasimhan and Rosakis. The J integral was evaluated at various points on the crack front by means of the domain integral approach (Chapter 11); J(z) is normalized by the nominal J for this specimen, $\hat{J}$, which is related to the area under the load-displacement curve (Eq. 3.39). The ratio $J(z)/\hat{J}$ is slightly greater than 1.0 at the mid plane, and decreases rapidly near the free surface. The relative magnitude of the through-thickness variation increases with $\hat{J}$. Figure 3.37 illustrates the through-thickness variation of CTOD, which is not as large as the J variation.

The through-thickness variation of J and CTOD in actual test specimens is probably not as great as in Figs. 3.36 and 3.37. The Narasimhan and Rosakis analysis assumed a straight crack front, while cracks in laboratory specimens are slightly bowed. In the latter case, the cracks are usually introduced by fatigue; the crack takes on a shape that tends to equalize the stress intensity factor along the crack front.

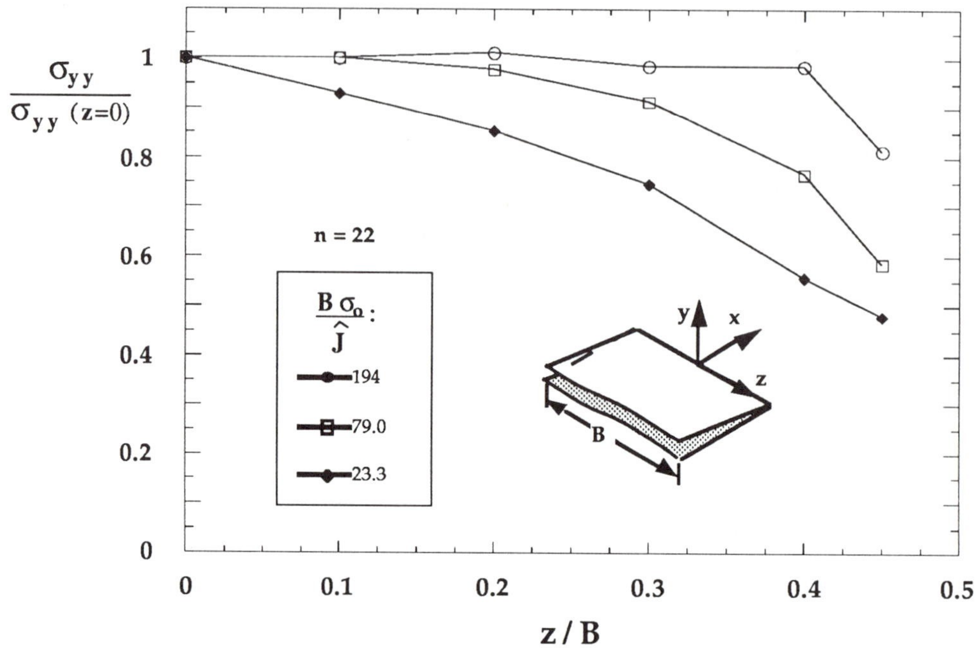

FIGURE 3.35 Variation of σ_{yy} through the thickness of an SENB specimen [28]. $\hat{J}$ is the nominal (average) value of J in the specimen.

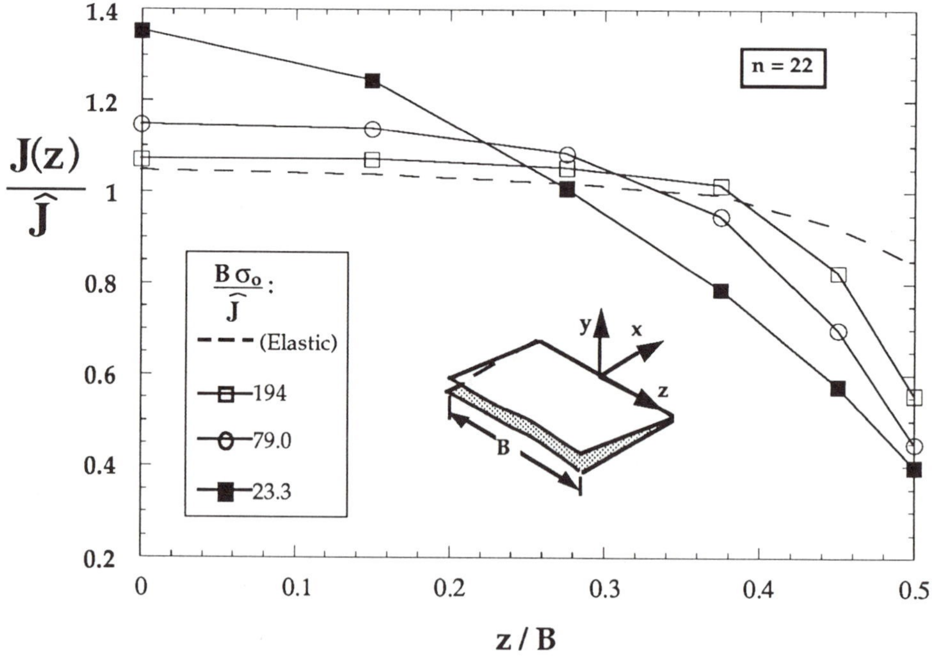

FIGURE 3.36 Through-thickness variation of J in an SENB specimen, where $\hat{J}$ is the nominal J [28].

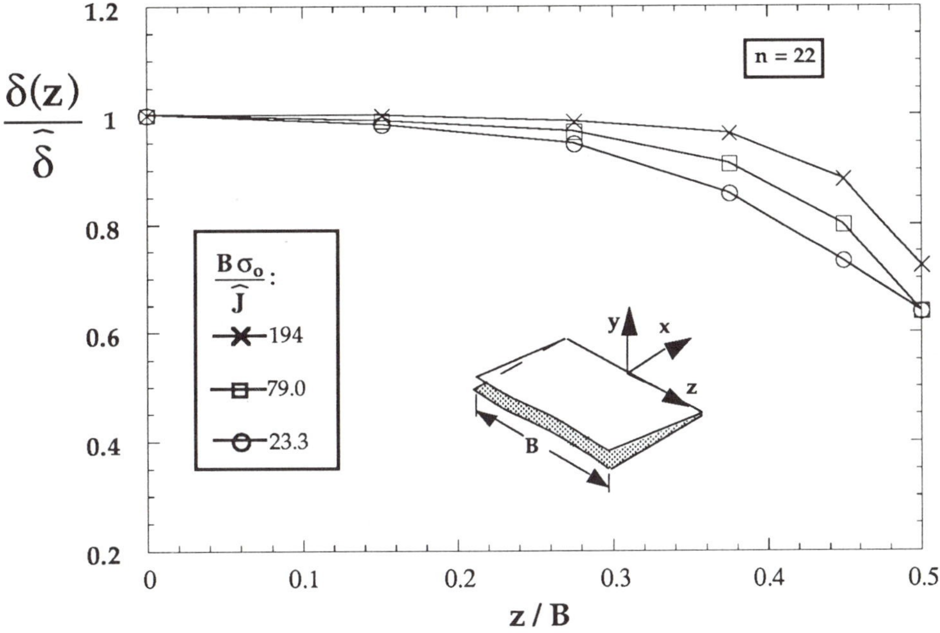

FIGURE 3.37 Through-thickness variation of CTOD in an SENB specimen, where $\hat{\delta}$ is the CTOD at midthickness [28].

The semi-elliptical surface crack is an important configuration in fracture mechanics because it is representative of many structural flaws. Three dimensional effects have a pronounced effect on the behavior of surface cracks. Figure 2.20, for example, shows that K_I varies along the crack front; the maximum K_I occurs at the deepest point along the crack, while the minimum value is at the free surface. Figure 3.38 shows the variation of J along the crack front for the case where the crack depth is a finite fraction of the plate thickness [29]. The J variation shows trends that are similar to the K variation under linear elastic conditions.

The normal stress distribution around the circumference of the surface flaw behaves in an usual manner, as illustrated in Fig. 3.39 [29]. The local stress reaches a peak at approximately 30° from the free surface. This effect is most pronounced at the highest σ^{∞}. The stress is probably lower a the maximum crack depth because of relaxation to the back surface of the plate; plane stress conditions occur at the free surface, resulting in lower stresses. The peak stress in Fig. 3.39 probably coincides to the location of the maximum stress triaxiality.

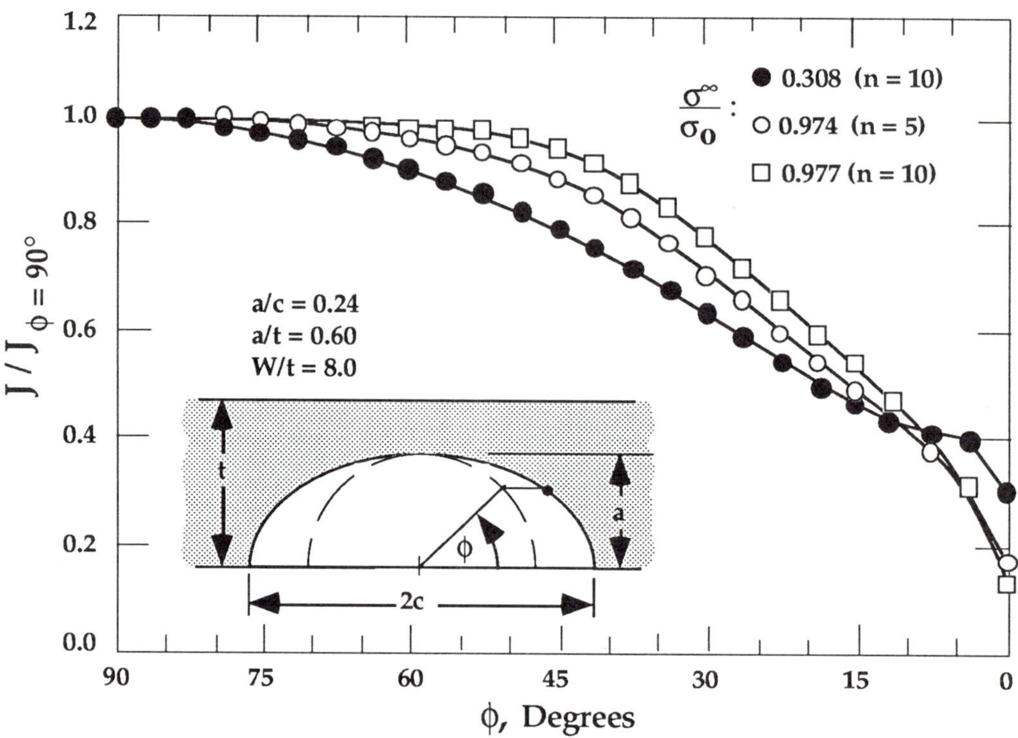

FIGURE 3.38 Computed variation of J along the crack front of a surface flaw [29].

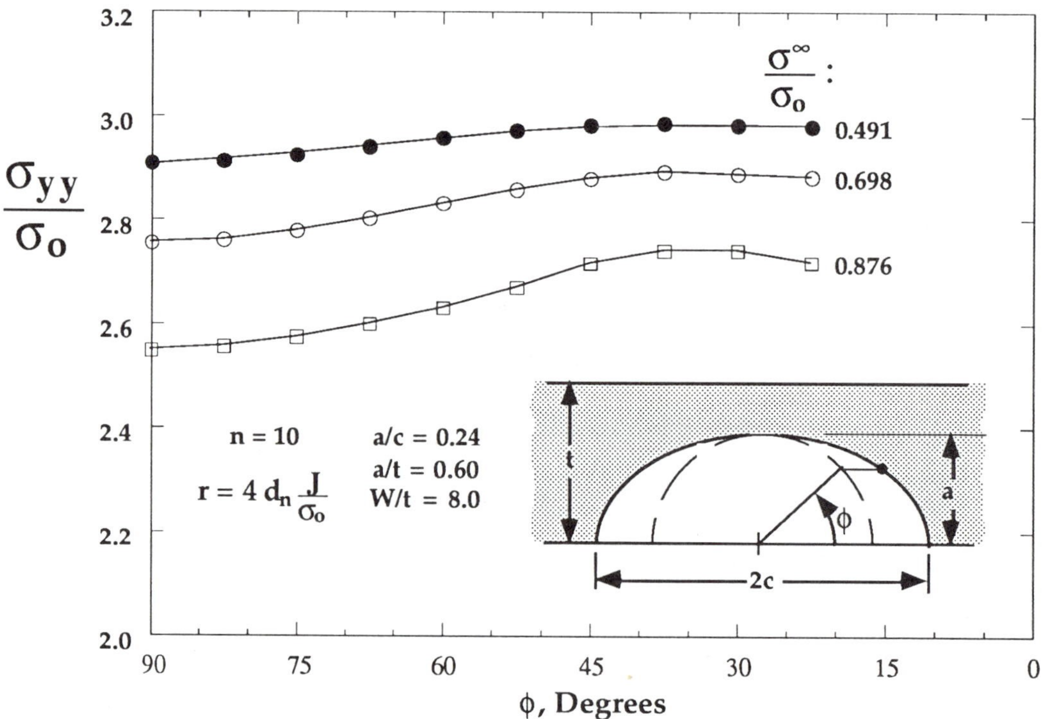

FIGURE 3.39 Computed variation of normal stress along the crack front of a surface flaw [29].

Although the apparent driving force for fracture (J) in a surface flaw is at a maximum at $\phi = 90°$, the maximum effective driving force occurs at $\phi \approx 30°$. The photograph in Fig. 3.40 shows unstable propagation and arrest of a semielliptical surface crack [30]. The maximum crack growth occurs at approximately 30° from the free surface, where the finite element analysis indicates that the maximum stress occurs; the crack grew at the highest rate where the effective driving force was at a maximum.

Three-dimensional effects in fracture mechanics are complex phenomena that are the subject of ongoing research. In most practical situations, it is necessary to idealize the problem as two-dimensional, because the tools necessary to incorporate three dimensional effects are not yet available. Laboratory specimens are often designed to correspond to a nearly two-dimensional situation. Side-grooves, for example, can minimize through-thickness variations of stress and driving force in test specimens, and produce nearly plane strain conditions. In the case of structural flaws, however, the shape of the crack is imposed by nature; it is obviously not possible to design the cracked configuration to match two-dimensional assumptions.

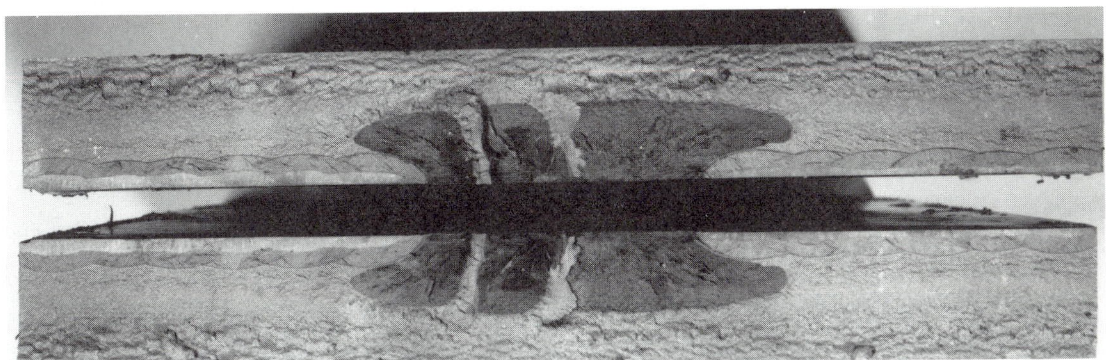

FIGURE 3.40 Arrested crack propagation of a surface flaw in a thick steel panel [30]. The dark heat-tinted region indicates the shape of the arrested crack. (Photograph provided by J.G. Merkle.)

3.6.3 Size Dependence of R Curves

The subject of the size and geometry dependence of J-R curves, and how best to deal with this effect, has been hotly debated in recent years. Seemingly contradictory experimental data has added a great deal of confusion. This issue is important to the nuclear power industry, who need to perform accurate ductile tearing analyses that involve very large structures and extensive stable crack growth.

Figure 3.41 schematically illustrates two types of behavior that have been observed in ductile materials. The J-R curves in part (a) exhibit a size dependence, when analyzed in terms of the deformation J, but the geometry dependence is apparently removed by applying the modified J. (See Section 3.4.2 for definitions of J_D and J_M.) The J_D-R curves agree at small amounts of crack growth but diverge at larger amounts of growth; the smallest specimens diverge first, and the R curve slope decreases from that in the larger specimens. In some cases, the R curves in small specimens have negative slopes.

Figure 3.41(b) illustrates behavior that is virtually the opposite of that in Fig. 3.41(a). The R curve slope decreases with specimen size in this case. The size dependence is actually worse for the J_M-R curve, because the application of Eq. (3.64) has the effect of raising the R curves for the smallest specimens. Hiser [31] has published fracture toughness data that display the trends in Fig. 3.41(b), while Ernst [16] showed an example of data that behave in a matter similar to Fig. 3.41(a).

The opposite trends in data with specimen size probably reflect two competing mechanisms. Constraint loss is the most likely explanation for the behavior in Fig. 3.41(b), while the size dependence illustrated by Fig. 3.41(a) is probably caused by finite ligament effects that are inherent in the definition of J_D, as discussed below.

It is well known that constraint loss tends to increase the apparent toughness. This effect is evident in the cleavage fracture toughness data in Figs. 3.32 and 3.34. In the case of ductile fracture, higher toughness implies an elevation of both the J at initiation and the R curve slope. The behavior in Fig. 3.41(b) indicates that the small specimens have less triaxiality than larger specimens made from the same material.

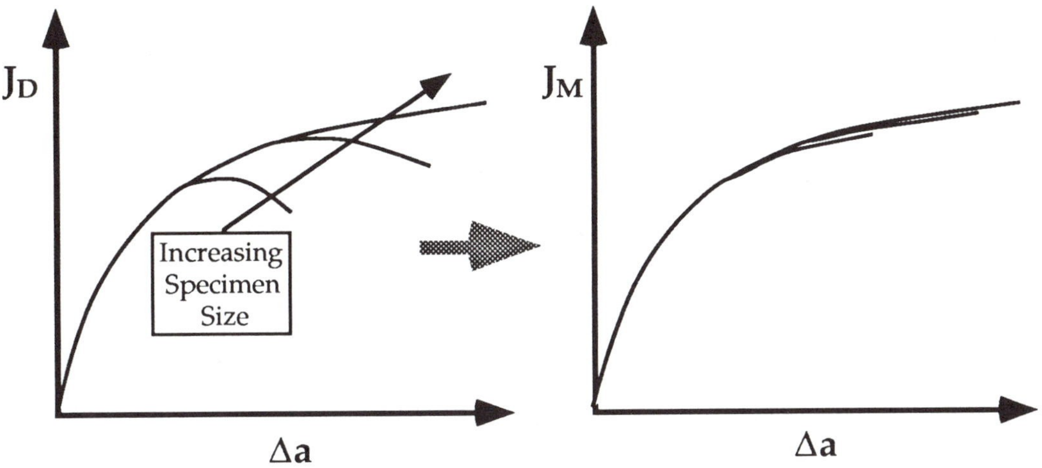

(a) R curve slope increases with specimen size.

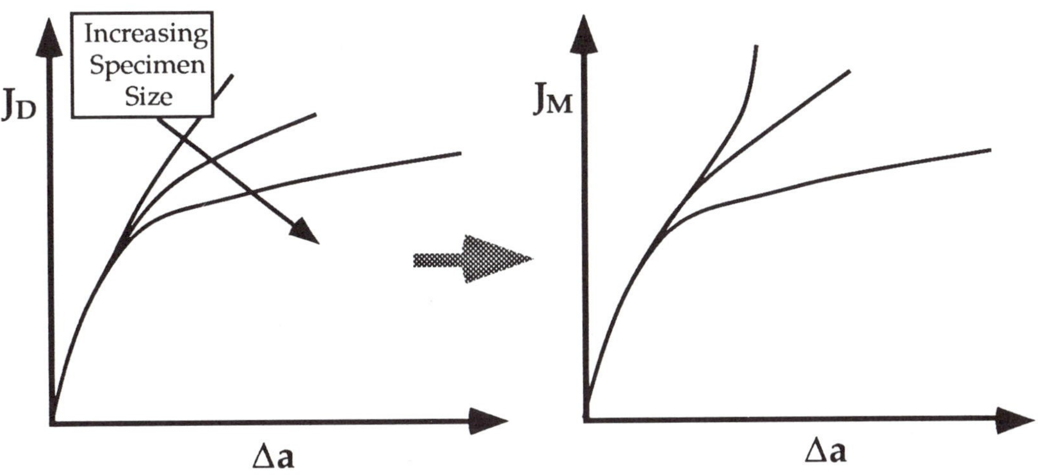

(b) R curve slope decreases with specimen size.

FIGURE 3.41 Two types of size dependence that have been observed in J-R curves for ductile materials.

The way in which J_D is defined leads to the type of size dependence illustrated in Fig. 3.41(a). Refer to Eq. (3.62), in which an incremental change in J_D is related to four partial derivatives. The incremental J_M is obtained by subtracting the last term in Eq. (3.62) from dJ_D. Applying Eq. (3.56(b)) leads to the the following relationship [16]:

$$\left(\frac{\partial J_p}{\partial a}\right)_P = \gamma \frac{J_p}{b} \tag{3.75}$$

where γ is a constant that depends on η_p (see Appendix 7 for a derivation of γ). Substituting this result into Eq. (3.63) and solving for the R curve slope gives

$$\frac{dJ_D}{da} = \frac{dJ_M}{da} - \gamma\frac{J_p}{b} \tag{3.76}$$

The relative contribution of the second term increases as the ligament length decreases in size. As the growing crack approaches the back surface of the specimen, the second term may become larger than the first term, resulting in a negative J_D-R curve slope (Fig. 3.41(a)).

It is important to note that although application of J_M removes the finite ligament effect, it does not correct for constraint loss. Figure 3.42 schematically compares the R curve behavior of a finite specimen with the small scale yielding limit. The R curve for the finite specimen lies above the curve for the small scale yielding case, because the finite specimen has less constraint at the crack tip. When the crack tip in the finite specimen approaches the back surface, the J_D-R curve slope decreases and eventually becomes negative, but the J_M-R curve continues to rise. Neither curve, however, agrees with the small scale yielding limit.

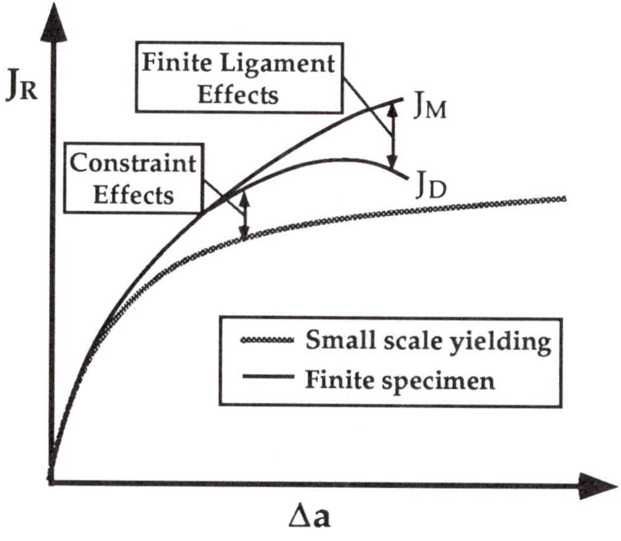

FIG 3.42 Schematic illustration of size and geometry effects on R curves.

Figure 3.43 shows experimental R curves that exhibit both constraint and finite ligament effects [32]. The R curve scatter band for the small (10 mm x 10 mm) specimens initially lies above the R curves for larger specimens, but the R curve slope for the small specimens decreases rapidly as the crack growth approaches specimen boundaries.

Additional experimental data that illustrate the effect of constraint on J-R curves are shown in Figs. 3.44 and 3.45 [33,34]. In the former case, all specimen dimensions were fixed except the crack length; specimens with shallow cracks have the lowest constraint and exhibit the steepest R curves. Figure 3.45 compares J-R curves for bending and tensile loading; the tensile panel has low constraint (Fig. 3.26(d)), which is reflected in the R curve.

This discussion in this section has been largely qualitative because the effect of specimen dimensions and loading on J-R curves in ductile materials has yet to be quantified.

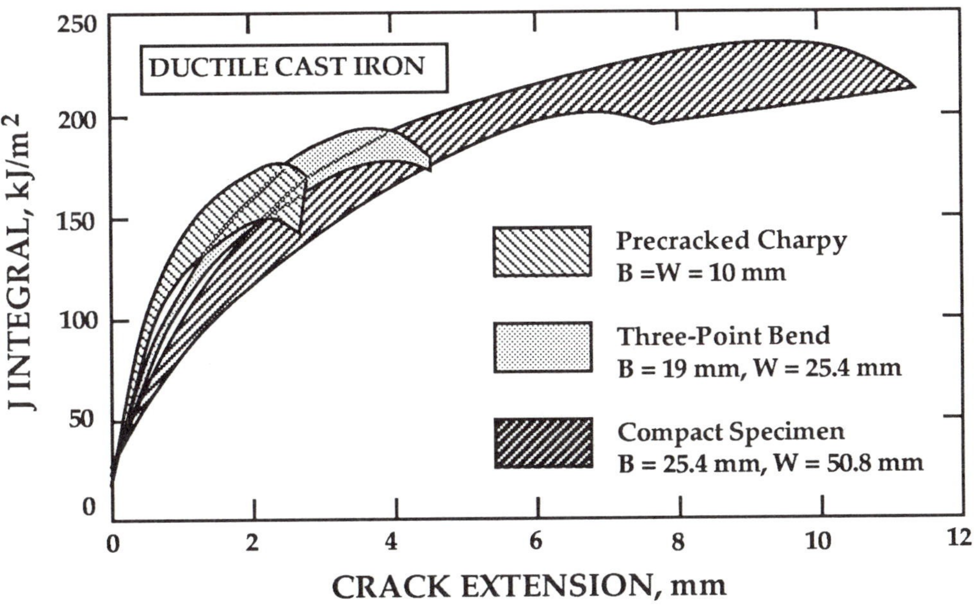

FIG. 3.43 Effect of specimen size and geometry on J-R curves for ductile iron [32]. Note that both constraint and finite ligament effects occur in this material.

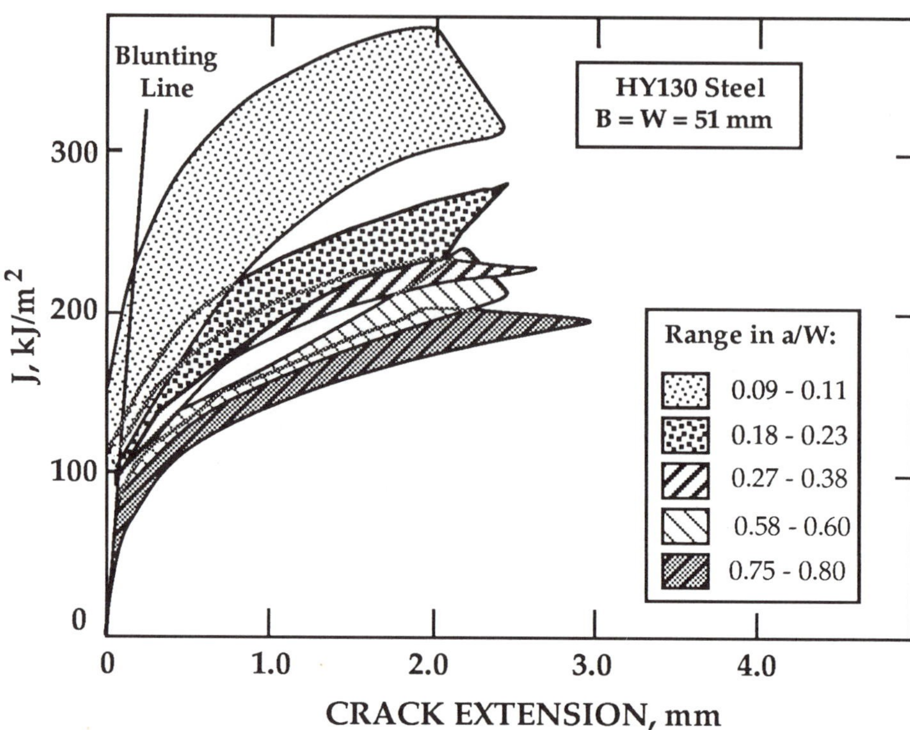

FIG. 3.44 Effect of crack length/specimen width ratio on J-R curves for HY130 steel three-point bend specimens [33].

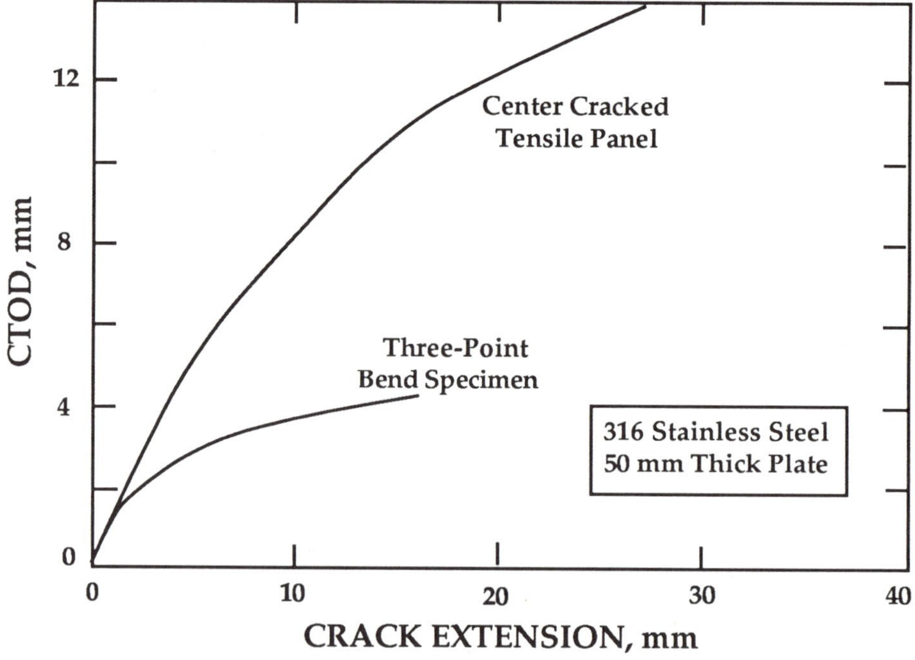

FIG. 3.45 Comparison of CTOD resistance curves for bending and tensile loading in 316 austenitic steel [34].

REFERENCES

1. Wells, A.A., "Unstable Crack Propagation in Metals: Cleavage and Fast Fracture." *Proceedings of the Crack Propagation Symposium,* Vol 1, Paper 84, Cranfield, UK, 1961.

2. Irwin, G.R., "Plastic Zone Near a Crack and Fracture Toughness." *Sagamore Research Conference Proceedings,* Vol. 4, 1961.

3. Burdekin, F.M. and Stone, D.E.W., "The Crack Opening Displacement Approach to Fracture Mechanics in Yielding Materials." *Journal of Strain Analysis,* Vol. 1, 1966, pp. 145-153.

4. Rice, J.R. " A Path Independent Integral and the Approximate Analysis of Strain Concentration by Notches and Cracks." *Journal of Applied Mechanics,* Vol. 35, 1968, pp. 379-386.

5. BS 5762: 1979, "Methods for Crack Opening Displacement (COD) Testing." British Standards Institution, London, 1979.

6. E 1290-89 "Standard Test Method for Crack Tip Opening Displacement Testing." American Society for Testing and Materials, Philadelphia, 1989.

7. Hutchinson, J.W., "Singular Behavior at the End of a Tensile Crack Tip in a Hardening Material." *Journal of the Mechanics and Physics of Solids,* Vol. 16, 1968, pp. 13-31.

8. Rice, J.R. and Rosengren, G.F., "Plane Strain Deformation near a Crack Tip in a Power-Law Hardening Material. *Journal of the Mechanics and Physics of Solids,* Vol. 16, 1968, pp. 1-12.

9. McMeeking, R.M. and Parks, D.M., "On Criteria for J-Dominance of Crack Tip Fields in Large-Scale Yielding." ASTM STP 668, American Society for Testing and Materials, Philadelphia, 1979, pp. 175-194.

10. Read, D.T., "Applied J-Integral in HY130 Tensile Panels and Implications for Fitness for Service Assessment." Report NBSIR 82- 1670, National Bureau of Standards, Boulder, CO, 1982.

11. Begley, J. A. and Landes, J.D., "The J-Integral as a Fracture Criterion." ASTM STP 514, American Society for Testing and Materials, Philadelphia, 1972, pp. 1-20.

12. Landes, J.D. and Begley, J.A., "The Effect of Specimen Geometry on J_{Ic}." ASTM STP 514, American Society for Testing and Materials, Philadelphia, 1972, pp. 24-29.

13. Rice, J.R., Paris, P.C., and Merkle, J.G., "Some Further Results of J-Integral Analysis and Estimates." ASTM STP 536, American Society of Testing and Materials, Philadelphia, 1973, pp. 231-245.

14. Shih, C.F. "Relationship between the J-Integral and the Crack Opening Displacement for Stationary and Extending Cracks." *Journal of the Mechanics and Physics of Solids*, Vol 29, 1981, pp. 305-326.

15. Rice, J.R., Drugan, W.J., and Sham, T.-L., "Elastic-Plastic Analysis of Growing Cracks." ASTM STP 700, American Society of Testing and Materials, Philadelphia, 1980, pp. 189-221.

16. Ernst, H.A., "Material Resistance and Instability Beyond J-Controlled Crack Growth." ASTM STP 803, American Society for Testing and Materials, Philadelphia, 1983, pp. I-191 - I-213.

17. Joyce, J.A., Davis, D.A., Hackett, E.M., and Hays, R.A., "Application of the J-Integral and the Modified J-Integral to Cases of Large Crack Extension." ASTM STP 1074, American Society for Testing and Materials, Philadelphia, (In press).

18. E1152-87 "Standard Test Method for Determining J-R Curves." American Society for Testing and Materials, Philadelphia, 1987.

19. Weatherby, J.R., "Finite Element Analysis of a Crack Growing in Inelastic Media." Ph.D. Dissertation, Texas A&M University, College Station, TX, May 1986.

20. McClintock, F.A., "Plasticity Aspects of Fracture." *Fracture: An Advanced Treatise*, Vol. 3, Academic Press, New York, 1971, pp. 47-225.

21. Dodds, R.H. Jr., Anderson T.L. and Kirk, M.T., "A Framework to Correlate a/W Effects on Elastic-Plastic Fracture Toughness (J_c)." to be published in *International Journal of Fracture.*

22. Anderson, T.L. and Dodds, R.H., Jr., "Specimen Size Requirements for Fracture Toughness Testing in the Ductile-Brittle Transition Region." *Journal of Testing and Evaluation*, Vol. 19, 1991, pp. 123-134.

23. Green, A.P. and Hundy, B.B., "Initial Plastic Yielding in Notch Bend Bars." *Journal of the Mechanics and Physics of Solids*, Vol. 4, 1956, pp.128-149.

24. E 813-87, "Standard Test Method for J_{Ic}, a Measure of Fracture Toughness." American Society for Testing and Materials, Philadelphia, 1987.

25. Shih, C.F. and German, M.D., "Requirements for a One Parameter Characterization of Crack Tip Fields by the HRR Singularity." *International Journal of Fracture*, Vol. 17, 1981, pp. 27-43.

26. Sorem,W.A., "The Effect of Specimen Size and Crack Depth on the Elastic-Plastic Fracture Toughness of a Low-Strength High-Strain Hardening Steel." Ph.D. Dissertation, The University of Kansas, Lawrence, KS, May 1989.

27. Anderson, T.L., "Ductile and Brittle Fracture Analysis of Surface Flaws Using CTOD." *Experimental Mechanics*, June 1988, pp. 188-193.

28. Narasimhan, R. and Rosakis A.J., "Three Dimensional Effects Near a Crack Tip in a Ductile Three Point Bend Specimen - Part I: A Numerical Investigation." California Institute of Technology, Division of Engineering and Applied Science, Report SM 88-6, Pasadena, CA, January 1988.

29. Parks, D.M. and Wang, Y.-Y., "Elastic-Plastic Analysis of Part-Through Surface Cracks." *Analytical, Numerical, and Experimental Aspects of Three-Dimensional Fracture Processes,* ASME AMD-91, American Society of Mechanical Engineers, New York, 1988, pp. 19-32.

30. Corwin, W.R., "Heavy Section Steel Technology Program Semiannual Progress Report for April-September 1987, U.S. Nuclear Regulatory Commission Report NUREG/CR-4219", Vol. 4, No. 2, October, 1987.

31. Hiser, A.L. and Terrell, J.B., "Size Effects on J-R Curves for a A 302-B plate." Nuclear Regulatory Commission Report NUREG/CR-5265, January, 1989.

32. Cayard, M.S. and Bradley, W.L., "A Comparison of Several Analytical Techniques for Calculating J-R Curves from Load-Displacement Data and Their Relation to Specimen Geometry." *Engineering Fracture Mechanics,* Vol. 33, 1989, pp. 121-132.

33. Towers, O.L. and Garwood, S.J., "Influence of Crack Depth on Resistance Curves for Three-Point Bend Specimens in HY130." ASTM STP 905 American Society for Testing and Materials, Philadelphia, 1986, pp. 454-484.

34. Anderson, T.L., Gordon, J.R., and Garwood, S.J., "On the Application of R-Curves and Maximum Load Toughness to Structures." ASTM STP 969, American Society for Testing and Materials, Philadelphia, 1988, pp. 291-317.

35. Westergaard, H.M., "Bearing Pressures and Cracks." *Journal of Applied Mechanics,* Vol. 6, 1939, pp. 49-53.

36. Bilby, B.A., Cottrell, A.H., and Swindon, K. H., "The Spread of Plastic Yield from a Notch." *Proceedings, Royal Society of London,* Vol. A-272, 1963, pp. 304-314,

37. Smith, E., "The Spread of Plasticity from a Crack: an Approach Based on the Solution of a Pair of Dual Integral Equations." CEGB Research Laboratories, Lab. Note No. RD/L/M31/62, July 1962.

38. Rice, J.R. and Tracey, D.M., *Journal of the Mechanics and Physics of Solids,* Vol. 17, 1969, pp. 201-217.

39. Budiansky, B. "A Reassessment of Deformation Theories of Plasticity." *Journal of Applied Mechanics.* Vol. 81, 1959, pp. 259-264.

APPENDIX 3: MATHEMATICAL FOUNDATIONS OF ELASTIC-PLASTI FRACTURE MECHANICS

(Selected Results)

A3.1 DETERMINING CTOD FROM THE STRIP YIELD MODEL

Burdekin and Stone [3] applied the Westergaard [35] complex stress function approach to the strip yield model. They derived an expression for CTOD by superimposing a stress function for closure forces on the crack faces in the strip yield zone. Their result was similar to previous analyses based on the strip yield model performed by Bilby, et al. [36] and Smith [37].

Recall from Appendix 2.3 that the Westergaard approach expresses the in-plane stresses (in a limited number of cases) in terms of Z:

$$\sigma_{xx} = \mathrm{Re}Z - y\mathrm{Im}Z' \tag{A3.1a}$$

$$\sigma_{yy} = \mathrm{Re}Z + y\mathrm{Im}Z' \tag{A3.1b}$$

$$\tau_{xy} = -\, y\mathrm{Re}Z' \tag{A3.1c}$$

where Z is an analytic function of the complex variable z = x + iy, and the prime denotes a first derivative with respect to z. By invoking the equations of elasticity for the plane problem, it can be shown that the displacement in the y direction is as follows:

$$u_y = \frac{1}{E}\left[2\,\mathrm{Im}\bar{Z} - y\,(1+\nu)\,\mathrm{Re}Z\right] \quad \text{for plane stress} \tag{A3.2a}$$

and

$$u_y = \frac{1}{E}\left[2\,(1-\nu^2)\,\mathrm{Im}\bar{Z} - y\,(1+\nu)\,\mathrm{Re}Z\right] \quad \text{for plane strain} \tag{A3.2b}$$

where $\bar{Z}$ is the integral of Z with respect to z, as discussed in Appendix 2. For a through crack of length $2a_1$ in an infinite plate under biaxial tensile stress σ, the Westergaard function is given by

$$Z = \frac{\sigma z}{\sqrt{z^2 - a_1^2}} \tag{A3.3}$$

where the origin is defined at the crack center.

The stress function for a pair of splitting forces, P, at ± x within a crack of length 2 a_1 (see Fig. 2.30) is given by

$$Z = \frac{2 P z \sqrt{a_1^2 - x^2}}{\pi \sqrt{z^2 - a_1^2}\,(z^2 - a_1^2)} \tag{A3.4}$$

For a uniform compressive stress = σ_{YS} along the crack surface between a and a_1 (Fig. A3.1), the Westergaard stress function is obtained by substituting $P = -\sigma_{YS}\,dx$ into Eq. (A3.4) and integrating:

$$Z = -\int_a^{a_1} \frac{2 \sigma_{YS} z \sqrt{a_1^2 - x^2}}{\pi \sqrt{z^2 - a_1^2}\,(z^2 - a_1^2)} dx$$

$$= -\frac{2\sigma_{YS}}{\pi}\left[\frac{z}{\sqrt{z^2 - a_1^2}} \cos^{-1}\left(\frac{a}{a_1}\right) - \cot^{-1}\left(\frac{a}{z}\sqrt{\frac{z^2 - a_1^2}{a_1^2 - a^2}}\right)\right] \tag{A3.5}$$

The stress functions of Eqs. (A3.3) and (A3.5) can be superimposed, resulting in the strip yield solution for the though crack. Recall from Section 2.8.2 that the size of the strip yield zone was chosen so that the stresses at the tip would be finite. Thus

$$k \equiv \frac{a}{a_1} = \cos\left(\frac{\pi \sigma}{\sigma_{YS}}\right) \tag{A3.6}$$

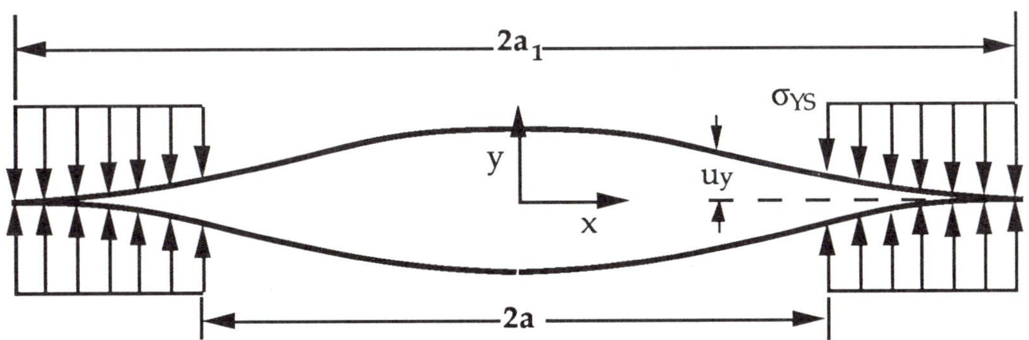

FIGURE A3.1 Strip yield model for a through crack.

When Eq. (A3.6) is substituted into Eq. (A3.5) and Eq. (A3.3) is superimposed, the first term in Eq. (A3.5) cancel with Eq. (A3.3), which leads to

$$Z = \frac{2\,\sigma_{YS}}{\pi}\left[\frac{k}{z}\sqrt{\frac{z^2 - a_1{}^2}{1 - k^2}}\right] \tag{A3.7}$$

Integrating Eq. (A3.7) gives

$$\bar{Z} = \frac{2\,\sigma_{YS}}{\pi}[z\,\omega_1 - a\,\omega_2] \tag{A3.8}$$

where

$$\omega_1 = \cot^{-1}\frac{\sqrt{1 - \left(\frac{a_1}{a}\right)^2}}{\sqrt{\frac{1}{k^2} - 1}}$$

and

$$\omega_2 = \cot^{-1}\sqrt{\frac{z^2 - a_1{}^2}{1 - k^2}}$$

On the crack plane, $y = 0$ and the displacement in the y direction (Eq. (A3.2)) reduces to

$$u_y = \frac{2}{E}\,\mathrm{Im}\bar{Z} \tag{A3.9}$$

for plane stress. Solving for the imaginary part of Eq. (A3.8) gives

$$u_y = \frac{4\,\sigma_{YS}}{\pi E}\left[a \coth^{-1}\left(\frac{1}{a_1}\sqrt{\frac{a_1^2 - z^2}{1 - k^2}}\right) - z \coth^{-1}\left(\frac{k}{z}\sqrt{\frac{a_1^2 - z^2}{1 - k^2}}\right)\right]$$

for $|z| \leq a_1$. Setting z = a leads to

$$\delta = 2u_y = \frac{8\,\sigma_{YS}\,a}{\pi E} \ln\left(\frac{1}{k}\right) \tag{A3.10}$$

which is identical to Eq. (3.5).

Recall the J-CTOD relationship (Eq. (3.44)) derived from the strip yield model. Let us define an effective stress intensity for elastic-plastic conditions in terms of the J integral:

$$K_{eff} \equiv \sqrt{J E} \tag{A3.11}$$

Combining Eqs. (3.44) (A3.10) and (A3.11) gives

$$K_{eff} = \sigma_{YS}\sqrt{\pi a}\left[\frac{8}{\pi^2} \ln \sec\left(\frac{\pi \sigma}{2\,\sigma_{YS}}\right)\right]^{1/2} \tag{A3.12}$$

which is the strip yield plastic zone correction given in Eq. (2.7(' and plotted in Fig. 2.31. Thus the strip yield correction to K_I is equi lent to a J-based approach for a nonhardening material in plane stress.

A3.2 THE J CONTOUR INTEGRAL

Rice [4] presented a mathematical proof of the path independence of the J contour integral. He began by evaluating J along a closed contour, Γ* (Fig A3.2):

$$J^* = \int_{\Gamma^*}\left(w dy - T_i \frac{\partial u_i}{\partial x} ds\right) \tag{A3.13}$$

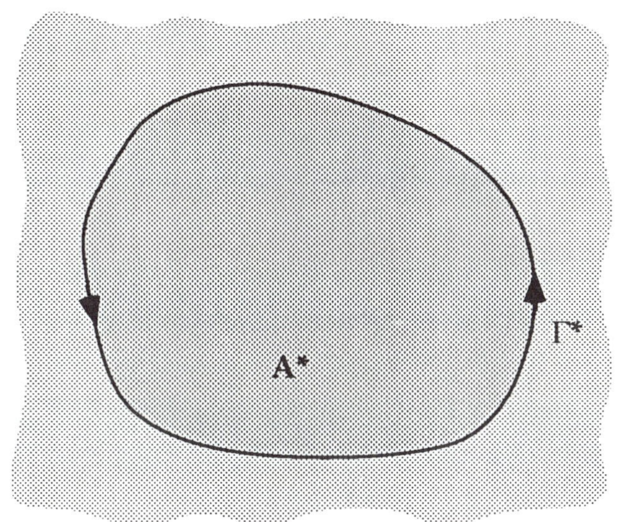

FIGURE A3.2 Closed contour, Γ*, in a two-dimensional solid.

where the various terms in this expression are defined in Section 3.2.2. Rice then invoked the divergence theorem to convert Eq. (A3.13) into an area integral:

$$J^* = \int_{A^*} \left[\frac{\partial w}{\partial x} - \frac{\partial}{\partial x_j}\left(\sigma_{ij} \frac{\partial u_i}{\partial x} \right) \right] dx\, dy \tag{A3.14}$$

where A* is the area enclosed by Γ*. By invoking the definition of strain energy density given by Eq. (3.20), we can evaluate the first term in square brackets in Eq. (A3.14):

$$\frac{\partial w}{\partial x} = \frac{\partial w}{\partial \varepsilon_{ij}} \frac{\partial \varepsilon_{ij}}{\partial x} = \sigma_{ij} \frac{\partial \varepsilon_{ij}}{\partial x} \tag{A3.15}$$

Note that Eq. (A3.15) applies only when w exhibits the properties of an elastic potential. Applying the strain-displacement relationship (for small strains) to Eq. (A3.15) gives

$$\frac{\partial w}{\partial x} = \frac{1}{2} \sigma_{ij} \left[\frac{\partial}{\partial x}\left(\frac{\partial u_i}{\partial x_j} \right) + \frac{\partial}{\partial x}\left(\frac{\partial u_j}{\partial x_i} \right) \right]$$

$$= \sigma_{ij} \frac{\partial}{\partial x_j}\left(\frac{\partial u_i}{\partial x}\right) \tag{A3.16}$$

since $\sigma_{ij} = \sigma_{ji}$. Invoking the equilibrium condition:

$$\frac{\partial \sigma_{ij}}{\partial x_j} = 0 \tag{A3.17}$$

leads to

$$\sigma_{ij} \frac{\partial}{\partial x_j}\left(\frac{\partial u_i}{\partial x}\right) = \frac{\partial}{\partial x_j}\left(\sigma_{ij} \frac{\partial u_i}{\partial x}\right) \tag{A3.18}$$

which is identical to the second term in square bracket in Eq. (A3.14). Thus the integrand in Eq. (A3.14) vanishes and J = 0 for any closed contour.

Consider now two arbitrary contours, Γ_1 and Γ_2 around a crack tip, as illustrated in Fig. A3.3. If Γ_1 and Γ_2 are connected by segments along the crack face (Γ_3 and Γ_4), a closed contour is formed. The total J along the closed contour is equal to the sum of contributions from each segment:

$$J = J_1 + J_2 + J_3 + J_4 = 0 \tag{A3.19}$$

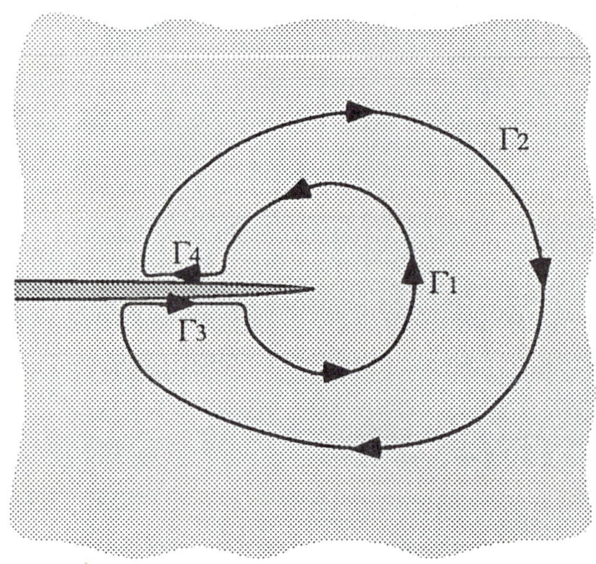

FIGURE A3.3 Two arbitrary contours, Γ_1 and Γ_2, around the tip of a crack. When these contours are connected by Γ_3 and Γ_4, a closed contour is formed, and the total J = 0.

On the crack face, $T_i = dy = 0$. Thus, $J_3 = J_4 = 0$ and $J_1 = - J_2$. Therefore, any arbitrary (counter-clockwise) path around a crack will yield the same value of J; *J is path-independent.*

A3.3 J AS A NONLINEAR ELASTIC ENERGY RELEASE RATE

Consider a two-dimensional cracked body bounded by the curve Γ' (Fig. A3.4). Let A' denote the area of the body. The coordinate axis is attached to the crack tip. Under quasistatic conditions and in the absence of body forces, the potential energy is given by

$$\Pi = \int_{A'} w \, dA - \int_{\Gamma''} T_i u_i \, ds \tag{A3.20}$$

where Γ" is the portion of the contour on which tractions are defined. Let us now consider the change in potential energy resulting from a virtual extension of the crack:

$$\frac{d\Pi}{da} = \int_{A'} \frac{dw}{da} dA - \int_{\Gamma'} T_i \frac{du_i}{da} ds \tag{A3.21}$$

The line integration in Eq. (A3.21) can be performed over the entire contour, Γ', because $du_i/da = 0$ over the region where displacements are specified; also, $dT_i/da = 0$ over the region the tractions are specified. When the crack grows, the coordinate axis moves. Thus a derivative with respect to crack length can be written as

$$\frac{d}{da} = \frac{\partial}{\partial a} + \frac{\partial x}{\partial a}\frac{\partial}{\partial x} = \frac{\partial}{\partial a} - \frac{\partial}{\partial x} \tag{A3.22}$$

since $\partial x/\partial a = -1$. Applying this result to Eq. (A3.21) gives

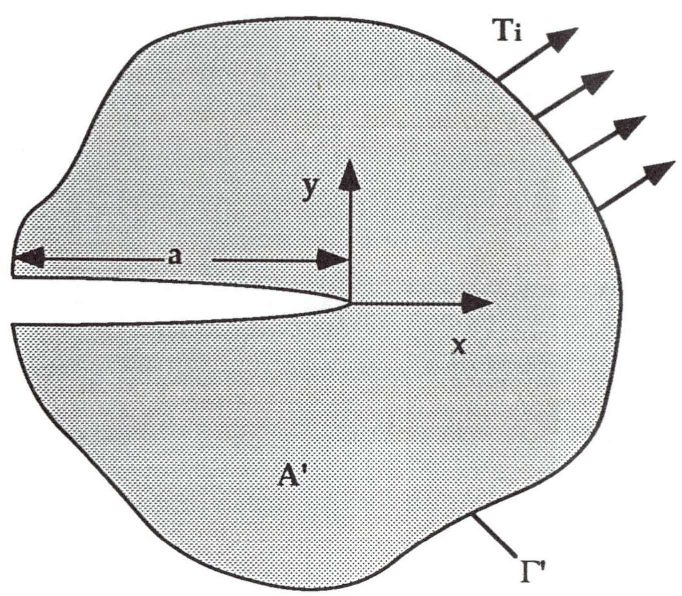

FIGURE A3.4 A two-dimensional cracked body bounded by the curve Γ'.

$$\frac{d\Pi}{da} = \int\limits_{A'}\left(\frac{\partial w}{\partial a} - \frac{\partial w}{\partial x}\right)dA - \int\limits_{\Gamma'} T_i\left(\frac{\partial u_i}{\partial a} - \frac{\partial u_i}{\partial x}\right)ds \qquad \text{(A3.23)}$$

By applying the same assumptions as in Eqs. (A3.15) and (A3.16), we obtain:

$$\frac{\partial w}{\partial a} = \frac{\partial w}{\partial \varepsilon_{ij}}\frac{\partial \varepsilon_{ij}}{\partial a} = \sigma_{ij}\frac{\partial}{\partial x_j}\left(\frac{\partial u_i}{\partial a}\right) \qquad \text{(A3.24)}$$

Invoking the principle of virtual work gives

$$\int\limits_{A'} \sigma_{ij}\frac{\partial}{\partial x_j}\left(\frac{\partial u_i}{\partial a}\right)dA = \int\limits_{\Gamma'} T_i\frac{\partial u_i}{\partial a}ds \qquad \text{(A3.25)}$$

which cancels with one of the terms in the line integral in Eq. (A3.23), resulting in the following:

$$\frac{d\Pi}{da} = \int_{\Gamma'} T_i \frac{\partial u_i}{\partial x} ds - \int_{A'} \frac{\partial w}{\partial x} dA \qquad \text{(A3.26)}$$

Applying the divergence theorem and multiplying both sides by -1 leads to

$$-\frac{d\Pi}{da} = \int_{\Gamma'} \left(w\, n_x - T_i \frac{\partial u_i}{\partial x} \right) ds$$

$$= \int_{\Gamma'} \left(w dy - T_i \frac{\partial u_i}{\partial x} ds \right) \qquad \text{(A3.26)}$$

since n_x ds = dy. Therefore, the J contour integral is equal to the energy release rate for a linear or nonlinear elastic material under quasistatic conditions.

A3.4 THE HRR SINGULARITY

Hutchinson [7] and Rice and Rosengren [8] independently evaluated the character of crack tip stress fields in the case of power-law hardening materials. Hutchinson evaluated both plane stress and plane strain, while Rice and Rosengren considered only plane strain conditions. Both articles, which were published in the same issue of the *Journal of the Mechanics and Physics of Solids,* argued that stress times strain varies as $1/r$ near the crack tip, although only Hutchinson was able to provide a mathematical proof of this relationship.

The Hutchinson analysis is outlined below. Some of the details are omitted for brevity. We focus instead on his overall approach and the ramifications of this analysis.

Hutchinson began by defining a stress function, **Φ**, for the problem. The governing differential equation for deformation plasticity theory for a plane problem in a Ramberg-Osgood material is more complicated than the linear elastic case:

$$\nabla^4\Phi + \gamma(\Phi; \sigma_e; r; n; \alpha) = 0 \tag{A3.27}$$

where the function γ differs for plane stress and plane strain. For the Mode I crack problem, Hutchinson chose to represent Φ in terms of an asymptotic expansion in the following form:

$$\Phi = C_1(\theta)\, r^s + C_2(\theta)\, r^t + \ldots \tag{A3.28}$$

where C_1 and C_2 are constants that depend on θ, the angle from the crack plane. Equation (A3.28) is analogous to the Williams expansion for the linear elastic case (Appendix 2.3). If $s < t$, and t is less than all subsequent exponents on r, then the first term dominates as $r \rightarrow 0$. If the analysis is restricted to the region near the crack tip, then the stress function can be expressed as follows:

$$\Phi = \kappa\, \sigma_o\, r^s\, \tilde{\Phi}(\theta) \tag{A3.29}$$

where κ is the amplitude of the stress function and $\tilde{\Phi}$ is a dimensionless function of θ. Although Eq. (A3.27) is different from the linear elastic case, the stresses can still be derived from Φ through Eqs. (A2.6) or (A2.13). Thus the stresses, in polar coordinates, are given by

$$\begin{aligned}
\sigma_{rr} &= \kappa\, \sigma_o\, r^{s-2}\, \tilde{\sigma}_{rr}(\theta) = \kappa\, \sigma_o\, r^{s-2}\, (s\, \tilde{\Phi} + \tilde{\Phi}'') \\
\sigma_{\theta\theta} &= \kappa\, \sigma_o\, r^{s-2}\, \tilde{\sigma}_{\theta\theta}(\theta) = \kappa\, \sigma_o\, r^{s-2}\, s\, (s-1)\tilde{\Phi} \\
\sigma_{r\theta} &= \kappa\, \sigma_o\, r^{s-2}\, \tilde{\sigma}_{r\theta}(\theta) = \kappa\, \sigma_o\, r^{s-2}\, (1-s)\tilde{\Phi}'
\end{aligned} \tag{A3.30}$$

$$\sigma_e = \kappa\, \sigma_o\, r^{s-2}\, \tilde{\sigma}_e(\theta) = \kappa\, \sigma_o\, r^{s-2}\, (\tilde{\sigma}_{rr}^2 + \tilde{\sigma}_{\theta\theta}^2 - \tilde{\sigma}_{rr}\, \tilde{\sigma}_{\theta\theta} + 3\, \tilde{\sigma}_{r\theta}^2)^{1/2}$$

The boundary conditions for the crack problem are as follows:

$$\tilde{\Phi}(\pm\pi) = \tilde{\Phi}'(\pm\pi) = 0$$

In the region close to the crack tip where Eq. (A3.29) applies, elastic strains are negligible compared to plastic strains; only the second term in Eq. (A3.27) is relevant in this case. Hutchinson substituted the boundary conditions and Eq. (A3.29) into Eq. (A3.27) and obtained a nonlinear eigenvalue equation for s. He then solved this equation numerically for a range of n values. The numerical analysis indicated that s could be described quite accurately (for both plane stress and plane strain) by a simple formula:

$$s = \frac{2n+1}{n+1} \tag{A3.31}$$

which implies that the strain energy density varies as $1/r$ near the crack tip. This numerical analysis also yielded relative values for the angular functions $\tilde{\sigma}_{ij}$. The amplitude, however, cannot be obtained without connecting the near-tip analysis with the remote boundary conditions. The J contour integral provides a simple means for making this connection in the case of small scale yielding. Moreover, by invoking the path-independent property of J, Hutchinson was able to obtain a direct proof of the validity of Eq. (A3.31).

Consider two circular contours of radius r_1 and r_2 around the tip of a crack in small scale yielding, as illustrated in Fig. A3.5. Assume that r_1 is in the region described by the elastic singularity, while r_2 is well inside of the plastic zone, where the stresses are described by Eq. (A3.30). When the stresses and displacements in Tables 2.1 and 2.2 are inserted into Eq. (A3.26), and the J integral is evaluated along r_1, one finds that $J = K_I^2/E'$, as expected from the previous section. Since the connection between K_I and global boundary conditions is well established for a wide range of configurations, and J is path-independent, the near-tip problem for small scale yielding can be solved by evaluating J at r_2 and relating J to the amplitude (κ).

Solving for the integrand in the J integral at r_2 leads to

$$w = \alpha\,\sigma_o\,\varepsilon_o\,\kappa^{n+1}\frac{n}{n+1}\,r^{(n+1)(s-2)}\,\tilde{\sigma}_e^{\,n+1} \tag{A3.32a}$$

and

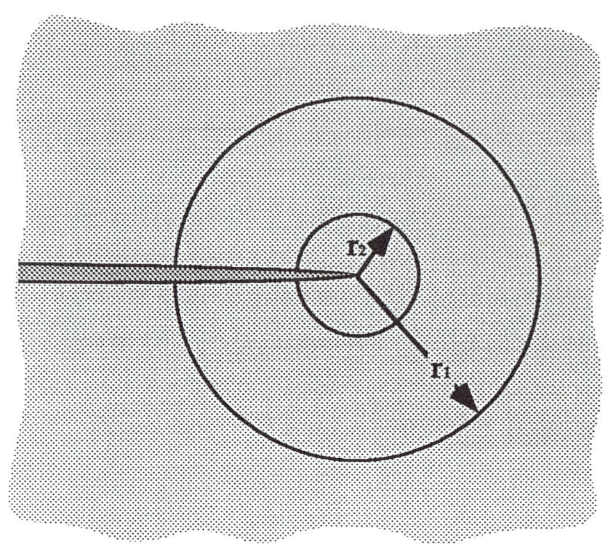

FIGURE A3.5 Two circular contours around the crack tip. r_1 is in the zone dominated by the elastic singularity, while r_2 is in the plastic zone where the leading term of the Hutchinson asymptotic expansion dominates.

$$T_i \frac{\partial u_i}{\partial x} = \alpha\, \sigma_o\, \varepsilon_o\, \kappa^{n+1}\, r^{(n+1)(s-2)} \{\sin\theta\, [\tilde{\sigma}_{rr}\, (\tilde{u}_\theta - \tilde{u}'_r) - \tilde{\sigma}_{r\theta}\, (\tilde{u}_r + \tilde{u}'_\theta)] + \cos\theta\, [n\,(s-2)+1]\, [\tilde{\sigma}_{rr}\, \tilde{u}_r + \tilde{\sigma}_{\theta\theta}\, \tilde{u}_\theta]\} \tag{A3.32b}$$

where $\tilde{u}_r$ and $\tilde{u}_\theta$ are dimensionless displacements, defined by

$$u_r = \alpha\, \varepsilon_o\, \kappa^n\, r^{n(s-2)+1}\, \tilde{u}_r(\theta)$$
$$u_\theta = \alpha\, \varepsilon_o\, \kappa^n\, r^{n(s-2)+1}\, \tilde{u}_\theta(\theta) \tag{A3.33}$$

u_r and u_θ can be derived from the strain-displacement relationships. Evaluating the J integral at r_2 gives

$$J = \alpha\, \kappa^{n+1}\, r_2^{(n+1)(s-2)+1}\, I_n \tag{A3.34}$$

where I_n is an integration constant, given by

$$I_n = \int_{-\pi}^{+\pi} \left\{\frac{n}{n+1}\, \tilde{\sigma}_e^{n+1} \cos\theta - [\sin\theta\, \{\tilde{\sigma}_{rr}\, (\tilde{u}_\theta - \tilde{u}'_r) - \tilde{\sigma}_{r\theta}\, (\tilde{u}_r + \tilde{u}'_\theta)\}\right.$$

$$+ \cos\theta\,(n\,(s-2)+1)\,(\tilde{\sigma}_{rr}\,\tilde{u}_r + \tilde{\sigma}_{\theta\theta}\,\tilde{u}_\theta)]\Big\}\,d\theta \tag{A3.35}$$

In order for J to be path independent, it cannot depend on r_2, which was defined arbitrarily. The radius vanishes in Eq. (A3.34) only when $(n+1)(s-2)+1 = 0$, or

$$s = \frac{2n+1}{n+1}$$

which is identical to the result obtained numerically (Eq. (A3.31)). Thus the amplitude of the stress function is given by

$$\kappa = \left(\frac{J}{\alpha\,\sigma_o\,\varepsilon_o\,I_n}\right)^{\frac{1}{n+1}} \tag{A3.36}$$

Substituting Eq. (A3.36) into Eq. (A3.30) yields the familiar form of the HRR singularity:

$$\sigma_{ij} = \sigma_o\left(\frac{E\,J}{\alpha\,\sigma_o^2\,I_n\,r}\right)^{\frac{1}{n+1}} \tilde{\sigma}_{ij}(n,\theta) \tag{A3.37}$$

since $\varepsilon_o = \sigma_o/E$. The integration constant is plotted in Fig. 3.10 for both plane stress and plane strain, while Fig. 3.11 shows the angular variation of $\tilde{\sigma}_{ij}$ for $n = 3$ and $n = 13$.

Rice and Rosengren [8] obtained essentially identical results to Hutchinson (for plane strain), although they approached the problem in a somewhat different manner. Rice and Rosengren began with a heuristic argument for the 1/r variation of strain energy density, and then introduced an Airy stress function of the form of Eq. (A3.29) with the exponent on r given by Eq. (A3.31). They computed stresses, strains and displacements in the vicinity of the crack tip by applying the appropriate boundary conditions.

The HRR singularity was an important result because it established J as a stress amplitude parameter within the plastic zone, where the linear elastic solution is invalid. The analyses of Hutchinson, Rice and Rosengren demonstrated that the stresses in the plastic zone are much higher in plane strain than in plane stress; recall that the elastic solution predicts identical in-plane stresses for both cases. These results provided a theoretical explanation for empirically observed thickness effects in fracture toughness tests.

One must bear in mind the limitations of the HRR solution. Since the singularity is merely the leading term in an asymptotic expansion, and elastic strains were assumed to be negligible, this analysis is only valid near the crack tip, well within the plastic zone. For very small r values, however, the HRR solution is invalid because it neglects finite geometry changes at the crack tip. An implicit assumption in the analysis is that the loading is proportional. This assumption allows the stresses and strains to be derived from a stress function, which in turn leads to the single parameter description. The large geometry changes at the crack tip cause local nonproportional loading, thereby eliminating the possibility of a single parameter description of stresses and strains.

A3.5 ANALYSIS OF STABLE CRACK GROWTH IN SMALL SCALE YIELDING

A3.5.1 The Rice-Drugan-Sham Analysis

Rice, Drugan and Sham (RDS) [15] performed an asymptotic analysis of a growing crack in an elastic-plastic solid in small-scale yielding. They assumed crack extension at a constant crack opening angle, and predicted the shape of J resistance curves. They also speculated about the effect of large scale yielding on the crack growth resistance behavior.

Small scale yielding

Rice et al. analyzed the local stresses and displacements at a growing crack by modifying the classical Prandtl slip line field to account for elastic unloading behind the crack tip. They assumed small scale yielding conditions and a nonhardening material; the details of the derivation are omitted for brevity. The RDS crack growth analysis resulted in the following expression:

$$\dot{\delta} = \alpha \frac{\dot{J}}{\sigma_o} + \beta \frac{\sigma_o}{E} \dot{a} \ln\left(\frac{R}{r}\right) \qquad \text{for } r \rightarrow 0 \tag{A3.38}$$

where $\dot{\delta}$ is the rate of crack opening displacement a distance R behind the crack tip, $\dot{J}$ is the rate of change in the J integral, $\dot{a}$ is the crack growth rate, and α, β, and R are constants[6]. The asymptotic analysis indicated that $\beta = 5.083$ for $\nu = 0.3$ and $\beta = 4.385$ for $\nu = 0.5$. The other constants, α and R, could not be inferred from the asymptotic analysis. Rice, et. al. [15] performed elastic-plastic finite element analysis of a growing crack and found that R, which has units of length, scales approximately with plastic zone size, and can be estimated by

$$R = \frac{\lambda E J}{\sigma_o^2} \qquad \text{where } \lambda \approx 0.2 \tag{A3.39}$$

The dimensionless constant α can be estimated by considering a stationary crack ($\dot{a} = 0$):

$$\delta = \alpha \frac{J}{\sigma_o} \tag{A3.40}$$

Referring to Eq. (3.48), α obviously equals d_n when δ is defined by the 90° intercept method. The finite element analysis performed by Rice, et al. indicated that α for a growing crack is nearly equal to the stationary crack case.

Rice et. al. performed an asymptotic integration of (Eq. A3.38) for the case where crack length increases continuously with J, which led to

$$\delta = \frac{\alpha r}{\sigma_o} \frac{dJ}{da} + \beta r \frac{\sigma_o}{E} \ln\left(\frac{e R}{r}\right) \tag{A3.41}$$

where δ, in this case, is the crack opening displacement at a r from the crack tip, and e (= 2.718) is the natural logarithm base. Equation (A3.41) can be rearranged to solve for the nondimensional tearing modulus:

---

[6]The constant α in the RDS analysis should not be confused with the dimensionless constant in the Ramberg-Osgood relationship (Eq. 3.22), for which the same symbol is used.

$$T \equiv \frac{E}{\sigma_o^2}\frac{dJ}{da} = \frac{E\,\delta}{\alpha\,\sigma_o\,r} - \frac{\beta}{\alpha}\ln\left(\frac{e\,R}{r}\right) \tag{A3.42}$$

Rice, et al. proposed a failure criterion that corresponds approximately to crack extension at a constant crack tip opening angle (CTOA). Since $d\delta/dr = \infty$ at the crack tip, CTOA is undefined, but an approximate CTOA can be inferred a finite distance from the tip. Figure A3.6 illustrates the RDS crack growth criterion. They postulated that crack growth occurs at a critical crack opening displacement, δ_c, at a distance r_m behind the crack tip. That is,

$$\frac{\delta_c}{r_m} = \frac{\alpha}{\sigma_o}\frac{dJ}{da} + \beta\frac{\sigma_o}{E}\ln\left(\frac{e\,R}{r_m}\right) = \text{constant} \tag{A3.43}$$

Rice, et al. found that it was possible to define the micromechanical failure parameters, δ_c and r_m, in terms of global parameters that are easy to obtain experimentally. Setting $J = J_{IC}$ and combining Eqs. (A3.39), (A3.42), and (A3.43) gives

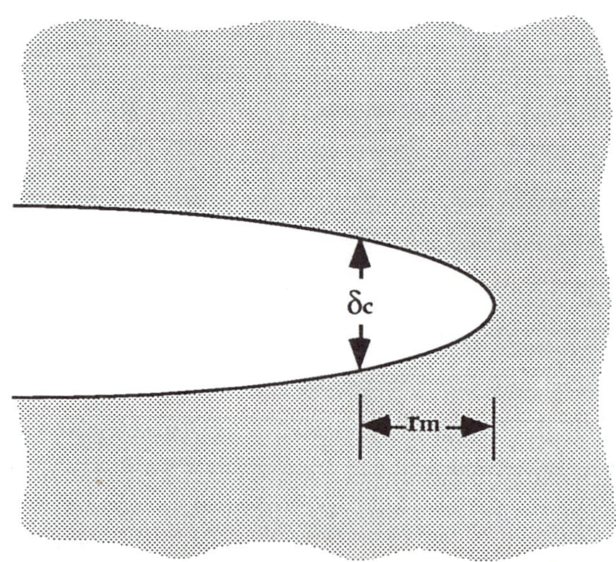

FIGURE A3.6 **The RDS crack growth criterion. The crack is assumed to extend with a constant opening displacement, δ_c, at distance r_m behind the crack tip. This criterion corresponds approximately to crack extension at a constant crack tip opening angle (CTOA).**

$$T_o = \frac{E\,\delta_c}{\alpha\,\sigma_o\,r_m} - \frac{\beta}{\alpha}\ln\left(\frac{e\,\lambda\,E\,J_{IC}}{r_m\,\sigma_o^2}\right) \tag{A3.44}$$

where T_o is the initial tearing modulus. Thus for, $J > J_{IC}$, the tearing modulus is given by

$$T = T_o - \frac{\beta}{\alpha}\ln\left(\frac{J}{J_{IC}}\right) \tag{A3.45}$$

Rice, et al. computed normalized R curves (J/J_{IC} v. $\Delta a/R$) for a range of T_o values and found that $T = T_o$ in the early stages of crack growth, but the R curve slope decrease until steady state plateau is reached. The steady state J can easily be inferred from Eq. (A3.45) by setting $T = 0$:

$$J_{ss} = J_{IC}\exp\left(\frac{\alpha\,T_o}{\beta}\right) \tag{A3.46}$$

Large scale yielding

Although the RDS analysis was derived for small scale yielding conditions, Rice, et al. speculated that the form of Eq. (A3.38) might also be valid for fully plastic conditions. The numerical values of some of the constants, however, probably differ for the large scale yielding case.

The most important difference between small scale yielding and fully plastic conditions is the value of R. Rice, et. al. argued that R would no longer scale with plastic zone size, but should be proportional to the ligament length. They made a rough estimate of $R \sim b/4$ for the fully plastic case.

The constant α depends on crack tip triaxiality and thus may differ for small scale yielding and fully yielded conditions. For highly constrained configurations, such as bend specimens, α for the two cases should be similar.

In small scale yielding, the definition of J is unambiguous, since it is related to the elastic stress intensity factor. The J integral for a growing crack under fully plastic conditions can be computed in a number of ways, however, and not all definitions of J are appropriate in the large scale yielding version of Eq. (A3.38).

Assume that the crack growth resistance behavior is to be characterized by a J-like parameter, J_x. Assuming J_x depends on crack length and displacement, the rate of change in J_x should be linearly related to displacement rate and $\dot{a}$:

$$\dot{J}_x = \xi \dot{\Delta} + \chi \dot{a} \tag{A3.47}$$

where ξ and χ are functions of displacement and crack length. Substituting Eq. (A3.47) into Eq. (A3.38) gives

$$\dot{\delta} = \frac{\alpha}{\sigma_o} \xi \dot{\Delta} + \left[\beta \frac{\sigma_o}{E} \ln\left(\frac{R}{r}\right) + \frac{\alpha}{\sigma_o} \chi \right] \dot{a} \tag{A3.48}$$

In the limit of a rigid-ideally plastic material, $\sigma_o/E = 0$. Also, the local crack opening rate must be proportional to the global displacement rate for a rigid-ideally plastic material:

$$\dot{\delta} = \psi \dot{\Delta} \tag{A3.49}$$

Therefore, the term in square brackets in Eq. (A3.48) must vanish, which implies that $\chi = 0$, at least in the limit of a rigid-ideally plastic material. Thus in order for the RDS model to apply to large scale yielding, the rate of change in the J-like parameter must not depend on the crack growth rate:

$$\dot{J}_x \neq \dot{J}_x(\dot{a}) \tag{A3.50}$$

Rice, et al. showed that neither the deformation theory J or the far-field J satisfy Eq. (A3.50) for all configurations. Ernst [16] defined a modified J that does satisfy Eq. (A3.50) for rigid, nonhardening materials.

Satisfying Eq. (A3.50) does not necessarily imply that a J_x-R curve is geometry independent. The RDS model suggests that a resistance curve obtained from a fully yielded specimen will not, in general, agree with the small scale yielding R curve for the same material. Assuming R = b/4 for the fully plastic case, the RDS model predicts the following tearing modulus:

$$T = \left[T_o - \frac{\beta}{\alpha_{ssy}} \ln \left(\frac{b/4}{\lambda \, E \, J_{IC}/\sigma_o^2} \right) \right] \frac{\alpha_{ssy}}{\alpha_{fy}} \tag{A3.51}$$

where the subscripts ssy and fy denote small scale yielding and fully yielded conditions, respectively. According to Eq. (A3.51), the crack growth resistance curve under fully yielded conditions has a constant initial slope, but this slope is not equal to T_o (the initial tearing modulus in small scale yielding) unless $\alpha_{fy} = \alpha_{ssy}$ and $b = 4\lambda \, E \, J_{IC}/\sigma_o^2$. Equation (A3.51) does not predict a steady state limit where $T = 0$; rather this relationship predicts that T actually increases as the ligament becomes smaller. An increasing tearing modulus has been observed experimentally when the R curve is characterized by Ernst's modified J [17].

The forgoing analysis implies that crack growth resistance curves obtained from specimens with fully yielded ligaments are suspect. One should exercise extreme caution when applying experimental results from small specimens to predict the behavior of large structures.

A3.5.2 Steady State Crack Growth

The RDS analysis, which assumed a local failure criterion based on crack opening angle, indicated crack growth in small scale yielding reaches a steady state, where $dJ/da \rightarrow 0$. The derivation that follows shows that the steady state limit is a general result for small scale yielding; the R curve must eventually reach a plateau in an infinite body, regardless of the failure mechanism.

<u>*Generalized Damage Integral*</u>

Consider a material element a small distance from a crack tip, as illustrated in Fig. A3.7. This material element will fail when it is deformed beyond its capacity. The crack will grow as consecutive material elements at the tip fail. Let us define a generalized damage integral, Θ, which characterizes the severity of loading at the crack tip:

$$\Theta = \int_0^{\varepsilon_{eq}} \Omega(\sigma_{ij}; \varepsilon_{ij})\, d\varepsilon_{eq} \tag{A3.52}$$

where ε_{eq} is the equivalent (von Mises) plastic strain and Ω is a function of the stress and strain tensors (σ_{ij} and ε_{ij}, respectively). The above integral is sufficiently general that it can depend on the current values of all stress and strain components, as well as the entire deformation history. Referring to Fig. A3.7, the material element will fail at a critical value of Θ. At the moment of crack initiation or during crack extension, material near the crack tip will be close to the point of failure. At a distance r* from the crack tip, where r* is arbitrarily small, we can assume that $\Theta = \Theta_c$.

The precise form of the damage integral depends on the micromechanism of fracture. For example, a modified Rice and Tracey [38] model for ductile hole growth (see Chapter 5) can be used to characterize ductile fracture in metals:

$$\Theta = \ln\left(\frac{R}{R_o}\right) = 0.283 \int_0^{\varepsilon_{eq}} \exp\left(\frac{1.5\,\sigma_m}{\sigma_e}\right) d\varepsilon_{eq} \tag{A3.53}$$

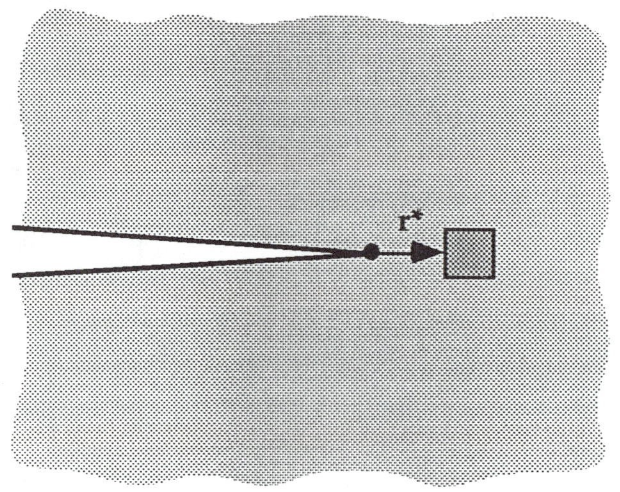

FIGURE A3.7 Material point a distance r* from the crack tip.

where R is the void radius, R_o is the initial radius, σ_m is the mean (hydrostatic) stress, and σ_e is the effective (von Mises) stress. Failure, in this case, is assumed when the void radius reaches a critical value.

Stable crack growth

Consider an infinite body[7] that contains a crack that is growing in a stable, self-similar, quasistatic manner. If the crack has grown well beyond the initial blunted tip, dimensional analysis indicates that the local stresses and strains are uniquely characterized by the far-field J integral, as stated in Eq. (3.69). In light of this single parameter condition, the integrand of Eq. (A3.52) becomes

$$\Omega = \Omega\left(\frac{J}{\sigma_o r}, \theta\right) \quad \text{(A3.54)}$$

We can restrict this analysis to $\theta = 0$ by assuming that the material on the crack plane fails during Mode I crack growth. For a given material point on the crack plane, r decreases as the crack grows, and the plastic strain increases. If strain increases monotonically as this material point approaches the crack tip, Eq. (A3.54) permits writing Ω as a function of the von Mises strain:

$$\Omega = \Omega(\varepsilon_{eq}) \quad \text{(A3.55)}$$

Therefore the local failure criterion is given by

$$\Theta_c = \int_0^{\varepsilon^*} \Omega(\varepsilon_{eq})\, d\varepsilon_{eq} \quad \text{(A3.56)}$$

where ε^* is the critical strain (i.e., the von Mises strain at $r = r^*$). Since the integrand is a function only of ε_{eq}, the integration path is the same for all material points ahead of the crack tip, and ε^* is constant during crack growth. That is, the equivalent plastic strain at r^* will always

[7]In practical terms "infinite" means that external boundaries are sufficiently far from the crack tip so that the plastic zone is embedded within an elastic singularity zone.

equal ε^* when the crack is growing. Based on Eqs. (3.69) and (A3.54), ε^* is a function only of r^* and the applied J:

$$\varepsilon^* = \varepsilon^*(J, r^*) \tag{A3.57}$$

Solving for the differential of ε^* gives

$$d\varepsilon^* = \frac{\partial \varepsilon^*}{\partial J} dJ + \frac{\partial \varepsilon^*}{\partial r^*} dr^* \tag{A3.58}$$

Since ε^* and r^* are both fixed, $d\varepsilon^* = dr^* = dJ = 0$.

Thus the J integral remains constant during crack extension ($dJ/da = 0$) when Eq. (3.69) is satisfied. Steady state crack growth is usually not observed experimentally because large scale yielding in finite sized specimens precludes characterizing a growing crack with J. Also, a significant amount of crack growth may be required before a steady state is reached (Fig. 3.25); the crack tip in a typical laboratory specimen approaches a free boundary well before the crack growth is sufficient to be unaffected by the initial blunted tip.

A3.6 NOTES ON THE APPLICABILITY OF DEFORMATION PLASTICITY TO CRACK PROBLEMS

Since elastic-plastic fracture mechanics is based on deformation plasticity theory, it may be instructive to take a closer look at this theory and assess its validity for crack problems.

Let us begin with the plastic portion of the Ramberg-Osgood equation for uniaxial deformation, which can be expressed in the following form:

$$\varepsilon_p = \alpha \left(\frac{\sigma}{\sigma_o}\right)^{n-1} \frac{\sigma}{E} \tag{A3.59}$$

Differentiating Eq. (A3.59) gives

$$d\varepsilon_p = \alpha\, n \left(\frac{\sigma}{\sigma_o}\right)^{n-2} \frac{\sigma}{E} \frac{d\sigma}{\sigma_o} \tag{A3.60}$$

for an increment of plastic strain. For the remainder of this section, the subscript on strain is suppressed for brevity; only plastic strains are considered, unless stated otherwise.

Equations (A3.59) and (A3.60) represent the deformation and incremental flow theories, respectively, for uniaxial deformation in a Ramberg-Osgood material. In this simple case, there is no difference between the incremental and deformation theories, provided no unloading occurs. Equation (A3.60) can obviously be integrated to obtain Eq. (A3.59). Stress is uniquely related to strain when both increase monotonically. It does not necessarily follow that deformation and incremental theories are equivalent in the case of three-dimensional monotonic loading, but there are many cases where this is a good assumption.

Equation (A3.59) can be generalized to three dimensions by assuming deformation plasticity and isotropic hardening:

$$\varepsilon_{ij} = \frac{3}{2}\alpha\left(\frac{\sigma_e}{\sigma_o}\right)^{n-1}\frac{S_{ij}}{E} \tag{A3.61}$$

where σ_e is the effective (von Mises) stress and S_{ij} is the deviatoric component of the stress tensor, defined by

$$S_{ij} = \sigma_{ij} - \frac{1}{3}\sigma_{kk}\,\delta_{ij} \tag{A3.62}$$

where δ_{ij} is the Kronecker delta. Equation (A3.61) is the deformation theory *flow rule* for a Ramberg-Osgood material. The corresponding flow rule for incremental plasticity theory is given by

$$d\varepsilon_{ij} = \frac{3}{2}\alpha\, n\left(\frac{\sigma_e}{\sigma_o}\right)^{n-2}\frac{S_{ij}}{E}\frac{d\sigma_e}{\sigma_o} \tag{A3.63}$$

By comparing Eqs. (A3.61) and (A3.63), one sees that the deformation and incremental theories of plasticity coincide only if the latter equation can be integrated to obtain the former. If the deviatoric stress components are proportional to the effective stress:

$$S_{ij} = \omega_{ij}\,\sigma_e \tag{A3.64}$$

where ω_{ij} is a constant tensor that does not depend on strain, then integration of Eq. (A3.63) results in Eq. (A3.61). Thus deformation and incremental theories of plasticity are identical when the loading is proportional in the deviatoric stresses. Note that the *total* stress components need not be proportional in order for the two theories to coincide; the flow rule is not influenced by the hydrostatic portion of the stress tensor.

Proportional loading of the deviatoric components does not necessarily mean that deformation plasticity theory is rigorously correct; it merely implies that deformation theory is no more objectionable than incremental theory. Classical plasticity theory, whether based on incremental strain or total deformation, contains simplifying assumptions about material behavior. Both Eqs. (A3.61) and (A3.63) assume that the yield surface expands symmetrically and that its radius does not depend on hydrostatic stress. For monotonic loading ahead of a crack in a metal, these assumptions are probably reasonable; the assumed hardening law is of little consequence for monotonic loading, and hydrostatic stress effects on the yield surface are relatively small for most metals.

Budiansky [39] showed that deformation theory is still acceptable when there are modest deviations from proportionality. Low work hardening materials are the least sensitive to nonproportional loading.

Since most of classical fracture mechanics assumes either plane stress or plane strain, it is useful to examine plastic deformation in the two-dimensional case, and determine under what conditions the the requirement of proportional deviatoric stresses is at least approximately satisfied. Consider, for example, plane strain. When elastic strains are negligible, the in-plane deviatoric normal stresses are given by

$$S_{xx} = \frac{\sigma_{xx} - \sigma_{yy}}{2} \quad \text{and} \quad S_{yy} = \frac{\sigma_{yy} - \sigma_{xx}}{2} \tag{A3.65}$$

assuming incompressible plastic deformation, where $\sigma_{zz} = (\sigma_{xx} + \sigma_{yy})/2$. The expression for von Mises stress in plane strain reduces to

$$\sigma_e = \frac{1}{\sqrt{2}}[3\,S_{xx}^2 + 6\,S_{xy}^2]^{1/2} \tag{A3.66}$$

where $S_{xy} = \tau_{xy}$. Alternatively, σ_e can be written in terms of principal normal stresses:

$$\sigma_e = \frac{\sqrt{3}}{2}[\sigma_1 - \sigma_2] \quad \text{where } \sigma_1 > \sigma_2$$
$$= \sqrt{3}\,S_1 \tag{A3.67}$$

Therefore, the principal deviatoric stresses are proportional to σ_e in the case of plane strain. It can easily be shown that the same is true for plane stress. If the principal axes are fixed, S_{xx}, S_{yy}, and S_{xy} must also be proportional to σ_e. If, however, the principal axes rotate during deformation, the deviatoric stress components defined by a fixed coordinate system will not increase in proportion to one another.

In the case of Mode I loading of a crack, τ_{xy} is always zero on the crack plane, implying that the principal directions on the crack plane are always parallel to the x-y-z coordinate axis. Thus, deformation and incremental plasticity theories should be equally valid on the crack plane, well inside the plastic zone (where elastic strains are negligible). At finite angles from the crack plane, the principal axes may rotate with deformation, which will produce nonproportional deviatoric stresses. If this effect is small, deformation plasticity theory should be adequate to analyze stresses and strains near the crack tip in either plane stress or plane strain.

The validity of deformation plasticity theory does not automatically guarantee that the crack tip conditions can be characterized by a single parameter, such as J or K. Single parameter fracture mechanics requires that the *total* stress components be proportional near the crack tip[8], a much more severe restriction. Proportional total stresses imply that the deviatoric stresses are proportional, but the reverse is not necessarily true. In both the linear elastic case (Appendix 2.3) and the non-

---

[8]The proportional loading region need not extend all the way to the crack tip, but the nonproportional zone at the tip must be embedded within the proportional zone in order for a single loading parameter to characterize crack tip conditions.

linear case (Appendix 3.4) the stresses near the crack tip were derived from a stress function of the form

$$\Phi = \kappa f(r) g(\theta) \tag{A3.68}$$

where κ is a constant, and f and g are dimensionless functions of r and θ. The form of Eq. (A3.68) guarantees that all stress components are proportional to κ, and thus proportional to one another. Therefore any monotonic function of κ uniquely characterizes the stress fields in the region where Eq. (A3.68) is valid. Nonproportional loading automatically invalidates Eq. (A3.68) and the single parameter description that it implies.

As stated earlier, the deviatoric stresses are proportional on the crack plane, well within the plastic zone. The hydrostatic stress may not be proportional to σ_e, however. For example, the loading is highly nonproportional in the large strain region, as Fig. 3.12 indicates. Consider a material point at a distance x from the crack tip, where x is in the current large strain region. At earlier stages of deformation the loading on this point was proportional, but σ_{yy} reached a peak when the ratio $x\,\sigma_o/J$ was approximately unity, and the normal stress decreased with subsequent deformation. Thus the most recent loading on this point was nonproportional, but the deviatoric stresses are still proportional to σ_e.

When the crack grows, material behind the crack tip unloads elastically and deformation plasticity theory is no longer valid. Deformation theory is also suspect near the elastic-plastic boundary. Equations (A3.65) to (A3.67) were derived assuming the elastic strains were negligible, which implies $\sigma_{zz} = (\sigma_{xx} + \sigma_{yy})/2$ in plane strain. At the onset of yielding, however, $\sigma_{zz} = \nu\,(\sigma_{xx} + \sigma_{yy})$, and the proportionality constants between σ_e and the deviatoric stress components are different than for the fully plastic case. Thus when elastic and plastic strains are of comparable magnitude, the deviatoric stresses are nonproportional, as ω_{ij} (Eq. (A3.64)) varies from its elastic value to the fully plastic limit. The errors in deformation theory that may arise from the transition from elastic to plastic behavior should not be appreciable in crack problems, because the strain gradient ahead of the crack tip is relatively steep, and the transition zone is small.

4. DYNAMIC AND TIME-DEPENDENT FRACTURE

In certain fracture problems, time is an important variable. At high loading rates, for example, inertia effects and material rate dependence can be significant. Metals and ceramics also exhibit rate-dependent deformation (creep) at temperatures that are close to the melting point of the material. The mechanical behavior of polymers is highly sensitive to strain rate, particularly above the glass transition temperature. In each of these cases, linear elastic and elastic-plastic fracture mechanics, which assume quasistatic, rate-independent deformation, are inadequate.

Early fracture mechanics researchers considered dynamic effects, but only for the special case of linear elastic material behavior. More recently, fracture mechanics has been extended to include time-dependent material behavior such as viscoplasticity and viscoelasticity. Most of these newer approaches are based on generalizations of the J contour integral.

This chapter gives an overview of time-dependent fracture mechanics. The treatment of this subject is far from exhaustive, but should serve as an introduction to a complex and rapidly developing field. The reader is encouraged to consult the published literature for further background.

4.1 DYNAMIC FRACTURE AND CRACK ARREST

As any undergraduate engineering student knows, dynamics is more difficult than statics. Problems become more complicated when the equations of equilibrium are replaced by the equations of motion.

In the most general case, dynamic fracture mechanics contains three complicating features that are not present in LEFM and elastic-plastic fracture mechanics: inertia forces, rate-dependent material behavior, and reflected stress waves. Inertia effects are important when the load changes abruptly or the crack grows rapidly; a portion of the work that is applied to the specimen is converted to kinetic energy.

Most metals are not sensitive to moderate variations in strain rate near ambient temperature, but the flow stress can increase appreciably when strain rate increases by several orders of magnitude. The effect of rapid loading is even more pronounced in rate sensitive materials such as polymers. When the load changes abruptly or the crack grows rapidly, stress waves propagate through the material and reflect off of free surfaces, such as the specimen boundaries and the crack plane. Reflecting stress waves influence the local crack tip stress and strain fields, which, in turn, affect the fracture behavior.

In certain problems, one or more of the above effects can be ignored. If all three effects are neglected, the problem reduces to the quasistatic case.

The dynamic version of LEFM is termed *elastodynamic fracture mechanics*, where nonlinear material behavior is neglected, but inertia forces and reflected stress waves are incorporated when necessary. The theoretical framework of elastodynamic fracture mechanics is fairly well established, and practical applications of this approach are becoming more common. Extensive reviews of this subject have been published by Freund [1-5], Kanninen and Poplar [6], Rose [7], and others. Elastodynamic fracture mechanics has limitations, but is approximately valid in many cases. When the plastic zone is restricted to a small region near the crack tip in a dynamic problem, the stress intensity approach, with some modifications, is still applicable.

Dynamic fracture analyses that incorporate nonlinear, time-dependent material behavior are a relatively recent innovation. A number of researchers have generalized the J integral to account for inertia and viscoplasticity [8-13].

There are two major classes of dynamic fracture problems: (1) fracture initiation as a result of rapid loading, and (2) rapid propagation of a crack. In the latter case, the crack propagation may initiate either by quasistatic or rapid application of a load; the crack may arrest after some amount of unstable propagation. Dynamic initiation, propagation, and crack arrest are discussed below.

4.1.1 Rapid Loading of a Stationary Crack

Rapid loading of a structure can come from a number sources, but most often occurs as the result of impact with a second object (e.g. a ship colliding with an offshore platform or a missile striking its target). Impact loading is often applied in laboratory tests when a high strain rate is desired. The Charpy test [14], where a pendulum dropped from a fixed height fractures a notched specimen, is probably the most common dynamic mechanical test. Dynamic loading of a fracture mechanics specimen can be achieved through impact loading [15,16], a controlled explosion near the specimen [17], or servohydraulic testing machines that are specially designed to impart high displacement rates. Chapter 7 describes some of the practical aspects of high rate fracture testing.

Figure 4.1 schematically illustrates a typical load-time response for dynamic loading. The load tends to increase with time, but oscillates at a particular frequency that depends on specimen geometry and material properties. Note that the loading rate is finite; i.e., a finite time is required to reach a particular load. The amplitude of the oscillations decreases with time, as kinetic energy is absorbed by the specimen. Thus inertia effects are most significant at short times, and are minimal after sufficiently long times, where the behavior is essentially quasistatic.

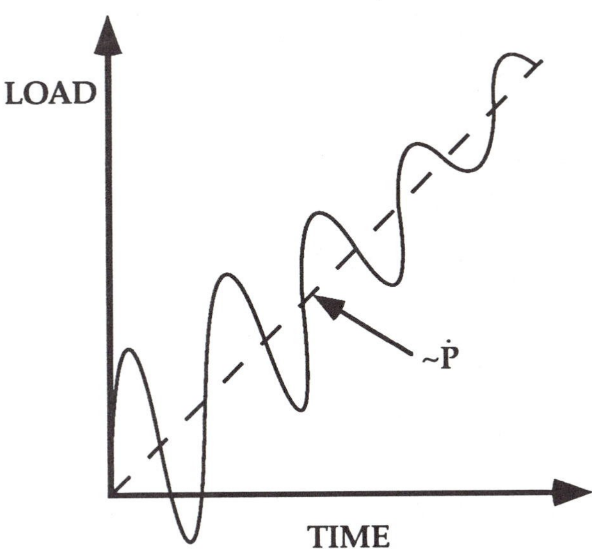

FIGURE 4.1 Schematic load-time response of a rapidly loaded structure.

Determining a fracture characterizing parameter, such as the stress intensity factor or the J integral, for rapid loading can be very difficult. Consider the case where the plastic zone is confined to a small region surrounding the crack tip. The near-tip stress fields for high rate Mode I loading are given by

$$\sigma_{ij} = \frac{K_I(t)}{\sqrt{2\pi r}} f_{ij}(\theta) \tag{4.1}$$

where (t) denotes a function of time. The angular functions, f_{ij}, are identical to the quasistatic case and are given in Table 2.1. The stress intensity factor, which characterizes the amplitude of the elastic singularity, varies erratically in the early stages of loading. Reflecting stress waves pass that through the specimen constructively and destructively interfere with one another, resulting in a highly complex time-dependent stress distribution. The instantaneous K_I depends on the magnitude of the discrete stress waves that pass through the crack tip region at that particular moment in time. When the discrete waves are significant, it is not possible to infer K_I from the remote loads.

Recent work by Nakamura et al. [18,19] quantified inertia effects in laboratory specimens and showed that these effects can be neglected in many cases. They observed that the behavior of a dynamically loaded specimen can be characterized by a short-time response, dominated by discrete waves, and a long-time response that is essentially quasistatic. At intermediate times, global inertia effects are significant but local oscillations at the crack are small, because kinetic energy is absorbed by the plastic zone. To distinguish short-time response from long-time response, Nakamura et al. defined a *transition time*, t_τ, when the kinetic energy and the deformation energy (the energy absorbed by the specimen) are equal. Inertia effects dominate prior to the transition time, but the deformation energy dominates at times significantly greater than t_t. In the latter case, a J-dominated field should exist near the crack tip and quasistatic relationships can be used to infer J from global load and displacement.

Since it is not possible to measure kinetic and deformation energies separately during a fracture mechanics experiment, Nakamura et al. developed a simple model to estimate kinetic energy and transition time in a three-point bend specimen (Fig. 4.2). This model was based on the Bernoulli-Euler beam theory and assumed that the kinetic energy at early times was dominated by the elastic response of the specimen. Incorporating the known relationship between load line displacement and strain energy in a three-point bend specimen leads to an approximate relationship for the ratio of kinetic to deformation energy:

$$\frac{E_k}{U} = \left(\Lambda \frac{W\, \dot{\Delta}(t)}{c_o\, \Delta(t)} \right)^2 \tag{4.2}$$

where E_k is the kinetic energy, U is the deformation energy, W is the specimen width, Δ is the load line displacement, $\dot{\Delta}$ is the displacement rate, c_o is the longitudinal wave speed (i.e. the speed of sound) in a one-dimensional bar, and Λ is a geometry factor, which for the bend specimen is given by

$$\Lambda = \sqrt{\frac{S\,B\,E\,C}{W}} \tag{4.3}$$

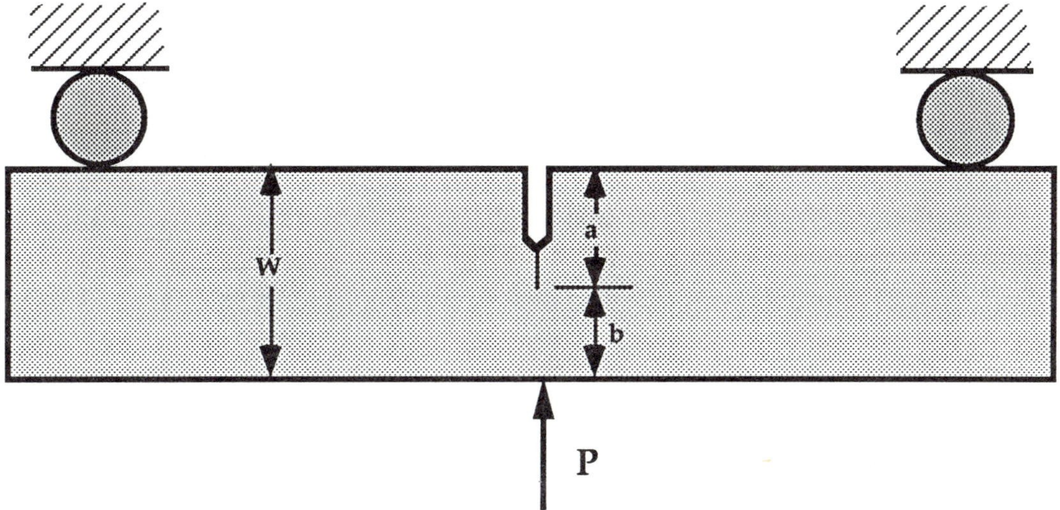

FIGURE 4.2 Three-point bend specimen.

where S is the span of the specimen. The advantage of Eq. (4.2) is that the displacement and displacement rate can be measured experimentally. The transition time is defined at the moment in the test when the ratio $E_k/U = 1$. In order to obtain an explicit expression for t_τ, it is convenient to introduce a dimensionless displacement coefficient, D:

$$D = \left. \frac{t\,\dot{\Delta}(t)}{\Delta(t)} \right|_{t_\tau} \tag{4.4}$$

If, for example, the displacement varies with time as a power law: $\Delta = \beta t^\gamma$, then $D = \gamma$. Combining Eqs. (4.2) and (4.4) and setting $E_k/U = 1$ leads to

$$t_\tau = D\,\Lambda \frac{W}{c_o} \tag{4.5}$$

Nakamura et al. [18,19] performed dynamic finite element analysis on a three point bend specimen in order to evaluate the accuracy of Eqs. (4.2) and (4.5). Figure 4.3 compares the E_k/U ratio computed from finite element analysis with that determined from experiment and Eq. (4.2). The horizontal axis is a dimensionless time scale, and c_1 is the longitudinal wave speed in an unbounded solid. The ratio W/c_1 is an estimate of the time required for a stress wave to traverse the width of the specimen. Based on Eq. (4.2) and experiment, $t_\tau\, c_1/W \approx 28$ (or $t_\tau\, c_o/H \approx 24$), while the finite element analysis estimated $t_\tau\, c_1/W \approx 27$. Thus the simple model agrees quite well with a more detailed analysis.

The simple model was based on the global kinetic energy and did not consider discrete stress waves. Thus the model is only valid after stress waves have traversed the width of the specimen several times. This limitation does not affect the analysis of transition time, since stress waves have made approximately 27 passes when t_τ is reached. Note, in Fig. 4.3, that the simple model agrees very well with the finite element analysis when $t\, c_1/W > 20$.

When $t >> t_\tau$, inertia effects are negligible and quasistatic models should apply to the problem. Consequently, the J integral for a deeply cracked bend specimen at long times can be estimated by

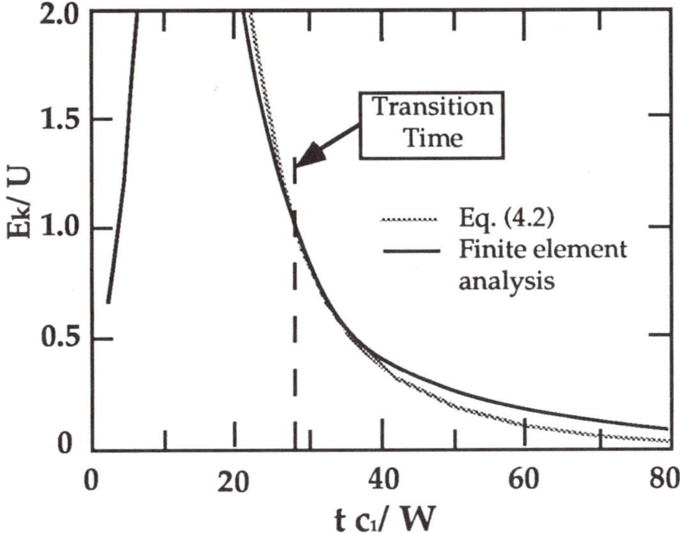

FIGURE 4.3 Ratio of kinetic to stress work energy in a dynamically loaded three-point bend specimen [19].

$$J_{dc} = \frac{2}{B\,b} \int_{0}^{\Omega(t^*)} M(t)\, d\Omega(t) \tag{4.6}$$

where B is the plate thickness, b is the uncracked ligament length, M is the applied moment on the ligament, Ω is the angle of rotation, and t* is the current time. Equation (4.6), which was originally published by Rice, et al. [20], is derived in Section 3.2.5.

Nakamura, et al. [19] performed a three-dimensional dynamic elastic-plastic finite element analysis on a three-point bend specimen in order to determine the range of applicability of Eq. (4.6). They evaluated a dynamic J integral (see Section 4.1.3) at various thickness positions and observed a through-thickness variation of J that is similar to Fig. 3.36. They computed a nominal J that averaged the through-thickness variations and compared this value with J_{dc}. The results of this exercise are plotted in Fig. 4.4. At short times, the average dynamic J is significantly lower than the J computed from the quasistatic relationship. For $t > 2t_\tau$, the J_{dc}/J_{ave} reaches a constant value that is slightly greater than 1. The modest discrepancy between J_{dc} and J_{ave} at long times is probably due to three-dimensional effects rather than dynamic effects (Eq. (4.6) is essentially a two-dimensional formula).

According to Fig. 4.4, Eq. (4.6) provides a good estimate of J in a high rate test at times greater than approximately twice the transition time. It follows that if fracture initiation occurs after $2t_\tau$, the critical value of J obtained from Eq. (4.6) is a measure of fracture toughness for high rate loading. If small scale yielding assumptions apply, the critical J can be converted to an equivalent K_{IC} through Eq. (3.18).

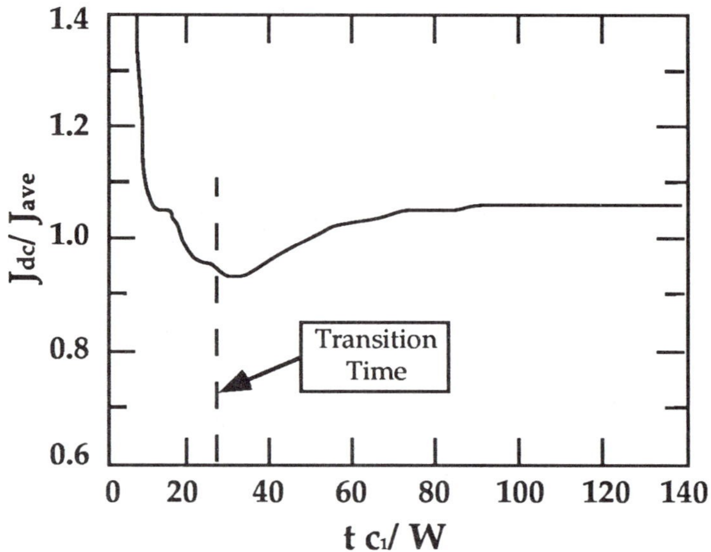

FIGURE 4.4 Ratio of J computed from Eq. (4.6) to the through-thickness average J computed from three-dimensional dynamic finite element analysis.

Given the difficulties associated with defining a fracture parameter in the presence of inertia forces and reflected stress waves, it is obviously preferable to apply Eq. (4.6) whenever possible. For a three point bend specimen with W = 50 mm, the transition time is approximately 300 μs [19]. Thus the quasistatic formula can be applied as long as fracture occurs after ~ 600 μs. This requirement is relatively easy to meet in impact tests on ductile materials [15,16]. For more brittle materials, the transition time requirement can be met by decreasing the displacement rate or the width of the specimen.

The transition time concept can be applied to other configurations by adjusting the geometry factor in Eq. (4.2). Duffy and Shih [17] have applied this approach to dynamic fracture toughness measurement in notched round bars. Small round bars have proved to be suitable for dynamic testing of brittle materials such as ceramics, where the transition time must be small.

If the effects of inertia and reflected stress waves can be eliminated, one is left with the rate-dependent material response. The transition time approach allows material rate effects to be quantified independent of inertia effects. High strain rates tend to elevate the flow stress of the material. The effect of flow stress on fracture toughness depends on the failure mechanism. High strain rates tend to decrease cleavage resistance, which is stress controlled. Materials whose fracture mechanisms are strain controlled often see an increase in toughness at high loading rates because more energy is required to reach a given strain value.

Figure 4.5 shows fracture toughness data for a structural steel at three loading rates [21]. The critical K_I values were determined from quasistatic relationships. For a given loading rate, fracture toughness increases rapidly with temperature at the onset of the ductile-brittle transition. Note that increasing the loading rate has the effect of shifting the transition to higher temperatures. Thus at a constant temperature, fracture toughness is highly sensitive to strain rate.

The effect of loading rate on fracture behavior of a structural steel on the upper shelf of toughness is illustrated in Fig. 4.6. In this instance, strain rate has the opposite effect from Fig. 4.5, because ductile fracture of metals is primarily strain controlled. The J integral at a given amount of crack extension is elevated by high strain rates.

4.1.2 Rapid Crack Propagation and Arrest

When the driving force for crack extension exceeds the material resistance, the structure is unstable, and rapid crack propagation occurs. Figure 4.7 illustrates a simple case, where the (quasistatic) energy release rate increases linearly with crack length and the material resistance is constant. Since the first law of thermodynamics must be obeyed even by an unstable system, the excess energy, denoted by the shaded area in Fig. 4.7, does not simply disappear, but is converted into

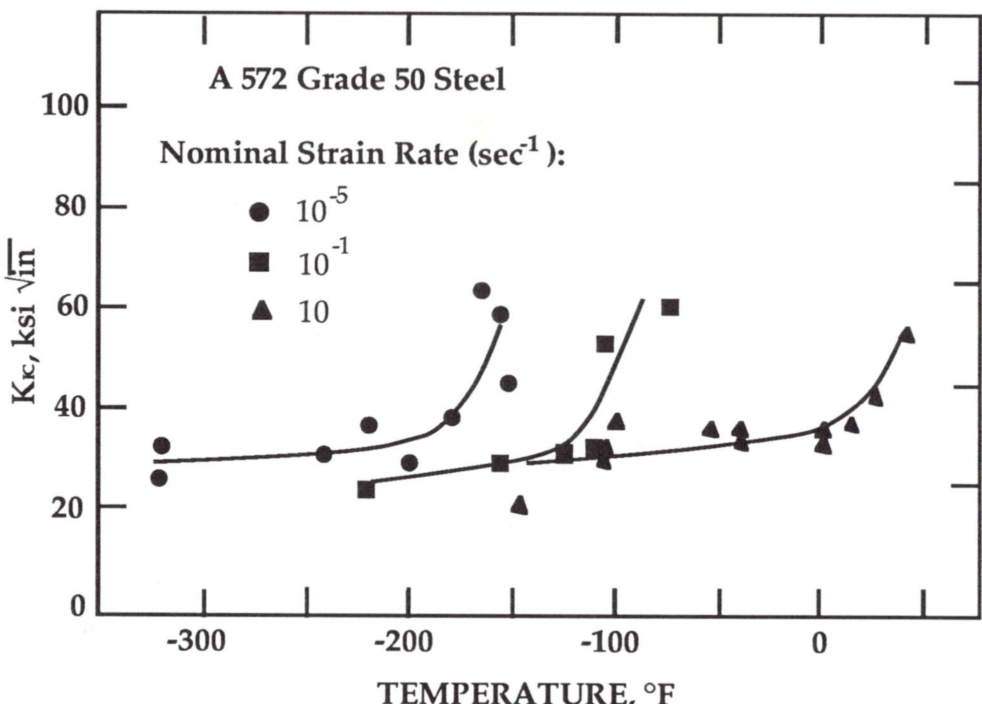

FIGURE 4.5 Effect of loading rate on the cleavage fracture toughness of a structural steel [21].

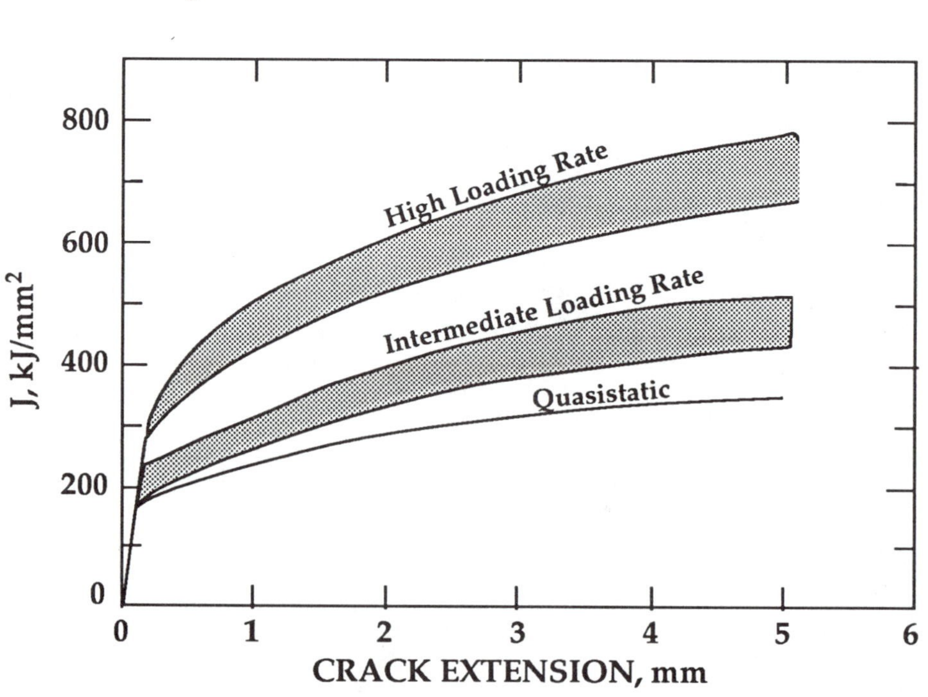

FIGURE 4.6 Effect of loading rate on the J-R curve behavior of HY80 steel [15].

kinetic energy. The magnitude of the kinetic energy dictates the crack speed.

In the quasistatic case, a crack is stable if the driving force is less than or equal to the material resistance. Similarly, if the energy available for an incremental extension of a rapidly propagating crack falls below the material resistance, the crack arrests. Figure 4.8 illustrates a simplified scenario for crack arrest. Suppose that cleavage fracture initiates when $K_I = K_{IC}$. The resistance encountered by a rapidly propagating cleavage crack is less than for cleavage initiation, because plastic deformation at the moving crack tip is suppressed by the high local strain rates. If the structure has a falling driving force curve, it eventually crosses the resistance curve. Arrest does not occur at this point, however, because the structure contains kinetic energy that can be converted to fracture energy. Arrest occurs below the resistance curve, after most of the available energy has been dissipated. The apparent arrest toughness, K_{Ia}, is less than the true material resistance, K_{IA}. The difference between K_{Ia} and K_{IA} is governed by the kinetic energy created during crack propagation; K_{IA} is a material property, but K_{Ia} depends on geometry.

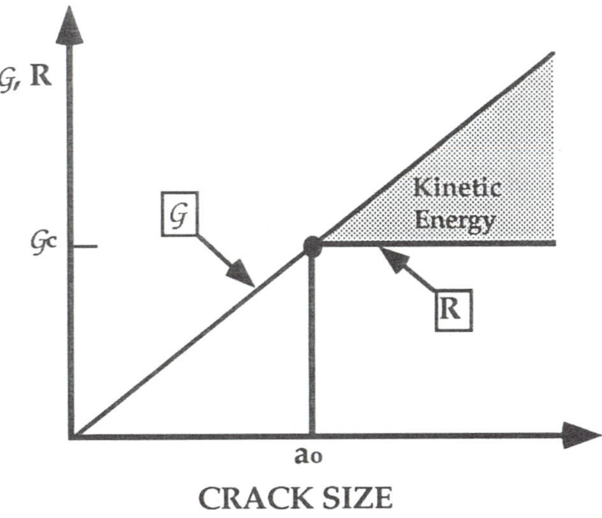

FIGURE 4.7 Unstable crack propagation, which results in the generation of kinetic energy.

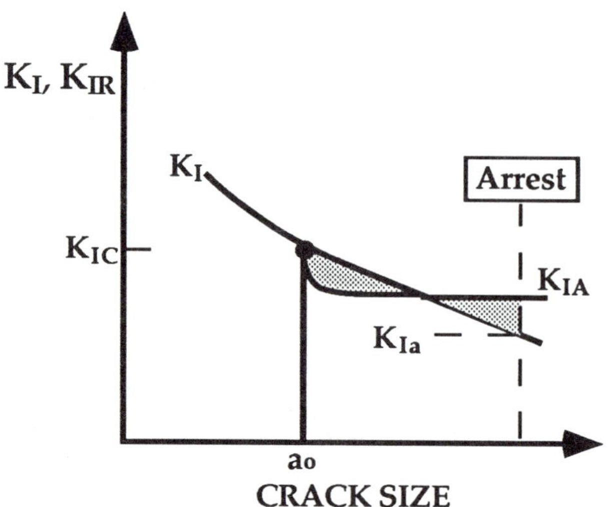

FIGURE 4.8 Unstable crack propagation and arrest with a falling driving force curve. The apparent arrest toughness, K_{Ia}, is slightly below the true material resistance, K_{IA}, due to excess kinetic energy.

Figures 4.7 and 4.8 compare material resistance with *quasistatic* driving force curves. That is, these curves represent K_I and $\mathcal{G}$ values computed with the procedures described in Chapter 2. Early researchers [22-26] realized that the crack driving force should incorporate the effect of kinetic energy. The Griffith-Irwin energy balance (Sections 2.3 and 2.4) can be modified to include kinetic energy, resulting in a dynamic definition of energy release rate:

$$\mathcal{G}(t) = \frac{dF}{d\mathcal{A}} - \frac{dU}{d\mathcal{A}} - \frac{dE_k}{d\mathcal{A}} \tag{4.7}$$

where F is the work done by external forces and $\mathcal{A}$ is the crack area. Equation (4.7) is consistent with the original Griffith approach, which is based on the first law of thermodynamics. The kinetic energy must be included in a general statement of the first law; Griffith implicitly assumed that $E_k = 0$.

Crack speed

Mott [22] applied dimensional analysis to a propagating crack in order to estimate the relationship between kinetic energy and crack speed. For a through crack of length 2a in an infinite plate in tension, the displacements must be proportional to crack size, since a is the only relevant length dimension. Assuming the plate is elastic, displacements must also be proportional to the nominal applied strain; thus

$$u_x = \alpha_x \, a \frac{\sigma}{E} \qquad \text{and} \qquad u_y = \alpha_y \, a \frac{\sigma}{E} \tag{4.8}$$

where α_x and α_y are dimensionless constants. (Note that quantitative estimates for α_x and α_y near the crack tip in the quasistatic case can be obtained by applying the relationships in Table 2.2.) The kinetic energy is equal to one half the mass times the velocity squared. Therefore, E_k for the cracked plate (assuming unit thickness) is given by

$$E_k = \frac{1}{2} \rho \, a^2 \, V^2 \left(\frac{\sigma}{E}\right)^2 \int\!\!\int (\alpha_x^2 + \alpha_y^2) \, dx \, dy \tag{4.9}$$

where ρ is the mass density of the material and V (= $\dot{a}$) is the crack speed. Assuming the integrand depends only on position[1], E_k can be written in the following form:

$$E_k = \frac{1}{2} k \, \rho \, a^2 \, V^2 \left(\frac{\sigma}{E}\right)^2 \tag{4.10}$$

where k is a constant. Applying the modified Griffith energy balance (Eq. (4.7)) gives

$$\mathcal{G}(t) = \frac{d}{da}\left(\frac{1}{2} k \, \rho \, a^2 \, V^2 \left(\frac{\sigma}{E}\right)^2 - \frac{\pi \sigma^2 a^2}{E}\right) = 2w_f \tag{4.11}$$

where w_f is the work of fracture, defined in Chapter 2; in the limit of an ideally brittle material, $w_f = \gamma_s$, the surface energy. Note that Eq. (4.11) assumes a flat R curve (constant w_f). At initiation, the kinetic energy term is not present, and the initial crack length, a_o, can be inferred from Eq. (2.22):

---

[1]In a rigorous dynamic analysis, α_x and α_y , and thus k, depend on crack speed.

$$a_o = \frac{2 E w_f}{\pi \sigma^2} \tag{4.12}$$

Substituting Eq. (4.12) into Eq. (4.11) and solving for V leads to

$$V = \sqrt{\frac{2\pi}{k}}\, c_o \left(1 - \frac{a}{a_o}\right) \tag{4.13}$$

where $c_o = \sqrt{E/\rho}$, the speed of sound for one-dimensional wave propagation. Mott [22] actually obtained a somewhat different relationship from Eq. (4.13), because he solved Eq. (4.11) by making the erroneous assumption that $dV/da = 0$. Dulaney and Brace [27] and Berry [28] later corrected the Mott analysis and derived Eq. (4.13).

Roberts and Wells [29] obtained an estimate for k by applying the Westergaard stress function (Appendix 2.3) for this configuration. After making a few assumptions, they showed that $\sqrt{2\pi/k} \approx 0.38$.

According to Eq. (4.13) and the Roberts and Wells analysis, the crack speed reaches a limiting value of 0.38 c_o when $a >> a_o$. This estimate compares favorably with measured crack speeds in metals, which typically range from 0.2 to 0.4 c_o [30].

Freund [2-4] performed a more detailed numerical analysis of a dynamically propagating crack in an infinite body and obtained the following relationship

$$V = c_r \left(1 - \frac{a}{a_o}\right) \tag{4.14}$$

where c_r is the Raleigh (surface) wave speed. For Poisson's ratio = 0.3, the c_r/c_o ratio = 0.57. Thus the Freund analysis predicts a larger limiting crack speed than the Roberts and Wells analysis. The limiting crack speed in Eq. (4.14) can be argued on physical grounds [26]. For the special case where $w_f = 0$, a propagating crack is merely a disturbance on a free surface, which must move at the Raleigh wave velocity. In both Eq. (4.13) and (4.14), the limiting velocity is independent of fracture energy; thus the maximum crack speed should be c_r for all w_f.

Experimentally observed crack speeds do not usually reach c_r. Both the simple analysis that resulted in Eq. (4.13) and Freund's more detailed dynamic analysis assumed that the fracture energy does not depend on crack length or crack speed. The material resistance actually increases with crack speed, as discussed below. The good agreement between experimental crack velocities and the Roberts and Wells estimate of 0.38 c_o is largely coincidental.

Elastodynamic crack tip parameters

The governing equation for Mode I crack propagation under elastodynamic conditions can be written as

$$K_I(t) = K_{ID}(V) \tag{4.15}$$

where K_I is the instantaneous stress intensity and K_{ID} is the material resistance to crack propagation, which depends on crack velocity. In general, $K_I(t)$ is not equal to the static stress intensity factor, as defined in Chapter 2. A number of researchers [8-10,31-33] have obtained a relationship for the dynamic stress intensity of the form

$$K_I(t) = k(V)\, K_I(0) \tag{4.16}$$

where k is a universal function of crack speed and $K_I(0)$ is the static stress intensity factor. The function $k(V) = 1.0$ when $V = 0$, and decreases to zero as V approaches the Raleigh wave velocity. An approximate expression for k was obtained by Rose [34]:

$$k(V) \approx \left(1 - \frac{V}{c_r}\right) \sqrt{1 - hV} \tag{4.17}$$

where h is a function of the elastic wave speeds and can be approximated by

$$h \approx \frac{2}{c_1}\left(\frac{c_2}{c_r}\right)^2 \left[1 - \left(\frac{c_2}{c_1}\right)\right]^2 \tag{4.18}$$

where c_1 and c_2 are the longitudinal and shear wave speeds, respectively.

Equation (4.16) is valid only at short times or in infinite bodies. This relationship neglects reflected stress waves, which can have a significant effect on the local crack tip fields. Since the crack speed is proportional to the wave speed, Eq. (4.16) is valid as long as the length of crack propagation ($a - a_o$) is small compared to specimen dimensions, because reflecting stress waves will not have had time to reach the crack tip (Example 4.1). In finite specimens where stress waves reflect back to the propagating crack tip, the dynamic stress intensity must be determined experimentally or numerically on a case-by-case basis.

EXAMPLE 4.1

Rapid crack propagation initiates in a deeply-notched specimen with initial ligament b_o (Fig. 4.9). Assuming the average crack speed = 0.2 c_1, estimate how far the crack will propagate before it encounters a reflected longitudinal wave.

Solution: At the moment the crack encounters the first reflected wave, the crack has traveled a distance Δa, while the wave has traveled $2\ b_o - \Delta a$. Equating travel times gives

$$\frac{\Delta a}{0.2\ c_1} = \frac{2\ b_o - \Delta a}{c_1}$$

Thus

$$\Delta a = \frac{b_o}{3}$$

Equation (4.16) is valid in this case as long as the crack extension is less than $b_o/3$ and the plastic zone is small compared to b_o.

For an infinite body or short times, Freund [10] showed that the dynamic energy release rate could be expressed in the following form:

$$\mathcal{G}(t) = g(V)\ G(0) \tag{4.19}$$

where g is a universal function of crack speed that can be approximated by

$$g(V) \approx 1 - \frac{V}{c_r} \tag{4.20}$$

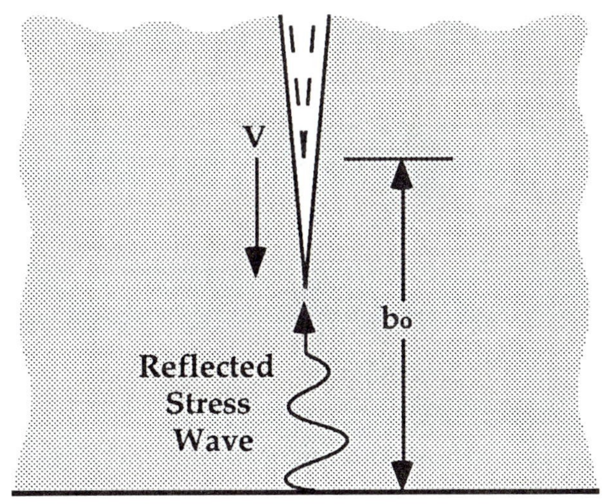

FIGURE 4.9 **Propagating crack encountering a reflected stress wave.**

Combining Eqs. (4.16) to (4.20) with Eq. (2.51) gives

$$\mathcal{G}(t) = A(V)\frac{K_I^2(t)}{E'} \tag{4.21}$$

where

$$A(V) \approx \left[\left(1 - \frac{V}{c_r}\right)(1 - hV)\right]^{-1} \tag{4.22}$$

Thus the relationship between K_I and $\mathcal{G}$ depends on crack speed. A more accurate (and more complicated) relationship for A(V), is given in Appendix 4.1.

When the plastic zone ahead of the propagating crack is small, $K_I(t)$ uniquely defines the crack tip stress, strain, and displacement fields, but the angular dependence of these quantities is different from the quasistatic case. For example, the stresses in the elastic singularity zone are given by [32,33,35]

$$\sigma_{ij} = \frac{K_I(t)}{\sqrt{2\pi r}} f_{ij}(\theta, V) \tag{4.23}$$

The function f_{ij} reduces to the quasistatic case (Table 2.1) when V = 0. Appendix 4.1 outlines the derivation of Eq. (4.23) and gives specific relationships for f_{ij} in the case of rapid crack propagation. The displacement functions also display an angular dependence that varies with V. Consequently, α_x and α_y in Eq. (4.9) must depend on crack velocity as well as position, and the Mott analysis is not rigorously correct for dynamic crack propagation.

Dynamic toughness

As Eq. (4.15) indicates, the dynamic stress intensity is equal to K_{ID}, the dynamic material resistance, which depends on crack speed. This equality permits experimental measurements of K_{ID}.

Dynamic propagation toughness can be measured as a function of crack speed by means of high speed photography and optical methods, such as photoelasticity [36,37] and the method of caustics [38]. Figure 4.10 shows photoelastic fringe patterns for dynamic crack propagation in Homalite 100 [37]. Each fringe corresponds to a contour of maximum shear stress. Sanford and Dally [36] describe procedures for inferring stress intensity from photoelastic patterns.

Figure 4.11 illustrates the typical variation of K_{ID} with crack speed. At low speeds, K_{ID} is relatively insensitive to V, but K_{ID} increases asymptotically as V approaches a limiting value. Figure 4.12 shows K_{ID} data for 4340 steel published by Rosakis and Freund [39].

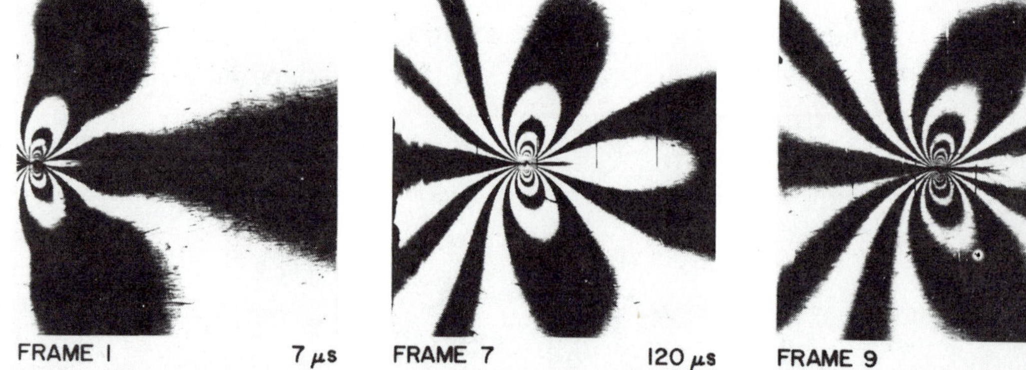

FIGURE 4.10 Photoelastic fringe patterns for a rapidly propagating crack in Homalite 100 [37]. (Photograph provided by R. Chona)

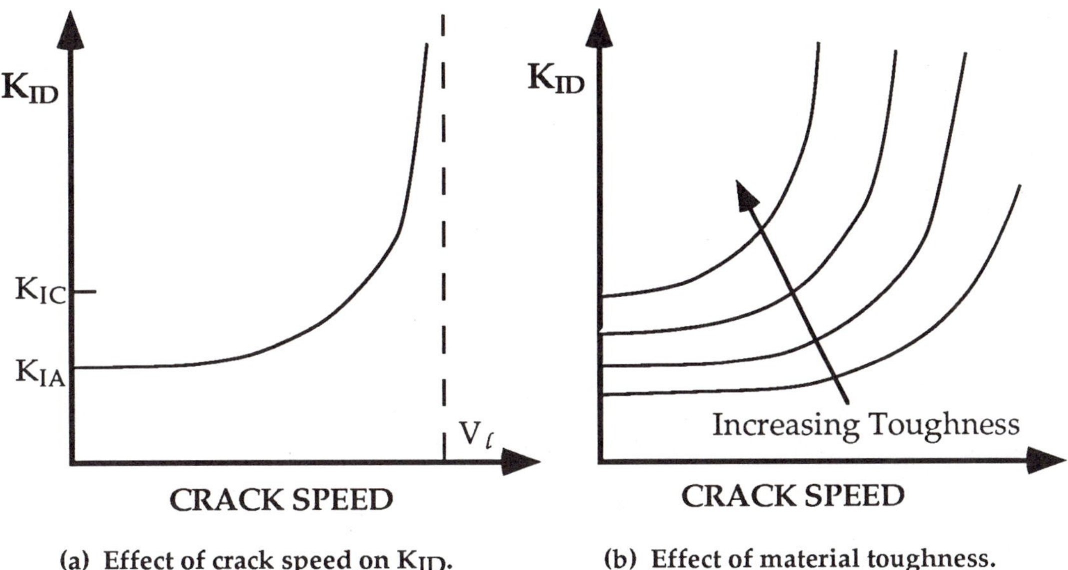

(a) Effect of crack speed on K_{ID}. **(b) Effect of material toughness.**

FIGURE 4.11 Schematic K_{ID}-crack speed curves.

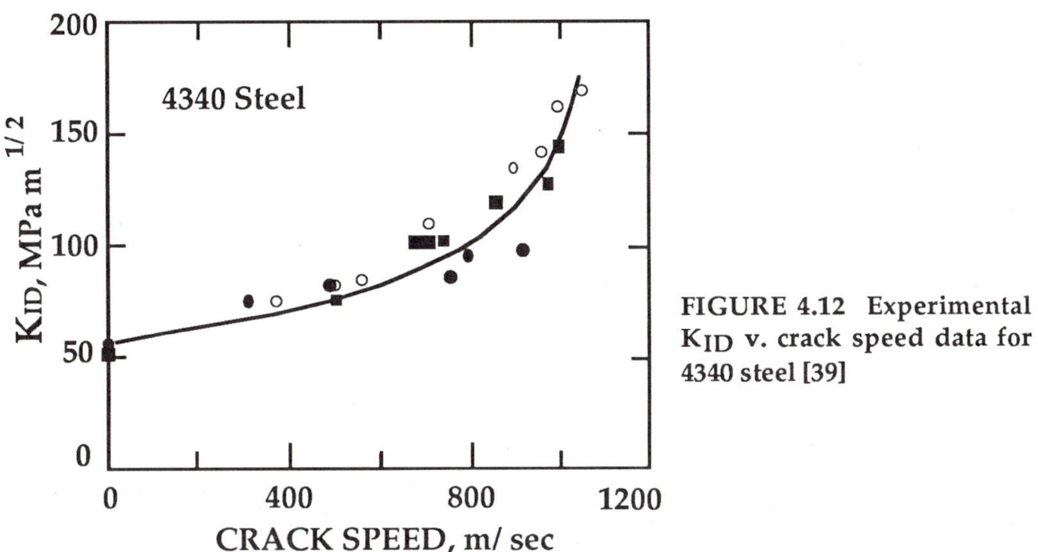

FIGURE 4.12 Experimental K_{ID} v. crack speed data for 4340 steel [39]

In the limit of V = 0, $K_{ID} = K_{IA}$, the arrest toughness of the material. In general, $K_{IA} < K_{IC}$, the quasistatic initiation toughness. When a stationary crack in an elastic-plastic material is loaded monotonically, the crack tip blunts and a plastic zone forms. A propagating crack, however, tends to be sharper and has a smaller plastic zone than a stationary crack. Consequently, more energy is required to initiate fracture from a stationary crack than is required to maintain propagation of a sharp crack.

The crack speed dependence of K_{ID} can be represented by an empirical equation of the form

$$K_{ID} = \frac{K_{IA}}{1 - \left(\frac{V}{V_\ell}\right)^m} \tag{4.24}$$

where V_ℓ is the limiting crack speed in the material and m is an experimentally determined constant. As Fig. 4.11(b) illustrates, K_{IA} increases and V_ℓ decreases with increasing material toughness. The trends in Figs. 4.11(a) and 4.11(b) have not only been observed experimentally, but have also been obtained by numerical simulation [40,41]. The upturn in propagation toughness at high crack speeds is apparently caused by local inertia forces in the plastic zone.

Crack arrest

Equation (4.15) defines the conditions for rapid crack advance. If, however, $K_I(t)$ falls below the minimum K_{ID} value for a finite length of time, propagation cannot continue, and the crack arrests. There are a number of situations that might lead to crack arrest. Figure 4.8 illustrates one possibility: if the driving force decreases with crack extension, it may eventually be less than the material resistance. Arrest is also possible when material resistance increases with crack extension. For example, a crack that initiates in a brittle region of a structure, such as a weld, may arrest when it reaches a material with higher toughness. A temperature gradient in a material that exhibits a ductile-brittle transition is another case where the toughness can increase with position: a crack may initiate in a cold region of the structure and arrest when it encounters warmer material with a higher toughness. An example of this latter scenario is a pressurized thermal shock event in a nuclear pressure vessel [42].

In many instances, it is not possible to guarantee with absolute certainty that an unstable fracture will not initiate in a structure. Transient loads, for example, may occur unexpectedly. In such instances crack arrest can be the second line of defence. Thus the crack arrest toughness, K_{IA}, is an important material property.

Based on Eq. (4.16), one can argue that $K_I(t)$ at arrest is equivalent to the quasistatic value, since V = 0. Thus it should be possible to infer K_{IA} from a quasistatic calculation based on the load and crack length at arrest. This quasistatic approach to arrest is actually quite common, and it is acceptable in many practical situations. Chapter 7 describes a standardized test method for measuring crack arrest toughness that is based on quasistatic assumptions.

The quasistatic arrest approach must be used with caution, however. Recall that Eq. (4.16) is only valid for infinite structures or short crack jumps, where reflected stress waves do not have sufficient time to return to the crack tip. When reflected stress wave effects are significant, Eq. (4.16) is no longer valid, and a quasistatic analysis tends to give misleading estimates of the arrest toughness. Quasistatic estimates of arrest toughness are sometimes given the designation K_{Ia}; for short crack jumps, $K_{Ia} = K_{IA}$.

The effect of stress waves on the apparent arrest toughness (K_{Ia}) was demonstrated dramatically by Kalthoff, et al. [43], who performed dynamic propagation and arrest experiments on wedge loaded double cantilever beam (DCB) specimens. Recall from Example 2.3 that the DCB specimen exhibits a falling driving force curve in displacement control. Kalthoff, et al. varied the K_I at initiation by varying the notch root radius. When the crack was sharp, fracture initiated slightly above K_{IA} and arrested after a short crack jump; the length of crack jump increased with notch tip radius.

Figure 4.13 is a plot of the Kalthoff, et al. results. For the shortest crack jump, the true arrest toughness and the apparent quasistatic value coincide, as expected. As the length of crack jump increases, the discrepancy between the true arrest and the quasistatic estimate increases, with $K_{IA} > K_{Ia}$. Note that K_{IA} appears to be a material constant but K_{Ia} varies with the length of crack propagation. Also note that the dynamic stress intensity during crack growth is considerably different from the quasistatic estimate of K_I. Kobayashi, et al. [43] obtained similar results.

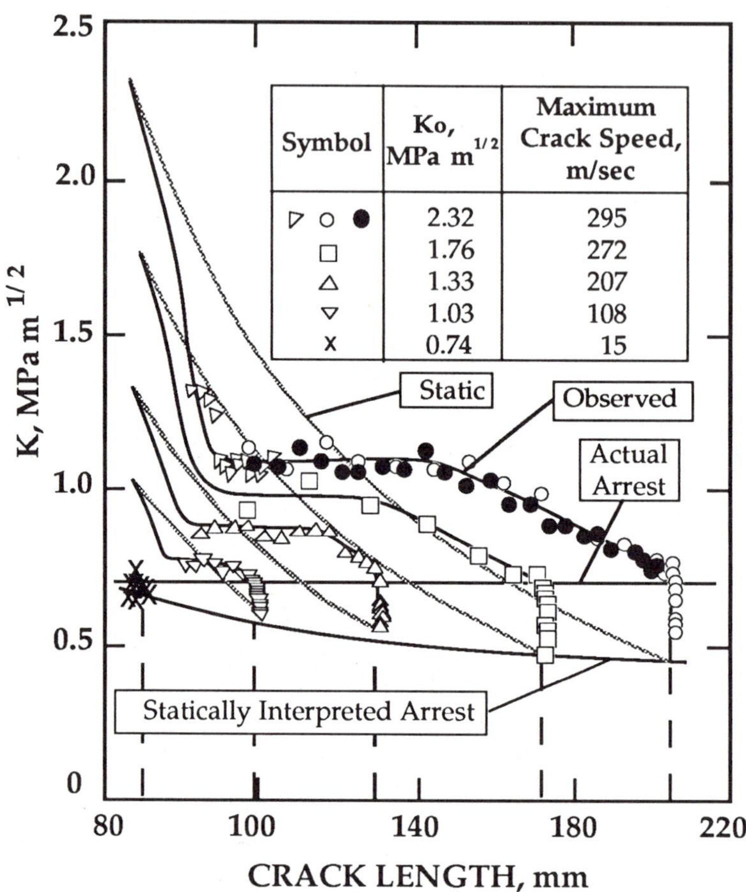

FIGURE 4.13 Crack arrest experiments on wedge-loaded DCB Araldite B specimens [43]. The statically interpreted arrest toughness underestimates the true K_{IA} of the material; this effect is most pronounced for long crack jumps.

A short time after arrest, the applied stress intensity reaches K_{Ia}, the quasistatic value. Figure 4.14 shows the variation of K_I after arrest in one of the Kalthoff, et al. experiments. When the crack arrests, $K_I = K_{IA}$, which is greater than K_{Ia}. Figure 4.14 shows that the specimen "rings down" to K_{Ia} after ~2000 μs. The quasistatic value, however, is not indicative of the true material arrest properties.

Recall the schematic in Fig. 4.8, where it was argued that arrest, when quantified by the quasistatic stress intensity, would occur below the true arrest toughness, K_{IA}, because of kinetic energy in the specimen. This argument is a slight oversimplification, but it leads to the correct qualitative conclusion.

The DCB specimen provides an extreme example of reflected stress wave effects; the specimen design is such that stress waves can traverse the width of the specimen and return to the crack tip in a very short time. In many structures, the quasistatic approach is approximately valid, even for relatively long crack jumps. In any case, K_{Ia} gives a lower bound estimate of K_{IA}, and thus is conservative in most instances.

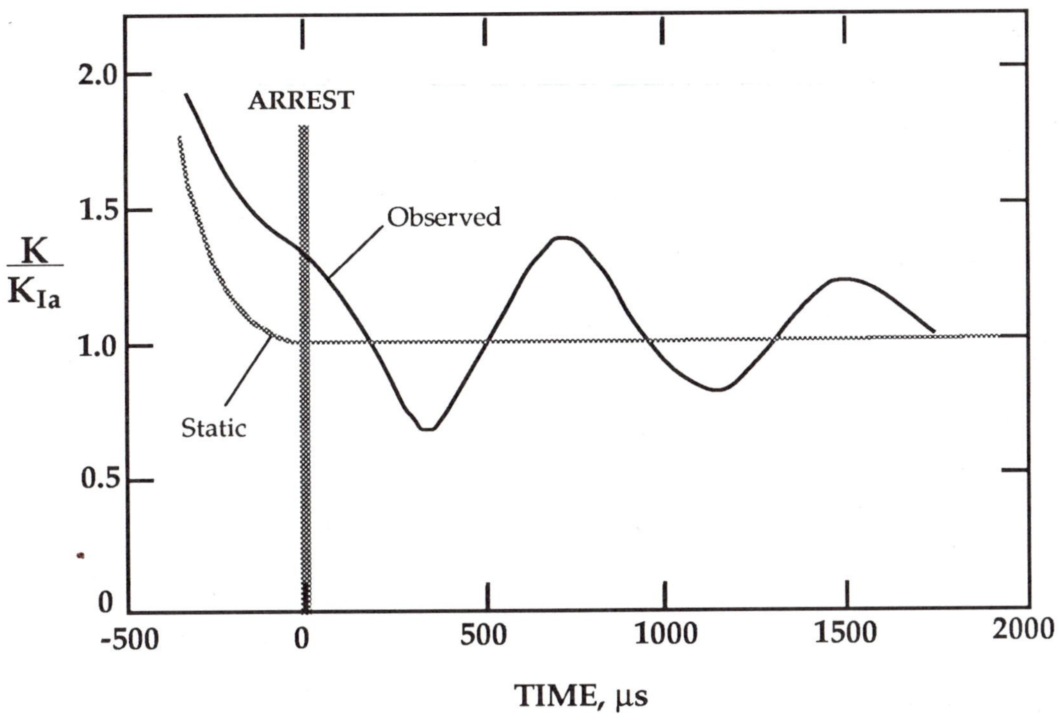

FIGURE 4.14 Comparison of dynamic measurements of stress intensity with static calculations for a wedge loaded DCB Araldite B specimen [43].

4.1.3 Dynamic Contour Integrals

The original formulation of the J contour integral is equivalent to the nonlinear elastic energy release rate for quasistatic deformation. By invoking a more general definition of energy release rate, it is possible to incorporate dynamic effects and time-dependent material behavior into the J integral.

The energy release rate is usually defined as the energy released from the body per unit crack advance. A more precise definition [11] involves the work input into the crack tip. Consider a vanishingly small contour, Γ, around the tip of a crack in a two-dimensional solid (Fig. 4.15). The energy release rate is equal to the energy flux into the crack tip, divided by the crack speed:

$$J = \frac{\mathcal{F}}{V} \tag{4.25}$$

where $\mathcal{F}$ is the energy flux into the area bounded by Γ. The generalized energy release rate, including inertia effects, is given by

$$J = \lim_{\Gamma \to 0} \int_{\Gamma} \left[(w + T)\, dy + \sigma_{ij}\, n_j \frac{\partial u_i}{\partial x}\, ds \right] \tag{4.26}$$

where w and T are the stress work and kinetic energy densities defined as

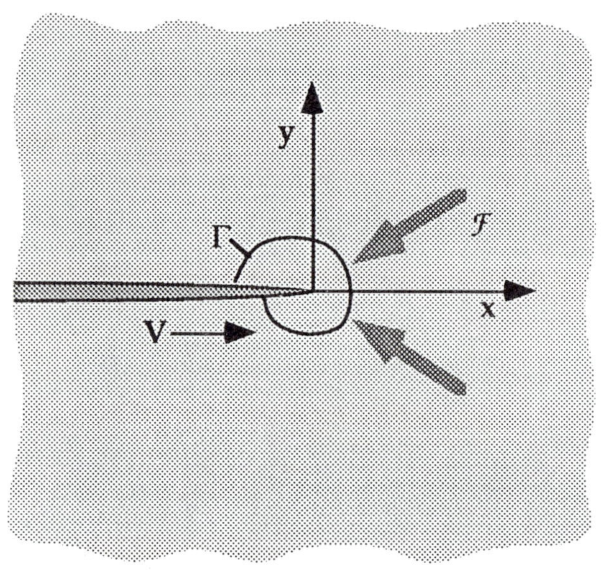

FIGURE 4.15 Energy flux into a small contour at the tip of a propagating crack.

$$w = \int_0^{\varepsilon_{ij}} \sigma_{ij}\, d\varepsilon_{ij} \tag{4.27}$$

and

$$T = \frac{1}{2}\rho \frac{\partial u_i}{\partial t}\frac{\partial u_i}{\partial t} \tag{4.28}$$

Equation (4.26) has been published in a variety of forms by several researchers [8-12]. Appendix 4.2 gives a derivation of this relationship.

Equation (4.26) is valid for time-dependent as well as history-dependent material behavior. When evaluating J for a time-dependent material, it may be convenient to express w in the following form:

$$w = \int_{t_o}^{t} \sigma_{ij}\, \dot{\varepsilon}_{ij}\, dt \tag{4.29}$$

where $\dot{\varepsilon}_{ij}$ is the strain rate.

Unlike the conventional J integral, the contour in Eq. (4.26) cannot be chosen arbitrarily. Consider, for example, a dynamically loaded cracked body with stress waves reflecting off of free surfaces. If the integral in Eq. (4.26) were computed at two arbitrary contours a finite distance from the crack tip, and a stress wave passed through one contour but not the other, the values of these integrals would normally be different for the two contours. Thus the generalized J integral is not path independent, except in the immediate vicinity of the crack tip. If, however, $T = 0$ at all points in the body, the integrand in Eq. (4.26) reduces to the form of the original J integral. In the latter case, the path-independent property of J is restored if w displays the property of an elastic potential (see Appendix 4.2).

The form of Eq. (4.26) is not very convenient for numerical calculations, since it is extremely difficult to obtain adequate numerical precision from a contour integration very close to the crack tip. Fortunately, Eq. (4.26) can be expressed in a variety of other forms that are more conducive to numerical analysis. The energy release rate can also be generalized to three dimensions. The results in Figs. 4.3 and 4.4 are obtained from finite element analysis that utilized alternate forms of Eq. (4.26). Chapter 11 discusses numerical calculations of J for both quasistatic and dynamic loading.

4.2 CREEP CRACK GROWTH

Components that operate at high temperatures relative to the melting point of the material may fail by slow, stable extension of a macroscopic crack. Traditional approaches to design in the creep regime apply only when creep and material damage are uniformly distributed. Time-dependent fracture mechanics approaches are required when creep failure is controlled by a dominant crack in the structure.

Figure 4.16 illustrates the typical creep response of a material subject to constant stress. Deformation at high temperatures can be divided into four regimes: instantaneous (elastic) strain, primary creep, secondary (steady state) creep, and tertiary creep. The elastic strain occurs immediately upon application of the load. As discussed in the previous section on dynamic fracture, the elastic stress-strain response of a material is not instantaneous (i.e., it is limited by the speed of sound in the material), but it can be viewed as such in creep problems, where the time scale is usually measured in hours. Primary creep dominates at short times after application of the load; the strain rate decreases with time, as the material strain hardens. In the secondary creep stage, the deformation reaches a steady state, where strain hardening and strain softening are balanced; the creep rate is constant in the secondary stage. In the tertiary stage, the creep rate accelerates, as the material approaches ultimate failure. Microscopic failure mechanisms, such as grain boundary cavitation, nucleate in this final stage of creep.

During growth of a macroscopic crack at high temperatures, all four types of creep response can occur simultaneously in the most general case (Fig. 4.17). The material at the tip of growing crack is in the tertiary stage of creep, since the material is obviously failing locally. The material may be elastic remote from the crack tip, and in the primary and secondary stages of creep at moderate distances from the tip.

Most analytical treatments of creep crack growth assume limiting cases, where one or more of these regimes are not present or are confined to a small portion of the component. If, for example, the component is predominantly elastic, and the creep zone is confined to a small region near the crack tip, the crack growth can be characterized by the stress intensity factor. In the other extreme, when the component deforms globally in steady state creep, elastic strains and tertiary creep can be disregarded. A parameter that applies to the latter case is described below, followed by a brief discussion of approaches that consider the transition from elastic to steady state creep behavior.

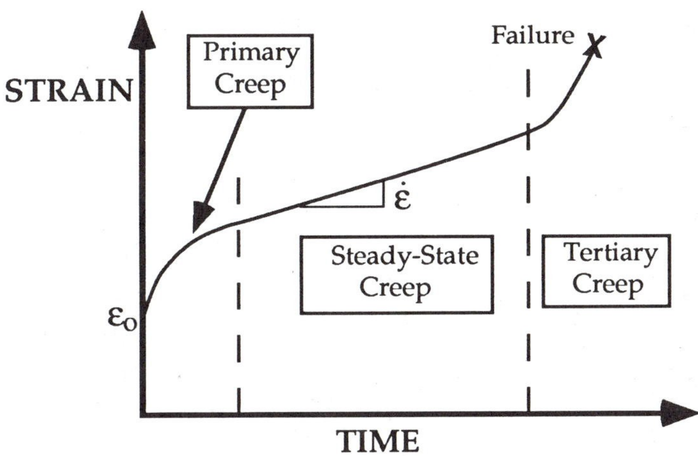

FIGURE 4.16 Schematic creep behavior of a material subject to a constant stress.

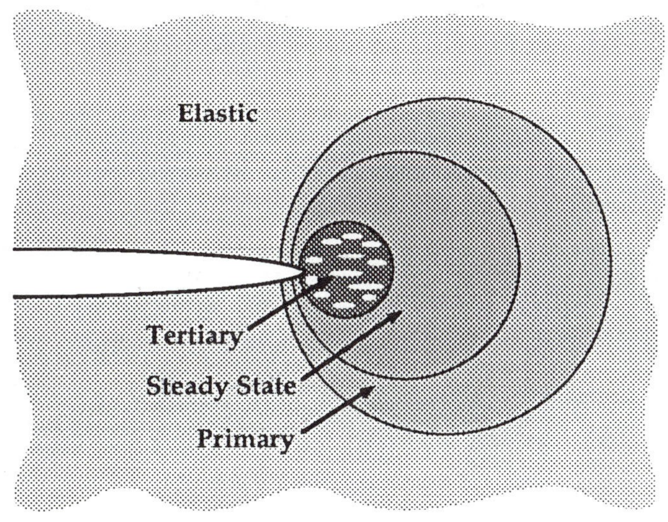

FIGURE 4.17 Creep zones at the tip of a crack.

4.2.1 The C* Integral

A formal fracture mechanics approach to creep crack growth was developed soon after the J integral was established as an elastic-plastic fracture parameter. Landes and Begley [45], Ohji, et al. [46], and Nikbin, et al. [47] independently proposed what became known as the C* integral to characterize crack growth in a material undergoing steady state creep. They applied Hoff's analogy [48], which states that if there exists a nonlinear elastic body that obeys the relationship $\varepsilon_{ij} = f(\sigma_{ij})$ and a viscous body that is characterized by $\dot{\varepsilon}_{ij} = f(\sigma_{ij})$, where the function of stress is the same for both, then both bodies develop identical stress distributions when the same load is applied. Hoff's analogy can be applied to steady state creep, since the creep rate is a function only of the applied stress.

The C* integral is defined by replacing strains with strain rates, and displacements with displacement rates in the J contour integral:

$$C^* = \int_{\Gamma} \left(\dot{w}\, dy + \sigma_{ij} n_j \frac{\partial \dot{u}_i}{\partial x} ds \right) \tag{4.30}$$

where $\dot{w}$ is the stress work rate (power) density, defined as

$$\dot{w} = \int_0^{\dot{\varepsilon}_{kl}} \sigma_{ij}\, d\dot{\varepsilon}_{ij} \tag{4.31}$$

Hoff's analogy implies that the C* integral is path-independent, because J is path-independent. Also, if secondary creep follows a power law:

$$\dot{\varepsilon}_{ij} = A \sigma_{ij}^{n} \tag{4.32}$$

where A and n are material constants, then it is possible to define an HRR-type singularity for stresses and strain rates near the crack tip:

$$\sigma_{ij} = \left(\frac{C^*}{A\, I_n\, r}\right)^{\frac{1}{n+1}} \tilde{\sigma}_{ij}(n,\theta) \tag{4.33a}$$

and

$$\dot{\varepsilon}_{ij} = \left(\frac{C^*}{A\, I_n\, r}\right)^{\frac{n}{n+1}} \tilde{\varepsilon}_{ij}(n,\theta) \tag{4.33b}$$

where the constants I_n, $\tilde{\sigma}_{ij}$, and $\tilde{\varepsilon}_{ij}$ are identical to the corresponding parameters in the HRR relationship (Eq. (3.24)). Note that in the present case, n is a creep exponent rather than a strain hardening exponent.

Just as the J integral characterizes the crack tip fields in an elastic or elastic-plastic material, the C* integral uniquely defines crack tip conditions in a viscous material. Thus the time-dependent crack growth rate in a viscous material should depend only on the value of C*. Experimental studies [45-49] have shown that creep crack growth rates correlate very well with C*, provided steady state creep is the dominant deformation mechanism in the specimen. Figure 4.18 shows typical creep crack growth data. Note that the crack growth rate follows a power law:

$$\dot{a} = \gamma (C^*)^m \tag{4.34}$$

where γ and m are material constants. In many materials, $m \approx n/(n+1)$, a result that is predicted by grain boundary cavitation models [49].

Experimental measurements of C* take advantage of analogies with the J integral. Recall that J is usually measured by invoking the energy release rate definition:

$$J = -\frac{1}{B}\left(\frac{\partial}{\partial a}\int_0^{\Delta} P\, d\Delta\right)_{\Delta} \tag{4.35}$$

where P is the applied load and Δ is the load line displacement. Similarly, C* can be defined in terms of a power release rate:

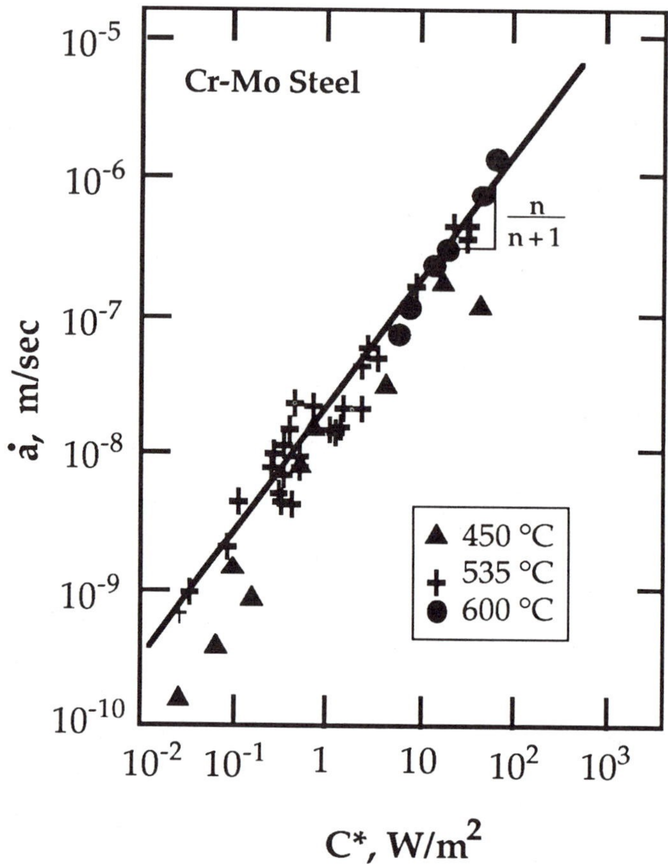

FIGURE 4.18 Creep crack growth data in a Cr-Mo Steel at three temperatures [49].

$$C^* = -\frac{1}{B}\left(\frac{\partial}{\partial a}\int_0^{\dot{\Delta}} P\, d\dot{\Delta}\right)_{\dot{\Delta}} \tag{4.36}$$

The J integral can be related to the energy absorbed by a laboratory specimen, divided by the ligament area[2]:

---

[2]The load line displacement, Δ, in Eqs. (4.37) to (4.39) corresponds to the portion of the displacement due to the presence of the crack, as discussed in Section 3.2.5. This distinction is not necessary in Eqs. (4.35) and (4.36), because the displacement component attributed to the uncracked configuration vanishes when differentiated with respect to a.

$$J = \frac{\eta}{B\,b} \int_0^{\Delta} P\, d\Delta \tag{4.37}$$

where η is a dimensionless constant that depends on geometry. Therefore, C* is given by

$$C^* = \frac{\eta}{B\,b} \int_0^{\dot{\Delta}} P\, d\dot{\Delta} \tag{4.38}$$

For a material that creeps according to a power law (Eq. (4.32)), the displacement rate is proportional to P^n, assuming global creep in the specimen. In this case Eq. (4.38) reduces to

$$C^* = \frac{n}{n+1} \frac{\eta}{B\,b} P \dot{\Delta} \tag{4.39}$$

The geometry factor η has been determined for a variety of test specimens. For example $\eta = 2.0$ for a deeply notched bend specimen (Eqs. (3.37) and (4.6)).

4.2.2 Short Time Versus Long Time Behavior

The C* parameter only applies to crack growth in the presence of global steady state creep. Stated another way, C* applies to long time behavior, as discussed below.

Consider a stationary crack in a material that is susceptible to creep deformation. If a remote load is applied to the cracked body, the material responds almost immediately with the corresponding elastic strain distribution. Assuming the loading is pure Mode I, the stresses and strains exhibit a $1/\sqrt{r}$ singularity near the crack tip and are uniquely defined by K_I. Large scale creep deformation does not occur immediately, however. Soon after the load is applied, a small creep zone, analogous to a plastic zone, forms at the crack tip. The crack tip conditions can be characterized by K_I as long as the creep zone is embedded within the singularity dominated zone. The creep zone grows with time, eventually invalidating K_I as a crack tip parameter. At long times, the creep zone spreads throughout the entire structure.

When the crack grows with time, the behavior of the structure depends on the crack growth rate relative to the creep rate. In brittle materials, the crack growth rate is so fast that it overtakes the creep zone; crack growth can be characterized by K_I because the creep zone at the tip of the growing crack remains small. At the other extreme, if the crack growth is sufficiently slow that the creep zone spreads throughout the structure, C* is the appropriate characterizing parameter.

Riedel and Rice [50] analyzed the transition from short time elastic behavior to long time viscous behavior. They assumed a simplified stress-strain rate law that neglects primary creep:

$$\dot{\varepsilon} = \frac{\dot{\sigma}}{E} + A\,\sigma^n \tag{4.40}$$

for uniaxial tension. If a load is suddenly applied and then held constant, a creep zone gradually develops in an elastic singularity zone, as discussed above. Riedel and Rice argued that the stresses well within the creep zone can be described by

$$\sigma_{ij} = \left(\frac{C(t)}{A\,I_n\,r}\right)^{\frac{1}{n+1}} \tilde{\sigma}_{ij}(n,\theta) \tag{4.41}$$

where C(t) is a parameter that characterizes the amplitude of the local stress singularity in the creep zone; C(t) varies with time and is equal to C* in the limit of long time behavior. If the remote load is fixed, the stresses in the creep zone relax with time, as creep strain accumulates in the crack tip region. For small scale creep conditions, C(t) decays as 1/t according to the following relationship:

$$C(t) = \frac{K_I^2\,(1 - \nu^2)}{(n + 1)\,E\,t} \tag{4.42}$$

And the approximate size of the creep zone is given by

$$r_c(\theta, t) = \frac{1}{2\pi}\left(\frac{K_I}{E}\right)^2 \left[\frac{(n+1)\, A\, I_n\, E^n\, t}{2\pi\,(1-\nu^2)}\right]^{\frac{2}{n-1}} \tilde{r}_c(\theta, n) \qquad (4.43)$$

At $\theta = 90°$, $\tilde{r}_c$ is a maximum and ranges from 0.2 to 0.5, depending on n. As r_c increases in size, C(t) approaches the steady state value C*. Riedel and Rice defined a characteristic time for the transition from short time to long time behavior:

$$t_1 = \frac{K_I^2\,(1-\nu^2)}{(n+1)\,E\,C^*} \qquad (4.44a)$$

or

$$t_1 = \frac{J}{(n+1)\,C^*} \qquad (4.44b)$$

When significant crack growth occurs over time scales much less than t_1, the behavior can be characterized by K_I, while C* is the appropriate parameter when significant crack growth requires times >> t_1. Based on finite element analysis, Riedel [51] suggested the following simple formula to interpolate between small scale creep and extensive creep (short and long time behavior, respectively):

$$C(t) \approx C^*\left(\frac{t_1}{t} + 1\right) \qquad (4.45)$$

Note the similarity to the transition time concept in dynamic fracture (Section 4.1.1). In both instances, a transition time characterizes the interaction between two competing phenomena.

The C_t parameter

Unlike K_I and C*, direct experimental measurement of C(t) under transient conditions is usually not possible. Consequently Saxena [52] defined an alternate parameter, C_t, which was originally intended as an approximation of C(t). The advantage of C_t is that it can be measured relatively easily.

Saxena began by separating global displacement into instantaneous elastic and time-dependent creep components:

$$\Delta = \Delta_e + \Delta_t \tag{4.46}$$

The creep displacement, Δ_t, increases with time as the creep zone grows. Also, if load is fixed, $\dot{\Delta}_t = \dot{\Delta}$. The C_t parameter is defined as the creep component of the power release rate:

$$C_t = -\frac{1}{B}\left(\frac{\partial}{\partial a}\int_0^{\dot{\Delta}_t} P\, d\dot{\Delta}_t\right)_{\dot{\Delta}_t} \tag{4.47}$$

Note the similarity between Eqs. (4.36) and (4.47).

For small scale creep (ssc) conditions, Saxena defined an effective crack length, analogous to the Irwin plastic zone correction described in Chapter 2:

$$a_{eff} = a + \beta r_c \tag{4.48}$$

where $\beta \approx 1/3$ and r_c is defined at $\theta = 90°$. The displacement due to the creep zone is given by

$$\Delta_t = \Delta - \Delta_e = P\frac{dC}{da}\beta r_c \tag{4.49}$$

where C is the elastic compliance, defined in Chapter 2. Saxena showed that the small scale creep limit for C_t can be expressed as follows

$$(C_t)_{ssc} = \left(\frac{f'(a/W)}{f(a/W)}\right)\frac{P\dot{\Delta}_t}{B\ W} \tag{4.50}$$

where $f(a/W)$ is the geometry correction factor for Mode I stress intensity (see Table 2.4):

$$f(a/W) = \frac{K_I B\sqrt{W}}{P}$$

and f' is the first derivative of f. Equation (4.50) predicts that $(C_t)_{ssc}$ is proportional to K_I^4; thus C_t does not coincide with C(t) in the limit of small scale creep (Eq. (4.42)).

Saxena proposed the following interpolation between small scale creep and extensive creep:

$$C_t = (C_t)_{ssc}\left(1 - \frac{\dot{\Delta}}{\dot{\Delta}_t}\right) + C^* \tag{4.51}$$

where C* is determined from Eq. (4.38) using the *total* displacement rate. In the limit of long time behavior, $C^*/C_t = 1.0$, but this ratio is less than unity for small scale creep and transient behavior.

Bassani, et al. [53] applied the C_t parameter to experimental data with various C^*/C_t ratios and found that C_t characterized crack growth rates much better than C* or K_I. They state that C_t, when defined by Eqs. (4.50) and (4.51), characterizes experimental data better than C(t), as defined by Riedel's approximation (Eq. (4.45)).

Although C_t was originally intended as an approximation of C(t), it has become clear that these two parameters are distinct from one another. The C(t) parameter characterizes the stresses ahead of a stationary crack, while C_t is related to the rate of expansion of the creep zone. The latter quantity appears to be better suited to materials that experience relatively rapid creep crack growth. Both parameters approach C* in the limit of steady-state creep.

Primary creep

The analyses introduced so far do not consider primary creep. Referring to Fig. 4.17, which depicts the most general case, the outer ring of the creep zone is in the primary stage of creep. Primary creep may have an appreciable effect on the crack growth behavior if the size of the primary zone is significant.

Recently, researchers have begun to develop crack growth analyses that include the effects of primary creep. One such approach [54] considers a strain hardening model for the primary creep deformation, resulting in the following expression for total strain rate:

$$\dot{\varepsilon} = \frac{\dot{\sigma}}{E} + A_1 \sigma^n + A_2 \sigma^{m(1+p)} \varepsilon^{-p} \tag{4.51}$$

Riedel [54] introduced a new parameter, C_h^*, which is the primary creep analog to C*. The characteristic time that defines the transition from primary to secondary creep is defined as

$$t_2 = \left(\frac{C_h^*}{(1+p)\, C^*}\right)^{\frac{p+1}{p}} \tag{4.52}$$

The stresses within the steady state creep zone are still defined by Eq. (4.41), but the interpolation scheme for C(t) is modified when primary creep strains are present [54]:

$$C(t) \approx \left[\frac{t_1}{t} + \left(\frac{t_2}{t}\right)^{\frac{p+1}{p}} + 1\right] C^* \tag{4.53}$$

Equation (4.53) has been applied to experimental data in a limited number of cases. This relationship appears to give a better description of experimental data than Eq. (4.45), where the primary term is omitted.

Chun-Pok and McDowell [55] have recently incorporated the effects of primary creep into the estimation of the C_t parameter.

4.3 VISCOELASTIC FRACTURE MECHANICS

Polymeric materials have seen increasing service in structural applications in recent years. Consequently, the fracture resistance of these materials has become an important consideration. Much of the fracture mechanics methodology that was developed for metals is not directly transferable to polymers, however, because the latter behave in a viscoelastic manner.

Theoretical fracture mechanics analyses that incorporate viscoelastic material response are relatively new, and practical applications of viscoelastic fracture mechanics are rare, as of this writing. Most current applications to polymers utilize conventional, time-independent fracture mechanics methodology (see Chapters 6 and 8). Approaches that incorporate time dependence should become more widespread, however, as the methodology is developed further and is validated experimentally.

This section introduces viscoelastic fracture mechanics and outlines a number of recent advances in this area. The work of Schapery [56-61] is emphasized, because he has formulated the most complete theoretical framework, and his approach is related to the J and C* integrals, which were introduced earlier in this text.

4.3.1 Linear Viscoelasticity

Viscoelasticity is perhaps the most general (and complex) type of time-dependent material response. From a continuum mechanics viewpoint, viscoplastic creep in metals is actually a special case of viscoelastic material behavior. While creep in metals is generally considered permanent deformation, the strains can recover with time in viscoelastic materials. In the case of polymers, time-dependent deformation and recovery is a direct result of their molecular structure, as discussed in Chapter 6.

Let us introduce the subject by considering linear viscoelastic material behavior. In this case, *linear* implies that the material meets two conditions: superposition and proportionality. The first condition requires that stresses and strains at time t be additive. For example, consider two uniaxial strains, ε_1 and ε_2, at time t, and the corresponding stresses, $\sigma(\varepsilon_1)$ and $\sigma(\varepsilon_2)$. Superposition implies

$$\sigma[\varepsilon_1(t)] + \sigma[\varepsilon_2(t)] = \sigma[\varepsilon_1(t) + \varepsilon_2(t)] \tag{4.54}$$

If each stress is multiplied by a constant, the proportionality condition gives

$$\lambda_1 \sigma[\varepsilon_1(t)] + \lambda_2 \sigma[\varepsilon_2(t)] = \sigma[\lambda_1 \varepsilon_1(t) + \lambda_2 \varepsilon_2(t)] \tag{4.55}$$

If a uniaxial constant stress creep test is performed on a linear viscoelastic material, such that $\sigma = 0$ for $t < 0$ and $\sigma = \sigma_o$ for $t > 0$, the strain increases with time according to

$$\varepsilon(t) = D(t)\,\sigma_o \tag{4.56}$$

where D(t) is the creep compliance. The loading in this case can be represented more compactly as σ_o H(t), where H(t) is the Heaviside step function, defined as

$$H(t) \equiv \begin{cases} 0 & \text{for } t < 0 \\ 1 & \text{for } t > 0 \end{cases}$$

In the case of a constant uniaxial strain, i.e., $\varepsilon = \varepsilon_o H(t)$, the stress is given by

$$\sigma(t) = E(t)\, \varepsilon_o \tag{4.57}$$

where E(t) is the relaxation modulus. When ε_o is positive, the stress relaxes with time. Figure 4.19 schematically illustrates creep at a constant stress, and stress relaxation at a fixed strain.

When stress and strain both vary, the entire deformation history must be taken into account. The strain at time t is obtained by summing strain increments from earlier times. The incremental strain at time τ, where $0 < \tau < t$, that results from an incremental stress $d\sigma\, H(t - \tau)$ is given by

$$d\varepsilon(\tau) = D(t - \tau)\, d\sigma(\tau) \tag{4.58}$$

Integrating this expression with respect to time t gives

$$\varepsilon(t) = \int_0^t D(t - \tau) \frac{d\sigma(\tau)}{d\tau}\, d\tau \tag{4.59}$$

where it is assumed that $\varepsilon = \sigma = 0$ at $t = 0$. In order to allow for a discontinuous change in stress at $t = 0$, the lower integration limit is assumed to be 0^-, an infinitesimal time before $t = 0$. Relationships such as Eq. (4.59) are called *hereditary integrals* because the conditions at time t depend on prior history. The corresponding hereditary integral for stress in given by the inverse of Eq. (4.59):

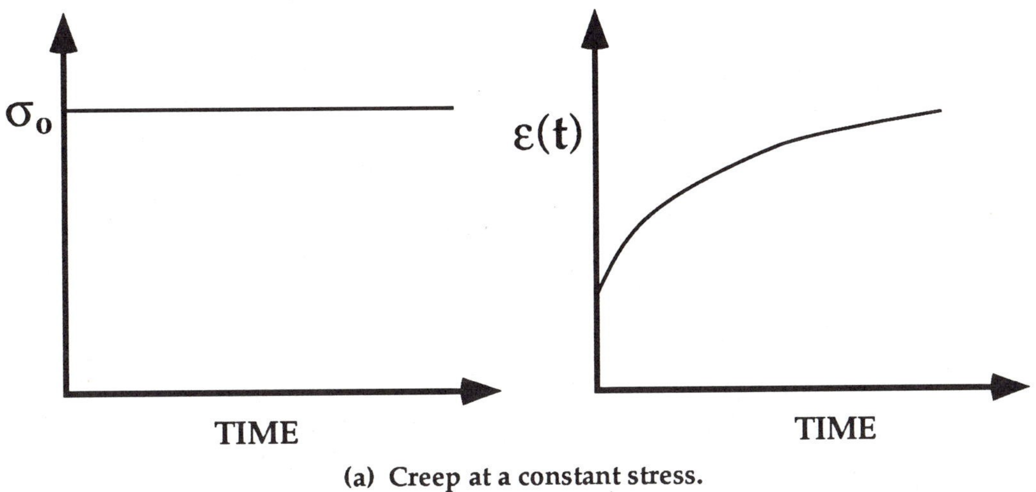

(a) Creep at a constant stress.

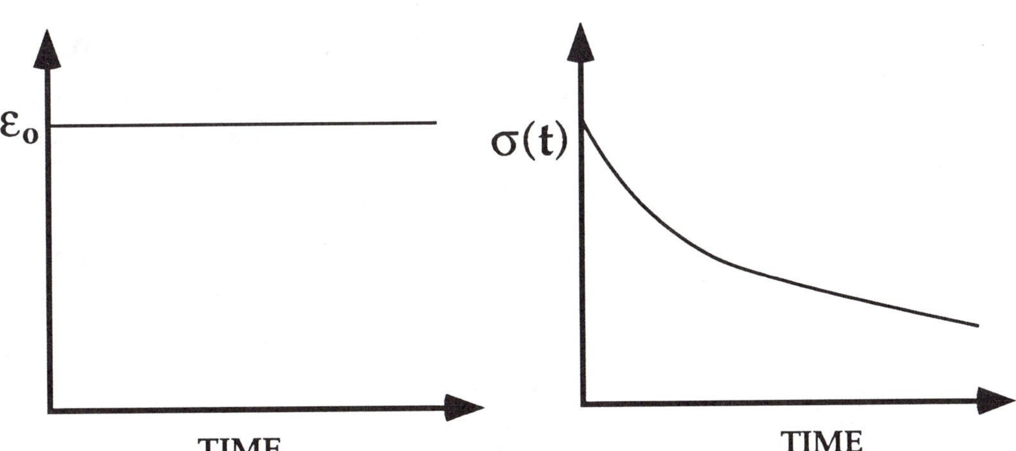

(b) Stress relaxation at a constant strain

FIGURE 4.19 Schematic uniaxial viscoelastic deformation.

$$\sigma(t) = \int_0^t E(t - \tau) \frac{d\varepsilon(\tau)}{d\tau} d\tau \tag{4.60}$$

By performing a Laplace transform on Eqs. (4.59) and (4.60), it can be shown that the creep compliance and the relaxation modulus are related as follows:

$$\int_{\tau_o}^{t} E(t-\tau)\,\frac{dD(\tau-\tau_o)}{d\tau}\,d\tau = H(t-\tau_o) \tag{4.61}$$

For deformation in three dimensions, the generalized hereditary integral for strain is given by

$$\varepsilon_{ij}(t) = \int_{0}^{t} D_{ijkl}(t-\tau)\,\frac{d\sigma_{kl}(\tau)}{d\tau}\,d\tau \tag{4.62}$$

but symmetry considerations reduce the number of independent creep compliance constants. In the case of a linear viscoelastic isotropic material, there are two independent constants, and the mechanical behavior can be described by E(t) or D(t), which are uniquely related, plus $\nu_c(t)$, the Poisson's ratio for creep.

Following an approach developed by Schapery [59], it is possible to define a pseudo elastic strain, which for uniaxial conditions is given by

$$\varepsilon^e(t) = \frac{\sigma(t)}{E_R} \tag{4.63}$$

where E_R is a reference modulus. Substituting Eq. (4.63) into Eq. (4.59) gives

$$\varepsilon(t) = E_R \int_{0}^{t} D(t-\tau)\,\frac{d\varepsilon^e(\tau)}{d\tau}\,d\tau \tag{4.64}$$

The pseudo strains in three dimensions are related to the stress tensor through Hooke's law, assuming isotropic material behavior:

$$\varepsilon_{ij}^{e} = E_R^{-1}\,[(1+\nu)\,\sigma_{ij} - \nu\,\sigma_{kk}\,\delta_{ij}] \tag{4.65}$$

where δ_{ij} is the Kronecker delta, and the standard convention of summation on repeated indices is followed. If $\nu_c = \nu$ = constant with time, it can be shown that the three-dimensional generalization of Eq. (4.64) is given by

$$\varepsilon_{ij}(t) = E_R \int_0^t D(t-\tau) \frac{d\varepsilon_{ij}^e(\tau)}{d\tau} d\tau \tag{4.66}$$

and the inverse of Eq. (4.66) is as follows.

$$\varepsilon_{ij}^e(t) = E_R^{-1} \int_0^t E(t-\tau) \frac{d\varepsilon_{ij}(\tau)}{d\tau} d\tau \tag{4.67}$$

The advantage of introducing pseudo strains is that they can be related to stresses through Hooke's law. Thus if a linear elastic solution is known for a particular geometry, it is possible to determine the corresponding linear viscoelastic solution through a hereditary integral. Given two identical configurations, one made from a linear elastic material and the other made from a linear viscoelastic material, the stresses in both bodies must be identical, and the strains are related through Eqs. (4.66) or (4.67), provided both configurations are subject to the same applied loads. This is a special case of a *correspondence principle,* which is discussed in more detail below; note the similarity to Hoff's analogy for elastic and viscous materials (Section 4.2).

4.3.2 The Viscoelastic J Integral

Constitutive Equations

Schapery [59] developed a generalized J integral that is applicable to a wide range of viscoelastic materials. He began by assuming a nonlinear viscoelastic constitutive equation in the form of a hereditary integral:

$$\varepsilon_{ij}(t) = E_R \int_0^t D(t - \tau, t) \frac{\partial \varepsilon_{ij}^e(\tau)}{\partial \tau} d\tau \tag{4.68}$$

where the lower integration limit is taken as 0^-. The pseudo elastic strain, ε_{ij}^e, is related to stress through a linear or nonlinear elastic constitutive law. The similarity between Eqs. (4.66) and (4.68) is obvious, but the latter relationship also applies to certain types of nonlinear viscoelastic behavior. The creep compliance, D(t), has a somewhat different interpretation for the nonlinear case.

The pseudo strain tensor and reference modulus in Eq. (4.68) are analogous to the linear case. In the previous section, these quantities were introduced to relate a linear viscoelastic problem to a reference elastic problem. This idea is generalized in the present case, where the nonlinear viscoelastic behavior is related to a reference nonlinear elastic problem through a correspondence principle, as discussed below.

The inverse of Eq. (4.68) is given by

$$\varepsilon_{ij}^e(t) = E_R^{-1} \int_0^t E(t - \tau, t) \frac{\partial \varepsilon_{ij}(\tau)}{\partial \tau} d\tau \tag{4.69}$$

Since hereditary integrals of the form of Eqs. (4.68) and (4.69) are used extensively in the remainder of this discussion, it is convenient to introduce an abbreviated notation:

$$\{D\, df\} \equiv E_R \int_0^t D(t - \tau, t) \frac{\partial f}{\partial \tau} d\tau \tag{4.70a}$$

and

$$\{E\, df\} \equiv E_R^{-1} \int_0^t E(t - \tau, t) \frac{\partial f}{\partial \tau} d\tau \tag{4.70b}$$

where f is a function of time. In each case, it is assumed that integration begins at 0^-. Thus Eqs. (4.68) and (4.69) become, respectively:

$$\varepsilon_{ij}(t) = \{D\, d\varepsilon_{ij}^{e}\} \qquad \text{and} \qquad \varepsilon_{ij}^{e} = \{E\, d\varepsilon_{ij}\}$$

Correspondence Principle

Consider two bodies with the same instantaneous geometry, where one material is elastic and the other is viscoelastic and is described by Eq. (4.68). Assume that at time t, a surface traction $T_i = \sigma_{ij} n_j$ is applied to both configurations along the outer boundaries. If the stresses and strains in the elastic body are σ_{ij}^e and ε_{ij}^e, respectively, while the corresponding quantities in the viscoelastic body are σ_{ij} and ε_{ij}, the stresses, strains, and displacements are related as follows [59]:

$$\sigma_{ij} = \sigma_{ij}^{e} \qquad \varepsilon_{ij} = \{D\, d\varepsilon_{ij}^{e}\} \qquad u_i = \{D\, du_i^{e}\} \tag{4.71}$$

Equation (4.71) defines a correspondence principle, introduced by Schapery [59], which allows the solution to a viscoelastic problem to be inferred from a reference elastic solution. This correspondence principle stems from the fact that the stresses in both bodies must satisfy equilibrium, and the strains must satisfy compatibility requirements in both cases. Also, the stresses are equal on the boundaries by definition:

$$T_i = \sigma_{ij} n_j = \sigma_{ij}^{e} n_j$$

Schapery [59] gives a rigorous proof of Eq. (4.71) for viscoelastic materials that satisfy Eq. (4.68).

Applications of correspondence principles in viscoelasticity, where the viscoelastic solution is related to a corresponding elastic solution, usually involve performing a Laplace transform on a hereditary integral in the form of Eq. (4.62), which contains actual stresses and strains. The introduction of pseudo quantities makes the connection between viscoelastic and elastic solutions more straightforward.

Generalized J integral

The correspondence principle in Eq. (4.71) makes it possible to define a generalized time-dependent J integral by forming an analogy with the nonlinear elastic case:

$$J_v = \int_\Gamma \left(w^e \, dy + \sigma_{ij} n_j \frac{\partial u_i^e}{\partial x} ds \right) \tag{4.72}$$

where w^e is the pseudo strain energy density:

$$w^e = \int \sigma_{ij} \, d\varepsilon_{ij}^e \tag{4.73}$$

The stresses in Eq. (4.72) are the actual values in the body, but the strains and displacements are pseudo elastic values. The actual strains and displacements are given by Eq. (4.71). Conversely, if ε_{ij} and u_i are known, J_v can be determined by computing pseudo values, which are inserted into Eq. (4.73). The pseudo strains and displacements are given by

$$\varepsilon_{ij}^e = \{E \, d\varepsilon_{ij}\} \qquad \text{and} \qquad u_i^e = \{E \, du_i\} \tag{4.74}$$

Consider a simple example, where the material exhibits steady state creep at $t > t_o$. The hereditary integrals for strain and displacement reduce to

$$\varepsilon_{ij}^e = \dot{\varepsilon}_{ij} \qquad \text{and} \qquad u_i^e = \dot{u}_i$$

By inserting the above results into Eq. (4.73), we see that $J_v = C^*$. Thus C^* is a special case of J_v. The latter parameter is capable of taking account of a wide range of time-dependent material behavior, and includes viscous creep as a special case.

Near the tip of the crack, the stresses and pseudo strains are characterized by J_v through an HRR-type relationship in the form of Eq. (4.33). The viscoelastic J can also be determined through a pseudo energy release rate:

$$J_v = -\frac{1}{B}\left(\frac{\partial}{\partial a}\int_0^{\Delta^e} P\, d\Delta^e\right)_{\Delta^e} \tag{4.75}$$

where Δ^e is the pseudo displacement in the loading direction, which is related to the actual displacement by

$$\Delta = \{D\, d\Delta^e\} \tag{4.76}$$

Finally, for Mode I loading of a linear viscoelastic material in plane strain, J_v is related to the stress intensity factor as follows:

$$J_v = \frac{K_I^2\,(1 - \nu^2)}{E_R} \tag{4.77}$$

The stress intensity factor is related to specimen geometry, applied loads, and crack dimensions through the standard equations outlined in Chapter 2.

Crack initiation and growth

When characterizing crack initiation and growth, it is useful to relate J_v to physical parameters such as CTOD and fracture work, which can be used as local failure criteria. Schapery [59] derived simplified relationships between these parameters by assuming a strip yield-type failure zone ahead of the crack tip, where a closure stress σ_m acts over ρ, as illustrated in Fig. 4.20. While material in the failure zone may be severely damaged and contain voids and other discontinuities, it is assumed that the surrounding material can be treated as a continuum. If σ_m does not vary with x, applying Eq. (3.44) gives

$$J_v = \sigma_m \delta^e \tag{4.78}$$

where δ^e is the pseudo crack tip opening displacement, which is related to the actual CTOD through a hereditary integral of the form of Eq. (4.77). Thus the CTOD is given by

$$\delta = \{D\, d(J_v/\sigma_m)\} \tag{4.79}$$

Although σ_m was assumed to be independent of x at time t, Eq. (4.79) permits σ_m to vary with time. The CTOD can be utilized as a local failure criterion: if crack initiation occurs at δ_i, the J_v at initiation can be inferred from Eq. (4.79). If δ_i is assumed to be constant, the critical J_v would, in general, depend on the strain rate. A more general version of Eq. (4.79) can be derived by allowing σ_m to vary with x.

An alternative local failure criterion is the fracture work, w_f. Equating the work input to the crack tip to the energy required to advance the crack tip by da results in the following energy balance at initiation:

$$\int_0^{\delta_i} \sigma_m \, d\delta = 2\, w_f \tag{4.80}$$

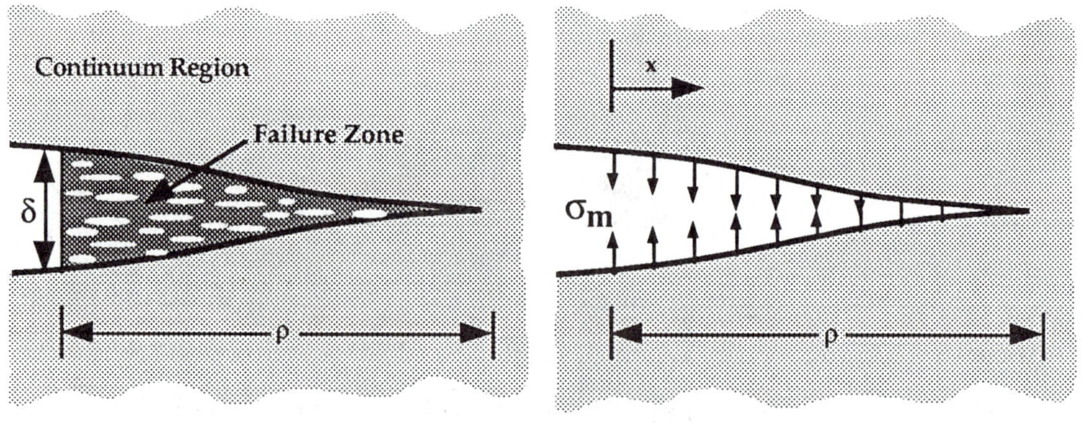

(a) Failure zone (b) Strip yield model

FIGURE 4.20 Failure zone at the crack tip in a viscoelastic material. This zone is modeled by surface tractions within $0 < x < \rho$.

assuming unit thickness and Mode I loading. This energy balance can also be written in terms of a time integral:

$$\int_0^{t_i} \sigma_m \frac{\partial \delta}{\partial t} dt = 2 w_f \tag{4.81}$$

Inserting Eq. (4.79) into Eq. (4.81) gives

$$\int_0^{t_i} \sigma_m \frac{\partial \{D \, d(J_v/\sigma_m)\}}{\partial t} dt = 2 w_f \tag{4.82}$$

If σ_m is independent of time, it cancels out of Eq. (4.82), which then simplifies to

$$E_R \int_0^{t_i} D(t_i - \tau, t_i) \frac{\partial J_v}{\partial \tau} d\tau = 2 w_f \tag{4.83}$$

For an elastic material, $D = E_R^{-1}$, and $J_v = 2 w_f$. If the failure zone is viscoelastic and the surrounding continuum is elastic, J_v may vary with time. If the surrounding continuum is viscous, $D = (t_v E_R)^{-1} (t - \tau)$, where t_v is a constant with units of time. Inserting this latter result into Eq. (4.83) and integrating by parts gives

$$t_v^{-1} \int_0^{t_i} J_v \, dt = 2 w_f \tag{4.84}$$

4.3.3 Transition from Linear to Nonlinear Behavior

Typical polymers are linear viscoelastic at low stresses and nonlinear at high stresses. A specimen that contains a crack may have a zone of nonlinearity at the crack tip, analogous to a plastic zone, that is surrounded by linear viscoelastic material. The approach described in the previous section applies only when one type of behavior (linear or nonlinear) dominates.

Schapery [61] has recently modified the J_v concept to cover the transition from small stress to large stress behavior. He introduced a modified constitutive equation, where strain is given by the sum of two hereditary integrals: one corresponding to linear viscoelastic strains and the other describing nonlinear strains. For the latter term, he assumed power-law viscoelasticity. For the case of uniaxial constant tensile stress, σ_o, the creep strain in this modified model is given by

$$\varepsilon(t) = E_R\, D(t) \left(\frac{\sigma_o}{\sigma_{ref}}\right)^n + D_L(t)\,\sigma_o \tag{4.85}$$

where D and D_L are the nonlinear and linear creep compliance, respectively, and σ_{ref} is a reference stress.

At low stresses and short times, the second term in Eq. (4.85) dominates, while the nonlinear term dominates at high stresses or long times. In the case of a viscoelastic body with a stationary crack at a fixed load, the nonlinear zone is initially small but normally increases with time, until the behavior is predominantly nonlinear. Thus there is a direct analogy between the present case and the transition from elastic to viscous behavior described in Section 4.2.

Close to the crack tip, but outside of the failure zone, the stresses are related to a pseudo strain through a power law:

$$\varepsilon^e = \left(\frac{\sigma_o}{\sigma_{ref}}\right)^n \tag{4.86}$$

In the region dominated by Eq. (4.86), the stresses are characterized by J_v, regardless of whether the global behavior is linear or nonlinear:

$$\sigma_{ij} = \sigma_{ref} \left(\frac{J_v}{\sigma_{ref}\, I_n\, r}\right)^{\frac{1}{n+1}} \tilde{\sigma}_{ij}(n,\theta) \tag{4.87}$$

If the global behavior is linear, there is a second singularity further away from the crack tip:

$$\sigma_{ij} = \frac{K_I}{\sqrt{2\pi r}} f_{ij}(\theta) \tag{4.88}$$

Let us define a pseudo strain tensor that, when inserted into the path-independent integral of Eq. (4.72), yields a value J_L. Also suppose that this pseudo strain tensor is related to the stress tensor by means of linear and power law pseudo complementary strain energy density functions (w_{cl} and w_{cn}, respectively):

$$\varepsilon_{ij}^{eL} = \frac{\partial}{\partial \sigma_{ij}} (f\, w_{cn} + w_{cl}) \tag{4.89}$$

where f(t) is an as yet unspecified aging function, and the complementary strain energy density is defined by

$$w_c = \int \varepsilon_{ij}^{eL}\, d\sigma_{ij}$$

For uniaxial deformation, Eq. (4.89) reduces to

$$\varepsilon^{eL} = f\left(\frac{\sigma}{\sigma_{ref}}\right)^n + \frac{\sigma}{E_R} \tag{4.90}$$

Comparing Eqs (4.85) and (4.90), it can be seen that

$$f = \frac{D(t)}{D_L(t)} \qquad \text{if} \qquad \varepsilon^{eL} \equiv \frac{\varepsilon(t)}{E_R\, D_L(t)} \qquad \text{for constant stress creep.}$$

The latter relationship for pseudo strain agrees with the conventional definition in the limit of linear behavior.

Let us now consider the case where the inner and outer singularities, Eqs. (4.87) and (4.88), exist simultaneously. For the outer singularity, second term in Eq. (4.90) dominates, the stresses are given by Eq. (4.88), and J_L is related to K_I as follows:

$$J_L = \frac{K_I^2 (1 - \nu^2)}{E_R} \tag{4.91}$$

Closer to the crack tip, the stresses are characterized by J_v through Eq. (4.87), but J_L is not necessarily equal to J_v, because f appears in the first term of the modified constitutive relationship (Eq. (4.90)), but not in Eq. (4.86). These two definitions of J coincide if σ_{ref} in Eq. (4.90) is replaced with $\sigma_{ref}\, f^{\,1/n}$. Thus, the near-tip singularity in terms of J_L is given by

$$\sigma_{ij} = \sigma_{ref} \left(\frac{J_L}{f\, \sigma_{ref}\, I_n\, r} \right)^{\frac{1}{n+1}} \tilde{\sigma}_{ij}(n,\theta) \tag{4.92}$$

Therefore,

$$J_v = \frac{J_L}{f} \tag{4.93}$$

Schapery showed that f = 1 in the limit of purely linear behavior; thus J_L is the limiting value of J_v when the nonlinear zone is negligible. The function f is indicative of the extent of nonlinearity. In most cases, f increases with time, until J_v reaches J_n, the limiting value when the specimen is dominated by nonlinear viscoelasticity. Schapery also confirmed that

$$f = \frac{D(t)}{D_L(t)} \tag{4.94}$$

for small scale nonlinearity. Equations (4.93) and (4.94) provides a reasonable description of the transition to nonlinear behavior. Schapery defined a transition time by setting $J_v = J_n$ in Eq. (4.93):

$$J_n = \frac{J_L}{f(t_\tau)} \tag{4.95a}$$

or

$$t_\tau = f^{-1}\left(\frac{J_L}{J_n}\right) \tag{4.95b}$$

For the special case of linear behavior followed by viscous creep, Eq. (4.95b) becomes

$$t_\tau = \frac{J}{(n + 1)\, C^*} \tag{4.96}$$

which is identical to the transition time defined by Riedel and Rice [50].

REFERENCES

1. Freund, L.B., "Dynamic Crack Propagation." *The Mechanics of Fracture*, American Society of Mechanical Engineers, New York, 1976, pp. 105-134.

2. Freund, L.B., "Crack Propagation in an Elastic Solid Subjected to General Loading--I. Constant Rate of Extension." *Journal of the Mechanics and Physics of Solids*, Vol. 20, 1972, pp. 129-140.

3. Freund, L.B., "Crack Propagation in an Elastic Solid Subjected to General Loading--II. Non-Uniform Rate of Extension." *Journal of the Mechanics and Physics of Solids*, Vol. 20, 1972, pp. 141-152.

4. Freund, L.B., "Crack Propagation in an Elastic Solid Subjected to General Loading--III. Stress Wave Loading." *Journal of the Mechanics and Physics of Solids*, Vol. 21, 1973, pp. 47-61.

5. Freund, L.B., *Dynamic Fracture Mechanics*, Cambridge University Press, Cambridge, UK, 1990.

6. Kanninen, M.F. and Poplar C.H., *Advanced Fracture Mechanics*, Oxford University Press, New York, Oxford, 1985.

7. Rose, L.R.F., "Recent Theoretical and Experimental Results on Fast Brittle Fracture." *International Journal of Fracture*, Vol. 12, 1976, pp. 799-813.

8. Atkinson, C. and Eshlby, J.D., "The Flow of Energy into the Tip of a Moving Crack." *International Journal of Fracture Mechanics*, Vol. 4, 1968, pp. 3-8.

9. Sih, G.C., "Dynamic Aspects of Crack Propagation." *Inelastic Behavior of Solids*, McGraw-Hill, New York, 1970, pp. 607-633.

10. Freund. L.B., "Energy Flux into the Tip of an Extending Crack in an Elastic Solid." *Journal of Elasticity*, Vol. 2, 1972, pp. 341-349.

11. Moran, B. and Shih, C.F., "A General Treatment of Crack Tip Contour Integrals." *International Journal of Fracture*, Vol. 35, 1987, pp. 295-310.

12. Atluri, S.N., "Path-Independent Integrals in Finite Elasticity and Inelasticity, with Body Forces, Inertia, and Arbitrary Crack Face Conditions." *Engineering Fracture Mechanics*, Vol. 16, 1982, pp. 341-369.

13. Kishimoto, K., Aoki, S., and Sakata, M., "On the Path-Independent J Integral." *Engineering Fracture Mechanics,* Vol. 13, 1980, pp. 841-850.

14. E 23-88, "Standard Test Methods for Notched Bar Impact Testing of Metallic Materials." American Society for Testing and Materials, Philadelphia, 1988.

15. Joyce J.A. and Hacket, E.M., "Dynamic J-R Curve Testing of a High Strength Steel Using the Multispecimen and Key Curve Techniques." ASTM STP 905, American Society of Testing and Materials, Philadelphia, 1984, pp. 741-774.

16. Joyce J.A. and Hacket, E.M., "An Advanced Procedure for J-R Curve Testing Using a Drop Tower." ASTM STP 995, American Society of Testing and Materials, Philadelphia, 1989, 298-317.

17. Duffy, J and Shih, C.F., "Dynamic Fracture Toughness Measurement Methods for Brittle and Ductile Materials." *Advances in Fracture Research: Seventh International Conference on Fracture,* Pergamon Press, Oxford, 1989, pp. 633-642.

18. Nakamura, T., Shih, C.F. and Freund, L.B., "Analysis of a Dynamically Loaded Three-Point-Bend Ductile Fracture Specimen." *Engineering Fracture Mechanics,* Vol. 25, 1986, pp. 323-339.

19. Nakamura, T., Shih, C.F. and Freund, L.B., "Three-Dimensional Transient Analysis of a Dynamically Loaded Three-Point-Bend Ductile Fracture Specimen." ASTM STP 995, Vol. I, American Society of Testing and Materials, Philadelphia, 1989, pp. 217-241.

20. Rice, J.R., Paris, P.C., and Merkle, J.G., "Some Further Results of J-Integral Analysis and Estimates." ASTM STP 536, American Society of Testing and Materials, Philadelphia, 1973, pp. 231-245.

21. Barsom, J.M., "Development of the AASHTO Fracture Toughness Requirements for Bridge Steels." *Engineering Fracture Mechanics,* Vol. 7, 1975, pp. 605-618.

22. Mott, N.F., "Fracture of Metals: Theoretical Considerations." *Engineering,* Vol. 165, 1948, pp. 16-18.

23. Yoffe, E.H., "The Moving Griffith Crack." *Philosophical Magazine,* Vol. 42, 1951, pp. 739-750.

24. Broberg, K.B., "The Propagation of a Brittle Crack." *Arkvik for Fysik,* Vol 18, 1960, pp. 159-192

25. Craggs, J.W., "On the Propagation of a Crack in an Elastic-Brittle Material." *Journal of the Mechanics and Physics of Solids,* Vol. 8, 1960, pp. 66-75.

26. Stroh, A.N., "A Simple Model of a Propagating Crack." *Journal of the Mechanics and Physics of Solids.,* Vol. 8, 1960, pp. 119-122.

27. Dulaney, E.N. and Brace, W.F., "Velocity Behavior of a Growing Crack." *Journal of Applied Physics*, Vol. 31, 1960, pp. 2233-2236.

28. Berry, J.P., "Some Kinetic Considerations of the Griffith Criterion for Fracture." *Journal of the Mechanics and Physics of Solids.*, Vol. 8, 1960, pp. 194-216.

29. Roberts, D.K. and Wells, A.A., "The Velocity of Brittle Fracture." *Engineering*, Vol. 178, 1954, pp. 820-821.

30. Bluhm, J.I., "Fracture Arrest." *Fracture: An Advanced Treatise*, Vol. V, Academic Press, New York, 1969.

31. Rice, J.R., "Mathematical Analysis in the Mechanics of Fracture." *Fracture: An Advanced Treatise*, Vol. II, Academic Press, New York, 1968, p. 191.

32. Freund, L.B. and Clifton, R.J., "On the Uniqueness of Plane Elastodynamic Solutions for Running Cracks." *Journal of Elasticity*, Vol. 4, 1974, pp. 293-299.

33. Nillson, F. "A Note on the Stress Singularity at a Non-Uniformly Moving Crack Tip." *Journal of Elasticity*, Vol. 4, 1974, pp. 293-299.

34. Rose, L.R.F., "An Approximate (Wiener-Hopf) Kernel for Dynamic Crack Problems in Linear Elasticity and Viscoelasticity." *Proceedings, Royal Society of London*, Vol. A-349, 1976, pp. 497-521.

35. Sih, G.C., "Some Elastodynamic Problems of Cracks." *International Journal of Fracture Mechanics*, Vol 4., 1968, p. 51-68.

36. Sanford, R.J. and Dally, J.W., "A General Method for Determining Mixed-Mode Stress Intensity Factors from isochromatic Fringe Patterns." *Engineering Fracture Mechanics*, Vol. 11, 1979, pp. 621-633.

37. Chona, R. , Irwin, G.R., and Shukla, A., "Two and Three Parameter Representation of Crack Tip Stress Fields." *Journal of Strain Analysis*, Vol. 17, 1982, pp. 79-86.

38. Kalthoff, J.F., Beinart, J., Winkler, S., and Klemm, W., "Experimental Analysis of Dynamic Effects in Different Crack Arrest Test Specimens." ASTM STP 711, American Society for Testing and Materials, Philadelphia, 1980, pp. 109-127.

39. Rosakis, A.J. and Freund L.B., "Optical Measurement of the Plane Strain Concentration at a Crack Tip in a Ductile Steel Plate." *Journal of Engineering Materials Technology*, Vol. 104, 1982, pp. 115-120.

40. Freund, L.B. and Douglas, A.S., "The Influence of Inertia on Elastic-Plastic Antiplane Shear Crack Growth." *Journal of the Mechanics and Physics of Solids*, Vol. 30, 1982, pp. 59-74.

41. Freund, L.B., "Results on the Influence of Crack-Tip Plasticity During Dynamic Crack Growth." ASTM STP 1020, American Society for Testing and Materials, Philadelphia, 1989, pp. 84-97.

42. Corwin, W.R., "Heavy Section Steel Technology Program Semiannual Progress Report for April-September 1987", U.S. Nuclear Regulatory Commission Report NUREG/CR-4219, Vol. 4, No. 2, October, 1987.

43. Kalthoff, J.F., Beinart, J. and Winkler, S. "Measurement of Dynamic Stress Intensity Factors for Fast Running and Arresting Cracks in Double-Cantilever Beam Specimens." ASTM STP 627, American Society for Testing and Materials, Philadelphia, 1977, pp. 161-176.

44. Kobayashi, A.S., Seo, K.K., Jou, J.Y, and Urabe, Y. "A Dynamic Analysis of Modified Compact Tension Specimens Using Homolite-100 and Polycarbonate Plates." *Experimental Mechanics*. Vol. 20, 1980, pp. 73-79.

45. Landes, J.D. and Begley, J.A., "A Fracture Mechanics Approach to Creep Crack Growth." ASTM STP, 590, American Society for Testing and Materials, Philadelphia, 1976, pp. 128-148.

46. Ohji, K., Ogura, K., and Kubo, S., *Transactions, Japanese Society of Mechanical Engineers*, Vol. 42, 1976, pp. 350-358.

47. Nikbin, K.M., Webster, G.A., and Turner, C.E., ASTM STP 601, American Society for Testing and Materials, Philadelphia, 1976, pp. 47-62.

48. Hoff, N.J., "Approximate Analysis of Structures in the Presence of Moderately Large Creep Deformations." *Quarterly of Applied Mathematics*, Vol. 12, 1954, pp. 49-55.

49. Riedel, H., "Creep Crack Growth." ASTM STP 1020, American Society for Testing and Materials, Philadelphia, 1989, pp. 101-126.

50. Riedel, H. and Rice, J.R., "Tensile Cracks in Creeping Solids." ASTM STP 700, American Society for Testing and Materials, Philadelphia, 1980, pp. 112-130.

51. Ehlers, R. and Riedel, H. "A Finite Element Analysis of Creep Deformation in a Specimen Containing a Macroscopic Crack." *Proceedings, 5th International Conference on Fracture*, Pergamon Press, Oxford, 1981, pp. 691-698.

52. Saxena, A., "Creep Crack Growth under Non-Steady-State Conditions." ASTM STP 905, American Society for Testing and Materials, Philadelphia, 1986, pp. 185-201.

53. Bassani, J.L., Hawk, D.E., and Saxena, A., "Evaluation of the C_t Parameter for Characterizing Creep Crack Growth Rate in the Transient Regime." ASTM STP 995, Vol. I, American Society for Testing and Materials, Philadelphia, 1990, pp. 112-130.

54. Riedel, H., "Creep Deformation at Crack Tips in Elastic-Viscoplastic Solids." *Journal of the Mechanics and Physics of Solids*, Vol 29, 1981, pp. 35-49.

55. Chun-Pok, L. and McDowell, D.L., "Inclusion of Primary Creep in the Estimation of the C_t Parameter." *International Journal of Fracture,* Vol. 46, 1990, pp. 81-104.

56. Schapery, R.A. "A Theory of Crack Initiation and Growth in Viscoelastic Media--I. Theoretical Development." *International Journal of Fracture,* Vol 11, 1975, pp. 141-159.

57. Schapery, R.A. "A Theory of Crack Initiation and Growth in Viscoelastic Media--II. Approximate Methods of Analysis." *International Journal of Fracture,* Vol 11, 1975, pp. 369-388.

58. Schapery, R.A. "A Theory of Crack Initiation and Growth in Viscoelastic Media--III. Analysis of Continuous Growth." *International Journal of Fracture,* Vol 11, 1975, pp. 549-562.

59. Schapery, R.A., "Correspondence Principles and a Generalized J Integral for Large Deformation and Fracture Analysis of Viscoelastic Media." *International Journal of Fracture,* Vol. 25, 1984, pp. 195-223.

60. Schapery, R.A., "Time-Dependent Fracture: Continuum Aspects of Crack Growth." *Encyclopedia of Materials Science and Engineering,* Pergamon Press, Oxford, 1986, pp. 5043-5054.

61. Schapery, R.A., "On Some Path Independent Integrals and Their Use in Fracture of Nonlinear Viscoelastic Media." *International Journal of Fracture,* Vol. 42, 1990, pp. 189-207.

62. Irwin, G.R., "Constant Speed Semi-Infinite Tensile Crack Opened by a Line Force." Lehigh University Memorandum, 1967.

APPENDIX 4: DYNAMIC FRACTURE ANALYSIS

(Selected Results)

A4.1 ELASTODYNAMIC CRACK TIP FIELDS

Rice [31], Sih [35], and Irwin [62] each derived expressions for the stresses ahead of a crack propagating at a constant speed. They found that the moving crack retained the $1/\sqrt{r}$ singularity, but that the angular dependence of the stresses, strains and displacements depends on crack speed. Freund and Clifton [32] and Nilsson [33] later showed that the solution for a constant speed crack was valid in general; the near-tip quantities depend only on instantaneous crack speed. The following derivation presents the more general case, where the crack speed is allowed to vary.

For dynamic problems, the equations of equilibrium are replaced by the equations of motion, which, in the absence of body forces, are given by

$$\frac{\partial \sigma_{ji}}{\partial x_j} = \rho\, \ddot{u}_i \tag{A4.1}$$

where x_j denotes the orthogonal coordinates and each dot indicates a time derivative. For quasistatic problems, the term on the right side of Eq. (A4.1) vanishes. For a linear elastic material, it is possible to write the equations of motion in terms of displacements and elastic constants by invoking the strain-displacement and stress-strain relationships:

$$\mu \frac{\partial^2 u_i}{\partial x_j^2} + (\lambda + \mu) \frac{\partial^2 u_j}{\partial x_i \partial x_j} = \rho\, \ddot{u}_i \tag{A4.2}$$

where μ and λ are the Lame constants; μ is the shear modulus and

$$\lambda = \frac{2\mu\nu}{1 - 2\nu}$$

Consider rapid crack propagation in a body subject to plane strain loading. Let us define a fixed coordinate axis, X-Y, with an origin on the crack plane at a(t) = 0, as illustrated in Fig. A4.1. It is convenient at this point to introduce two displacement potentials, defined by

$$u_X = \frac{\partial \psi_1}{\partial X} + \frac{\partial \psi_2}{\partial Y} \qquad u_Y = \frac{\partial \psi_1}{\partial Y} - \frac{\partial \psi_2}{\partial X} \tag{A4.3}$$

Substituting Eq. (A4.3) into Eq. (A4.2) leads to

$$\frac{\partial^2 \psi_1}{\partial X^2} + \frac{\partial^2 \psi_1}{\partial Y^2} = \frac{1}{c_1^2}\ddot{\psi}_1$$

and (A4.4)

$$\frac{\partial^2 \psi_2}{\partial X^2} + \frac{\partial^2 \psi_2}{\partial Y^2} = \frac{1}{c_2^2}\ddot{\psi}_2$$

since the wave speeds are given by

$$c_1^2 = \frac{\lambda + \mu}{\rho} \qquad c_2^2 = \frac{\mu}{\rho}$$

for plane strain. Thus ψ_1 and ψ_2 are the longitudinal and shear wave potentials, respectively. The stresses can be written in terms of ψ_1 and ψ_2 by invoking Eqs. (A2.1) and (A2.2):

$$\sigma_{XX} + \sigma_{YY} = 2(\lambda + \mu)\left(\frac{\partial^2 \psi_1}{\partial X^2} + \frac{\partial^2 \psi_1}{\partial Y^2}\right)$$

$$\sigma_{XX} - \sigma_{YY} = 2\,\mu\left(\frac{\partial^2 \psi_1}{\partial X^2} - \frac{\partial^2 \psi_1}{\partial Y^2} + 2\frac{\partial^2 \psi_2}{\partial X \partial Y}\right) \tag{A4.5}$$

$$\tau_{XY} = \mu\left(\frac{\partial^2 \psi_2}{\partial Y^2} - \frac{\partial^2 \psi_2}{\partial X^2} + 2\,\frac{\partial^2 \psi_1}{\partial X \partial Y}\right)$$

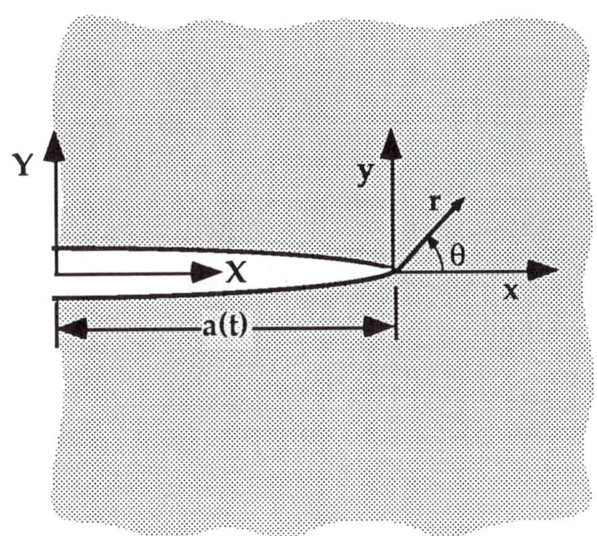

FIGURE A4.1 Definition of coordinate axes for a rapidly propagating crack. The X-Y axes are fixed in space and the x-y axes are attached to the crack tip.

Let us now introduce a moving coordinate system, x-y, attached to the crack tip, where x = X - a(t) and y = Y. The rate of change of each wave potential can be written as

$$\frac{d\psi_i}{dt} = \frac{\partial\psi_i}{\partial t} - V\frac{\partial\psi_i}{\partial x} \quad (i = 1,2) \tag{A4.6}$$

where V (= - dx/dt) is the crack speed. Differentiating Eq. (4.6) with respect to time gives

$$\ddot{\psi}_i = V^2\frac{\partial^2\psi_i}{\partial x^2} - 2V\frac{\partial^2\psi_i}{\partial x\,\partial t} + \frac{\partial^2\psi_i}{\partial t^2} - \dot{V}\frac{\partial\psi_i}{\partial x} \tag{A4.7}$$

According to Eq. (A4.5) the first term on the right-hand side of Eq. (4.7) is proportional to the stress tensor. This term should dominate close to the crack tip, assuming there is a stress singularity. Substituting the first term of Eq. (A4.7) into Eq. (A4.4) leads to

$$\beta_1{}^2\frac{\partial^2\psi_1}{\partial x^2} + \frac{\partial^2\psi_1}{\partial y^2} = 0$$

and (A4.8)

$$\beta_2^2 \frac{\partial^2 \psi_2}{\partial x^2} + \frac{\partial^2 \psi_2}{\partial y^2} = 0$$

where

$$\beta_1^2 = 1 - \left(\frac{V}{c_1}\right)^2 \quad \text{and} \quad \beta_2^2 = 1 - \left(\frac{V}{c_2}\right)^2$$

Note that the governing equations depend only on instantaneous crack speed; the term that contains crack acceleration in Eq. (A4.7) is negligible near the crack tip.

If we scale y by defining new coordinates, $y_1 = \beta_1 y$ and $y_2 = \beta_2 y$, Eq. (A4.8) becomes the Laplace equation. Freund and Clifton [32] applied a complex variable method to solve Eq. (A4.8). The general solutions to the wave potentials are as follows:

$$\psi_1 = \text{Re}\,[F(z_1)]$$

and (A4.9)

$$\psi_2 = \text{Im}\,[G(z_2)]$$

where F and G are as yet unspecified complex functions, $z_1 = x + iy_1$, and $z_2 = x + iy_2$.

The boundary conditions are the same as for a stationary crack: $\sigma_{yy} = \tau_{xy} = 0$ on the crack surfaces. Freund and Clifton showed that these boundary conditions can be expressed in terms of second derivatives for F and G at y = 0 and x < 0:

$$(1 + \beta_2^2)\,[F''(x)_+ + F''(x)_-] + 2\,\beta_2\,[G''(x)_+ + G''(x)_-] = 0$$

(A4.10)

$$2\,\beta_1\,[F''(x)_+ - F''(x)_-] + (1 + \beta_1^2)\,[G''(x)_+ - G''(x)_-] = 0$$

where the subscripts + and - correspond to the upper and lower crack surfaces, respectively. The following functions satisfy the boundary conditions and lead to integrable strain energy density and finite displacement at the crack tip:

$$F''(z_1) = \frac{C}{\sqrt{z_1}} \qquad G''(z_2) = \frac{-2\,\beta_2\,C}{(1 + \beta_2^2)\sqrt{z_2}} \qquad \text{(A4.11)}$$

where C is a constant. Making the substitution $z_1 = r_1 e^{i\theta_1}$ and $z_2 = r_2 e^{i\theta_2}$ leads to the following expressions for the Mode I crack tip stress fields:

$$\sigma_{xx} = \frac{K_I(t)}{\sqrt{2\pi r}} \frac{1+\beta_2^2}{D(t)} \left[(1 + 2\beta_1^2 - \beta_2^2) \cos\left(\frac{\theta_1}{2}\right)\sqrt{\frac{r}{r_1}} - \frac{4\beta_1\beta_2}{1+\beta_2^2} \cos\left(\frac{\theta_2}{2}\right)\sqrt{\frac{r}{r_2}} \right]$$

$$\sigma_{yy} = \frac{K_I(t)}{\sqrt{2\pi r}} \frac{1+\beta_2^2}{D(t)} \left[-(1 + \beta_2^2) \cos\left(\frac{\theta_1}{2}\right)\sqrt{\frac{r}{r_1}} + \frac{4\beta_1\beta_2}{1+\beta_2^2} \cos\left(\frac{\theta_2}{2}\right)\sqrt{\frac{r}{r_2}} \right]$$

$$\tau_{xy} = \frac{K_I(t)}{\sqrt{2\pi r}} \frac{2\beta_1(1+\beta_2^2)}{D(t)} \left[\sin\left(\frac{\theta_1}{2}\right)\sqrt{\frac{r}{r_1}} - \sin\left(\frac{\theta_2}{2}\right)\sqrt{\frac{r}{r_2}} \right] \quad \text{(A4.12)}$$

where

$$D(t) = 4\beta_1\beta_2 - (1+\beta_2^2)^2$$

Equation (4.12) reduces to the quasistatic relationship (Table 2.1) when $V = 0$.

Craggs [25] and Freund [10] obtained the following relationship between $K_I(t)$ and energy release rate for crack propagation at a constant speed:

$$\mathcal{G} = A(V)\frac{K_I^2(1-\nu^2)}{E} \quad \text{(A4.13)}$$

for plane strain, where

$$A(V) = \frac{V^2\beta_1}{(1-\nu)c_2^2 D(t)}$$

It can be shown that

$$\lim_{V \to 0} A = 1$$

and Eq. (A4.13) reduces to the quasistatic result. Equation (A4.13) can be derived by substituting the dynamic crack tip solution (Eq. (4.12) and the corresponding relationships for strain and displacement) into the generalized contour integral given by Eq. (4.26).

The derivation that led to Eq. (A4.12) implies that Eq. (A4.13) is a general relationship that applies to accelerating cracks as well as constant speed cracks.

A4.2 DERIVATION OF THE GENERALIZED ENERGY RELEASE RATE

Equation (4.26) will now be derived. The approach closely follows that of Moran and Shih [11], who applied a general balance law to derive a variety of contour integrals, including the energy release. Other authors [8-10] have derived equivalent expressions using slightly different approaches.

Beginning with the equation of motion, Eq. (A4.1), taking an inner product of both sides with displacement rate, $\dot{u}_i$, and rearranging gives

$$\frac{\partial(\sigma_{ji}\,\dot{u}_i)}{\partial x_j} = \rho\,\ddot{u}_i + \sigma_{ji}\frac{\partial(\dot{u}_i)}{\partial x_j}$$

$$= \dot{T} + \dot{w} \tag{A4.14}$$

where T and w are the kinetic energy and stress work densities, respectively, as defined in Eqs. (4.27) to (4.29). Equation (A4.14) is a general balance law that applies to all material behavior. Integrating this relationship over an arbitrary volume, and applying the divergence and transport theorems gives

$$\int_{\partial \mathcal{V}} \sigma_{ji}\, \dot{u}_i\, m_j\, dS = \frac{d}{dt} \int_{\mathcal{V}} (w + T)\, d\mathcal{V} - \int_{\partial \mathcal{V}} (w + T)\, V_j\, m_j\, dS \qquad \text{(A4.15)}$$

where $\mathcal{V}$ is volume, m_j is the outward normal to the surface $\partial \mathcal{V}$, and V_i is the instantaneous velocity of $\partial \mathcal{V}$.

Consider now the special case of a crack in a two-dimensional body, where the crack is propagating along the x axis and the origin is attached to the crack tip. (Fig. A4.2). Let us define a contour, C_o, fixed in space, that contains the propagating crack and bounds the area $\mathcal{A}$. The crack tip is surrounded by a small contour, Γ, that is fixed in size and moves with the crack. The balance law in Eq. (A4.15) becomes

$$\int_{C_o} \sigma_{ji}\, \dot{u}_i\, m_j\, dC = \frac{d}{dt} \int_{\mathcal{A}} (w + T)\, d\mathcal{A} - \int_{\Gamma} \left[(w + T)\, V\, \delta_{1j} + \sigma_{ji}\, \dot{u}_i\right] m_j\, d\Gamma \qquad \text{(A4.16)}$$

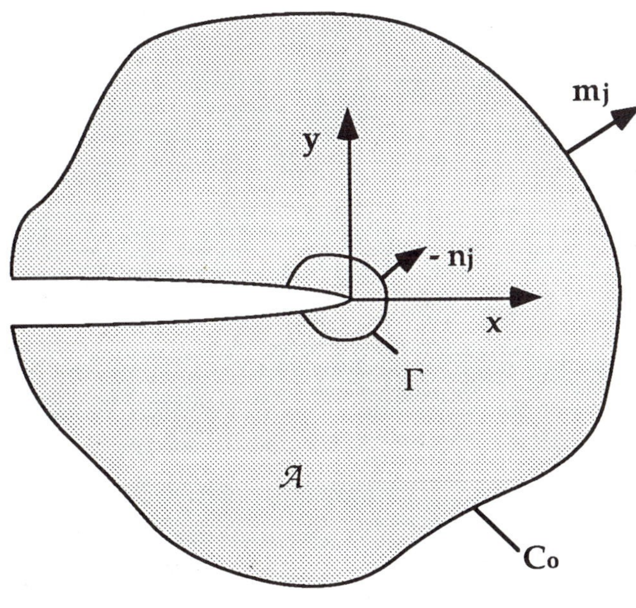

FIGURE A4.2 Conventions for the energy balance for a propagating crack. The outer contour, C_o, is fixed in space, and the inner contour, Γ, and the x-y axes are attached to the moving crack tip.

where V is the crack speed. The integral on the left side of Eq. (A4.16) is the rate at which energy is input into the body. The first term on the right side of this relationship is the rate of increase in internal energy in the body. Consequently, the second integral on the right side of Eq. (A4.16) corresponds to the rate at which energy is lost from the body due to flux through Γ. By defining $n_j = -m_j$ on Γ, we obtain the following expression for the energy flux into Γ:

$$F(\Gamma) = \int_{\Gamma} \left[(w + T)\, V\, \delta_{1j} + \sigma_{ji}\, \dot{u}_i \right] n_j \, d\Gamma \tag{A4.17}$$

In the limit of a vanishingly small contour, the flux is independent of the shape of Γ. Thus the energy flux to the crack tip is given by

$$\mathcal{F} = \lim_{\Gamma \to 0} \int_{\Gamma} \left[(w + T)\, V\, \delta_{1j} + \sigma_{ji}\, \dot{u}_i \right] n_j \, d\Gamma \tag{A4.18}$$

In an increment of time dt, the crack extends by da = V dt and the energy expended is $\mathcal{F}$ dt. Thus the energy release rate is given by

$$J = \frac{\mathcal{F}}{V} \tag{A4.19}$$

Substituting Eq. (A4.18) into Eq. (A4.19) will yield a generalized expression for the J integral. First, however, we must express displacement rate in terms of crack speed. By analogy with Eq. (A4.6), displacement rate can be written as

$$\dot{u}_i = -V \frac{\partial u_i}{\partial x} + \frac{\partial u_i}{\partial t} \tag{A4.20}$$

Under steady state conditions, the second term in Eq. (A4.20) vanishes; the displacement at a fixed distance from the propagating crack tip remains constant. Close to the crack tip, displacement changes rapidly with position (at a fixed time) and the first term in Eq. (A4.20) dominates in all cases. Thus the J integral is given by

$$J = \lim_{\Gamma \to 0} \int_{\Gamma} \left[(w + T)\, \delta_{1j} - \sigma_{ji} \frac{\partial u_i}{\partial x} \right] n_j \, d\Gamma$$

$$= \lim_{\Gamma \to 0} \int_{\Gamma} \left[(w + T)\, dy - \sigma_{ji}\, n_j \frac{\partial u_i}{\partial x} \, d\Gamma \right] \tag{A4.21}$$

Equation (A4.21) applies to all types of material response (e.g. elastic, plastic, viscoplastic, and viscoelastic behavior), because it was derived from a generalized energy balance.[3] In the special case of an elastic material (linear or nonlinear), w is the strain energy density, which displays the properties of an elastic potential:

$$\sigma_{ij} = \frac{\partial w}{\partial \varepsilon_{ij}} \tag{A4.22}$$

Recall from Appendix 3 that Eq. (A4.22) is necessary to demonstrate path independence of J in the quasistatic case. In general, Eq. (A4.21) is not path independent except in local region near the crack tip. For an elastic material, however, J is path independent in the dynamic case when the crack propagation is steady state ($\partial u_i/\partial t = 0$) [8].

Although Eq. (A4.22) is, in principle, applicable to all types of material response, special care must be taken when J is evaluated for a growing crack. Figure A4.3(a) illustrates a growing crack under small scale yielding conditions. A small plastic zone (or process zone) is embedded within an elastic singularity zone. The plastic zone leaves behind a wake as it sweeps through the material. Unrecoverable work is performed on material inside the plastic wake, as Fig. A4.3(b)

---

[3]Since the divergence and transport theorems were invoked, there is an inherent assumption that the material behaves as a continuum with smoothly varying displacement fields.

illustrates. The work necessary to form the plastic wake comes from the energy flux into the contour Γ. In an ideally elastic body, the energy flux is released from the body through the crack tip, but in an elastic-plastic material, the majority of this energy is dissipated in the wake.

Recall the modified Griffith model (Section 2.3.2), where the work required to increase the crack area a unit amount is equal to $2(\gamma_s + \gamma_p)$, where γ_s is the surface energy and γ_p is the plastic work. The latter term corresponds to the energy dissipated in the plastic wake (Fig. 2.6(b)).

The energy release rate computed from Eq. (A4.21) must therefore be interpreted as the energy flow to the plastic zone and plastic wake, rather than to the crack tip. That is, Γ cannot shrink to zero; rather, the contour must have a small, but finite radius. The J integral is path-independent as long as Γ is defined within the elastic singularity zone, but J becomes path-dependent when the contour is taken inside the plastic zone. In the limit as Γ shrinks to the crack tip, the computed energy release rate would approach zero (in a continuum analysis), since the calculation would exclude the work dissipated by the plastic wake. The actual energy flow to the crack tip is not zero, since a portion of the energy is required to break bonds at the tip. In all but the most brittle materials, however, the bond energy (γ_s) is a small fraction of the total fracture energy.

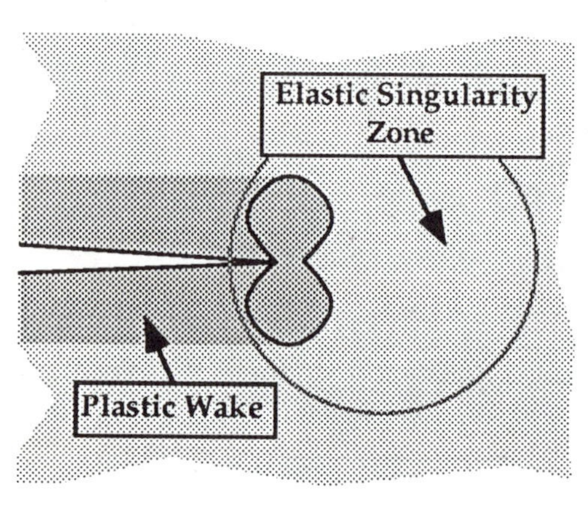

(a) Growing crack

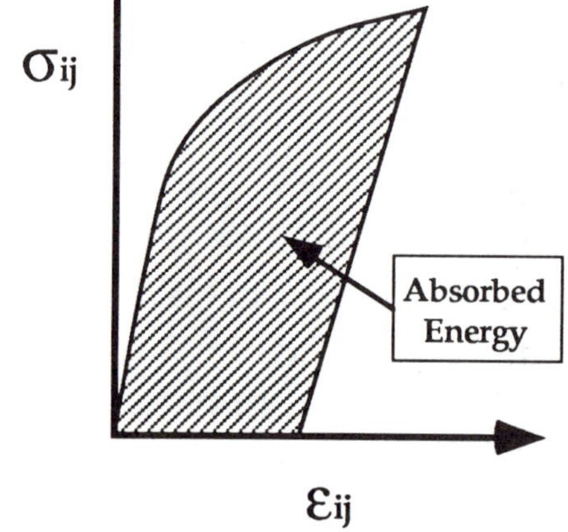

(b) Schematic stress-strain curve for material in the plastic wake

FIGURE A4.3 Crack growth in small scale yielding. The plastic wake, which forms behind the growing crack, dissipates energy

As long as the plastic zone or processes zone is embedded within an elastic singularity, the energy release rate can be defined unambiguously for a growing crack. In large scale yielding conditions, however, J is path dependent. *Consequently, an unambiguous definition of energy release rate does not exist for a crack growing in an elastic-plastic or fully plastic body.* Recall from Chapter 3 that there are several definitions of J for growing cracks. The so-called deformation J, which is based on a pseudo energy release rate concept, is the most common methodology. The deformation J is not, in general, equal to the J integral inferred from a contour integration.

PART III: MATERIAL BEHAVIOR

Chapters 5 and 6 give an overview of the micromechanisms of fracture in various material systems. This subject is of obvious importance to materials scientists, because an understanding of microstructural events that lead to fracture is essential to the development of materials with optimum toughness. Those who approach fracture from a solid mechanics viewpoint, however, often sidestep microstructural issues and consider only continuum models.

In certain cases, classical fracture mechanics provides some justification for disregarding microscopic failure mechanisms. Just as it is not necessary to understand dislocation theory to apply tensile data to design, it *may* not be necessary to consider the microscopic details of fracture when applying fracture mechanics on a global scale. When a single parameter (i.e., K, J, or CTOD) uniquely characterizes crack tip conditions, a critical value of this parameter is a material constant that is transferable from a test specimen to a structure made from the same material (see Sections 2.10 and 3.5). A laboratory specimen and a flawed structure experience identical crack tip conditions at failure when the single parameter assumption is valid, and it is not necessary to delve into the details of microscopic failure to characterize global fracture.

The situation becomes considerably more complicated when the single parameter assumption ceases to be valid. A fracture toughness test on a small scale laboratory specimen is no longer a reliable indicator of how a large structure will behave. The fracture toughness of the structure and test specimen are likely to be different, and the two configurations may even fail by different mechanisms. A number of researchers are currently attempting to develop alternatives to single parameter fracture mechanics. Such approaches cannot succeed with continuum theory alone, but must also consider microscopic fracture mechanisms (see Appendix 5.1). Thus the next two chapters should be of equal value to materials scientists and solid mechanicians.

5. FRACTURE MECHANISMS IN METALS

Figure 5.1 schematically illustrates three of the most common fracture mechanisms in metals and alloys. (A fourth mechanism, fatigue, is discussed in Chapter 10.) Ductile materials (Fig. 5.1(a)) usually fail as the result of nucleation, growth and coalescence of microscopic voids that initiate at inclusions and second phase particles. Cleavage fracture (Fig. 5.1(b)) involves separation along specific crystallographic planes. Note that the fracture path is transgranular. Although cleavage is often called brittle fracture, it can be preceded by large scale plasticity and ductile crack growth. Intergranular fracture (Fig. 5.1(c)), as its name implies, occurs when the grain boundaries are the preferred fracture path in the material.

5.1 DUCTILE FRACTURE

Figure 5.2 schematically illustrates the uniaxial tensile behavior in a ductile metal. The material eventually reaches an instability point, where strain hardening cannot keep pace with loss in cross sectional area, and a necked region forms beyond the maximum load. In very high purity materials, the tensile specimen may neck down to a sharp point, resulting in extremely large local plastic strains and nearly 100% reduction in area. Materials that contain impurities, however, fail at much lower strains. Microvoids nucleate at inclusions and second phase particles; the voids grow together to form a macroscopic flaw, which leads to fracture.

The commonly observed stages in ductile fracture are [1-5]:

(1) Formation of a free surface at an inclusion or second phase particle by either interface decohesion or particle cracking.

(2) Growth of the void around the particle, by means of plastic strain and hydrostatic stress.

(3) Coalescence of the growing void with adjacent voids.

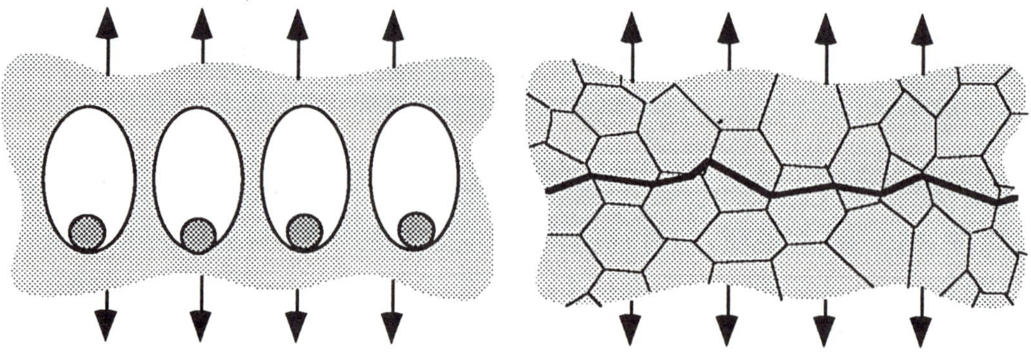

(a) Ductile fracture.

(b) Cleavage

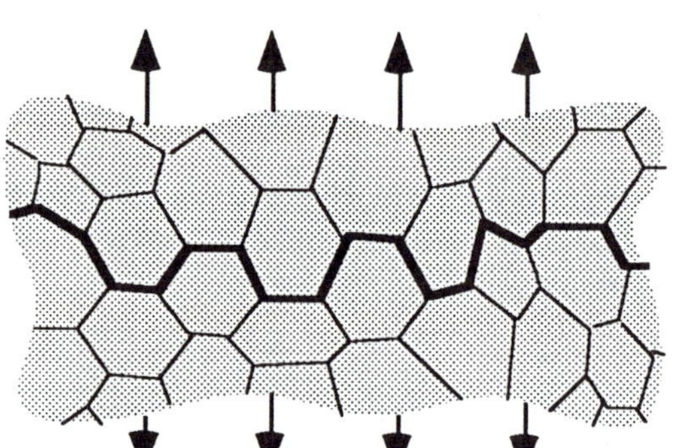

(c) Intergranular fracture.

FIGURE 5.1 Three micromechanisms of fracture in metals.

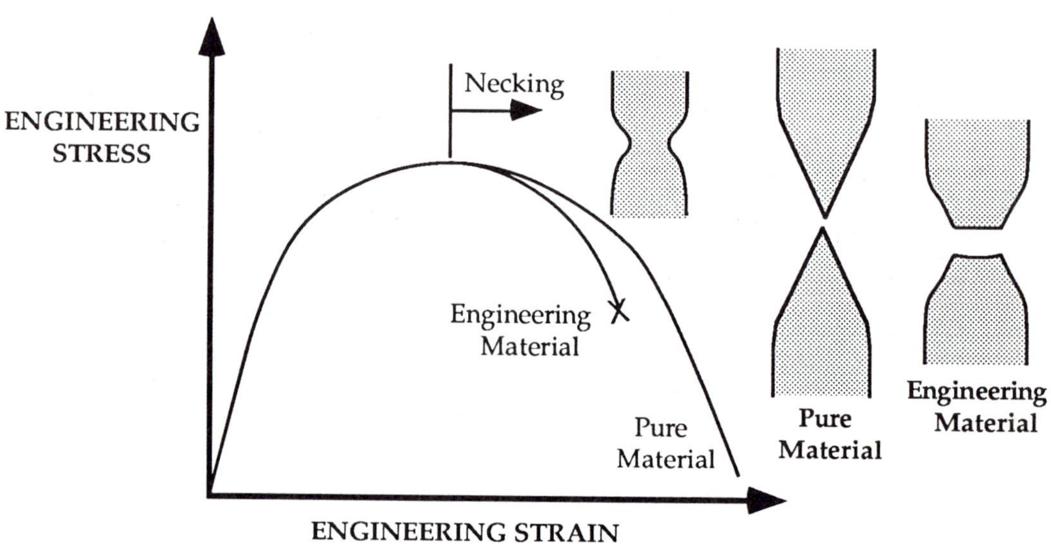

FIGURE 5.2 Uniaxial tensile deformation of ductile materials.

In materials where the second phase particles and inclusions are well bonded to the matrix, void nucleation is often the critical step; fracture occurs soon after the voids form. When void nucleation occurs with little difficulty, the fracture properties are controlled by the growth and coalescence of voids; the growing voids reach a critical size, relative to their spacing, and a local plastic instability develops between voids, resulting in failure.

5.1.1 Void Nucleation

A void forms around a second phase particle or inclusion when sufficient stress is applied to break the interfacial bonds between the particle and the matrix. A number of models for estimating void nucleation stress have been published, some of which are based on continuum theory [6,7] while others incorporate dislocation-particle interactions [8,9]. The latter models are required for particles < 1 μm in diameter.

The most widely used continuum model for void nucleation is due to Argon, et al. [6]. They argued that the interfacial stress at a cylindrical particle is approximately equal to the sum of the mean (hydrostatic) stress and the effective (von Mises) stress. The decohesion stress is defined as a critical combination of these two stresses:

$$\sigma_c = \sigma_e + \sigma_m \tag{5.1}$$

where σ_e is the effective stress, given by

$$\sigma_e = \frac{1}{\sqrt{2}}\left[(\sigma_1 - \sigma_2)^2 + (\sigma_1 - \sigma_3)^2 + (\sigma_3 - \sigma_2)^2\right]^{1/2} \tag{5.2}$$

σ_m is the mean stress, defined as

$$\sigma_m = \frac{\sigma_1 + \sigma_2 + \sigma_3}{3} \tag{5.3}$$

and σ_1, σ_2, and σ_3 are the principal normal stresses. According to the Argon, et al model, the nucleation strain decreases as the hydrostatic stress increases. That is, void nucleation occurs more readily in a triaxial tensile stress field, a result that is consistent with experimental observations.

The Beremin research group in France [7] applied the Argon et al. criterion to experimental data for a carbon manganese steel, but found that the following semi-empirical relationship gave better predictions of void nucleation at MnS inclusions that were elongated in the rolling direction:

$$\sigma_c = \sigma_m + C\,(\sigma_e - \sigma_{YS}) \tag{5.4}$$

where σ_{YS} is the yield strength and C is a fitting parameter that is approximately 1.6 for longitudinal loading and 0.6 for loading transverse to the rolling direction.

Goods and Brown [9] have developed a dislocation model for void nucleation at submicron particles. They estimated that dislocations near the particle elevate the stress at the interface by the following amount:

$$\Delta\sigma_d = 5.4\,\alpha\,\mu\sqrt{\frac{\varepsilon_1 b}{r}} \tag{5.5}$$

where α is a constant that ranges from 0.14 to 0.33, μ is the shear modulus, ε_1 is the maximum remote normal strain, b is the magnitude of the Burger's vector, and r is the particle radius. The total maximum interface stress is equal to the maximum principal stress plus $\Delta\sigma_d$. Void nucleation occurs when the sum of these stresses reaches a critical value:

$$\sigma_c = \Delta\sigma_d + \sigma_1 \tag{5.6}$$

An alternative but equivalent expression can be obtained by separating σ_1 into deviatoric and hydrostatic components:

$$\sigma_c = \Delta\sigma_d + S_1 + \sigma_m \tag{5.7}$$

where S_1 is the maximum deviatoric stress.

The Goods and Brown dislocation model indicates that the local stress concentration increases with decreasing particle size; void nucleation is more difficult with larger particles. The continuum models (Eqs. (5.1) and (5.4)), which apply to particles with $r > 1\ \mu m$, imply that σ_c is independent of particle size.

Experimental observations usually differ from both continuum and dislocation models, in that void nucleation tends to occur more readily at large particles [10]. Recall, however, that these models only considered nucleation by particle-matrix debonding. Voids can also be nucleated when particles crack. Larger particles are more likely to crack in the presence of plastic strain, because they are more likely to contain small defects which can act like Griffith cracks (see Section 5.2). In addition, large nonmetallic inclusions, such as oxides and sulfides, are often damaged during fabrication; some of these particles may be cracked or debonded prior to plastic deformation, making void nucleation relatively easy. Further research is obviously needed to develop void nucleation models that are more in line with experiment.

5.1.2 Void Growth and Coalescence

Once voids form, further plastic strain and hydrostatic stress cause the voids to grow and eventually coalesce. Figures 5.3 and 5.4 are scanning electron microscope (SEM) fractographs which show dimpled fracture surfaces that are typical of microvoid coalescence. Figure 5.4 shows an inclusion that nucleated a void.

Figure 5.5 schematically illustrates the growth and coalescence of microvoids. If the initial volume fraction of voids is low (< 10%), each void can be assumed to grow independently; upon further growth, neighboring voids interact. Plastic strain is concentrated along a sheet of voids, and local necking instabilities develop. The orientation of the fracture path depends on the stress state [11].

Many materials contain a bimodal or trimodal distribution of particles. For example, a precipitation-hardened aluminum alloy may contain relatively large intermetallic particles, together with a fine dispersion of submicron second phase precipitates. These alloys also contain micron-size dispersoid particles for grain refinement. Voids form much more readily in the inclusions, but the smaller particles can contribute in certain cases. Bimodal particle distributions can lead to so-called "shear" fracture surfaces, as described below.

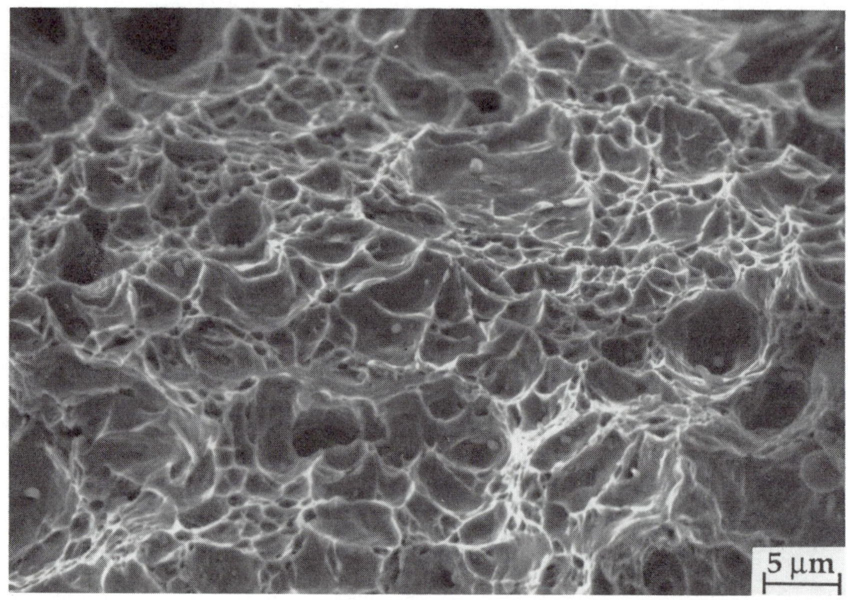

FIGURE 5.3 Scanning electron microscope (SEM) fractograph which shows ductile fracture in a low carbon steel. (Photograph provided by Mr. Sun Yongqi.)

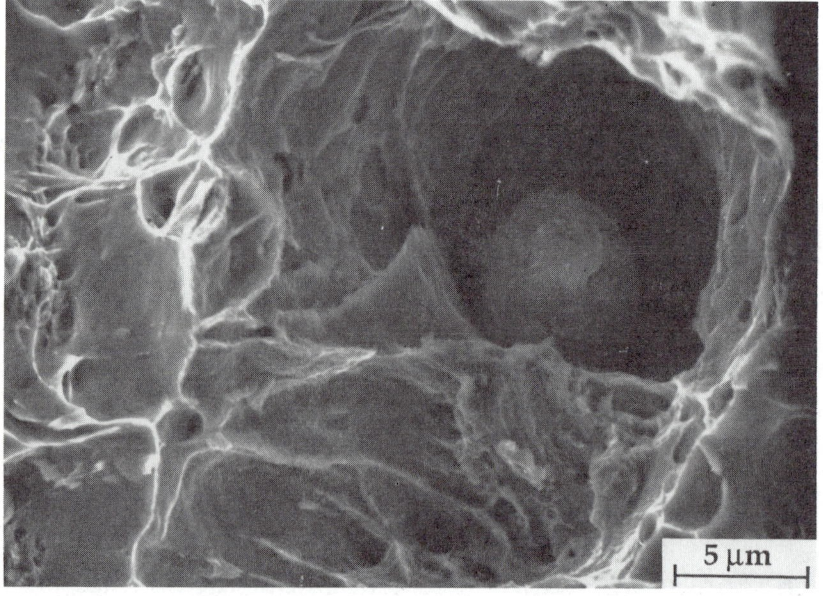

FIGURE 5.4 High magnification fractograph of the steel ductile fracture surface. Note the spherical inclusion which nucleated a microvoid. (Photograph provided by Mr. Sun Yongqi.)

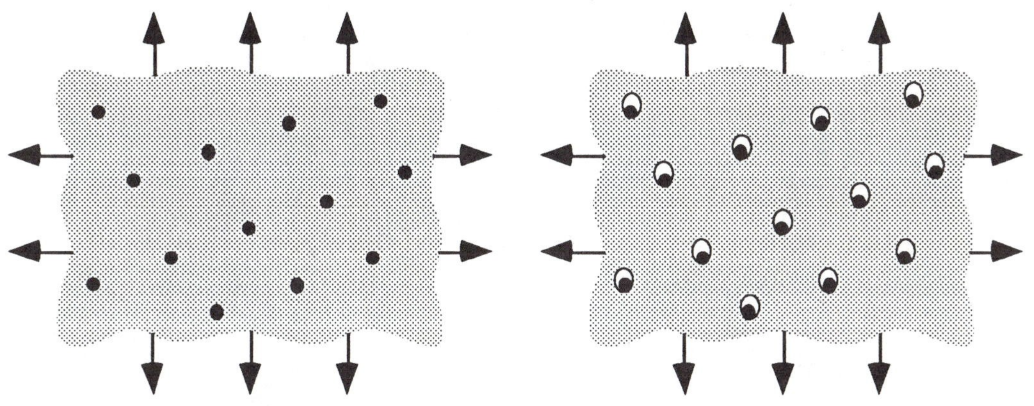

(a) Inclusions in a ductile matrix. (b) Void nucleation.

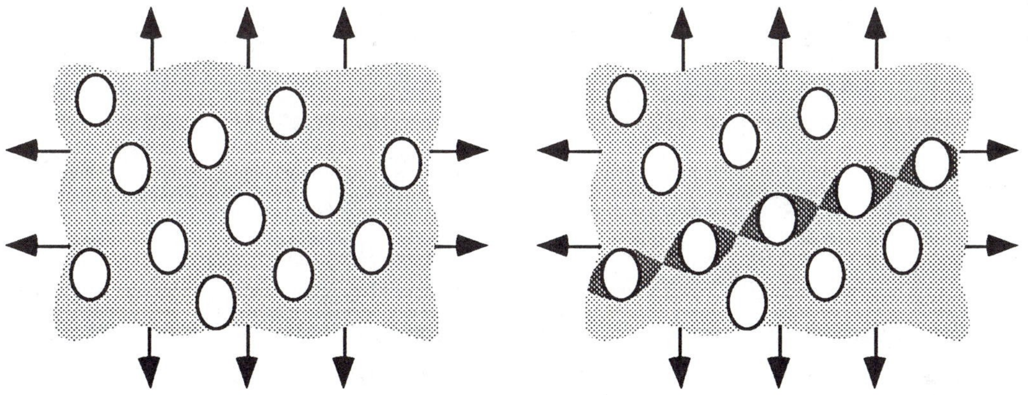

(c) Void growth. (d) Strain localization between voids.

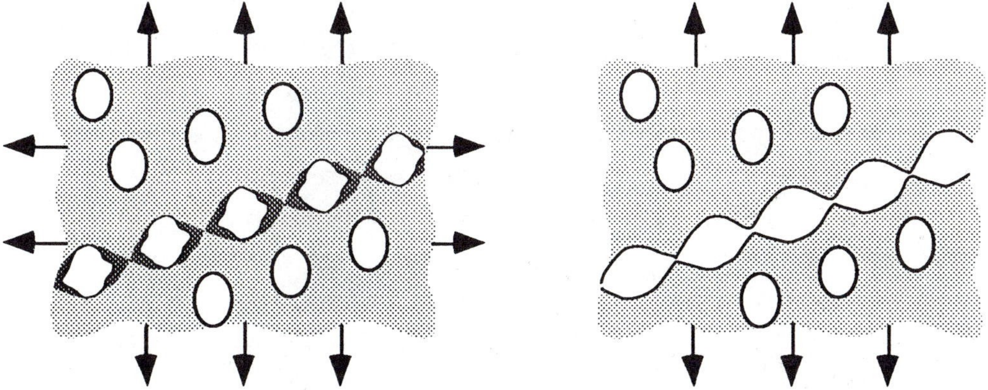

(e) Necking between voids. (f) Void coalescence and fracture.

FIGURE 5.5 Void nucleation, growth, and coalescence in ductile metals.

Figure 5.6 illustrates the formation of the "cup and cone" fracture surface that is commonly observed in uniaxial tensile tests. The neck produces a triaxial stress state in the center of the specimen, which promotes void nucleation and growth in the larger particles. Upon further strain, the voids coalesce, resulting in a penny-shaped flaw. The outer ring of the specimen contains relatively few voids, because the hydrostatic stress is lower than in the center. The penny-shaped flaw produces deformation bands at 45° from the tensile axis. This concentration of strain provides sufficient plasticity to nucleate voids in the smaller more numerous particles. Since the small particles are closely spaced, an instability occurs soon after these smaller voids form, resulting in total fracture of the specimen and the cup and cone appearance of the matching surfaces. The central region of the fracture surface has a fibrous appearance at low magnifications, but the outer region is relatively smooth. Because the latter surface is oriented 45° from the tensile axis and there is little evidence (at low magnifications) of microvoid coalescence, many refer to this type of surface as shear fracture. Although there is a certain amount of shear on the microscopic level, the axisymmetric geometry of a tensile specimen precludes Mode II or Mode III displacements on a global scale [11]. This so-called shear fracture is actually a tensile fracture, as Fig. 5.6 illustrates.

Figure 5.7 is a photograph of the cross section of a fractured tensile specimen; note the high concentration of microvoids in the center of the necked region, compared with the edges of the necked region.

Figure 5.8 shows SEM fractographs of a cup and cone fracture surface. The central portion of the specimen exhibits a typical dimpled appearance, but the outer region appears to be relatively smooth, particularly at low magnification (Fig. 5.8(a)). At somewhat higher magnification (Fig. 5.8(b)), a few widely spaced voids are evident in the outer region. Figure 5.9 shows a representative fractograph at higher magnification of the "shear" surface. Note the dimpled appearance, that is characteristic of microvoid coalescence. The average void size and spacing, however, are much smaller than in the central region of the specimen.

There are a number of mathematical models for void growth and coalescence. The two most widely referenced models were published by Rice and Tracey [12] and Gurson [13]. The latter approach was actually based on the work of Berg [14], but it is commonly known as the Gurson model. Both the Gurson and Rice and Tracey models have been modified in more recent investigations [15,16].

Rice and Tracey considered a single void in an infinite solid, as illustrated in Fig. 5.10. The void is subject to remote normal stresses σ_1, σ_2, σ_3, and remote normal strain rates $\dot{\varepsilon}_1$, $\dot{\varepsilon}_2$, $\dot{\varepsilon}_3$. The initial void is assumed to be spherical, but it becomes ellipsoidal as it deforms. Rice and Tracey analyzed both rigid plastic material behavior and linear strain hardening. They showed that the rate of change of radius in each principal direction has the form:

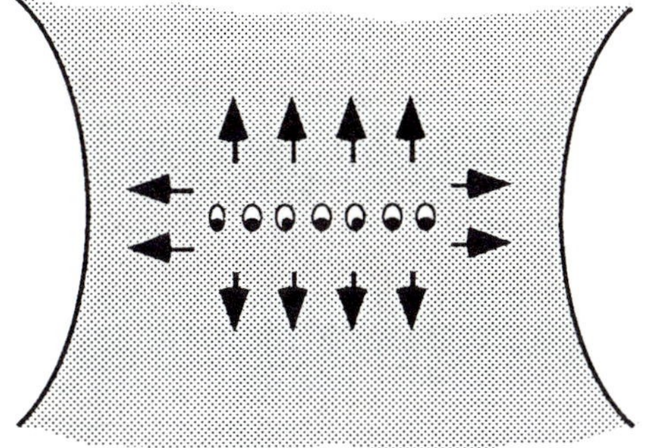

(a) Void growth in a triaxial stress state.

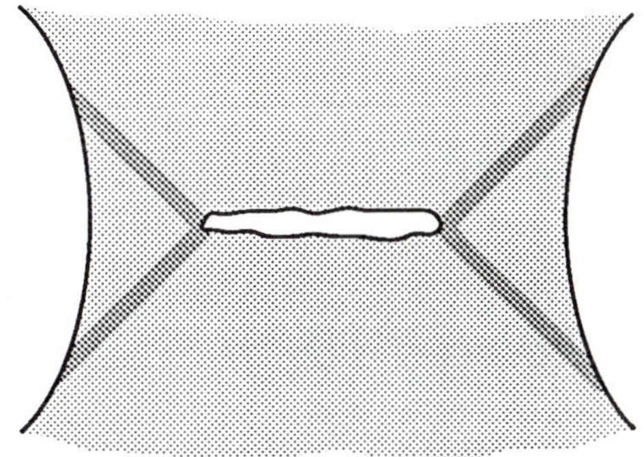

(b) Crack and deformation band formation.

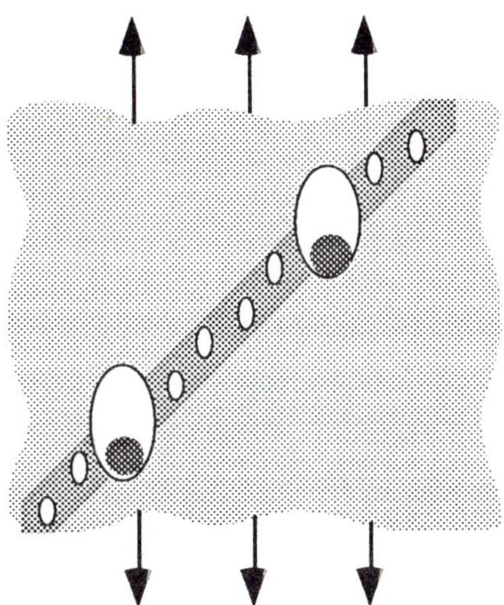

(c) Nucleation at smaller particles along the deformation bands.

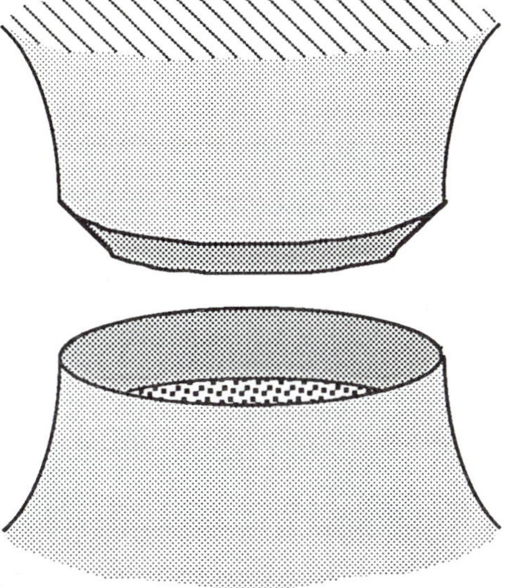

(d) Cup and cone fracture.

FIGURE 5.6 Formation of the cup and cone fracture surface in uniaxial tension.

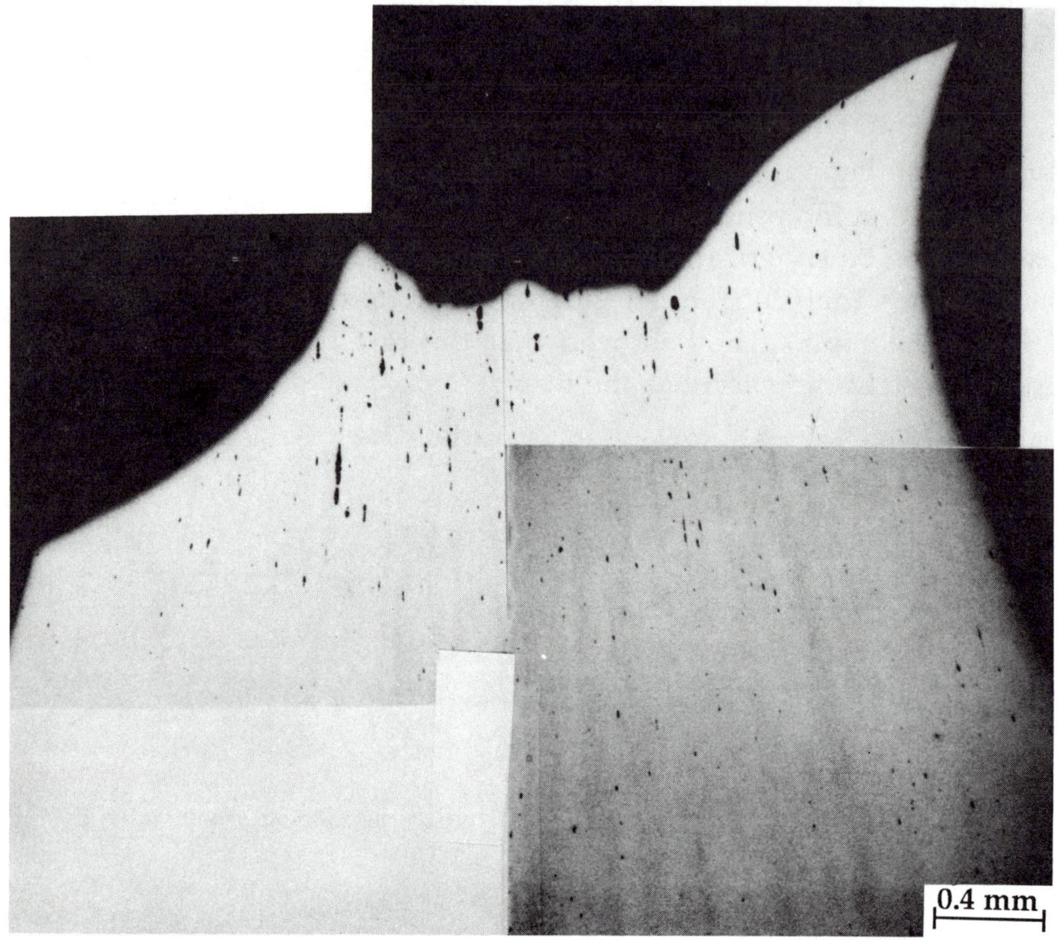

FIGURE 5.7 Metallographic cross section (unetched) of a ruptured austenitic stainless steel tensile specimen. The dark areas in the necked region are microvoids. (Photograph provided by P.T. Purtscher.)

$$\dot{R}_i = \left[(1 + G)\,\dot{\varepsilon}_i + \sqrt{\frac{2}{3}\dot{\varepsilon}_j\dot{\varepsilon}_j}\;\; D \right] R_o \qquad (i, j = 1, 2, 3) \tag{5.8}$$

where D and G are constants that depend on stress state and strain hardening, and R_o is the radius of the initial spherical void. The standard notation, where repeated indices implies summation, is followed here. Invoking the incompressibility condition ($\dot{\varepsilon}_1 + \dot{\varepsilon}_2 + \dot{\varepsilon}_3 = 0$) reduces the number of independent principal strain rates to two. Rice

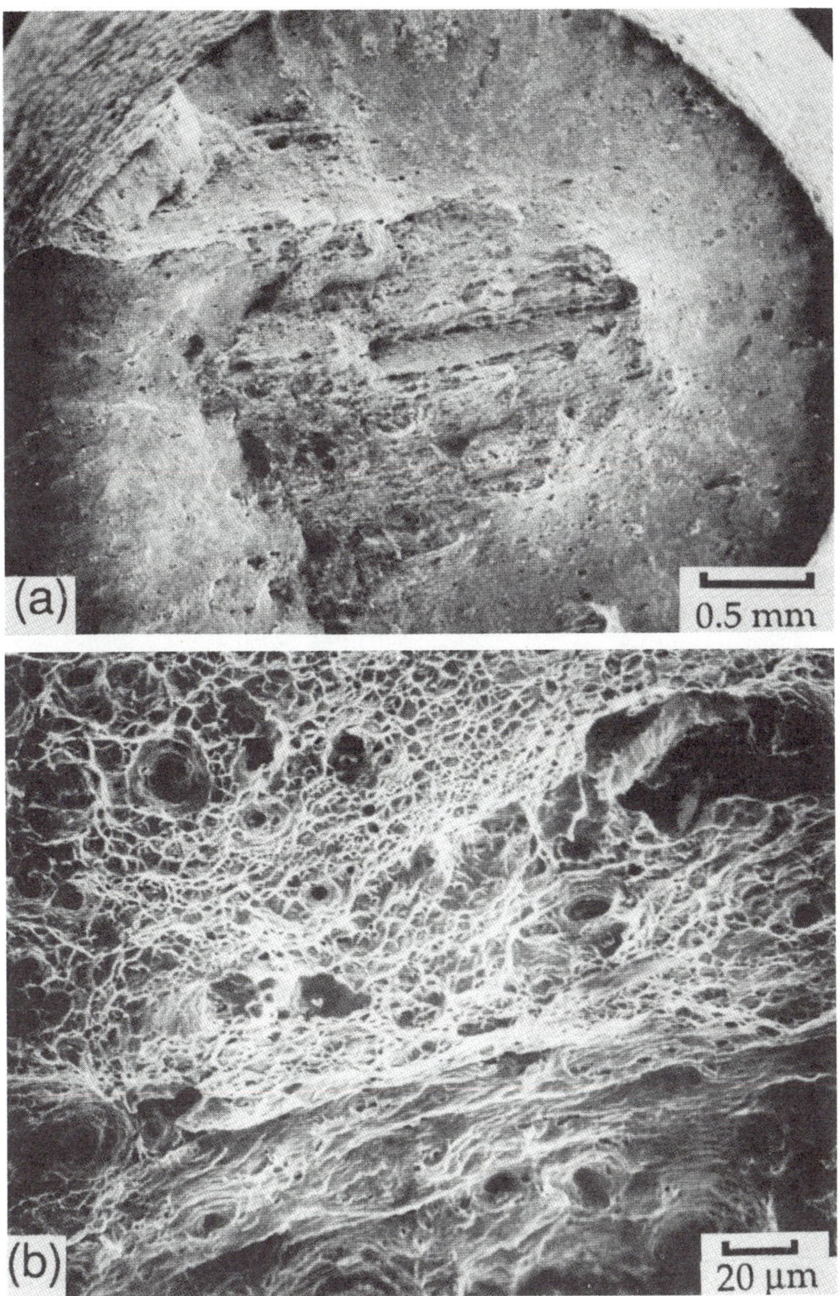

FIGURE 5.8 Cup and cone fracture in an austenitic stainless steel [17]. (Photographs provided by P.T. Purtscher.)

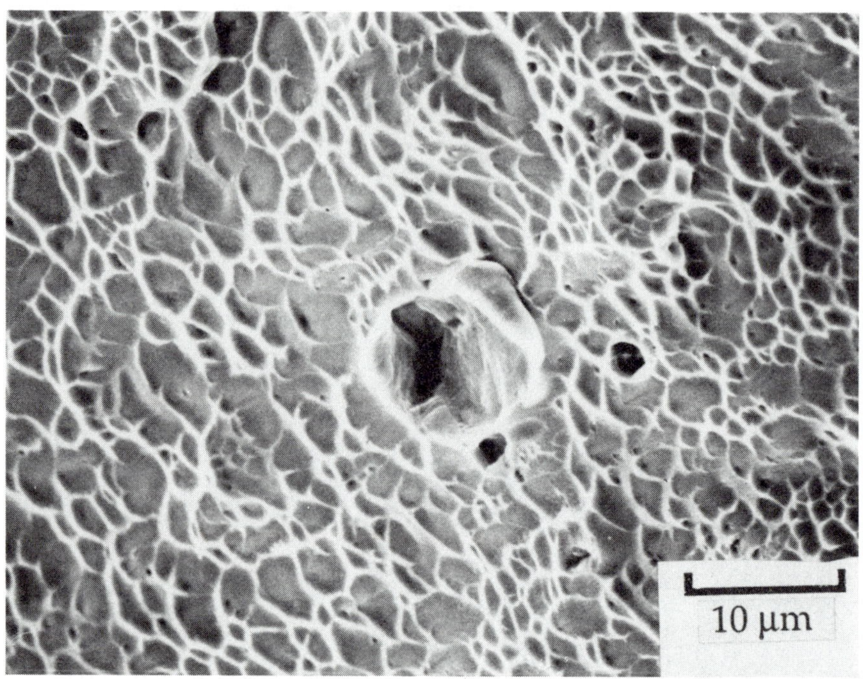

FIGURE 5.9 High magnification fractograph of the "shear" region of a cup and cone fracture surface in austenitic stainless steel [17]. (Photograph provided by P.T. Purtscher.)

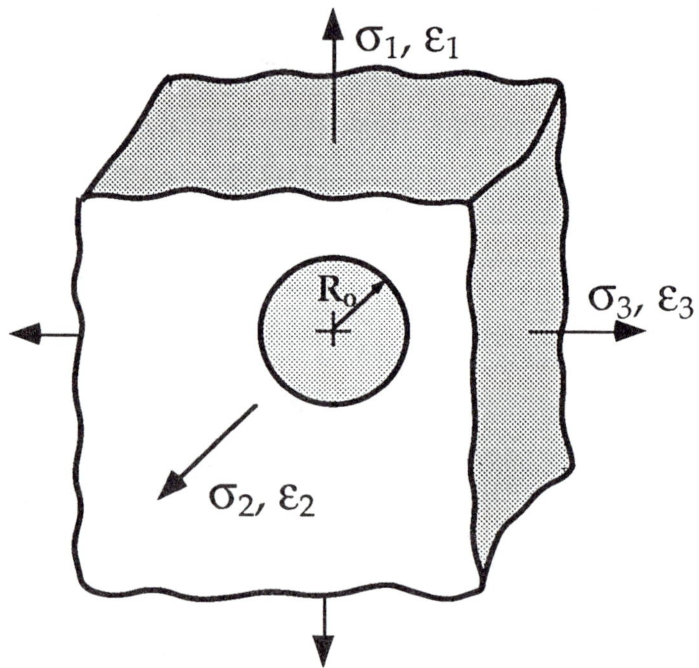

FIGURE 5.10 Spherical void in a solid, subject to a triaxial stress stress state.

and Tracey chose to express $\dot{\varepsilon}_2$ and $\dot{\varepsilon}_3$ in terms of $\dot{\varepsilon}_1$ and a second parameter:

$$\dot{\varepsilon}_2 = \frac{-2\phi}{3+\phi}\dot{\varepsilon}_1$$
$$\dot{\varepsilon}_3 = \frac{\phi - 3}{3+\phi}\dot{\varepsilon}_1 \tag{5.9}$$

where

$$\phi = -\frac{3\dot{\varepsilon}_2}{\dot{\varepsilon}_1 - \dot{\varepsilon}_3}$$

Substituting Eq. (5.9) into Eq. (5.8) and making a few simplifying assumptions leads to the following expressions for the radial displacements of the ellipsoidal void:

$$R_1 = \left(A + \frac{(3+\phi)}{2\sqrt{\phi^2+3}}B\right)R_o$$
$$R_2 = \left(A - \frac{\phi B}{\sqrt{\phi^2+3}}\right)R_o \tag{5.10}$$
$$R_3 = \left(A + \frac{(\phi - 3)B}{2\sqrt{\phi^2+3}}\right)R_o$$

where

$$A = \exp\left(\frac{2\sqrt{\phi^2+3}}{(3+\phi)}D\,\varepsilon_1\right)$$

$$B = \frac{(1+F)(A-1)}{D}$$

and ε_1 is the total strain, integrated from the undeformed configuration to the current state.

Rice and Tracey solved Eq. (5.10) for a variety of stress states and found that the void growth in all cases could be approximated by the following semi-empirical relationship:

$$\ln\left(\frac{\bar{R}}{R_o}\right) = 0.283 \int_0^{\varepsilon_{eq}} \exp\left(\frac{1.5\,\sigma_m}{\sigma_{YS}}\right) d\varepsilon_{eq} \tag{5.11}$$

where $\bar{R} = (R_1 + R_2 + R_3)/3$ and ε_{eq} is the equivalent (von Mises) plastic strain. Subsequent investigators found that Eq. (5.11) could be approximately modified for strain hardening by replacing the yield strength with σ_e, the effective stress [18].

Since the Rice and Tracey model is based on a single void, it does not take account of interactions between voids, nor does it predict ultimate failure. A separate failure criterion must be applied to characterize microvoid coalescence. For example, one might assume that fracture occurs when the nominal void radius reaches a critical value.

The Gurson model [13] analyzes plastic flow in a porous medium by assuming that the material behaves as a continuum. Voids appear in the model indirectly through their influence on the global flow behavior. The effect of the voids is averaged through the material, which is assumed to be continuous and homogeneous (Fig. 5.11). The main difference between the Gurson model and classical plasticity is that the yield surface in the former exhibits a weak hydrostatic stress dependence, while classical plasticity assumes that yielding is independent of hydrostatic stress. This modification to conventional plasticity theory has the effect of introducing a strain softening term.

Unlike the Rice and Tracey model, the Gurson model contains a failure criterion. Ductile fracture is assumed to occur as the result of a plastic instability that produces a band of localized deformation. Such an instability occurs more readily in a Gurson material because of the strain softening induced by hydrostatic stress. However, because the model does not consider discrete voids, it is unable to predict necking instability between voids.

The original Gurson model describes the yield surface as follows:

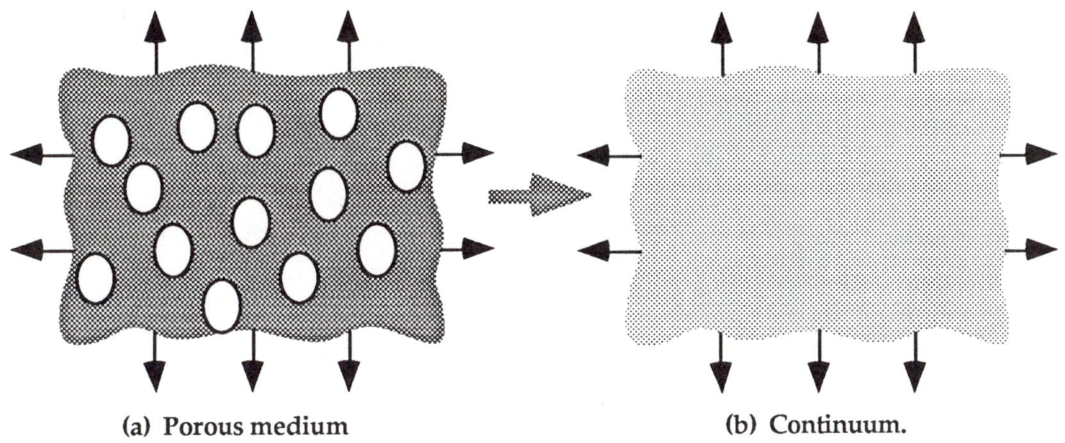

FIGURE 3.11 The continuum assumption for modeling a porous medium. The material is assumed to be homogeneous, and the effect of the voids is averaged through the solid.

$$\Phi = \frac{3}{2}\frac{S_{ij}\,S_{ij}}{\sigma_{YS}{}^2} + 2\,f\cosh\left(\frac{3}{2}\frac{\sigma_m}{\sigma_{YS}}\right) - (1 + f^2) = 0 \tag{5.12}$$

where f is the void volume fraction, S_{ij} is the deviatoric stress, defined as

$$S_{ij} = \sigma_{ij} - \sigma_m\,\delta_{ij} \tag{5.13}$$

and δ_{ij} is the Kronecker delta:

$$\delta_{ij} = \begin{cases} 1 & i = j \\ 0 & i \neq j \end{cases} \tag{5.14}$$

When f = 0, Eq. (5.12) reduces to the classical von Mises yield surface with isotropic hardening. Equation (5.12) greatly overpredicts failure strains in real materials. Tvergaard [15] attempted to correct the Gurson model by adding two adjustable parameters, q_1 and q_2:

$$\Phi = \frac{3}{2}\frac{S_{ij}\,S_{ij}}{\sigma_{YS}{}^2} + 2\,q_1\,f\cosh\left(\frac{3}{2}\frac{q_2\,\sigma_m}{\sigma_{YS}}\right) - [1 + (q_1\,f)^2] = 0 \tag{5.15}$$

Tvergaard calibrated the revised equation with experimental data and found that reasonable predictions of failure could be obtained when $q_1 = 2$ and $q_2 = 1$. This modification has the effect of amplifying the influence of hydrostatic stress at *all* strain levels. In real materials, the behavior deviates only slightly from classical plasticity theory through most of the deformation; at incipient failure, the deviation is rather abrupt.

Tvergaard and Needleman [16] have modified the Gurson model further by replacing f with an effective void volume fraction, f*:

$$f^* = \begin{cases} f & \text{for } f \le f_c \\ f_c - \dfrac{f_u^* - f_c}{f_F - f_c}(f - f_c) & \text{for } f > f_c \end{cases} \tag{5.16}$$

where f_c, f_u^* and f_F are fitting parameters. This most recent modification introduces an abrupt failure point, which more closely matches experimental observation. The effect of hydrostatic stress is amplified when $f > f_c$, which accelerates the onset of a plastic instability. A major disadvantage of the revised Gurson model is that it contains numerous adjustable parameters.

Although the Gurson model (and its subsequent modifications) may adequately characterize plastic flow in the early stages of the ductile fracture process, it does not provide a good description of the events that lead to final failure. Ductile failure results from local necking instabilities between voids. Since the Gurson model does not contain discrete voids, it is incapable of predicting void interactions that lead to failure.

Thomason [11] developed a simple limit load model for internal necking between microvoids. This model states that failure occurs when the net section stress between voids reaches a critical value, $\sigma_{n(c)}$. Figure 5.12 illustrates a two-dimensional case, where cylindrical voids are growing in a material subject to plane strain loading ($\varepsilon_3 = 0$). If the in-plane dimensions of the voids are 2a and 2b, and the spacing between voids is 2d, the row of voids illustrated in Fig. 5.12 is stable if

$$\sigma_{n(c)} \frac{d}{d + b} > \sigma_1 \tag{5.17a}$$

and fracture occurs when

$$\sigma_{n(c)} \frac{d}{d + b} = \sigma_1 \tag{5.17b}$$

where σ_1 is the maximum remote principal stress. Thomason applied the Rice and Tracey void growth model to predict the size and shape of growing voids, and utilized Eq. (5.17b) as a failure criterion. He predicted failure strains that were relatively close to experimental observations and were an order of magnitude lower than estimated by Eq. (5.12).

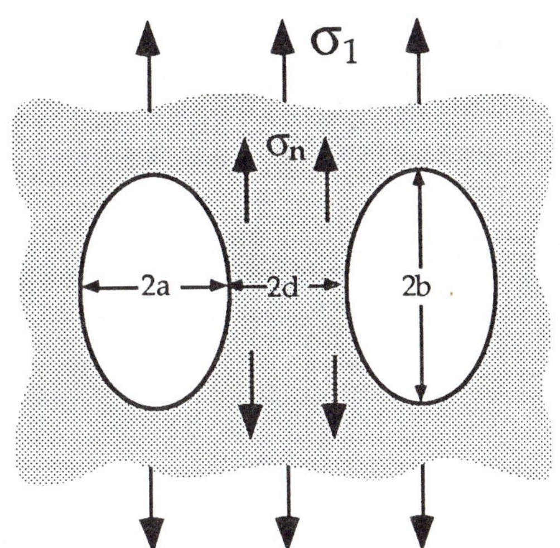

FIGURE 5.12 The limit load model for void instability. Failure is assumed to occur when the net section stress between voids reaches a critical value.

5.1.3 Ductile Crack Growth

Figure 5.13 schematically illustrates microvoid initiation, growth, and coalescence at the tip of a pre-existing crack. As the cracked structure is loaded, local strains and stresses at the crack tip become sufficient to nucleate voids. These voids grow as the crack blunts, and they eventually link with the main crack. As this process continues, the crack grows.

Figure 5.14 is a plot of stress and strain near the tip of a blunted crack [19]. The strain exhibits a singularity near the crack tip, but the stress reaches a peak at approximately two times the crack tip opening displacement (CTOD)[1]. In most materials, the triaxiality ahead of the crack tip provides sufficient stress elevation for void nucleation; thus growth and coalescence of microvoids are usually the critical steps in ductile crack growth. Nucleation typically occurs when a particle is ~ 2δ from the crack tip, while most of the void growth occurs much closer to the crack tip, relative to CTOD. (Note that although a void remains approximately fixed in absolute space, its distance from the crack tip, relative to CTOD, decreases as the crack blunts; the *absolute* distance from the crack tip also deceases as the crack grows)

Ductile crack growth is usually stable because it produces a rising resistance curve, at least during the early stages of crack growth. Stable crack growth and R curves are discussed in detail in Chapter 3.

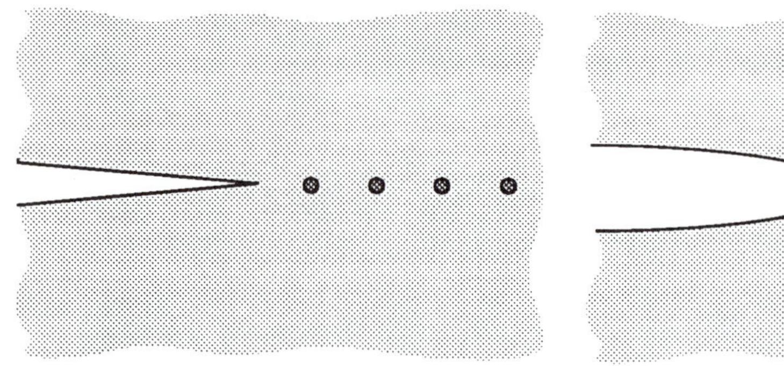

(a) Initial state.

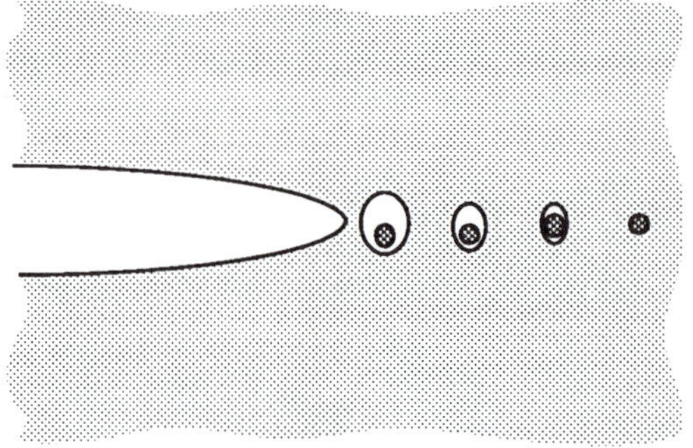

(b) Void growth at the crack tip.

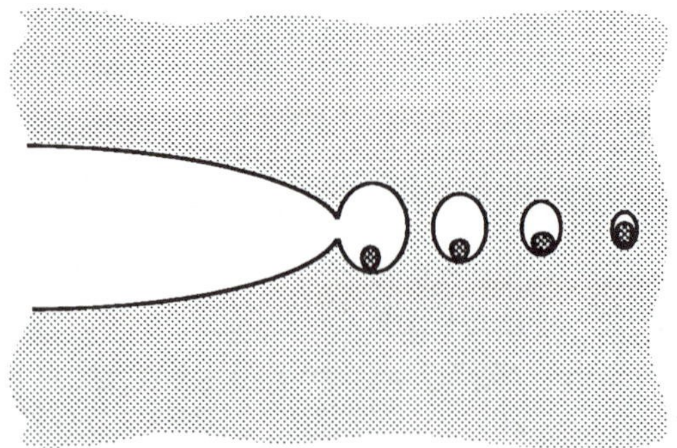

(c) Coalescence of voids with the crack tip.

FIGURE 5.13 Mechanism for ductile crack growth.

1Finite element analysis and slip line analysis of blunted crack tips predict a stress singularity very close to the crack tip (~0.1 CTOD), but it is not clear whether or not this actually occurs in real materials because the continuum assumptions break down at such fine scales.

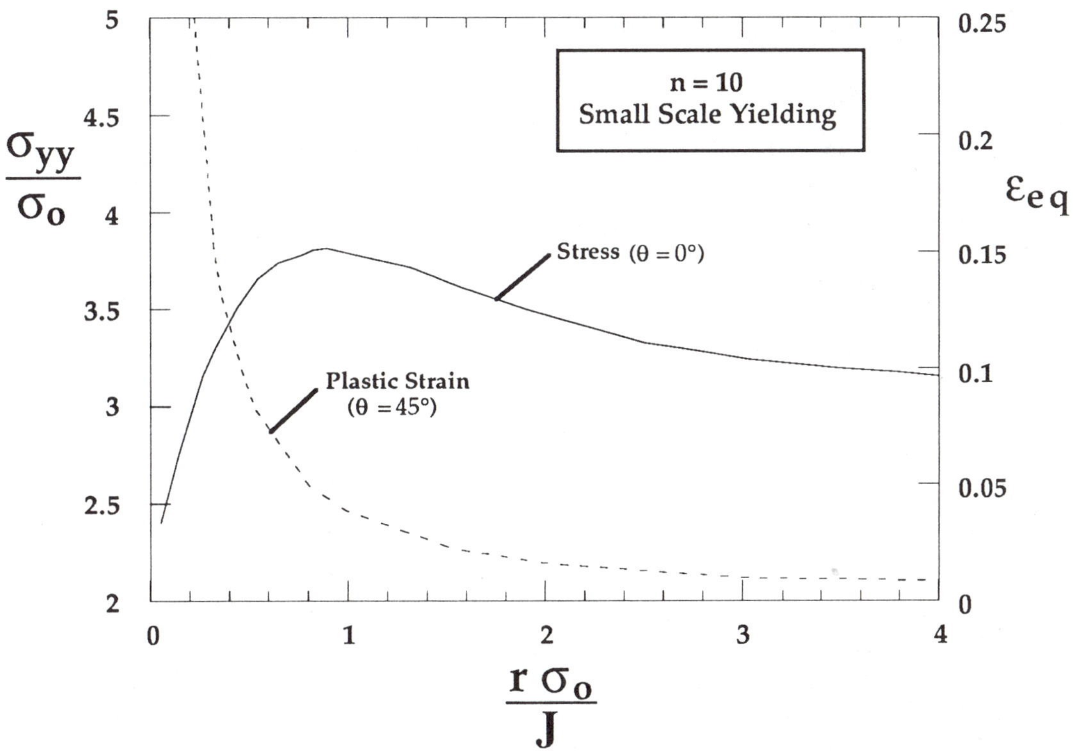

FIGURE 5.14 Stress and strain ahead of a blunted crack tip, determined by finite element analysis [19].

When an edge crack in a plate grows by microvoid coalescence, the crack exhibits a *tunneling* effect, where it grows faster in the center of the plate, due to the higher stress triaxiality. The through-thickness variation of triaxiality also produces *shear lips,* where the crack growth near the free surface occurs at a 45° angle from the maximum principal stress, as illustrated in Fig. 5.15. The shear lips are very similar to the cup and cone features in fractured tensile specimens. The growing crack in the center of the plate produces deformation bands which nucleate voids in small particles (Fig. 5.6). Thus the so-called shear lips are caused by a tensile (Mode I) fracture, despite the fact that the preferred fracture path is not perpendicular to the tensile axis.

Plane strain crack growth in the center of a plate appears to be relatively flat, but closer examination reveals a more complex structure. For a crack subject to plane strain Mode I loading, the maximum plastic strain occurs at 45° from the crack plane, as illustrated in Fig. 5.16(a). On a local level, this angle is the preferred path for void coalescence, but global constraints require that the crack propagation remain in its original plane. One way to reconcile these competing requirements is for the crack to grow in a zig-zag pattern (Fig. 5.16(b)), such that the crack appears flat on a global scale, but oriented ± 45° from the crack propagation direction when viewed at higher magnification. This zig-zag pattern is often observed in ductile materials (20,21). Figure 5.17 shows a metallographic cross section of a growing crack that exhibits this behavior.

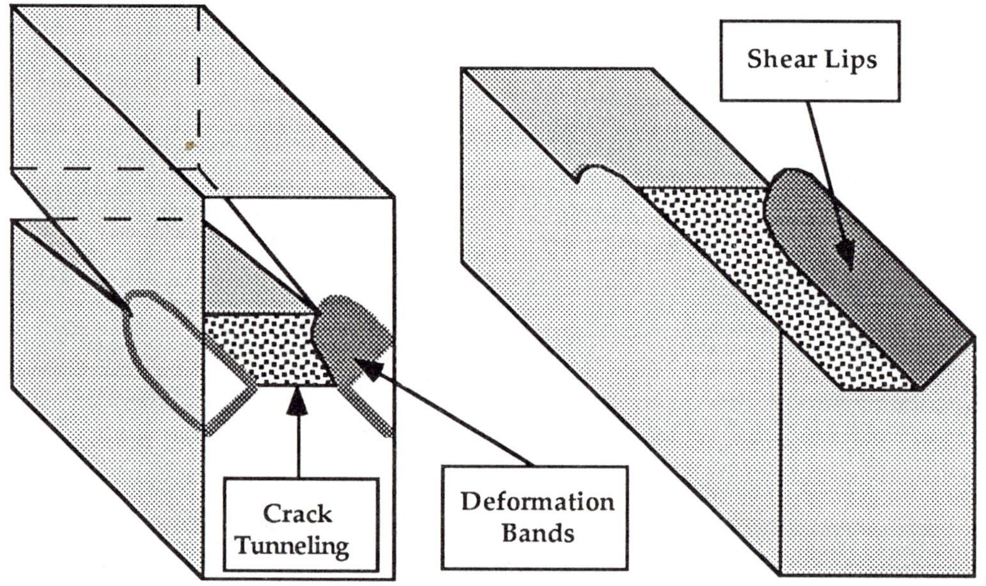

FIGURE 5.15. Ductile growth of an edge crack. The so-called shear lips are produced by the same mechanism as the cup and cone in uniaxial tension (Fig. 5.7).

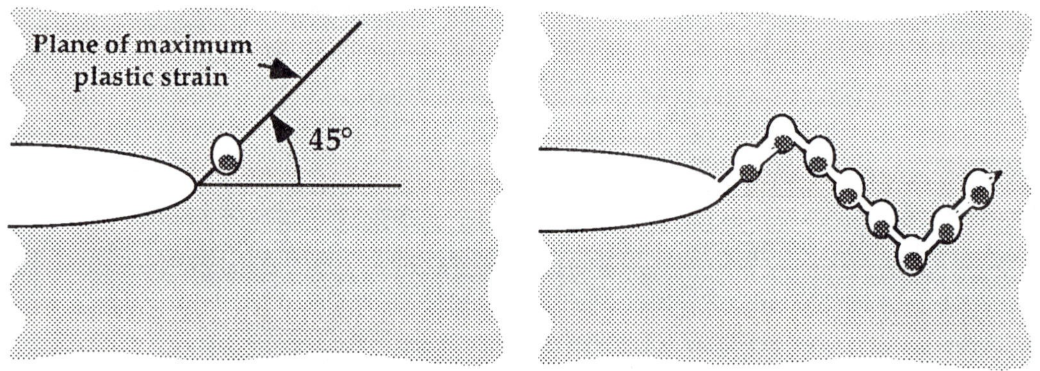

FIGURE 5.16. Ductile crack growth in a 45° zig-zag pattern.

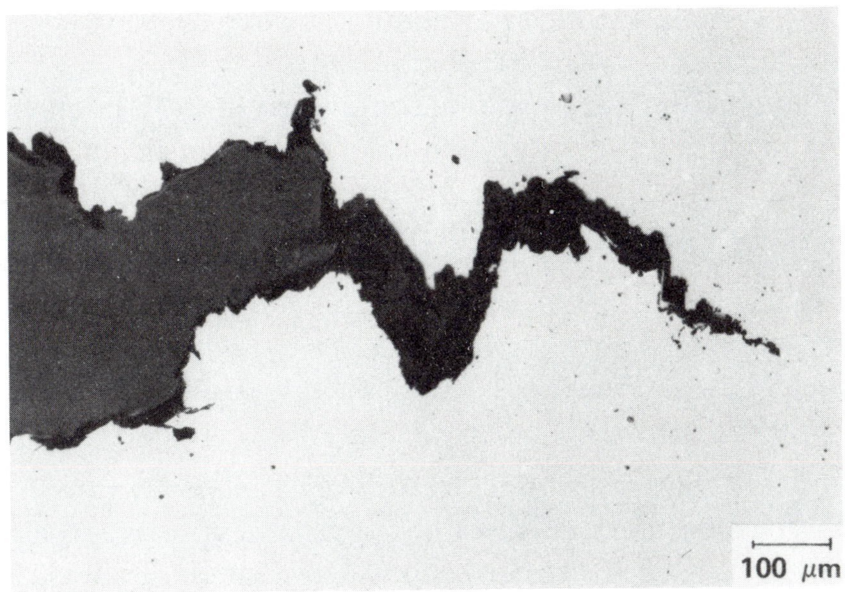

FIGURE 5.17 Optical micrograph (unetched) of ductile crack growth in an A 710 high strength-low alloy steel [21]. (Photograph provided by J.P. Gudas.)

5.2 CLEAVAGE

Cleavage fracture can be defined as rapid propagation of a crack along a particular crystallographic plane. Cleavage may be brittle, but it can be preceded by large scale plastic flow and ductile crack growth (see Section 5.3) The preferred cleavage planes are those with the lowest packing density, since fewer bonds must be broken and the spacing between planes is greater. In the case of body centered cubic (BCC) materials, cleavage occurs on {100} planes. The fracture path is transgranular in polycrystalline materials, as Fig. 5.1(b) illustrates. The propagating crack changes direction each time it crosses a grain boundary; the crack seeks the most favorably oriented cleavage plane in each grain. The nominal orientation of the cleavage crack is perpendicular to the maximum principal stress.

Cleavage is most likely when plastic flow is restricted. Face centered cubic (FCC) metals are usually not susceptible to cleavage because there are ample slip systems for ductile behavior at all temperatures. At low temperatures, BCC metals fail by cleavage because there are a limited number of active slip systems. Polycrystalline hexagonal close packed (HCP) metals, which have only three slip systems per grain, are also susceptible to cleavage fracture.

This section and Section 5.3 focus on ferritic steel, because it is the most technologically important (and the most extensively studied) material that is subject to cleavage fracture. This class of materials has a BCC crystal structure, which undergoes a ductile-brittle transition with decreasing temperature. Many of the mechanisms described below also operate in other material systems that fail by cleavage.

5.2.1 Fractography

Figure 5.18 shows SEM fractographs of cleavage fracture in a low alloy steel. The multifaceted surface is typical of cleavage in a polycrystalline material; each facet corresponds to a single grain. The "river patterns" on each facet are also typical of cleavage fracture. These markings are so named because multiple lines converge to a single line, much like tributaries to a river.

Figure 5.19 illustrates how river patterns are formed. A propagating cleavage crack encounters a grain boundary, where the nearest cleavage plane in the adjoining grain is oriented at a finite twist angle from the current cleavage plane. Initially, the crack accommodates the twist mismatch by forming on several parallel planes. As the multiple cracks propagate, they are joined by tearing between planes. Since this process consumes more energy than crack propagation on a single plane, there is a tendency for the multiple cracks to converge into a single crack. Thus the direction of crack propagation can be inferred from river patterns. Figure 5.20 shows a fractograph of river patterns in a low alloy steel, where tearing between parallel cleavage planes is evident.

5.2.2 Mechanisms of Cleavage Initiation

Since cleavage involves breaking bonds, the local stress must be sufficient to overcome the cohesive strength of the material. In Chapter 2, we learned that the theoretical fracture strength of a crystalline solid is approximately E/π. Figure 5.14, however, indicates that the maximum stress achieved ahead of the crack tip is 3 to 4 times the

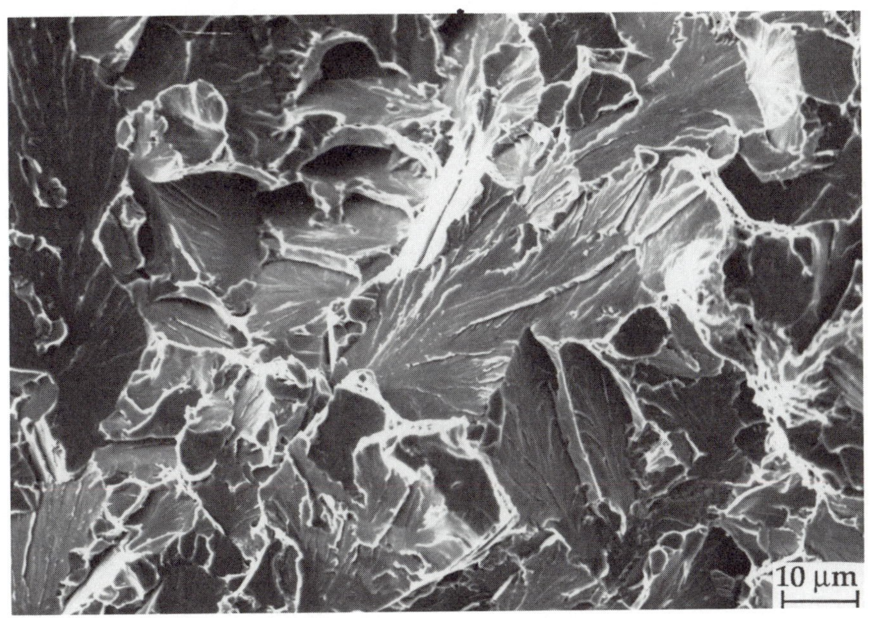

(a)

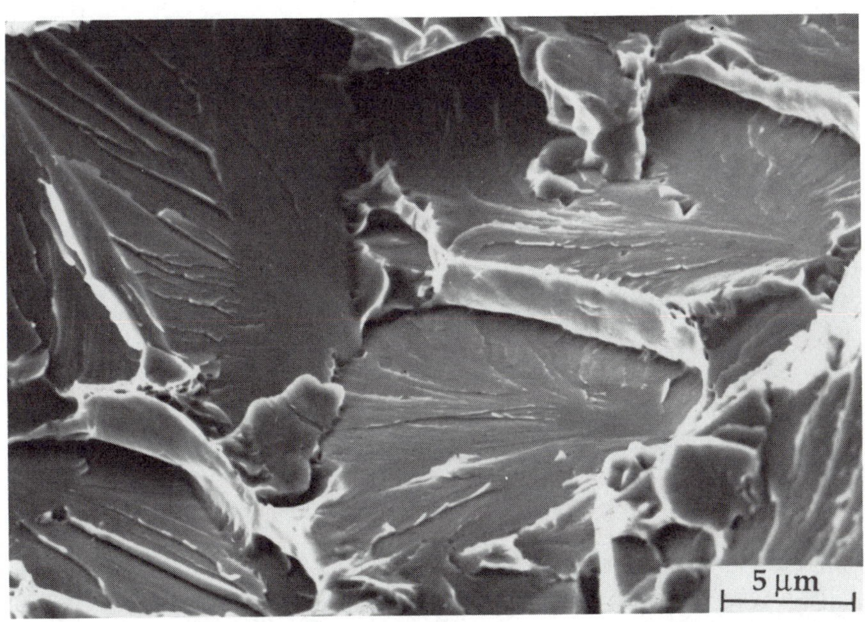

(b)

FIGURE 5.18 SEM fractographs of cleavage in an A 508 Class 3 alloy. (Photographs provided by Mr. Sun Yongqi.)

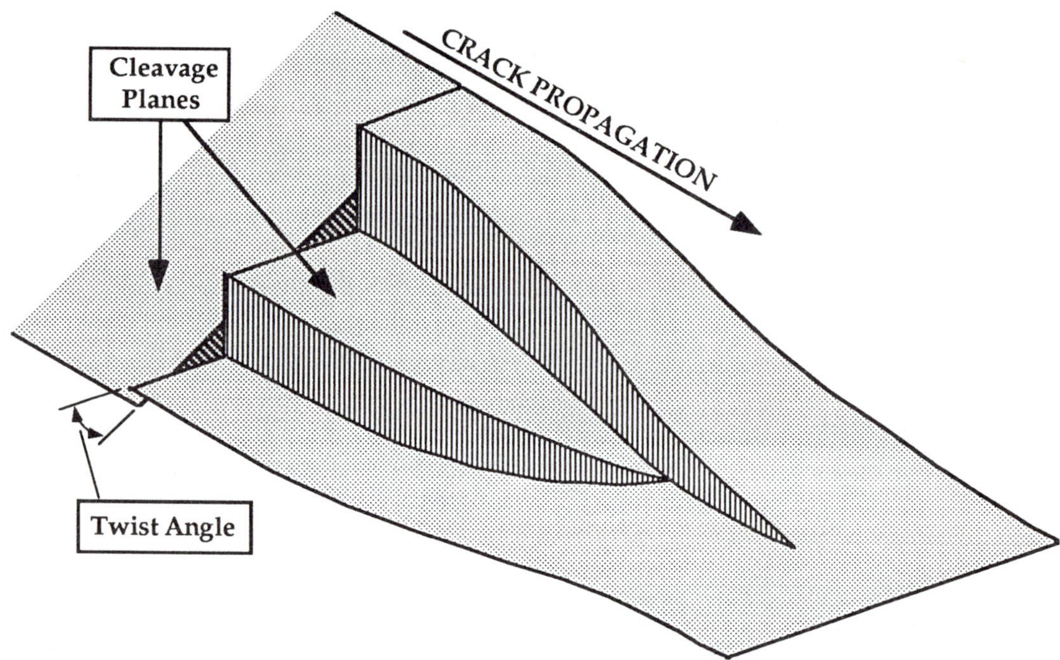

FIGURE 5.19 Formation of river patterns, as a result of a cleavage crack crossing a twist boundary between grains.

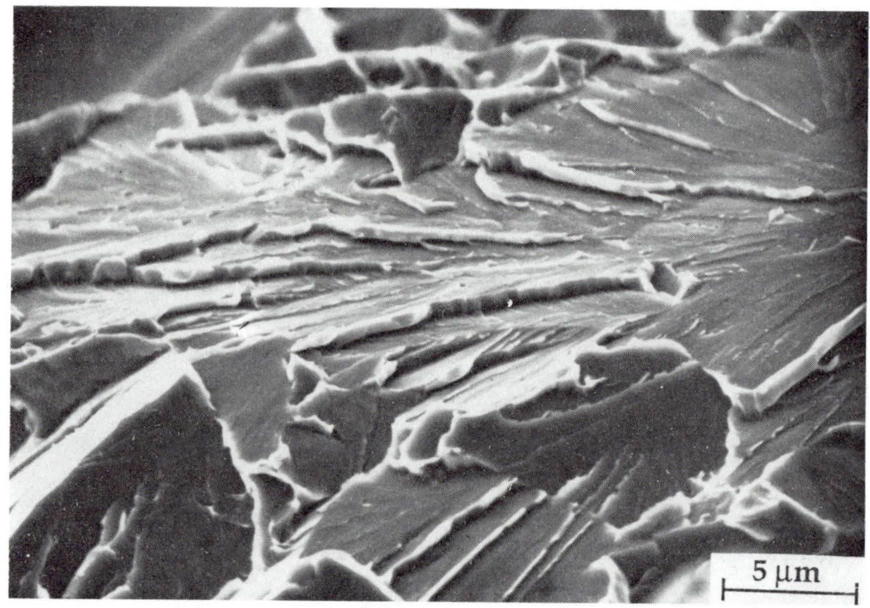

FIGURE 5.20 River patterns in an A 508 Class 3 steel. Note the tearing (light areas) between parallel cleavage planes. (Photograph provided by Mr. Sun Yongqi.)

yield strength. For a steel with σ_{YS} = 400 MPa and E = 210,000 MPa, the cohesive strength would be ~50 times higher than the maximum stress achieved ahead of the crack tip. Thus a macroscopic crack provides insufficient stress concentration to exceed the bond strength.

In order for cleavage to initiate, there must be a local discontinuity ahead of the macroscopic crack that is sufficient to exceed the bond strength. A sharp microcrack is one way to provide sufficient local stress concentration. Cottrell [22] postulated that microcracks form at intersecting slip planes by means of dislocation interaction. A far more common mechanism for microcrack formation in steels, however, involves inclusions and second phase particles [1,23,24].

Figure 5.21 illustrates the mechanism of cleavage nucleation in ferritic steels. The macroscopic crack provides a local stress and strain concentration. A second phase particle, such as a carbide or inclusion, cracks because of the plastic strain in the surrounding matrix. At this point the microcrack can be treated as a Griffith crack (Section 2.3). If the stress ahead of the macroscopic crack is sufficient, the microcrack propagates into the ferrite matrix, causing failure by cleavage. For example, if the particle is spherical and it produces a penny-shaped crack, the fracture stress is given by

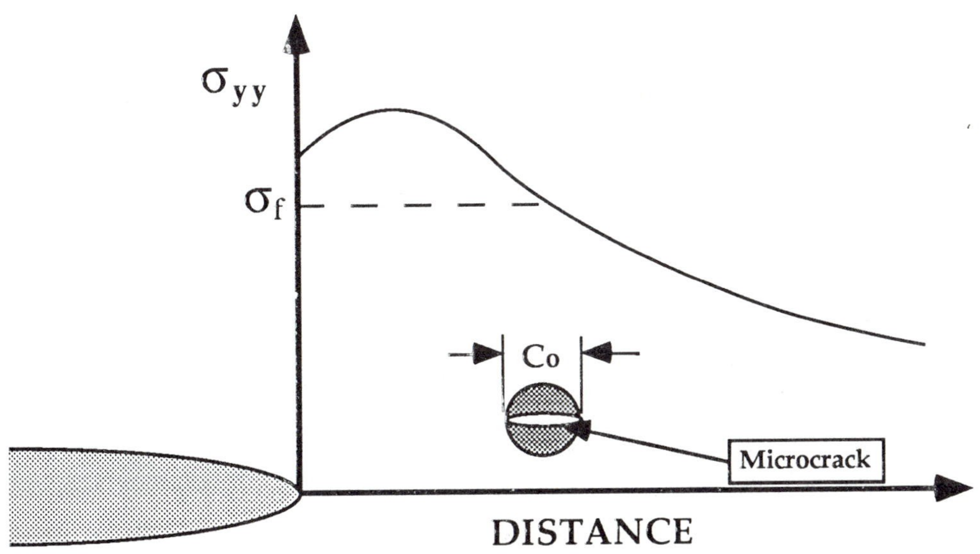

FIGURE 5.21 Initiation of cleavage at a microcrack that forms in a second phase particle ahead of a macroscopic crack.

$$\sigma_f = \left(\frac{\pi \, E \, \gamma_p}{(1 - \nu^2) \, C_o} \right)^{1/2} \tag{5.18}$$

where γ_p is the plastic work required to create a unit area of fracture surface in the ferrite and C_o is the particle diameter. It is assumed that $\gamma_p >> \gamma_s$, the surface energy (c.f. Eq. (2.21)). Note that the stress ahead of the macrocrack is treated as a remote stress in this case.

Consider the hypothetical material described earlier, where σ_{YS} = 400 MPa and E = 210,000 MPa. Knott [1] has estimated γ_p = 14 J/m^2 for ferrite. Setting $\sigma_f = 3\,\sigma_{YS}$ and solving for critical particle diameter yields C_o = 7.05 μm. Thus the Griffith criterion can be satisfied with relatively small particles.

The nature of the microstructural feature that nucleates cleavage depends on the alloy and heat treatment. In mild steels, cleavage usually initiates at grain boundary carbides [1,23,24]. In quenched and tempered alloy steels, the critical feature is usually either a spherical carbide or an inclusion [1,25]. Various models have been developed to explain the relationship between cleavage fracture stress and microstructure; most of these models resulted in expressions similar to Eq. (5.18). Smith [24] proposed a model for cleavage fracture that considers stress concentration due to a dislocation pile-up at a grain boundary carbide. The resulting failure criterion is as follows:

$$C_o \, \sigma_f^2 + k_y^2 \left[\frac{1}{2} + \frac{2 \, \tau_i \sqrt{C_o}}{\pi \, k_y} \right]^2 = \frac{4 \, E \, \gamma_p}{\pi \, (1 - \nu^2)} \tag{5.19}$$

where C_o, in this case, is the carbide thickness, and τ_i and k_y are the friction stress and pile-up constant, respectively, as defined in the Hall-Petch equation:

$$\tau_y = \tau_i + k_y \, d^{-1/2}$$

where τ_y is the yield strength in shear. The second term on the left side of Eq. (5.19) contains the dislocation contribution to cleavage initiation. If this term is removed, Eq. (5.19) reduces to the Griffith relationship for a grain boundary microcrack.

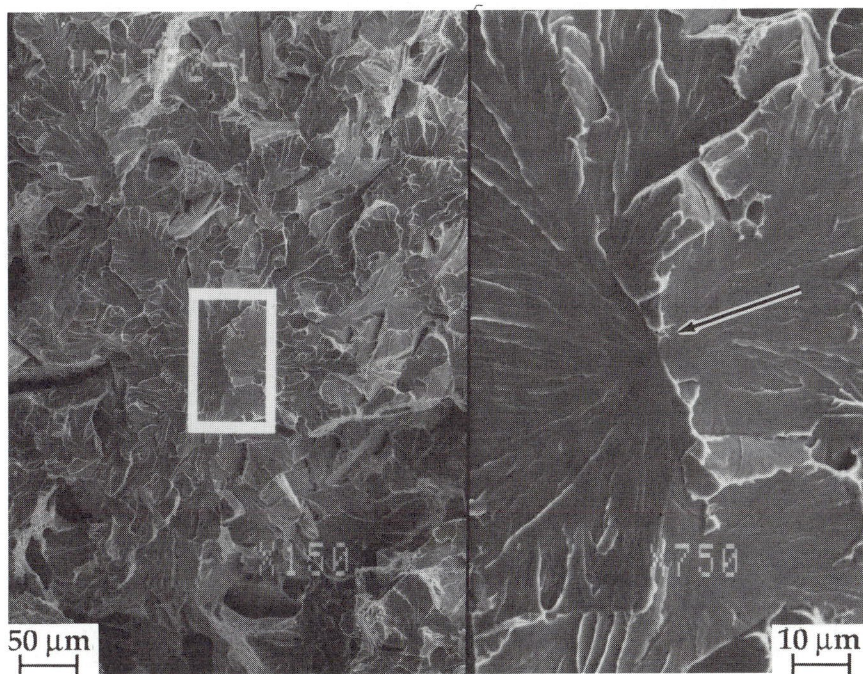

(a) Initiation at a grain boundary carbide.

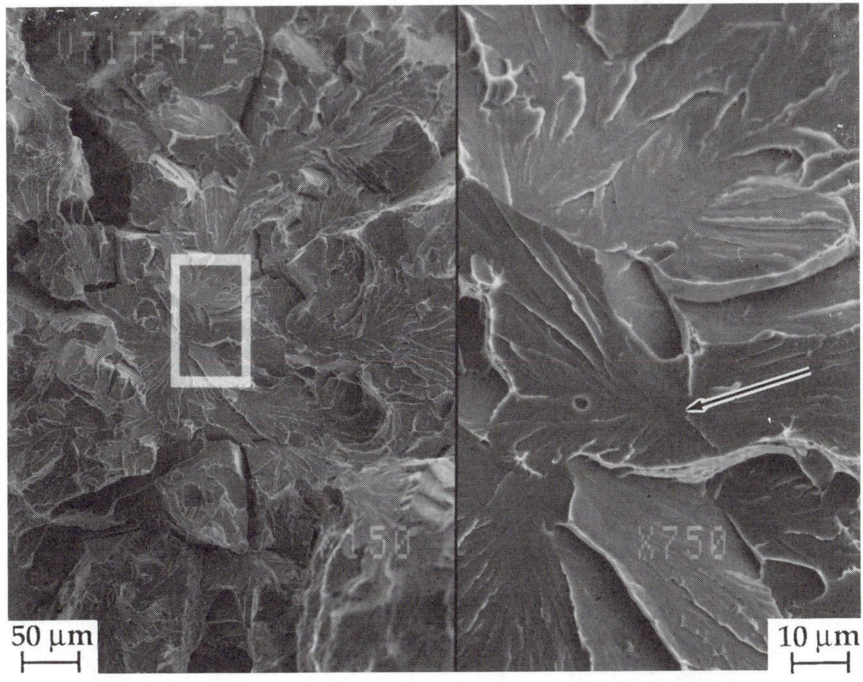

(b) Initiation at an inclusion near the center of a grain.

FIGURE 5.22 SEM fractographs of cleavage initiation in an A 508 Class 3 alloy. (Photographs provided by M.T. Miglin.)

Figure 5.22 shows SEM fractographs which give examples of cleavage initiation from a grain boundary carbide (a) and an inclusion at the interior of a grain (b). In both cases, the fracture origin was located by following river patterns on the fracture surface.

Susceptibility to cleavage fracture is enhanced by almost any factor that increases the yield strength, such as low temperature, a triaxial stress state, radiation damage, high strain rate, and strain aging. Grain size refinement increases the yield strength but also increases σ_f. There are a number of reasons for the grain size effect. In mild steels, a decrease in grain size implies an increase in grain boundary area, which leads to smaller grain boundary carbides and an increase in σ_f. In fine grained steels, the critical event may be propagation of the microcrack across the first grain boundary it encounters. In such cases the Griffith model implies the following expression for fracture stress:

$$\sigma_f = \left(\frac{\pi E \gamma_{gb}}{(1 - \nu^2) d} \right)^{1/2} \tag{5.20}$$

where γ_{gb} is the plastic work per unit area required to propagate into the adjoining grain. Since there tends to be a high degree of mismatch between grains in a polycrystalline material, $\gamma_{gb} > \gamma_p$. Equation (5.20) assumes an equiaxed grain structure. For martensitic and bainitic microstructures, Dolby and Knott [26] derived a modified expression for σ_f based on the packet diameter.

In some cases cleavage nucleates, but total fracture of the specimen or structure does not occur. Figure 5.23 illustrates three examples of unsuccessful cleavage events. Part (a) shows a microcrack that has arrested at the particle/matrix interface. The particle cracks due to strain in the matrix, but the crack is unable to propagate because the applied stress is less than the required fracture stress. This microcrack does not re-initiate because subsequent deformation and dislocation motion in the matrix causes the crack to blunt. Microcracks must remain sharp in order for the stress on the atomic level to exceed the cohesive strength of the material. If a microcrack in a particle propagates into the ferrite matrix, it may arrest at the grain boundary, as illustrated in Fig. 5.23(b). This corresponds to a case where Eq. (5.20) governs cleavage. Even if a crack successfully propagates into the surrounding grains, it may still arrest if there is a steep stress gradient ahead of the macroscopic crack (Fig. 5.23(c)). This tends to occur at low applied K_I values. Locally, the stress is sufficient to satisfy Eqs. (5.18) and (5.20) but there is insufficient global driving force to continue crack propagation. Figure 5.24 shows an example of arrested cleavage cracks in front of a macroscopic crack in a spherodized 1008 steel [27].

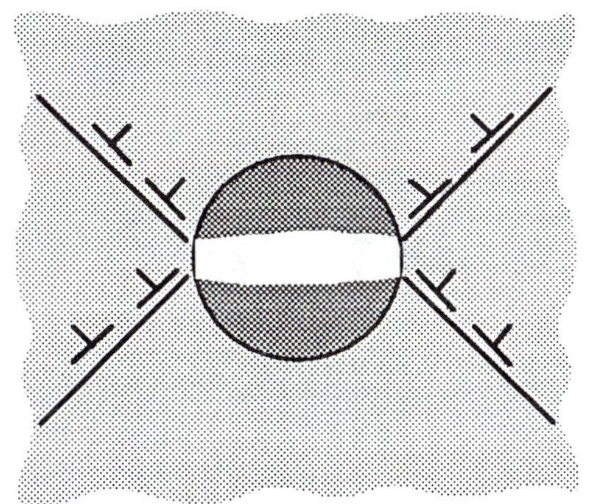

(a) Arrest at particle/matrix interface

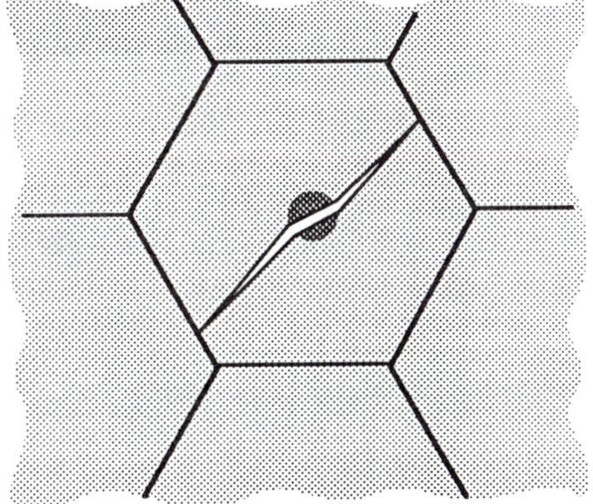

(b) Arrest at a grain boundary.

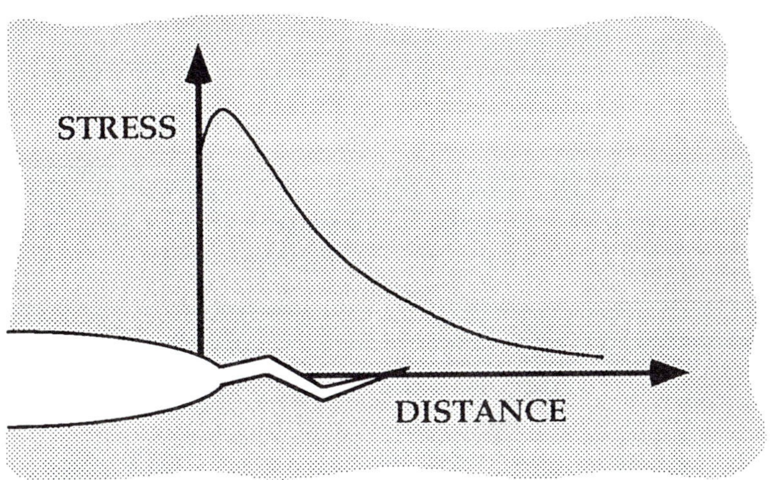

(c) Arrest due to a steep stress gradient.

FIGURE 5.23 Examples of unsuccessful cleavage events.

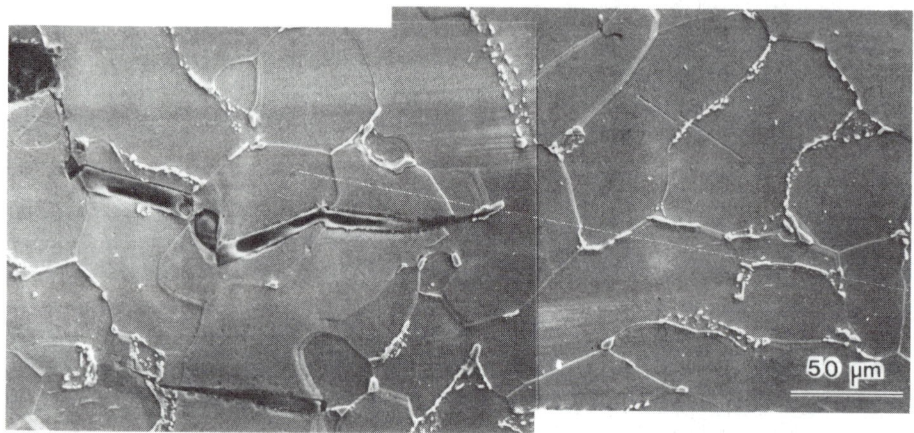

FIGURE 5.24 **Arrested cleavage cracks ahead of a macroscopic crack in a spherodized 1008 steel [27].**

5.2.3 Mathematical Models of Cleavage Fracture Toughness

A difficulty emerges when trying to predict fracture toughness from Eqs. (5.18) to (5.20). The maximum stress ahead of a macroscopic crack occurs at approximately 2δ from the crack tip, but the absolute value of this stress is constant in small scale yielding (Fig. 5.14); the distance from the crack tip at which this stress occurs increases with increasing K, J, and δ. Thus if attaining a critical fracture stress were a sufficient condition for cleavage fracture, the material might fail upon application of an infinitesimal load, because the stresses would be high near the crack tip. Since ferritic materials have finite toughness, attainment of a critical stress ahead of the crack tip is apparently necessary but not sufficient.

Ritchie, Knott and Rice (RKR) [28] introduced a simple model to relate fracture stress to fracture toughness, and to explain why steels did not spontaneously fracture upon application of minimal load. They postulated that cleavage failure occurs when the stress ahead of the crack tip exceeds σ_f over a characteristic distance, as illustrated in Fig.

5.25. They inferred σ_f in a mild steel from blunt notched four-point bend specimens and measured K_{IC} with conventional fracture toughness specimens. They inferred the crack tip stress field from a finite element solution published by Rice and Tracey [29]. They found that the characteristic distance was equal to two grain diameters for the material they tested. Ritchie, et al. argued that if fracture initiates in a grain boundary carbide and propagates into a ferrite grain, the stress must be sufficient to propagate the cleavage crack across the opposite grain boundary and into the next grain; thus σ_f must be exceeded over 1 or 2 grain diameters. Subsequent investigations [25,30,31], however, revealed no consistent relationship between the critical distance and grain size.

Curry and Knott [32] provided a statistical explanation for the RKR critical distance. A finite volume of material must be sampled ahead of the crack tip in order to find a particle that is sufficiently large to nucleate cleavage. Thus a critical sample volume, over which $\sigma_{YY} \geq \sigma_f$, is required for failure. The critical volume, which can easily be related to a critical distance, depends on the average spacing of cleavage nucleation sites.

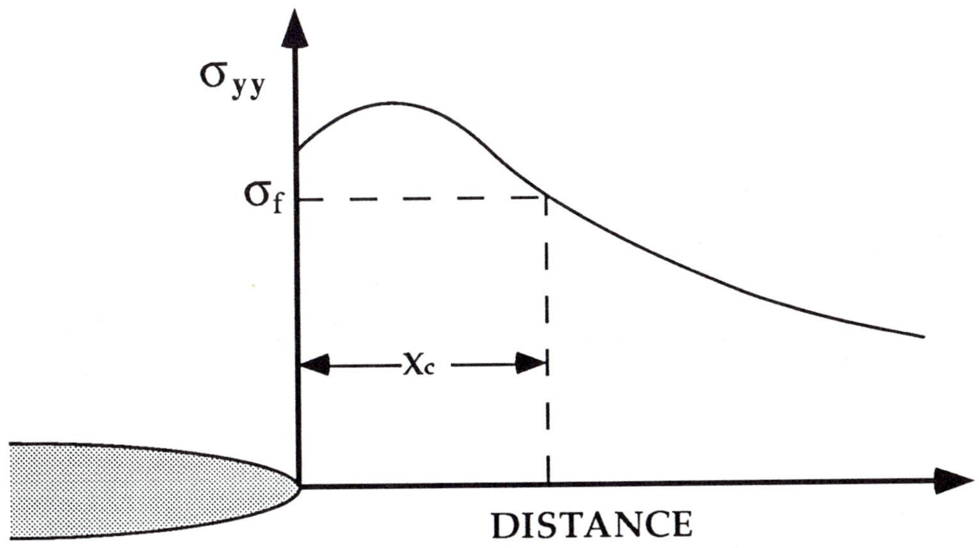

FIGURE 5.25 The Ritchie-Knott-Rice model for cleavage fracture [28]. Failure is assumed to occur when the fracture stress is exceeded over a characteristic distance.

The statistical argument also explains why cleavage fracture toughness data tend to be widely scattered. Two nominally identical specimens made from the same material may display vastly different toughness values because the location of the critical fracture-triggering particle is random. If one specimen samples a large fracture-triggering particle near the crack tip, while the fracture trigger in the other specimen is further from the crack tip, the latter specimen will display a higher fracture toughness, because a higher load is required to elevate the stress at the particle to a critical value. The statistical nature of fracture also leads to an apparent thickness effect on toughness. A thicker specimen is more likely to sample a large fracture trigger along the crack front and therefore, will have a lower toughness that a thin specimen, on average [33-35].

The Curry and Knott approach was followed by more formal statistical models for cleavage [27,35-38]. These models all treated cleavage as a weakest link phenomenon, where the probability of failure is equal to the probability of sampling at least one critical fracture-triggering particle. For a volume of material V, with ρ critical particles per unit volume, the probability of failure can be inferred from the Poisson distribution:

$$F = 1 - \exp(-\rho V) \tag{5.21a}$$

The second term is the probability of finding zero critical particles in V, so F is the probability of sampling one or critical more particles. The Poisson distribution can be derived from the binomial distribution by assuming that ρ is small and V is large, an assumption that is easily satisfied in the present problem.[2] Since the critical particle size depends on stress, which varies ahead of the crack tip, ρ must vary with position. Therefore, for crack problems, the failure probability must be integrated over individual volume elements ahead of the crack tip:

---

[2]For a detailed discussion of the Poisson assumption, consult any textbook on probability and statistics.

$$F = 1 - \exp\left[- \int_V \rho \, dV \right] \tag{5.21b}$$

Assuming ρ depends only on the locally applied stress, and the crack tip conditions are uniquely defined by K or J, it can be shown (Appendix 5.2) that critical values of K and J follow a characteristic distribution when failure is controlled by a weakest link mechanism[3]:

$$F = 1 - \exp\left[- \left(\frac{K_{IC}}{\Theta_K} \right)^4 \right] \tag{5.22a}$$

or

$$F = 1 - \exp\left[- \left(\frac{J_c}{\Theta_J} \right)^2 \right] \tag{5.22b}$$

where Θ_K and Θ_J are material properties that depend on microstructure and temperature. Equations (5.22a) and (5.22b) have the form of a two-parameter Weibull distribution [39]. The Weibull shape parameter, which is sometimes called the Weibull slope, is equal to 4.0 for K_{IC} data and (because of the relationship between K and J) 2.0 for J_c values for cleavage. The Weibull scale parameters, Θ_K and Θ_J, are the 63rd percentile values of K_{IC} and J_c, respectively. If Θ_K or Θ_J are known, the entire fracture toughness distribution can be inferred from Eq. (5.22a) or (5.22b).

The prediction of a fracture toughness distribution that follows a two parameter Weibull function with a known slope is an important result. The Weibull slope is a measure of the relative scatter; a prior knowledge of the Weibull slope enables the relative scatter to be predicted a priori, as Example 5.1 illustrates.

---

[3]Equations (5.22a) and (5.22b) apply only when the thickness (i.e. the crack front length) is fixed. The weakest link model predicts a thickness effect, which is described in Appendix 5.2 but is omitted here for brevity.

EXAMPLE 5.1

Determine the relative size of the 90% confidence bounds of K_{IC} and J_c data, assuming Eqs. (5.22a) and (5.22b) describe the respective distributions.

Solution: The median, 5% lower bound and 95% upper bound values are obtained by setting F = 0.5, 0.05 and 0.95, respectively, in Eqs. (5.22a) and (5.22b). Both equations have the form:

$$F = 1 - \exp(-\lambda)$$

Solving for λ at each probability level gives

$$\lambda_{0.50} = 0.693 \qquad \lambda_{0.05} = 0.0513 \quad \lambda_{0.95} = 2.996$$

The width of the 90% confidence band in K_{IC} data, normalized by the median, is given by

$$\frac{K_{0.95} - K_{0.05}}{K_{0.50}} = \frac{(2.996)^{0.25} - (0.0513)^{0.25}}{(0.693)^{0.25}} = 0.920$$

and the relative width of the J_c scatter band is

$$\frac{J_{0.95} - J_{0.05}}{J_{0.50}} = \frac{\sqrt{2.996} - \sqrt{0.0513}}{\sqrt{0.693}} = 1.81$$

Note that Θ_K and Θ_J cancel out of the above results and the relative scatter depends only on the Weibull slope.

There are two major problems with the weakest link model that leads to Eqs. (5.22a) and (5.22b). First, these equations predict zero as the minimum toughness in the distribution. Intuition suggests that such a prediction is incorrect, and more formal arguments can be made for a nonzero threshold toughness. A crack cannot propagate in a material unless there is sufficient energy available to break bonds and perform plastic work. If the material is a polycrystal, additional work must be performed when the crack crosses randomly oriented grains. Thus one can make an estimate of threshold toughness in terms of energy release rate:

$$\mathcal{G}_{c(min)} \approx 2\,\gamma_p\,\phi \tag{5.23}$$

where ϕ is a grain misorientation factor. If the global driving force is less than $\mathcal{G}_{c(min)}$, the crack cannot propagate. The threshold toughness can also be viewed as a crack arrest value: a crack cannot propagate if $K_I < K_{IA}$.

A second problem with Eqs. (5.22a) and (5.22b) is that they tend to overpredict the experimental scatter. That is, scatter in experimental cleavage fracture toughness data is usually less severe than predicted by the weakest link model.

According to the weakest link model, failure is controlled by the initiation of cleavage in the ferrite as the result of cracking of a critical particle; i.e a particle that satisfies Eq. (5.18) or (5.19). While weakest link initiation is necessary, it is apparently not sufficient for total failure. A cleavage crack, once initiated, must have sufficient driving force to propagate. Recall Fig. 5.22, which gives examples of unsuccessful cleavage events.

Both problems, threshold toughness and scatter, can be addressed by incorporating a conditional probability of propagation into the statistical model [40,41]. Figure 5.26 is a probability tree for cleavage initiation and propagation. When a flawed structure is subject to an applied K, a microcrack may or may not initiate, depending on the temperature as well as the location of the eligible cleavage triggers. Initiation of cleavage cracks should be governed by a weakest link mechanism, because the process involves searching for a large enough trigger to propagate a microcrack into the first ferrite grain. Once cleavage initiates, the crack may either propagate in an unstable fashion or arrest (as in Fig. 5.23(b) and (c)). Initiation is governed by the local stress at the critical particle, while propagation is controlled by the orientation of the neighboring grains and the global driving force. The overall probability of failure is equal to the probability of initiation times the conditional probability of propagation.

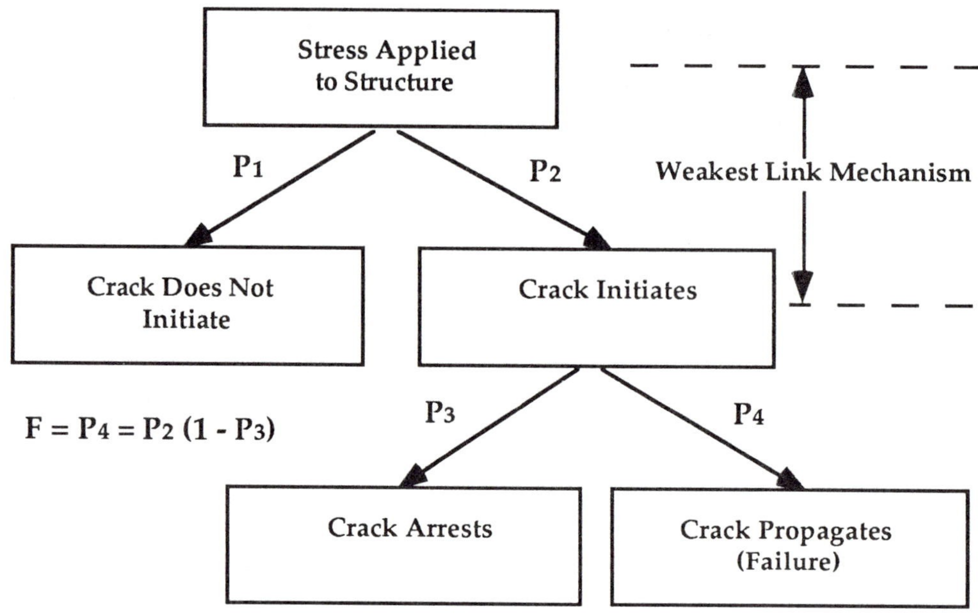

FIGURE 5.26 Probability tree for cleavage initiation and propagation.

This model assumes that if a microcrack arrests, it does not contribute to subsequent failure. This is a reasonable assumption, since only a rapidly propagating crack is sufficiently sharp to give the stress intensification necessary to break bonds. Once a microcrack arrests, it is blunted by local plastic flow.

Consider the case where the conditional probability of propagation is a step function:

$$P_{pr} = \begin{cases} 0 & K_I < K_o \\ 1 & K_I \geq K_o \end{cases}$$

That is, assume that all cracks arrest when $K_I < K_o$ and that a crack propagates if $K_I \geq K_o$ at the time of initiation. This assumption implies that the material has a crack arrest toughness that is single valued. It can be shown (see Appendix 5.2) that such a material exhibits the following fracture toughness distribution on K values:

$$F = 1 - \exp\left\{-\left[\left(\frac{K_{IC}}{\Theta_K}\right)^4 - \left(\frac{K_o}{\Theta_K}\right)^4\right]\right\} \quad \text{for } K_I > K_o \qquad (5.24a)$$

$$F = 0 \qquad \text{for } K_I \leq K_o \tag{5.24b}$$

Equation (5.24) is a *truncated Weibull* distribution; Θ_K can no longer be interpreted as the 63rd percentile K_{IC} value. Note that a threshold has been introduced, which removes one of the shortcomings of the weakest link model. Equation (5.24) also exhibits less scatter than the two parameter distribution (Eq. 5.22a), thereby removing the other objection to the weakest link model.

The threshold is obvious in Eq. (5.24), but the reduction in relative scatter is less so. The latter effect can be understood by considering the limiting cases of Eq. (5.24). If $K_o/\Theta_K >> 1$, there are ample initiation sites for cleavage, but the microcracks cannot propagate unless $K_I > K_o$. Once K_I exceeds K_o, the next microcrack to initiate will cause total failure. Since initiation events are frequent in this case, K_{IC} values will be clustered near K_o, and the scatter will be minimal. On the other hand, if $K_o/\Theta_K << 1$, Eq. (5.24) reduces to the weakest link case. Thus the relative scatter decreases as K_o/Θ_K increases.

Equation (5.24) is an oversimplification, because it assumes a single-valued crack arrest toughness. In reality, there is undoubtedly some degree of randomness associated with microscopic crack arrest. When a cleavage crack initiates in a single ferrite grain, the probability of propagation into the surrounding grains depends in part on their relative orientation; a high degree of mismatch increases the likelihood of arrest at the grain boundary. Stienstra and Anderson [41] recently performed a probabilistic simulation of microcrack propagation and arrest in a polycrystalline solid. Initiation in a single grain ahead of the crack tip was assumed, and the tilt and twist angles at surrounding grains were allowed to vary randomly (within the geometric constrains imposed by assuming {100} cleavage planes). An energy-based propagation criterion, suggested by the work of Gell and Smith [42], was applied. The conditional probability of propagation was estimated over a range of applied K_I values. The results fit an offset power law expression:

$$P_{pr} = \alpha(K_I - K_o)^\beta \tag{5.25}$$

where α and β are material constants. Incorporating Eq. (5.25) into the overall probability analysis leads to a complicated distribution function that is very difficult to apply to experimental data (see Appendix 5.2). Stienstra and Anderson found, however, that this new function could be approximated by a three-parameter Weibull distribution:

$$F = 1 - \exp\left[-\left(\frac{K_{IC} - \Theta_o}{\Theta_K}\right)^4\right] \tag{5.26}$$

where Θ_o is the Weibull location parameter. Stienstra and Anderson showed that when experimental data are fit to Eq. (5.26), Θ_o gives a conservative estimate of K_o, the true threshold toughness of the material.

Figure 5.27 shows experimental cleavage fracture toughness data for a low alloy steel. Critical J values measured experimentally were converted to equivalent K_{IC} data. The data were corrected for constraint loss through an analysis developed by Anderson and Dodds [43] (see Section 3.6.1). Equations (5.22a), (5.24), and (5.26) were fit to to the

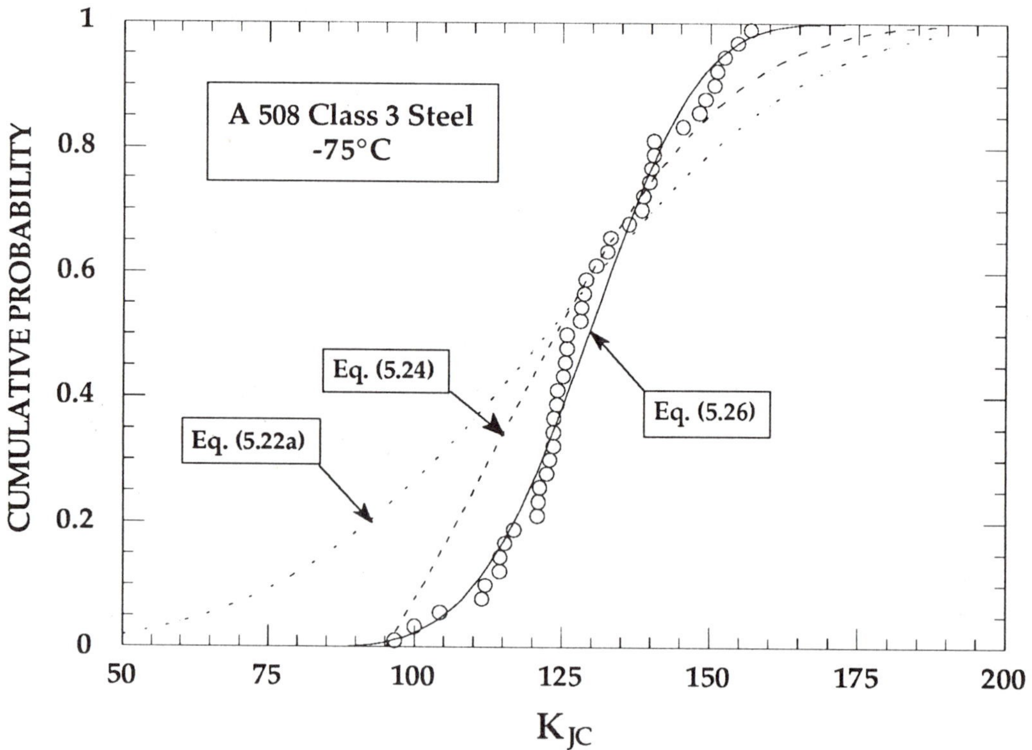

FIGURE 5.27 Cleavage fracture toughness data for an A 508 Class 3 steel at -75°C [41]. The data have been fit to various statistical distributions.

experimental data. The three parameter Weibull distribution obviously gives the best fit. The weakest link model (Eq. (5.22a)) overestimates the scatter, while the truncated Weibull distribution does not follow the data in the lower tail, presumably because the assumption of a single valued arrest toughness is incorrect.

5.3 THE DUCTILE-BRITTLE TRANSITION

The fracture toughness of ferritic steels can change drastically over a small temperature range, as illustrated in Fig 5.28. At low temperatures, steel is brittle and fails by cleavage. At high temperatures, the material is ductile and fails by microvoid coalescence. Ductile fracture initiates at a particular toughness value, as indicated by the dashed line in Fig. 5.28. The crack grows as load is increased. Eventually, the specimen fails by plastic collapse or tearing instability. In the transition region between ductile and brittle behavior, both micromechanisms of fracture can occur in the same specimen. In the lower transition region, the fracture mechanism is pure cleavage, but the toughness increases rapidly with temperature as cleavage becomes more difficult. In the upper transition region, a crack initiates by microvoid coalescence but ultimate failure occurs by cleavage. On initial loading in the upper transition region, cleavage does not occur because there are no critical particles near the crack tip. As the crack grows by ductile tearing, however, more material is sampled. Eventually, the growing crack samples a critical particle and cleavage occurs. Because fracture toughness in the transition region is governed by these statistical sampling effects, the data tend to be highly scattered. Wallin [44] has developed a statistical model for the transition region which incorporates the effect of prior ductile tearing on the cleavage probability.

Recent work by Heerens and Read [25] demonstrates the statistical sampling nature of cleavage fracture in the transition region. They performed a large number of fracture toughness tests on a quenched and tempered alloy steel at several temperatures in the transition region. As expected, the data at a given temperature were highly scattered. Some specimens failed without significant stable crack growth while other specimens sustained high levels of ductile tearing prior to cleavage. Heerens and Read examined the fracture surface of each specimen to determine the site of cleavage initiation. The measured distance from the initiation site to the original crack tip correlated very well with the measured fracture toughness. In specimens that exhibited low toughness, this distance was small; a critical nucleus was available near the crack tip. In the specimens that exhibited high toughness, there were no critical particles near the crack tip; the crack had to grow and sample additional material before a critical cleavage nucleus was found. Figure 5.29 is a plot of fracture toughness versus the critical distance, r_c, which Heerens and Read measured from the fracture surface; r_c is defined as the distance from the fatigue crack tip to the cleavage initiation site. The resistance curve for ductile crack growth is also shown on this plot. In every case, cleavage initiated near the location of the maximum tensile stress (c.f. Fig. 5.14). Similar fractographic studies by Wantanabe et al. [31] and Rosenfield and Shetty [45] also revealed a correlation between J_c, Δa, and r_c.

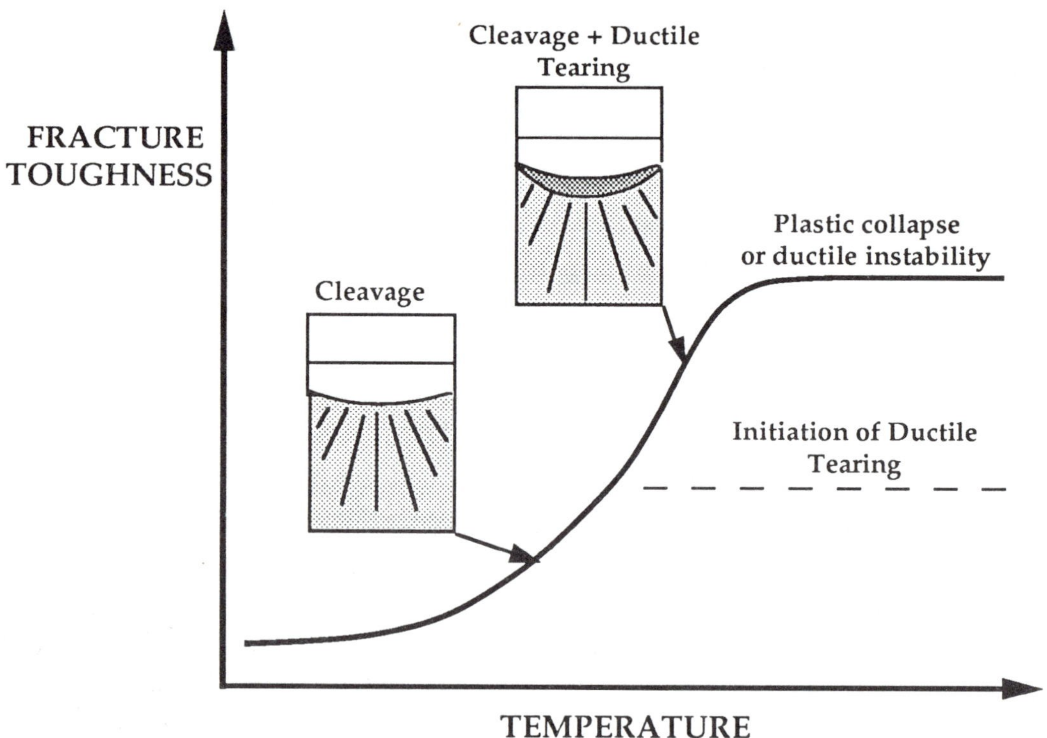

FIGURE 5.28 The ductile-brittle transition in ferritic steels. The fracture mechanism changes from cleavage to microvoid coalescence as temperature increases.

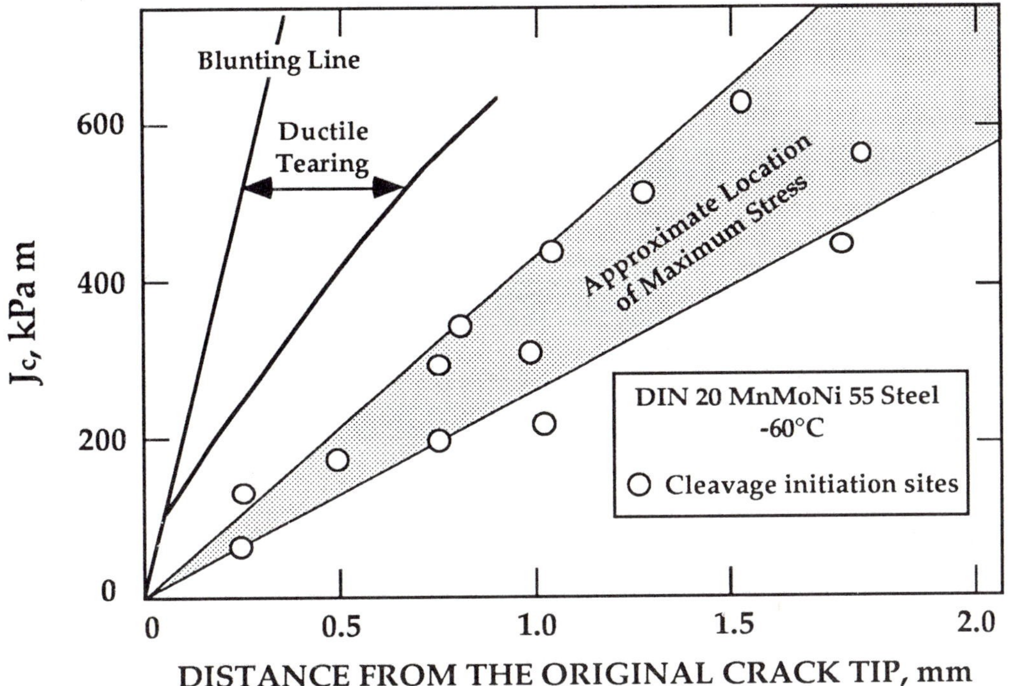

FIGURE 5.29 Relationship between cleavage fracture toughness and the distance between the crack tip and the cleavage trigger [25].

Cleavage propagation in the upper transition region often displays isolated islands of ductile fracture [21,46]. When specimens with arrested macroscopic cleavage cracks are studied metallographically, unbroken ligaments are sometimes discovered behind the arrested crack tip. These two observations imply that a propagating cleavage crack in the upper transition region encounters barriers, such as highly misoriented grains or particles, through which the crack cannot propagate. The crack is diverted around these obstacles, leaving isolated unbroken ligaments in its wake. As the crack propagation continues and the crack faces open, the ligaments that are well behind the crack tip rupture. Figure 5.30 schematically illustrates this postulated mechanism. The energy required to rupture the ductile ligaments may provide the majority of the propagation resistance a cleavage crack experiences. The concentration of ductile ligaments on a fracture surface increases with temperature [46], which may explain why crack arrest toughness (K_{Ia}) exhibits a steep brittle-ductile transition, much like K_{IC} and J_c.

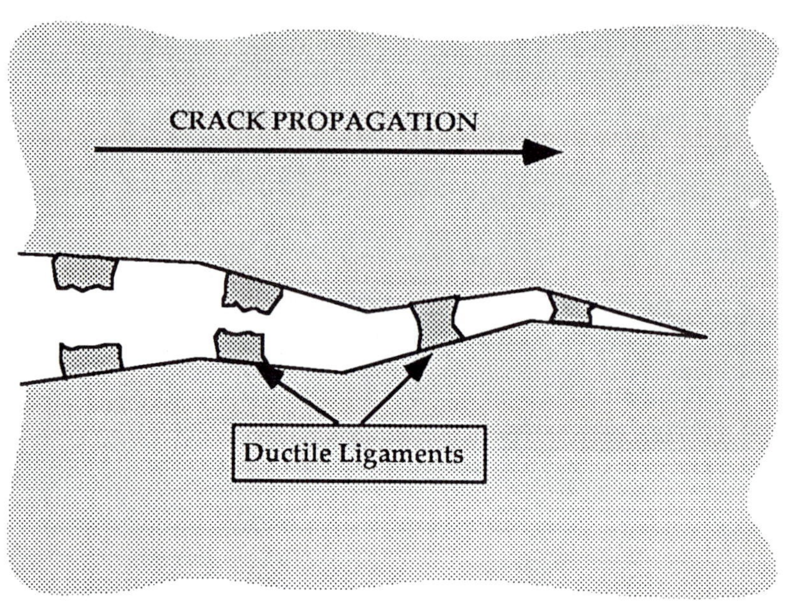

FIGURE 5.30 Schematic illustration of cleavage crack propagation in the ductile-brittle transition region. Ductile ligaments rupture behind the crack tip, resulting in increased propagation resistance.

5.4 INTERGRANULAR FRACTURE

In most cases metals do not fail along grain boundaries. Ductile metals usually fail by coalescence of voids formed at inclusions and second phase particles, while brittle metals typically fail by transgranular cleavage. Under special circumstances, however, cracks can form and propagate along grain boundaries.

There is no single mechanism for intergranular fracture. Rather, there are a variety of situations that can lead to cracking on grain boundaries, including:

(1) Precipitation of a brittle phase on the grain boundary.

(2) Hydrogen embrittlement and liquid metal embrittlement.

(3) Environmental assisted cracking.

(4) Intergranular corrosion.

(5) Grain boundary cavitation and cracking at high temperatures.

Space limitations preclude discussing each of these mechanisms in detail. A brief description of the intergranular cracking mechanisms is given below.

Brittle phases can be deposited on grain boundaries of steel through improper tempering [47]. Tempered martensite embrittlement, which results from tempering near 350°C, and temper embrittlement, which occurs when an alloy steel is tempered at ~ 550°C, both apparently involve segregation of impurities, such as phosphorous and sulphur, to prior austenite grain boundaries. These thin layers of impurity atoms are not resolvable on the fracture surface, but can be detected with surface analysis techniques such as Auger electron spectroscopy. Segregation of aluminum nitride particles on grain boundaries during solidification is a common embrittlement mechanism in cast steels [47]. Aluminum nitride, if present in sufficient quantity, can also contribute to degradation of toughness resulting from temper embrittlement in wrought alloys.

Hydrogen can severely degrade the toughness of an alloy, and much has been written on this subject in the last 50 years [48,49]. Although the precise mechanism of hydrogen embrittlement is not completely understood, atomic hydrogen apparently bonds with the metal atoms and reduces the cohesive strength at grain boundaries. Hydrogen can come from a number of sources, including moisture, hydrogen containing compounds such as H_2S, and hydrogen gas. A common problem in steel weldments is moisture adsorption during welding, which leads to cracking in the heat affected zone. Liquid metals, when slightly above their melting temperature, can embrittle a second metal with a higher melting point. Steel, for example, can be embrittled when placed in contact with molten metals with low melting points, such as lithium and sodium. The mechanism for liquid metal embrittlement is believed to be similar to hydrogen embrittlement.

Environmental assisted cracking is related to hydrogen embrittlement, in that hydrogen plays a role in the cracking process. High strength alloys are most susceptible to environmental assisted cracking, and deleterious environments include H_20-NaCl solutions, H_2S, ammonia, and gaseous hydrogen. The cracking is time-dependent and usually follows grain boundaries. Figure 5.31 shows the fracture surface of an ammonia tanker that experienced environmental assisted cracking. Note the smooth "rock candy" appearance of the intergranular fracture. The chemical and transport processes that lead to environmental assisted cracking are as follows [51]:

(1) Transport of the deleterious environment to the crack tip.

(2) Reactions of the environment with the crack surfaces, resulting in localized dissolution and production of hydrogen.

(3) Hydrogen absorption into the alloy.

(4) Diffusion of the hydrogen to an embrittlement site ahead of the crack tip.

(5) Hydrogen-metal interactions leading to embrittlement and crack propagation.

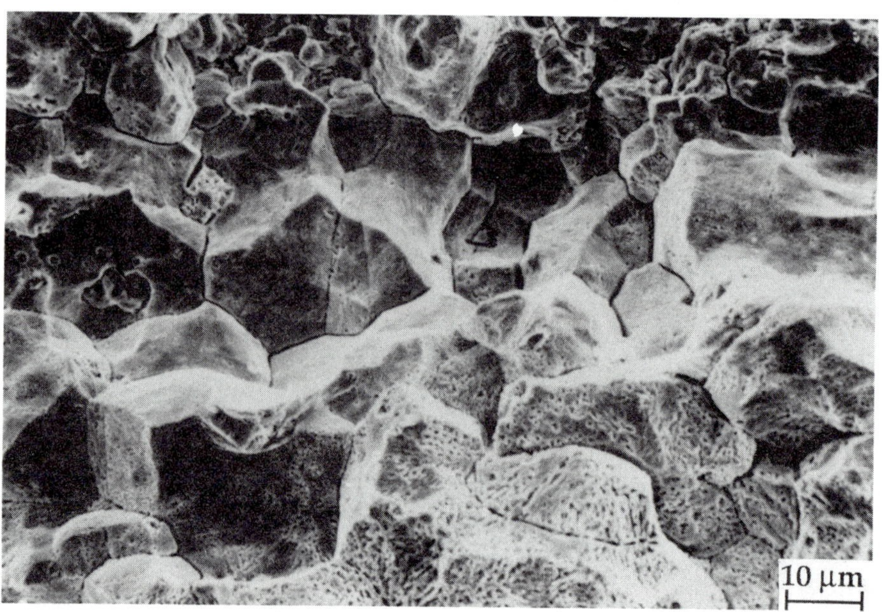

FIGURE 5.31 Intergranular fracture in a steel ammonia tank. (Photograph provided by W.L. Bradley)

Intergranular corrosion involves preferential attack of the grain boundaries, as opposed to general corrosion, where the material is dissolved relatively uniformly across the surface. Intergranular attack is different from environmental assisted cracking, in that there is no embrittlement mechanism associated with the grain boundary corrosion.

At high temperatures, grain boundaries are weak relative to the matrix, and a significant portion of creep deformation is accommodated by grain boundary sliding. In such cases void nucleation and growth (at second phase particles) is concentrated at the crack boundaries, and cracks form as grain boundary cavities grow and coalesce. Grain boundary cavitation is the dominant mechanism of creep crack growth in metals [51], and it can be characterized with time dependent parameters such as the C* integral (Chapter 4).

REFERENCES

1. Knott, J.F., "Micromechanisms of Fracture and the Fracture Toughness of Engineering Alloys." *Fracture 1977*, Vol 1, ICF4, Waterloo Canada, June 1977, pp. 61-91.

2. Knott, J.F. "Effects of Microstructure and Stress-State on Ductile Fracture in Metallic Alloys." In: *Advances in Fracture Research, Proceedings of the Seventh International Conference on Fracture (ICF7)*, K. Salama, et. al., Eds., Pergamon Press, Oxford, UK, 1989, pp. 125-138.

3. Wilsforf, H.G.F., "The Ductile Fracture of Metals: a Microstructural Viewpoint." *Materials Science and Engineering*, Vol 59, 1983, pp. 1-19.

4. Garrison, W.M. Jr. and Moody, N.R., "Ductile Fracture." *Journal of the Physics and Chemistry of Solids*, Vol. 48, 1987, pp. 1035-1074.

5. Knott, J.F., "Micromechanisms of Fibrous Crack Extension in Engineering Alloys." *Metal Science*, Vol. 14, 1980, pp. 327-336.

6. Argon, A.S., Im, J., Safoglu, R., "Cavity Formation from Inclusions in Ductile Fracture." *Metallurgical Transactions*, Vol. 6A, 1975, pp. 825-837.

7. Berimin, F.M., "Cavity Formation from Inclusions in Ductile Fracture of A 508 Steel." *Metallurgical Transactions*, Vol 12A, 1981, pp. 723-731.

8. Brown, L.M. and Stobbs, W.M., "The Work-Hardening of Copper-Silica v. Equilibrium Plastic Relaxation by Secondary Dislocations." *Philosophical Magazine*, 1976, Vol. 34, pp. 351-372.

9. Goods, S.H. and Brown, L.M., "The Nucleation of Cavities by Plastic Deformation." *Acta Metallurgica*, Vol. 27, 1979, pp. 1-15.

10. Van Stone, R.H., Cox, T.B., Low, J.R. Jr., and Psioda, P.A., "Microstructural Aspects of Fracture by Dimpled Rupture." *International Metallurgical Reviews*, Vol. 30, 1985, pp. 157-179.

11. Thomason, P.F., *Ductile Fracture of Metals*, Pergamon Press, Oxford, UK, 1990.

12. Rice, J.R. and Tracey, D.M., "On the Ductile Enlargement of Voids in Triaxial Stress Fields." *Journal of the Mechanics and Physics of Solids*, Vol. 17, 1969, pp. 201-217.

13. Gurson, A.L., "Continuum Theory of Ductile Rupture by Void Nucleation and Growth: Part 1—Yield Criteria and Flow Rules for Porous Ductile Media." *Journal of Engineering Materials and Technology*, Vol. 99, 1977, pp. 2-15.

14. Berg, C.A., "Plastic Dilation and Void Interaction." *Inelastic Behavior of Solids*, McGraw-Hill, New York, 1970, pp. 171-210.

15. Tvergaard, V., "On Localization in Ductile Materials Containing Spherical Voids." *International Journal of Fracture*, Vol. 18, 1982, pp. 237-252.

16. Tvergaard, V and Needleman, A., "Analysis of the Cup-Cone Fracture in a Round Tensile Bar." *Acta Metallurgica*, Vol 32, 1984, pp. 157-169.

17. Purtscher, P.T., "Micromechanisms of Ductile Fracture and Fracture Toughness in a High Strength Austenitic Stainless Steel." Ph.D. Dissertation, Colorado School of Mines, Golden, CO, April, 1990.

18. d'Escata, Y. and Devaux, J.C., "Numerical Study of Initiation, Stable Crack Growth, and Maximum Load with a Ductile Fracture Criterion Based on the Growth of Holes." ASTM STP 668, American Society of Testing and Materials, Philadelphia, 1979, pp. 229-248.

19. McMeeking, R.M. and Parks, D.M., "On Criteria for J-Dominance of Crack-Tip Fields in Large-Scale Yielding." ASTM STP 668, American Society of Testing and Materials, Philadelphia, 1979, pp. 175-194.

20. Beachem, C.D. and Yoder, G.R., "Elastic-Plastic Fracture by Homogeneous Microvoid Coalescence Tearing Along Alternating Shear Planes." Metallurgical Transactions, Vol. 4A, 1973, pp. 1145-1153.

21. Gudas, J.P. "Micromechanisms of Fracture and Crack Arrest in Two High Strength Steels." Ph.D. Dissertation, Johns Hopkins University, Baltimore, MD, 1985.

22. Cottrell, A.H., "Theory of Brittle Fracture in Steel and Similar Metals." *Transactions of the ASME*, Vol 212, 1958, pp. 192-203.

23. McMahan, C.J. Jr and Cohen, M., "Initiation of Cleavage in Polycrystalline Iron." *Acta Metallurgica*, Vol 13, 1965, pp. 591-604.

24. Smith, E., "The Nucleation and Growth of Cleavage Microcracks in Mild Steel." *Proceedings of the Conference on the Physical Basis of Fracture,* Institute of Physics and Physics Society, 1966, pp. 36-46.

25. Heerens, J. and Read, D.T., "Fracture Behavior of a Pressure Vessel Steel in the Ductile-to-Brittle Transition Region." NISTIR 88-3099, National Institute for Standards and Technology, Boulder, CO, December, 1988.

26. Dolby, R.E. and Knott, J.F., "Toughness of Martensitic and Martensitic-Bainitic Microstructures with Particular Reference to Heat-Affected Zones." *Journal of the Iron and Steel Institute,* Vol 210, 1972, p. 857-865.

27. Lin, T., Evans, A.G. and Ritchie, R.O., "Statistical Model of Brittle Fracture by Transgranular Cleavage." *Journal of the Mechanics and Physics of Solids,* Vol 34, 1986, pp. 477-496.

28. Ritchie, R.O., Knott, J.F., and Rice, J.R. "On the Relationship between Critical Tensile Stress and Fracture Toughness in Mild Steel." *Journal of the Mechanics and Physics of Solids,* Vol. 21, 1973, pp. 395-410.

29. Rice, J.R. and Tracey, D.M. "Computational Fracture Mechanics." *Numerical Computer Methods in Structural Mechanics,* Academic Press, New York, 1973, pp. 585-623.

30. Curry, D.A. and Knott, J.F., "Effects of Microstructure on Cleavage Fracture Stress in Steel." *Metal Science,* 1978, pp. 511-514.

31. Watanabe, J., Iwadate, T., Tanaka, Y., Yokoboro, T. and Ando, K., "Fracture Toughness in the Transition Region." *Engineering Fracture Mechanics,* Vol. 28, 1987, pp. 589-600.

32. Curry D.A. and Knott, J.F., "Effect of Microstructure on Cleavage Fracture Toughness in Mild Steel." *Metal Science,* Vol. 13, 1979, pp. 341-345

33. Landes J.D. and Shaffer, D.H., "Statistical Characterization of Fracture in the Transition Region." ASTM STP 700, American Society of Testing and Materials, Philadelphia, 1980, pp. 368-372.

34. Anderson, T.L. and Williams, S., "Assessing the Dominant Mechanism for Size Effects in the Ductile-to-Brittle Transition Region.", ASTM STP 905, American Society of Testing and Materials, Philadelphia, 1986, pp. 715-740.

35. Anderson, T.L. and Stienstra, D., "A Model to Predict the Sources and Magnitude of Scatter in Toughness Data in the Transition Region." *Journal of Testing and Evaluation,* Vol. 17, 1989, pp. 46-53.

36. Evans, A.G.. "Statistical Aspects of Cleavage Fracture in Steel.", *Metallurgical Transactions,* Vol. 14A, 1983, pp. 1349-1355.

37. Wallin, K., Saario, T., and Törrönen, K., "Statistical Model for Carbide Induced Brittle Fracture in Steel." *Metal Science*, Vol. 18, 1984, pp. 13-16.

38. Beremin, F.M., "A Local Criterion for Cleavage Fracture of a Nuclear Pressure Vessel Steel." *Metallurgical Transactions*, Vol. 14A, 1983, pp. 2277-2287.

39. Weibull, W., "A Statistical Distribution Function of Wide Applicability." *Journal of Applied Mechanics*, Vol. 18, 1953, pp. 293-297.

40. Stienstra, D.I.A., "Stochastic Micromechanical Modeling of Cleavage Fracture in the Ductile-Brittle Transition Region." Ph.D. Dissertation, Texas A&M University, College Station, TX, August, 1990.

41. Stienstra, D.I.A. and Anderson, T.L., to be published.

42. Gell, M and Smith, E., "The Propagation of Cracks Through Grain Boundaries in Polycrystalline 3% Silicon-Iron." *Acta Metallurgica*, Vol. 15, 1967, pp. 253-258.

43. Anderson, T.L. and Dodds, R.H., Jr., "Specimen Size Requirements for Fracture Toughness Testing in the Ductile-Brittle Transition Region." *Journal of Testing and Evaluation*, to appear, 1991.

44. Wallin, K., "Fracture Toughness Testing in the Ductile-Brittle Transition Region." In: *Advances in Fracture Research, Proceedings of the Seventh International Conference on Fracture (ICF7)*, K. Salama, et. al., Eds., Pergamon Press, Oxford, UK, 1989, pp. 267-276.

45. Rosenfield, A.R. and Shetty, D.K., "Cleavage Fracture in the Ductile-Brittle Transition Region." ASTM STP 856, American Society for Testing and Materials, Philadelphia, 1985, pp. 196-209.

46. Hoagland, R,G,., Rosenfield, A.R., and Hahn, G.T., "Mechanisms of Fast Fracture and Arrest in Steels." *Metallurgical Transactions*, Vol. 3, 1972, pp. 123-136.

47. Krauss, G., *Principles of Heat Treatment of Steel*. American Society for Metals, Metals Park, OH, 1980.

48. Thompson, A.W. and Bernstein, I.M. (Eds.), *Effect of Hydrogen on Behavior of Materials*. TMS-AIME, Warrendale, 1976.

49. Anon, *Hydrogen Damage*. American Society for Metals, Metals Park, OH, 1977.

50. Wei, R.P. and Gangloff, R.P., "Environmentally Assisted Crack Growth in Structural Alloys: Perspectives and New Directions." ASTM STP 1020, American Society for Testing and Materials, Philadelphia, 1989, pp. 233-264.

51. Riedel, H., "Creep Crack Growth." ASTM STP 1020, American Society for Testing and Materials, Philadelphia, 1989, pp. 101-126.

52. Bain, L.J., *Statistical Analysis of Reliability and Life-Testing Models*, Marcel Dekker, Inc. New York, 1978.

APPENDIX 5: MICROMECHANICAL MODELING OF FRACTURE IN METALS

A5.1 THE EFFECT OF MICROMECHANISM ON THE SIZE DEPENDENCE OF FRACTURE TOUGHNESS

The single parameter assumption of classical fracture mechanics is only strictly valid for an infinite structure. In finite structures, this assumption may be approximately valid, but there is always a danger that the crack tip conditions are influenced by external boundaries and the measured fracture toughness is geometry dependent. The sensitivity of fracture toughness to specimen size and geometry depends strongly on the micromechanism of failure, as illustrated below.

Consider a material that obeys the HRR singularity (Chapter 3) in the limit of small scale yielding. The stress normal to the crack plane in this case can be written in the following form:

$$\sigma_{yy}^{(ssy)} = C_1 \left(\frac{J}{r}\right)^{\frac{1}{n+1}} \tag{A5.1}$$

were r is the radial distance from the crack tip, n is the Ramberg-Osgood strain hardening exponent, and C_1 is a constant that contains all of the remaining parameters in Eq. (3.24a). Consider now a finite specimen, where σ_{yy} is relaxed from the small scale yielding limit by the factor ϕ_σ:

$$\sigma_{yy} = \phi_\sigma C_1 \left(\frac{J}{r}\right)^{\frac{1}{n+1}} \tag{A5.2}$$

where $\phi_\sigma \leq 1$. Assume that fracture occurs when σ_{yy} reaches a critical value, σ_f, at a distance r* from the crack tip (i.e., the RKR model). The critical value of J under small scale yielding conditions is given by

$$J_c^{(ssy)} = r^* \left(\frac{\sigma_f}{C_1}\right)^{n+1} \tag{A5.3a}$$

and the critical J in the finite specimen is given by

$$J_c = r^* \left(\frac{\sigma_f}{\phi_\sigma C_1} \right)^{n+1} \tag{A5.3b}$$

Therefore,

$$\frac{J_c}{J_c^{(ssy)}} = \left(\frac{1}{\phi_\sigma} \right)^{n+1} \tag{A5.4}$$

The fracture toughness is very sensitive to slight deviations from the small scale yielding limit in this case. For example, if a material with n = 10 has stresses on the crack plane that are 10% below the HRR prediction (ϕ = 0.90), the elevation in fracture toughness can be estimated as follows:

$$\frac{J_c}{J_c^{(ssy)}} = \left(\frac{1}{0.90} \right)^{11} = 3.19$$

Thus a 10% deviation in stress results in more than a three-fold increase in the measured fracture toughness.

If failure is controlled by the strain ahead of the crack tip rather than the stress, the fracture toughness is much less sensitive to deviations from small scale yielding. The HRR solution for the crack tip strain fields is given by

$$\varepsilon_{ij}^{(ssy)} = C_2 \left(\frac{J}{r} \right)^{\frac{n}{n+1}} \tag{A5.5}$$

If we introduce a strain ratio, ϕ_ε, analogous to the stress ratio in Eq. (A5.2), the fracture toughness ratio for strain controlled fracture is given by

$$\frac{J_c}{J_c^{(ssy)}} = \left(\frac{1}{\phi_\varepsilon} \right)^{\frac{n+1}{n}} \tag{A5.6}$$

For n = 10 and $\phi_\varepsilon = 0.90$, the fracture toughness ratio = 1.12. Thus strain controlled fracture is much less sensitive to deviations from the small scale yielding limit. Other failure criteria would produce varying levels of sensitivity to crack tip constraint.

Since cleavage fracture is stress controlled, it is highly sensitive to the crack tip triaxiality. Figures 3.29 to 3.31 illustrate the predicted size and geometry dependence of cleavage fracture toughness, assuming the the maximum principal stress distribution controls failure [43]. In order to obtain a nearly size independent fracture toughness value in deeply notched bend specimens, the following requirement must be met:

$$B, b \geq \frac{200\, J_c}{\sigma_Y} \tag{A5.7}$$

where B and b are the thickness and ligament length, respectively, and σ_Y is the flow stress.

Experimental studies have shown that ductile fracture initiation toughness (J_{IC}) is nearly size independent when the following size requirement is met:

$$B, b \geq \frac{25\, J_{IC}}{\sigma_Y} \tag{A5.8}$$

which is eight times less restrictive than Eq. (A5.7).

The relative insensitivity to specimen size for ductile initiation toughness can be understood by considering the micromechanism of fracture. Refer to Eq. (5.11), which is a semi-empirical result obtained from the Rice and Tracey hole growth model [12]. Note that the rate of void growth, and presumably the ductile fracture toughness, depend on the hydrostatic stress and the equivalent plastic strain. When a finite sized specimen is loaded beyond small scale yielding, the stresses are typically lower than they would be in an infinite body, while the strains are somewhat higher [19]. Thus the relaxation in hydrostatic stress tends to inhibit ductile fracture, while the increase in plastic strain has the opposite effect; the stress relation is partially balanced by higher strains.

In summary, one must always consider the micromechanism of failure when assessing the validity of the single parameter assumption of fracture mechanics. Some fracture mechanisms, such as cleavage, are highly sensitive to specimen size; the single parameter assumption breaks down early in such materials. In other cases, such as ductile fracture, the single parameter assumption is valid to relatively large J values.

A5.2 STATISTICAL MODELING OF CLEAVAGE FRACTURE

When one assumes that fracture occurs by a weakest link mechanism under J controlled conditions, it is possible to derive a closed-form expression for the fracture toughness distribution. When weakest link initiation is necessary but not sufficient for cleavage fracture, the problem becomes somewhat more complicated, but it is still possible to describe the cleavage process mathematically.

A5.2.1 Weakest Link Fracture

As discussed in Section 5.2, the weakest link model for cleavage assumes that failure occurs when at least one critical fracture-triggering particle is sampled by the crack tip. Equation (5.21) describes the failure probability in this case.[4] Since cleavage is stress controlled, the microcrack density (i.e., the number of critical microcracks per unit volume) should depend only on the maximum principal stress[5]:

$$\rho = \rho(\sigma_1) \tag{A5.9}$$

---

[4]It turns out that Eq. (5.21) is valid even when the Poisson assumption is not [40]; the quantity ρ is not the microcrack density in such cases but ρ is *uniquely related* to microcrack density. Thus, the derivation of the fracture toughness distribution presented in this section does not hinge on the Poisson assumption.

[5]Although this derivation assumes that the maximum principal stress at a point controls the incremental cleavage probability, the same basic result can be obtained by inserting any stress component Eq. (A5.9). For example one might assume that the tangential stress, $\sigma_{\theta\theta}$, governs cleavage.

This quantity must be integrated over the volume ahead of the crack tip. In order to perform this integration, it is necessary to relate the crack tip stresses to the volume sampled at each stress level.

Recall Section 3.5, where dimensional analysis indicated that the stresses ahead of the crack tip in the limit of small scale yielding are given by

$$\frac{\sigma_1}{\sigma_o} = f\left(\frac{J}{r\,\sigma_o}, \theta\right) \qquad \text{(A5.10)}$$

assuming Young's modulus is fixed in the material (and thus does not need to be included in the dimensional analysis). Equation (A5.10) can be inverted to solve for the distance ahead of the crack tip (at a given angle) which corresponds to a particular stress value:

$$r(\sigma_1/\sigma_o, \theta) = \frac{J}{\sigma_o} g(\sigma_1/\sigma_o, \theta) \qquad \text{(A5.11)}$$

By fixing σ_1 and varying θ from $-\pi$ to $+\pi$, we can construct a contour of constant principal stress, as illustrated in Fig. A5.1. The area inside this contour is given by

$$A(\sigma_1/\sigma_o) = \frac{J^2}{\sigma_o^2} h(\sigma_1/\sigma_o) \qquad \text{(A5.12)}$$

where h is a dimensionless shape factor:

$$h(\sigma_1/\sigma_o) = \frac{1}{2}\int_{-\pi}^{+\pi} g(\sigma_1/\sigma_o, \theta)\, d\theta \qquad \text{(A5.13)}$$

For plane strain conditions in an edge cracked test specimen, the volume sampled at a given stress value is simply B A, where B is the specimen thickness. Therefore, the incremental volume at a fixed J and σ_o is given by

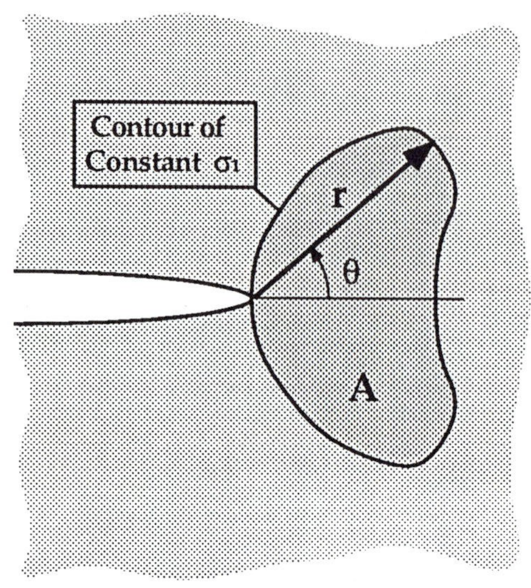

FIGURE A5.1 Definition of r, θ, and area for a principal stress contour.

$$dV(\sigma_1) = \frac{B\,J^2}{\sigma_o^2}\frac{\partial h}{\partial \sigma_1}\,d\sigma_1 \tag{A5.14}$$

Inserting Eqs. (A5.9) and (A5.14) into Eq. (5.21) gives

$$F = 1 - \exp\left(-\frac{B\,J^2}{\sigma_o^2}\int_{\sigma_u}^{\sigma_{max}} \rho(\sigma_1)\frac{\partial h}{\partial \sigma_1}\,d\sigma_1 \right) \tag{A5.15}$$

where σ_{max} is the peak value of stress that occurs ahead of the crack tip and σ_u is the threshold fracture stress, which corresponds to the largest fracture-triggering particle the material is likely to contain.

Note that J appears outside of the integral in Eq. (A5.15). By setting $J = J_c$ in Eq. (A5.15), we obtain an expression for the statistical distribution of critical J values, which can be written in the following form:

$$F = 1 - \exp\left[-\frac{B}{B_o}\left(\frac{J_c}{\Theta_J}\right)^2 \right] \tag{A5.16}$$

where B_o is a reference thickness, which is added to nondimensionalize B. When $B = B_o$, Θ_J is the 63rd percentile J_c value. Equation (A5.16) has the form of a two parameter Weibull distribution, as discussed in Section 5.2.3. Invoking the relationship between K and J for small scale yielding gives

$$F = 1 - \exp\left[-\frac{B}{B_o}\left(\frac{K_{IC}}{\Theta_K}\right)^4\right] \tag{A5.17}$$

Equations (A5.16) and (A5.17) both predict a thickness effect on toughness. The average toughness is proportional to $1/\sqrt{B}$ for critical J values and $B^{-0.25}$ for K_{IC} data. The average toughness does not increase indefinitely with thickness, however. There are limits to the validity of the weakest link model, as discussed in the next section.

All of the above relationships are only valid when weakest link failure occurs under J controlled conditions; i.e., the single parameter assumption must apply. When constraint relaxes, critical J values no longer follow a Weibull distribution with a specific slope, but the effective small scale yielding J values, J_{ssy}, (see Section 3.5) follow Eq. (A5.16) if a weakest link mechanism controls failure. Actual J_c values would be more scattered than J_{ssy} values, however, because the ratio J/J_{ssy} increases with J, as Figs. 3.29 to 3.31 illustrate.

A5.2.2 Incorporating a Conditional Probability of Propagation

In many materials, weakest link initiation of cleavage appears to be necessary but not sufficient. Figure 5.26 schematically illustrates a probability tree for cleavage initiation and propagation. This diagram is a slight oversimplification, because the cumulative failure probability must be computed incrementally.

Modifying the statistical cleavage model to account for propagation requires that the probability be expressed in terms of a *hazard function* [52], which defines the instantaneous risk of fracture. For a random variable T, the hazard function, H(T), and the cumulative probability are related as follows:

$$F = 1 - \exp\left(- \int_{T_o}^{T} H(T)\, dT \right) \qquad \text{(A5.18)}$$

where T_o is the minimum value of T. By comparing Eqs. (A5.17) and (A5.18), it can easily be shown that the hazard function for weakest link initiation, in terms of stress intensity, is given by

$$H(K) = \frac{4\,K^3}{\Theta_K{}^4} \qquad \text{(A5.19)}$$

assuming $B = B_o$. The hazard function for *total failure* is equal to Eq. (5.19) times the conditional probability of failure:

$$H(K) = P_{pr} \frac{4\,K^3}{\Theta_K{}^4} \qquad \text{(A5.20)}$$

Thus the overall probability of failure is given by

$$F = 1 - \exp\left(- \int_{0}^{K} P_{pr} \frac{4\,K^3}{\Theta_K{}^4}\, dK \right) \qquad \text{(A5.21)}$$

Consider the case where P_{pr} is a constant; i.e., it does not depend on the applied K. Suppose, for example, that half of the carbides of a critical size have a favorable orientation with respect to a cleavage plane in a ferrite grain. The failure probability becomes:

$$F = 1 - \exp\left[- 0.5 \left(\frac{K}{\Theta_K} \right)^4 \right] \qquad \text{(A5.22)}$$

In this instance, the finite propagation probability merely shifts the 63rd percentile toughness to a higher value:

$$\Theta_K{}^* = 2^{0.25}\,\Theta_K = 1.19\,\Theta_K$$

The shape of the distribution is unchanged, and the fracture process still follows a weakest link model. In this case, the weak link is defined as a particle that greater than the critical size *that is also oriented favorably*.

Deviations from the weakest link distribution occur when P_{pr} depends on the applied K. If the conditional probability of propagation is a step function:

$$P_{pr} = \begin{cases} 0 & K_I < K_o \\ 1 & K_I \geq K_o \end{cases}$$

the fracture toughness distribution becomes a truncated Weibull (Eq. 5.24); failure can only occur when $K > K_o$. The introduction of a threshold toughness also reduces the relative scatter, as discussed in Section 5.2.3.

Equation (5.24) implies that the arrest toughness is single valued; a microcrack always propagates above K_o, but always arrests at or below K_o. Experimental data, however, indicate that arrest can occur over a range of K values. The data in Fig. 5.27 exhibit a sigmoidal shape, while the truncated Weibull is nearly linear near the threshold.

A computer simulation of cleavage propagation in a polycrystalline material [40,41] resulted in a prediction P_{pr} as a function of the applied K; these results fit an offset power law expression (Eq. (5.25)). The absolute values obtained from the simulation are questionable, but the predicted trend is reasonable. Inserting Eq. (5.25) into Eq. (A5.21) gives

$$F = 1 - \exp\left(- \int_{K_o}^{K} \alpha (K - K_o)^{\beta} \frac{4\,K^3}{\Theta_K{}^4}\, dK \right) \tag{A5.22}$$

The integral in Eq. (A5.22) has a closed-form solution, but it is rather lengthy. The above distribution exhibits a sigmoidal shape, much like the experimental data in Fig. 5.27. Unfortunately, it is very difficult to fit experimental data to Eq. (A5.22). Note that there are four fitting parameters in this distribution: α, β, K_o, and Θ_K. Even with fewer unknown parameters, the form of Eq. (A5.22) is not conducive to curve fitting because it cannot be linearized.

Equation (A5.22) can be approximated with a conventional three parameter Weibull distribution with the slope fixed at 4 (Eq. 5.26). The latter expression also gives a reasonably good fit of experimental data (Fig. 5.27). The three parameter Weibull distribution is sufficiently flexible to model a wide range of behavior. The advantage of Eq. (5.26) is that there are only two parameters to fit (the Weibull shape parameter is fixed at 4.0) and it can be linearized. The apparent threshold, Θ_0, obtained by curve fitting tends to be a conservative estimate of the true threshold toughness.

6. FRACTURE MECHANISMS IN NONMETALS

Traditional structural materials, such as steel and aluminum, are being replaced with plastics, ceramics, and composites in a number of applications. Engineering plastics have a number of advantages, including low cost, ease of fabrication, and corrosion resistance. Ceramics provide superior wear resistance and creep strength. Composites offer high strength/weight ratios, and enable engineers to design materials with specific elastic and thermal properties.

None of these newer structural materials, however, is immune to fracture. Recall from Chapter 1 the example of pinch clamping of polyethylene pipe that led to time-dependent fracture. The so-called high toughness ceramics that have been developed in recent years (Section 6.2), have lower toughness than even the most brittle steels. Relatively minor impact (e.g. an airplane mechanic accidentally dropping his wrench on a wing) can cause microscale damage in a composite material, which can adversely affect subsequent performance.

Compared with fracture of metals, research into the fracture behavior of nonmetals is in its infancy. Much of the necessary theoretical framework is not yet fully developed for nonmetals, and there are many instances where fracture mechanics concepts that apply to metals have been misapplied to other materials.

This chapter gives a brief overview of the current state of understanding of fracture and failure mechanisms in selected nonmetallic structural materials. Although the coverage of the subject is far from complete, this chapter should enable the reader to gain an appreciation of the diverse fracture behavior that various materials can exhibit. The references listed at the end of the chapter provide a wealth of information to those who desire a more in-depth understanding of a particular material system. The reader should also refer to Chapter 8, which describes current methods for fracture toughness measurements in nonmetallic materials.

Section 6.1 outlines the molecular structure and mechanical properties of polymeric materials, and describes how these properties influence the fracture behavior. This section also includes a discussion of the fracture mechanisms in polymer matrix composites. Section 6.2 considers fracture in ceramic materials, including the newest generation of ceramic composites.

This chapter does not specifically address metal matrix composites, but these materials have many features in common with polymer and ceramic matrix composites [1]. Also, the metal matrix in these materials should exhibit the fracture mechanisms described in Chapter 5.

6.1 ENGINEERING PLASTICS

The fracture behavior of polymeric materials has only recently become a major concern, as engineering plastics have begun to appear in critical structural applications. In most consumer products made from polymers (e.g.. toys, garbage bags, ice chests, lawn furniture, etc.), fracture may be an annoyance, but it is not a significant safety issue. Fracture in plastic natural gas piping systems or aircraft wings, however, can have dire consequences.

Several books devoted solely to fracture and fatigue of plastics have been published in recent years [2-5]. These reference proved invaluable to the author in preparing Chapters 6 and 8.

Let us begin the discussion of fracture in plastics by reviewing some of the basic principles of polymeric materials.

6.1.1 Structure and Properties of Polymers

A polymer is defined as the union of two or more compounds called mers. The *degree of polymerization* is a measure of the number of these units in a given molecule. Typical engineering plastics consist of very long chains, with the degree of polymerization on the order of several thousand.

Consider polyethylene, a polymer with a relatively simple molecular structure. The building block in this case is ethylene (C_2H_4), which consists of two carbon atoms joined by a double bond, with two hydrogen atoms attached to each carbon atom. If sufficient energy is applied to this compound, the double bond can be broken, resulting in two free radicals which can react with other ethylene groups:

$$\begin{array}{c} H\ \ H \\ |\ \ \ | \\ C{=}C \\ |\ \ \ | \\ H\ \ H \end{array} + \text{Energy} \rightarrow \begin{array}{c} H\ \ H \\ |\ \ \ | \\ -C-C- \\ |\ \ \ | \\ H\ \ H \end{array}$$

The degree of polymerization (i.e., the length of the chain) can be controlled by the heat input, catalyst, as well as reagents that may be added to aid the polymerization process.

Molecular Weight

The molecular weight is a measure of the length of a polymer chain. Since there is typically a distribution of molecule sizes in a polymer sample, it is convenient to quantify an average molecular weight, which can be defined in one of two ways. The *number average* molecular weight is the total weight divided by the number of molecules:

$$\overline{M}_n = \frac{\sum_{i=1}^{n} N_i M_i}{\sum_{i=1}^{n} N_i} \tag{6.1}$$

where N_i is the number of molecules with molecular weight M_i. The number average molecular weight attaches equal importance to all molecules, while the *weight average* molecular weight reflects the actual average weight of molecules by placing additional emphasis on the larger molecules:

$$\overline{M}_w = \frac{\sum_{i=1}^{n} N_i M_i^2}{\sum_{i=1}^{n} N_i M_i} \tag{6.2}$$

These two measures of molecular weight are obviously identical if all molecules in the sample are the same size, but the number average is usually lower that the weight average molecular weight. The *polydispersity* is defined as the ratio of these two quantities:

$$PD = \frac{\overline{M}_W}{\overline{M}_n} \tag{6.3}$$

A narrow distribution of molecular weights implies a PD close to 1, while PD can be greater than 20 in materials with broadly distributed molecule sizes. Both measures of molecular weight, as well as the PD, influence the mechanical properties of a polymer.

Molecular Structure

The structure of polymer chains also has a significant effect on the mechanical properties. Figure 6.1 illustrates three general classifications of polymer chains: linear, branched, and cross-linked. *Linear polymers* are not actual straight lines; rather, the carbon atoms in a linear molecule form a single continuous path from one end of the chain to the other. A *branched polymer* molecule, as the name suggests, contains a serious of smaller chains that branch off from a main "backbone". A *cross-linked polymer*, consists of a network structure rather than linear chains. A highly cross-linked structure is typical of *thermoset polymerss*, while *thermoplastics* consist of linear and branched chains. *Elastomers* typically have lightly cross-linked structures and are capable of large elastic strains.

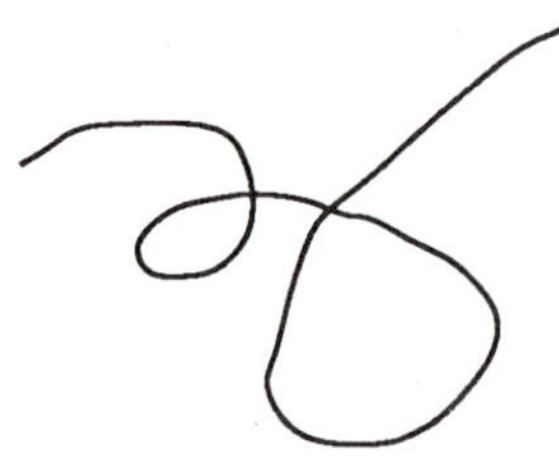

(a) Linear polymer.

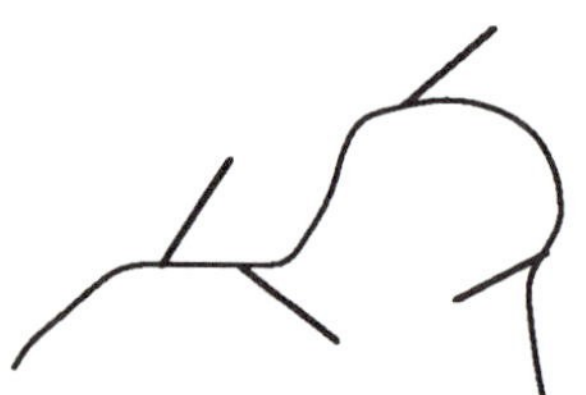

(b) Branched polymer.

(c) Cross-linked polymer

FIGURE 6.1 Three types of polymer chains.

Epoxies are the most common example of thermoset polymers. Typically, two compounds that are in the liquid state at ambient temperature are mixed together to form an epoxy resin, which solidifies into a cross-linked lattice upon curing. This process is irreversible; a thermoset cannot be formed into another shape once it solidifies.

Thermomechanical processes in thermoplastics are reversible, because these materials do not form cross-linked networks. Thermoplastics become viscous upon heating (see below), where they can be formed into the desired shape.

Crystalline and Amorphous Polymers

Polymer chains can be packed tightly together in a regular pattern, or they can form random entanglements. Materials that display the former configuration are called crystalline polymers, while the disordered state corresponds to amorphous (glassy) polymers. Figure 6.2 schematically illustrates crystalline and amorphous arrangements of polymer molecules.

The term *crystalline* does not have the same meaning for polymers as for metals and ceramics. A crystal structure in a metal or ceramic is a regular array of atoms with three-dimensional symmetry; all atoms in a crystal have identical surroundings (except atoms that are adjacent to a defect, such as a dislocation or vacancy). The degree of symmetry in a crystalline polymer, however, is much lower, as Fig. 6.2 illustrates.

Figure 6.3 illustrates the volume-temperature relationships in crystalline and amorphous thermoplastics. As a crystalline polymer cools from the liquid state, an abrupt decrease in volume occurs at the melting temperature, T_m, and the molecular chains pack efficiently in response to the thermodynamic drive to order into a crystalline state. The volume discontinuity at T_m resembles the behavior of crystalline metals and ceramics. An amorphous polymer bypasses T_m upon cooling, and remains in a viscous state until it reaches the glass transition temperature, T_g, at which time the relative motion of the molecules becomes restricted. An amorphous polymer contains more free volume than the same material in the crystalline state, and thus has a lower density. The glass transition temperature is sensitive to cooling rate; rapid heating or cooling tends to increase T_g, as Fig. 6.3 indicates.

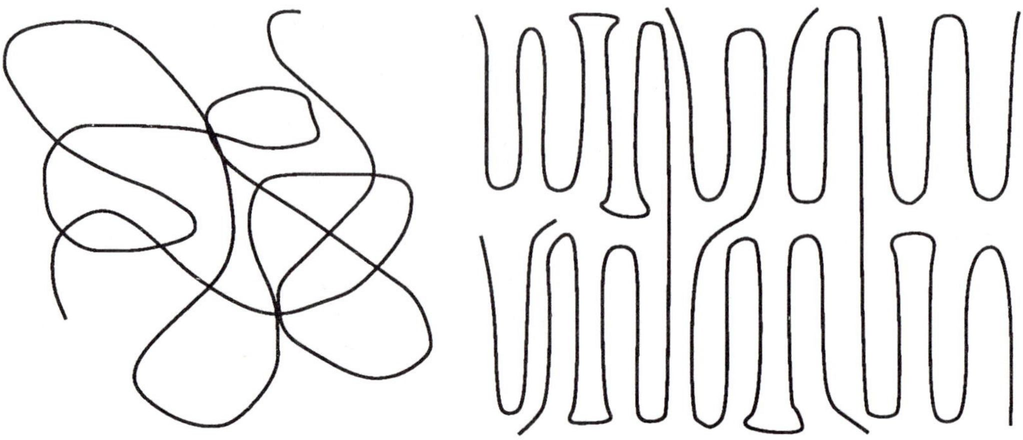

(a) Amorphous polymer. (b) Crystalline polymer.

FIGURE 6.2 Amorphous and crystalline polymers.

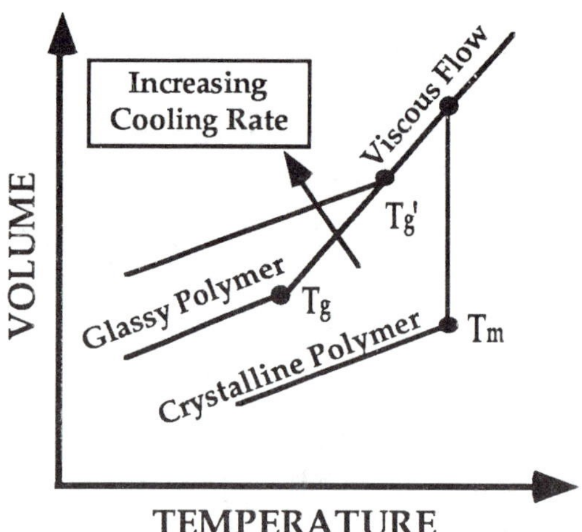

FIGURE 6.3 Volume-temperature relationships for amorphous (glassy) and crystalline polymers.

Semicrystalline polymers contain both crystalline and glassy regions. The relative fraction of each state depends on a number of factors, including molecular structure and cooling rate. Slow cooling provides more time for the molecules to arrange themselves in an equilibrium crystal structure.

Viscoelastic Behavior

Polymers exhibit rate-dependent viscoelastic deformation, which is a direct result of their molecular structure. Figure 6.4 gives a simplified view of viscoelastic behavior on the molecular level. Two neighboring molecules, or different segments of a single molecule that is folded back upon itself, experience weak attractive forces called *Van der Waals bonds*. These secondary bonds resist any external force that attempts to pull the molecules apart. The elastic modulus of a typical polymer is significantly lower than Young's modulus for metals and ceramics, because the Van der Waals bonds are much weaker than primary bonds. Deforming a polymer requires cooperative motion among molecules. The material is relatively compliant if the imposed strain rate is sufficiently low to provide molecules sufficient time to move. At faster strain rates, however, the forced molecular motion produces friction, and a higher stress is required to deform the material. If the load is removed, the material attempts to return to its original shape, but molecular entanglements prevent instantaneous elastic recovery. If the strain is sufficiently large, yielding mechanisms occur, such as crazing and shear deformation (see Section 6.6.2 below), and much of the induced strain is essentially permanent.

Section 4.3 introduced the relaxation modulus, E(t), and the creep compliance, D(t), which describe the time-dependent response of viscoelastic materials. The relaxation modulus and creep compliance can be obtained experimentally by fixing strain and stress, respectively:

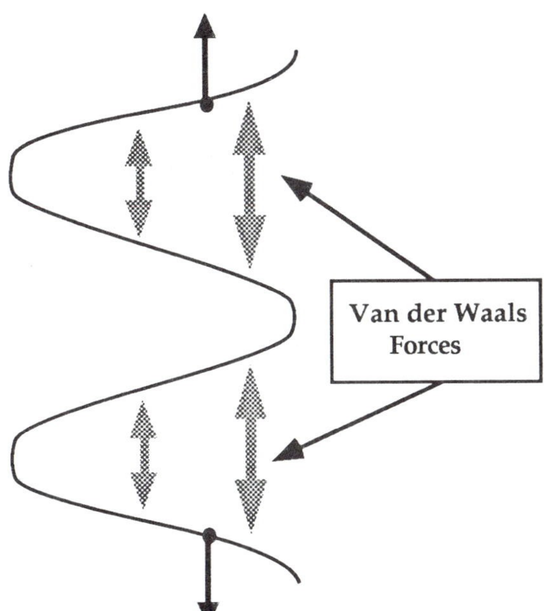

FIGURE 6.4 Schematic deformation of a polymer chain. Secondary Van der Waals bonds between chain segments resist forces that try to extend the molecule.

$$E(t) = \frac{\sigma(t)}{\varepsilon_o} \qquad D(t) = \frac{\varepsilon(t)}{\sigma_o} \tag{6.4}$$

See Fig. 4.19 for a schematic illustration of stress relaxation and creep experiments. For linear viscoelastic materials[1], E(t) and D(t) are related through a hereditary integral (Eq. (4.61)).

Figure 6.5(a) is a plot of relaxation modulus versus temperature at a fixed time for a thermoplastic. Below T_g, the modulus is relatively high, as molecular motion is restricted. At around T_g, the modulus (at a fixed time) decreases rapidly, and the polymer exhibits a "leathery" behavior. At higher temperatures, the modulus reaches a lower plateau, and the polymer is in a rubbery state. Natural and synthetic rubbers are merely materials whose glass transition temperature is below room temperature[2]. If the temperature is sufficiently high, linear polymers loose virtually all load-carrying capacity and behave like a viscous fluid. Highly cross-linked polymers, however, maintain a modulus plateau.

Figure 6.5(b) shows a curve with the same characteristic shape as Fig. 6.5(a), but with fixed temperature and varying time. At short times, the polymer is glassy, but exhibits leathery, rubbery, and liquid behavior at sufficiently long times. Of course, *short time* and *long time* are relative terms that depend on temperature. A polymer significantly below T_g might remain in a glassy state during the time frame of a stress relaxation test, while a polymer well above T_g may pass through this state so rapidly that the glassy behavior cannot be detected.

The equivalence between high temperature and long times (i.e., the time-temperature superposition principle) led Williams, Landel, and Ferry [6] to develop a semiempirical equation that collapses data at different times onto a single modulus-temperature master curve. They defined a time shift factor, a_T, as follows:

$$\log a_T = \log \frac{t_T}{t_{T_o}} = \frac{C_1 (T - T_o)}{C_2 + T - T_o} \tag{6.5}$$

---

[1]Linear viscoelastic materials do not, in general, have linear stress-strain curves (since the modulus is time dependent), but display other characteristics of linear elasticity such as superposition. See Section 4.3.1 for a definition of linear viscoelasticity.

[2]To demonstrate the temperature dependence of viscoelastic behavior, try blowing up a balloon after it has been in a freezer for an hour.

where t_T and t_{T_o} are the times to reach a specific modulus at temperatures T and T_o, respectively, T_o is a reference temperature (usually defined at T_g), and C_1 and C_2 are fitting parameters that depend on material properties. Equation (6.5), which is known as the WLF relationship, typically is valid in the range $T_g < T < T_g + 100°C$. Readers familiar with creep in metals may recognize an analogy with the Larson-Miller parameter [7], which assumes a time-temperature equivalence for creep rupture.

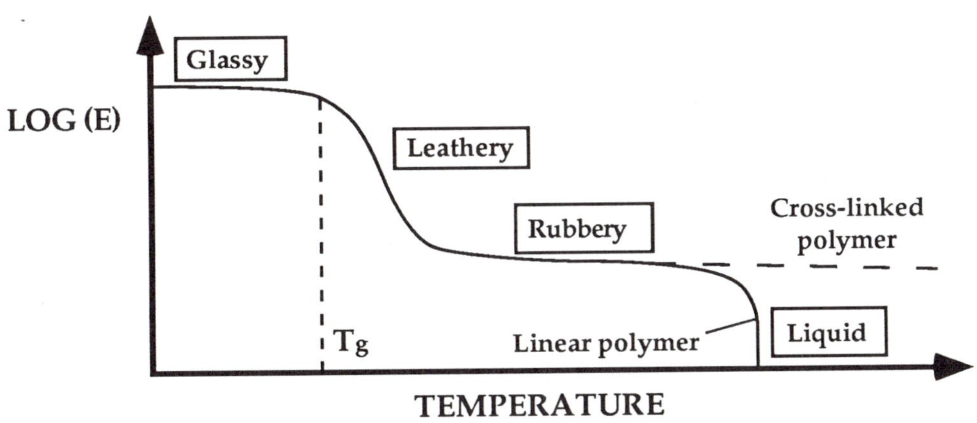

(a) Modulus versus temperature at a fixed time.

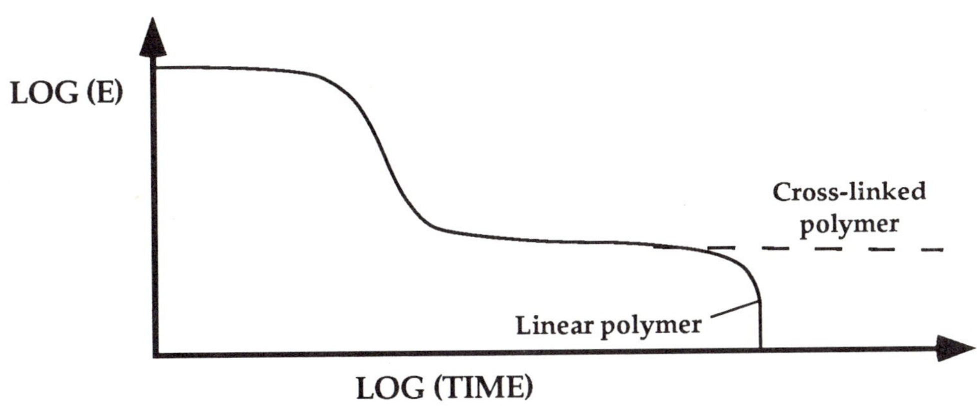

(b) Modulus versus time at a fixed temperature.

FIGURE 6.5 Effect of temperature an time on the modulus of an amorphous polymer.

Mechanical Analogs

Simple mechanical models are useful for understanding the viscoelastic response of polymers. Three such models are illustrated in Fig. 6.6. The Maxwell model (Fig. 6.6(a)) consists of a spring and a dashpot in series, where a dashpot is a moving piston in a cylinder of viscous fluid. The Voight model (Fig. 6.6(b)) contains a spring and a dashpot in parallel. Figure 6.6(c) shows a combined Maxwell-Voight model. In each case, the stress-strain response in the spring is instantaneous:

$$\varepsilon = \frac{\sigma}{E} \tag{6.6}$$

while the dashpot response is time-dependent:

$$\dot{\varepsilon} = \frac{\sigma}{\eta} \tag{6.7}$$

where $\dot{\varepsilon}$ is the strain rate and η is the fluid viscosity in the dashpot. The temperature dependence of η can be described by an Arrhenius rate equation:

$$\eta = \eta_o \exp\left(\frac{Q}{RT}\right) \tag{6.8}$$

where Q is the activation energy for viscous flow (which may depend on temperature), T is the absolute temperature, and R is the gas constant (= 8.314 J/(mole °K)).

In the Maxwell model, the stresses in the spring and dashpot are equal, and the strains are additive. Therefore,

$$\dot{\varepsilon} = \frac{\sigma}{\eta} + \frac{1}{E}\frac{d\sigma}{dt} \tag{6.9}$$

For a stress relation experiment (Fig. 4.19(b)), the strain is fixed at ε_o, and $\dot{\varepsilon} = 0$. Integrating stress with respect to time for this case leads to

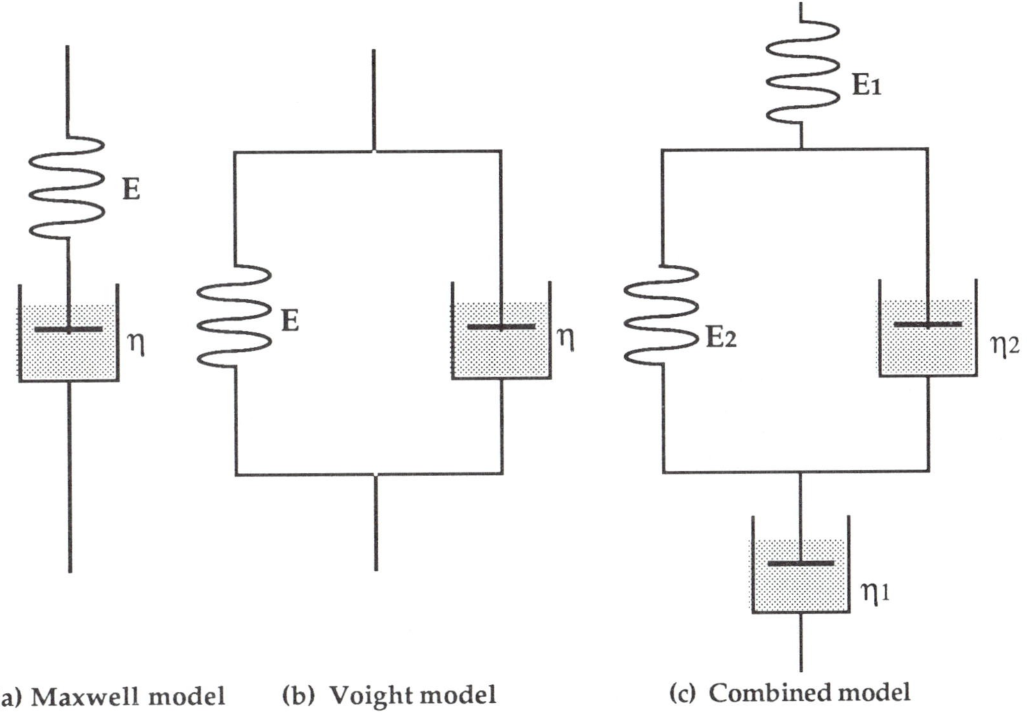

FIGURE 6.6 Mechanical analogs for viscoelastic deformation in polymers.

$$\sigma(t) = \sigma_o \exp^{-t/t_R} \tag{6.10}$$

where σ_o is the stress at $t = 0$, and $t_R \equiv \eta/E$ is the relaxation time.

When the spring and dashpot are in parallel (the Voight model) the strains are equal and the stresses are additive:

$$\sigma(t) = E\varepsilon + \eta\dot{\varepsilon} \tag{6.11}$$

For a constant stress creep test, Eq. (6.11) can be integrated to give:

$$\varepsilon(t) = \frac{\sigma_o}{E}\left(1 - \exp^{-t/t_R}\right) \tag{6.12}$$

Note that the limiting value of creep strain in this model is σ_o/E, which corresponds to zero stress on the dash pot. If the stress is removed, the strain recovers with time:

$$\varepsilon(t) = \varepsilon_0 \exp^{-t/t_R} \tag{6.13}$$

where ε_0 is the strain at t = 0, and zero time is defined at the moment the load is removed.

Neither model describes all types of viscoelastic response. For example, the Maxwell model does not account for viscoelastic recovery, because strain in the dashpot is not reversed when the stress is removed. The Voight model cannot be applied to the stress relaxation case, because when strain is fixed in Eq. (6.11), all of the stress is carried by the spring; the problem reduces to simple static loading, where both stress and strain remain constant.

If we combine the two models, however, we obtain a more realistic and versatile model of viscoelastic behavior. Figure 6.6(c) illustrates the combined Maxwell-Voight model. In this case, the strains in the Maxwell and Voight contributions are additive, and the stress carried by the Maxwell spring and dashpot is divided between the Voight spring and dashpot. For a constant stress creep test, combining Eqs. (6.9) and (6.13) gives

$$\varepsilon(t) = \frac{\sigma_0}{E_1} + \frac{\sigma_0}{E_2}\left(1 - \exp^{-t/t_{R(2)}}\right) + \frac{\sigma_0 t}{\eta_1} \tag{6.14}$$

All three models are oversimplifications of actual polymer behavior, but are useful for approximating different types of viscoelastic response.

6.1.2 Yielding and Fracture in Polymers

In metals, fracture and yielding are competing failure mechanisms. Brittle fracture occurs in materials in which yielding is difficult. Ductile metals, by definition, experience extensive plastic deformation before they eventually fracture. Low temperatures, high strain rates, and triaxial tensile stresses tend to suppress yielding and favor brittle fracture.

From a global point of view, the forgoing also applies to polymers, but the microscopic details of yielding and fracture in plastics are different from metals. Polymers do not contain crystallographic planes, dislocations, and grain boundaries; rather, they consist of long molecular chains. Section 2.1 states that fracture on the atomic level involves breaking bonds, and polymers are no exception. A complicating feature for polymers, however, is that two types of bond govern the mechanical response: the covalent bonds between carbon atoms and the secondary van der Waals forces between molecule segments. Ultimate fracture normally requires breaking the latter, but the secondary bonds often play a major role in the deformation mechanisms that lead to fracture.

The factors that govern the toughness and ductility of polymers include strain rate, temperature, and molecular structure. At high rates or low temperatures (relative to T_g) polymers tend to be brittle, because there is insufficient time for the material to respond to stress with large-scale viscoelastic deformation or yielding. Highly cross-linked polymers are also incapable of large scale viscoelastic deformation. The mechanism illustrated in Fig. 6.4, where molecular chains overcome van der Waals forces, does not apply to to cross-linked polymers; *primary* bonds between chain segments must be broken for these materials to deform.

Chain Scission and Disentanglement

Fracture, by definition, involves material separation, which normally implies severing bonds. In the case of polymers, fracture on the atomic level is called *chain scission.*

Recall from Chapter 2 that the theoretical bond strength in most materials is several orders of magnitude larger than measured fracture stresses, but crack-like flaws can produce significant local stress concentrations. Another factor that aids chain scission in polymers is that molecules are not stressed uniformly. When a stress is applied to a polymer sample, certain chain segments carry a disproportionate amount of load, which can be sufficient to exceed the bond strength. The degree of nonuniformity in stress is more pronounced in amorphous polymers, while the limited degree of symmetry in crystalline polymers tends to distribute stress more evenly.

Free radicals form when covalent bonds in polymers are severed. Consequently, chain scission can be detected experimentally by means of electron spin resonance (ESR) and infrared spectroscopy [8,9].

In some cases, fracture occurs by *chain disentanglement,* where molecules separate from one another intact. The likelihood of chain disentanglement depends on the length of molecules and the degree to which they are interwoven.[3]

Chain scission can occur at relatively low strains in cross-linked or highly aligned polymers, but the mechanical response of isotropic polymers with low cross link density is governed by secondary bonds at low strains. At high strains, many polymers yield before fracture, as discussed below.

Shear Yielding and Crazing

Most polymers, like metals, yield at sufficiently high stresses. While metals yield by dislocation motion along slip planes, polymers can exhibit either shear yielding or crazing.

Shear yielding in polymers resembles plastic flow in metals, at least from a continuum mechanics viewpoint. Molecules slide with respect to one another when subjected to a critical shear stress. Shear yielding criteria can either be based on the maximum shear stress or the octahedral shear stress [10,11]:

$$\tau_{max} = \tau_o - \mu_s \sigma_m \tag{6.15a}$$

or

$$\tau_{oct} = \tau_o - \mu_s \sigma_m \tag{6.15b}$$

where σ_m is the hydrostatic stress and μ_s is a material constant that characterizes the sensitivity of the yield behavior to σ_m. When $\mu_s = 0$, Eqs. (6.15a) and (6.15b) reduce to the Tresca and von Mises yield criteria, respectively.

Glassy polymers subject to tensile loading often yield by crazing, which is a highly localized deformation that leads to cavitation (void formation) and strains on the order of 100% [12,13]. On the macroscopic level, crazing appears as a *stress-whitened region,* due to a low

[3]An analogy that should be familiar to most Americans is the process of disentangling Christmas tree lights that have been stored in a box for a year. For those who are not acquainted with this holiday ritual, a similar example is a large mass of tangled strands of string; pulling on a single strand will either free it (chain disentanglement) or cause it to break (chain scission).

refractive index. The craze zone usually forms perpendicular to the maximum principal normal stress.

Figure 6.7 illustrates the mechanism for crazing in homogeneous glassy polymers. At sufficiently high strains, molecular chains form aligned packets called fibrils. Microvoids form between the fibrils due to an incompatibility of strains in neighboring fibrils. The aligned structure enables the fibrils to carry very high stresses relative to the undeformed amorphous state, because covalent bonds are much stronger and stiffer than the secondary bonds. The fibrils elongate by incorporating additional material, as Fig. 6.7 illustrates. Figure 6.8 shows an SEM fractograph of a craze zone.

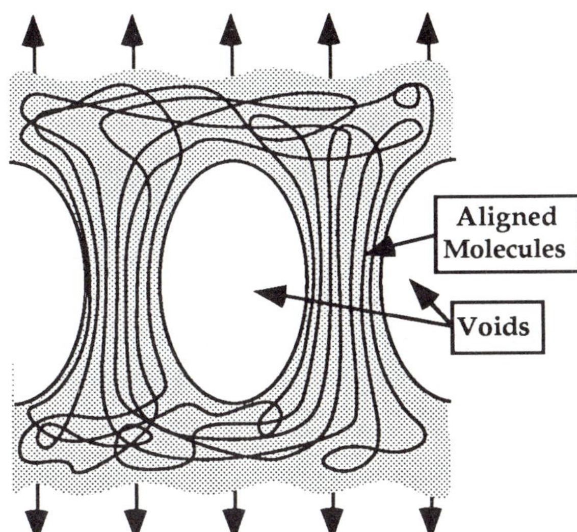

FIGURE 6.7 Craze formation in glassy polymers. Voids form between fibrils, which are bundles of aligned molecular chains. The craze zone grows by drawing additional material into the fibrils.

Oxborough and Bowden [14] proposed the following craze criterion:

$$\varepsilon_1 = \frac{\beta(t, T)}{E} + \frac{\gamma(t, T)}{3\,\sigma_m} \tag{6.16}$$

where ε_1 is the maximum principal normal strain, and β and γ are parameters that are time- and temperature-dependent. According to this model, the critical strain for crazing decreases with increasing modulus and hydrostatic stress.

Fracture occurs in a craze zone when individual fibrils rupture. This process can be unstable if, when a fibril fails, the redistributed stress is sufficient to rupture one or more neighboring fibrils. Fracture in a craze zone usually initiates from inorganic dust particles that are entrapped in the polymer [15]. There are a number of ways to neutralize the detrimental effects of these impurities, including the addition of soft second-phase particles (see below).

Crazing and shear yielding are competing mechanisms; the dominant yielding behavior depends on molecular structure, stress state and temperature. A large hydrostatic tensile component in the stress tensor is conducive to crazing, while shear yielding favors a large deviatoric stress component. Each yielding mechanism displays a different temperature dependence; thus the dominant mechanism may change with temperature.

Crack Tip Behavior

As with metals, a yield zone typically forms at the tip of a crack in polymers. In the case of shear yielding, the damage zone resembles the plastic zone in metals, because slip in metals and shear in polymers are governed by similar yield criteria. Craze yielding, however, produces a Dugdale-type strip yield zone ahead of the crack tip. Of the two yielding mechanisms in polymers, crazing is somewhat more likely ahead of a crack tip, because of the triaxial tensile stress state. Shear yielding, however, can occur at crack tips in some materials, depending on the temperature and specimen geometry [16].

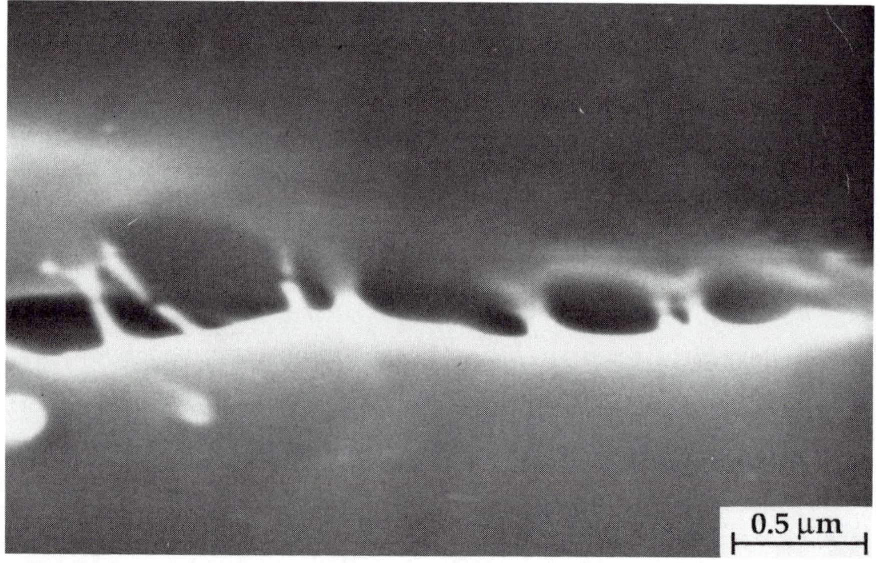

FIGURE 6.8 Craze zone in polypropylene. (Photograph provided by M. Cayard.)

Figure 6.9 illustrates a craze zone ahead of a crack tip. If the craze zone is small compared to specimen dimensions[4], we can estimate its length, ρ, from the Dugdale-Barenblatt [17,18] strip yield model:

$$\rho = \frac{\pi}{8}\left(\frac{K_I}{\sigma_c}\right)^2 \tag{6.17}$$

which is a restatement of Eq. (2.24), except that we have replaced the yield strength with σ_c, the crazing stress. Figure 6.10 is a photograph of a crack tip craze zone [16], which exhibits a typical stress whitening appearance.

The crack advances when the fibrils at the trailing edge of the craze rupture. In other words, cavities in the craze zone coalesce with the crack tip. Figure 6.11 is an SEM fractograph of the surface of a polypropylene fracture toughness specimen that has experienced craze crack growth. Note the similarity to fracture surfaces for microvoid coalescence in metals (Figs. 5.3 and 5.8).

Craze crack growth can either be stable or unstable, depending on the relative toughness of the material. Some polymers with intermediate toughness exhibit sporadic, so-called *stick/slip* crack growth: at a critical crack tip opening displacement, the entire craze zone ruptures, the crack arrests, and the craze zone reforms at the new crack tip [3]. Stick/slip crack growth can also occur in materials that exhibit shear yield zones.

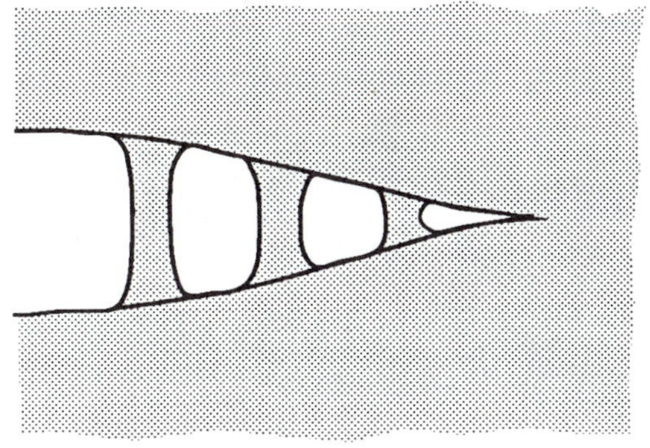

FIGURE 6.9 Schematic crack tip craze zone.

[4]Another implicit assumption of Eq. (6.17) is that the global material behavior is linear elastic or linear viscoelastic. Chapter 8 discusses the requirements for the validity of the stress intensity factor in polymers.

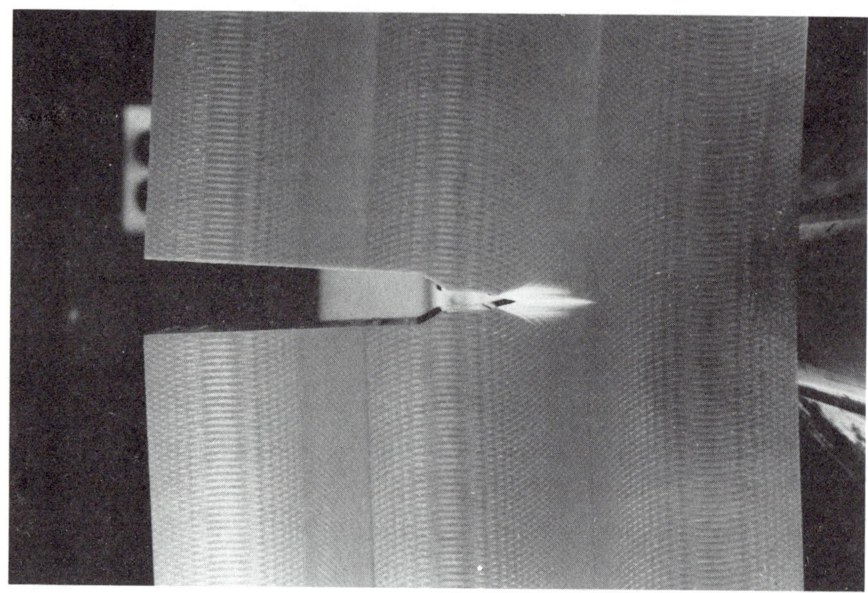

FIGURE 6.10 Stress-whitened zone ahead of a crack tip, which indicates crazing. (Photograph provided by M. Cayard.)

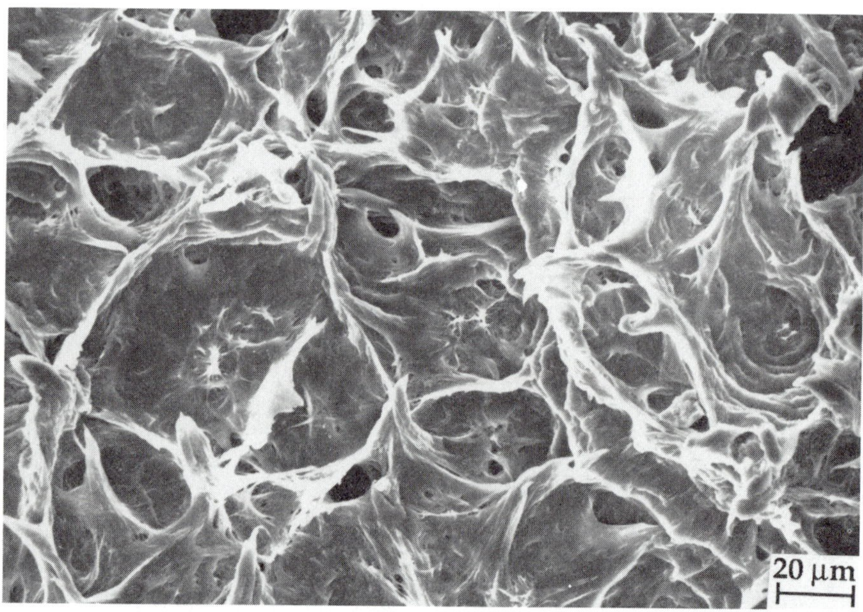

FIGURE 6.11 Fracture surface of craze crack growth in polypropylene. (Photograph provided by Mr. Sun Yongqi.)

Rubber Toughening

As stated earlier, rupture of fibrils in a craze zone can lead to unstable crack propagation. Fracture initiates at inorganic dust particles in the polymer when the stress exceeds a critical value. It is possible to increase the toughness of a polymer by lowering the crazing stress to well below the critical fracture stress.

The addition of rubbery second-phase particles to a polymer matrix significantly increases toughness by making craze initiation easier [15]. The low modulus particles provide sites for void nucleation, thereby lowering the stress required for craze formation. The detrimental effect of the dust particles is largely negated, because the stress in the fibrils tends to be well below that required for fracture. Figure 6.12 is an SEM fractograph that shows crack growth in a rubber-toughened polymer. Note the high concentration of voids, compared to the fracture surface in Fig. 6.11.

Of course there is a trade-off with rubber toughening, in that the increase in toughness and ductility comes at the expense of yield strength. A similar trade-off between toughness and strength often occurs in metals and alloys.

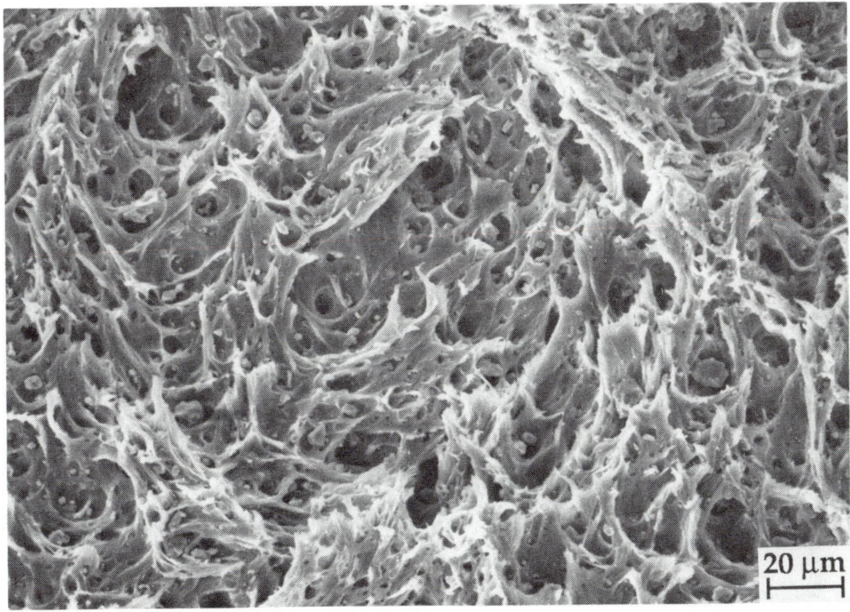

FIGURE 6.12 Fracture surface of a rubber-toughened polyvinyl chloride (PVC). Note the high concentration of microvoids. (Photograph provided by Mr Sun Yongqi.)

Fatigue

Time-dependent crack growth in the presence of cyclic stresses is a problem in virtually all material systems. Two mechanisms control fatigue in polymers: chain scission and hysteresis heating [5].

Crack growth by chain scission occurs in brittle systems, where crack tip yielding is limited. A finite number of bonds are broken during each stress cycle, and measurable crack advance takes place after sufficient cycles.

Tougher materials exhibit significant viscoelastic deformation and yielding at the crack tip. Figure 6.13 illustrates the stress-strain behavior of a viscoelastic material for a single load-unload cycle. Unlike elastic materials, where the unloading and loading paths coincide and the strain energy is recovered, a viscoelastic material displays a hysteresis loop in the stress-strain curve; the area inside this loop represents energy that remains in the material after it is unloaded. When a viscoelastic material is subject to multiple stress cycles, a significant amount of work is performed on the material. Much of this work is converted to heat, and the temperature in the material rises. The crack tip region in a polymer subject to cyclic loading may rise to well above T_g, resulting in local melting and viscous flow of the material. The rate of crack growth depends on the temperature at the crack tip, which is governed by the loading frequency and the rate of heat conduction away from the crack tip. Fatigue crack growth data from small laboratory coupons may not be applicable to structural components because heat transfer properties depend on the size and geometry of the sample.

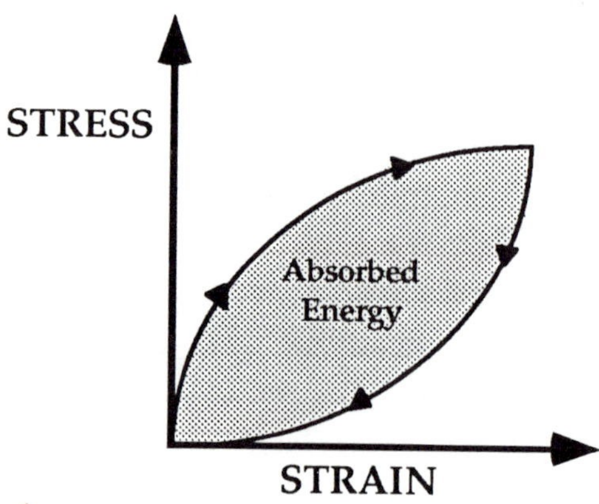

FIGURE 6.13 Cyclic stress-strain curve in a viscoelastic material. Hysteresis results in absorbed energy, which is converted to heat.

6.1.3 Fiber-Reinforced Plastics

This section focuses on the fracture behavior of continuous fiber-reinforced plastics, as opposed to other types of polymer composites. The latter materials tend to be isotropic on the macroscopic scale, and their behavior is often similar to homogeneous materials. Continuous fiber-reinforced plastics, however, have orthotropic mechanical properties which lead to unique failure mechanisms such as delamination and microbuckling.

The combination of two or more materials can lead to a third material with highly desirable properties. Precipitation-hardened aluminum alloys and rubber-toughened plastics are examples of materials whose properties are superior to those of the parent constituents. While these materials form "naturally" through careful control of chemical composition and thermal treatments, the manufacture of *composite materials* normally involves somewhat more heavy-handed human intervention. The constituents of a composite material are usually combined on a macroscopic scale through physical rather than chemical means [19]. The distinction between composites and multiphase materials is somewhat arbitrary, since many of the same strengthening mechanisms operate in both classes of material.

Composite materials usually consist of a matrix and a reinforcing constituent. The matrix is often soft and ductile compared to the reinforcement, but this is not always the case (see Section 6.2). Various types of reinforcement are possible, including continuous fibers, chopped fibers, whiskers, flakes, and particulates [19].

When a polymer matrix is combined with a strong, high modulus reinforcement, the resulting material can have superior strength/weight and stiffness/weight ratios compared to steel and aluminum. Continuous fiber-reinforced plastics tend to give the best overall performance (compared to other types of polymer composites), but can also exhibit troubling fracture and damage behavior. Consequently, these materials have been the subject of extensive research over the past 20 years.

A variety of fiber-reinforced polymer composites are commercially available. The matrix material is usually a thermoset polymer (i.e., an epoxy), although thermoplastic composites have become increasingly popular in recent years. Two of the most common fiber materials are carbon, in the form of graphite, and aramid (also known by the trade name, Kevlar[5]), which is a high modulus polymer. Polymers reinforced by continuous graphite or Kevlar fibers are intended for high performance applications such as fighter planes, while *fiberglass* is an example of a polymer composite that appears in more down-to-earth applications. The latter material consists of randomly oriented chopped glass fibers in a thermoset matrix.

Figure 6.14 illustrates the structure of a fiber-reinforced composite. Consider a single ply (Fig. 6.14(a)). The material has high strength and stiffness in the fiber direction, but has relatively poor mechanical properties when loaded transverse to the fibers. In the latter case, the strength and stiffness are controlled by the properties of the matrix. When the composite is subject to biaxial loading, several plies with differing fiber orientations can be bonded to form a laminated composite (Fig. 6.14(b)). The individual plies interact to produce complex elastic properties in the laminate. The desired elastic response can be achieved through the appropriate choice of the fiber and matrix material, the fiber volume, and the lay-up sequence of the plies. The fundamentals of orthotropic elasticity and laminate theory are well established [20].

Overview of Failure Mechanisms

Many have attempted to apply fracture mechanics to fiber-reinforced composites, and have met with mixed success. Conventional fracture mechanics methodology assumes a single dominant crack that grows in a self-similar fashion; i.e. the crack increases in size (either through stable or unstable growth), but its shape and orientation remain the same. Fracture of a fiber-reinforced composite, however, is often controlled by numerous microcracks distributed throughout the material, rather than a single macroscopic crack. There are situations where fracture mechanics is appropriate for composites, but it is important to recognize the limitations of theories that were intended for homogeneous materials.

---

[5]Kevlar is a trademark of the E.I. Dupont Company.

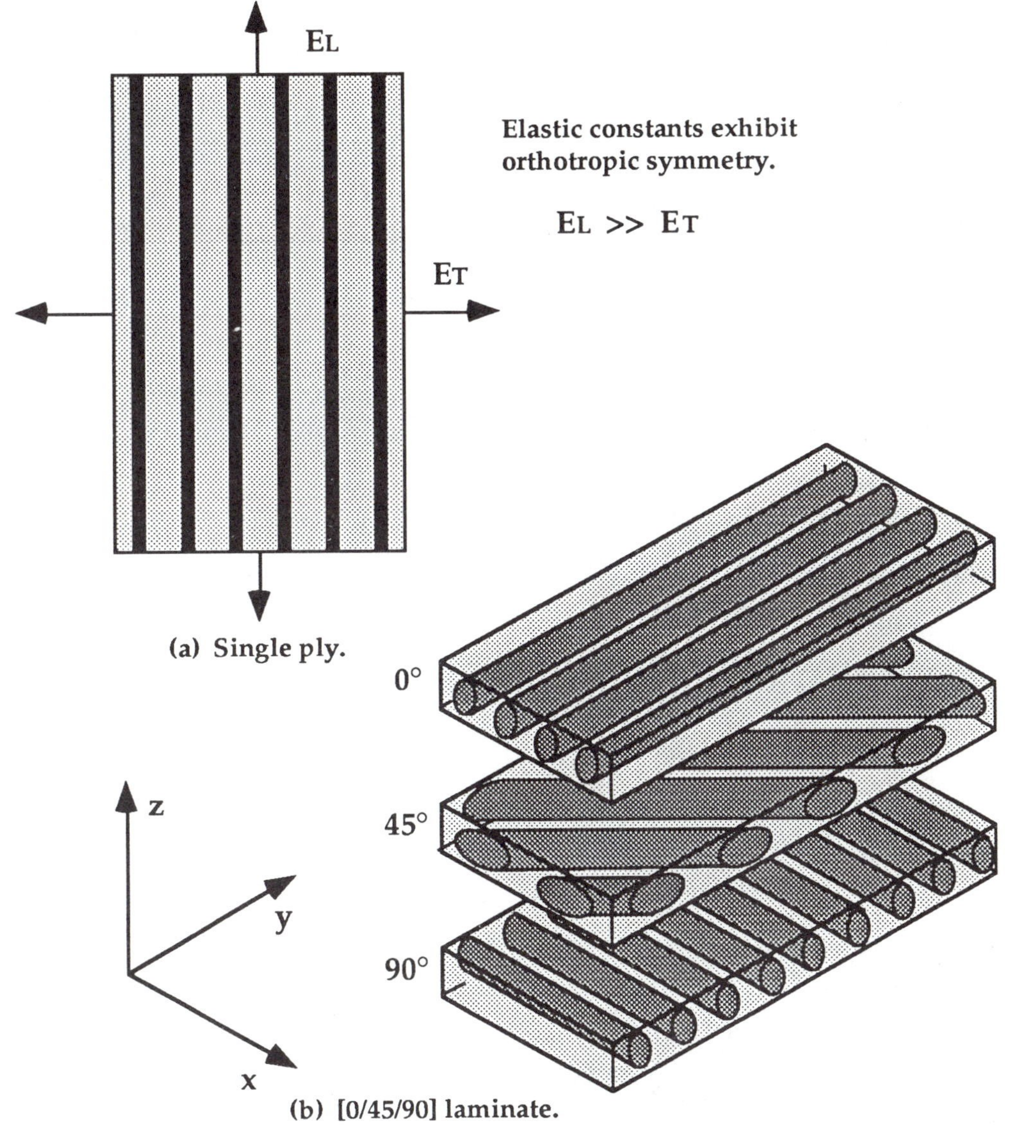

FIGURE 6.14 Schematic structure of fiber-reinforced composites.

Figure 6.15 illustrates various failure mechanisms in fiber-reinforced composites. One advantage of composite materials is that fracture seldom occurs catastrophically without warning, but tends to be progressive, with subcritical damage widely dispersed through the material. Tensile loading (Fig. 6.15 (a)) can produce matrix cracking, fiber bridging, fiber rupture, fiber pullout and fiber/matrix debonding. Ultimate tensile failure of a fiber-reinforced composite often involves several of these mechanisms. Out-of-plane stresses can lead to delamination (Fig. 6.15 (b)), because the fibers do not contribute significantly to strength in this direction. Compressive loading can produce microbuckling of fibers (Fig. 6.15 (c)); since the polymer matrix is soft compared to the fibers, the fibers are unstable in compression. Compressive loading can also lead to macroscopic delamination buckling (Fig. 6.15 (d)), particularly if the material contains a pre-existing delaminated region.

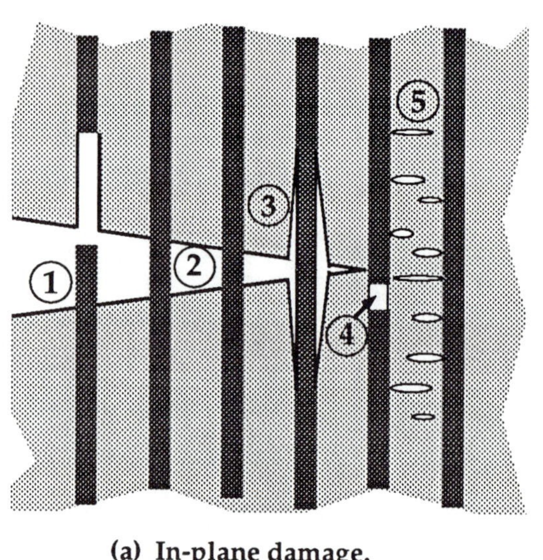

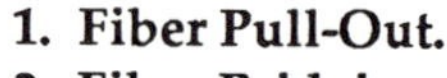

(a) In-plane damage.

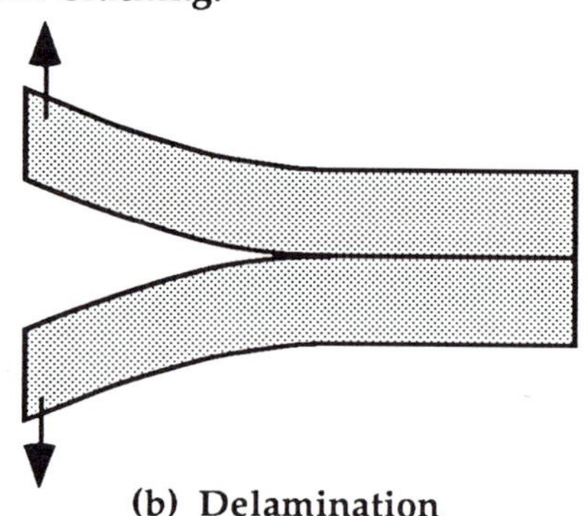

(b) Delamination

(c) Microbuckling

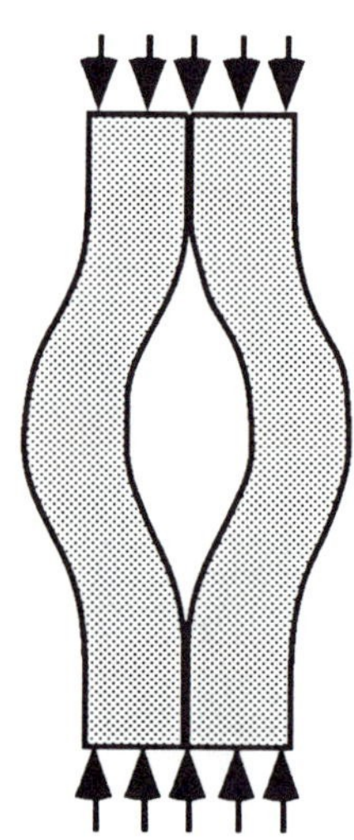

(d) Buckling delamination.

FIGURE 6.15 Examples of damage and fracture mechanisms in fiber-reinforced composites.

Delamination

Out-of-plane tensile stresses can cause failure between plies, as Fig. 6.15 (b) illustrates. Stresses that lead to delamination could result from the structural geometry, such as if two composite panels are joined in a "T" configuration. Out-of-plane stresses, however, also arise from an unexpected source. Mismatch in Poisson ratios between plies results in shear stresses in the x-y plane near the ply interface. These shear stresses produce a bending moment that is balanced by a stress in the z direction. For some lay-up sequences, substantial out-of-plane tensile stresses occur at the edge of the panel, which can lead to the formation of a delamination crack. Figure 6.16 shows a computed σ_Z distribution for a particular lay-up [21].

Although the assumption of self-similar growth of a dominant crack often does not apply to failure of composite materials, such an assumption is appropriate in the case of delamination. Consequently fracture mechanics has been very successful in characterizing this failure mechanism.

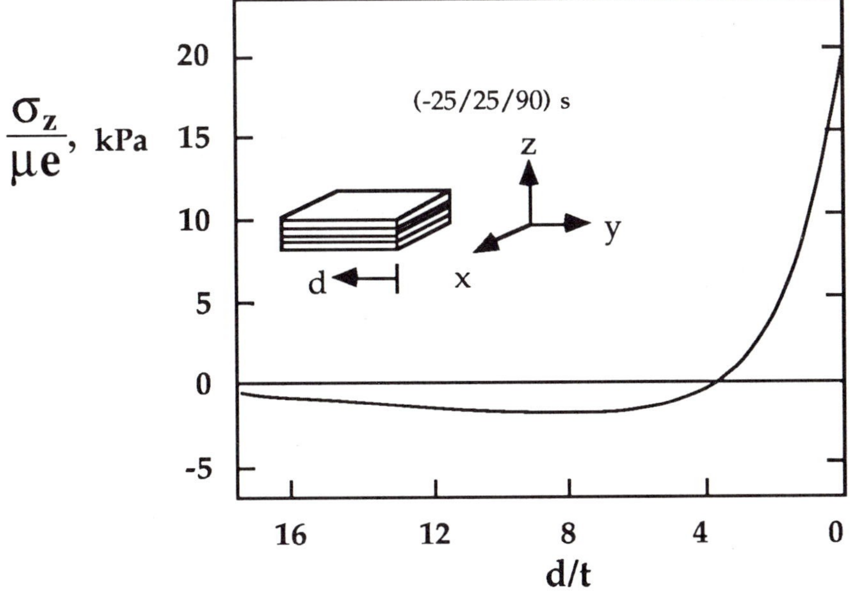

FIGURE 6.16. Out-of-plane stress at mid-thickness in a composite laminate, normalized by nominal strain and shear modulus [21]. The distance from the free edge, d, is normalized by the ply thickness, t.

Delamination can occur in both Mode I and Mode II. The interlaminar fracture toughness, which is usually characterized by a critical energy release rate (see Chapter 8), is related to to the fracture toughness of the matrix material. The matrix and composite toughness are seldom equal, however, due to the influence of the fibers in the latter.

Figure 6.17 is a compilation of $\mathcal{G}_{IC}$ values for various matrix materials, compared with the interlaminar toughness of the corresponding composite [22]. For brittle thermosets, the composite has higher toughness than the neat resin, but the effect is reversed for high toughness matrices. Attempts to increase the composite toughness through tougher resins have yielded disappointing results; only a fraction of the toughness of a high ductility matrix is transferred to the composite.

Let us first consider the reasons for the high relative toughness of composites with brittle matrices. Figure 6.18 shows the fracture surface in a composite specimen with a brittle epoxy resin. The crack followed the fibers, implying that fiber/matrix debonding was the crack growth mechanism in this case. The fracture surface has a "corrugated roof" appearance; more surface area was created in the composite experiment, which apparently resulted in higher fracture energy. Another contributing factor in the composite toughness in this case is fiber bridging. In some instances, the crack grows around a fiber, which then bridges the crack faces, and adds resistance to further crack growth.

With respect to fracture of tough matrices, one possible explanation for the lower relative toughness of the composite is that the latter is limited by the fiber/matrix bond, which is weaker than the matrix material. Experimental observations, however, indicate that fiber constraint is a more likely explanation [23]. In high toughness polymers, a shear or craze damage zone forms ahead of the crack tip. If the toughness is sufficient for the size of the damage zone to exceed the fiber spacing, the fibers restrain the crack tip yielding, resulting in a smaller zone than in the neat resin. The smaller damage zone leads to a lower fracture energy between plies.

Delamination in Mode II loading is possible, but $\mathcal{G}_{IIC}$ is typically 2 to 10 times higher than the corresponding $\mathcal{G}_{IC}$ [23]. The largest disparity between Mode I and Mode II interlaminar toughness occurs in brittle matrices.

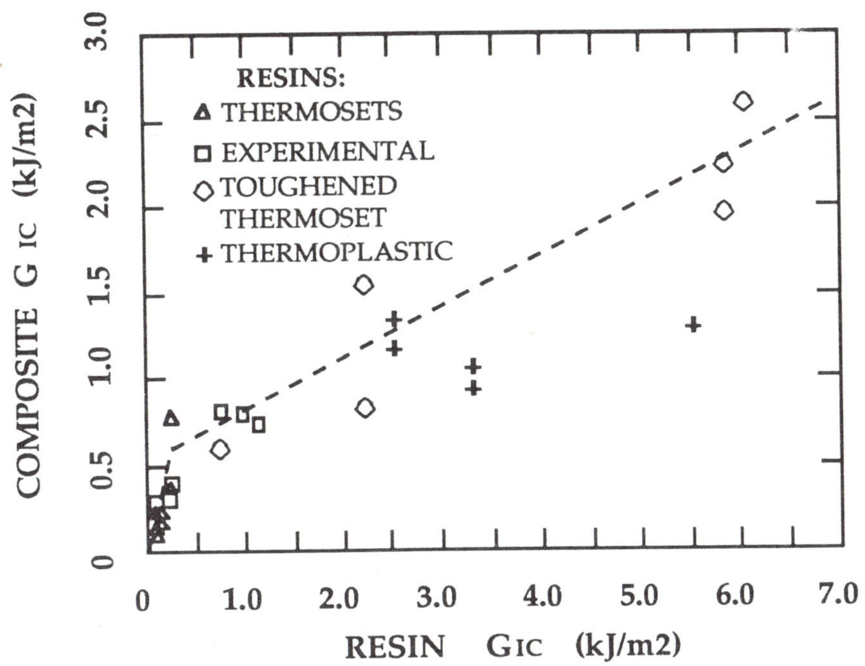

FIGURE 6.17 Compilation of interlaminar fracture toughness data, compared with the toughness of the corresponding neat resin [22].

FIGURE 6.18 Fracture surface resulting from Mode I delamination of a graphite-epoxy composite with a brittle resin [23]. (Photograph provided by W.L. Bradley.)

In-situ fracture experiments in an SEM enable one to view the fracture process during delamination [23-25]. Long, slender damage zones containing numerous microcracks form ahead of the crack tip during Mode II loading. Figure 6.19 shows a sequence of SEM fractographs of a Mode II damage zone ahead of a interlaminar crack in a brittle resin; the same region was photographed at different damage states. Note that the microcracks are oriented approximately 45° from the main crack, which is subject to Mode II shear. Thus the microcracks are oriented perpendicular to the maximum normal stress, and are actually Mode I cracks. As loading progresses, these microcracks coalesce with the main crack tip. The high relative toughness in Mode II results from energy dissipation in this damage zone.

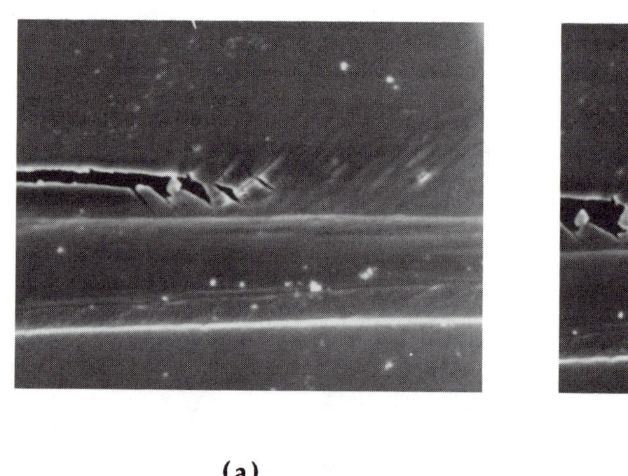

(a)

(b)

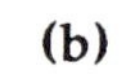

(c)

FIGURE 6.19 Sequence of photographs which show microcrack coalescence in a Mode II delamination experiment. (Photographs provided by Mr. Sun Yongqi.)

In more ductile matrices, the appearance of the Mode II damage zone is similar to the Mode I case, and the difference between $\mathcal{G}_{IC}$ and $\mathcal{G}_{IIC}$ is not as large as for brittle matrices [23].

Compressive Failure

High modulus fibers provide excellent strength and stiffness in tension, but are of limited value for compressive loading. According to the Euler buckling equation, a column of length L with a cross section moment of inertia I, subject to a compressive force P becomes unstable when

$$P \geq \frac{\pi^2 E I}{L^2} \tag{6.18}$$

assuming the loading is applied on the central axis of the column and the ends are unrestrained. Thus a long, slender fiber has very little load-carrying capacity in compression.

Equation (6.18) is much too pessimistic for composites, because the fibers are supported by matrix material. Early attempts [26] to model fiber buckling in composites incorporated an elastic foundation into the Euler bucking analysis, as Fig. 6.20 illustrates. This led to the following compressive failure criterion for unidirectional composites:

$$\sigma_c = \mu_{LT} + \pi^2 E_f V_f \left(\frac{r}{L}\right)^2 \tag{6.19}$$

where μ_{LT} is the longitudinal-transverse shear modulus of the matrix and E_f is Youngs modulus of the fibers. This model overpredicts the actual compressive strength of composites by a factor of ~ 4.

One problem with Eq. (6.19) is that it assumes that the response of the material remains elastic; matrix yielding is likely for large lateral displacements of fibers. Another shortcoming of this simple model is that it considers global fiber instability, while fiber buckling is a local phenomenon; microscopic *kink bands* form, usually at a free edge, and propagate across the panel [27,28].[6] Figure 6.21 is a photograph of local fiber buckling in a graphite-epoxy composite.

An additional complication in real composites is *fiber waviness.* Fibers are seldom perfectly straight, rather they tend to have a sine wave-like profile, as Fig. 6.22 illustrates [29]. Such a configuration is less stable in compression than a straight column.

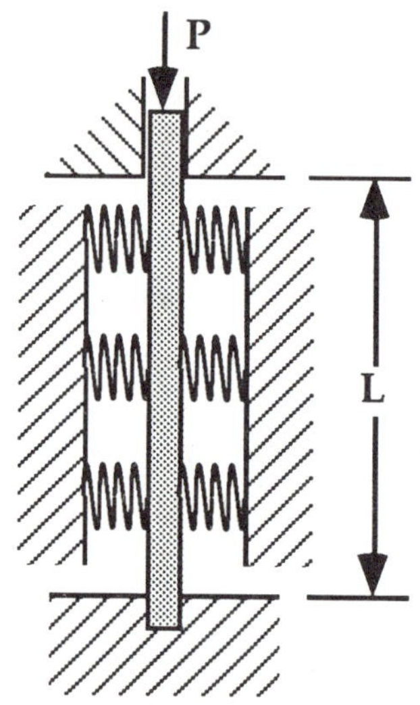

FIGURE 6.20 Compressive loading of a column that is supported laterally by an elastic foundation.

Recent investigators [29-31] have incorporated the effects of matrix nonlinearity and fiber waviness into failure models. Most failure models are based on continuum theory and thus do not address the localized nature of microbuckling. Guynn [31], however, has recently performed detailed numerical simulations of compression loading of fibers in a nonlinear matrix.

Microbuckling is not the only mechanism for compressive failure. Figure 6.15 (d) illustrates buckling delamination, which is a macroscopic instability. This type of failure is common in composites that have been subject to impact damage, which produces microcracks and delamination flaws in the material. Delamination buckling induces Mode I loading, which causes the delamination flaw to propagate at sufficiently high loads. This delamination growth can be characterized with fracture mechanics methodology [32]. A *compression after impact*

---

[6]The long, slender appearance of the kink bands led several investigators [27,28] to apply the Dugadale-Barenblatt strip yield model to the problem. This model has been moderately successful in quantifying the size of the compressive damage zones.

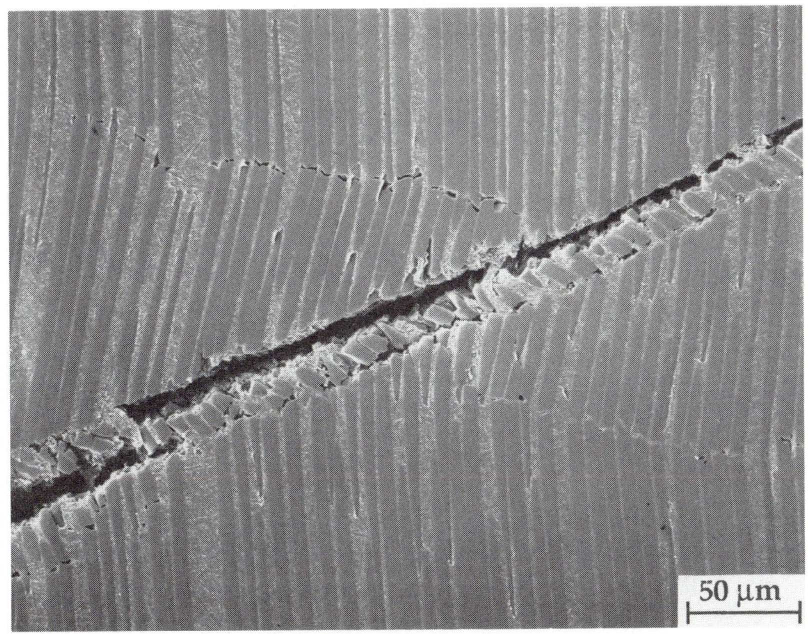

FIGURE 6.21 Kink band formation in a graphite-epoxy composite [31]. (Photograph provided by E.G. Guynn.)

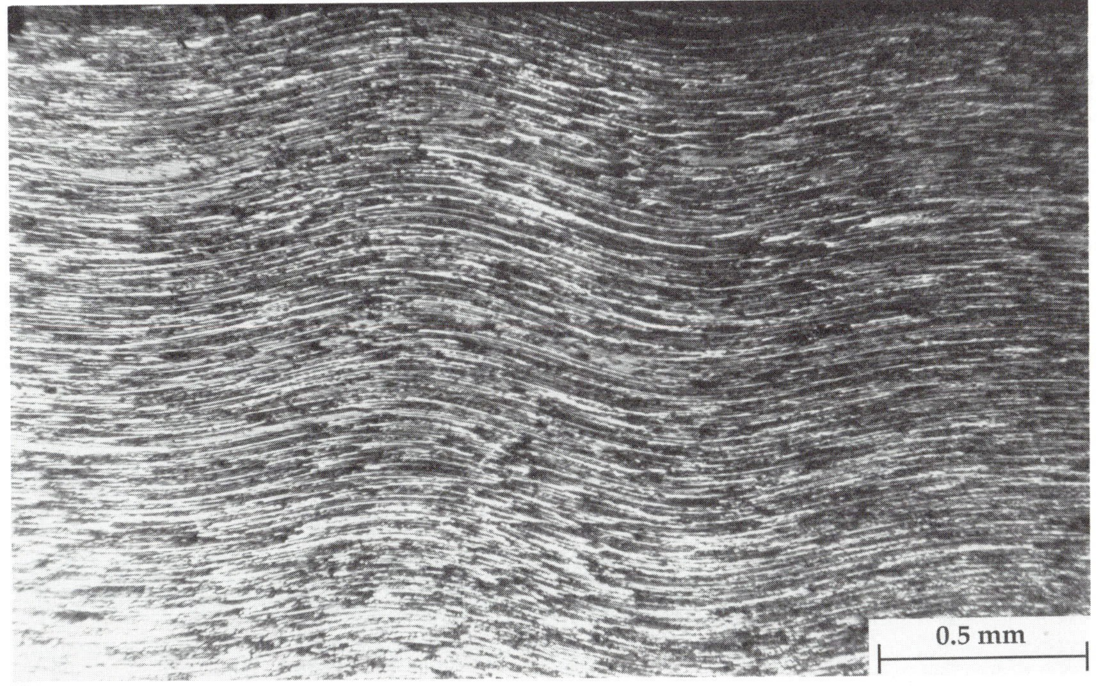

FIGURE 6.22 Fiber waviness in a graphite-epoxy composite [29]. (Photograph provided by A.L. Highsmith.)

test is a common screening criterion for assessing the ability of a material to withstand impact loading without sustaining significant damage.

Notch Strength

The strength of a composite laminate that contains a hole or a notch is less than the unnotched strength, because of the local stress concentration effect. A circular hole in an isotropic plate has a stress concentration factor (SCF) of 3.0, and the SCF can be much higher for a elliptical notch (Section 2.2). If a composite panel with a circular hole fails when the maximum stress reaches a critical value, the strength should be independent of hole size, since SCF does not depend on radius. Actual strength measurements, however, indicate a hole size effect, where strength decreases with increasing hole size [33].

Figure 6.23 illustrates the elastic stress distributions ahead of a large hole and a small hole. Although the peak stress is the same for both holes, the stress concentration effects of the large hole act over a wider distance. Thus the *volume* over which the stress acts appears to be important.

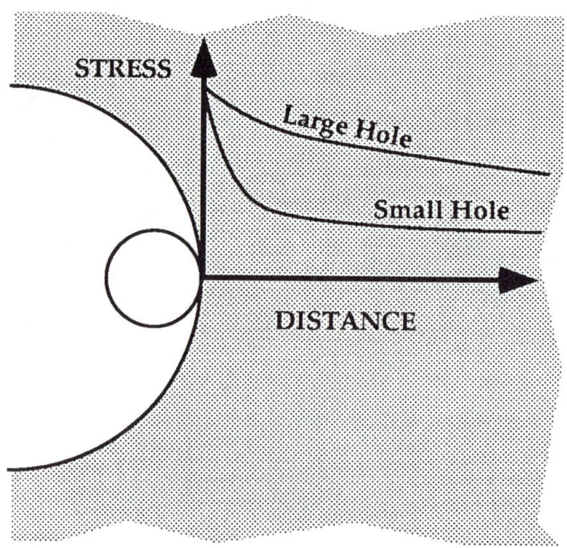

FIGURE 6.23 Effect of hole size on local stress distribution

Whitney and Nuismer [34] proposed a simple model for notch strength, where failure is assumed to occur when the stress exceeds the unnotched strength over a critical distance.[7] This distance is a fitting parameter that must be obtained by experiment. Subsequent modifications to this model, including the work of Pipes, et al. [35], yielded additional fitting parameters, but did not result in a better understanding of the failure mechanisms.

Figure 6.24 shows the effect of notch length on the strength panels that contain elliptical center notches [33]. These experimental data actually apply to a boron-aluminum composite, but polymer composites exhibit a similar trend. The simple Whitney and Nuismer criterion gives a reasonably good fit of the data in this case.

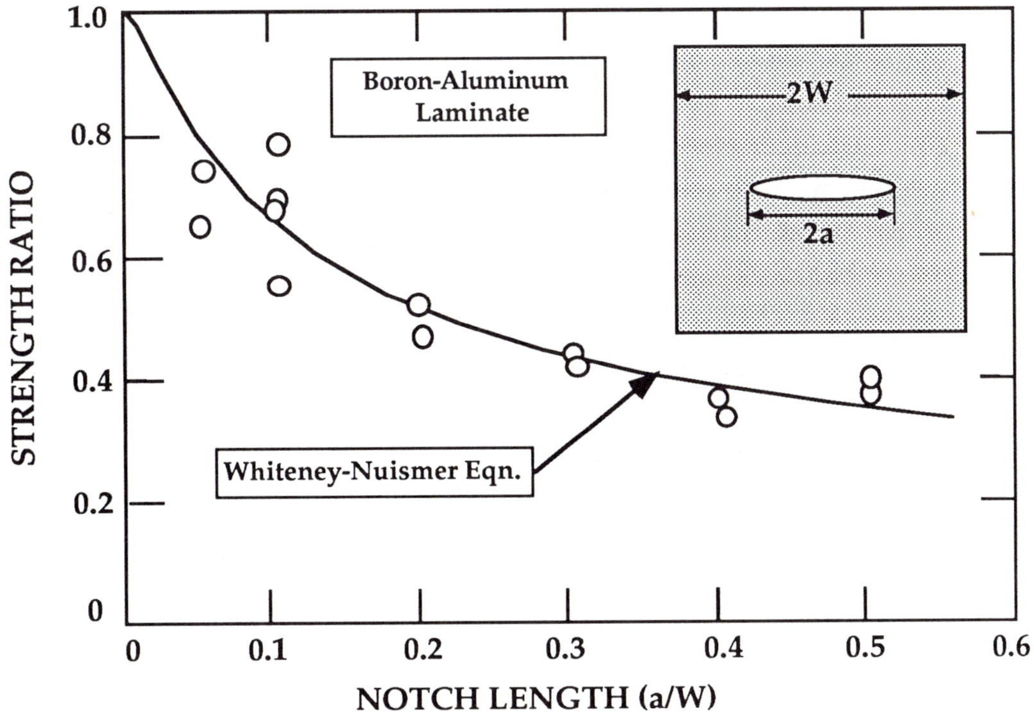

FIGURE 6.24 **Strength of center-notched composite laminates, relative to the unnotched strength [33].**

---

[7]Note the similarity to the Ritchie-Knott-Rice model for cleavage fracture (Chapter 5).

Some researchers [36] have applied fracture mechanics concepts to the failure of composites panels that contain holes and notches. They assume failure at a critical K, which is usually modified with a plastic zone correction to account for subcritical damage. Some of these models are capable of fitting experimental data such as that in Fig. 6.24, because the plastic zone correction is an adjustable parameter. The physical basis of these models is dubious, however. Fracture mechanics formalism gives these models the illusion of rigor, but they have no more theoretical basis than the simple strength-of-materials approaches such as the Whitney-Nuismer criterion.

That linear elastic fracture mechanics is invalid for circular holes and blunt notches in composites should be self evident, since LEFM theory assumes sharp cracks. If, however, a sharp slit is introduced into a composite panel (Fig. 6.25), the validity (or lack of validity) of fracture mechanics is less obvious. This issue is explored below.

Recall Chapter 2, which introduced the concept of a singularity zone, where the stress and strain vary as $1/\sqrt{r}$ from the crack tip. Outside of the singularity zone, higher order terms, which are geometry dependent, become significant. For K to define uniquely the crack tip conditions and be a valid failure criterion, all nonlinear material behavior must be confined to a small region inside the singularity zone. This theory is based entirely on continuum mechanics. While metals, plastics and ceramics are often heterogeneous, the scale of microstructural constituents is normally small compared to the size of the singularity zone; thus the continuum assumption is approximately valid.

For LEFM to be valid for a sharp crack in a composite panel, the following conditions must be met:

(1) The fiber spacing must be small compared to the size of the singularity zone. Otherwise, the continuum assumption is invalid.

(2) Nonlinear damage must be confined to a small region within the singularity zone.

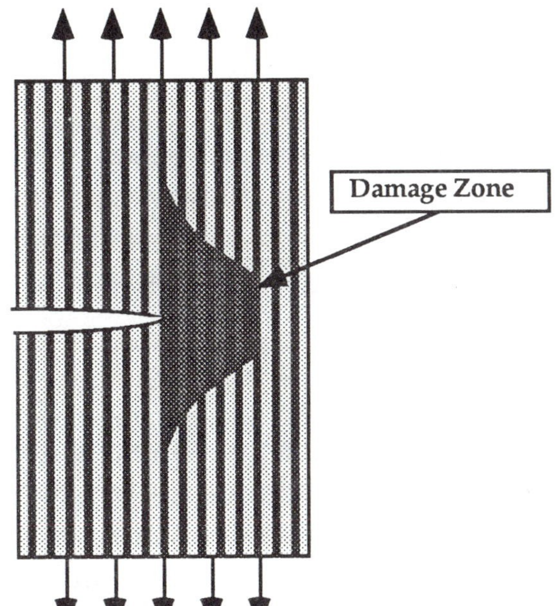

FIGURE 6.25 Sharp notch artificially introduced into a composite panel.

Harris and Morris [37] showed that K characterizes the *onset* of damage in cracked specimens, but not ultimate failure, because damage spreads throughout the specimen before failure, and K no longer has any meaning. Figure 6.25 illustrates a typical damage zone in a specimen with a sharp macroscopic notch. The damage, which includes fiber/matrix debonding and matrix cracking, actually propagates perpendicular to the macrocrack. Thus the crack does not grow in a self-similar fashion.

One of the most significant shortcomings of tests on composite specimens with narrow slits is that defects of this type do not occur naturally in fiber-reinforced composites; therefore, the geometry in Fig. 6.25 is of limited practical concern. Holes and blunt notches may be unavoidable in a design, but a competent design engineer would not be foolish enough to include a sharp notch in a load-bearing member of a structure.

Fatigue Damage

Cyclic loading of composite panels produces essentially the same type of damage as monotonic loading. Fiber rupture, matrix cracking, fiber/matrix debonding, and delamination all occur in response to fatigue loading. Fatigue damage reduces the strength and modulus of a composite laminate, and eventually leads to total failure.

Figures 6.26 and 6.27 show the effect of cyclic stresses on the residual strength and modulus of graphite/epoxy laminates [38]. Both strength and modulus decrease rapidly after relatively few cycles, but remain approximately constant up to around 80% of the fatigue life. Near the end of the fatigue life, strength and modulus decrease further.

6.2 CERAMICS AND CERAMIC COMPOSITES

A number of technological initiatives have been proposed, whose implementation depend on achieving major advances in materials technology. For example, the National Aerospace Plane will re-enter the Earth's atmosphere at speeds of up to Mach 25, creating extremes of both temperature and stress. Also, the Advanced Turbine Technology Applications Program (ATTAP) has the stated goal of developing heat engines which have a service life of 3000 h at 1350 °C. Additional applications are on the horizon that will require materials that can perform at temperatures in excess of 2000 °C. All metals, including cobalt-based superalloys, are inadequate at these temperatures. Only ceramics possess adequate creep resistance above 1000°C.

Ceramic materials include oxides, carbides, sulfides, and intermetallic compounds, which are joined either by covalent or ionic bonds. Most ceramics are crystalline but, unlike metals, they do not have close-packed planes on which dislocation motion can occur. Therefore, ceramic materials tend to be very brittle compared to metals.

Typical ceramics have very high melting temperatures, which explains their good creep properties. Also, many of these materials have superior wear resistance, and have been used for bearings and machine tools. Most ceramics, however, are too brittle for critical load-bearing applications. Consequently, a vast amount of research has been devoted to improving the toughness of ceramics.

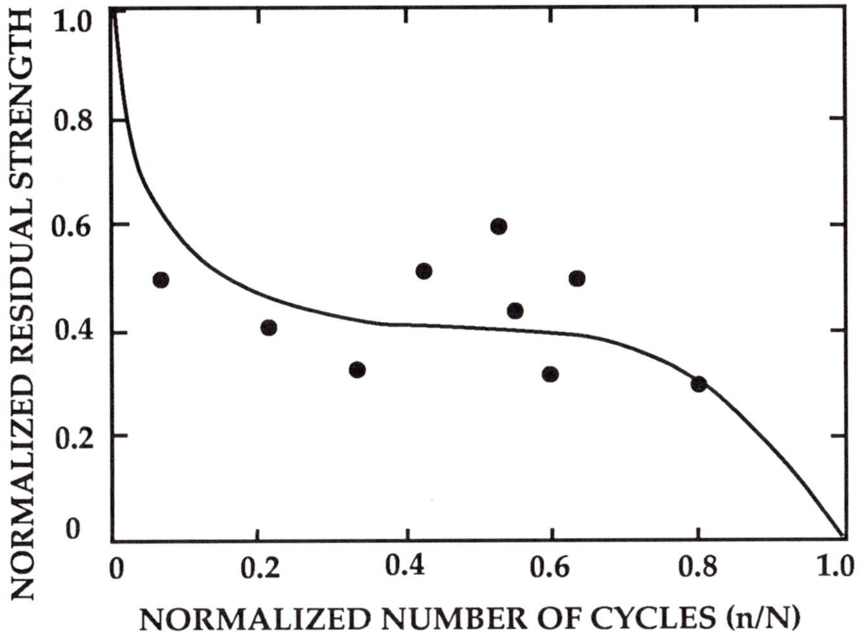

FIGURE 6.26 Residual strength after fatigue damage in a graphite-epoxy laminate [38].

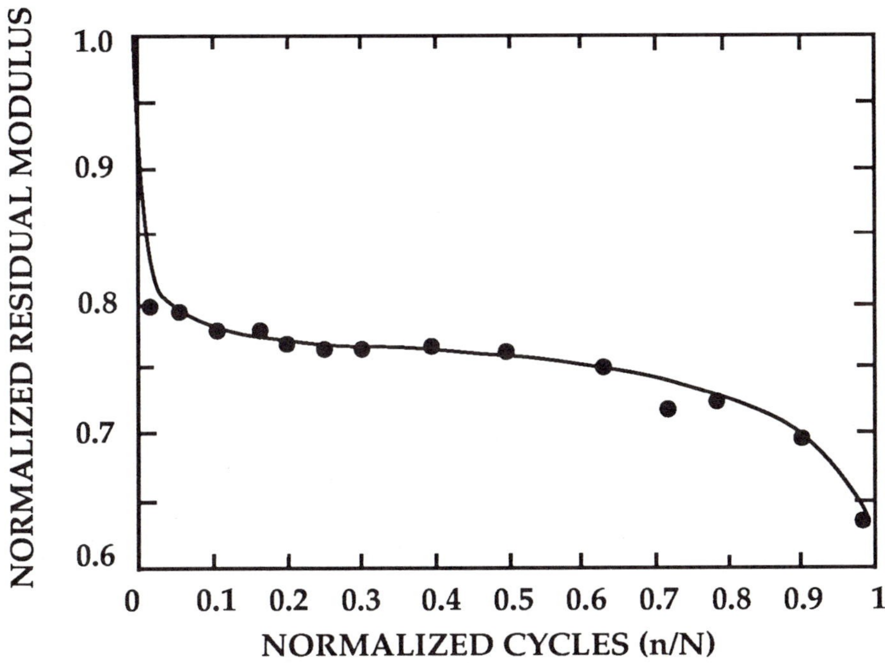

FIGURE 6.27 Residual modulus after fatigue damage in a graphite-epoxy laminate [38].

Most traditional ceramics are monolithic (single phase) and have very low fracture toughness. Because they do not yield, monolithic ceramics behave as ideally brittle materials (Fig. 2.6(a)), and a propagating crack need only overcome the surface energy of the material. The new generation of ceramics, however, includes multiphase materials and ceramic composites that have vastly improved toughness. Under certain conditions, two brittle solids can be combined to produce a material that is significantly tougher than either parent material.

The micromechanisms that lead to improved fracture resistance in modern ceramics include microcrack toughening, transformation toughening, ductile phase toughening, fiber toughening, and whisker toughening. Table 6.1 lists the dominant toughening mechanism in several materials, along with the typical fracture toughness values [39]. Fiber toughening, the most effective mechanism, produces toughness values around 20 MPa $\sqrt{m}$, which is below the lower shelf toughness of steels but is significantly higher than most ceramics.

Evans [39] divides toughening mechanisms for ceramics into two categories: process zone formation and bridging. Both mechanisms involve energy dissipation at the crack tip. A third mechanism, crack deflection, elevates toughness by increasing the area of the fracture surface (Fig. 2.6(c)).

The process zone mechanism for toughening is illustrated in Fig. 6.28. Consider a material that forms a process zone at the crack tip (Fig. 6.28(a)). When this crack propagates, it leaves a wake behind the crack tip. The critical energy release rate for propagation is equal to the work required to propagate the crack from a to a + da, divided by da:

$$\mathcal{G}_R = 2\int_0^h \left[\int_0^{\varepsilon_{ij}} \sigma_{ij}\, d\varepsilon_{ij}\right] dy + 2\gamma_s \tag{6.20}$$

where h is the half width of the process zone and γ_s is the surface energy. The integral in the square brackets is the strain energy density, which is simply the area under the stress-strain curve in the case of uniaxial loading. Figure 6.28(b) compares the stress-strain curve of brittle and toughened ceramics. The latter material is capable of higher strains, and absorbs more energy prior to failure.

Many toughened ceramics contain second-phase particles that are capable of nonlinear deformation, and are primarily responsible for the elevated toughness. Figure 6.28(c) illustrates the process zone for such a material. Assuming the particles provide all of the energy dissipation in the process zone, and the strain energy density in this region does not depend on y, the fracture toughness is given by

$$\mathcal{G}_R = 2\,h\,f \int_0^{\varepsilon_{ij}} \sigma_{ij}\, d\varepsilon_{ij} + 2\,\gamma_s \tag{6.21}$$

Table 6.1 Ceramics with enhanced toughness [39].

Toughening Mechanism	Material	Maximum Toughness, MPa $\sqrt{m}$
Fiber reinforced	LAS/SiC	~20
	Glass/C	~20
	SiC/SiC	~20
Whisker reinforced	Al_2O_3/SiC(0.2)	10
	Si_3N_4/SiC(0.2)	14
Ductile network	Al_2O_3/Al(0.2)	12
	B_4C/Al(0.2)	14
	WC/Co(0.2)	20
Transformation toughened	PSZ	18
	TZP	16
	ZTA	10
Microcrack toughened	ZTA	7
	Si_3N_4/SiC	7

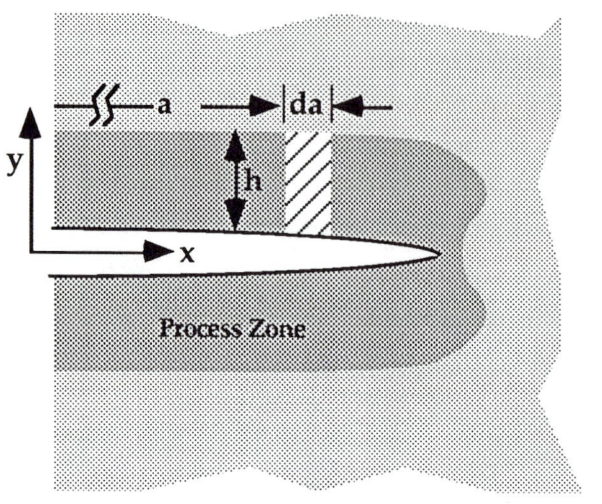

(a) Process zone formed by growing crack

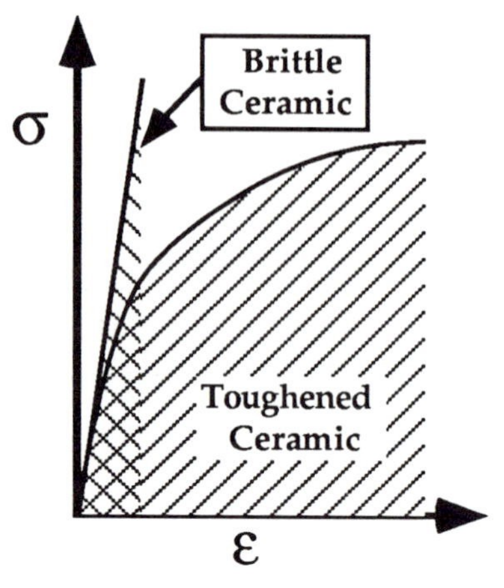

(b) Schematic stress-strain behavior

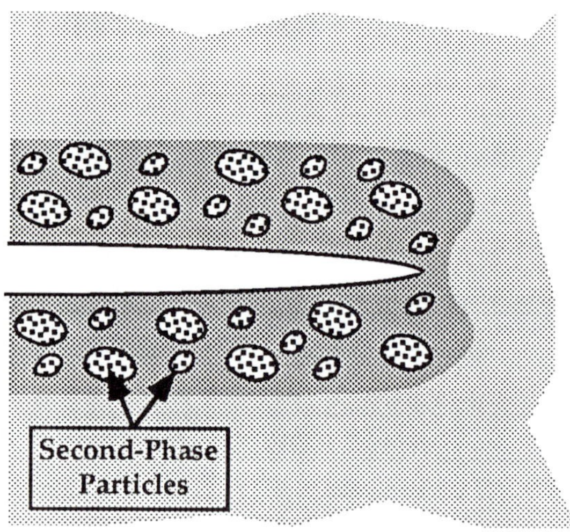

(c) Nonlinear deformation of second-phase particles

FIGURE 6.28 The process zone mechanism for ceramic toughening.

where f is the volume fraction of second-phase particles. Thus the toughness is controlled by the width of the process zone, the concentration of second-phase particles, and the area under the stress-strain curve. Recall the delamination of composites with tough resins (Section 6.1.3), where the fracture toughness of the composite was not as great as the neat resin because the fibers restricted the size of the process zone (h).

The process zone mechanism often results in a rising R curve, as Fig. 6.29 illustrates. The material resistance increases with crack growth, as the width of the processes zone grows. Eventually, h and $\mathcal{G}_R$ reach steady-state values.

Figure 6.30 illustrates the crack bridging mechanism, where the propagating crack leaves fibers or second-phase particles intact. The unbroken fibers or particles exert a traction force on the crack faces, much like the Dugdale-Barenblatt strip yield model [17,18]. The fibers eventually rupture when the stress reaches a critical value. According to Eqs. (3.42) and (3.43), the critical energy release rate for crack propagation is given by

$$\mathcal{G}_C = J_c = f \int_0^{\delta_C} \sigma_{yy} \, d\delta \tag{6.21}$$

The sections that follow outline several specific toughening mechanisms in modern ceramics.

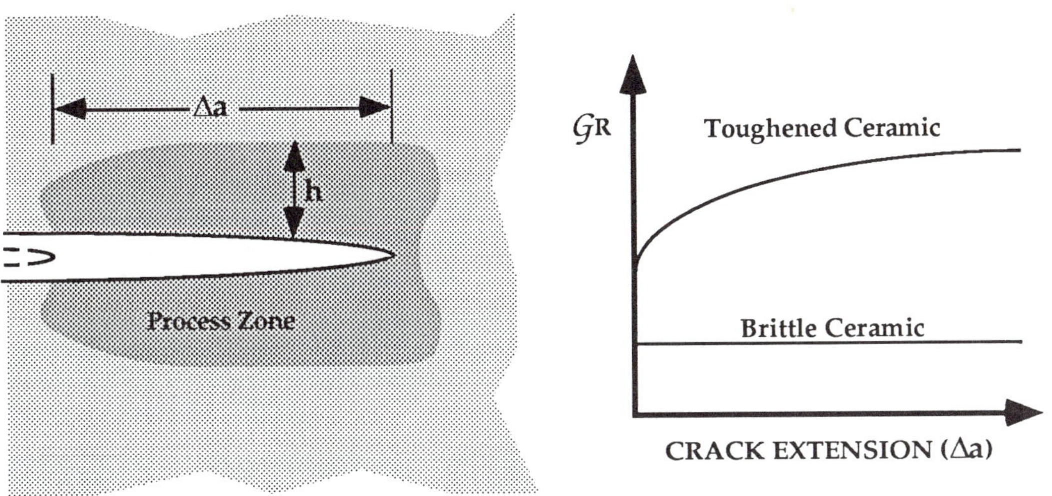

FIGURE 6.29 The process zone toughening mechanism usually results in a rising R curve.

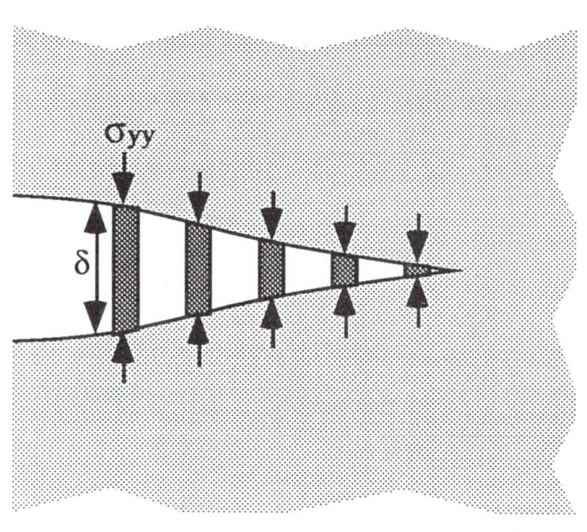

FIGURE 6.30 The fiber bridging mechanism for ceramic toughening.

6.2.1 Microcrack Toughening

Although the formation of cracks in a material is generally considered deleterious, microcracking can sometimes lead to improved toughness. Consider a material sample of volume V that forms N microcracks when subject to a particular stress. If these cracks are penny shaped with an average radius a, the total work required to form these microcracks is equal to the surface energy times the total area created:

$$W_c = 2 N \pi a^2 \gamma_s \tag{6.22}$$

The formation of microcracks releases strain energy from the sample, which results in an increase in compliance. If this change in compliance is gradual, as existing microcracks grow and new cracks form, a nonlinear stress-strain curve results. The change in strain energy density due to the microcrack formation is given by

$$\Delta w = 2 \rho \pi a^2 \gamma_s \tag{6.23}$$

where $\rho \equiv N/V$ is the microcrack density. For a macroscopic crack that produces a process zone of microcracks, the increment of toughening due to microcrack formation can be inferred by inserting Eq. (6.23) into Eq. (6.21).

A major problem with the above scenario is that *stable* microcrack growth does not usually occur in a brittle solid. Pre-existing flaws in the material remain stationary until they satisfy the Griffith criterion, at which time they become unstable. Stable crack advance normally requires a either a rising R curve, where the fracture work (w_f, see Fig. 2.6) increases with crack extension, or physical barriers in the material that inhibit crack growth.

Certain multiphase ceramics possess the latter characteristic, and therefore have the potential for microcrack toughening. Figure 6.31 schematically illustrates this toughening mechanism [39]. Second-phase particles often are subject to residual stress due to thermal expansion mismatch or transformation. If the residual stress in the particle is tensile and the local stress in the matrix is compressive[8], the particle cracks. If the signs on the stresses are reversed, the matrix material cracks at the interface. In both cases there is a residual opening of the microcracks, which leads to an increase in volume in the sample. Figure 6.31 (b) illustrates the stress-strain response of such a material. The material begins to crack at a critical stress, σ_c, and the stress-strain curve becomes nonlinear, due to a combination of compliance increase and dilatational strain. If the material is unloaded prior to total failure, the relative contributions of dilatational effects (residual microcrack opening) and modulus effects (due to the release of strain energy) are readily apparent.

A number of multiphase ceramic materials exhibit trends in toughness with particle size and temperature that are consistent with the microcracking mechanism, but this phenomena has only been directly observed in aluminum oxide toughened with monoclinic zirconium dioxide [40].

This mechanism is relatively ineffective, as Table 6.1 indicates. Moreover, the degree of microcrack toughening is temperature dependent. Thermal mismatch and the resulting residual stresses tend to be lower at elevated temperatures, which implies less dilatational strain. Also, lower residual stresses may not prevent the microcracks from becoming unstable and propagating through the particle/matrix interface.

---

[8]The residual stresses in the matrix and particle must balance in order to satisfy equilibrium.

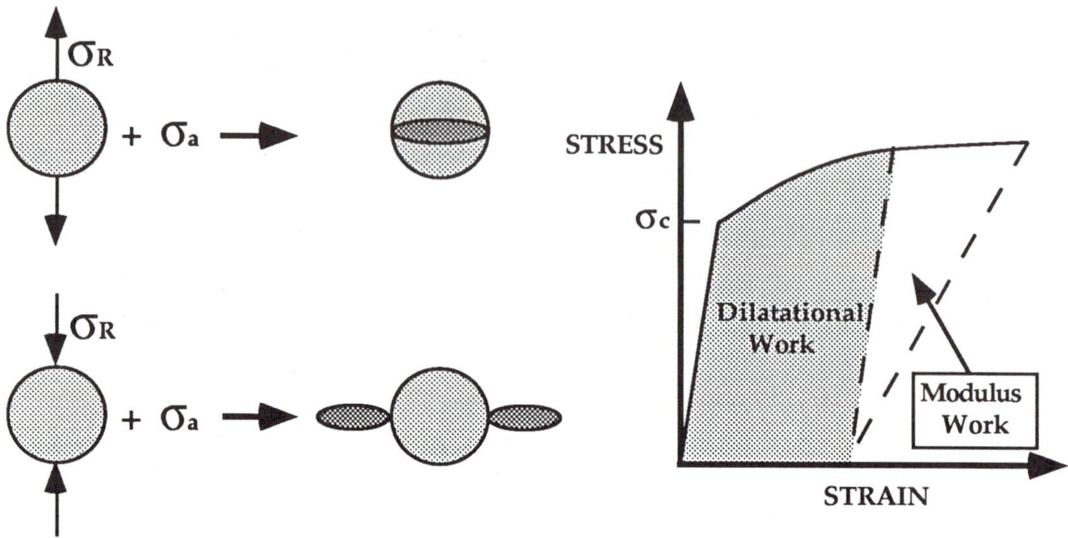

FIGURE 6.31 The microcrack toughening mechanism [39]. The formation of microcracks in or near second-phase particles results in release of strain energy (modulus work) and residual microcrack opening (dilatational work).

6.2.2 Transformation Toughening

Some ceramic materials experience a stress-induced martensitic transformation that results in shear deformation and a volume change (i.e., a dilatational strain). Ceramics that contain second-phase particles that transform often have improved toughness. Zirconium dioxide (ZrO_2) is the most widely studied material that exhibits a stress-induced martensitic transformation [41].

Figure 6.32 illustrates the typical stress-strain behavior for a martensitic transformation [41]. At a critical stress, the material transforms, resulting in both dilatational and shear strains. Figure 6.33(a) shows a crack tip process zone, where second-phase particles have transformed.

The toughening mechanism for such a material can be explained in terms of the work argument: energy dissipation in the process zone results in higher toughness. An alternative explanation is that of *crack tip shielding*, where the transformation lowers the local crack driving force [41]. Figure 6.33(b) shows the stress distribution ahead of the crack with a transformed process zone. Outside of this zone, the stress field is defined by the global stress intensity, but the stress field in the process zone is lower, due to dilatational effects. The crack tip work and shielding explanations are consistent with one another; more work is required for crack extension when the local stresses are reduced. Crack tip shielding due to the martensitic transformation is analogous to the stress redistribution that accompanies plastic zone formation in metals (Chapter 2).

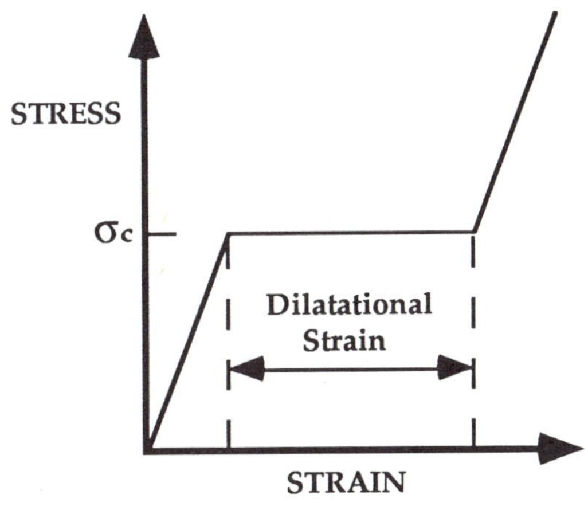

FIGURE 6.32 Schematic stress-strain response of a material that exhibits a martensitic transformation at a critical stress.

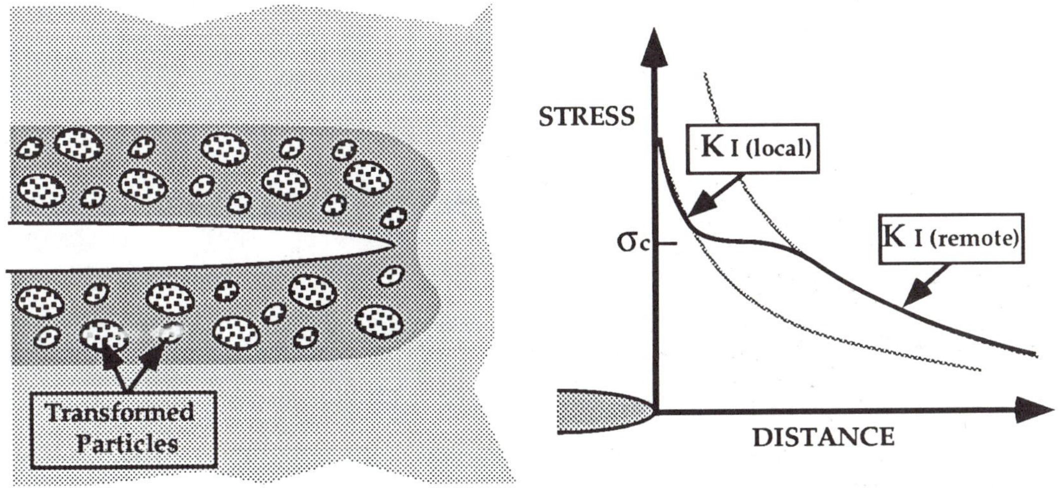

(a) Process zone.

(b) Crack tip stress field

FIGURE 6.33 The martensitic toughening mechanism. Transformation of particles near the crack tip results in a nonlinear process zone (a) and crack tip shielding (b).

The transformation stress and the dilatational strain are temperature dependent. These quantities influence the size of the process zone, h, and the strain energy density within this zone. Consequently, the effectiveness of the transformation toughening mechanism also depends on temperature. Below M_S, the martensite start temperature, the transformation occurs spontaneously, and the transformation stress is essentially zero. Thermally transformed martensite does not cause crack tip shielding, however [41]. Above M_S, the transformation stress increases with temperature. When this stress becomes sufficiently large, the transformation toughening mechanism is no longer effective.

6.2.3 Ductile Phase Toughening

Ceramics alloyed with ductile particles exhibit both bridging and process zone toughening, as Fig. 6.34 illustrates. Plastic deformation of the particles in the process zone contributes toughness, as does the ductile rupture of the particles that intersect the crack plane. Figure 6.35 is an SEM fractograph of bridging zones in Al_2O_3 reinforced with aluminum [39]. Residual stresses in the particles can also add to the material's toughness. The magnitude of the bridging and process zone toughening depends on the volume fraction and flow properties of the particles. The process zone toughening also depends on the particle size, with small particles giving the highest toughness [39].

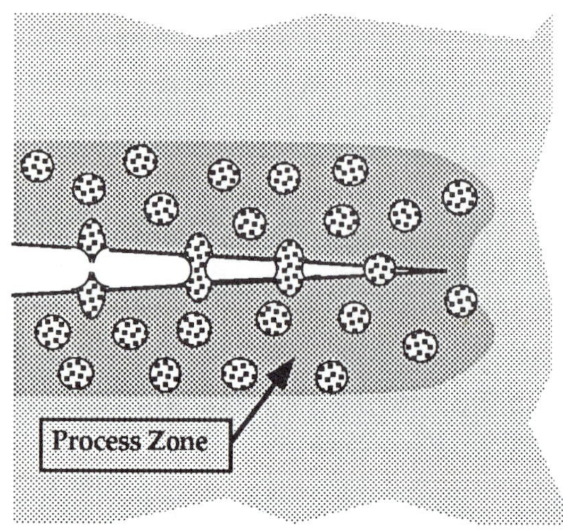

FIGURE 6.34 Ductile phase toughening. Ductile second-phase particles increase the ceramic toughness by plastic deformation in the process zone, as well as by a bridging mechanism.

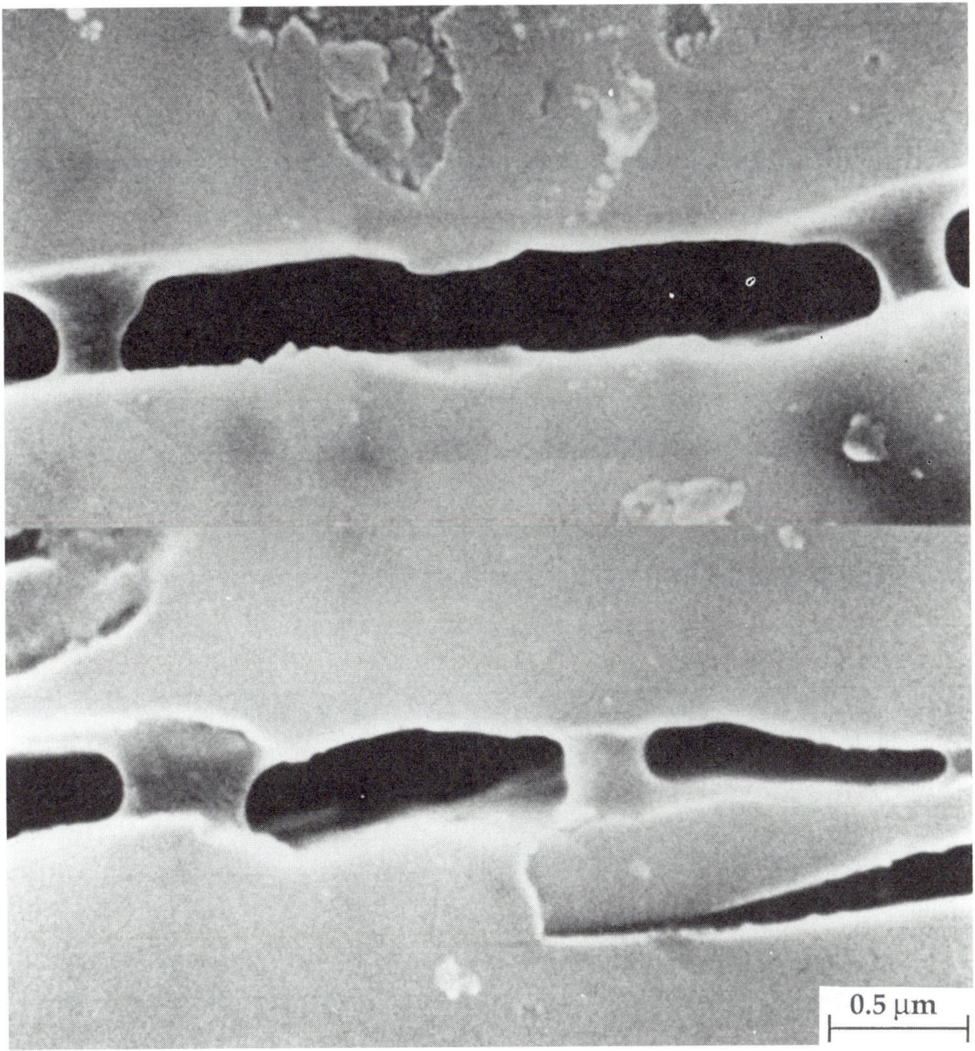

FIGURE 6.35 Ductile-phase bridging in Al_2O_3/Al [39]. (Photograph provided by A.G. Evans.)

This toughening mechanism is temperature dependent, since the flow properties of the metal particles vary with temperature. Ductile phase ceramics are obviously inappropriate for applications above the melting temperature of the metal particles.

6.2.4 Fiber and Whisker Toughening

One of the most interesting features of ceramic composites is that the combination of a brittle ceramic matrix with brittle ceramic fibers or whiskers can result in a material with relatively high toughness (Table 6.1). The secret to the high toughness of ceramic composite lies in the bond between the matrix and the fibers or whiskers. Having a *brittle* interface leads to higher toughness than a strong interface. Thus ceramic composites defy intuition: a brittle matrix bonded to a brittle fiber by a brittle interface results in a tough material.

A weak interface between the matrix and reinforcing material aids the bridging mechanism. When a matrix crack encounters a fiber/matrix interface, this interface experiences Mode II loading; debonding occurs if the fracture energy of the interface is low (Fig. 6.36(a)). If the extent of debonding is sufficient, the matrix crack bypasses the fiber, leaving it intact. Mathematical models [42] of fiber/matrix debonding predict crack bridging when the interfacial fracture energy is an order of magnitude smaller than the matrix toughness. If the interfacial bond is strong, matrix cracks propagate through the fiber, and the composite toughness obeys a rule of mixtures; but bridging increases the composite toughness (Fig. 6.36(c)).

An alternate model [42-44] for bridging in fiber-reinforced ceramics assumes that the fibers are not bonded, but that friction between the fibers and the matrix restrict the crack opening (Fig. 6.36(b)). The model that considers Mode II debonding [42] neglects friction effects, and predicts that the length of the debond controls the crack opening.

Both models predict steady-state cracking, where the matrix cracks at a constant stress that does not depend on the initial flaw distribution in the matrix. Experimental data support the steady-state cracking theory. Because the cracking stress is independent of flaw size, fracture toughness measurements (e.g., K_{IC} and $\mathcal{G}_c$) have little or no meaning.

Figure 6.37 illustrates the stress-strain behavior of a fiber-reinforced ceramic. The behavior is linear elastic up to σ_c, the steady-state cracking stress in the matrix. Once the matrix has cracked, the load is carried by the fibers. The fibers do not fail simultaneously, because the fiber strength is subject to statistical variability [45]. Consequently, the material exhibits quasiductility, where damage accumulates gradually until final failure.

Not only is fiber bridging the most effective toughening mechanism for ceramics (Table 6.1), it is also effective at high temperatures [46,47]. Consequently applications that require load-bearing capability at

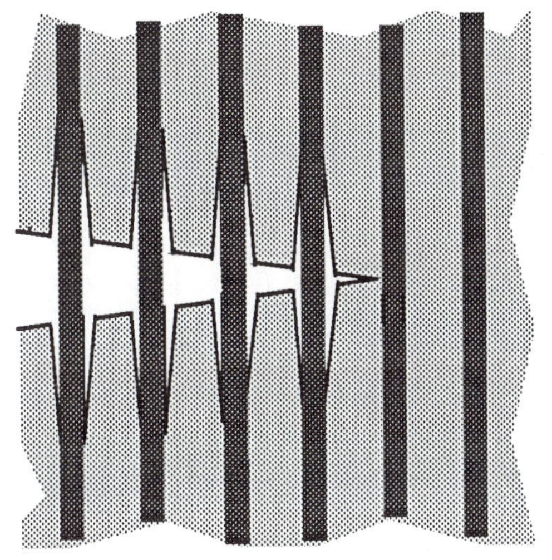

(a) Fiber/matrix debonding.

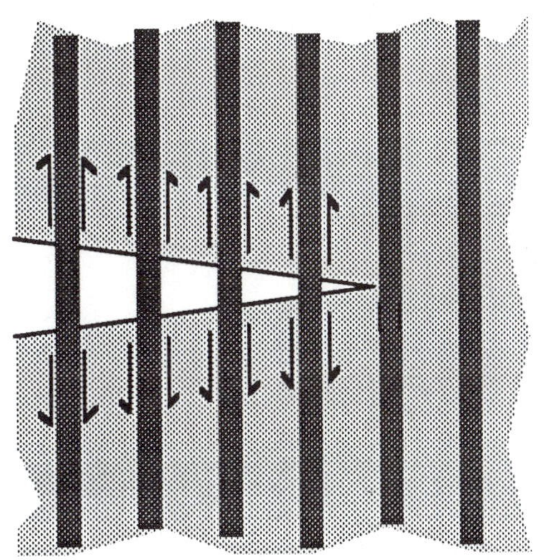

(b) Frictional sliding along interfaces.

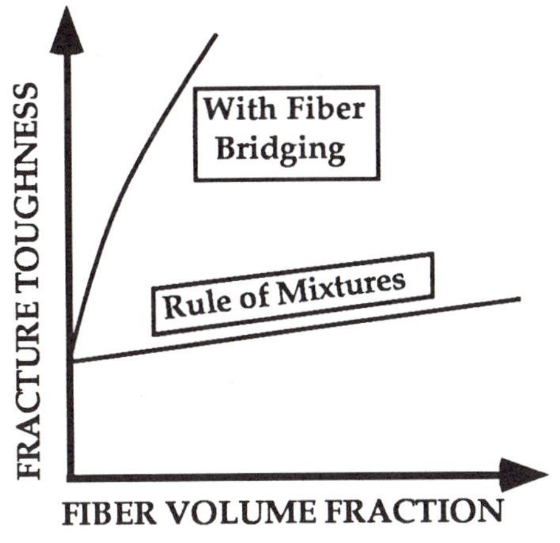

(c) Effect of bridging on toughness.

FIGURE 6.36 Fiber bridging in ceramic composites. Mathematical models treat bridging either in terms of fiber/matrix debonding (a) or frictional sliding (b). This mechanism provides composite toughness well in excess of that predicted by the rule of mixtures (c).

temperatures above 1000°C will undoubtedly utilize fiber-reinforced ceramics.

Whisker-reinforced ceramics possess reasonably high toughness, although whisker reinforcement is not as effective as continuous fibers. The primary failure mechanism in whisker composites appears to be bridging [48]; crack deflection also adds an increment of toughness. Figure 6.38 is a micrograph that illustrates crack bridging in an Al_2O_3 ceramic reinforced by SiC whiskers.

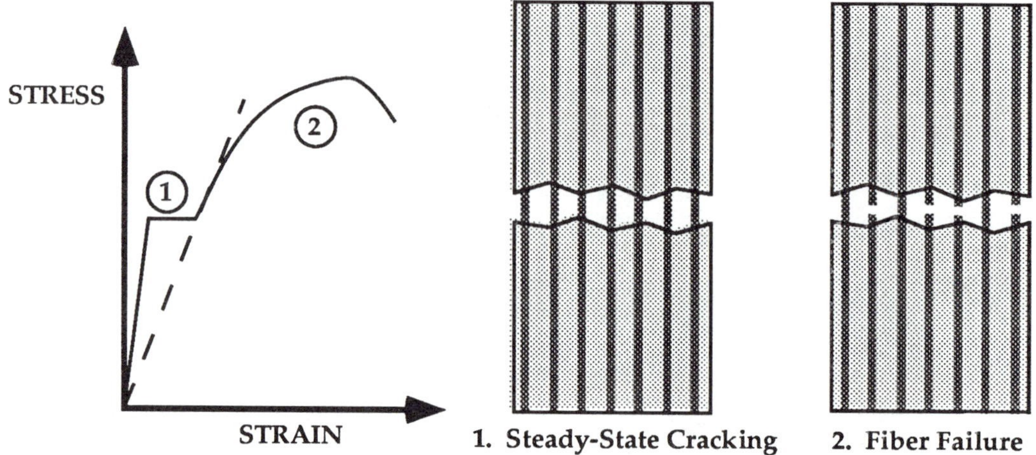

FIGURE 6.37 Stress-strain behavior of fiber-reinforced ceramic composites.

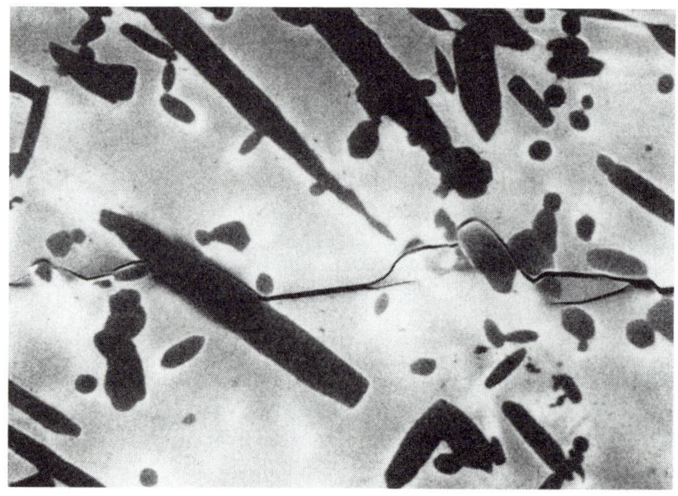

FIGURE 6.38 Crack bridging in Al_2O_3 reinforced with SiC whiskers [39]. (Photograph provided by A.G. Evans.)

REFERENCES

1. Johnson, W.S., ed., *Metal Matrix Composites: Testing, Analysis, and Failure Modes*. ASTM STP 1032, American Society for Testing and Materials, Philadelphia, 1989.

2. Williams, J.G. *Fracture Mechanics of Polymers*, , Halsted Press, John Wiley & Sons, New York, 1984.

3. Kinloch, A.J. and Young, R.J., *Fracture Behavior of Polymers,* Elsevier Applied Science Publishers, London, 1983.

4. Brostow, W. and Corneliussen, R.D., eds., *Failure of Plastics,* Hanser Publishers, Munich, 1986.

5. Hertzberg, R.W. and Manson, J.A., *Fatigue of Engineering Plastics,* Academic Press, New York, 1980.

6. Ferry, J.D., Landel, R.F., and Williams, M.L., "Extensions of the Rouse Theory of Viscoelastic Properties to Undiluted Linear Polymers." *Journal of Applied Physics,* Vol 26, pp. 359-362, 1955.

7. Larson, F.R. and Miller, J., "A Time-Temperature Relationship for Rupture and Creep Stresses", *Transactions of the American Society for Mechanical Engineers,* Vol 74, pp. 765-775, 1952.

8. Kausch, H.H., *Polymer Fracture,* Springer, Heidelberg-New York, 1978.

9. Zhurkov, S.N. and Korsukov, V.E., "Atomic Mechanism of Fracture of Solid Polymers," *Journal of Polymer Science: Polymer Physics Edition,* Vol. 12, pp. 385-398, 1974.

10. Ward, I.M., *Mechanical Properties of Solid Polymers.* John Wiley & Sons Ltd., New York, 1971.

11. Sternstein, S.S. and Ongchin, L., "Yield Criteria for Plastic Deformation of Glassy High Polymers in General Stress Fields." *American Chemical Society, Polymer Preprints,* Vol. 10, pp. 1117-1124, 1969.

12. Bucknall, C.B., *Toughened Plastics,* Applied Science Publishers, London, 1977.

13. Donald, A.M. and Kramer, E.J., "Effect of Molecular Entanglements on Craze Microstructure in Glassy Polymers." *Journal of Polymer Science: Polymer Physics Edition,* Vol. 27, pp. 899-909, 1982.

14. Oxborough, R.J. and Bowden, P.B., "A General Critical-Strain Criterion for Crazing in Amorphous Glassy Polymers.", *Philosophical Magazine,* Vol. 28, 1973, pp. 547-559.

15. Argon, A.S., "The Role of Heterogeneities in Fracture." ASTM STP 1020, American Society for Testing and Materials, Philadelphia, 1989, pp. 127-148.

16. Cayard, M., "Fracture Toughness Testing of Polymeric Materials." Ph.D. Dissertation, Texas A&M University, College Station, TX, September, 1990.

17. Dugdale, D.S., "Yielding in Steel Sheets Containing Slits." *Journal of the Mechanics and Physics of Solids,* Vol 8, pp. 100-104.

18. Barenblatt, G.I., "The Mathematical Theory of Equilibrium Cracks in Brittle Fracture." *Advances in Applied Mechanics*, Vol VII, Academic Press, 1962, pp. 55-129.

19. *Engineered Materials Handbook, Volume 1: Composites*. ASM International, Metals Park, OH, 1987.

20. Vinson, J.R. and Sierakowski, R.L., *The Behavior of Structures Composed of Composite Materials*. Marinus Nijhoff, Dordrecht, The Netherlands, 1987.

21. Wang, A.S.D., "An Overview of the Delamination Problem in Structural Composites." *Key Engineering Materials*, Vol. 37, 1989, pp. 1-20.

22. Hunston, D. and Dehl, R., "The Role of Polymer Toughness in Matrix Dominated Composite Fracture." Paper EM87-355, Society of Manufacturing Engineers, Deerborn, MI, 1987.

23. Bradley, W.L., "Understanding the Translation of Neat Resin Toughness into Delamination Toughness in Composites." *Key Engineering Materials*, Vol. 37, 1989, pp. 161-198.

24. Jordan, W.M, and Bradley, W.L., "Micromechanisms of Fracture in Toughened Graphite-Epoxy Laminates." ASTM STP 937, American Society for Testing and Materials, Philadelphia, 1987, pp 95-114.

25. Hibbs, M.F., Tse, M. K., and Bradley, W.L., "Interlaminar Fracture Toughness and Real-Time Fracture Mechanisms of Some Toughened Graphite/Epoxy Composites." ASTM STP 937, American Society for Testing and Materials, Philadelphia, 1987, pp 115-130.

26. Rosen, B.W., "Mechanics of Composite Strengthening." *Fiber Composite Materials*, American Society for Metals, Metals Park, OH, 1965, pp. 37-75.

27. Guynn, E.G., Bradley, W.L. and Elber, W., "Micromechanics of Compression Failures in Open Hole Composite Laminates." ASTM STP 1012, American Society for Testing and Materials, Philadelphia, 1989, pp 118-136.

28. Soutis, C., Fleck, N.A., and Smith, P.A., 'Failure Prediction Technique for Compression Loaded Carbon Fibre-Epoxy Laminate with Open Holes." *Submitted to Journal of Composite Materials*, 1990.

29. Highsmith, A.L. and Davis, J., "The Effects of Fiber Waviness on the Compressive Response of Fiber-Reinforced Composite Materials." Progress Report for NASA Research Grant NAG-1-659, NASA Langley Research Center, Hampton, VA, January 1990.

30. Wang, A.S.D., "A Non-Linear Microbuckling Model Predicting the Compressive Strength of Unidirectional Composites." ASME Paper 78-WA/Aero-1, American Society for Mechanical Engineers, New York, 1978.

31. Guynn, E.G., "Experimental Observations and Finite Element Analysis of the Initiation of Fiber Microbuckling in Notched Composite Laminates." Ph.D. Dissertation, Texas A&M University, College Station, TX, December 1990.

32. Whitcomb, J.D., "Finite Element Analysis of Instability Related Delamination Growth." *Journal of Composite Materials,* Vol. 15, 1981, pp. 403-425.

33. Awerbuch, J. and Madhukar, M.S., "Notched Strength of Composite Laminates: Predictions and Experiments–A Review." *Journal of Reinforced Plastics and Composites,* Vol. 4, 1985, pp. 3-159.

34. Whitney, J.M. and Nuismer, R.J., "Stress Fracture Criteria for Laminated Composites Containing Stress Concentrations." *Journal of Composite Materials,* Vol. 8, 1974, pp. 253-265.

35. Pipes, R.B., Wetherhold, R.C., and Gillespie, J.W., Jr., "Notched Strength of Composite Materials." *Journal of Composite Materials,* Vol. 12, 1979, pp. 148-160.

36. Waddoups, M.E., Eisenmann, J.R., and Kaminski, B.E., "Macroscopic Fracture Mechanics of Advanced Composite Materials." *Journal of Composite Materials,* Vol. 5, 1971, pp. 446-454.

37. Harris, C.E. and Morris, D.H., "A Comparison of the Fracture Behavior of Thick Laminated Composites Utilizing Compact Tension, Three-Point Bend, and Center-Cracked Tension Specimens." ASTM STP 905, American Society for Testing and Materials, Philadelphia, 1986, pp. 124-135.

38. Charewicz, A. and Daniel, I.M., "Damage Mechanisms and Accumulation in Graphite/Epoxy Laminates." ASTM STP 907, American Society for Testing and Materials, Philadelphia, 1986, pp. 274-297.

39. Evans, A.G., "The New High Toughness Ceramics." ASTM STP 907, American Society for Testing and Materials, Philadelphia, 1989, pp. 267-291.

40. Hutchinson, J.W., "Crack Tip Shielding by Micro Cracking in Brittle Solids", *Acta Metallurgica,* Vol. 35, 1987, p. 1605-1619.

41. A.G. Evans, ed., *Fracture in Ceramic Materials: Toughening Mechanisms, Machining Damage, Shock.* Noyes Publications, Park Ridge, NJ, 1984.

42. Budiansky, B., Hutchinson, J.W., and Evans, A.G., "Matrix Fracture in Fiber-Reinforced Ceramics." *Journal of the Mechanics and Physics of Solids,* Vol. 34, 1986, pp. 167-189.

43. Aveston, J., Cooper G.A., and Kelly, A., *The Properties of Fiber Composites,* 1971, pp 15-26.

44. Marshall, D.B., Cox, B.N. and Evans, A.G., "The Mechanics of Matrix Cracking in Brittle-Matrix Fiber Composites." *Acta Metallurgica,* Vol 33, 1985, pp. 2013-2021.

45. Marshall, D.B. and Ritter, J.E., "Reliability of Advanced Structural Ceramics and Ceramic Matrix Composites—A Review." *Ceramic Bulletin*, Vol. 68, 1987, pp. 309-317.

46. Mah, T., Mendiratta, M.G., Katz, A.P., Ruh, R., and Mazsiyasni, K.S., "Room Temperature Mechanical Behavior of Fiber-Reinforced Ceramic Composites." *Journal of the American Ceramic Society*, Vol. 68, 1985, pp. C27-C30.

47. Mah, T., Mendiratta, M.G., Katz, A.P., Ruh, R., and Mazsiyasni, K.S., "High-Temperature Mechanical Behavior of Fiber-Reinforced Glass-Ceramic-Matrix Composites." *Journal of the American Ceramic Society.*, Vol. 68, 1985, pp. C248-C251.

48. Ruhle, M., Dalgleish, B.J., and Evans, A.G., "On the Toughening of Ceramics by Whiskers." *Scripta Metallurgica*, Vol 21, pp. 681-686.

PART IV: APPLICATIONS

7. FRACTURE TOUGHNESS TESTING OF METALS

A fracture toughness test measures the resistance of a material to crack extension. Such a test may yield either a single value of fracture toughness or a resistance curve, where a toughness parameter such as K, J, or CTOD is plotted against crack extension. A single toughness value is usually sufficient to describe a test that fails by cleavage, because this fracture mechanism is typically unstable. The situation is similar to the schematic in Fig 2.10(a), which illustrates a material with a flat R curve. Cleavage fracture actually has a falling resistance curve, as Fig. 4.8 illustrates. Crack growth by microvoid coalescence, however, usually yields a rising R curve, such as that shown in Fig. 2.10(b); ductile crack growth can be stable, at least initially. When ductile crack growth initiates in a test specimen, that specimen seldom fails immediately. Therefore, one can quantify ductile fracture resistance either by the initiation value or by the entire R curve.

A variety of organizations throughout the world publish standardized procedures for fracture toughness measurements, including the American Society for Testing and Materials (ASTM), the British Standards Institution (BSI), and the Japan Society of Mechanical Engineers (JSME). The first standards for K and J testing were developed by ASTM in 1970 and 1981, respectively, while BSI published the first CTOD test method in 1979.

Existing fracture toughness standards include procedures for K_{IC}, K-R curve, J_{IC}, J-R curve, CTOD, and K_{Ia} testing. This chapter focuses primarily on ASTM standards, since they are the most widely used throughout the world. Standards produced by other organizations, however, are generally consistent with the ASTM procedures, and usually differ only in minute details. A few of the more substantive differences between alternative procedures are discussed briefly in this chapter.

The ongoing work of the various standardizing bodies is also discussed. The existing standards are continuously evolving, as the technology improves and more experience is gained. Also, there are still some important applications, such as weldment testing, that present standards do not address. The fracture testing community has, however, obtained substantial experience in some of these areas, and draft standards are currently being prepared.

The reader should not rely on this chapter alone for guidance on conducting fracture toughness tests, but should consult the relevant standards. Also, the reader is strongly encouraged to review Chapters 2 and 3 in order to gain an understanding of the fundamental basis of K, J, and CTOD, as well as the limitations of these parameters.

7.1 GENERAL CONSIDERATIONS

Virtually all fracture toughness tests have several common features. The design of test specimens is similar in each of the standards, and the orientation of the specimen relative to symmetry directions in the material is always an important consideration. The cracks in test specimens are introduced by fatigue in each case, although the requirements for fatigue loads varies from one standard to the next. The basic instrumentation required to measure load and displacement is common to all fracture mechanics tests, but some tests require addition instrumentation to monitor crack growth.

7.1.1 Specimen Configurations

There are five types of specimens that are permitted in ASTM standards that characterize fracture initiation and crack growth, although no single standard allows all five configurations, and the design of a particular specimen type may vary between standards. The configurations that are currently standardized include the compact specimen, the single edge notched bend (SENB) geometry, the arc-shaped specimen, the disk specimen, and the middle tension (MT) panel. Figure 7.1 shows a drawing of each specimen type.

An additional configuration, the compact crack arrest specimen, is used for K_{Ia} measurements and is described in Section 7.6. Specimens for qualitative toughness measurements, such as Charpy and drop weight tests, are discussed in Section 7.9. Chevron notched specimens, which are usually applied to brittle materials, are discussed in Chapter 8.

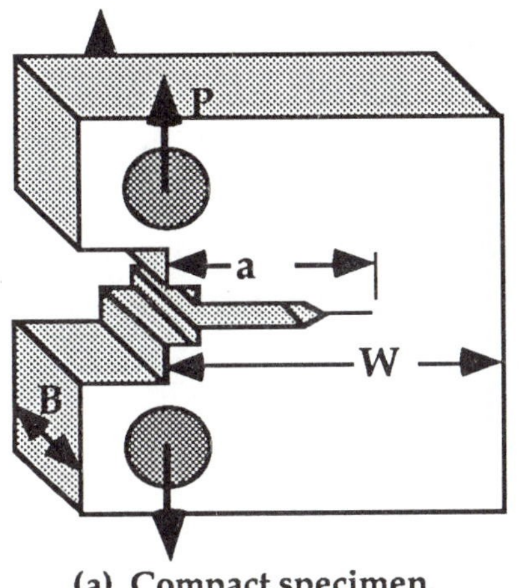

(a) Compact specimen.

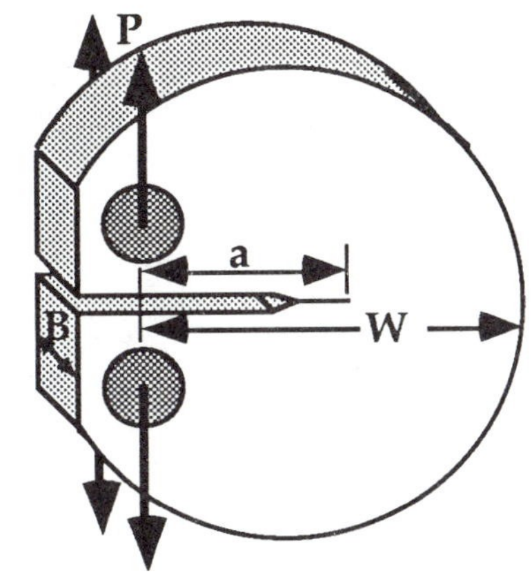

(b) Disk shaped compact specimen.

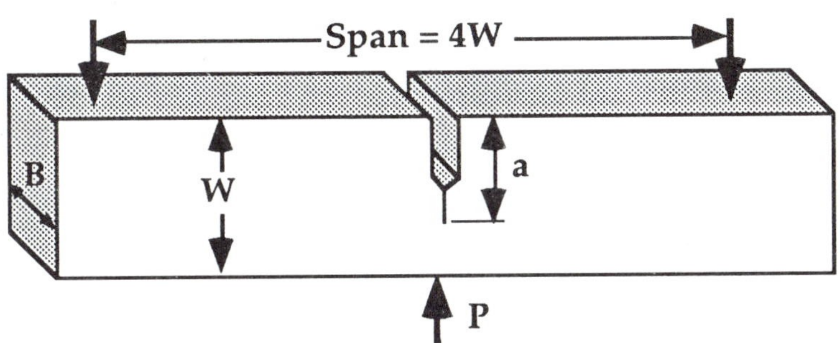

(c) Single edge notched bend (SENB) specimen.

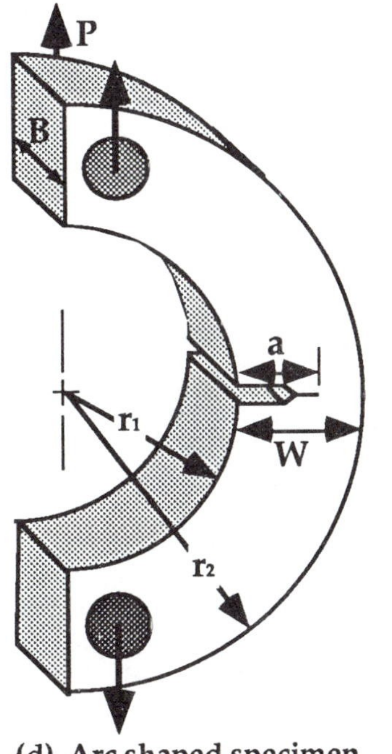

(d) Arc shaped specimen

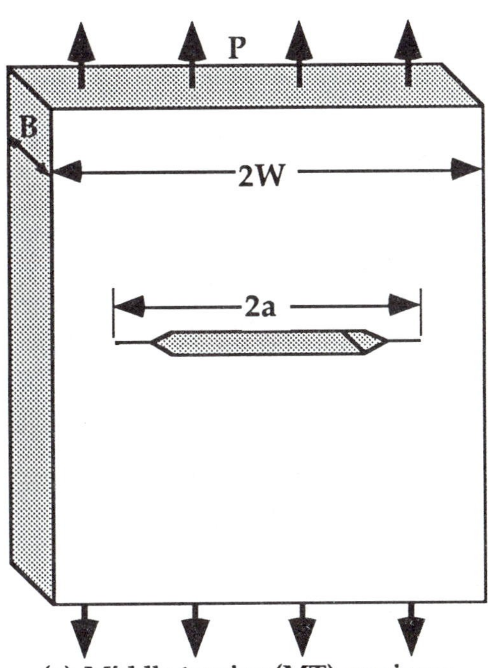

(e) Middle tension (MT) specimen.

FIGURE 7.1 Standardized fracture mechanics test specimens.

Each specimen configuration has three important characteristic dimensions: the crack length (a), the thickness (B) and the width (W). In most cases, W = 2 B and a/W ≈ 0.5, but there are exceptions which are discussed later in this chapter.

There are number of specimen configurations that are used in research, but have yet to be standardized. Some of the more common nonstandard configurations include the single edge notch tensile panel, the double edge notched tensile panel, the axisymmetric notched bar, and the double cantilever beam specimen.

The vast majority of fracture toughness tests are performed on either compact or SENB specimens. Figure 7.2 illustrates the profiles of these two specimen types, assuming the same characteristic dimensions (B, W, a). It is obvious from Fig. 7.2 that the compact geometry consumes less material, but this specimen requires extra material in the width direction, due to the holes. If one is testing plate material or a forging, the compact specimen is more economical, but the SENB configuration may be preferable for weldment testing, because less weld metal is consumed in some orientations (Section 7.7).

The compact specimen is pin-loaded by special clevises, as illustrated in Fig. 7.3. Compact specimens are usually machined in a limited number of sizes, because a separate test fixture must be fabricated for each specimen size. Specimen size is usually scaled geometrically; standard sizes include: $^1/_2$T, 1T, 2T and 4T, where the nomenclature refers to the thickness in inches[1]. For example, a standard 1T compact specimen has the dimensions B = 1 in (25.4 mm) and W = 2 in (50.8 mm). Although ASTM has converted to SI units, the above nomenclature for compact specimen sizes persists.

The SENB specimen is more flexible with respect to size. The standard loading span for SENB specimens is 4W. If the fixture is designed properly, the span can be adjusted continuously to any value that is within its capacity. Thus SENB specimens with a wide range of thicknesses can be tested with a single fixture. An apparatus for three-point bend testing is shown in Fig. 7.4.

1An exception to this interpretation of the nomenclature occurs in thin sheet specimens, as discussed in Section 7.3.

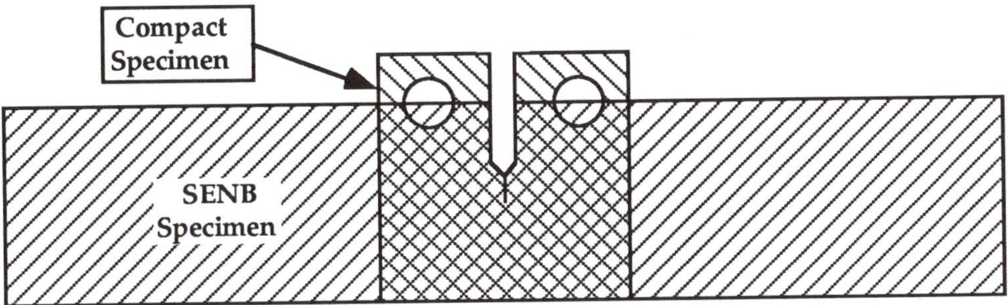

FIGURE 7.2 Comparison of the profiles of compact and SENB specimens with the same in-plane characteristic dimensions (W and a).

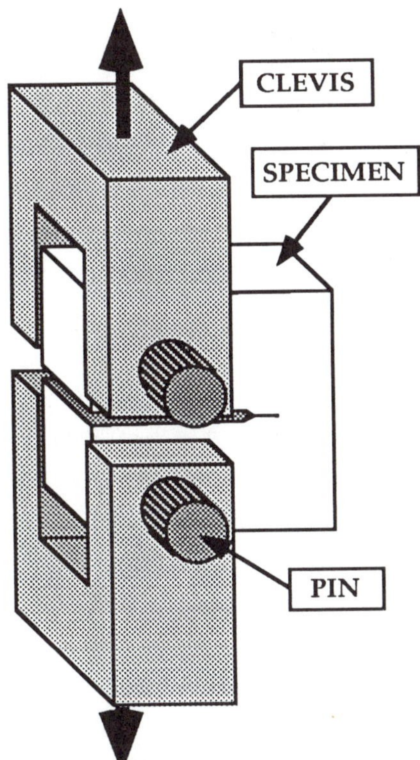

FIGURE 7.3 Apparatus for testing compact specimens.

7.1.2 Specimen Orientation

Engineering materials are seldom homogeneous and isotropic. Microstructure, and thus mechanical properties, are often sensitive to direction. The sensitivity to orientation is particularly pronounced in fracture toughness measurements, because a microstructure with a preferred orientation may contain planes of weakness, where crack propagation is relatively easy. Since specimen orientation is such an important variable in fracture toughness measurements, all ASTM fracture testing standards require that the orientation be reported along with the measured toughness; ASTM has adopted a notation for this purpose [1].

Figure 7.5 illustrates the ASTM notation for fracture specimens extracted from a rolled plate or forging. When the specimen is aligned with the axes of symmetry in the plate, there are six possible orientations. The letters L, T, and S denote the longitudinal, transverse, and short transverse directions, respectively, relative to the rolling direction or forging axis. Note that two letters are required to identify the orientation of a fracture mechanics specimen; the first letter indicates the direction of the principal tensile stress (which is always perpendicular to the crack plane in Mode I tests) and the second letter denotes the direction of crack propagation. For example, the L-T orientation corresponds to loading in the longitudinal direction and crack propagation in the transverse direction.

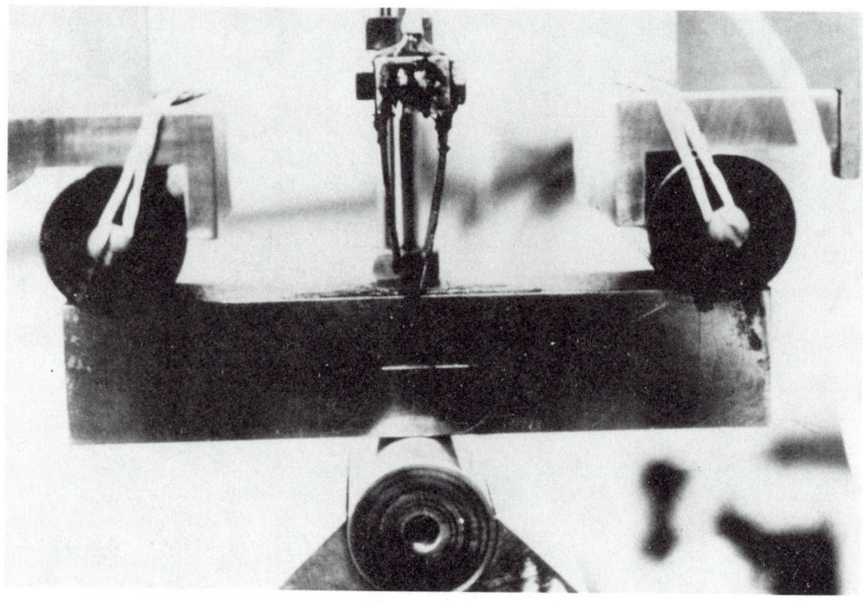

FIGURE 7.4 Three-point bending apparatus for testing SENB specimens.

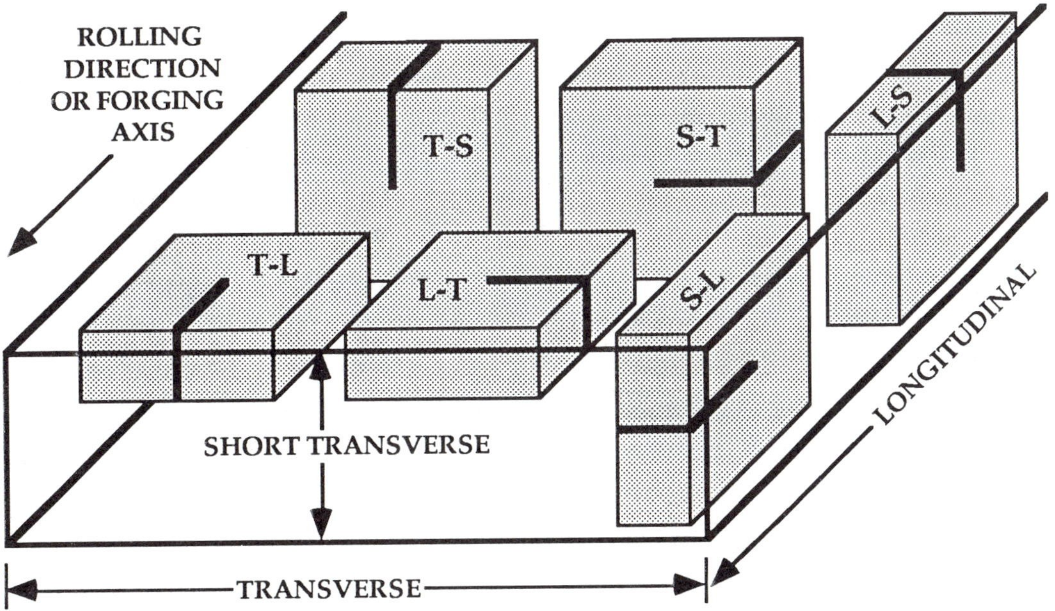

FIGURE 7.5 ASTM notation for specimens extracted from rolled plate and forgings [1].

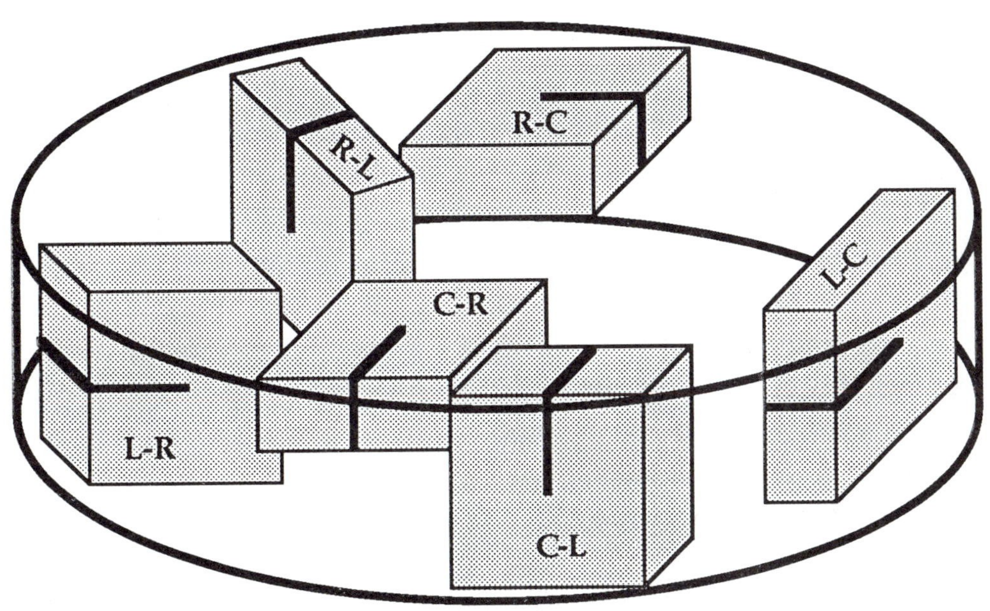

FIGURE 7.6 ASTM notation for specimens extracted from disks and hollow cylinders [1].

A similar notation applies to round bars and hollow cylinders, as Fig. 7.6 illustrates. The symmetry directions in this case are circumferential, radial, and longitudinal (C, R, and L, respectively).

Ideally, one should measure the toughness of a material in several orientations, but this is often not practical. When choosing an appropriate specimen orientation, one should bear in mind the purpose of the test, as well as geometrical constraints imposed by the material. A low toughness orientation, where the crack propagates in the rolling direction (T-L or S-L), should be adopted for general material characterization or screening. When the purpose of the test is to simulate conditions in a flawed structure, however, the crack orientation should match that of the structural flaw. Geometrical constraints may preclude testing some configurations; the S-L and S-T orientations, for example, are only practical in thick sections. The T-S and L-S orientations may limit the size of compact specimen that can be extracted from a rolled plate.

7.1.3 Fatigue Precracking

Fracture mechanics theory applies to cracks that are infinitely sharp prior to loading. While laboratory specimens invariably fall short of this ideal, it is possible to introduce cracks that are sufficiently sharp for practical purposes. The most efficient way to produce such a crack is through cyclic loading.

Figure 7.7 illustrates the precracking procedure in a typical specimen, where a fatigue crack initiates at the tip of a machined notch and grows to the desired size through careful control of the cyclic loads. Modern servohydraulic test machines can be programmed to produce sinusoidal loading, as well as a variety of other wave forms. Dedicated fatigue precracking machines that cycle at a high frequency are also available.

The fatigue crack must be introduced in such a way as not to adversely influence the toughness value that is to be measured. Cyclic loading produces a crack of finite radius with a small plastic zone at the tip, which contains strain hardened material and a complicated residual stress distribution (see Chapter 10). In order for a fracture toughness to reflect true material properties, the fatigue crack must satisfy the following conditions:

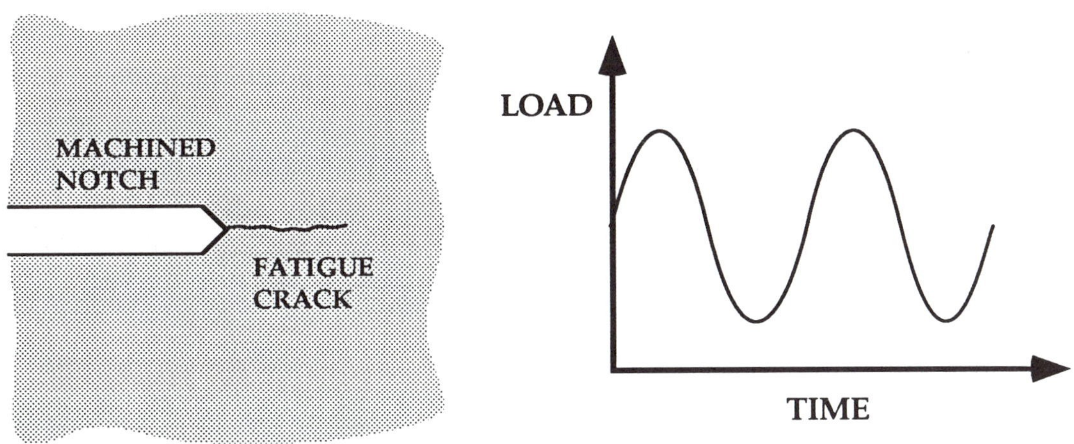

FIGURE 7.7 Fatigue precracking a fracture mechanics specimen. A fatigue crack is introduced at the tip of a machined notch by means of cyclic loading.

- The crack tip radius at failure must be much larger than the initial radius of the fatigue crack.
- The plastic zone produced during fatigue cracking must be small compared to the plastic zone at fracture.

Each of the various fracture testing standards contains restrictions on fatigue loads, which are designed to satisfy the above requirements. The precise guidelines depend on the nature of the test. In K_{IC} tests, for example, the maximum K during fatigue loading must be no greater than a particular fraction of K_{IC}. In J and CTOD tests, where the test specimen is typically fully plastic at failure, the maximum fatigue load is defined as a fraction of the load at ligament yielding. Of course one can always perform fatigue precracking well below the allowable loads in order to gain additional assurance of the validity of the results, but the time required to produce the crack (i.e., the number of cycles) increases rapidly with decreasing fatigue loads.

7.1.4 Instrumentation

At a minimum, the applied load and a characteristic displacement on the specimen must be measured during a fracture toughness test.

Additional instrumentation is applied to some specimens, in order to monitor crack growth or to measure more than one displacement.

Measuring load during a conventional fracture toughness test is relatively straightforward, since nearly all test machines are equipped with load cells. The most common displacement transducer in fracture mechanics tests is the clip gage [2], which is illustrated in Fig. 7.8. The clip gage, which attaches to the mouth of the crack, consists of four resistance strain gages bonded to a pair of cantilever beams. Deflection of the beams results in a change in voltage across the strain gages, which varies linearly with displacement. A clip gage must be attached to sharp knife edges in order to ensure that the ends of each beam are free to rotate. The knife edges can either be machined into the specimen or attached to the specimen at the crack mouth.

A *linear variable differential transformer* (LVDT) provides an alternative means for inferring displacements in fracture toughness tests. Figure 7.9 schematically illustrates the underlying principal of an LVDT. A steel rod is placed inside a hollow cylinder that contains a pair of tightly wound coils of wire. When a current passes through the first coil, the core becomes magnetized and induces a voltage in the second core. When the rod moves, the voltage drop in the second coil changes; the change in voltage varies linearly with displacement of the rod. The LVDT is useful for measuring displacements on a test specimen at locations other than the crack mouth.

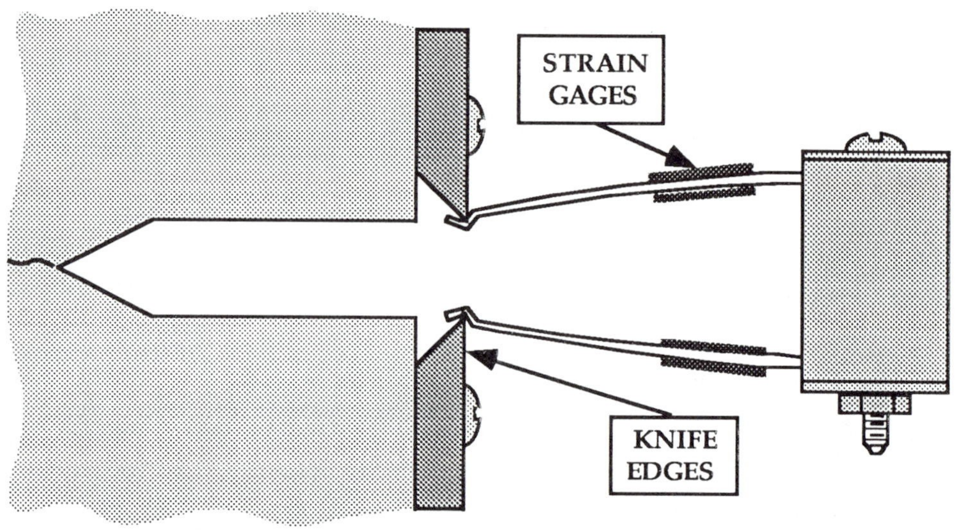

FIGURE 7.8. Measurement of the crack mouth opening displacement with a clip gage.

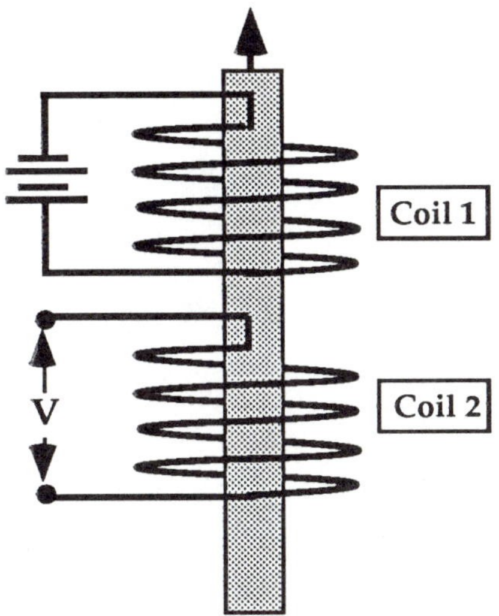

FIGURE 7.9 Schematic of a linear variable differential transformer (LVDT). Electric current in the first coil induces a magnetic field, which produces a voltage in the second coil. Displacement of the central core causes a variation in the output voltage.

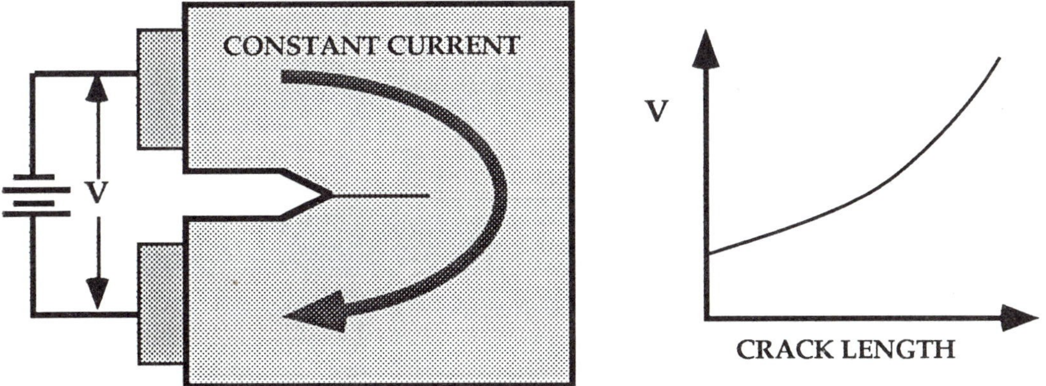

FIGURE 7.10 Potential drop method for monitoring crack growth. As the crack grows and the net cross sectional area decreases, the effective resistance increases, resulting in an increase in voltage (V).

The *potential drop* technique utilizes a voltage change to infer crack growth, as illustrated in Fig. 7.10. If a constant current passes through the uncracked ligament of a test specimen, the voltage must increase as the crack grows, because the electrical resistance increases and the net cross-sectional area decreases. The potential drop method can use either DC or AC current. See Refs. [3] and [4] for examples of this technique.

The disadvantage of the potential drop technique is that it requires additional instrumentation. The *unloading compliance* technique

[5,6], however, allows crack growth to be inferred from the load and displacement transducers that are part of any standard fracture mechanics test. A specimen can be partially unloaded at various points during the test in order to measure the elastic compliance, which can be related to the crack length. Section 7.4 describes the unloading compliance technique in more detail.

In some cases it is necessary to measure more than one displacement on a test specimen. For example, one may want to measure both the crack mouth opening displacement (CMOD) and the displacement along the loading axis. A compact specimen can be designed such that the load line displacement and the CMOD are identical, but these two displacements do not coincide in an SENB specimen. Figure 7.11 illustrates simultaneous CMOD and load line displacement measurement in an SENB specimen. The CMOD is inferred from a clip gage attached to knife edges; the knife edge height must be taken into account when computing the relevant toughness parameter (see Section 7.5). The load line displacement can be inferred by a number of methods, including the comparison bar technique [7,8] that is illustrated in Fig. 7.11. A bar is attached to the specimen at two points which are aligned with the outer loading points. The outer coil of an LVDT is attached to the comparison bar, which remains fixed during deformation, while the central rod is free to move as the specimen deflects.

7.1.5 Side Grooving

In certain cases, grooves are machined into the sides of a fracture toughness specimen [9], as illustrated in Fig. 7.12. The primary purpose of side grooving is to maintain close to plane strain conditions across the entire crack front. A specimen without side grooves is subject to crack tunneling and shear lip formation (Fig. 5.15) because the material near the outer surfaces is in a state of low stress triaxiality. Side grooves remove the free surfaces, where plane stress conditions prevail. Typical side grooved fracture toughness specimens have a net thickness that is approximately 80% of the gross thickness. If the side grooves are too deep, they produce lateral singularities, which cause the crack to grow more rapidly at the outer edges.

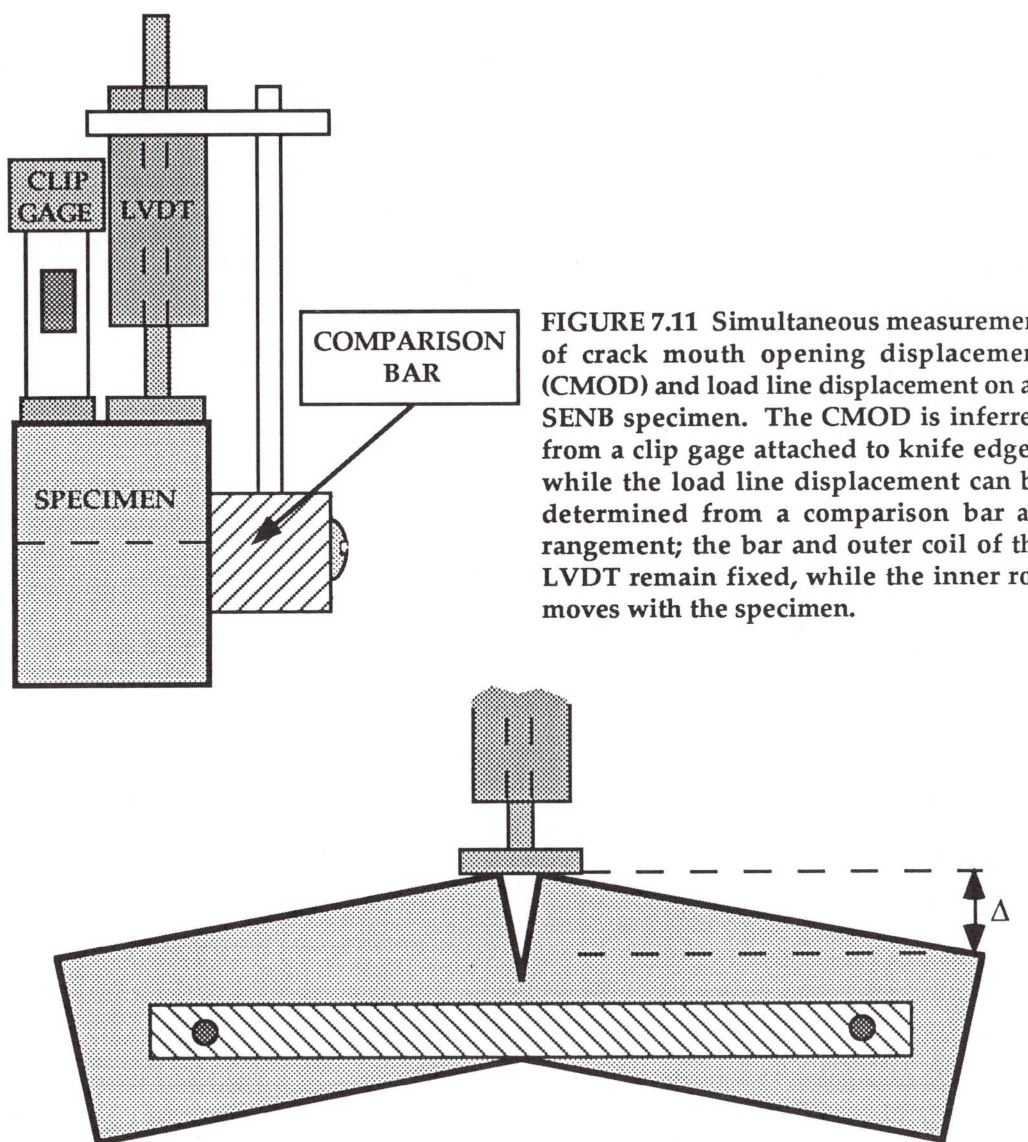

FIGURE 7.11 Simultaneous measurement of crack mouth opening displacement (CMOD) and load line displacement on an SENB specimen. The CMOD is inferred from a clip gage attached to knife edges, while the load line displacement can be determined from a comparison bar arrangement; the bar and outer coil of the LVDT remain fixed, while the inner rod moves with the specimen.

7.2 K_{IC} TESTING

When a material behaves in a linear elastic manner prior to failure, such that the plastic zone is small compared to specimen dimensions, a critical value of the the stress intensity factor, K_{IC}, may be an appropriate fracture parameter. Standard methods for K_{IC} testing include ASTM E 399 [2] and BS 5447 [10], the latter of which was published by the British Standards Institution. There are few substantive differences between these two standards.

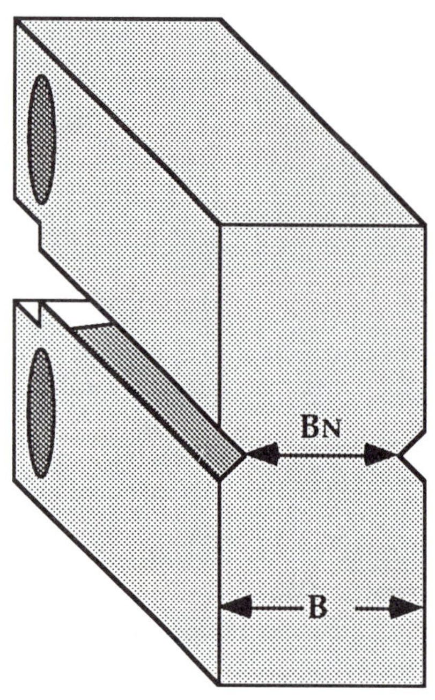

FIGURE 7.12 Side grooves in a fracture mechanics test specimen.

The ASTM standard E 399 was first published in 1970, and has been revised several times since then. The title, *"Standard Test Method for Plane Strain Fracture Toughness of Metallic Materials,"* is somewhat misleading. Although plane strain is a necessary condition for a valid K_{IC} test, it is not sufficient; a specimen must also behave in a linear elastic manner. The validity requirements in this standard are very stringent because even a relatively small amount of plastic deformation invalidates the assumptions of K theory (see Chapter 2).

Four specimen configurations are permitted by the current version of E 399: the compact, SENB, arc-shaped, and disk-shaped specimens. Specimens for K_{IC} tests are usually fabricated with the width, W, equal to twice the thickness, B. They are fatigue precracked so that the crack length/width ratio (a/W) lies between 0.45 and 0.55. Thus the specimen design is such that all the critical dimensions, a, B, and W-a, are approximately equal. This design results in efficient use of material, since each of these dimensions must be large compared to the plastic zone.

Most standardized mechanical tests (fracture toughness and otherwise), lead to valid results as long as the technician follows all of the procedures outlined in the standard. The K_{IC} test, however, often produces invalid results through no fault of the technician. If the plastic zone at fracture is too large, it is not possible to obtain a valid K_{IC}, regardless of how skilled the technician is.

Because of the strict size requirements, ASTM E 399 recommends that the user perform a preliminary validity check to determine the appropriate specimen dimensions. The size requirements for a valid K_{IC} are as follows:

$$B, (W\text{-}a) \geq 2.5 \left(\frac{K_{IC}}{\sigma_{YS}}\right)^2 \tag{7.1}$$

In order to determine the required specimen dimensions, the user must make a rough estimate of the anticipated K_{IC} for the material. Such an estimate can come from data for similar materials. If such data are not available, the ASTM standard provides a table of recommended thicknesses for various strength levels. Although there is a tendency for toughness to decrease with increasing strength, there is not a unique relationship between K_{IC} and σ_{YS} in metals. Thus the strength-thickness table in E 399 should only be used when better data are not available.

During the initial stages of fatigue precracking, the peak value of stress intensity in a single cycle, K_{max}, should be no larger than 0.8 K_{IC}, according to ASTM E 399. As the crack approaches its final size, K_{max} should be less than 0.6 K_{IC}. If the specimen is fatigued at one temperature (T_1) and tested at a different temperature (T_2), the final K_{max} must be $\leq 0.6(\sigma_{YS(1)}/\sigma_{YS(2)})K_{IC}$. The fatigue load requirements are less stringent at initiation because the final crack tip is remote from any damaged material that is produced in the early part of precracking. The maximum stress intensity during fatigue must always be less than K_{IC}, however, in order to avoid premature failure of the specimen.

Of course, one must know K_{IC} in order to determine the maximum allowable fatigue loads. The user must specify fatigue loads based on the anticipated toughness of the material. If he or she is conservative and selects low loads, precracking could take a very long time. On the other hand, if precracking is conducted at high loads, the user risks an invalid result, in which case the specimen and the technician's time are wasted.

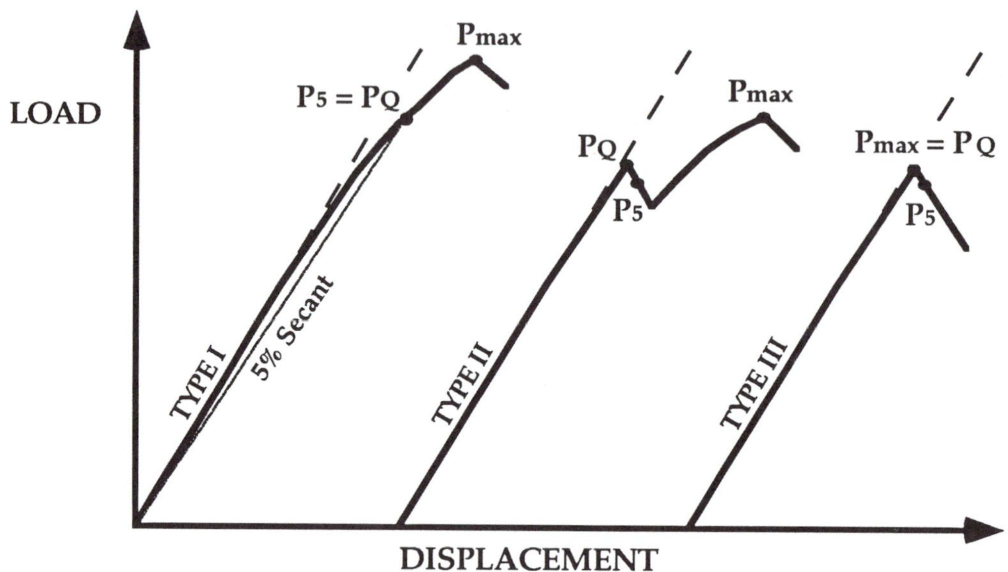

FIGURE 7.13 Three types of load-displacement behavior in a K_{IC} test.

When a precracked test specimen is loaded to failure, load and displacement are monitored. Three types of load-displacement curves are shown in Fig. 7.13. The critical load, P_Q, is defined in one of several ways, depending on the type of curve. One must construct a 5% secant line (i.e. a line from the origin with a slope equal to 95% of the initial elastic loading slope) to determine P_5. In the case of Type I behavior, the load-displacement curve is smooth and it deviates slightly from linearity before ultimate failure at P_{max}. This nonlinearity can be caused by plasticity, subcritical crack growth, or both. For a Type I curve, $P_Q = P_5$. With a Type II curve, a small amount of unstable crack growth (i.e. a pop-in) occurs before the curve deviates from linearity by 5%. In this case P_Q is defined at the pop-in. A specimen that exhibits Type III behavior fails completely before achieving 5% nonlinearity. In such cases, $P_Q = P_{max}$.

The crack length must be measured from the fracture surface. Since there is a tendency for the crack depth to vary through the thickness, the crack length is defined as the average of three evenly spaced measurements. Once P_Q and crack length are determined, a provisional fracture toughness, K_Q, is computed from the following relationship:

$$K_Q = \frac{P_Q}{B\sqrt{W}} f(a/W) \tag{7.2}$$

where f(a/W) is a dimensionless function of a/W. This function is given in polynomial form in the E 399 standard for the four specimen types. Individual values of f(a/W) are also tabulated in ASTM E399. (See Table 2.2 and Chapter 12 for K solutions for a variety of configurations.)

The K_Q value computed from Eq. (7.2) is a valid K_{IC} result only if all validity requirements in the standard are met, including

$$0.45 \leq a/W \leq 0.55 \tag{7.3a}$$

$$B, (W-a) \geq 2.5 \left(\frac{K_Q}{\sigma_{YS}}\right)^2 \tag{7.3b}$$

$$P_{max} \leq 1.10\, P_Q \tag{7.3c}$$

Additional validity requirements include the restrictions on fatigue load mentioned earlier, as well as limits on fatigue crack curvature. If the test meets all of the requirements of ASTM E399, then $K_Q = K_{IC}$.

Section 2.10 describes the limitations of stress intensity factor, and outlines the theoretical reasons for the strict size requirements for K_{IC}. Recall that Eqs. (7.3a) and (7.3b) ensure that the critical specimen dimensions, B, a, and (W-a), are at least ~50 times larger than the plane strain plastic zone. The third requirement, Eq. (7.3c), is necessary to correct a loophole in the K_{IC} test procedure, as discussed below.

The deviation from linearity in a load-displacement curve can be caused by crack growth, plastic zone effects, or both. In the absence of plastic deformation, 5% deviation from the initial slope of the load-displacement curve corresponds to crack growth through approximately 2% of the ligament in test specimens with $a/W \approx 0.5$; when a plastic zone forms, a 5% deviation from linearity can be viewed as 2% *apparent* crack growth. (Recall Section 2.8, where crack tip plasticity was modeled by pretending that the crack was slightly longer than the actual size.) If the nonlinearity in the load displacement curve is caused only by plasticity, a 5% deviation from linearity corresponds to a plastic zone size that is roughly 2% (i.e. 1/50) of the uncracked ligament. *Thus the plastic zone size at P_5 in a Type I test is approximately equal to its maximum allowable size, as defined by Eq. (7.3a).*

Consider a fracture toughness test that displays considerable plastic deformation prior to failure. Figure 7.14 schematically illustrates the load-displacement curve for such a test. Since this is a Type I curve, $P_Q = P_5$. A K_Q value computed from P_Q may just barely satisfy the size requirements of Eq. (7.3a) for reasons described in the previous paragraph. Such a quantity, however, would have little relevance to the fracture toughness of the material, since the specimen fails well beyond P_Q; the K_Q value in this case would grossly underestimate the true toughness of the material. Consequently the third validity requirement, Eq. (7.3c), is necessary to ensure that a K_{IC} value is indicative of the true toughness of the material.

Because the size requirements of ASTM E 399 are very stringent, it is very difficult and sometimes impossible to measure a valid K_{IC} in most structural materials, as Examples 7.1 and 7.2 illustrate. A material must either be relatively brittle or the test specimen must be very large for linear elastic fracture mechanics to be valid. In low- and medium strength structural steels, valid K_{IC} tests are normally only possible on the lower shelf of toughness; in the ductile-brittle transition and the upper shelf, elastic-plastic parameters such as the J integral and CTOD are required to characterize fracture.

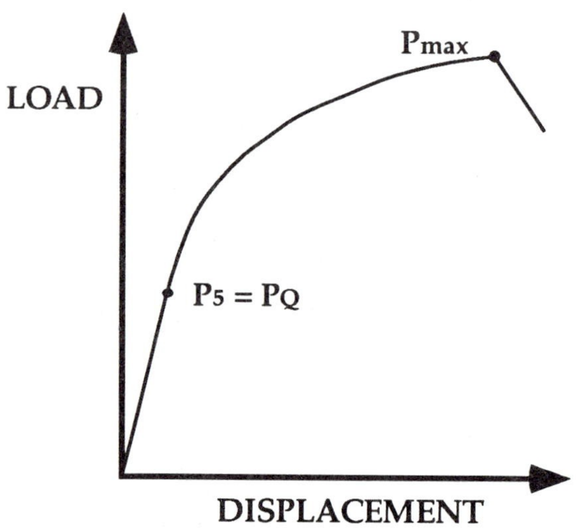

FIGURE 7.14 Load-displacement curve for an invalid K_{IC} test, where ultimate failure occurs well beyond P_Q.

Because of the strict validity requirements, the K_{IC} test is of limited value to structural metals. The toughness and thickness of most materials precludes a valid K_{IC} result. If, however, a valid K_{IC} test can be measured on a given material, it is probably too brittle for most structural applications.

EXAMPLE 7.1

Consider a structural steel with σ_{YS} = 350 MPa (51 ksi). Estimate the specimen dimensions required for a valid K_{IC} test. Assume that this material is on the upper shelf of toughness, where typical K_{IC} values for initiation of microvoid coalescence in these materials are around 200 MPa $\sqrt{m}$.

Solution: Inserting the yield strength and estimated toughness into Eq. (7.1) gives

$$B, a = 2.5\left(\frac{200 \text{ MPa} \sqrt{\text{m}}}{350 \text{ MPa}}\right)^2 = 0.816 \text{ m (32.1 in)}$$

Since $a/W \approx 0.5$, W = 1.63 m (64.2 in)! Thus a very large specimen would be required for a valid K_{IC} test. Materials are seldom available in such thicknesses. Even if a sufficiently large section thickness were fabricated, testing such a large specimen would not be practical; machining would be prohibitively expensive, and a special testing machine with a high load capacity would be needed.

EXAMPLE 7.2

Suppose that the material in Example 7.1 is fabricated in 25 mm (1 in) thick plate. Estimate the largest valid K_{IC} that can be measured on such a specimen.

EXAMPLE 7.2 (cont.)

Solution: For the L-T or T-L orientation, a test specimen with a standard design could be no larger than B = a = 25 mm and W = 50 mm. Inserting these dimensions and the yield strength into Eq. (7.1) and solving for K_{IC} gives

$$K_{IC} = 350 \text{ MPa} \sqrt{\frac{0.025 \text{ m}}{2.5}} = 35 \text{ MPa} \sqrt{\text{m}}$$

Figure 4.5 shows fracture toughness data for an A 572 Grade 50 steel. Note that the toughness level computed above corresponds to the lower shelf in this material. Thus valid K_{IC} tests on this material would only be possible at low temperatures, where the material is too brittle for most structural applications.

7.3 K-R CURVE TESTING

Some materials whose fracture behavior is predominantly linear elastic exhibit a rising R curve. The ASTM Standard E 561-86 [11] outlines a procedure for determining K versus crack growth curves in such materials. Unlike ASTM E 399, the K-R standard does not contain a minimum thickness requirement, and thus can be applied to thin sheets. This standard, however, is only appropriate when the plastic zone is small compared to the in-plane dimensions of the test specimen. This test method is often applied to high strength sheet materials, where the fracture behavior is plane stress linear elastic.

There is a common misconception about plane stress, plane strain, and R curves. A number of published articles and textbooks imply that a material in plane strain exhibits a single value of fracture toughness (K_{IC}), while the same material in plane stress displays a rising R curve. While this may occur in some cases, it is not a universal phenomenon. The shape of an R curve depends on the fracture mechanism as well as the stress state at the crack tip. Cleavage tends to exhibit a flat or falling R curve, while microvoid coalescence can produce a rising R curve. The slope of an R curve tends to decrease with increasing stress triaxiality, and the fracture mechanism (in steels) can change from ductile tearing to cleavage as the stress state ranges from plane stress to plane strain. This leads to the impression that plane stress conditions *always* produce a rising resistance curve while plane strain fracture can be described by a single toughness value (K_{IC}). It is possible, however, for cleavage, and thus a falling R curve, to occur in thin sheets. Similarly, rising R curves under plane strain conditions are common in ductile materials, as discussed in Section 7.4.

Figure 7.15 illustrates a typical K-R curve in a predominantly linear elastic material. The R curve is initially very steep, as little or no crack growth occurs with increasing K_I. As the crack begins to grow, K increases with crack growth until a steady state is reached, where the R curve becomes flat (see Section 3.5 and Appendix 3.5). It is possible to define a critical stress intensity, K_c, where the driving force is tangent to the R curve. This instability point is not a material property, however, because the point of tangency depends on the shape of the driving force curve, which is governed by the geometry of the cracked body. Thus K_c values obtained from laboratory specimens are not usually transferable to structures.

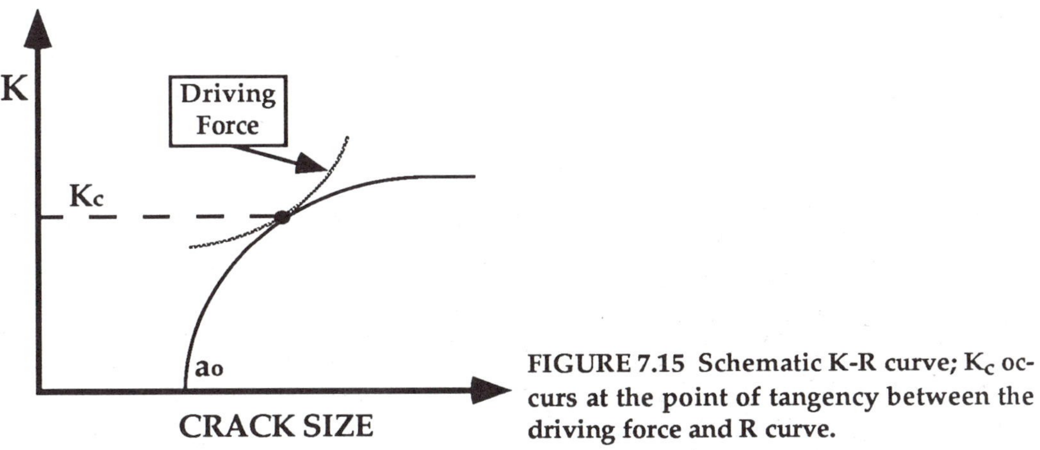

FIGURE 7.15 Schematic K-R curve; K_c occurs at the point of tangency between the driving force and R curve.

7.3.1 Specimen Design

The ASTM standard for K-R curve testing [11] permits three configurations of test specimen: the middle tension (MT) geometry, the conventional compact specimen, and a wedge loaded compact specimen. The latter configuration, which is similar to the compact crack arrest specimen discussed in Section 7.6, is the most stable of the three specimen types, and thus is suitable for materials with relatively flat R curves.

Since this test method is often applied to thin sheets, specimens do not usually have the conventional geometry, with the width equal to twice the thickness. The specimen thickness is normally fixed by the sheet thickness, and the width is governed by the anticipated toughness of the material, as well as the available test fixtures.

A modified nomenclature is applied to thin sheet compact specimens. For example, a specimen with W = 50 mm (2 in) is designated as a *1T plan* specimen, since the in-plane dimensions correspond to the conventional 1T compact geometry. Standard fixtures can be used to test thin sheet compact specimens, provided the specimens are fitted with spacers, as illustrated in Fig. 7.16.

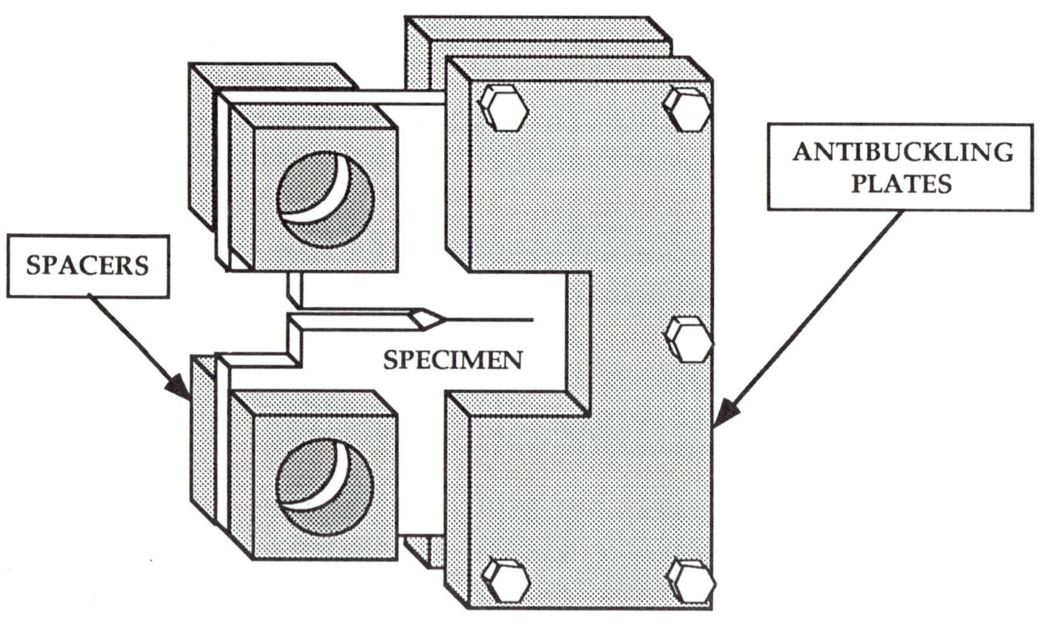

FIGURE 7.16 Antibuckling fixtures for testing thin compact specimens

One problem with thin sheet fracture toughness testing is that the specimens are subject to out-of-plane buckling, which leads to combined Mode I-Mode III loading of the crack. Consequently, an antibuckling device should be fitted to the specimen. Figure 7.16 illustrates a typical antibuckling fixture for thin sheet compact specimens. Plates on either side of the specimen prevent out-of-plane displacements. These plates should not be bolted too tightly together, because loads applied by the test machine should be carried by the specimen rather than the antibuckling plates. Some type of lubricant (e.g. Teflon sheet) is usually required to allow the specimen to slide freely through the two plates during the test.

7.3.2 Experimental Measurement of K-R Curves

The ASTM Standard E 561-86 outlines a number of alternative methods for computing both K_I and the crack extension in an R curve test; the most appropriate approach depends on the relative size of the plastic zone. Let us first consider the special case of negligible plasticity, which exhibits a load-displacement behavior that is illustrated in Figure 7.17. As the crack grows, the load-displacement curve deviates from its initial linear shape because the compliance continuously changes. If the specimen were unloaded prior to fracture, the curve would return to the origin, as the dashed lines indicate. The compliance at any point during the test is equal to the displacement divided by the load. The instantaneous crack length can be inferred from the compliance through relationships that are given in the ASTM standard. (See Chapter 12 for compliance-crack length equations for a variety of configurations.) The crack length can also be measured optically during tests on thin sheets, where there is negligible through-thickness variation of crack length. The instantaneous stress intensity is related to the current values of load and crack length:

$$K_I = \frac{P}{B\sqrt{W}} f(a/W) \tag{7.4}$$

Consider now the case where a plastic zone forms ahead of the growing crack. The nonlinearity in the load-displacement curve is caused by a combination of crack growth and plasticity, as Fig. 7.18 illustrates. If the specimen is unloaded prior to fracture, the load-displacement curve does not return to the origin; crack tip plasticity produces a finite amount of permanent deformation in the specimen. The physical crack length can be determined optically or from unloading compliance, where the specimen is partially unloaded, the elastic compliance is measured, the crack length is inferred from compliance. The stress intensity should be corrected for plasticity effects by determining an effective crack length. The ASTM standard suggests two alternative approaches for computing a_{eff}: the Irwin plastic zone correction and the secant method. According to the Irwin approach (Section 2.8.1), the effective crack length for plane stress is given by

$$a_{eff} = a + \frac{1}{2\pi}\left(\frac{K}{\sigma_{YS}}\right)^2 \tag{7.5}$$

The secant method consists of determining an effective crack size from the effective compliance, which is equal to the total displacement divided by the load (Fig. 7.18). The effective stress intensity factor for both methods is computed from the load and the effective crack length:

$$K_{eff} = \frac{P}{B\sqrt{W}} f\,(a_{eff}/W) \tag{7.6}$$

The Irwin correction requires an iterative calculation, where a first order estimate of a_{eff} is used to estimate K_{eff}, which is inserted into Eq. (7.5) to obtain a new a_{eff}; the process is repeated until the K_{eff} estimates converge.

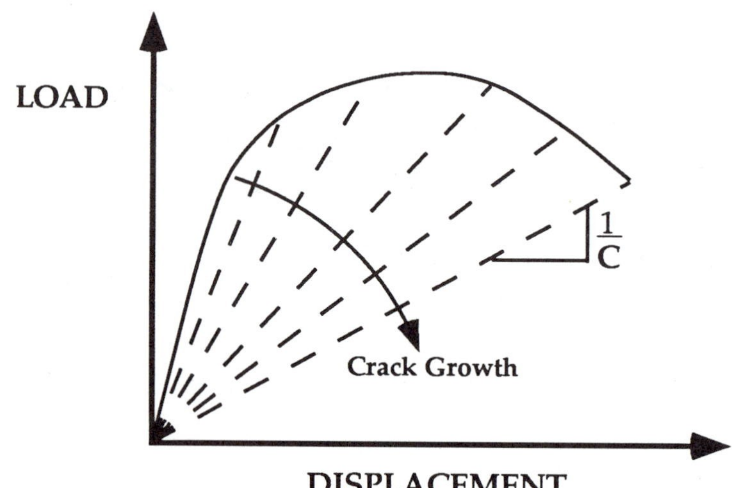

FIGURE 7.17 Load-displacement curve for crack growth in the absence of plasticity.

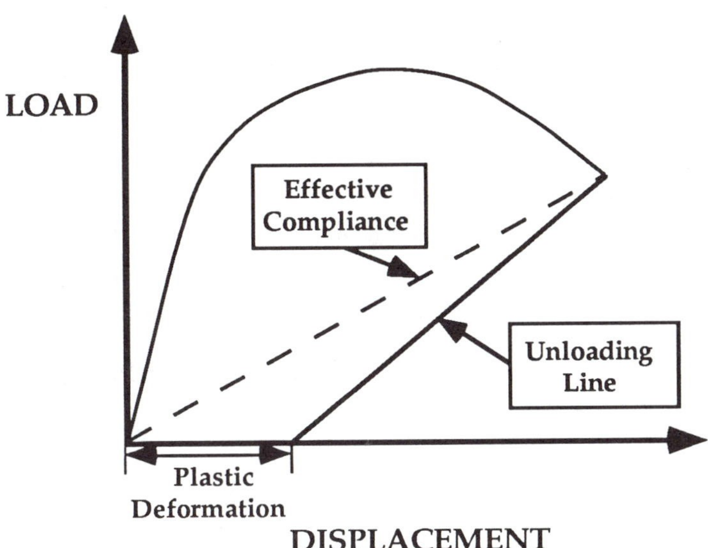

FIGURE 7.18 Load-displacement curve for crack growth with plasticity.

The choice of plasticity correction is left largely up to the user. When the plastic zone is small, ASTM E 561 suggests that the Irwin correction is acceptable, but recommends applying the secant approach when the crack tip plasticity is more extensive. Experimental data typically display less size dependence when the stress intensity is determined by the secant method [13].

The ASTM K-R curve standard requires that the stress intensity be plotted against the *effective* crack extension (Δa_{eff}). This practice is inconsistent with the J_{IC} and J-R curve approaches (Section 7.4), where J is plotted against the *physical* crack extension. The estimate of the instability point (K_c) should not be sensitive to the way in which crack growth in quantified, particularly when both the driving force and resistance curves are computed with a consistent definition of Δa.

The ASTM E 561-86 Standard does not contain requirements on specimen size or the maximum allowable crack extension; thus there is no guarantee that a K-R curve produced according to this standard will be a geometry-independent material property. As discussed in Section 2.10, application of LEFM to thin sections is acceptable as long as the specimen thickness matches the section thickness of the structure. The in-plane dimensions, however, must be large compared to the plastic zone in order for LEFM to be valid. Also, the growing crack must be remote from all external boundaries.

Unfortunately, the size dependence of R curves in high strength sheet materials has yet to be quantified, so it is not possible to recommend specific size and crack growth limits for this type of testing. The user must be aware of the potential for size dependence in K-R curves. Application of the secant approach reduces but does not eliminate the size dependence. The user should test wide specimens whenever possible in order to ensure that the laboratory test is indicative of the structure under consideration.

7.4 J TESTING OF METALS

Two ASTM standards currently address J testing. The J_{IC} standard, E 813 [5], which was first published in 1981 and revised in 1987, outlines a test method for estimating the critical J near initiation of ductile crack growth. A J-R curve testing standard, E 1152 [6], was first published in 1987. The Japan Society of Mechanical Engineers (JSME) has published a standard test method that utilizes scanning electron microscopic (SEM) examination of the fracture surfaces to infer a J_{IC} for initiation of ductile crack growth [13].

Because J integral test methods were originally developed by the nuclear power industry for applications well above ambient temperature, most existing J testing standards apply only to ductile fracture. However, the J integral can, in principal, be applied to cleavage fracture. Both ASTM and BSI [14,15] have produced draft standards that permit J testing on the lower shelf and in the ductile-brittle transition of steels (see Section 7.8).

The ASTM standards E 813-87 and E 1152-87 both produce a J-R curve, a plot of J versus crack extension. The E 1152 standard applies to the entire J-R curve, while E 813 is concerned only with J_{IC}, a single point on the R curve. The same test can be reported in terms of both standards. This is analogous to a tensile test, where one can report either the yield strength or the entire stress-strain curve. Both standards apply to compact and SENB specimens.

7.4.1 J_{IC} Measurements

The R curve for J_{IC} measurements can be generated by either multiple specimen or single specimen techniques. With the multiple specimen technique, a series of nominally identical specimens are loaded to various levels and then unloaded. Some stable crack growth occurs in most specimens. This crack growth is marked by heat tinting or fatigue cracking after the test. Each specimen is then broken open and the crack extension is measured.

The most common single specimen test technique is the unloading compliance method, which is illustrated in Fig. 7.19. The crack length is computed at regular intervals during the test by partially unloading the specimen and measuring the compliance. As the crack grows, the specimen becomes more compliant (less stiff). Both E 813 and E 1152 provide polynomial expressions that relate a/W to compliance. These standards require relatively deep cracks ($0.50 \leq a/W < 0.70$) because the unloading compliance technique is not sufficiently sensitive for $a/W < 0.5$. An alternative single specimen test method is the potential drop procedure (Fig. 7.10), yet to be standardized by ASTM, in which crack growth is monitored through the change in electrical resistance which accompanies a loss in cross sectional area. Both single specimen procedures are practical only in conjunction with a computer data acquisition and analysis system.

Regardless of the method for monitoring crack growth, a corresponding J value must be computed for each point on the R curve. For estimation purposes, ASTM E 813-87 divides J into elastic and plastic components:

$$J = J_{el} + J_{pl} \tag{7.7}$$

The elastic J is computed from the elastic stress intensity:

$$J_{el} = \frac{K^2(1-\nu^2)}{E} \tag{7.8}$$

where K is inferred from load and crack size through Eq. (7.4). If, however, side grooved specimens are used, the expression for K is modified:

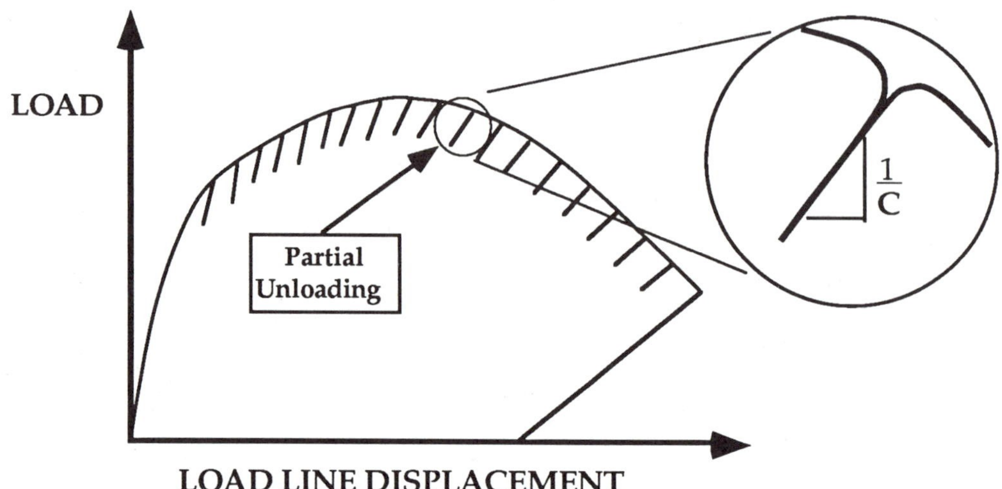

FIGURE 7.19. The unloading compliance method for monitoring crack growth.

$$K = \frac{P}{\sqrt{B\,B_N\,W}} f(a/W) \tag{7.9}$$

where B is the gross thickness and B_N is the net thickness (Fig 7.12). The ASTM J_{IC} standard enables the plastic J at each point on the R curve to be estimated from the plastic area under the load-displacement curve:[2]

$$J_{pl} = \frac{\eta\,A_{pl}}{B_N\,b_o} \tag{7.10}$$

where η is a dimensionless constant, A_{pl} is the plastic area under the load-displacement curve (see Fig. 7.20), and b_o is the initial ligament length. For an SENB specimen,

$$\eta = 2.0 \tag{7.11a}$$

and for a compact specimen,

$$\eta = 2 + 0.522\,b_o/W \tag{7.11b}$$

[2]Since J is defined in terms of the energy absorbed divided by the net cross sectional area, B_N appears in the denominator. For nonside grooved specimens $B_N = B$.

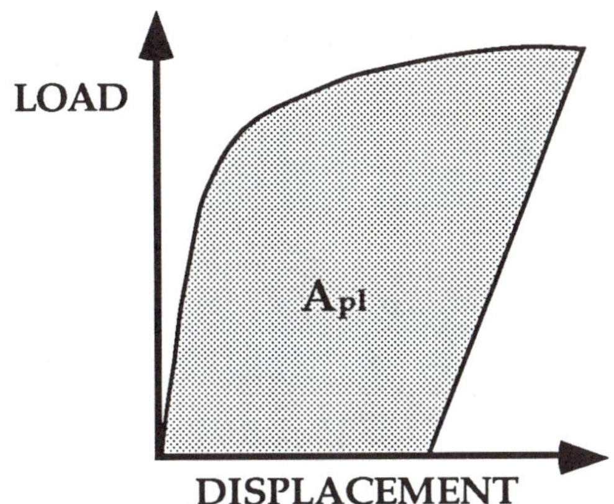

FIGURE 7.20 Plastic energy absorbed by a test specimen during a J_{IC} test.

Recall from Section 3.2.5 that Eq. (7.10) was derived from the energy release rate definition of J.

Note that Eqs. (7.10) and (7.11b) do not correct J for crack growth, but are based on the initial crack length. The ASTM J-R curve standard requires a more complicated procedure, in which J is computed incrementally with updated values of crack length and ligament length (see Section 7.4.2). Since E 813-87 is concerned with the initiation point rather than the shape of the R curve, the more detailed formula that corrects J for crack growth is optional in J_{IC} tests. In the limit of a stationary crack, both formulas give identical results. Thus the measured initiation toughness is insensitive to the choice of J equation.

The E 813-87 procedure for computing J_Q, a provisional J_{IC}, from the R curve is illustrated in Fig. 7.21. Exclusion lines are drawn at crack extension (Δa) values of 0.15 and 1.5 mm. These lines have a slope of $2\sigma_Y$, where σ_Y is the flow stress, defined as the average of the yield and tensile strengths. The slope of the exclusion lines corresponds approximately to the component of crack extension that is due to crack blunting, as opposed to ductile tearing. A horizontal exclusion line is defined at a maximum value of J:

$$J_{max} = \frac{b_o \sigma_Y}{15} \tag{7.12}$$

All data that fall within the exclusion limits are fit to a power law expression:

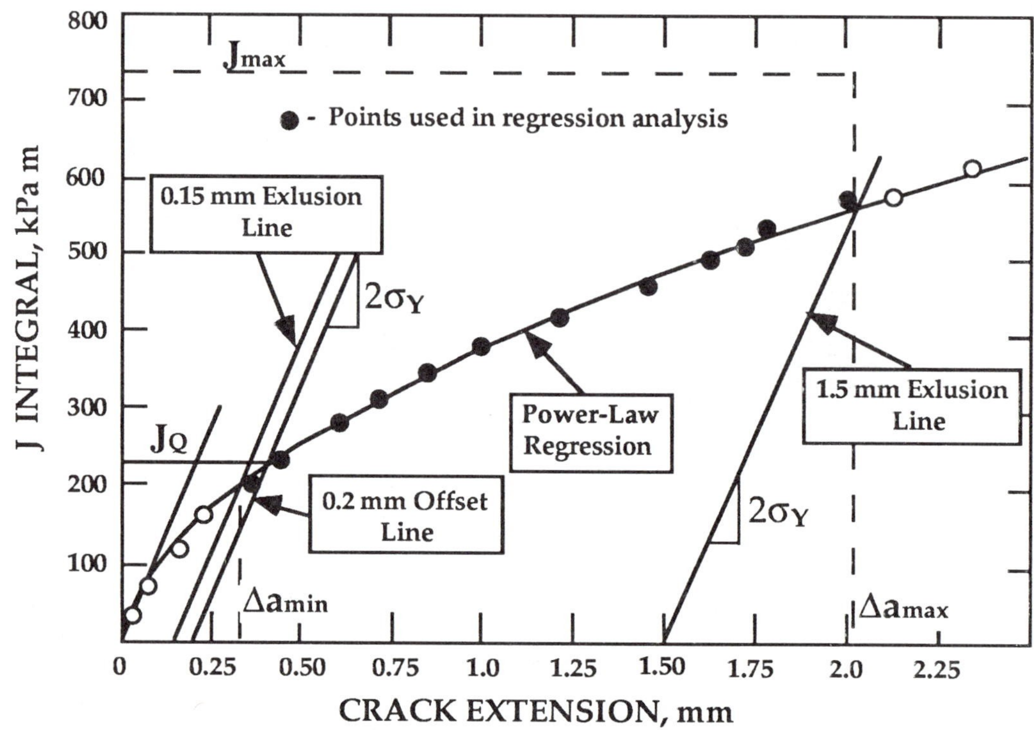

FIGURE 7.21 Determination of J_Q from a J-R curve [5].

$$J = C_1(\Delta a)^{C_2} \tag{7.13}$$

The J_Q is defined as the intersection between Eq. (7.13) and a 0.2 mm offset line. If all other validity criteria are met, $J_Q = J_{IC}$ as long the following size requirements are satisfied:

$$B, b_o \geq \frac{25\, J_Q}{\sigma_y} \tag{7.14}$$

EXAMPLE 7.3

Estimate the specimen size requirements for a valid J_{IC} test on the material in Example 7.1. Assume $\sigma_{TS} = 450$ MPa and E = 207,000 MPa.

EXAMPLE 7.3 (cont.)

Solution: First we must convert the K_{IC} value in Example 7.1 to an equivalent J_{IC}:

$$J_{IC} = \frac{K_{IC}^2 (1 - \nu^2)}{E} = \frac{(200 \text{ MPa} \sqrt{\text{m}})^2 (1 - 0.3^2)}{207000 \text{ MPa}} = 0.176 \text{ MPa m}$$

Substituting the above result into Eq. (7.14) gives,

$$B, b_o \geq \frac{(25)(0.176 \text{ MPa m})}{400 \text{ MPa}} = 0.0110 \text{ m} = 11.0 \text{ mm } (0.433 \text{ in})$$

which is nearly two orders of magnitude lower than the specimen dimension that ASTM E 399 requires for this material. Thus the J_{IC} size requirements are much more lenient than the K_{IC} requirements.

7.4.2 J-R Curve Testing

The ASTM standard E 1152-87 is applied when the entire J-R curve is of interest. Only single specimen unloading compliance tests are allowed by this standard. Test specimens should be side grooved in order to avoid tunneling and maintain a straight crack front.

There are a number of ways to compute J for a growing crack, as outlined in Section 3.4.2. The ASTM standard for J-R curve testing utilizes the *deformation theory* definition of J, which corresponds to the rate of energy dissipation by the growing crack (i.e., the energy release rate). Recall Fig. 3.22, which contrasts the actual loading path with the "deformation" path. The deformation J is related to the area under the load-displacement curve for a stationary crack, rather than the area under the actual load-displacement curve, where the crack length varies (see Eqs. (3.55) and (3.56)).

Since the crack length changes continuously during a J-R curve test, the J integral must be calculated incrementally. The most logical time to update the J value is at each unloading point, where the crack length is also updated. Consider an unloading compliance test with n unloading points. For a given unloading point i , where $1 \le i \le n$, the elastic and plastic components of J can be estimated from the following expressions (see Fig. 7.22)[3]:

$$J_{el(i)} = \frac{K_{(i)}^2 (1 - \nu^2)}{E} \tag{7.15a}$$

$$J_{pl(i)} = \left[J_{pl(i-1)} + \left(\frac{\eta_i}{B_N b_i} \right) \frac{(P_i + P_{i-1})(\Delta_{i(pl)} - \Delta_{i-1(pl)})}{2} \right] \times \left[1 - \gamma_{i-1} \frac{a_i - a_{i-1}}{b_{i-1}} \right] \tag{7.15b}$$

where $\Delta_{i(pl)}$ is the plastic load line displacement, $\gamma_i = 1.0$ for SENB specimens and $\gamma_i = 1 + 0.76\, b_i/W$; η_i is as defined in Eq. (7.11), except that b_o is replaced by b_i, the instantaneous ligament length. The instantaneous K is related to P_i and a_i/W through Eq. (7.9).

Equation (7.15b) gives an approximation of the plastic component of the deformation J. Appendix 7.1 explores the basis of this relationship, as well as its accuracy.

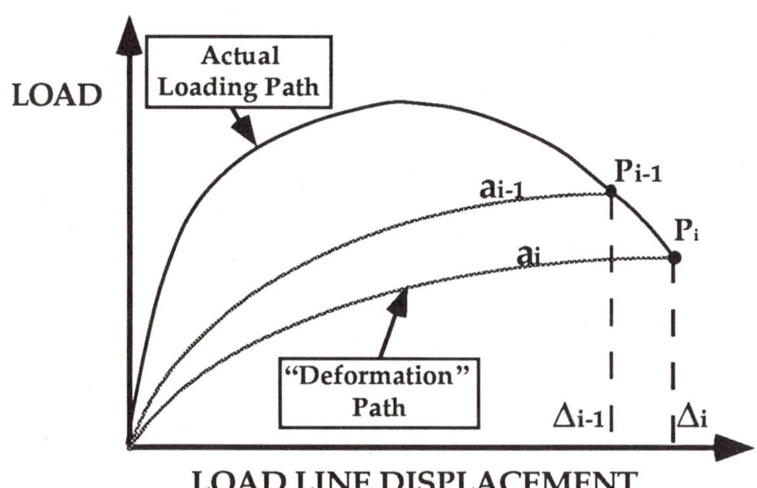

FIGURE 7.22 Schematic load-displacement curve for a J-R curve test.

---

[3]Equation (7.15b) contains different subscripts on b and γ from the corresponding equation in E 1152. The above equation is correct (see Appendix 7), and E 1152 apparently contains a typographical error. The differences in the computed J values between the two equations is minimal, however.

The ASTM J-R curve standard has the following limits on J and crack extension relative to specimen size:

$$B, b_o \geq \frac{20\, J_{max}}{\sigma_y} \tag{7.16}$$

and

$$\Delta a_{max} \leq 0.10\, b_o \tag{7.17}$$

Figure 7.23 shows a typical J-R curve with the E 1152-87 validity limits. The portion of the J-R curve that falls outside these limits is considered invalid. Note that the limit on J_{max} in ASTM E 1152-87 is more severe than the J_{max} limit of ASTM E 813-87 (Eq. (7.12)), but this limit is less restrictive than the J_{IC} size criterion (Eq. (7.14)).

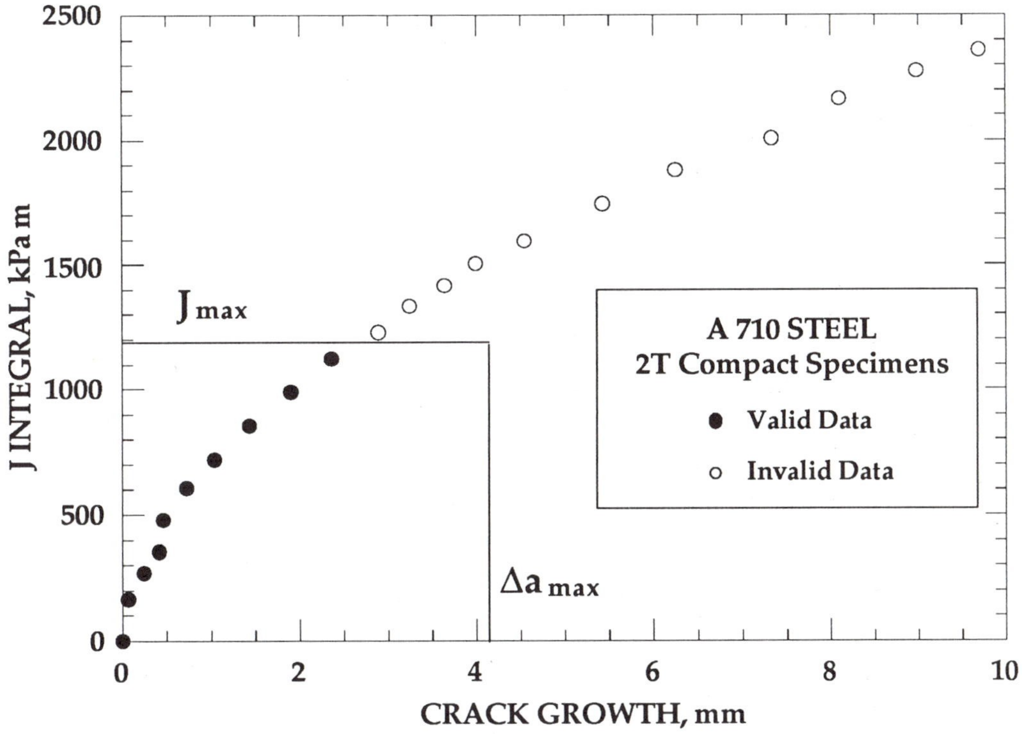

FIGURE 7.23 J-R curve for A 710 steel [16]. In this case, the data exceed the maximum J (Eq. 7.16) before the crack growth limit (Eq. 7.17).

Both the size and crack growth limits in E 1152 (Eqs. (7.16) and (7.17), respectively) are currently under review by ASTM. Many users would like to extend the crack growth limits, because ductile instability analyses of large structures typically require J-R curve data with large amounts of crack growth. Allowing the crack to grow through a significant portion of the ligament may, however, lead to a geometry-dependent R curve that is not transferable to structures. Refer to Sections 3.5 and 3.6 for further discussion of this subject.

7.4.3 Critical J Values for Cleavage Fracture

Although E 813-87 and E 1152-87 apply only to ductile fracture, there is, in principle, no reason why J cannot be applied to materials that fail by cleavage. The reasons for restricting J testing to ductile fracture are more historical than technical: existing J testing procedures were originally developed for nuclear pressure vessel steels, which are normally on the upper shelf at the service temperature.

Although the approach has yet to be standardized by ASTM or BSI, many researchers have measured critical J values and converted these values to equivalent K values through the following relationship:

$$K_{JC} = \sqrt{\frac{J_{crit}\,E}{(1 - \nu^2)}} \tag{7.18}$$

The K_{JC} values can be applied to structures that are elastically loaded, as discussed in Chapter 9. This approach is only valid if the critical J value is independent of specimen size. The test specimen must be sufficiently large that further increases in size have no effect on J_{crit}.

There are currently no standardized size requirements for J tests for cleavage, but recent research [17] indicates that the following criterion is sufficient to guarantee J-controlled cleavage:

$$B, b_o = \frac{200\,J_{crit}}{\sigma_Y} \tag{7.19}$$

which is eight times as severe as the size requirements for J_{IC} in ASTM E 813-87. The stricter requirements are necessary because cleavage is more sensitive to constraint loss than is ductile tearing, as discussed in

Appendix 5.1. Thus the material in Example 7.3 would require B, b_o = 88.0 mm (3.46 in) for a valid J test if the micromode of failure were cleavage.

The recommended size limits of Eq. (7.19) only apply to cleavage without significant prior stable crack growth. In the upper transition region of steels, cleavage is usually preceded by ductile tearing. Further research is necessary to understand upper transition fracture better in order to develop suitable J testing procedures for this region.

7.5 CTOD TESTING

Because of the strict limits on plastic deformation, the K_{IC} test can only be applied on the lower shelf of toughness in structural steels and welds. The ASTM J_{IC} and J-R curve test methods allow considerably more plastic deformation, but these tests are only valid on the upper shelf. The CTOD test is currently the only standardized method to measure fracture toughness in the ductile-brittle transition region.

The first CTOD test standard was published in Great Britain in 1979 [18]. ASTM recently published E 1290-89 [19], an American version of the CTOD standard. The British CTOD standard allows only the SENB specimen, while the ASTM standard provides for CTOD measurements on both the compact and SENB specimens. Both standards allow two configurations of SENB specimens: 1) a rectangular cross section with W = 2B, the standard geometry for K_{IC} and J_{IC} tests; and 2) a square cross section with W = B. The rectangular specimen is most useful with L-T or T-L orientations (Fig. 7.5); the square section is generally applied to the L-S or T-S orientations.

Experimental CTOD estimates are made by separating the CTOD into elastic and plastic components, similar to the J_{IC} and J-R tests. The elastic CTOD is obtained from the elastic K:

$$\delta_{el} = \frac{K^2(1-\nu^2)}{2\sigma_{YS} E} \tag{7.20}$$

The elastic K is related to applied load through Eq. (7.4). The above relationship assumes that d_n = 0.5 for linear elastic conditions (Eq. (3.48)). The plastic component of CTOD is obtained by assuming that the test specimen rotates about a plastic hinge. This concept is illustrated in

Fig. 7.24 for an SENB specimen. The plastic displacement at the crack mouth, V_p, is related to the plastic CTOD through a similar triangles construction:

$$\delta_{pl} = \frac{r_p\,(W-a)\,V_p}{r_p\,(W-a) + a + z} \tag{7.21}$$

where r_p is the plastic rotational factor, a constant between 0 and 1 that defines the relative position of the apparent hinge point. The mouth opening displacement is measured with a clip gage. In the case of an SENB specimen, knife edges must often be attached in order to hold the clip gage. Thus Eq. (7.21) must take account of the knife edge height, z. The compact specimen can be designed so that z = 0. The plastic component of V is obtained from the load-displacement curve by constructing a line parallel to the elastic loading line, as illustrated in Fig. 3.6. According to ASTM E 1290, the plastic rotational factor is given by

$$r_p = 0.44 \tag{7.22a}$$

for the SENB specimen and

$$r_p = 0.4\left\{1 + 2\left[\left(\frac{a}{b_o}\right)^2 + \frac{a}{b_o} + 0.5\right]^{1/2} - 2\left[\frac{a}{b_o} + 0.5\right]\right\} \tag{7.22b}$$

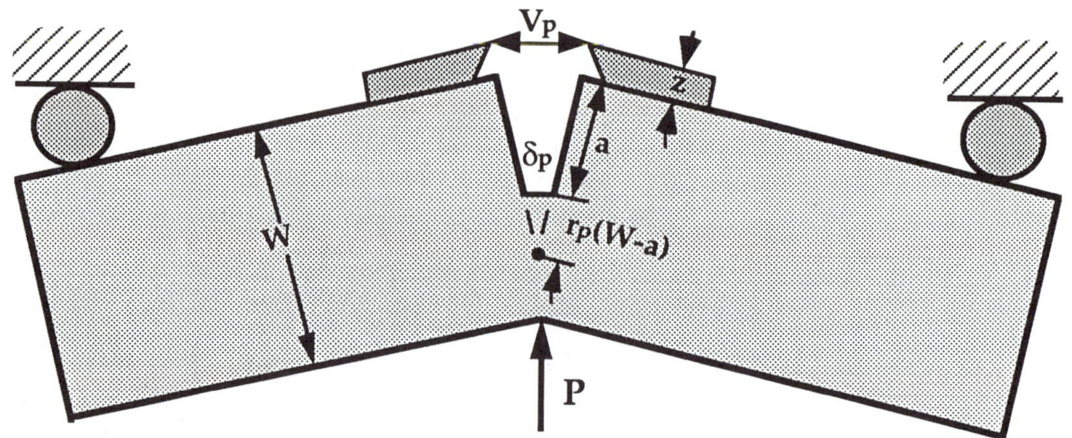

FIGURE 7.24 Hinge model for plastic displacements in an SENB specimen.

for the compact specimen. The original British standard for CTOD tests, BS 5762:1979 applied only to SENB specimens and specified r_p = 0.40.

The crack mouth opening displacement, V, on an SENB specimen is not the same as the load line displacement, Δ. The latter displacement measurement is required for J estimation because A_{pl} in Fig. 7.20 represents the plastic energy absorbed by the specimen. The CTOD standard utilizes V_p because this displacement is easier to measure in SENB specimens. If r_p is known, however, it is possible to infer J from a P-V curve or CTOD from a P-Δ curve [7,8]. The compact specimen simplifies matters somewhat because V = Δ as long as z = 0.

The ASTM CTOD standard test method can be applied to ductile and brittle materials, as well as steels in the ductile-brittle transition. This standard includes a notation for critical CTOD values that describes the fracture behavior of the specimen:

δ_c - Critical CTOD at the onset of unstable fracture with less than 0.2 mm of stable crack growth. This corresponds to the lower shelf and lower transition region of steels where the fracture mechanism is pure cleavage.

δ_u - Critical CTOD at the onset of unstable fracture which has been preceded by more than 0.2 mm of stable crack growth. In the case of ferritic steels, this corresponds to the "ductile thumbnail" observed in the upper transition region .

δ_i - CTOD near the initiation of stable crack growth. This measure of toughness is analogous to J_{IC}.

δ_m - CTOD at the first attainment of a maximum load plateau. This occurs on or near the upper shelf of steels.

Figure 7.25 is a series of schematic load-displacement curves that manifest each of the above failure scenarios. Curve (a) illustrates a test that results in a δ_c value; cleavage fracture occurs at P_c. Figure 7.25(b) corresponds to a δ_u result, where ductile tearing precedes cleavage. The ductile crack growth initiates at P_i. A test on or near the upper shelf produces a load-displacement curve like Fig. 7.25(c); a maximum load plateau occurs at P_m. The specimen is still stable after maximum load because the material has a rising R curve and the test is performed in displacement control. Three types of CTOD result, δ_c, δ_u and δ_m, are mutually exclusive; i.e, they cannot occur in the same test. It is possible, however, to measure a δ_i value in the same test as either a δ_m or δ_u result.

As Fig. 7.25 illustrates, there is usually no detectable change in the load-displacement curve at P_i. The only deviation in the load-displacement behavior is the reduced rate of increase in load as the crack grows. The maximum load plateau (Fig. 7.25(c)) occurs when the rate of strain hardening is exactly balanced by the rate of decrease in the cross section. However, the initiation of crack growth can not be detected from the load-displacement curve because the loss of cross section is gradual. Thus δ_i must be determined from an R curve.

As with the ASTM J_{IC} standard, the δ-R curve can be generated by either single or multiple specimen procedures, but the two standards vary in the way in which the initiation point is defined. Figure 7.26 illustrates the E 1290-89 procedure for δ_i determination from a δ-R curve. CTOD values are plotted against the physical crack extension, Δa_p. Vertical exclusion lines are drawn at 0.15 and 1.5 mm of crack extension. The data are then fit to an offset power law expression:

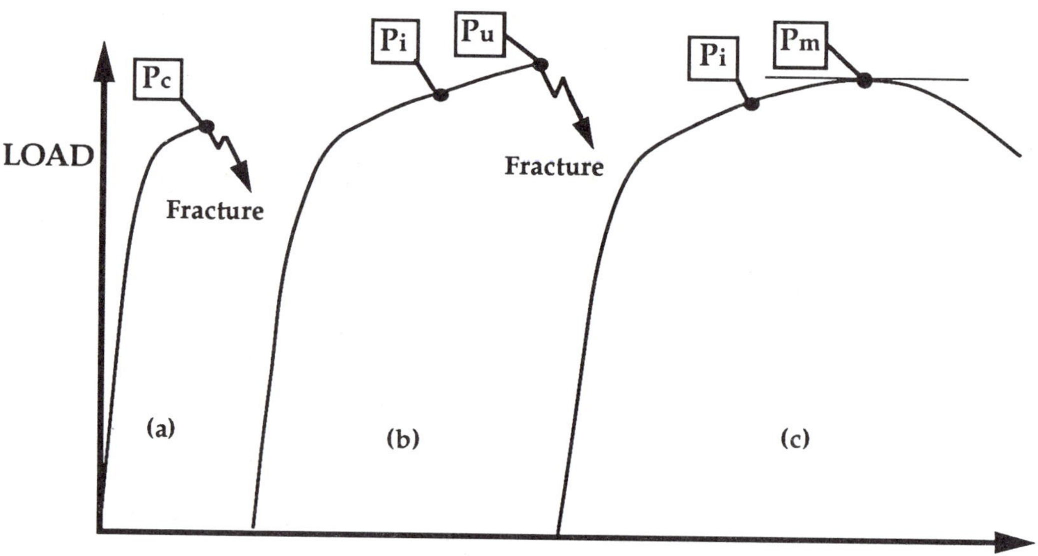

FIGURE 7.25 Various types of load-displacement curves from CTOD tests.

$$\delta = C_1 (C_2 + \Delta a_p)^{C_3} \tag{7,23}$$

The initiation toughness, δ_i, is then defined at the CTOD value corresponding to Δa_p = 0.2 mm. The offset power law is more cumbersome than a simple power law, because three constants are required to define the δ-R curve. At this writing, ASTM was considering replacing Eq. (7.23) with a simple power law (Eq. (7.13)) in order to make the determination of δ_i more consistent with J_{IC} measurements.

The only specimen size requirement of the British and ASTM CTOD standards is a recommendation to test full section thicknesses. For example, if a structure is to be made of 25 mm (1 in) thick plate, then B in the test specimens should be nominally 25 mm. If the specimen is notched from the surface (L-S or T-S orientations), a square section specimen is required for B to equal the plate thickness. The British CTOD standard allows a/W ratios ranging from 0.15 to 0.70, while the ASTM standard restricts the permissible a/W values to the range of 0.45 to 0.55. Shallow cracked specimens have certain advantages, particularly for weldment tests (see Section 7.7), but critical CTOD values from such tests are usually geometry dependent [17,20].

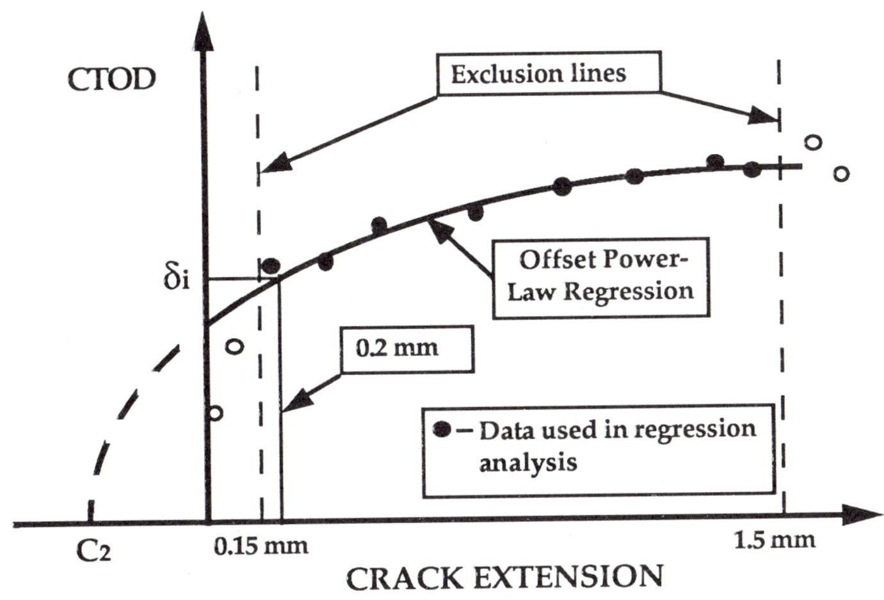

FIGURE 7.26 Schematic R curve for determining δ_i from CTOD tests.

Although existing CTOD standards do not contain size requirements, recent research [17] indicates that size independent CTOD values (for cleavage) can be obtained in deeply notched bend and compact specimens if the following criteria are met:

$$B, b_o \geq 300\, \delta_c \tag{7.24}$$

The size requirements for δ_i have not been established, but should be roughly eight times less strict than Eq. (7.24).

7.6 DYNAMIC AND CRACK ARREST TOUGHNESS

When a material is subject to a rapidly applied load or a rapidly propagating crack, the response of that material may be drastically different from the quasistatic case. When rapid loading or unstable crack propagation are likely to occur in practice, it is important to duplicate these conditions when measuring material properties in the laboratory.

The dynamic fracture toughness and the crack arrest toughness are two important material properties for many applications. The dynamic fracture toughness is a measure of the resistance of a material to crack propagation under rapid loading, while the crack arrest toughness quantifies the ability of a material to stop a rapidly propagating crack. In the latter case, the crack may *initiate* under either dynamic or quasistatic conditions, but unstable *propagation* is generally a dynamic phenomenon.

Dynamic fracture problems are often complicated by inertia effects, material rate dependence, and reflected stress waves. One or more of these effects can be neglected in some cases, however. Refer to Chapter 4 for additional discussion on this subject.

7.6.1 Rapid Loading in Fracture Testing

Aside from an Annex to E 399 [2], there are currently no ASTM Standards for high rate fracture testing. This type of testing is more difficult than conventional fracture toughness measurements, and requires considerably more instrumentation.

High loading rates can be achieved in the laboratory by a number of means, including a drop tower, a high rate testing machine, and explosive loading. With a drop tower, the load is imparted to the specimen through the force of gravity; a cross head of with a known weight is dropped onto the specimen from a specific height. A pendulum device such as a Charpy testing machine is a variation of this principle. Some servohydraulic machines are capable of high displacement rates. While conventional testing machines are *closed loop*, where the hydraulic fluid circulates through the system, high rate machines are *open loop*, where a single burst of hydraulic pressure is released over short time interval. For moderately high displacement rates, a closed-loop machine may be adequate. Explosive loading involves setting off a controlled charge which sends stress waves through the specimen [21].

The dynamic loads resulting from impact are often inferred from an instrumented tup. Alternatively, strain gages can be mounted directly on the specimen; the output can be calibrated for load measurements, provided the gages are placed in a region of the specimen that remains elastic during the test. Cross head displacements can be be measured directly through an optical device mounted to the cross head. If this instrumentation is not available, a load-time curve can be converted to a load-displacement curve through momentum transfer relationships.

Certain applications require more advanced optical techniques, such photoelasticity [22,23] and the method of caustics [24]. These procedures provide more detailed information about the deformation of the specimen, but are also more complicated than global measurements of load and displacement.

Because high rate fracture tests typically last only a few microseconds, conventional data acquisition tools are inadequate. A storage oscilloscope has traditionally been required to capture data in a high rate test; when a computer data acquisition system was used, the data were downloaded from the oscilloscope after the test. The newest generation of data acquisition cards for microcomputers removes the need for this two-step process. These cards are capable of collecting data at high rates, and enable the computer to simulate the functions of an oscilloscope.

Inertia effects can severely complicate measurement of the relevant fracture parameters. The stress intensity factor and J integral cannot be inferred from global loads and displacements when there is a significant kinetic energy component. Optical methods such as photoelasticity and caustics are necessary to measure J and K in such cases.

The transition time concept [25,26], which was introduced in Chapter 4, removes much of the complexity associated with J and K determination in high rate tests. Recall that the transition time, t_τ, is defined as the time at which the kinetic energy and deformation energy are approximately equal. At times much less than t_τ, inertia effects dominate, while inertia is negligible at times significantly greater than t_τ. The latter case corresponds to essentially quasistatic conditions, where conventional equations for J and K apply. According to Fig. 4.4, the quasistatic equation for J, based on the global load displacement curve, is accurate at times greater that 2 t_τ. Thus if the critical fracture event occurs after 2 t_τ, the toughness can be inferred from the conventional quasistatic relationships. For drop tower tests on ductile materials, the transition time requirement is relatively easy to meet [27,28]. For brittle materials (which fail sooner) or higher loading rates, the transition time can be shortened through specimen design.

7.6.2 K_{Ia} Measurements

In order to measure arrest toughness in a laboratory specimen, one must create conditions under which a crack initiates, propagates in an unstable manner, and then arrests. Unstable propagation followed by arrest can be achieved either through a rising R curve or a falling driving force curve. In the former case, a temperature gradient across a steel specimen produces the desired result; fracture can be initiated on the cold side of the specimen, where toughness is low, and propagate into warmer material where arrest is likely. A falling driving force can be obtained by loading the specimen in displacement control, as Example 2.3 illustrates.

The Robertson crack arrest test [29] was one of the earliest applications of the temperature gradient approach. This test is only qualitative, however, since the arrest temperature, rather than K_{Ia}, is determined from this test. The temperature at which a crack arrests in the Robertson specimen is only indicative of the relative arrest toughness of the material; designing above this temperature does not guarantee crack arrest under all loading conditions. The drop weight test developed by Pellini (see Section 7.9) is another qualitative arrest test that yields a critical temperature. In this case, however, arrest is accomplished through a falling driving force.

While most crack arrest tests are performed on small laboratory specimens, a limited number of experiments have been performed on larger configurations in order to validate the small scale data. An extreme example of large scale testing is the wide plate crack arrest experiments conducted at the National Institute of Standards and Technology (NIST)[4] in Gaithersburg, Maryland [30]. Figure 7.27 shows a photograph of the NIST testing machine and one of the crack arrest specimens. This specimen, which is a single edge notched tensile panel, is 10 m long by 1 m wide. A temperature gradient is applied across the width, such that the initial crack is at the cold end. The specimen is then loaded until unstable cleavage occurs. These specimens are heavily instrumented, so that a variety of information can be inferred from each test. The crack arrest toughness values measured from these tests is in broad agreement with small scale specimen data.

In 1988, ASTM published a standard for crack arrest testing, E 1221 [31]. This standard outlines a test procedure that is considerably more modest than the NIST experiments. A side grooved compact crack arrest specimen is wedge loaded until unstable fracture occurs. Because the specimen is held at a constant crack mouth opening displacement, the running crack experiences a falling K field. The crack arrest toughness, K_{Ia}, is determined from the mouth opening displacement and the arrested crack length.

The test specimen and loading apparatus for K_{Ia} testing are illustrated in Figs. 7.28 and 7.29. In most cases, a starter notch is placed in a brittle weld bead in order to facilitate fracture initiation. A wedge is driven through a split pin that imparts a displacement to the specimen. A clip gage measures the displacement at the crack mouth (Fig. 7.29).

Since the load normal to the crack plane is not measured in these tests, the stress intensity must be inferred from the clip gage displacement. The estimation of K is complicated, however, by extraneous displacements, such as seating of the wedge/pin assembly. Also, local yielding can occur near the starter notch prior to fracture initiation. The ASTM standard outlines a cyclic loading procedure for identifying

---

[4]NIST was formerly known as the National Bureau of Standards (NBS), which explains the initials on either end of the specimen in Fig. 7.27.

FIGURE 7.27 Photograph of a wide plate crack arrest test performed at NIST. [30]. (Photograph provided by J.G. Merkle.)

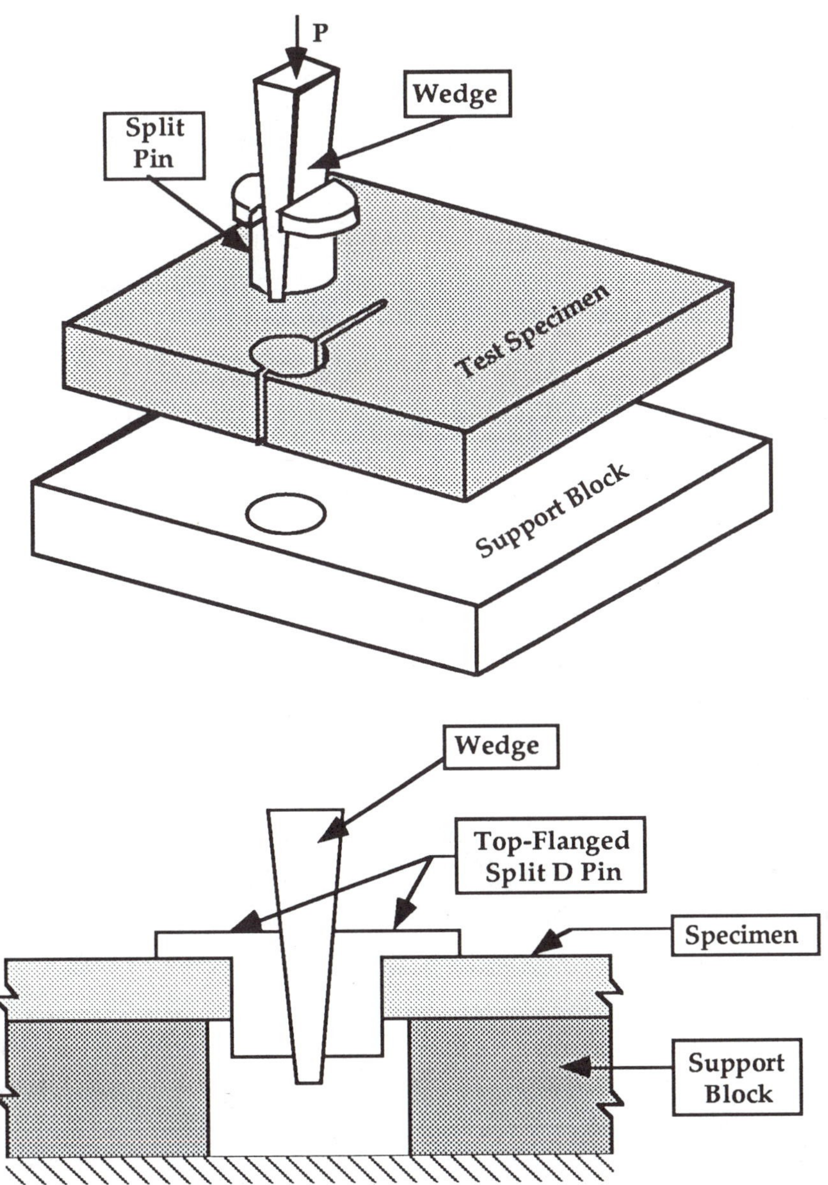

FIGURE 7.28 Apparatus for K_{Ia} tests [31].

these displacements; Fig. 7.30 shows a schematic load-displacement curve that illustrates this method. The specimen is first loaded to a predetermined displacement and, assuming the crack has not initiated, the specimen is unloaded. The displacement at zero load is assumed to represent the effects of fixture seating, and this component is subtracted from the total displacement when stress intensity is computed. The specimen is reloaded to a somewhat higher displacement and then unloaded; this process continues until fracture initiates. The zero load offset displacements that occur after the first cycle can be considered to be due to notch tip plasticity. The correct way to treat this displacement component in K calculations is unclear at present. Once the crack propagates through the plastic zone, the plastic displacement is largely recovered (i.e., converted into an elastic displacement), and thus may contribute to the driving force. It is not known whether or not there is sufficient time for this displacement component to exert an influence on the running crack. The ASTM standard takes the middle ground on this question, and requires that *half* of the plastic offset be included in the stress intensity calculations.

After the test, the specimen should be heat tinted at 250-350°C for 10 to 90 min to mark the crack propagation. When the specimen is broken open, the arrested crack length can then be measured on the fracture surface. The critical stress intensity at initiation, K_o, is computed from the initial crack size and the critical clip gage displacement. The provisional arrest toughness, K_a, is calculated from the *final* crack size, assuming constant displacement. These calculations assume quasistatic conditions. As discussed in Chapter 4, this assumption can lead to underestimates of arrest toughness. The ASTM standard, however, cites experimental evidence [32,33] that implies that the errors introduced by a quasistatic assumption are small in this case.

In order for the test to be valid, the crack propagation and arrest should occur under predominantly plane strain linear elastic conditions. The following validity requirements in ASTM E 1221-88 are designed to ensure that the plastic zone is small compared to specimen dimensions, and that the crack jump length is within acceptable limits:

$$W - a_a \geq 0.15\, W \tag{7.25a}$$

$$W - a_a \geq 1.25 \left(\frac{K_a}{\sigma_{Yd}} \right)^2 \tag{7.25b}$$

$$B \geq 1.0 \left(\frac{K_a}{\sigma_{Yd}} \right)^2 \tag{7.25c}$$

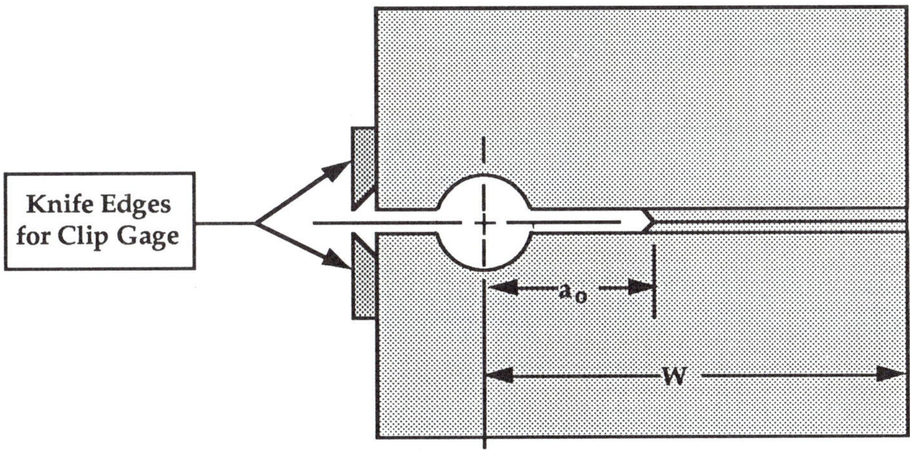

FIGURE 7.29 Side-grooved compact crack arrest specimen [31].

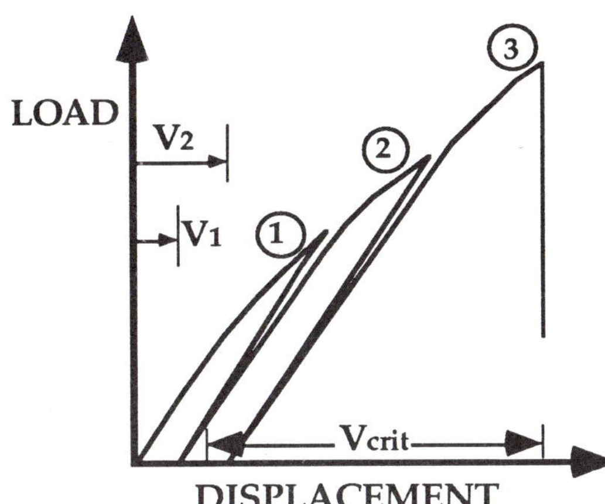

FIGURE 7.30 Schematic load-displacement curve for a K_{Ia} test [31], where V_1 and V_2 are zero load offset displacement. When computing V_{crit}, all of the first offset and half of the subsequent offsets are subtracted from the total displacement.

$$a_a - a_o \geq \frac{1}{2\pi}\left(\frac{K_a}{\sigma_{YS}}\right)^2 \qquad (7.25d)$$

where a_a is the arrested crack length, a_o is the initial crack length, and σ_{Yd} is the assumed dynamic yield strength, which the ASTM standard specifies at 205 MPa (30 ksi) above the quasistatic value. Since unstable crack propagation results in very high strain rates, the recommended estimate of σ_{Yd} is probably very conservative.

If the above validity requirements are satisfied and all other provisions of ASTM E 1221 are followed, $K_a = K_{Ia}$.

7.7 FRACTURE TESTING OF WELDMENTS

All of the test methods discussed so far are suitable for specimens extracted from uniform sections of homogeneous material. Welded joints, however, have decidedly heterogeneous microstructures and, in many cases, irregular shapes. Weldments also contain complex residual stress distributions. Existing fracture toughness testing standards do not address the special problems associated with weldment testing. The factors that make weldment testing difficult (i.e. heterogeneous microstructures, irregular shapes, and residual stresses) also tend to increase the risk of brittle fracture in welded structures. Thus, one cannot simply evaluate the regions of a structure where ASTM testing standards apply and ignore the fracture properties of weldments.

Although there are currently no fracture toughness testing standards for weldments, a number of laboratories and industries have significant of experience in this area. The Welding Institute in Cambridge, England, which probably has the most expertise, has recently published detailed recommendations for weldment testing [34]. The International Institute of Welding (IIW) has produced a similar document [35], although not as detailed. The American Petroleum Institute (API) has published guidelines for heat affected zone (HAZ) testing as part of a weld procedure qualification approach [36]. Committees within ASTM and BSI are currently drafting weldment test methods, relying heavily on 20 years of practical experience as well as the aforementioned documents.

Some of the general considerations and current recommendations for weldment testing are outlined below, with emphasis on the Welding Institute procedure [34] because it is the most complete document to date. Early drafts of both the ASTM and BSI guidelines incorporate many of the suggestions in The Welding Institute document.

When performing fracture toughness tests on weldments, a number of factors need special consideration. Specimen design and fabrication is more difficult because of the irregular shapes and curved surfaces associated with some welded joints. The heterogeneous microstructure of typical weldments requires special attention to the location of the notch in the test specimen. Residual stresses make fatigue precracking of weldment specimens more difficult. After the test, a weldment must often be sectioned and examined metallographically to determine whether or not the fatigue crack sampled the intended microstructure.

7.7.1 Specimen Design and Fabrication

The underlying philosophy of the Welding Institute [34] guidelines on specimen design and fabrication is that the specimen thickness should be as close to the section thickness as possible. Thicker specimens tend to produce more crack tip constraint, and hence lower toughness (See Chapters 2 and 3). Achieving nearly full thickness weldment specimens often requires sacrifices in other areas. For example if a specimen is to be extracted from a curved section such as a pipe, one can either produce a subsize rectangular specimen which meets the tolerances of the existing ASTM standards, or a full thickness specimen that is curved. The Welding Institute recommends the latter.

If curvature or distortion of a weldment is excessive, the specimen can be straightened by bending on either side of the notch to produce a "gull wing" configuration, which is illustrated in Fig. 7.31. The bending must be performed so that the three loading points (in an SENB specimen) are aligned.

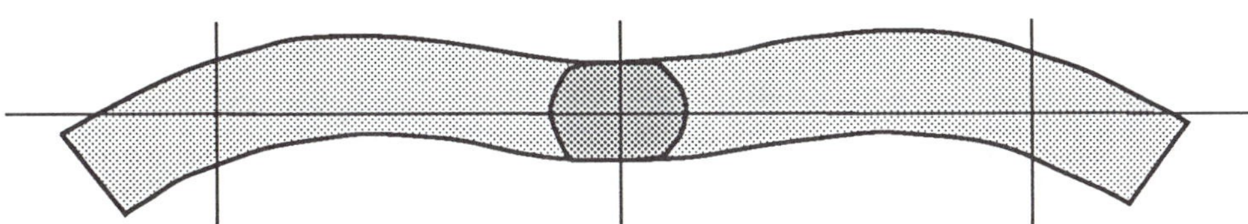

FIGURE 7.31 The gull-wing configuration for weldment specimens with excessive curvature [34].

Fabrication of either a compact or SENB weldment specimen is possible, but the SENB specimen is preferable in nearly every case. Although the compact specimen consumes less material (for a given B and W) in parent metal tests, it requires more weld metal in a through-thickness orientation (L-T or T-L) than an SENB specimen (Fig. 7.2). It is impractical to use a compact geometry for surface notched specimens (T-S or L-S); such a specimen would be greatly undersized with the standard B X 2B geometry.

The Welding Institute recommendations cover both the rectangular and square section SENB specimens. The appropriate choice of specimen type depends on the orientation of the notch.

7.7.2 Notch Location and Orientation

Weldments have a highly heterogeneous microstructure. Fracture toughness can vary considerably over relatively short distances. Thus it is important to take great care in locating the fatigue crack in the correct region. If the fracture toughness test is designed to simulate an actual structural flaw, then the fatigue crack must sample the same microstructure as the flaw. For a weld procedure qualification or a general assessment of a weldment's fracture toughness, location of the crack in the most brittle region may be desirable, but it is difficult to know in advance which region of the weld has the lowest toughness. In typical C-Mn structural steels, low toughness is usually associated with the coarse grained HAZ and the intercritically reheated HAZ. A microhardness survey can help identify low toughness regions because high hardness is often coincident with brittle behavior. The safest approach is to perform fracture toughness tests on a variety of regions in a weldment.

Once the microstructure of interest is identified, a notch orientation must be selected. The two most common alternatives are a through-thickness notch and a surface notch, which are illustrated in Fig. 7.32. Since full thickness specimens are desired, the surface notched specimen should be square section (BXB), while the through thickness notch will usually be in a rectangular (B X 2B) specimen.

For weld metal testing, the through-thickness orientation is usually preferable because a variety of regions in the weld are sampled. However, there may be cases where the surface notched specimen is the most suitable for testing the weld metal. For example, a surface notch can sample a particular region of the weld metal, such as the root or cap, or the notch can be located in a particular microstructure, such as unrefined weld metal.

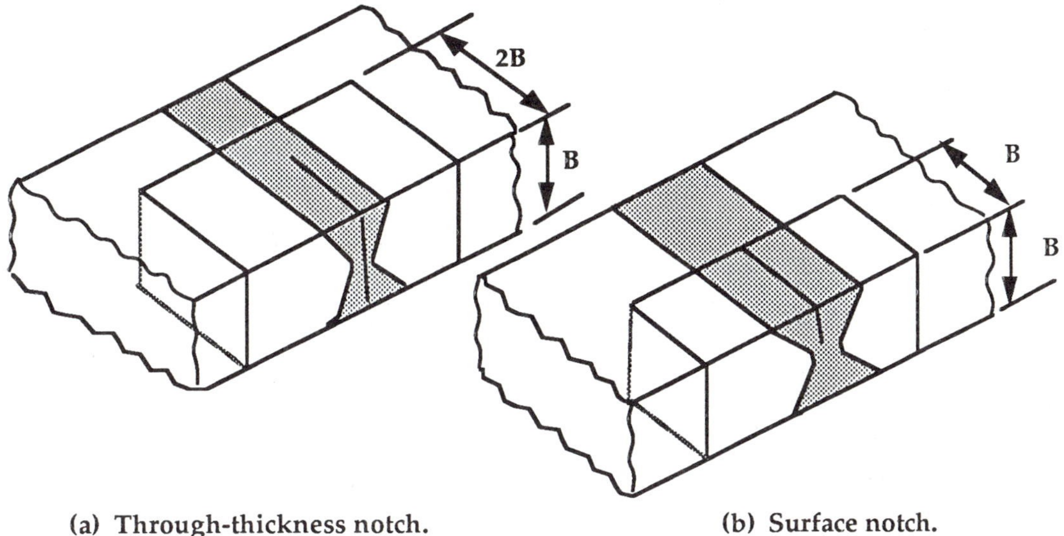

(a) Through-thickness notch. (b) Surface notch.

FIGURE 7.32 Notch orientation in weldment specimens [34].

Notch location in the HAZ often depends on the type of weldment. If welds are produced solely for mechanical testing, for example as part of a weld procedure qualification or a research program, the welded joint can be designed to facilitate HAZ testing. Figure 7.33 illustrates the K and half-K preparations, which simulate double-V and single-V welds, respectively. The plates should be tilted when these weldments are made, to have the same angle of attack for the electrode as in an actual single- or double-V joint. For fracture toughness testing, a through-thickness notch is placed in the straight side of the K or half-K HAZ.

In many instances, fracture toughness testing must be performed on an actual production weldment, where the joint geometry is governed by the structural design. In such cases, a surface notch is often necessary for the crack to sample sufficient HAZ material. The measured toughness is sensitive to the volume of HAZ material sampled by the crack tip because of the weakest link nature of cleavage fracture (see Chapter 5).

Another application of the surface notched orientation is the simulation of structural flaws. Figure 7.34 illustrates HAZ flaws in a structural weld and a surfaced notched fracture toughness specimen that models one of the flaws.

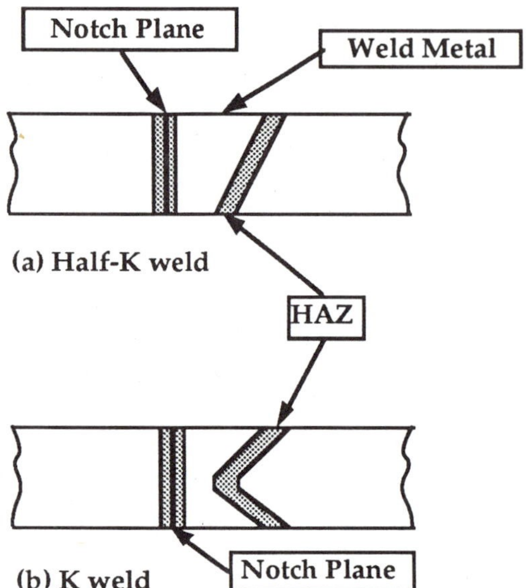

FIGURE 7.33 Special weld joint designs for fracture toughness testing of the heat-affected zone (HAZ) [34].

Figure 7.34 demonstrates the advantages of allowing a range of a/W ratios in surface notched specimens. A shallow notch is often required to locate a crack in the desired region, but existing ASTM standards do not allow a/W ratios less than 0.45. Shallow notched fracture toughness specimens tend to have lower constraint than deeply cracked specimens, as Figs 3.29 to 3.32 illustrate. Thus there is a conflict between the need to simulate a structural condition and the traditional fracture mechanic approach, where a toughness value is supposed to be a size independent material property. One way to resolve this conflict is through constraint corrections, such as that applied to the data in Fig. 3.32.

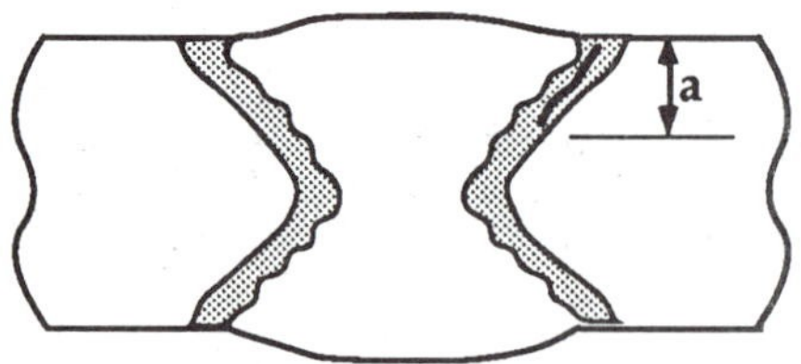

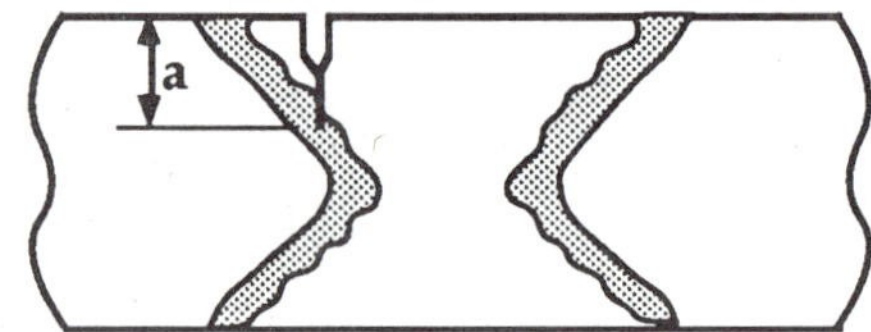

(a) Weldment with a flaw in the HAZ. **(b) Test specimen with simulated structural flaw.**

FIGURE 7.34 Test specimen with notch orientation and depth that matches a flaw in a structure [34].

7.7.3 Fatigue Precracking

Weldments that have not been stress relieved typically contain complex residual stress distributions that interfere with fatigue precracking of fracture toughness specimens. Tensile residual stresses accelerate fatigue crack initiation and growth, but compressive stresses retard fatigue. Since residual stresses vary through the cross section, fatigue crack fronts in as-welded samples are typical very nonuniform.

Towers and Dawes [37] evaluated the various methods for producing straight fatigue cracks in welded specimens, including reverse bending, high R ratio, and local compression.

The first method bends the specimen in the opposite direction to the normal loading configuration to produce residual tensile stresses along the crack front that counterbalance the compressive stresses. Although this technique gives some improvement, it does not produce acceptable fatigue crack fronts.

The R ratio in fatigue cracking is the ratio of the minimum stress to the maximum. A high R ratio minimizes the effect of residual stresses on fatigue, but also tends to increase the apparent toughness of the specimen. In addition, fatigue precracking at a high R ratio takes much longer than precracking at $R = 0.1$, the recommended R ratio of the various ASTM fracture testing standards.

The only method evaluated that produced consistently straight fatigue cracks was local compression, where the ligament is compressed to produce nominally 1% plastic strain through the thickness, mechanically relieving the residual stresses. However, local compression can reduce the toughness slightly. Towers and Dawes concluded that the benefits of local compression outweigh the disadvantages, particularly in the absence of a viable alternative.

7.7.4 Post-Test Analysis

Correct placement of a fatigue crack in weld metal is usually not difficult because this region is relatively homogeneous. The microstructure in the HAZ, however, can change dramatically over very small distances. Correct placement of a fatigue crack in the HAZ is often accomplished by trial and error. Because fatigue cracks are usually slightly bowed, the precise location of the crack tip in the center of a specimen cannot be inferred from observations on the surface of the specimen. Thus HAZ fracture toughness specimens must be examined metallographically after the test to determine the microstructure that initiated fracture. In certain cases, post-test examination may be required in weld metal specimens.

Figure 7.35 illustrates a procedure for sectioning surface notched and through-thickness notched specimens [34]. First, the origin of the fracture must be located by the chevron markings on the fracture surface. After marking the origin with a small spot of paint the specimen is sectioned perpendicular to the fracture surface and examined metallographically. The specimen should be sectioned slightly to one side of the origin and polished down to the initiation site. The spot of paint appears on the polished specimen when the origin is reached.

The API document RP2Z [36] outlines a post-test analysis of HAZ specimens which is more detailed and cumbersome than the procedure outlined above. In addition to sectioning the specimen, the amount of coarse-grained material at the crack tip must be quantified. For the test to be valid, at least 15% of the crack front must be in the coarse-grained HAZ. The purpose of this procedure is to prequalify steels with respect to HAZ toughness, identifying those that produce low HAZ toughness so that they can be rejected before fabrication.

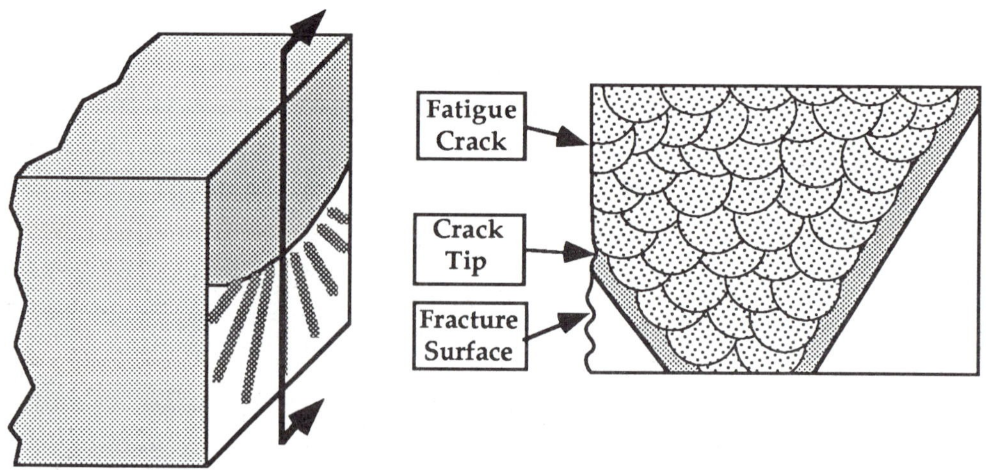

FIGURE 7.35 Post-test sectioning of a weldment fracture toughness specimen to identify the microstructural that caused fracture.

7.8 REMOVING INCONSISTENCY: DEVELOPMENT OF A COMMON TEST PROCEDURE

With existing fracture testing standards, one must decide in advance which standard to apply in a given situation. If the material does not behave as expected, the test is invalid and the specimen is wasted. For example, if one attempts to measure K_{IC} for a ductile material, large scale plasticity will invalidate the test. On the other hand, the J_{IC} and J-R curve standards cannot be applied to brittle materials.

The basic apparatus and specimens for the various types of quasistatic fracture toughness tests are similar from one standard to the next, but not identical. For example, the ASTM standards E 399 (K_{IC}), E 813 (J_{IC}), and E 1152 (J-R curve) all utilize compact specimens, but the E 399 design is slightly different from the other two. For the two J standards, the clip gage displacement is measured at the loading line, while the displacement is measured at the edge of the specimen in K_{IC} tests. (Figure 7.1a shows the design for J tests where a section is removed near the notch mouth to facilitated placement of the clip gage on the load line.)

Many of the differences in test procedures that exist between standards do not have a sound technical basis, but are the merely result of different committees producing standards at different times, without sufficient regard for consistency.

Both ASTM and BSI have recognized the disadvantages of the current situation. Each organization has drafted a standard [14,15] that combines several approaches into a single document. When these standards are officially adopted, the user will be able to apply a single testing procedure to most situations, and analyze the results in a manner that is appropriate to the material behavior. If the material is sufficiently brittle, for example, one could compute a K_{IC}; if the material fails the K_{IC} size requirements, it will still be possible to compute a critical J or CTOD.

The ASTM document [14] combines the K_{IC}, J_{IC}, J-R curve, and CTOD tests into a single standard test method. The procedure for determining CTOD near initiation of stable tearing (δ_i) is consistent with the J-based approach, unlike the current situation where E 813 and E 1290 define initiation differently. The ASTM draft standard offers a choice between a basic apparatus and more sophisticated instrumentation. The former case corresponds to simple loading to failure, while the latter applies to single specimen unloading compliance testing.

Both approaches permit J testing in the ductile-brittle transition region, where unstable cleavage is the ultimate failure mechanism. The BSI Draft standard includes guidelines for high rate testing. For the present time, crack arrest and K-R curve testing are not included in either document, since these methods are intended for special circumstances.

The individual testing standards, such as E 399 and E 813 will probably become obsolete soon after the combined standards are adopted.

7.9 QUALITATIVE TOUGHNESS TESTS

Before the development of formal fracture mechanics methodology, engineers realized the importance of material toughness in avoiding brittle fracture. In 1901 a French scientist named G. Charpy developed a pendulum test that measured the energy of separation in notched metallic specimens. This energy was believed to be indicative of the resistance of the material to brittle fracture. An investigation of the Liberty ship failures during World War II revealed that fracture was much more likely in steels with Charpy energy less than 20 J (15 ft-lb).

During the 1950s, when Irwin and his colleagues at the Naval Research Laboratory (NRL) were formulating the principles of linear elastic fracture mechanics, a metallurgist at NRL named W.S. Pellini developed the drop weight test, a qualitative measure of crack arrest toughness.

Both the Charpy test and the Pellini drop weight test are still widely applied today to structural materials. ASTM has standardized the drop weight tests, as well as a number of related approaches, including the Izod, drop weight tear and dynamic tear tests [38-41] (see below). Although these tests lack the mathematical rigor and predictive capabilities of fracture mechanics methods, these approaches provide a qualitative indication of material toughness. The advantage of these qualitative methods is that they are cheaper and easier to perform than fracture mechanics tests. These tests are suitable for material screening and quality control, but are not reliable indicators of structural integrity.

7.9.1 Charpy and Izod Impact Test

The ASTM Standard E 23-88 [38] covers Charpy and Izod testing. These tests both involve impacting a small notched bar with a pendulum and measuring the fracture energy. The Charpy specimen is a simple beam that is impacted in three-point bending, while the Izod specimen is a cantilever beam that is fixed at one end and impacted at the other. Figure 7.36 illustrates both types of test.

Charpy and Izod specimens are relatively small, and thus do not consume much material. The standard cross section of both specimens is 10 mm x 10 mm, and the lengths are 55 and 75 mm for Charpy and Izod specimens, respectively.

The pendulum device provides a simple but elegant method for quantifying fracture energy. As Fig. 7.37 illustrates, the pendulum is released from a height y_1 and swings through the specimen to a height y_2. Assuming negligible friction and aerodynamic drag, the energy absorbed by the specimen is equal to the height difference times the weight of the pendulum. A simple mechanical device on the Charpy machine converts the height difference to a direct read-out of absorbed energy.

A number of investigators [42-47] have attempted to correlate Charpy energy to fracture toughness parameters such as K_{IC}. These empirical correlations seem to work reasonably well in some cases, but are unreliable in general. There are several important differences between the Charpy test and fracture mechanics tests that preclude simple relationships between the qualitative and quantitative measures of toughness. The Charpy test contains a blunt notch, while fracture mechanics specimens have sharp fatigue cracks. The Charpy specimen is subsize, and thus has low constraint. In addition, the Charpy specimen experiences impact loading, while most fracture toughness tests are conducted under quasistatic conditions.

It is possible to obtain quantitative information from fatigue precracked Charpy specimens, provided the tup (i.e. the striker) is instrumented [48,49]. Such an experiment is essentially a miniature dynamic fracture toughness test.

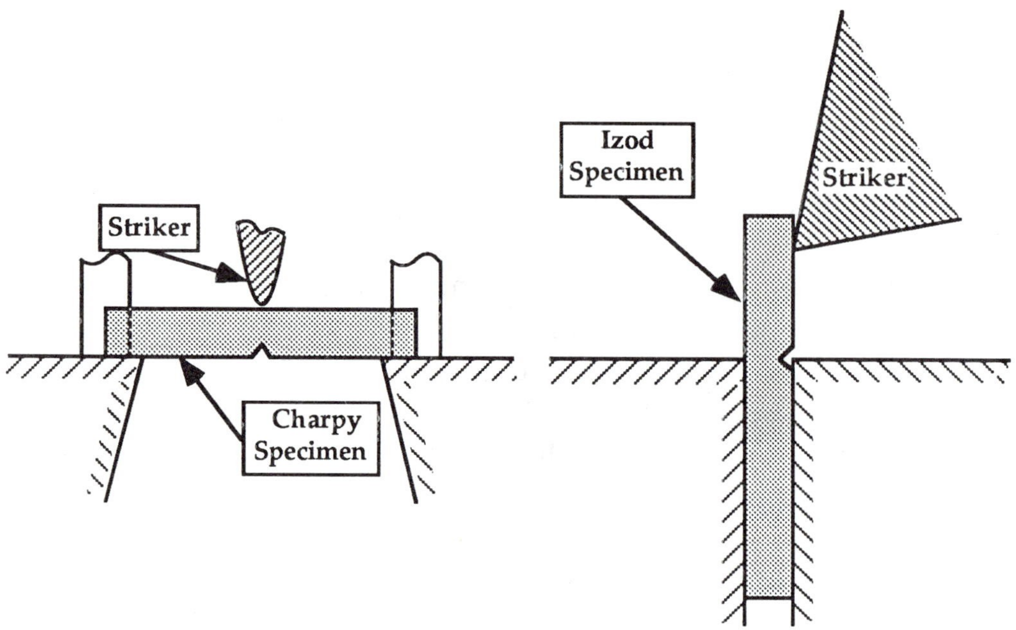

FIGURE 7.36 Charpy and Izod notched impact tests [38].

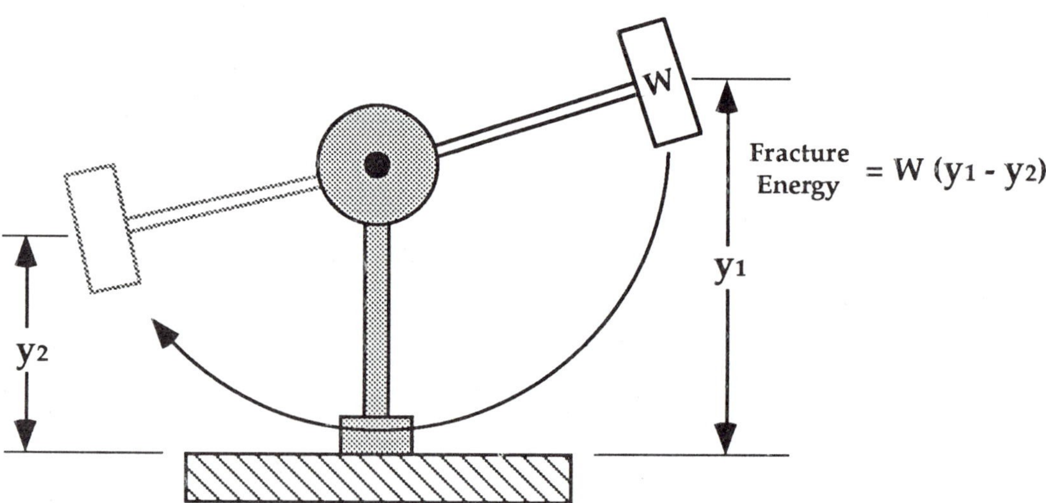

FIGURE 7.37 A pendulum device for impact testing. The energy absorbed by the specimen is equal to the weight of the crosshead, times the difference in height before and after impact.

7.9.2 Drop Weight Test

The ASTM standard E 208-87 [39] outlines the procedure for performing the Pellini drop weight test. A plate specimen with a starter notch in a brittle weld bead is impacted in three-point bending. A cleavage crack initiates in the weld bead and runs into the parent metal. If the material is sufficiently tough, the crack arrests; otherwise the specimen fractures completely.

Figure 7.38 illustrates the drop weight specimen and the testing fixture. The crosshead drops onto the specimen, causing it to deflect a predetermined amount. The fixture is designed with a deflection stop, which limits the displacement in the specimen. A crack initiates at the starter notch and either propagates or arrests, depending on the temperature and material properties. A "break" result is recorded when the running crack reaches at least one specimen edge. A "no-break" result is recorded if the crack arrests in the parent metal. Figure 7.39 gives examples of break and no-break results.

A nil-ductility transition temperature (NDTT) is obtained by performing drop weight tests over a range of temperatures, in 5°C or 10°F increments. When a no-break result is recorded, the temperature is decreased for the next test; test temperature is increased when a specimen fails. When break and no-break results are obtained at adjoining temperatures, a second test is performed at the no-break temperature. If this specimen fails, a test is performed at one temperature increment (5°C or 10°F) higher. The process is repeated until two no-break results are obtained at one temperature. The NDTT is defined as 5°C or 10°F below the lowest temperature where two no-breaks are recorded.

The nil-ductility transition temperature gives a qualitative estimate of the ability of a material to arrest a running crack. Arrest in structures is more likely to occur if the service temperature is above NDTT, but structures above NDTT are not immune to brittle fracture.

The ship building industry in the United States currently uses the drop weight test to qualify steels for ship hulls. The nuclear power industry relies primarily on quantitative fracture mechanics mechanics methodology, but uses the NDTT to index fracture toughness data for different heats of steel (see Chapter 9).

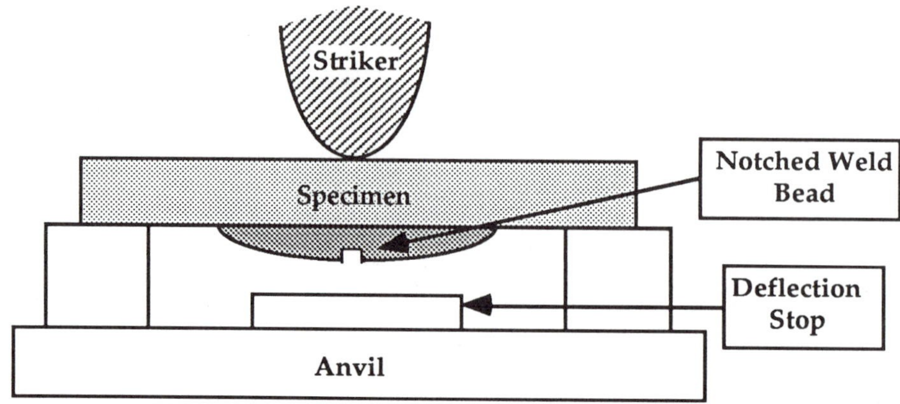

FIGURE 9.38 Apparatus for drop weight testing according to the ASTM E 208-87 [39]

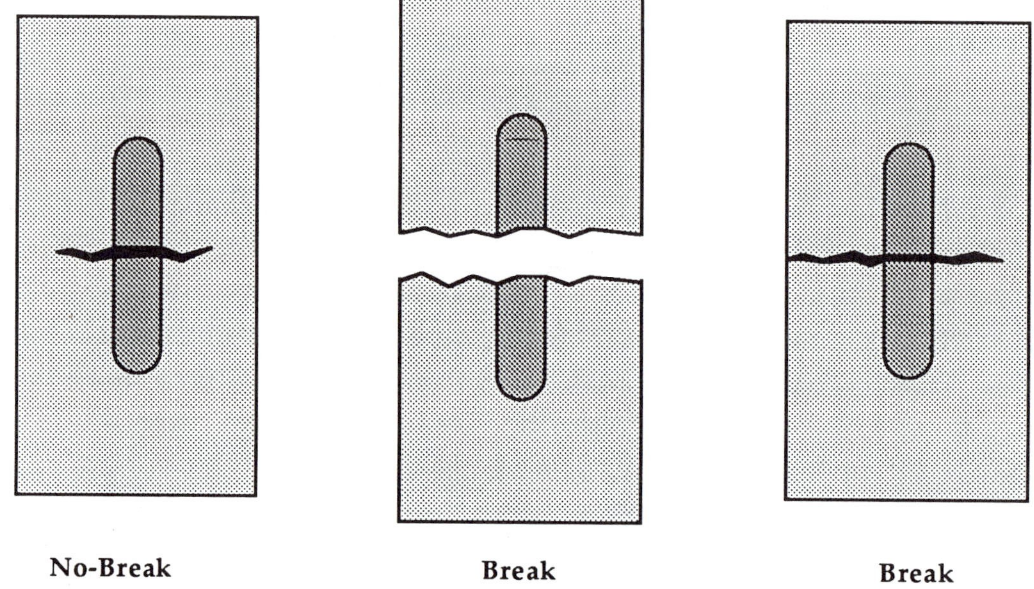

FIGURE 9.39 Examples of break and no-break behavior in drop weight tests. A break is recorded when the crack reaches at least one edge of the specimen.

7.9.3 Drop Weight Tear and Dynamic Tear Tests

Drop weight tear and dynamic tear tests are similar to the Charpy test, except that the former are performed on large specimens. The ASTM standards E 604-83 [41] and E 436-74 [40] cover drop weight tear and dynamic tear tests, respectively. Both test methods utilize three point bend specimens that are impacted in a drop tower or pendulum machine.

Drop weight tear specimens are 41 mm (1.6 in) wide, 16 mm (0.625 in) thick, and are loaded over a span of 165 mm (6.5 in). These specimens contain a sharp machined notch. A 0.13 mm (0.010 in) deep indentation is made at the tip of this notch. The fracture energy is measured in this test, much like the Charpy and Izod tests. Since drop weight specimens are significantly larger than Charpy specimens, the fracture energy is much greater, and the capacity of the testing machine must be scaled accordingly. If a pendulum machine is used, the energy can be determined in the same manner as in the the Charpy and Izod tests. A drop test must be instrumented, however; because only a portion of the potential energy is absorbed by the specimen; the remainder is transmitted through the foundation of the drop tower.

The dynamic tear test quantifies the toughness of steel through the appearance of the fracture surface. In the ductile-brittle transition region, a dynamic test produces a mixture of cleavage fracture and microvoid coalescence; the relative amount of each depends on the test temperature. The percent "shear" on the fracture surface is reported in dynamic tear tests, where the so-called shear fracture is actually microvoid coalescence (Chapter 5). Dynamic tear specimens are 76 mm (3 in) wide, 305 mm (12 in) long, and are loaded over a span of 254 mm (10 in). The specimen thickness is equal to the thickness of the plate under consideration. The notch is pressed into the specimen by indentation.

REFERENCES

1. E 616-89, "Terminology Relating to Fracture Testing." American Society for Testing and Materials, Philadelphia, 1989.

2. E 399-83, "Standard Test Method for Plane-Strain Fracture Toughness of Metallic Materials." American Society for Testing and Materials, Philadelphia, 1983.

3. Baker, A., "A DC Potential Drop Procedure for Crack Initiation and R Curve Measurements During Ductile Fracture Tests." ASTM STP 856, American Society of Testing and Materials, Philadelphia, 1985, pp. 394-410.

4. Schwalbe, K-H, Hellmann, D., Heerens, J., Knaack, J., Muller-Roos, J., "Measurement of Stable Crack Growth Including Detection of Initiation of Growth

Using the DC Potential Drop and the Partial Unloading Methods." ASTM STP 856, American Society of Testing and Materials, Philadelphia, 1985, pp. 338-362.

5. E 813-87, "Standard Test Method for J_{IC}, a Measure of Fracture Toughness." American Society for Testing and Materials, Philadelphia, 1987.

6. E 1152-87 "Standard Test Method for Determining J-R Curves." American Society for Testing and Materials, Philadelphia, 1987.

7. Dawes, M.G. "Elastic-Plastic Fracture Toughness Based on the COD and J-Contour Integral Concepts." ASTM STP 668, American Society of Testing and Materials, Philadelphia, 1979, pp. 306-333.

8. Anderson, T.L., McHenry, H.I. and Dawes, M.G., "Elastic-Plastic Fracture Toughness Testing with Single Edge Notched Bend Specimens." ASTM STP 856, American Society of Testing and Materials, Philadelphia, 1985. pp. 210-229.

9. Andrews, W.R. and Shih, C.F., "Thickness and Side-Groove Effects on J- and δ-Resistance Curves for A533-B Steel at 93°C." ASTM STP 668, American Society of Testing and Materials, Philadelphia, 1979, pp. 426-450.

10. BS 5447:1974 "Methods of Testing for Plane Strain Fracture Toughness (K_{IC}) of Metallic Materials." British Standards Institution, London, 1974.

11. E 561-86, "Standard Practice for R-Curve Determination." American Society of Testing and Materials, Philadelphia, 1986.

12. Stricklin, L.L., "Geometry Dependence of Crack Growth Resistance Curves in Thin Sheet Aluminum Alloys." Master of Science Thesis, Texas A&M University, College Station, TX, December 1988.

13. JSME S 001-1981, "Standard Method of Test for Elastic-Plastic Fracture Toughness, J_{Ic}." Japan Society of Mechanical Engineers, Tokyo, October 1981.

14. "Standard Method for Measurement of Fracture Toughness." Draft 9, American Society of Testing and Materials, Philadelphia, November 1989.

15. "Draft British Standard Fracture Mechanics Toughness Tests." British Standards Institution, London, May 1990.

16. Joyce, J.A. and Hackett, E.M., "Development of an Engineering Definition of the Extent of J Singularity Controlled Crack Growth." NUREG/CR-5238, U.S. Nuclear Regulatory Commission, Washington, D.C., May 1989.

17. Anderson, T.L. and Dodds, R.H., Jr., "Specimen Size Requirements for Fracture Toughness Testing in the Ductile-Brittle Transition Region." *Journal of Testing and Evaluation*, Vol. 19, 1991, pp. 123-134.

18. BS 5762: 1979, "Methods for Crack Opening Displacement (COD) Testing." British Standards Institution, London, 1979.

19. E 1290-89 "Standard Test Method for Crack Tip Opening Displacement Testing." American Society for Testing and Materials, Philadelphia, 1989.

20. Sorem, W.A., "The Effect of Specimen Size and Crack Depth on the Elastic-Plastic Fracture Toughness of a Low Strength High-Strain Hardening Steel." Ph.D. Dissertation, The University of Kansas, Lawrence, KS, May 1989.

21. Duffy, J. and Shih, C.F., "Dynamic Fracture Toughness Measurements for Brittle and Ductile Materials." *Advances in Fracture Research: Seventh International Conference on Fracture.*, Pergamon Press, Oxford, 1989, pp. 633-642.

22. Sanford, R.J. and Dally, J.W., "A General Method for Determining Mixed-Mode Stress Intensity Factors from Isochromatic Fringe Patterns." *Engineering Fracture Mechanics*, Vol. 11, 1979, pp. 621-633.

23. Chona, R., Irwin, G.R., and Shukla, A., "Two and Three Parameter Representation of Crack Tip Stress Fields." *Journal of Strain Analysis*, Vol. 17, 1982, pp. 79-86.

24. Kalthoff, J.F., Beinart, J., Winkler, S., and Klemm, W., "Experimental Analysis of Dynamic Effects in Different Crack Arrest Test Specimens." ASTM STP 711, American Society for Testing and Materials, Philadelphia, 1980, pp. 109-127.

25. Nakamura, T., Shih, C.F. and Freund, L.B., "Analysis of a Dynamically Loaded Three-Point-Bend Ductile Fracture Specimen." *Engineering Fracture Mechanics*, Vol. 25, 1986, pp. 323-339.

26. Nakamura, T., Shih, C.F. and Freund, L.B., "Three-Dimensional Transient Analysis of a Dynamically Loaded Three-Point-Bend Ductile Fracture Specimen." ASTM STP 995, Vol. I, American Society of Testing and Materials, Philadelphia, 1989, pp. 217-241.

27. Joyce J.A. and Hacket, E.M., "Dynamic J-R Curve Testing of a High Strength Steel Using the Multispecimen and Key Curve Techniques." ASTM STP 905, American Society of Testing and Materials, Philadelphia, 1984, pp. 741-774.

28. Joyce J.A. and Hacket, E.M., "An Advanced Procedure for J-R Curve Testing Using a Drop Tower." ASTM STP 995, American Society of Testing and Materials, Philadelphia, 1989, 298-317.

29. Robertson, T.S., "Brittle Fracture of Mild Steel." *Engineering*, Vol. 172, 1951, pp. 445-448.

30. Naus, D.J., Nanstad, R.K., Bass, B.R., Merkle, J.G., Pugh. C.E., Corwin, W.R., and Robinson G.C., "Crack-Arrest Behavior in SEN Wide Plates of Quenched and Tempered A 533 Grade B Steel Tested under Nonisothermal Conditions." NUREG/CR-4930, U.S. Nuclear Regulatory Commission, Washington, D.C., August 1987.

31. E 1221-88, "Standard Method for Determining Plane-Strain Crack-Arrest Toughness, K_{Ia}, of Ferritic Steels." American Society of Testing and Materials, Philadelphia, 1988.

32. Crosley, P.B., Fourney, W.L., Hahn, G.T., Hoagland, R.G., Irwin, G.R., and Ripling, E.J., "Final Report on Cooperative Test Program on Crack Arrest Toughness Measurements." NUREG/CR-3261, U.S. Nuclear Regulatory Commission, Washington, D.C., April 1983.

33. Barker, D.B., Chona, R., Fourney, W.L., and Irwin, G.R., "A Report on the Round Robin Program Conducted to Evaluate the Proposed ASTM Test Method of Determining the Crack Arrest Fracture Toughness, K_{Ia}, of Ferritic Materials." NUREG/CR-4996, January 1988.

34. Dawes, M.G., Pisarski, H.G. and Squirrell, H.G., "Fracture Mechanics Tests on Welded Joints" ASTM STP 995, American Society of Testing and Materials, Philadelphia, 1989, pp. II-191 - II-213.

35. Satok K. and Toyoda, M., "Guidelines for Fracture Mechanics Testing of WM/HAZ." Working Group on Fracture Mechanics Testing of Weld Metal/HAZ, International Institute of Welding, Commission X, IIW Document X-1113-86.

36. RP 2Z, "Recommended Practice for Preproduction Qualification of Steel Plates for Offshore Structures." American Petroleum Institute, 1987.

37. Towers O.L. and Dawes, M.G., "Welding Institute Research on the Fatigue Precracking of Fracture Toughness Specimens." ASTM STP 856, American Society of Testing and Materials, Philadelphia, 1985. pp. 23-46.

38. E 23-88, "Standard Test Methods for Notched Bar Impact Testing of Metallic Materials." American Society of Testing and Materials, Philadelphia, 1988.

39. E 208-87, "Standard Test Method for Conducting Drop-Weight Test to Determine Nil-Ductility Transition Temperature of Ferritic Steels." American Society of Testing and Materials, Philadelphia, 1987.

40. E 436-74, "Standard Method for Drop-Weight Tear Tests of Ferritic Steels." American Society of Testing and Materials, Philadelphia, 1974.

41. E 604-83, "Standard Test Method for Dynamic Tear Testing of Metallic Materials." American Society of Testing and Materials, Philadelphia, 1983.

42. Marandet, B. and Sanz, G., "Evaluation of the Toughness of Thick Medium Strength Steels by LEFM and Correlations Between K_{IC} and CVN." ASTM STP 631, American Society of Testing and Materials, Philadelphia, 1977, pp. 72-95.

43. Rolfe, S.T. and Novak, S.T., "Slow Bend KIC Testing of Medium Strength High Toughness Steels." ASTM STP 463, American Society of Testing and Materials, Philadelphia, 1970, pp. 124-159.

44. Barsom, J.M. and Rolfe, S.T., "Correlation Between K_{IC} and Charpy V Notch Test Results in the Transition Temperature Range." ASTM STP 466, American Society of Testing and Materials, Philadelphia, 1970, pp. 281-301.

45. Sailors, R.H. and Corten, H.T., "Relationship between Material Fracture Toughness Using Fracture Mechanics and Transition Temperature Tests." ASTM STP 514, American Society of Testing and Materials, Philadelphia, 1973, pp. 164-191.

46. Begley, J.A. and Logsdon, W.A., "Correlation of Fracture Toughness and Charpy Properties for Rotor Steels." Westinghouse Report, Scientific Paper 71-1E7, MSLRF-P1-1971.

47. Ito, T., Tanaka, K. and Sato, M. "Study of Brittle Fracture Initiation from Surface Notch in Welded Fusion Line." IIW Document X-704-730, September 1973.

48. Wullaert, R.A., "Applications of the Instrumented Charpy Impact Test." ASTM STP 466, American Society of Testing and Materials, Philadelphia, 1970, pp. 148-164.

49. Turner, C.E., "Measurement of Fracture Toughness by Instrumented Impact Test." ASTM STP 466, American Society of Testing and Materials, Philadelphia, 1970, pp. 93-114.

APPENDIX 7: EXPERIMENTAL ESTIMATES OF DEFORMATION J

This appendix presents a derivation of Eq. (7.15b), which estimates the plastic component of the deformation theory J for a growing crack. Only the plastic component need be considered here, because the elastic J at any point in the test can be computed directly from the current load and crack length.

Figure A7.1 shows a schematic load-plastic displacement curve for a growing crack. Consider two neighboring points on the curve, where the crack grows from a_1 to a_2, and the plastic component of the J integral varies from J_1 to J_2. Each of these J values can be computed from the area under the appropriate "deformation theory" curve. For example, J_1 is given by

$$J_1 = \frac{\eta_1 A_{pl(OABE)}}{B_N b_1} \tag{A7.1}$$

where $A_{pl(OABE)}$ is the area defined by the corresponding points in Fig. A7.1. It is convenient to separate J_2 into two components:

$$J_2 = \frac{\eta_2 A_{pl(OCDE)}}{B_N b_2} + \frac{\eta_2 A_{pl(EDFG)}}{B_N b_2}$$

$$= J_\alpha + J_\beta \tag{A7.2}$$

Assume that J_1 is known, and that we wish to compute J_2. First, let us determine the relationship between J_1 and J_α, the component of J_2 that corresponds to the displacement Δ_1. This relationship can be approximated by

$$J_\alpha = J_1 + \left(\frac{\partial J_{pl}}{\partial a}\right)_{\Delta_1} (a_2 - a_1) \tag{A7.3}$$

The partial derivative, for the general case, can be solved as follows:

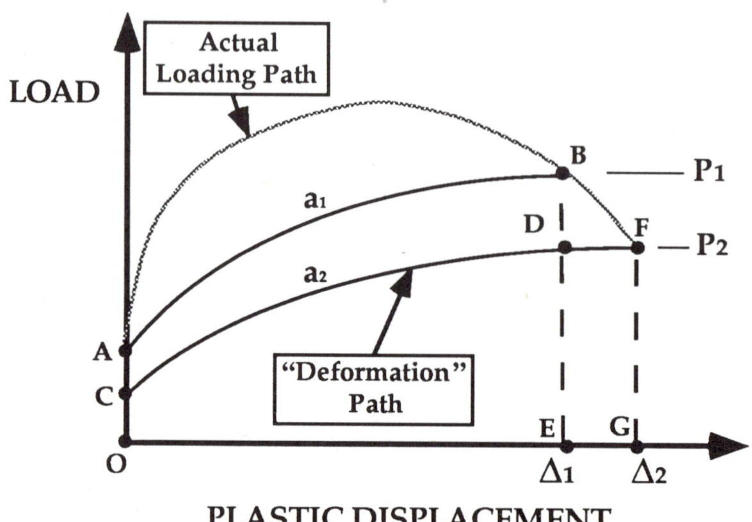

FIGURE A7.1 Schematic load-plastic displacement curve for a specimen in which the crack grows from a_1 to a_2.

$$\left(\frac{\partial J_{pl}}{\partial a}\right)_{\Delta_{pl}} = \left[\frac{\partial}{\partial a}\left(\frac{\eta A_{pl}}{B_N b}\right)\right]_{\Delta_{pl}}$$

$$= \frac{\eta}{B_N b}\left(\frac{\partial A_{pl}}{\partial a}\right)_{\Delta_{pl}} + \frac{\eta A_{pl}}{B_N b^2} + \frac{A_{pl}}{B_N b}\frac{d\eta}{da} \tag{A7.4}$$

Recall the energy release rate definition of J_{pl}:

$$J_{pl} = -\frac{1}{B_N}\left(\frac{\partial A_{pl}}{\partial a}\right)_{\Delta_{pl}} \tag{A7.5}$$

Substituting Eq. (A7.5) into Eq. (A7.4) leads to

$$\left(\frac{\partial J_{pl}}{\partial a}\right)_{\Delta_{pl}} = -\gamma \frac{J_{pl}}{b} \tag{A7.6}$$

where

$$\gamma = \left(\eta - 1 + \frac{1}{\eta}\frac{b}{W}\frac{d\eta}{d(b/W)}\right) \tag{A7.7}$$

Evaluating the derivative in Eq. (A7.3) at J_1 gives

$$J_\alpha = J_1\left[1 - \gamma_1 \frac{a_2 - a_1}{b_1}\right] \tag{A7.8}$$

For a deeply notched SENB specimen, $\eta = 2$; thus $\gamma = 1$, and Eq. (A7.8) reduces to

$$J_\alpha = J_1 \frac{b_2}{b_1} \tag{A7.9}$$

For a deeply notched compact specimen, η is approximately given by Eq. (7.11b). Substituting this result into Eq. (A7.7) gives

$$\gamma = 1 + 0.522\, b/W + \frac{0.522\, b/W}{2 + 0.522\, b/W}$$

which can be approximated by

$$\gamma \approx 1 + 0.76\, b/W \tag{A7.10}$$

for $0.50 \le a/W \le 0.70$.

The second term in Eq. (A7.2), J_β, can be estimated in a number of ways, including

$$J_\beta \approx \frac{\eta_2 P_2 (\Delta_2 - \Delta_1)}{B_N b_2} \tag{A7.11}$$

which is a good approximation, provided the load (on the deformation theory curve for a_2) does not vary significantly between Δ_1 and Δ_2. The ASTM Standard E 1152-87 chose the following estimate for J_β:

$$J_\beta \approx \left[\left(\frac{\eta_2}{B_N b_2}\right)\frac{(P_1 + P_2)(\Delta_2 - \Delta_1)}{2}\right]\left[1 - \gamma_1 \frac{a_2 - a_1}{b_1}\right] \tag{A7.12}$$

Summing J_α (Eq. (A7.8)) and the ASTM estimate for J_β (Eq. (A7.12) gives

$$J_2 = \left[J_1 + \left(\frac{\eta_2}{B_N b_2} \right) \frac{(P_1 + P_2)(\Delta_2 - \Delta_1)}{2} \right] \times \left[1 - \gamma_1 \frac{a_2 - a_1}{b_1} \right] \quad \text{(A7.13)}$$

which is essentially identical to Eq. (7.15b). The J-R curve can be computed by applying Eqs. (7.15a) and (7.15b) to successive increments of crack growth.

Figure A7.2 illustrates the numerical error that results from Eq. (A7.13). Note that this equation causes a slight underestimate of J_β. There may also be small errors in the estimate of J_α, since a partial derivative is applied to a finite change in crack size (Eq. (A7.3)). Equation (A7.9), however, is rigorously correct for a deeply notched bend specimen. This can be readily shown by applying the dimensional argument that was invoked in Section 3.2.5.

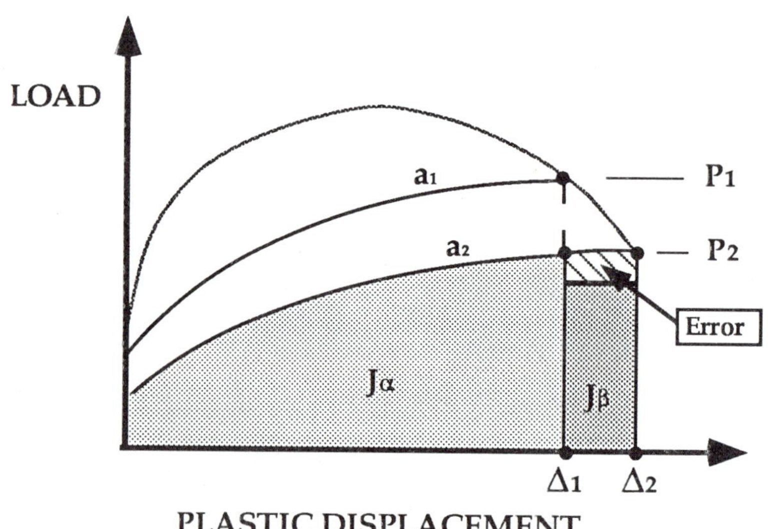

FIGURE A7.2 Schematic illustration of the error in J_{pl} that results from Eq. (A7.13).

8. FRACTURE TESTING OF NONMETALS

The procedures for fracture toughness testing of metals, which are described in Chapter 7, are fairly well established. Fracture testing of plastics, composites and ceramics is relatively new, however, and there are a number of unresolved issues.

Although many aspects of fracture toughness testing are similar for metals and nonmetals, there are several important differences. In some cases, metals fracture testing technology is inadequate on theoretical grounds. For example, the mechanical behavior of plastics can be highly rate dependent, and composites often violate continuum assumptions (see Chapter 6). There are also more pragmatic differences between fracture testing of metals and nonmetals. Ceramics, for instance, are typically very hard and brittle, which makes specimen fabrication and testing more difficult.

This chapter briefly summarizes the current procedures for measuring fracture toughness in plastics, fiber-reinforced composites, and ceramics. The reader should be familiar with the material in Chapter 7, since much of the same methodology (e.g., specimen design, instrumentation, fracture parameters) is currently being applied to nonmetals.

8.1 FRACTURE TOUGHNESS MEASUREMENTS IN ENGINEERING PLASTICS

Engineers and researchers who have attempted to measure fracture toughness of plastics have relied almost exclusively on metals testing technology. Existing experimental approaches implicitly recognize the potential for time-dependent deformation, but do not specifically address viscoelastic behavior in most instances. The recent work of Schapery [1,2], who developed a viscoelastic J integral (Chapter 4), has not seen widespread application to laboratory testing.

The Mode I stress intensity factor, K_I, and the (conventional) J integral were originally developed for time-independent materials, but may also be suitable for viscoelastic materials in certain cases. The restrictions on these parameters are explored below, followed by a summary of procedures for K and J testing on plastics. Section 8.1.5 briefly outlines possible approaches for taking account of viscoelastic behavior and time-dependent yielding in fracture toughness measurements.

8.1.1 The Suitability of K and J for Polymers

A number of investigators [3-6] have reported K_{IC}, J_{IC}, and J-R curve data for plastics. They applied testing and data analysis procedures that are virtually identical to metals approaches (Chapter 7). The validity of K and J is not guaranteed, however, when a material exhibits rate dependent mechanical properties. For example, neither J nor K are suitable for characterizing creep crack growth in metals (Section 4.2)[1]; an alternate parameter, C*, is required to account for the time-dependent material behavior. Schapery [1,2] has proposed an analogous parameter, J_v, to characterize viscoelastic materials (Section 4.3).

Let us examine the basis for applying K and J to viscoelastic materials, as well as the limitations on these parameters.

K-Controlled Fracture

In linear viscoelastic materials, remote loads and local stresses obey the same relationships as in the linear elastic case. Consequently, the stresses near the crack tip exhibit a $1/\sqrt{r}$ singularity:

$$\sigma_{ij} = \frac{K_I}{\sqrt{2\pi r}} f_{ij}(\theta) \tag{8.1}$$

and K_I is related to remote loads and geometry through the conventional linear elastic fracture mechanics (LEFM) equations introduced in Chapter 2. The strains and displacements depend on the viscoelastic properties, however. Therefore, the critical stress intensity factor for a viscoelastic material can be rate dependent; a K_{IC} value from a laboratory specimen is transferable to a structure only if the local crack tip strain histories of the two configurations are similar. Equation (8.1)

[1]The stress intensity factor is suitable for high temperature behavior in limited situations. At short times, when the creep zone is confined to a small region surrounding the crack tip, K uniquely characterizes crack tip conditions, while C* is appropriate for large scale creep.

only applies when yielding and nonlinear viscoelasticity are confined to a small region surrounding the crack tip.

Under plane strain linear viscoelastic conditions, K_I is related to the viscoelastic J integral, J_v, as follows [1]:

$$J_v = \frac{K_I^2 (1 - \nu^2)}{E_R} \tag{8.2}$$

where E_R is a reference modulus, which is sometimes defined as the short-time relaxation modulus.

Figure 8.1 illustrates a growing crack at times t_o and $t_o + t_\rho$.[2] Linear viscoelastic material surrounds a Dugdale strip yield zone, which is small compared to specimen dimensions. Consider a point A, which is at the leading edge of the yield zone at t_o and is at the trailing edge at $t_o + t_\rho$. The size of the yield zone and the crack tip opening displacement (CTOD) can be approximated as follows (see Chapters 2 & 3):

$$\rho_c = \frac{\pi}{8}\left(\frac{K_{IC}}{\sigma_{cr}}\right)^2 \tag{8.3}$$

and

$$\delta_c \approx \frac{K_{IC}^2}{\sigma_{cr}\, E(t_\rho)} \tag{8.4}$$

where σ_{cr} is the crazing stress. Assume that crack extension occurs at a constant CTOD. The time interval t_ρ is given by

$$t_\rho = \frac{\rho_c}{\dot{a}} \tag{8.5}$$

where $\dot{a}$ is the crack velocity. For many polymers, the time-dependence of the relaxation modulus can be represented by a simple power law:

$$E(t) = E_1\, t^{-n} \tag{8.6}$$

---

[2]This derivation, which was adapted from Marshall, et al. [7], is only heuristic and approximate. Schapery [8] performed a more rigorous analysis that led to a result that differs slightly from Eq. (8.9).

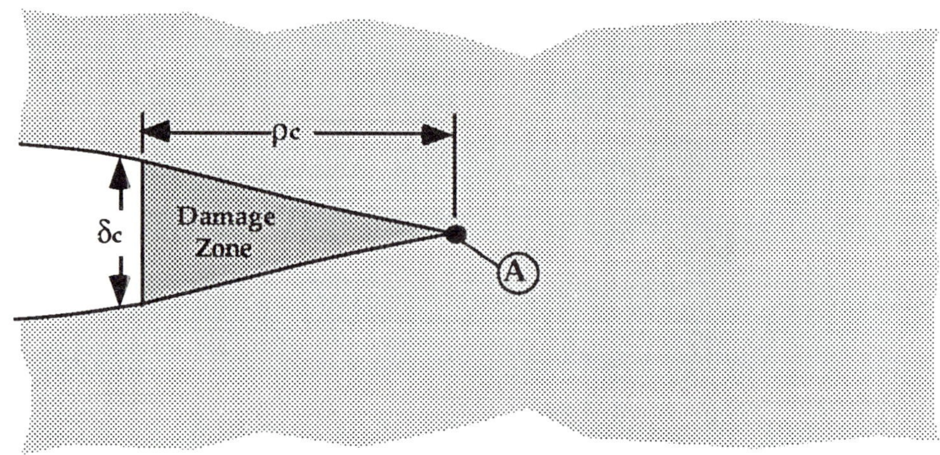

(a) Crack tip position at time t_o.

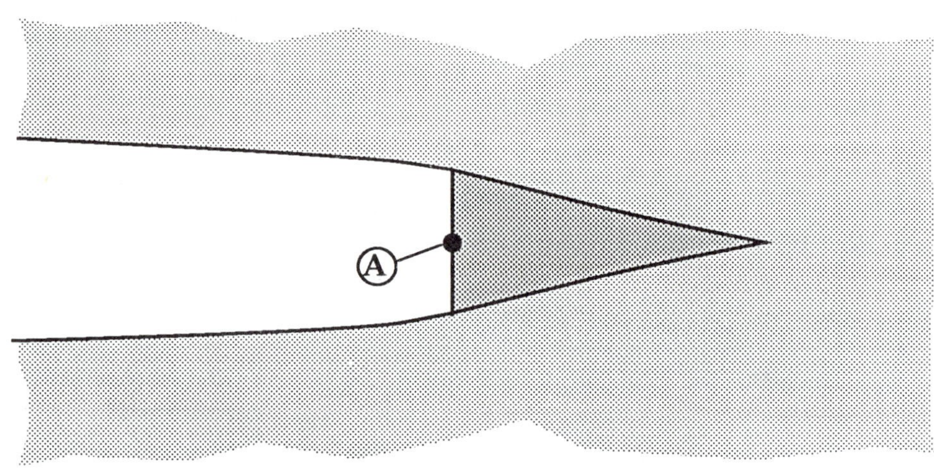

(b) Crack tip position at time $t_o + t_\rho$.

FIGURE 8.1 Crack growth at a constant CTOD in a linear viscoelastic material.

where E_1 and n are material constants that depend on temperature. If crazing is assumed to occur at a critical strain that is time-independent, the crazing stress is given by

$$\sigma_{cr} = E(t)\,\varepsilon_{cr} \tag{8.7}$$

Substituting Eqs. (8.5) to (8.7) into Eq. (8.4) leads to

$$K_{IC}^2 = \delta_c\,\varepsilon_{cr}\,E_1^2\,t^{-2n}$$

$$= \delta_c \, \varepsilon_{cr} \, E_1{}^2 \left(\frac{\rho_c}{\dot{a}}\right)^{-2n} \tag{8.8}$$

Solving for ρ_c and inserting the result in Eq. (8.8) gives

$$K_{IC} = \sqrt{\delta_c \, \varepsilon_{cr}} \left(\frac{8 \, \varepsilon_{cr}}{\pi \, \delta_c}\right)^n E_1 \, \dot{a}^n \tag{8.9}$$

Therefore, according to this analysis, fracture toughness is proportional to $\dot{a}^n$, and crack velocity varies as $K_I^{1/n}$. Several investigators have derived relationships similar to Eq. (8.9), including Marshall, et al. [7] and Schapery [8].

Figure 8.2 is a schematic plot of crack velocity versus K_I for various n values. In a time-independent material, n = 0; the crack remains stationary below K_{IC}, and becomes unstable when $K_I = K_{IC}$. In such materials, K_{IC} is a unique material property. Most metals and ceramics are nearly time independent at ambient temperature. When n > 0, crack propagation can occur over a range of K_I values. If, however, n is small, the crack velocity is highly sensitive to stress intensity, and the $\dot{a}$ - K_I curve exhibits a sharp knee. For example, if n = 0.1, the crack velocity is proportional to K_I^{10}. In typical polymers below T_g, n < 0.1.

Consider a short-time K_{IC} test on a material with $n \leq 0.1$, where K_I increases monotonically until the specimen fails. At low K_I values (i.e., in the early portion of the test), the crack growth would be negligible. The crack velocity would accelerate rapidly when the specimen reached the knee in the $\dot{a}$ - K_I curve. The specimen would then fail at a critical K_{IC} that would be relatively insensitive to rate. Thus if the knee in the crack velocity-stress intensity curve is sufficiently sharp, a short-time K_{IC} test can provide a meaningful material property.

One must be careful in applying a K_{IC} value to a polymer structure, however. While a statically loaded structure made from a time-independent material will not fail as long as $K_I < K_{IC}$, slow crack growth below K_{IC} does occur in viscoelastic materials. Recall from Chapter 1 the example of the polyethylene pipe that failed by time-dependent crack growth over a period of several years. The power law form of Eq. (8.9) enables long-time behavior to be inferred from short-time tests, as Example 8.1 illustrates.

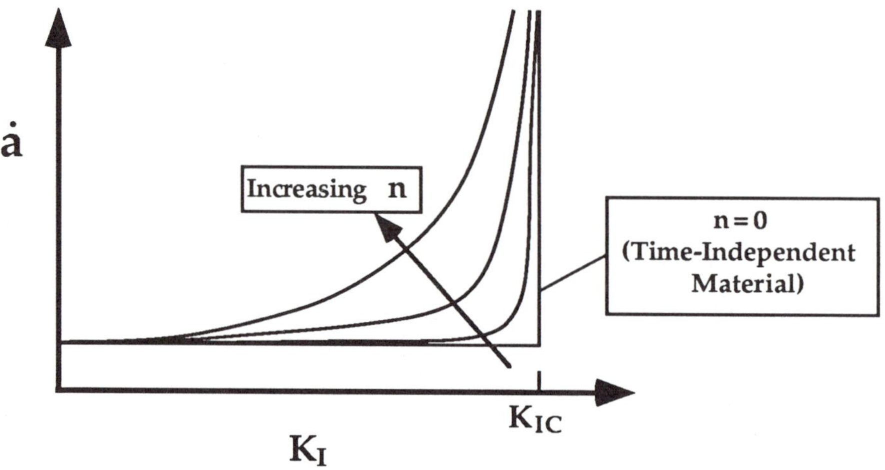

FIGURE 8.2 Effect of applied K_I on crack velocity for a variety of material responses.

EXAMPLE 8.1

Short-time fracture toughness tests on a polymer specimen indicate a crack velocity of 10 mm/s at K_{IC} = 5 MPa $\sqrt{m}$. If a pipe made from this material contains a flaw such that K_I = 2.5 MPa $\sqrt{m}$, estimate the crack velocity, assuming n = 0.08.

Solution: Since the crack velocity is proportional to $K_I^{12.5}$, the growth rate at 2.5 MPa $\sqrt{m}$ is given by

$$\dot{a} = 10\,\text{mm/s}\left(\frac{2.5\ \text{MPa}\sqrt{\text{m}}}{5\ \text{MPa}\sqrt{\text{m}}}\right)^{12.5} = 0.0017\ \text{mm/s} = 6.2\ \text{mm/hr}.$$

Equation (8.9) assumes that the critical CTOD for crack extension is rate-independent, which is a reasonable assumption for materials that are well below T_g. For materials near T_g, where E is highly sensitive to temperature and rate, the critical CTOD often exhibits a rate dependence [3].

J Controlled Fracture

Schapery [1,2] has introduced a viscoelastic J integral, J_V, that takes into account various types of linear and nonlinear viscoelastic behavior. For any material that obeys the assumed constitutive law, Schapery showed that J_V uniquely defines the crack tip conditions (Section 4.3.2). Thus J_V is a suitable fracture criteria for a wide range of time-dependent materials. Most practical applications of fracture mechanics to polymers, however, have considered only the conventional J integral, which does not account for time-dependent deformation.

Conventional J tests on polymers can provide useful information, but is important to recognize the limitations of such an approach. One way to assess the significance of critical J data for polymers is by evaluating the relationship between J and J_V. The following exercise considers a constant rate fracture test on a viscoelastic material.

Recall from Chapter 4 that strains and displacements in viscoelastic materials can be related to pseudo elastic quantities through hereditary integrals. For example, the pseudo elastic displacement, Δ^e is given by

$$\Delta^e = E_R^{-1} \int_0^t E(t - \tau) \frac{\partial \Delta}{\partial \tau} d\tau \tag{8.10}$$

where Δ is the actual load line displacement and τ is an integration variable. Equation (8.10) stems from the correspondence principle, and applies to linear viscoelastic materials for which Poisson's ratio is constant. This approach also applies to a wide range of *nonlinear* viscoelastic material behavior, although E(t) and E_R have somewhat different interpretations in the latter case.

For a constant displacement rate fracture test, Eq. (8.10) simplifies to

$$\Delta^e = \dot{\Delta} E_R^{-1} \int_0^t E(t - \tau) \, d\tau$$

$$= \Delta \frac{\bar{E}(t)}{E_R} \tag{8.11}$$

where $\dot{\Delta}$ is the displacement rate and $\bar{E}(t)$ is a time-average modulus, defined by

$$\bar{E}(t) = \frac{1}{t}\int_0^t E(t-\tau)\,d\tau \qquad (8.12)$$

Figure 8.3 schematically illustrates load-displacement and load-pseudo displacement curves for constant rate tests on viscoelastic materials. For a linear viscoelastic material (Fig. 8.3(a)), the P-Δ^e curve is linear, while the P-Δ curve is nonlinear due to time dependence. Evaluation of pseudo strains and displacements effectively removes the time dependence. When Δ^e is evaluated for a nonlinear viscoelastic material (Fig. 8.3(b)), the material nonlinearity can be decoupled from the time-dependent nonlinearity.

The viscoelastic J integral can be defined from the load-pseudo displacement curve:

$$J_v = -\frac{\partial}{\partial a}\left[\int_0^{\Delta^e} P\,d\Delta^e\right]_{\Delta^e} \qquad (8.14)$$

where P is the applied load in a specimen of unit thickness. Assume that the P-Δ^e curve obeys a power law:

$$P = M\,(\Delta^e)^N \qquad (8.15)$$

where M and N are time-independent parameters; N is a material property, while M depends on both the material and geometry. For a linear viscoelastic material, N = 1, and M is the elastic stiffness. Inserting Eq. (8.15) into Eq. (8.14) leads to

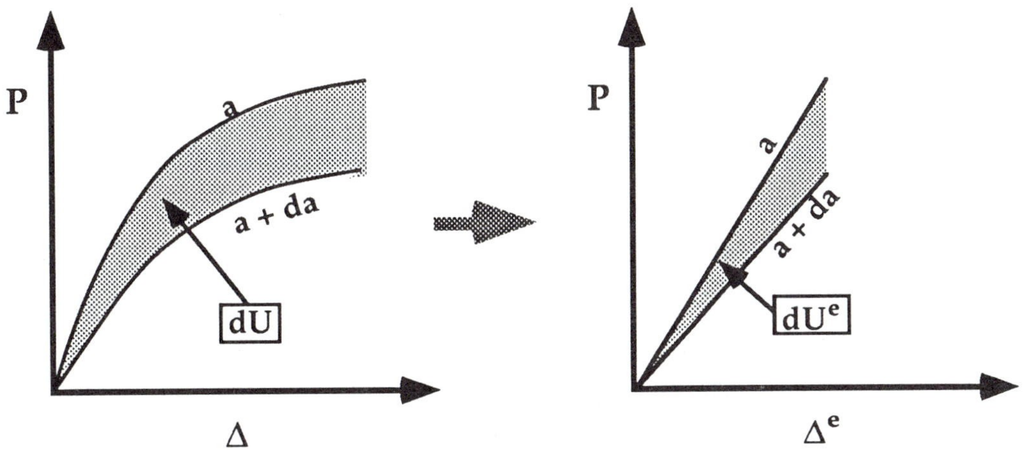

(a) Linear viscoelastic material.

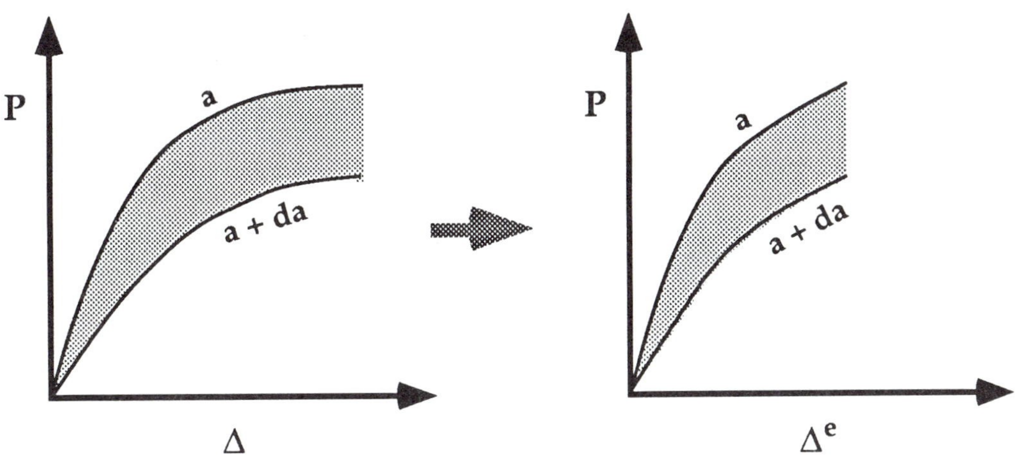

(b) Nonlinear viscoelastic material.

FIGURE 8.3 Load-displacement and load-pseudo displacement curves for viscoelastic materials.

$$J_V = -\frac{(\Delta^e)^{N+1}}{N+1}\left(\frac{\partial M}{\partial a}\right)_{\Delta^e} \tag{8.16}$$

Solving for J_V in terms of physical displacement (Eq. (8.11)) gives

$$J_V = -\frac{\Delta^{N+1}}{N+1}\left(\frac{\bar{E}(t)}{E_R}\right)^{N+1}\left(\frac{\partial M}{\partial a}\right)_{\Delta} \tag{8.17}$$

Let us now evaluate J from the same constant rate test:

$$J = -\frac{\partial}{\partial a}\left[\int_0^{\Delta} P\, d\Delta\right]_{\Delta} \tag{8.18}$$

The load can be expressed as a function of physical displacement by combining Eqs. (8.11) and (8.15):

$$P = M\,\Delta^N \left(\frac{\bar{E}(t)}{E_R}\right)^N \tag{8.19}$$

Substituting Eq. (8.19) into Eq. (8.18) leads to

$$J = -\,\Delta^{N+1}\left(\frac{\partial M}{\partial a}\right)_{\Delta} \frac{1}{t^{N+1}} \int_0^t \left(\frac{\bar{E}(\tau)}{E_R}\right)^N \tau^N\, d\tau \tag{8.20}$$

since $\Delta = \dot{\Delta}\, t$. Therefore,

$$J_v = J\,\phi(t) \tag{8.22}$$

where

$$\phi(t) = \frac{\left[t\,\bar{E}(t)\right]^{N+1}}{(N+1)\,E_R}\left\{\int_0^t [\bar{E}(\tau)]^N\, \tau^N\, d\tau\right\}^{-1} \tag{8.23}$$

Thus J and J_v are related through a dimensionless function of time in the case of a constant rate test. For a linear viscoelastic material in plane strain, the relationship between J and K_I is given by

$$J = \frac{K_I^2(1-\nu^2)}{E_R\,\phi(t)} \tag{8.24}$$

The conventional J integral uniquely characterizes the crack tip conditions in a viscoelastic material for a *given time.* A critical J value from a laboratory test is transferable to a structure, provided the failure times in the two configurations are the same.

A constant rate J test apparently provides a rational measure of fracture toughness in polymers, but applying such data to structural components may be problematic. Many structures are statically loaded at either a fixed load or remote displacement. Thus a constant load creep test or a load relaxation test on a cracked specimen might be more indicative of structural conditions than a constant displacement rate test. It is unlikely that the J integral would uniquely characterize viscoelastic crack growth behavior under all loading conditions. For example, in the case of viscous creep in metals, plots of J versus da/dt fail to exhibit a single trend, but C* (which is a special case of J_v) correlates crack growth data under different loading conditions (see Chapter 4).

Application of fracture mechanics to polymers presents additional problems for which both J and J_v may be inadequate. At sufficiently high stresses, polymeric materials typically experience irreversible deformation, such as yielding, microcracking, and microcrazing. This nonlinear material behavior exhibits a different time dependence than viscoelastic deformation; computing pseudo strains and displacements may not account for rate effects in such cases.

In certain instances, the J integral may be approximately applicable to polymers that exhibit large scale yielding. Suppose that there exists a quantity J_y that accounts for time-dependent yielding in polymers. A conventional J test will reflect material fracture behavior if J and J_y are related through a separable function of time [9]:

$$J_y = J\,\phi_y(t) \tag{8.25}$$

Section 8.1.5 outlines a procedure for determining J_y experimentally.

In metals, the J integral ceases to provide a single parameter description of crack tip conditions when the yielding is excessive. Critical J values become geometry dependent when the single parameter assumption is no longer valid (see Chapter 3). A similar situation undoubtedly exists in polymers: the single parameter assumption becomes invalid after sufficient irreversible deformation. Neither J nor J_y will give geometry independent measures of fracture toughness in such cases. Specimen size requirements for a single parameter description of fracture behavior in polymers have yet to be established, although there has been some research in this area (see Sections 8.1.3 and 8.1.4).

Crack growth presents further complications when the plastic zone is large. Material near the crack tip experiences nonproportional loading and unloading when the crack grows, and the J integral is no longer path-independent. The appropriate definition of J for a growing crack is unclear in metals (Section 3.4.2), and the problem is complicated further when the material is rate sensitive. The rate dependence of unloading in polymers is often different from that of loading.

In summary, the J integral can provide a rational measure of toughness for viscoelastic materials, but the applicability of J data to structural components is suspect. When the specimen experiences significant time-dependent yielding prior to fracture, J may give a reasonable characterization of fracture initiation from a stationary crack, as long as the extent of yielding does not invalidate the single-parameter assumption. Crack growth in conjunction with time-dependent yielding is a formidable problem that requires further study.

8.1.2 Precracking and Other Practical Matters

As with metals, fracture toughness tests on polymers require that the initial crack be sharp. Precracks in plastic specimens can be introduced by a number of methods including fatigue and razor notching.

Fatigue precracking in polymers can be very time consuming. The loading frequency must be kept low in order to minimize hysteresis heating, which can introduce residual stresses at the crack tip.

Because polymers are soft relative to metals, plastic fracture toughness specimens can be precracked by pressing a razor blade into a machined notch. Razor notching can produce a sharp crack in a fraction of the time required to grow a fatigue crack, and the measured toughness is not adversely affected if the notching is done properly [4].

Two types of razor notching are common: razor notch guillotine and razor sawing. In the former case, the razor blade is simply pressed into the material by a compressive force, while razor sawing entails a lateral slicing motion in conjunction with the compressive force.

Figures 8.4(a) and 8.4(b) are photographs of fixtures for the razor notch guillotine and razor sawing procedures, respectively.

In order to minimize material damage and residual stresses that result from razor notching, Cayard [4] recommends a three-step procedure: (1) fabrication of a conventional machined notch; (2) extension of the notch with a narrow slitting saw; and (3) final sharpening with a razor blade (by either of the techniques described above). Cayard found that such an approach produced very sharp cracks with minimal residual stresses. The notch tip radius is typically much smaller than the radius of the razor blade, apparently because a small pop-in propagates ahead of the razor notch.

While the relative softness of plastics aids the precracking process, it can cause problems during testing. The crack opening force that a clip gage applies to a specimen (Fig. 7.8) is negligible for metal specimens, but this load can be significant in plastic specimens. The conventional cantilever design may be too stiff for soft plastic specimens; a ring-shaped clip gage may be more suitable.

One may choose to infer specimen displacement from the crosshead displacement. In such cases it is necessary to correct for extraneous displacements due to indentation of the specimen by the test fixture. A displacement calibration can be inferred from a load-displacement curve for an unnotched specimen. If the calibration curve is linear, the correction to displacement is relatively simple:

$$\Delta = \Delta_{tot} - C_i P \tag{8.26}$$

where Δ_{tot} is the measured displacement and C_i is the compliance due to indentation. Since the deformation of the specimen is time-dependent, the crosshead rate in the calibration experiment should match that in the actual fracture toughness tests.

8.1.3 K_{IC} Testing

The American Society for Testing and Materials (ASTM) has published a number of standards for fracture testing of metals, which Chapter 7 describes. Committee D20 within ASTM is currently developing corresponding standards for plastics. A standard method for K_{IC} testing of plastics is in draft form, as of this writing [10].

(a) Razor notched guillotine.

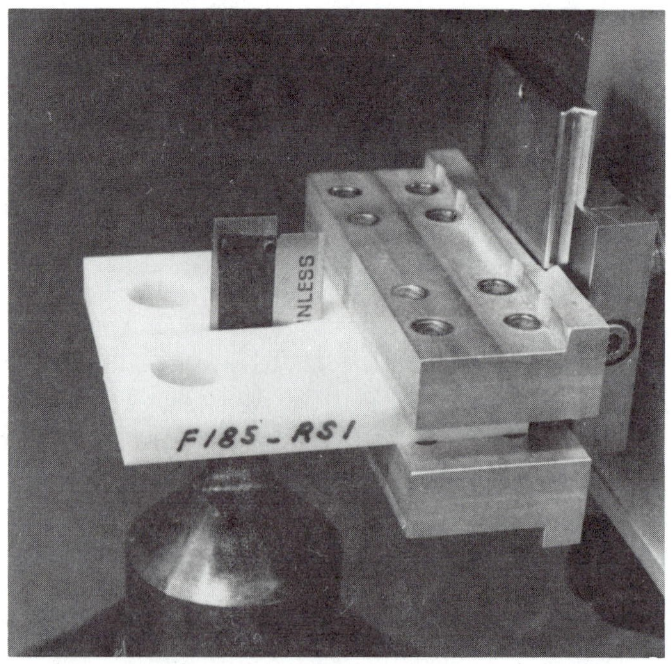

(b) Razor sawing.

FIGURE 8.4 Razor notching of polymer specimens. (Photographs provided by M. Cayard.)

The draft ASTM K_{IC} standard for plastics is very similar to E 399 [11], the ASTM K_{IC} standard for metals. Both test methods define an apparent crack initiation load, P_Q, by a 5% secant construction (Fig. 7.13). This load must be greater than 1.1 times the maximum load in the test for the result to be valid. The provisional fracture toughness, K_Q, must meet the following specimen size requirements:

$$B, a \geq 2.5\left(\frac{K_Q}{\sigma_{YS}}\right)^2 \tag{8.27a}$$

$$0.45 \leq \frac{a}{W} \leq 0.55 \tag{8.27b}$$

where B is the specimen thickness, a is the crack length, and W is the specimen width, as defined in Fig. 7.1.

The yield strength, σ_{YS}, is defined in a somewhat different manner for plastics. Figure 8.5 schematically illustrates a typical stress-strain curve for engineering plastics. When a polymer yields, it often experiences strain softening followed by strain hardening. The yield strength is defined at the peak stress prior to strain softening, as Fig. 8.5 illustrates. Because the flow properties are rate dependent, the draft ASTM K_{IC} standard for plastics requires that the time to reach σ_{YS} in a tensile test coincide with the time to failure in the fracture test to within ± 20%.

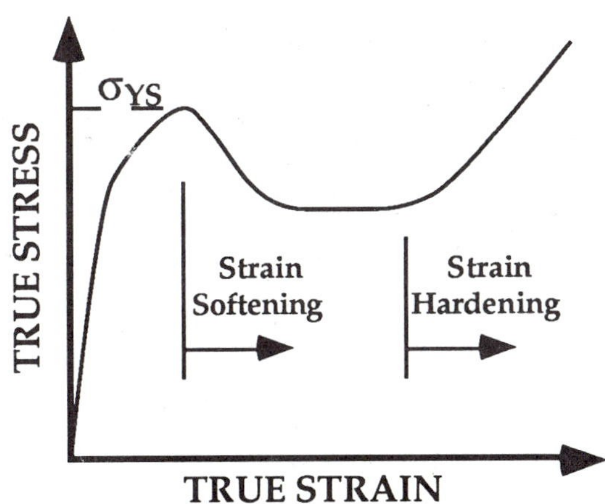

FIGURE 8.5 Typical stress-strain response of engineering plastics

The size requirements for metals (Eq. (8.27)) have been incorporated into the draft K_{IC} standard for plastics, apparently without assessing the suitability of these criteria for polymers. Recall from Chapters 2 and 7 the reasons for the K_{IC} size requirements:

- the plastic zone should be small compared to in-plane dimensions to ensure the presence of an elastic singularity zone ahead of the crack tip.

- the plastic zone should be small compared to the thickness to ensure predominantly plane strain conditions.

Because the yielding behavior of metals and plastics are different, one should not expect both materials to exhibit the same size limits for a valid K_{IC}. Even within a given material system, the sensitivity of toughness to specimen size is influenced by the micromechanism of fracture (see Appendix 5.1).

Cayard [4] has studied the size dependence of fracture toughness for a range of engineering plastics. The results for two typical materials are described below.

Figure 8.6 shows the effect of specimen size on K_Q values for a rigid polyvinyl chloride (PVC) and a polycarbonate (PC). In most cases, the specimens were geometrically similar, with W = 2B and a/W = 0.5. For specimen widths greater that 50 mm in the PC, the thickness was fixed at 25 mm, which corresponds to the plate thickness. Note that in the small specimens, $K_Q < K_{IC}$, because P_Q was defined from a 5% secant; in small specimens, this deviation in linearity depends on flow properties rather than fracture properties, as discussed in Section 7.2. The ASTM E 399 requirements for in-plane dimensions appear to be adequate for the PVC, but are nonconservative for the PC when the yield strength is defined by the peak stress (Fig. 8.5).

The different size dependence for the two polymer systems can be partially attributed to strain softening effects. Figure 8.7 shows the stress-strain curves for these two materials. Note that the PC exhibits significant strain softening, while the rigid PVC stress-strain curve is relatively flat after yielding. Significant strain softening probably increases the size of the yielded zone. If one defines σ_{YS} as the lower flow stress plateau, the size requirements are more restrictive

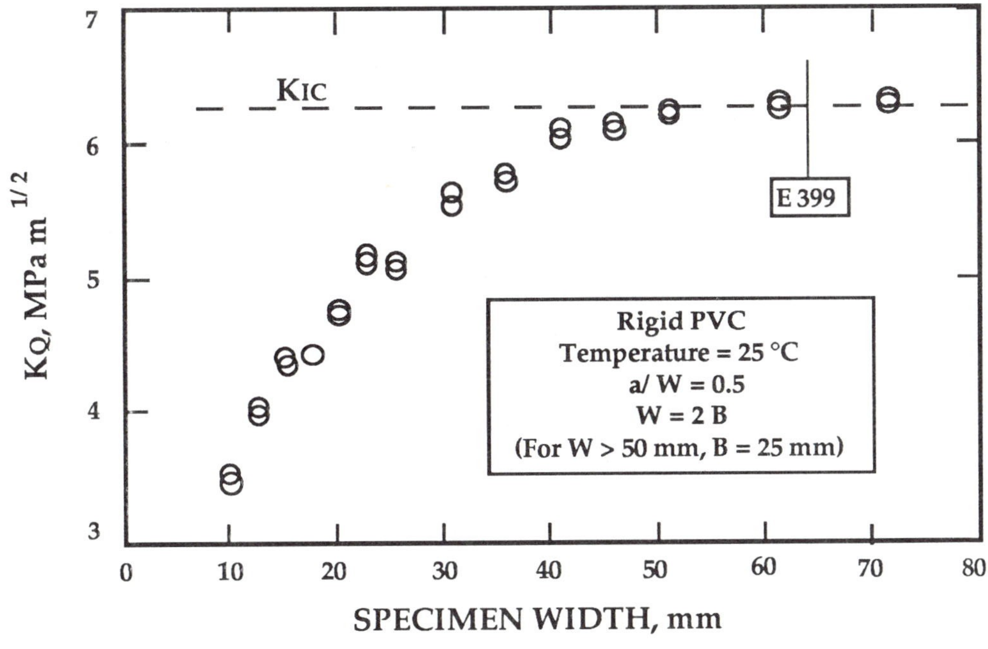

(a) Rigid PVC.

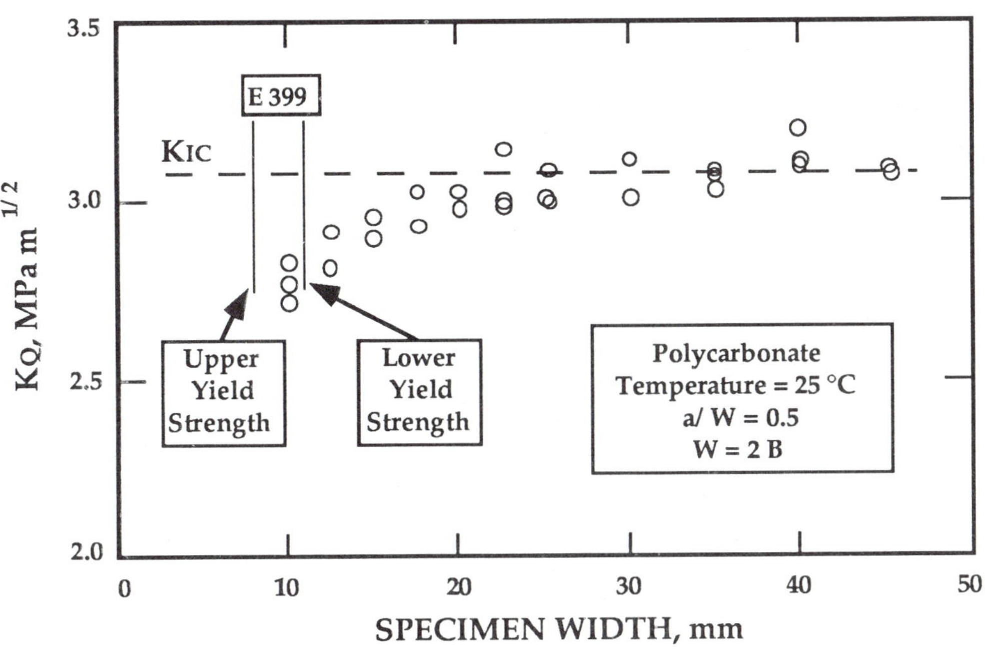

(b) Polycarbonate.

FIGURE 8.6 Effect of specimen size on K_Q in two engineering plastics [4].

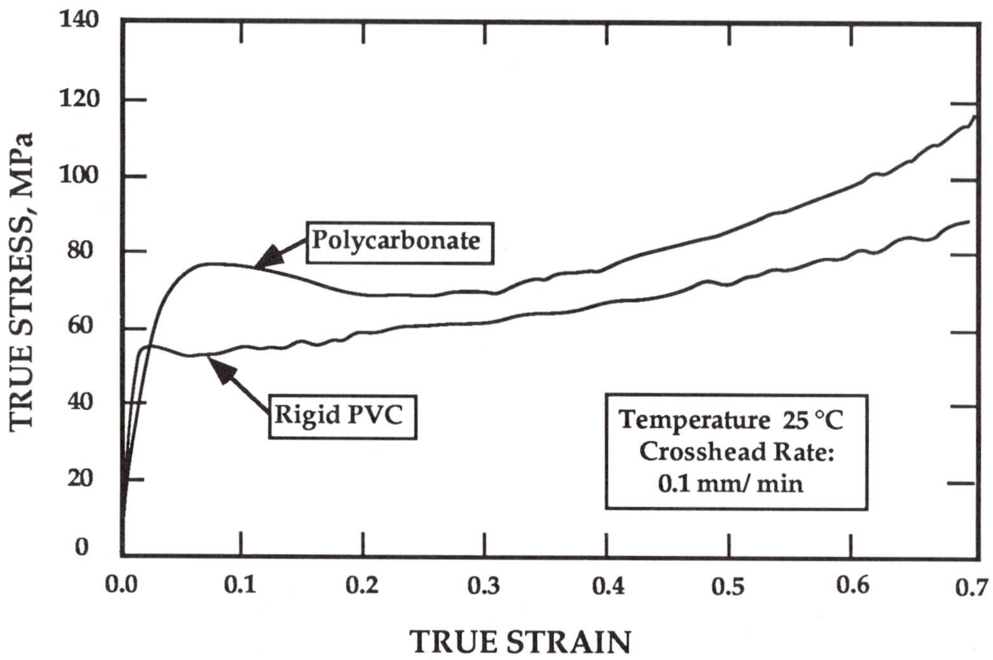

FIGURE 8.7 Stress-strain curves for the rigid PVC and polycarbonate [4].

for materials that strain soften. Figure 8.6(b) shows the E 399 in-plane requirements corresponding to the lower yield strength in the polycarbonate. Although this latter requirement is still nonconservative for this material, it represents a slight improvement over the approach in the draft ASTM standard for plastics.

Cayard [4] also examined the effect of thickness at constant in-plane dimensions. Figures 8.8(a) and 8.8(b) are plots of fracture toughness versus thickness for the PVC and the PC, respectively. Although all of the experimental data for the PVC are below the required thickness (according to Eq. (8.27)), these data do not exhibit a thickness dependence; Fig. 8.8(a) indicates that the E 399 thickness requirement is too severe for this material. In the case of the PC, most of the data are above the E 399 thickness requirement. These data also do not exhibit a thickness dependence, which implies that the E 399 requirement is at least adequate for this material. Further testing of thinner sections would be required to determine if the E 399 thickness requirement is overly conservative for the PC.

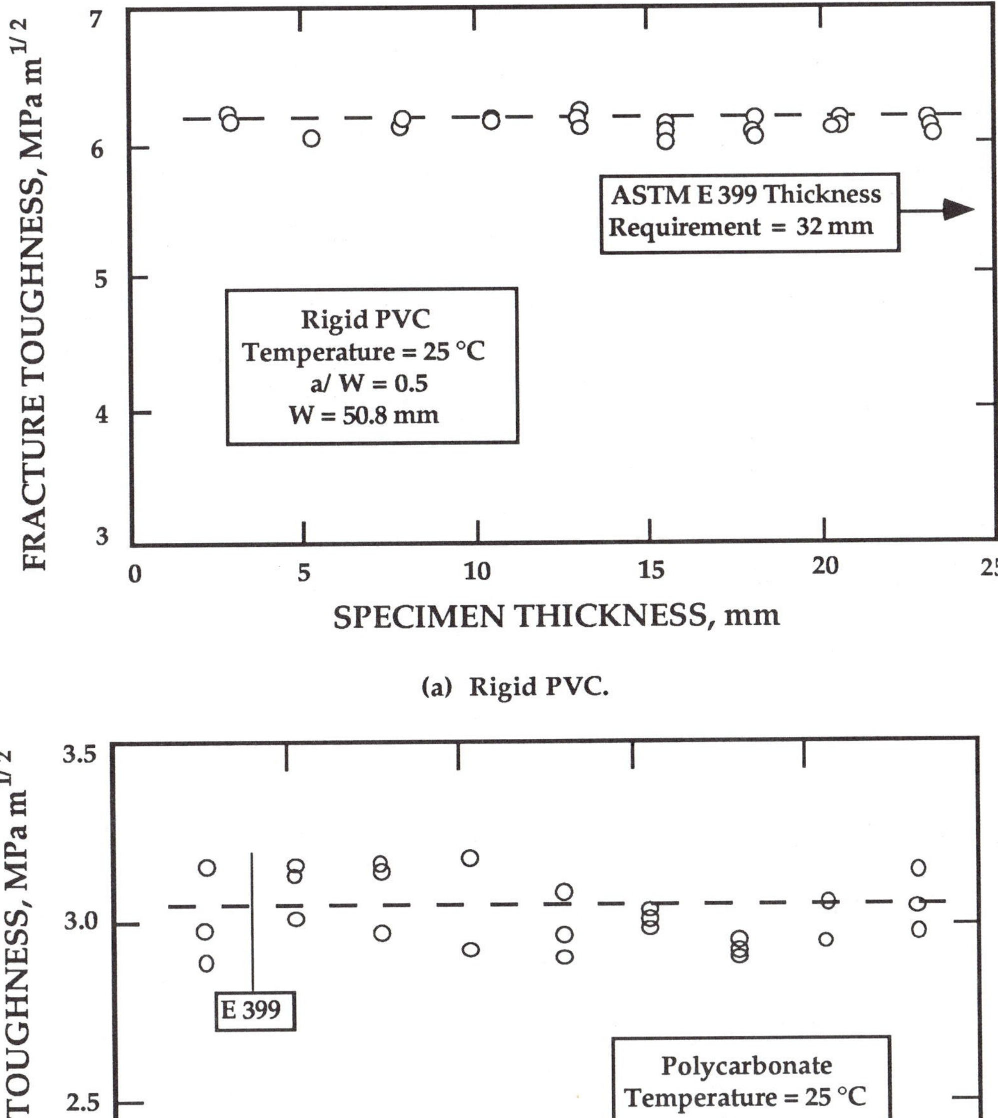

(a) Rigid PVC.

(b) Polycarbonate.

FIGURE 8.6 Effect of specimen thickness on fracture toughness of plastics [4].

The PVC material fails by crazing. Recall from Chapter 6 that a craze zone ahead of a crack tip contains a high concentration of voids. The material inside of the craze zone is subject essentially to plane stress loading, regardless of the specimen thickness. Consequently, the fracture toughness of materials that craze may be relatively insensitive to specimen thickness. Thickness effects are not necessarily absent from materials that craze, however. While the material in the craze zone is subject to plane stress, the surrounding material may be experience plane strain or mixed conditions; the stress state in the surrounding material could influence the toughness by dictating the size and shape of the craze zone.

Not all polymeric materials fracture by a crazing mechanism. For example, Cayard found no evidence of cavitation (void formation) on the fracture surfaces of the PC. The fracture toughness of this material may be more sensitive to specimen thickness than materials that craze, such as the PVC. The development of rational thickness requirements for polymers requires further study.

One final observation regarding the draft K_{IC} standard for plastics is that the procedure for estimating P_Q ignores time effects. Recall from Section 7.2 that nonlinearity in the load displacement curve from K_{IC} tests on metals can come from two sources: yielding and crack growth. In the case of polymers, viscoelasticity can also contribute to nonlinearity in the load-displacement curve. Consequently, at least a portion of the 5% deviation from linearity at P_Q could result from a decrease in the modulus during the test. Linear elastic fracture mechanics (LEFM) is valid for linear *viscoelastic* deformation, even when the load-displacement curve is nonlinear. The draft standard could be unduly restrictive in defining the critical load by a 5% secant, irrespective of the source of the nonlinearity.

For most practical situations, however, viscoelastic effects are probably negligible during K_{IC} tests. In order to obtain a valid K_{IC} result in most polymers, the test temperature must be well below T_g, where rate effects are minimal at short times. The duration of a typical K_{IC} test is on the order of several minutes, and the elastic properties *probably* will not change significantly prior to fracture. The rate sensitivity should be quantified, however, to evaluate the assumption that E does not change during the test.

8.1.4 J Testing

A number of researchers have applied J integral test methods to polymers [3-6], but a standard for measuring J_{IC} and J-R curves in plastics does not exist, as of this writing. A committee within ASTM is currently drafting a standard test method for J_{IC} measurements in plastics. The methodology that is being applied is very similar to that in E 813-87 [11], the ASTM standard for J_{IC} testing of metals.

The current formulas for estimating J from plastic test specimens are identical to those in E 813. For common test specimens, such as the compact and SENB geometries, J is related to the area under the load versus load line displacement curve:

$$J = \frac{\eta}{B\,b} \int_0^{\Delta} P \, d\Delta \tag{8.28}$$

where η is a dimensionless parameter that depends on geometry.

The most common approach for inferring the J-R curve for a polymer is the multiple specimen method. A set of nominally identical specimens are loaded to various displacements, unloaded, cooled to a low temperature, and then fractured. The initial crack length and stable crack growth are measured optically from each specimen, resulting in a series of data points on a J-Δa plot. The J_{IC} can then be inferred by fitting an equation, such as a power law or straight line, to the data. This latter exercise is described in Section 7.4.1 for J_{IC} testing of metals.

Single specimen techniques, such as unloading compliance, may also be applied to the measurement of J_{IC} and the J-R curve [6]. Time-dependent material behavior can complicate unloading compliance measurements, however. Figure 8.9 schematically illustrates the unload-reload behavior of a viscoelastic material. If rate effects are significant during the time frame of the unload-reload, the resulting curve can exhibit a hysteresis effect. One possible approach to account for viscoelasticity in such cases is to relate instantaneous crack length to *pseudo* elastic displacements (see Section 8.1.5).

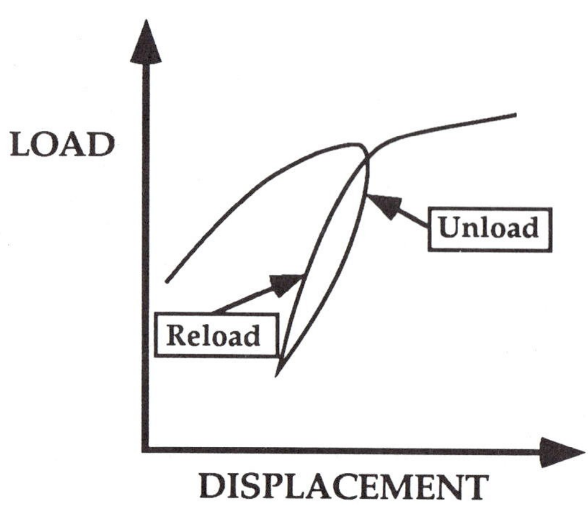

FIGURE 8.9 Schematic unloading behavior in a polymer. Hysteresis in the unload-reload curve complicates unloading compliance measurements.

Critical J values for polymers exhibit less size dependence than K_Q values. Figure 8.10 compares K_Q values for the polycarbonate with K_{JC} values, which were obtained by converting critical J values at fracture to an equivalent critical K through the following relationship:

$$K_{JC} = \sqrt{\frac{J_{crit} E}{1 - \nu^2}} \tag{8.29}$$

For sufficiently large specimens, where the global behavior is predominantly elastic, $K_{JC} = K_Q = K_{IC}$. Note that the K_{JC} values are independent of specimen size over the range of available data.

Crack growth resistance curves can be highly rate dependent. Figure 8.11 shows J-R curves for a polyethylene pipe material that was tested at three crosshead rates [3]. Increasing the crosshead rate from 0.254 mm/min to 1.27 mm/min (0.01 and 0.05 in/min, respectively) results in nearly a three-fold increase in J_{IC} in this case.

8.1.5 Experimental Estimates of Time-Dependent Fracture Parameters

While J_{IC} values may be indicative of a polymer's relative toughness, the existence of a unique correlation between J and crack growth rate is unlikely. Parameters such as J_v may be more suitable for some viscoelastic materials. For polymers that experience large scale yielding, neither J nor J_v may characterize crack growth.

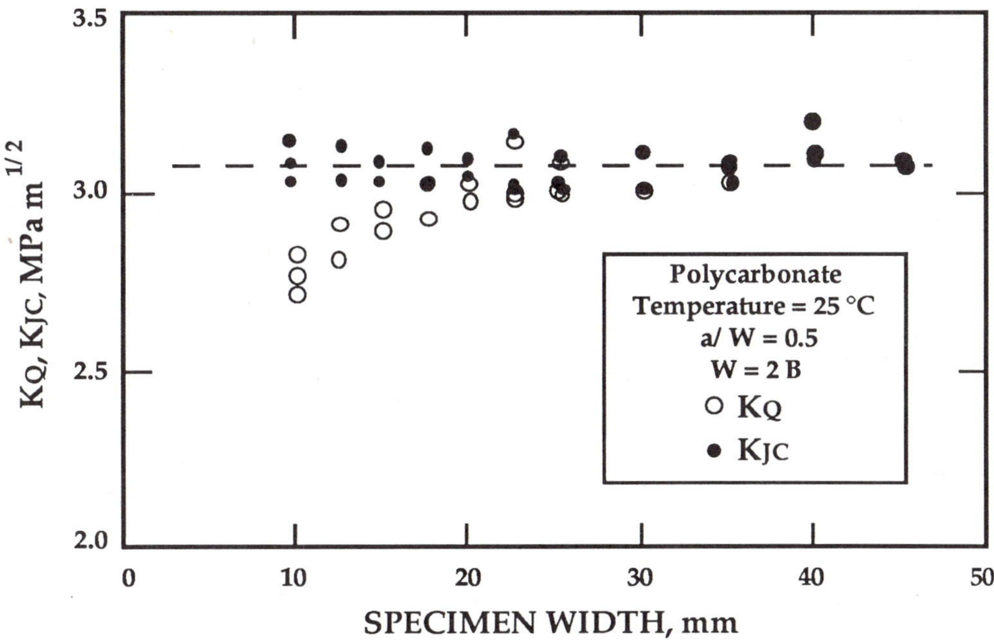

FIGURE 8.10 Size dependence of K_Q and J-based fracture toughness for PC [4].

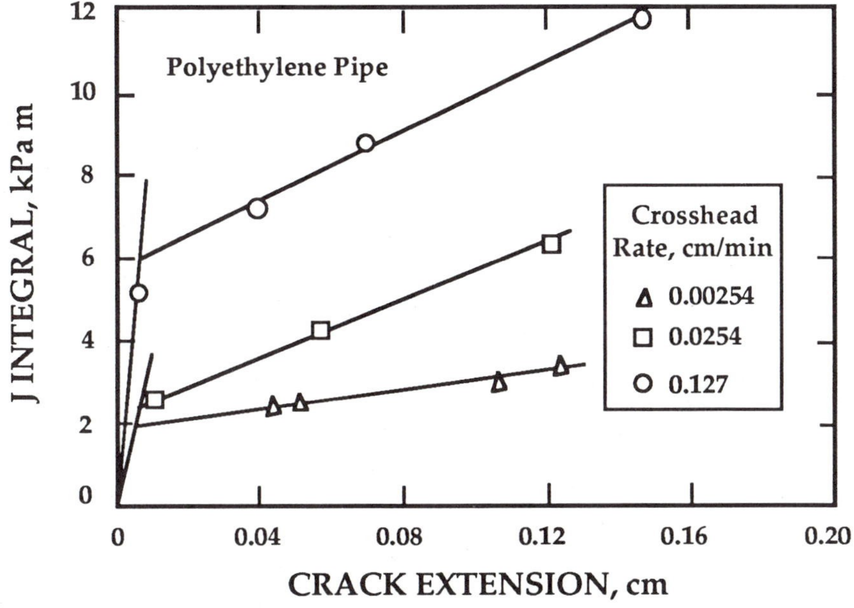

FIGURE 8.11 Crack growth resistance curves for polyethylene pipe at three crosshead rates [3].

The section outlines a few suggestions for inferring crack tip parameters that take into account the time-dependent deformation of engineering plastics. Since most of these approaches have yet to be validated experimentally, much of what follows contains an element of conjecture. These proposed methods, however, are certainly no worse than conventional J integral approaches, and may be considerably better for many engineering plastics.

The viscoelastic J integral, J_V, can be inferred by converting physical displacements to pseudo displacements. For a constant rate test, Eq. (8.11) gives the relationship between Δ and Δ^e. The viscoelastic J integral is given by Eq. (8.14); J_V can also be evaluated directly from the area under the P-Δ^e curve:

$$J_V = \frac{\eta}{b} \int_0^{\Delta^e} P \, d\Delta^e \tag{8.30}$$

for a specimen with unit thickness. If the load-pseudo displacement is a power-law (Eq. (8.19)), Eq. (8.30) becomes

$$J_V = \frac{\eta \, M \, (\Delta^e)^{N+1}}{b \, (N+1)} \tag{8.31}$$

Comparing Eqs. (8.30) and (8.16) leads to

$$\eta = -\frac{b}{M} \left(\frac{\partial M}{\partial a} \right)_{\Delta^e} \tag{8.32}$$

Since M does not depend on time, the dimensionless η factor is the same for both J and J_V.

Computing pseudo elastic displacements might also remove hysteresis effects in unloading compliance tests. If the unload-reload behavior is linear viscoelastic, the P-Δ^e unloading curves would be linear, and crack length could be correlated to the *pseudo elastic compliance*.

Determining pseudo displacements from Eq. (8.11) or the more general expression (Eq. (8.10)) requires a knowledge of E(t). A separate experiment to infer E(t) would not be particularly difficult, but such data would not be relevant if the material experienced large scale yielding in a fracture test. An alternative approach to inferring crack tip parameters that takes time effects into account is outlined below.

Schapery [9] has suggested evaluating a J-like parameter from isochronous (fixed time) load-displacement curves. Consider a series of fracture tests that are performed over a range of crosshead rates (Fig. 8.12(a)). If one selects a fixed time and determines the various combinations of load and displacement that correspond to this time, the resulting locus of points forms an isochronous load-displacement curve (Fig. 8.12(b)). Since the viscoelastic and yield properties are time-dependent, the isochronous curve represents the load displacement behavior for fixed material properties, as if time stood still while the test was performed. A fixed-time J integral can be defined as follows:

$$J_t = \frac{\eta}{b}\left(\int_0^{\Delta} P\, d\Delta\right)_{t\,=\,\text{constant}} \tag{8.33}$$

Suppose that the displacements at a given load are related by a separable function of time, such that it is possible to relate all displacements (at that particular load) to a reference displacement:

$$\Delta^R = \Delta\, \gamma(t) \tag{8.34}$$

The isochronous load-displacement curves could then be collapsed onto a single trend by multiplying each curve by $\gamma(t)$, as Fig. 8.12(c) illustrates. It would also be possible to define a reference J:

$$J^R = J_t\, \gamma(t) \tag{8.35}$$

Note the similarity between Eqs. (8.25) and (8.35).

The viscoelastic J is a special case of J^R. For a constant rate test, comparing Eqs. (8.11) and (8.34) gives

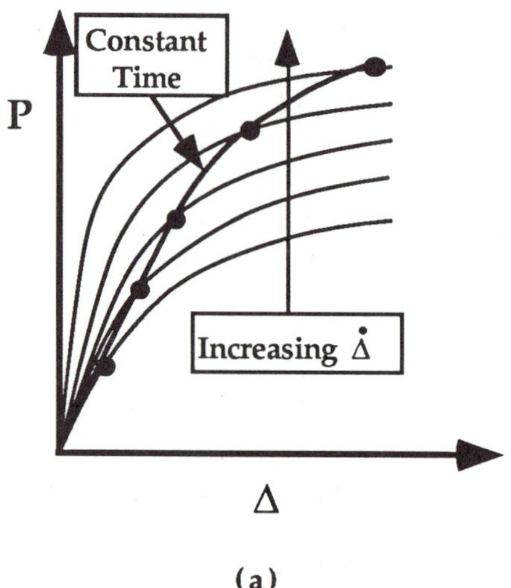

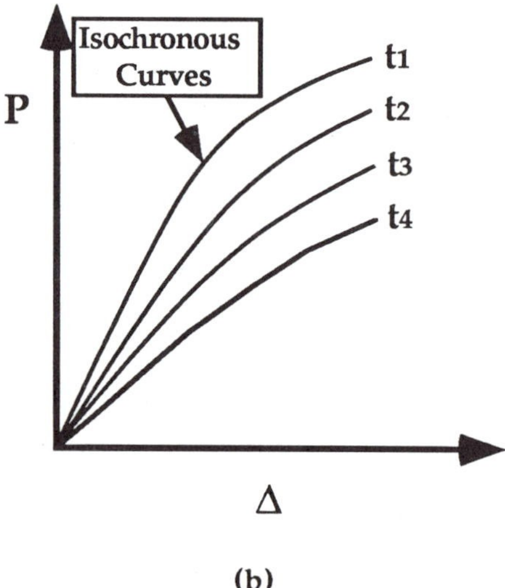

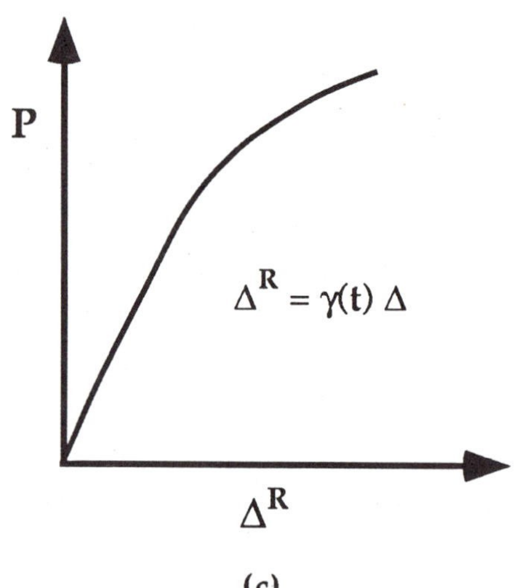

FIGURE 8.12 Proposed method for removing time dependence from load-displacement curves. First, a set of tests are performed over a range of displacement rates (a). Next, isochronous load displacement curves are inferred (b). Finally, the displacement axis of each curve is multiplied by a function $\gamma(t)$, resulting in a single curve (c).

$$J_v = J_t \frac{\bar{E}(t)}{E_R} \tag{8.36}$$

Thus for a linear viscoelastic material in plane strain,

$$J_t = \frac{K_I^2 (1 - \nu^2)}{\bar{E}(t)} \tag{8.37}$$

Isochronous load-displacement curves would be linear for a linear viscoelastic material, since the modulus is constant at a fixed time.

The parameter J^R is more general than J_v; the former may account for time dependence in cases where extensive yielding occurs in the specimen. The reference J should characterize crack initiation and growth in materials where Eq. (8.35) removes time dependence of displacement. Figure 8.13 schematically illustrates the postulated relationship between J_t, J_R, and crack velocity. The J_t-$\dot{a}$ curves should be parallel on a log-log plot, while a J^R-$\dot{a}$ plot should yield a unique curve. Even if it is not possible to produce a single J^R-$\dot{a}$ curve for a material, the Jt parameter should still characterize fracture at a fixed time.

Although J^R may characterize fracture initiation and the early stages of crack growth in a material that exhibits significant time-dependent yielding, this parameter would probably not be capable of characterizing extensive crack growth, since unloading and nonproportional loading occur near the growing crack tip. (See Section 8.1.1 above.)

8.1.6 Qualitative Fracture Tests on Plastics

The ASTM standard D 256-88 [13] describes impact testing of notched polymer specimens. This test method is currently the most common technique for characterizing the toughness of engineering plastics. The D 256 standard covers both Charpy and Izod tests (Fig. 7.36), but the plastics industry utilizes the Izod specimen in the vast majority of cases.

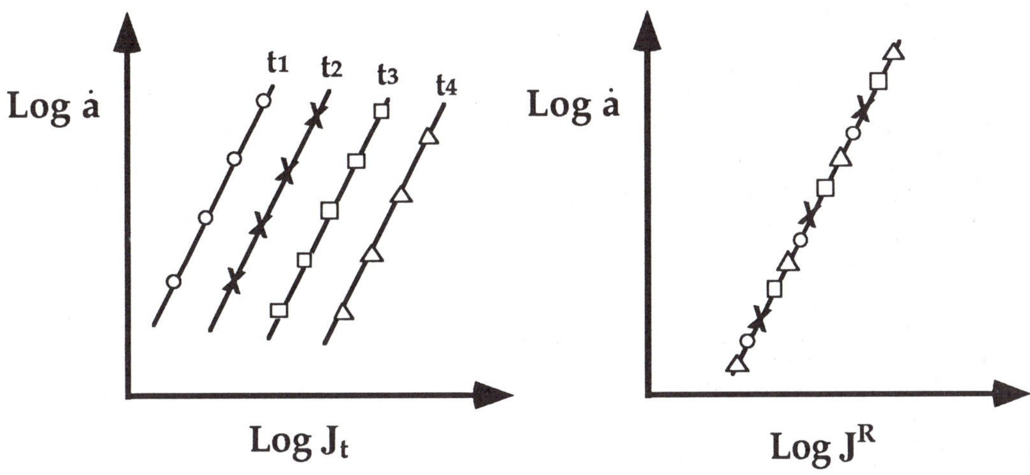

FIGURE 8.13 Postulated crack growth behavior in terms of J_t and J^R.

The procedure for impact testing of plastics is very similar to the metals approach, which is outlined in ASTM E 23 [14] (see Section 7.9). A pendulum strikes a notched specimen, and the energy required to fracture the specimen is inferred from the initial and final heights of the pendulum (Fig. 7.37). In the case of the Izod test, the specimen is a simple cantilever beam that is restrained at one end and struck by the pendulum at the other end. One difference between the metals and plastics test methods is that the absorbed energy is normalized by the net ligament area in plastics tests, while tests according to ASTM E 23 report only the total energy. The normalized fracture energy in plastics is known as the *impact strength*.

The impact test for plastics is pervasive throughout the plastics industry because it is a simple and inexpensive measurement. Its most common application is as a material screening criterion. The value of impact strength measurements is questionable, however.

One problem with this test method is that the specimens contain blunt notches. Figure 8.14 [15] shows Izod impact strength values for several polymers as a function of notch radius. As one might expect, the fracture energy decreases as the notch becomes sharper. The slope of the lines in Fig. 8.14 is a measure of the *notch sensitivity* of the material. Some materials are highly notch sensitive, while others are relatively insensitive to the radius of the notch. Note that the relative ordering of the materials' impact strengths in Fig. 8.14 changes with notch acuity. Thus a fracture energy for a particular notch radius may not be an appropriate criterion for ranking material toughness. Moreover, the notch strength is often not a reliable indicator of how the material will behave when it contains a sharp crack.

Since Izod and Charpy tests are performed under impact loading, the resulting fracture energy values are governed by the short-time material response. Many polymer structures, however, are loaded quasistatically and must be resistant to slow, stable crack growth. The ability of a material to resist crack growth at long times is not necessarily related to the fracture energy of a blunt-notched specimen in impact loading.

The British Standards Institution (BSI) specification for unplasticized polyvinyl chloride (PVC-U) pipe, BS 3505:1986 [16], contains a procedure for fracture toughness testing. Although the toughness test in

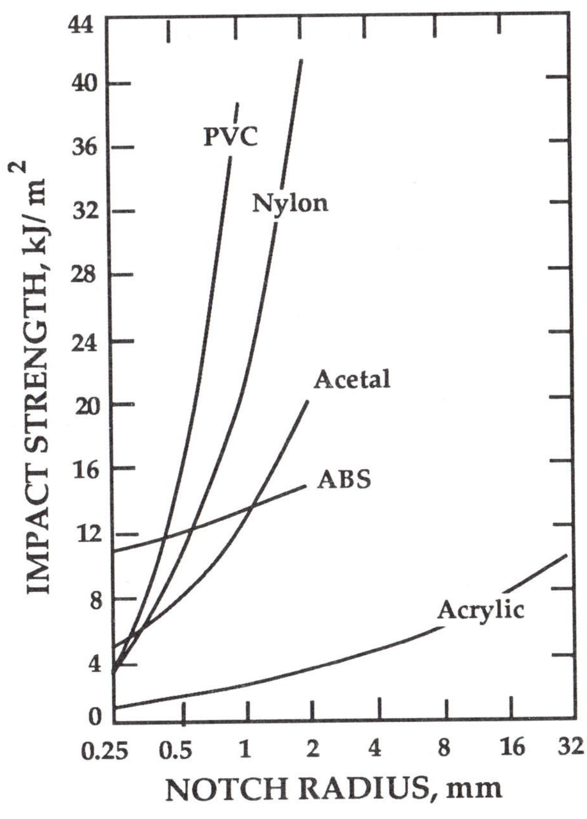

FIGURE 8.14 Effect of notch radius on the Izod impact strength of several engineering plastics [15].

BS 3506 is primarily a qualitative screening criterion, it is much more relevant to structural performance than the Izod impact test.

Appendices C and D of BS 3506 outline a procedure for inferring toughness of PVC-U pipe after exposure to an aggressive environment. A C-shaped section is removed from the pipe of interest and is submerged in dichloromethane liquid. After 15 min of exposure, the specimen is removed from the liquid and the surface is inspected for bleaching or whitening. A sharp notch is placed on the inner surface of the specimen which is then dead-loaded for 15 min or until cracking or total fracture is observed. Figure 8.15 is a schematic drawing of the testing apparatus. The loading is such that the notch region is subject to a bending moment. If the specimen cracks or fails completely during the test, the fracture toughness of the material can be computed from applied load and notch depth by means of standard K_I formulae. If no cracking is observed during the 15 min test, the toughness can be quantified by testing additional specimens at higher loads. The BS 3506 standard includes a semiempirical size correction for small pipes and high toughness materials that do not behave in an elastic manner.

8.2 INTERLAMINAR TOUGHNESS OF COMPOSITES

Chapter 6 outlined some of the difficulties in applying fracture mechanics to fiber-reinforced composites. The continuum assumption is often inappropriate, and cracks may not grow in a self-similar manner. The lack of a rigorous framework to describe fracture in composites has led to a number of qualitative approaches to characterize toughness.

Interlaminar fracture is one of the few instances where fracture mechanics formalism is applicable to fiber-reinforced composites on a global scale. A zone of delamination can be treated as a crack; the resistance of the material to the propagation of this crack is the fracture toughness. Since the crack typically is confined to the matrix material between plies, continuum theory is applicable, and the crack growth is self similar.

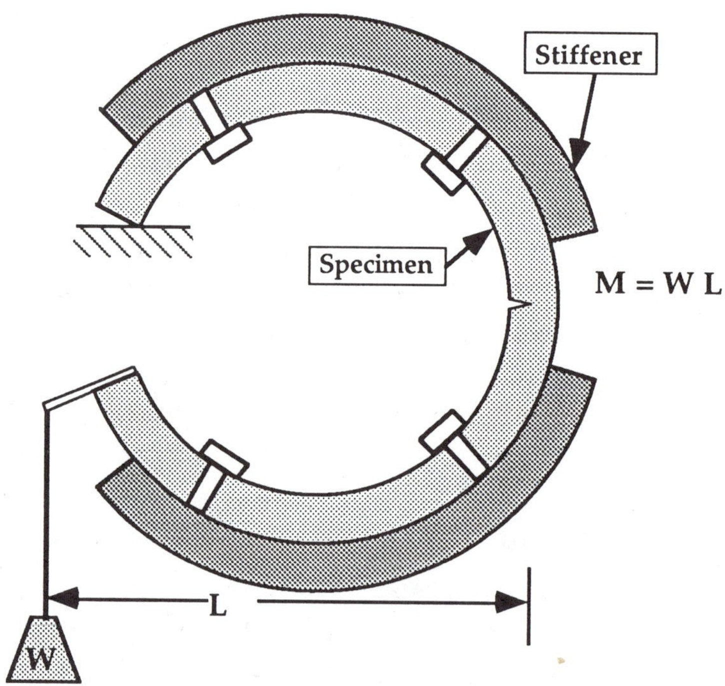

FIGURE 8.15 Loading apparatus for evaluating the toughness of PVC-U pipe according to BS 3506 [16].

A standard for interlaminar fracture toughness does not exist as of this writing, but ASTM, the European Group on Fracture (EGF) and Japanese Industrial Standards (JIS) are currently developing standardized test procedures for carbon/Epoxy and carbon/PEEK composites. The published literature contains a large amount of $\mathcal{G}_{IC}$ and $\mathcal{G}_{IIC}$ data for composites, but test methods differ widely between laboratories.

Figure 8.16 illustrates three common specimen configurations for interlaminar fracture toughness measurements. The double cantilever beam (DCB) specimen is probably the most common configuration for this type of test. One advantage of this specimen geometry is that it permits measurements of Mode I, Mode II or mixed mode fracture toughness. The end notched flexure (ENF) specimen has essentially the same geometry as the DCB specimen, but the latter is loaded in three-point bending, which imposes Mode II displacements of the crack faces. The edge delamination specimen simulates the conditions in an actual structure. Recall from Chapter 6 that tensile stresses normal to the ply are highest at the free edge (Fig. 6.16); thus delamination zones often initiate at the edges of a panel.

Procedures for measuring interlaminar toughness with DCB specimens are outlined below; analogous methods can be applied to other specimen configurations. The approaches that follow are not definitive test methods, but are representative of current practice [17-20].

The initial flaw in a DCB specimen is normally introduced by placing a thin film (e.g. aluminum foil) between plies prior to molding. The film should be coated with a release agent so that it can be removed prior to testing.

Figure 8.17 illustrates two common fixtures that facilitate loading the DCB specimen. The blocks or hinges are normally adhesively bonded to the specimen. These fixtures must allow free rotation of the specimen ends with a minimum of stiffening.

The DCB specimen can be tested in Mode I, Mode II, or mixed-mode conditions, as Fig. 8.18 illustrates. Recall from Chapter 2 that the energy release rate of this specimen configuration can be inferred from beam theory.

For pure Mode I loading (Fig. 8.18(a)), elastic beam theory leads to the following expression for energy release rate (see Example 2.2):

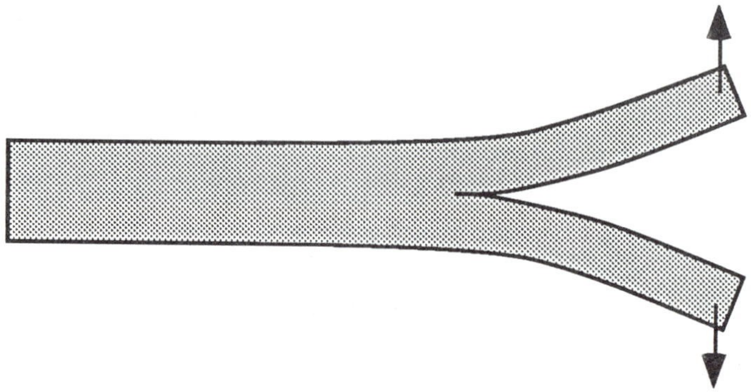

(a) Double cantilever beam specimen.

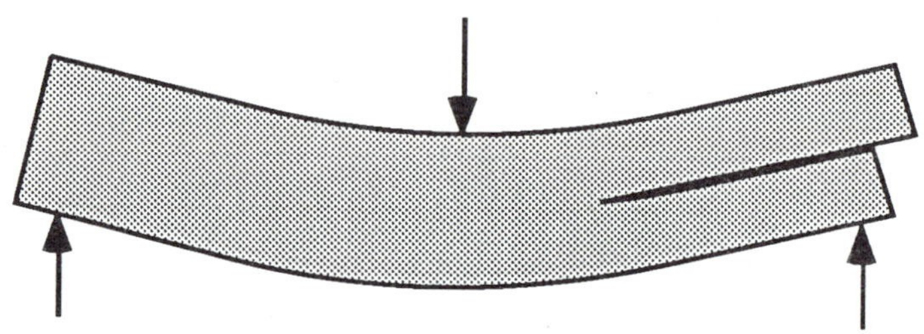

(b) End notched flexure specimen.

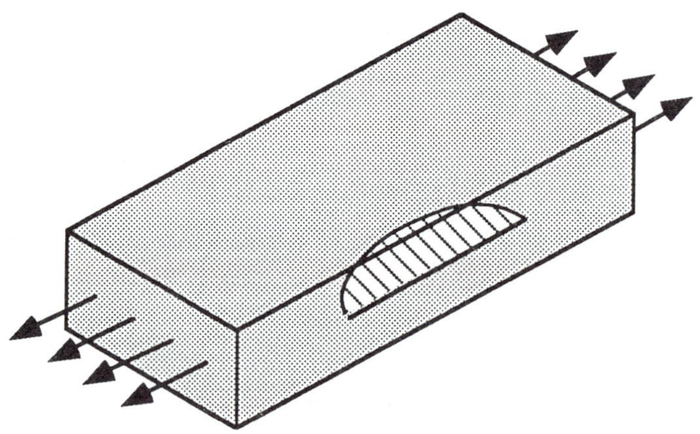

(c) Edge delamination specimen.

FIGURE 8.16 Common configurations for evaluating interlaminar fracture toughness.

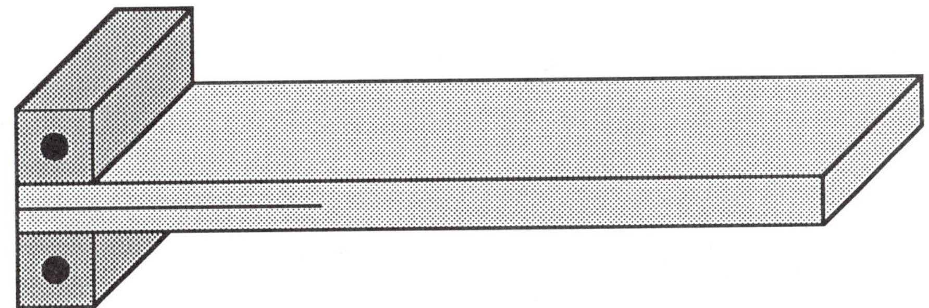

(a) End blocks

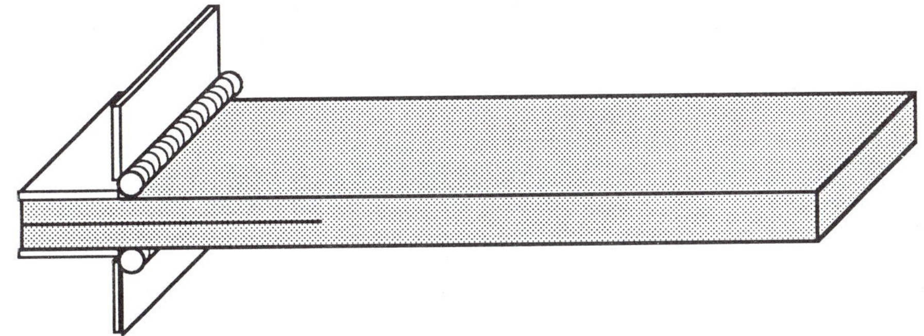

(b) Piano hinges

FIGURE 8.17 Loading fixtures for DCB specimens.

$$\mathcal{G}_I = \frac{P_I^2 a^2}{B E I} \tag{8.38}$$

where

$$E I = \frac{2 P_I a^3}{3 \Delta_I} \tag{8.39}$$

The corresponding relationship for Mode II (Fig 8.18(b)) is given by

$$\mathcal{G}_{II} = \frac{3 P_{II}^2 a^2}{4 B E I} \tag{8.40}$$

assuming linear beam theory. Mixed loading conditions can be achieved by unequal tensile loading of the upper and lower portions of the specimens, as Fig. 8.18(c) illustrates. The applied loads can be resolved into Mode I and Mode II components as follows:

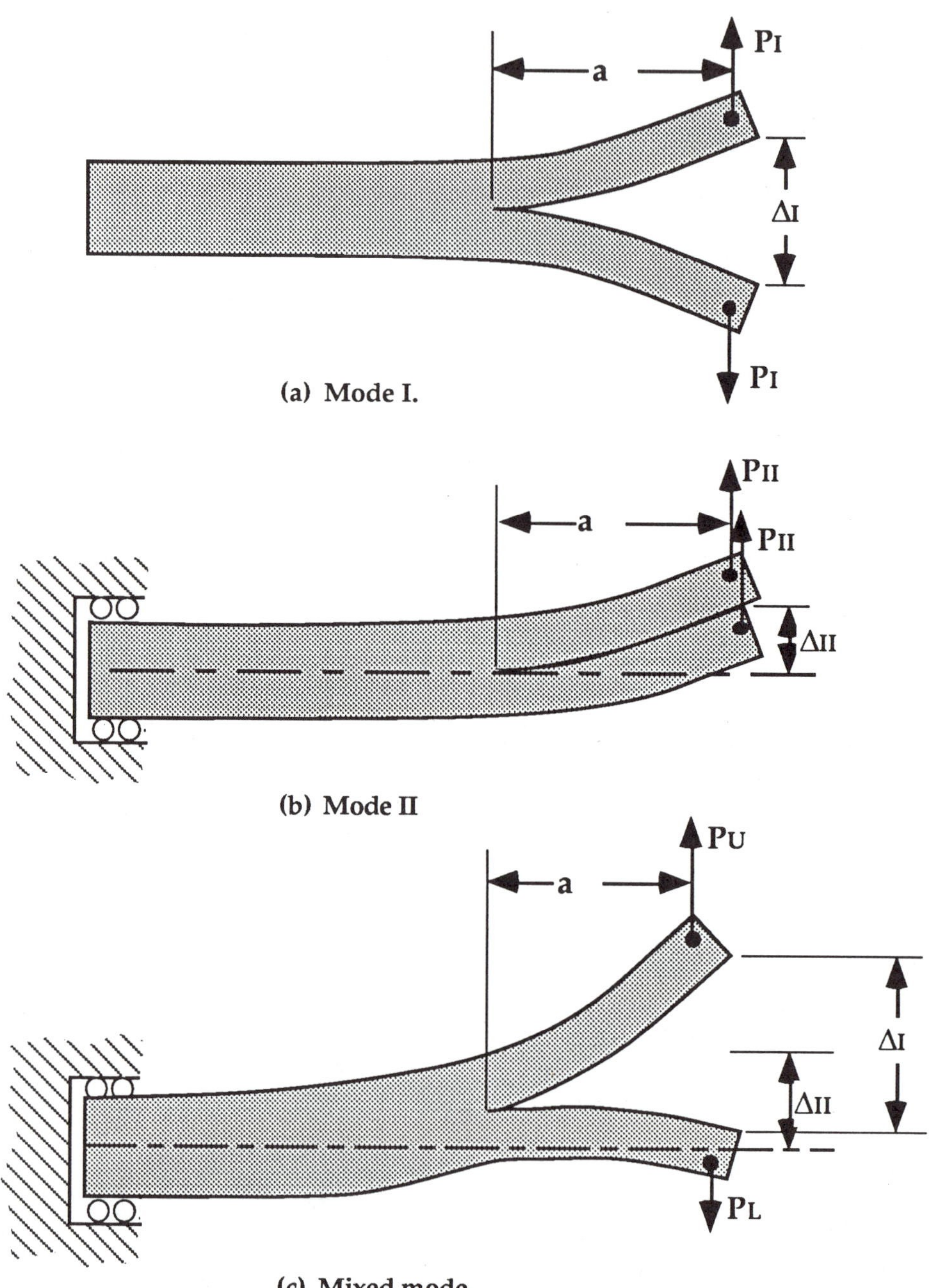

(a) Mode I.

(b) Mode II

(c) Mixed mode.

FIGURE 8.18 Mode I, II and mixed mode loading of DCB specimens.

$$P_I = |P_L| \tag{8.41a}$$

$$P_{II} = \frac{|P_U| - |P_L|}{2} \tag{8.41b}$$

where P_U and P_L are the upper and lower loads, respectively. The components of G can be computed by inserting P_I and P_{II} into Eqs. (8.38) and (8.40). Recall from Chapter 2 that Mode I and Mode II components of energy release rate are additive.

Linear beam theory may result in erroneous estimates of energy release rate, particularly when the specimen displacements are large. The area method [19-20] provides an alternative measure of energy release rate. Figure 8.19 schematically illustrates a typical load-displacement curve, where the specimen is periodically unloaded. The loading portion of the curve is typically nonlinear, but the unloading curve is usually linear and passes through the origin. The energy release rate can be estimated from the incremental area inside the load displacement curve, divided by the change in crack area:

$$\mathcal{G} = \frac{\Delta U}{B\, \Delta a} \tag{8.42}$$

The Mode I and Mode II components of G can be inferred from the P_I-Δ_I and P_{II}-Δ_{II} curves, respectively. The corresponding loads and displacements for Modes I and II are defined in Fig. 8.18 and Eq. (8.41).

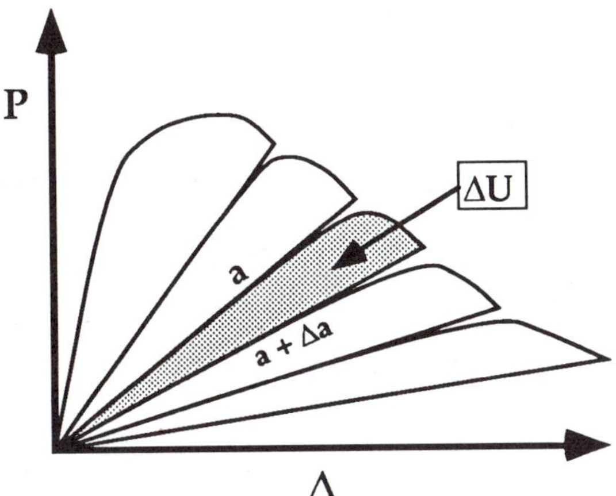

FIGURE 8.19 Schematic load-displacement curve for a delamination toughness measurement.

Figure 8.20 illustrates a typical delamination resistance curve for Mode I. After initiation and a small amount of growth, delamination occurs at a steady-state G_{IC} value, provided the global behavior of the specimen is elastic.

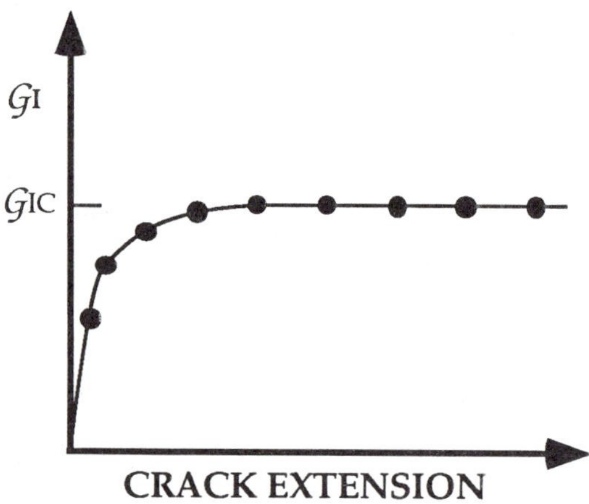

FIGURE 8.20 Schematic R curve inferred from a delamination experiment.

8.3 CERAMICS

Fracture toughness is usually the limiting property in ceramic materials. Ceramics tend to have excellent creep properties and wear resistance, but are excluded from many load-bearing applications because they are relatively brittle. The latest generation of ceramics (see Section 6.2) have enhanced toughness, but brittle fracture is still a primary area of concern in these materials.

Because toughness is a crucial property for ceramic materials, rational fracture toughness measurements are absolutely essential. Unfortunately, fracture toughness tests on ceramics can be very difficult and expensive. Specimen fabrication, for example, requires special grinding tools, since ordinary machining tools are inadequate. Precracking by fatigue is extremely time-consuming; some investigators have reported precracking times in excess of one week per specimen [21]. During testing, it is difficult to achieve stable crack growth with most specimen configurations and testing machines.

Several test methods have been developed to overcome some of the difficulties associated with fracture toughness measurements in ceramics. The chevron-notched specimen [22-24] eliminates the need for precracking, while the bridge indentation approach [21,25-29] is a novel method for introducing a crack without resorting to a lengthy fatigue precracking process.

8.3.1 Chevron-Notched Specimens

A chevron notch has a V-shaped ligament, such that the notch depth varies through the thickness, with the minimum notch depth at the center. Figure 8.21 shows two common configurations of chevron-notched specimens: the short bar and the short rod. In addition, single edge notched bend (SENB) and compact specimens (Fig. 7.1) are sometimes fabricated with chevron notches. The chevron notch is often utilized in conventional fracture toughness tests on metals because this shape facilitates initiation of the fatigue precrack. For fracture toughness tests on brittle materials, the unique properties of the chevron notch can eliminate the need for precracking altogether, as discussed below.

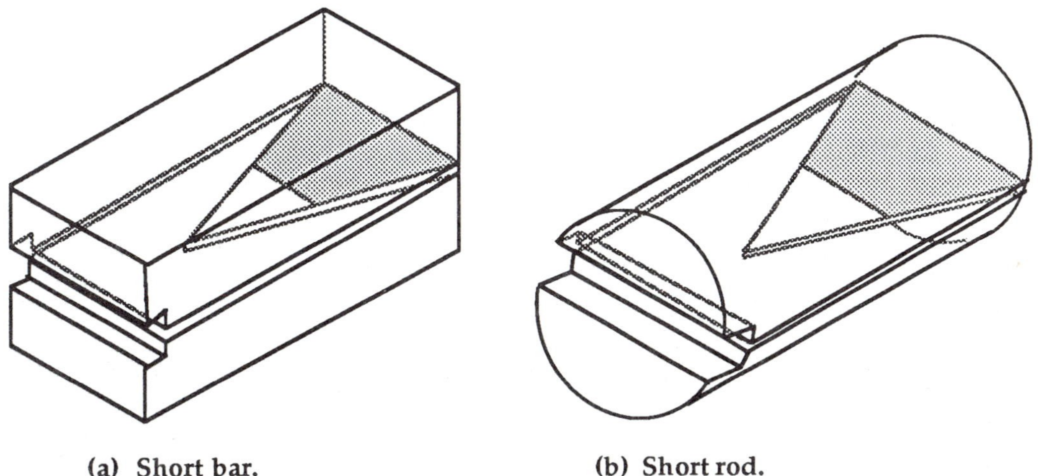

(a) Short bar. **(b) Short rod.**

FIGURE 8.21 Two common designs of chevron notched specimens [22].

Figure 8.22 schematically compares the stress intensity factor versus crack length for chevron and straight notch configurations. When the crack length = a_o, the stress intensity factor in the chevron-notched specimen is very high, because a finite load is applied over a very small net thickness. When $a \geq a_1$, the K_I values for the two notch configurations are identical, since the chevron notch no longer has an effect. The K_I for the chevron-notched specimen exhibits a minimum at a particular crack length, a_m, which is between a_o and a_1.

The K_I v. crack length behavior of the chevron-notched specimen makes this specimen particularly suitable for measuring the toughness in brittle materials. Consider a material in which the R curve reaches a steady-state plateau soon after the crack initiates (Fig. 8.23). The crack should initiate at the tip of the chevron upon application of a small load, since the local K_I is high. The crack is stable at this point, because the driving force decreases rapidly with crack advance; thus additional load is required to grow the crack further. The maximum load in the test, P_M, is achieved when the crack grows to a_m, the crack length corresponding to the minimum in the K_I-a curve. At this point, the specimen will be unstable if the test is conducted in load control, but stable crack growth may be possible beyond a_m if the specimen is subject to crosshead control. The point of instability in the latter case depends on the compliance of the testing machine, as discussed in Section 2.5.

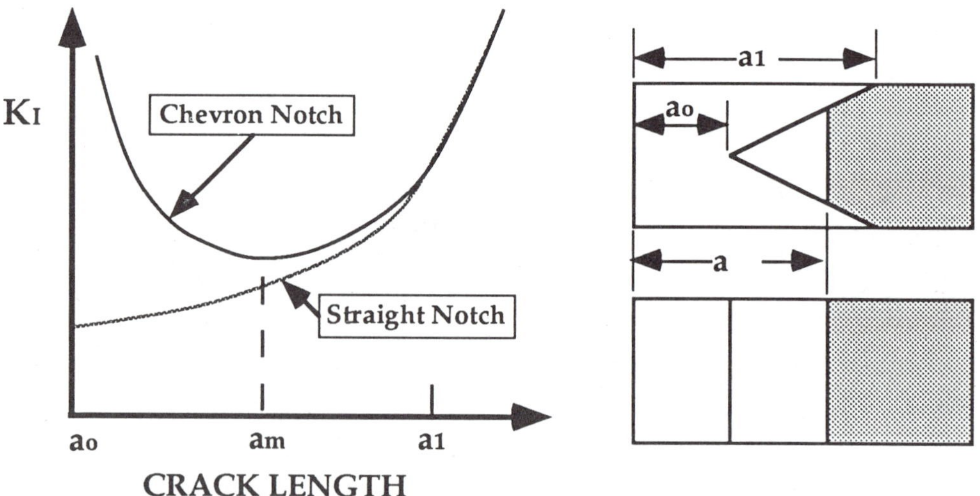

FIGURE 8.22 Comparison of stress intensity factors in specimens with chevron and straight notches. Note that K_I exhibits a minimum in the chevron notched specimens.

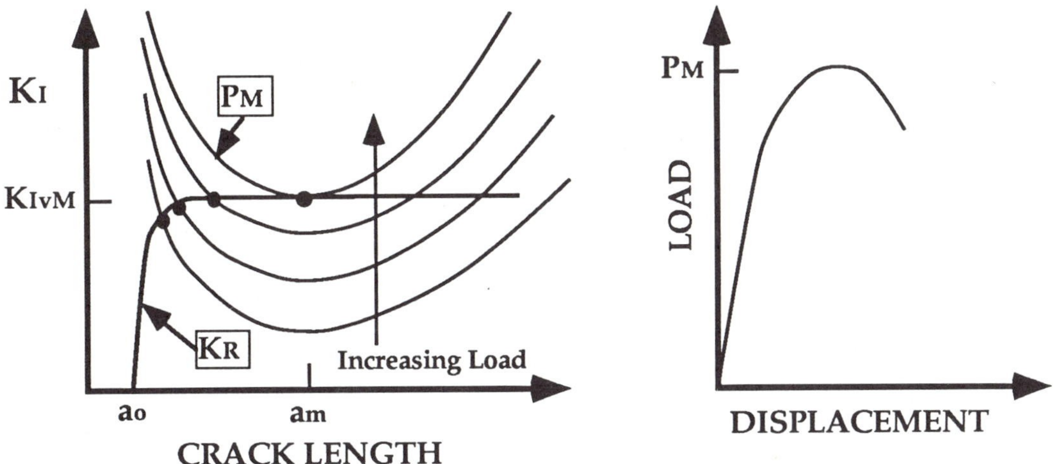

FIGURE 8.23 Fracture toughness testing of a material with a flat R curve. The maximum load in the test occurs when $a = a_m$.

Since the maximum load occurs at a_m, and a_m is known a priori (from the K_I v. crack length relationship), it is necessary only to measure the maximum load in this test. The fracture toughness is given by

$$K_{IvM} = \frac{P_M}{B\sqrt{W}} f(a_m/W) \tag{8.43}$$

where K_{IvM} is the chevron-notched toughness defined at maximum load, and f(a/W) is the geometry correction factor. Early researchers developed simple models to estimate f(a/W) for chevron-notched specimens, but more recent (and more accurate) estimates are based on three-dimensional finite element and boundary element analysis of this configuration [23].

The maximum load technique for inferring toughness does not work as well when the material exhibits a rising R curve, as Fig. 8.24 schematically illustrates. The point of tangency between the driving force and R curve may not occur at a_m in this case, resulting in an error in the stress intensity calculation. Moreover, the value of K_R at the point of tangency is geometry dependent when the R curve is rising.

If both load and crack length are measured throughout the test, it is possible to construct the R curve for the material under consideration. Optical observation of the growing crack is not usually feasible for a chevron-notched specimen, but the crack length can be inferred through an unloading compliance technique [22], in which the specimen is periodically unloaded and the crack length is computed from the elastic compliance.

An ASTM standard for chevron notched specimens, E 1304-90 [22], has recently been published. This standard actually applies to brittle metals, such as high strength aluminum alloys, but a corresponding standard for ceramics is currently under consideration. The E 1304 standard includes both the maximum load and compliance measures of fracture toughness, which are designated K_{IvM} and K_{Iv}, respectively. A number of researchers have measured fracture toughness of chevron-notched ceramic specimens with test techniques that are virtually identical to the provisions in ASTM E 1304.

The chevron-notched specimen has proved to be very useful in characterizing the toughness of brittle materials. The advantages of this test specimen include its compact geometry, the simple instrumentation requirements (in the case of the K_{IvM} measurement), and the fact that no precracking is required. One of the disadvantages of this specimen is its complicated design, which leads to higher machining costs. Also, this specimen is poorly suited to high temperature testing, and the K_{IvM} measurement is inappropriate for material with rising R curves.

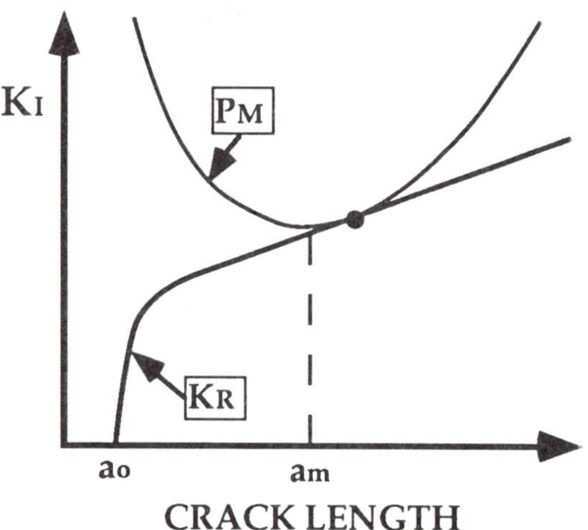

FIGURE 8.24 Application of the chevron-notched specimen to a material with a rising R curve.

8.3.2 Bend Specimens Precracked by Bridge Indentation

A novel technique for precracking ceramic SENB specimens has recently been developed in Japan [25]. A number of researchers [21,25-29] have adopted this method, which has been incorporated into an upcoming Japanese standard for fracture toughness testing of ceramics. Warren, et al. [26], who were among the first to apply this precracking technique have termed it the "bridge indentation" method.

Figure 8.25 is a schematic drawing of the loading fixtures for the bridge indentation method of precracking. A starter notch is introduced into an SENB specimen by means of a Vickers hardness indentation. The specimen is compressed between two anvils, as Fig. 8.25 illustrates. The top anvil is flat, while the bottom anvil has a gap in the center. This arrangement induces a local tensile stress in the specimen, which leads to a pop-in fracture. The fracture arrests because the propagating crack experiences a falling K field.

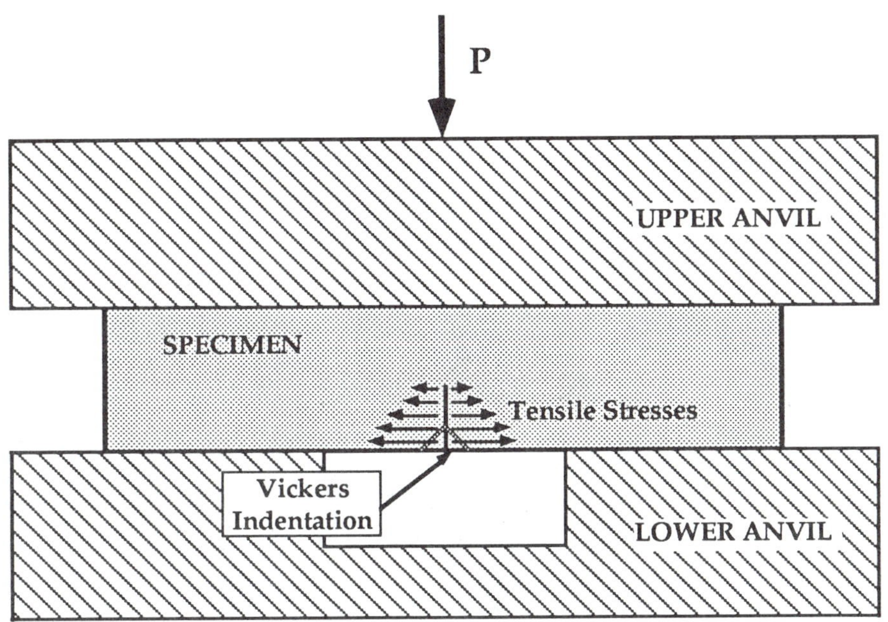

FIGURE 8.25 The bridge indentation method for precracking [21].

The bridge indentation technique is capable of producing highly uniform crack fronts in SENB specimens. After precracking, these specimens can be tested in three- or four-point bending with conventional fixtures. Nose and Fujii [21] showed that fracture toughness values obtained from bridge precracked specimens compared favorably with data from conventional fatigue precracked specimens.

Bar-On, et al. [27] investigated the effect of precracking variables on the size of the crack that is produced by this technique. Figure 8.26 shows that the length of the pop-in in alumina decreases with increasing Vickers indentation load. Also note that the pop-in load decreases with increasing indentation load. Large Vickers indentation loads produce significant initial flaws and tensile residual stresses, which enable the pop-in to initiate at a lower load; the crack arrests sooner at lower loads because there is less elastic energy available for crack propagation. Thus it is possible to control the length of the precrack though the Vickers indentation load.

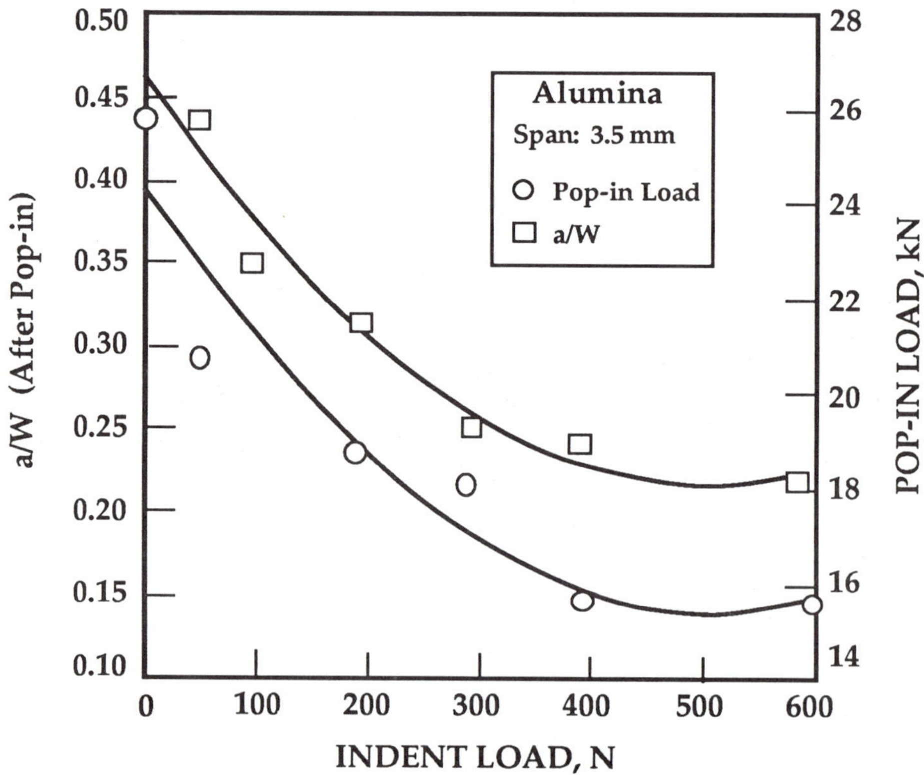

FIGURE 8.26 Effect of bridge indentation load on the crack length after pop-in [27].

The bridge indentation technique is obviously much more economical than fatigue precracking of ceramic specimens. The SENB configuration is simple, and therefore less expensive to fabricate. Also, three- and four-point bend fixtures are suitable for high temperature testing. One problem with the SENB specimen is that it consumes more material than the chevron notched specimens illustrated in Fig. 8.21; this is a major shortcoming when evaluating new materials, where only small samples are available. Another disadvantage of the beam configuration is that it tends to be unstable; most test machines are too compliant to achieve stable crack growth in brittle SENB specimens [28,29].

REFERENCES

1. Schapery, R.A., "Correspondence Principles and a Generalized J Integral for Large Deformation and Fracture Analysis of Viscoelastic Media." *International Journal of Fracture*, Vol. 25, 1984, pp. 195-223.

2. Schapery, R.A., "Time-Dependent Fracture: Continuum Aspects of Crack Growth." *Encyclopedia of Materials Science and Engineering*, Pergamon Press, Oxford, 1986, pp. 5043-5054.

3. Jones, R.E. and Bradley, W.L., "Fracture Toughness Testing of Polyethylene Pipe Materials." ASTM STP 995, Vol. 1, 1989, American Society for Testing and Materials, Philadelphia,PA, pp. 447-456.

4. Cayard, M., "Fracture Toughness Testing of Polymeric Materials." Ph.D. Dissertation, Texas A&M University, College Station, TX, September, 1990.

5. Williams, J.G. *Fracture Mechanics of Polymers*, Halsted Press, John Wiley & Sons, New York, 1984.

6. Letton, A., "The Use of Specimen Compliance in Predicting Crack Growth." Submitted to *Polymer Science and Engineering*, 1991.

7. Marshall, G.P., Coutts, L.H., and Williams, J.G., "Temperature Effects in the Fracture of PMMA." *Journal of Materials Science*, Vol. 13, 1974, pp. 1409-

8. Schapery, R.A. "A Theory of Crack Initiation and Growth in Viscoelastic Media--I. Theoretical Development." *International Journal of Fracture*, Vol 11, 1975, pp. 141-159.

9. Schapery, R.A., Private communication, 1990.

10. "Standard Test Method for Plane Strain Fracture Toughness and Strain Energy Release Rate of Plastic Materials." (Draft) American Society for Testing and Materials, Philadelphia, PA, August 1989.

11. E 399-83, "Standard Test Method for Fracture Toughness of Metallic Materials." American Society for Testing and Materials, Philadelphia, PA, 1983.

12. E 813-87, "Standard Test Method for J_{IC}, a Measure of Fracture Toughness." American Society for Testing and Materials, Philadelphia, PA, 1987.

13. D 256-88, "Impact Resistance of Plastics and Electrical Insulating Materials." American Society for Testing and Materials, Philadelphia, PA, 1988.

14. E 23-88, "Standard Test Methods for Notched Bar Impact Testing of Metallic Materials." American Society of Testing and Materials, Philadelphia, PA, 1988.

15. *Engineered Materials Handbook, Volume 2: Engineering Plastics.* ASM International, Metals Park, OH, 1988.

16. BS 3505:1986, "British Standard Specification for Unplasticized Polyvinyl Chloride (PVC-U) Pressure Pipes for Cold Potable Water." British Standards Institution, London, 1986.

17. Whitney, J.M., Browning, C.E., and Hoogsteden, W., "A Double Cantilever Beam Test for Characterizing Mode I Delamination of Composite Materials." *Journal of Reinforced Plastics and Composites,* Vol. 1, 1982, pp. 297-313.

18. Prel, Y.J., Davies, P., Benzeggah, M.L., and de Charentenay, F.-X., "Mode I and Mode II Delamination of Thermosetting and Thermoplastic Composites." ASTM STP 1012, American Society for Testing and Materials, Philadelphia, PA, 1989, pp. 251-269.

19. Corleto, C.R. and Bradley, W.L., "Mode II Delamination Fracture Toughness of Unidirectional Graphite/Epoxy Composites." ASTM STP 1012, American Society for Testing and Materials, Philadelphia, PA, 1989, pp. 201-221.

20. Hibbs, M.F., Tse, M.K., and Bradley, W.L., "Interlaminar Fracture Toughness and Real-Time Fracture Mechanism of Some Toughened Graphite/Epoxy Composites." ASTM STP 937, American Society for Testing and Materials, Philadelphia, PA, 1987, pp. 115-130.

21. Nose, T. and Fujii, T., "Evaluation of Fracture Toughness for Ceramic Materials by a Single-Edge-Precracked-Beam Method." *Journal of the American Ceramic Society,* Vol. 71, 1988, pp. 328-333.

22. E 1304-89, "Standard Test Method for Plane-Strain (Chevron Notch) Fracture Toughness of Metallic Materials." American Society for Testing and Materials, Philadelphia, PA, 1989.

23. Newman, J.C., "A Review of Chevron-Notched Fracture Specimens." ASTM STP 855, American Society for Testing and Materials, Philadelphia, PA, 1984, pp. 5-31.

24. Shannon, J.L., Jr. and Munz, D.G., "Specimen Size Effects on Fracture Toughness of Aluminum Oxide Measured with Short-Rod and Short Bar Chevron-Notched Specimens." ASTM STP 855, American Society for Testing and Materials, Philadelphia, PA, 1984, pp. 270-280.

25. Nunomura, S. and Jitsukawa, S. "Fracture Toughness for Bearing Steels by Indentation Cracking under Multiaxial Stress." (In Japanese) *Tetsu to Hagane,* Vol. 64, 1978.

26. Warren, R. and Johannsen, B. "Creation of Stable Cracks in Hard Metals Using 'Bridge' Indentation." *Powder Metallurgy,* Vol. 27, 1984, pp. 25-29.

27. Bar-On, I., Beals, J.T., Leatherman, G.L., and Murray, C.M., "Fracture Toughness of Ceramic Precracked Bend Bars." *Journal of the American Ceramic Society,* Vol. 73, 1990, pp. 2519-2522.

28. Barratta, F.I. and Dunlay, W.A., "Crack Stability in Simply Supported Four-Point and Three-Point Loaded Beams of Brittle Materials." *Proceedings of the Army Symposium on Solid Mechanics, 1989 - Mechanics of Engineered Materials and Applications,* U.S. Army Materials Technology Laboratory, Watertown, MA, 1989, pp. 1-11.

29. Underwood, J.H., Barratta, F.I., and Zalinka, J.J., "Fracture Toughness Tests and Displacement and Crack Stability Analyses of Round Bar Bend Specimens of Liquid-Phase Sintered Tungsten." *Proceedings of the 1990 SEM Spring Conference on Experimental Mechanics,* Albuquerque, NM, 1990, pp. 535-542.

9. APPLICATION TO STRUCTURES

Figure 9.1 illustrates the so-called fracture mechanics triangle. When designing a structure against fracture, there are three critical variables that must be considered: stress, flaw size, and toughness. Fracture mechanics provides a mathematical relationship between these quantities. In most cases there are two degrees of freedom (i.e., one equation and three unknowns); a knowledge of two quantities is required to compute the third. For example, if the stress is specified by the design and the material toughness is known, fracture mechanics relationships can predict the critical flaw size in the structure.

A number of relationships are available that attempt to quantify the critical relationship between stress, flaw size, and toughness, but each of these approaches is only suitable in limited situations. Linear elastic fracture models, for example, should not be applied to structures that exhibit significant plastic flow.

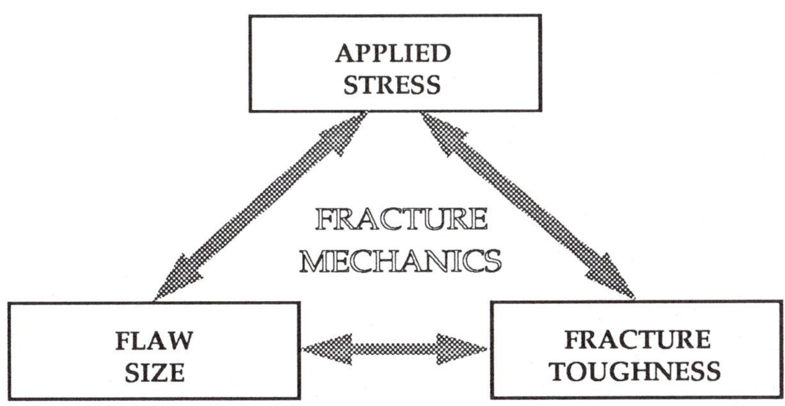

FIGURE 9.1 The fracture mechanics triangle, which identifies the three critical variables in fracture design.

The fracture design methodology should be selected based on the available data, material properties, environment, and the loading on the structure. If K_{IC} data are available and the design stress is low, linear elastic fracture mechanics (LEFM) may be appropriate. If the behavior of the structure is linear elastic but the laboratory fracture toughness tests behave in an elastic-plastic manner, it may be possible to convert a critical J value to the equivalent K_{IC} and analyze the structure with linear elastic relationships (Section 9.1.4). If the structure and test specimen both behave in an elastic-plastic fashion, J- and CTOD-based analyses are available (Section 9.3 to 9.7). Rapid loading may require special consideration, as will design for crack arrest. Time-dependent crack growth, such as fatigue, environmental assisted cracking, and creep crack growth may complicate the analysis further.

Most fracture analyses are deterministic. That is, the stress, flaw size, and toughness are assumed to be single-valued quantities. In practical situations, however, there is usually some degree of uncertainty associated with each of these variables. Consequently, it is usually not possible to predict the precise moment of failure. Probabilistic analyses (Section 9.9) can quantify the *risk* of failure, however.

This chapter focuses on fracture initiation and instability in structures made from linear elastic and elastic-plastic materials. A number of engineering approaches are discussed; the basis of these approaches and their limitations are explored. This chapter only covers quasistatic methodologies, but such approaches can be applied to rapid loading and crack arrest in certain circumstances (see Chapter 4). The analyses presented in this chapter do not address time-dependent crack growth. Chapter 10 considers fatigue crack growth in detail, and describes life predictions for all types of time-dependent crack growth.

9.1 LINEAR ELASTIC FRACTURE MECHANICS

Analyses based on LEFM apply to structures where crack tip plasticity is small. Chapter 2 introduced many of the fundamental concepts of LEFM. The fracture behavior of a linear elastic structure can be inferred by comparing the applied K (the driving force) to a critical K or a K-R curve (the fracture toughness). The elastic energy release rate, $\mathcal{G}$, is an alternative measure of driving force, and a critical value of $\mathcal{G}$ quantifies the material toughness.

For Mode I loading (Fig. 2.14), the stress intensity factor can be expressed in the following form:

$$K_I = C\sigma\sqrt{\pi a} \tag{9.1}$$

where C is a dimensionless geometry correction factor, σ is a characteristic stress, and a is a characteristic crack dimension. If the geometry factor is known, the applied K_I can be computed for any combination of σ and a. The applied stress intensity can then be compared to the appropriate material property, which could be a K_{IC} value, a K-R curve, environmental assisted cracking data, or, in the case of cyclic loading, fatigue crack growth data (see Chapter 10).

Fracture analysis of a linear elastic structure becomes relatively straightforward, once a K solution is obtained for the geometry of interest. Stress intensity solutions can come from a number of sources, including handbooks, the published literature, experiments, and numerical analysis.

A large number of stress intensity solutions have been published over the past 35 years. Several handbooks [1-3] contain compilations of solutions for a wide variety of configurations. The published literature contains many more solutions. It is usually possible to find a K solution for a geometry that is similar to the structure of interest.

When a published K solution is not available, or the accuracy of such a solution is in doubt, one can obtain the solution experimentally or numerically. (Deriving a closed-form solution is probably not a viable alternative, since this is only possible with simple geometries, and nearly all such solutions have already been published.) Experimental measurement of K is possible through optical techniques, such as photoelasticity [4,5] and the method of caustics [6], or by determining $\mathcal{G}$ from the rate of change in compliance with crack length (Eq. (2.30)) and computing K from $\mathcal{G}$ (Eq. (2.58)). Chapter 11 describes a number of computational techniques for deriving stress intensity.

An alternative is to utilize the principle of elastic superposition, which enables new K solutions to be constructed from known cases. Section 2.6.4 outlined this approach, and applied the principle of superposition to a pressure loaded semicircular surface crack (Example 2.5). Section 2.6.4 also briefly introduced the concept of influence coefficients, which stems directly from the principal of superposition.

The influence coefficient approach [7] enables stress intensity factors to be inferred for a given geometry with arbitrary loading, provided the crack is loaded in a single mode (Mode I, II, or III). Weight functions [8,9] provide an alternative technique that enables stress intensity factors to be inferred for arbitrary loading under mixed mode conditions. This approach, however, is somewhat more complicated than influence coefficients, which are based on elastic superposition. The influence approach is outlined below for the case of Mode I loading; several examples of this technique are included. Chapter 11 describes the weight function approach.

9.1.1 Obtaining K_I for Arbitrary Loading

Many structures contain regions where the stresses are highly nonuniform. Stress gradients, for example, occur near stress raisers such as holes and notches. Residual stresses produced by welding or other means are invariably nonuniform through the cross section. Most "handbook" stress intensity solutions apply only to simple loading, such as uniform tension or bending.

In the case of Mode I loading, an approximate method for accounting for stress gradients is to linearize the stress distribution and divide it into membrane (tensile) and bending components, as Fig. 9.2 illustrates. Separate stress intensity factors can be obtained for the membrane and bending stresses, and the two K_I values can then be added to obtain the total K_I. Consider a semielliptical surface flaw in a plate of finite thickness (Fig. 9.3). The Mode I stress intensity factor at $\phi = \pi/2$ for combined bending and membrane loading is given by

$$K_I = (\sigma_m + H\,\sigma_b)\,F\sqrt{\frac{\pi a}{Q}} \tag{9.2}$$

where Q is the flaw shape parameter, which is based on the solution of an elliptical integral of the second kind (see Fig. 2.19, Eq. 2.67, and Table 12.22), and F and H are geometry constants, which Newman and Raju [10] obtained from finite element analysis. The parameters F and H depend on a/c, a/t, and ϕ (see Fig. 9.3), and plate width. Table 12.22 gives polynomial expressions for F and H.

Equation (9.2) is reasonably flexible, since it can account for a range of stress gradients, and includes pure tension and pure bending as special cases. This equation, however, is actually a special case of the influence coefficient approach, an example of which is described below.

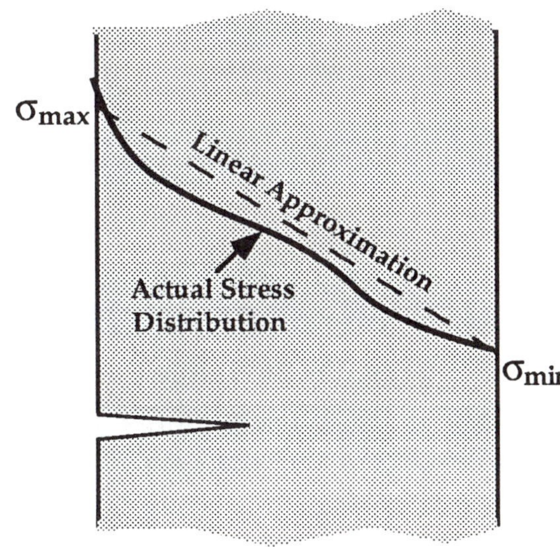

$$\sigma_m = \frac{\sigma_{max} + \sigma_{min}}{2}$$

$$\sigma_b = \frac{\sigma_{max} - \sigma_{min}}{2}$$

FIGURE 9.2 Approximating a nonuniform stress distribution as linear, and resolving the stresses into membrane and bending components.

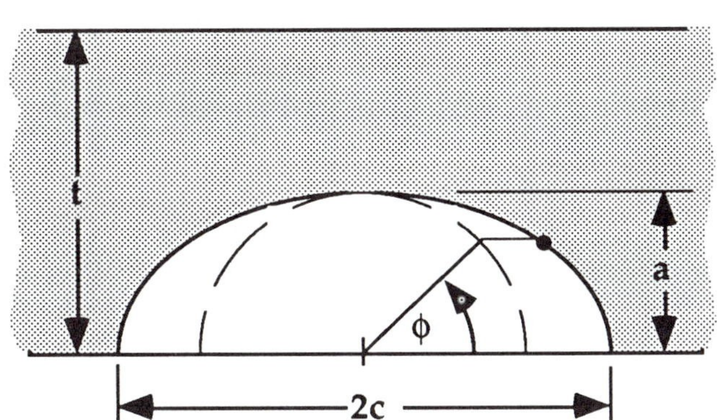

FIGURE 9.3 Semi-elliptical surface crack in a plate with finite thickness.

Suppose the remote stress normal to the crack plane can be represented by a cubic equation:

$$\sigma_{yy} = A_o + A_1 x + A_2 x^2 + A_3 x^3$$

$$= \sum_{j=0}^{3} A_j x^j \tag{9.3}$$

Figure 9.4 schematically illustrates this stress distribution, and defines the coordinate axes. The stress intensity factor that results from the above equation can be constructed by obtaining K_I solutions for power-law loading, where the exponent ranges from 0 to 3. The following dimensionless stress distributions can be applied to the crack face independently in a finite element model of the crack geometry:

$$\sigma_j = \left(\frac{x}{a}\right)^j \qquad \text{for } j = 0, 1, 2, \text{ or } 3 \tag{9.4}$$

Figure 9.5 illustrates application of a power-law stress distribution to the crack faces, which is equivalent to applying this stress field remote from the crack. Note that only the stresses that act in the range $0 \le x \le a$ need be considered; normal stresses at $x > a$ do not contribute to K_I. This last statement can be proven by invoking a superposition argument similar to that in Example 2.5.

Raju and Newman [7] applied power law stress distributions to a wide range of semi-elliptical surface flaws in cylinders, as Fig. 9.6 illustrates. They considered t/R_i ratios of 0 (flat plate), 0.10, and 0.25. Their analysis included both internal flaws and external flaws. For the stress distribution in Eq. (9.4), the stress intensity factor can be expressed as follows:

$$K_I = \sqrt{\frac{\pi a}{Q}}\, G_j\left(\frac{a}{c}, \frac{a}{t}, \frac{t}{R_i}, \phi\right) \tag{9.5}$$

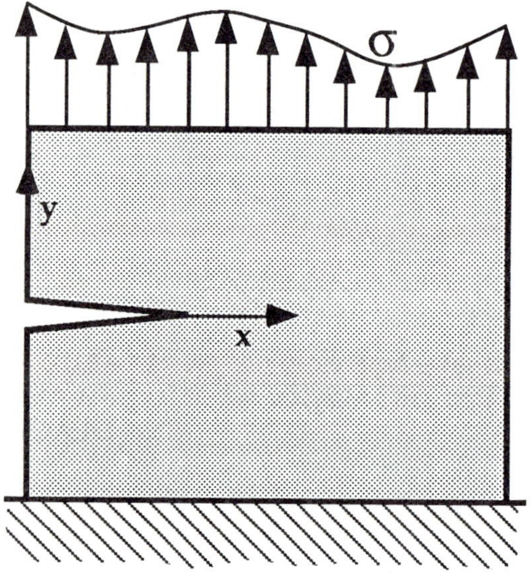

FIGURE 9.4 Arbitrary stress distribution which can be fit to a four-term polynomial (Eq. (9.3)).

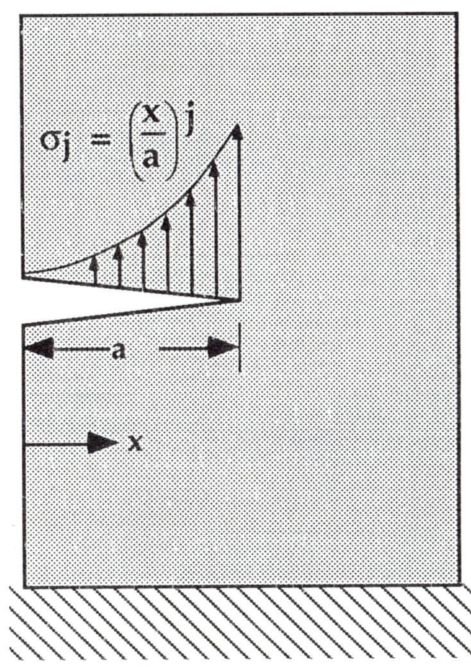

FIGURE 9.5 Power-law stress distribution applied to the crack face.

for j = 0, 1, 2, or 3, where G_j is an influence coefficient. Chapter 12 lists influence coefficients for various flaw geometries. For a given flaw shape and j value, G_j is relatively insensitive to the t/R_i ratio, which is indicative of the curvature of at the free surface. Thus it is not critical to match the curvature of a structure exactly; the influence coefficients for a surface flaw in a flat plate (Table 12.23) should give reasonable estimates of K_I in most structures.

When an arbitrary stress distribution is approximated by Eq. (9.3), the contribution of each term in the polynomial can be summed to obtain the total K_I for the crack. Equations (9.4) and (9.5), however, must be scaled to the actual stress distribution. Equation (9.3) can be rewritten in the following form:

$$\sigma_{yy} = = \sum_{j=0}^{3} A_j a^j \left(\frac{x}{a}\right)^j \tag{9.6}$$

By comparing Eqs. (9.4) and (9.6), we see that the scaling factor for each term in the polynomial is $A_j a^j$. Therefore, the stress intensity factor for a cubic polynomial stress distribution is given by

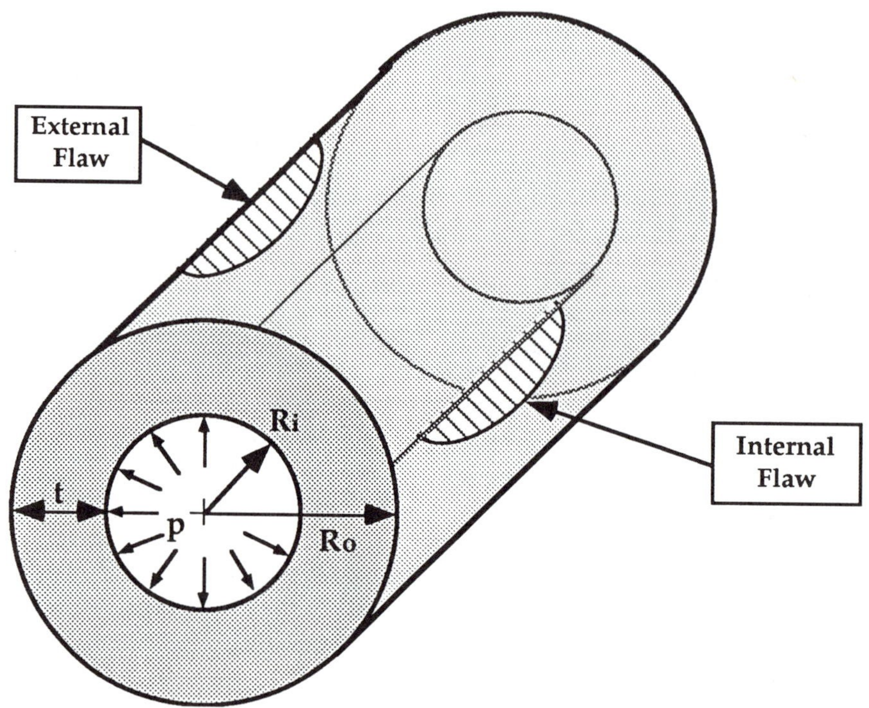

FIGURE 9.6 Internal and external axial surface flaws in a pressurized cylinder.

$$K_I = \sqrt{\frac{\pi a}{Q}} \sum_{j=0}^{3} G_j A_j a^j \tag{9.7}$$

Let us now consider the example of a pressurized cylinder with an internal axial surface flaw, as illustrated in Fig. 9.6. In the absence of the crack, the hoop stress in a thick wall pressure vessel is as follows [11]:

$$\sigma_{\theta\theta} = \frac{p R_i^2}{R_o^2 - R_i^2}\left[1 + \left(\frac{R_o}{r}\right)^2\right] \tag{9.8}$$

where p is the internal pressure and the other terms are defined in Fig. 9.6. If we define the origin at the inner wall ($x = r - R_i$) and perform a Taylor series expansion about $x = 0$, Eq. (9.8) becomes

$$\sigma_{\theta\theta} = \frac{p R_o^2}{R_o^2 - R_i^2}\left[1 + \left(\frac{R_o}{R_i}\right)^2 - 2\left(\frac{x}{R_i}\right) + 3\left(\frac{x}{R_i}\right)^2 - 6\left(\frac{x}{R_i}\right)^3 + \ldots\right] \tag{9.9}$$

The first four terms of this expansion give the desired cubic polynomial. An alternate approach would be to curve-fit a cubic polynomial to the stress field. This latter method is necessary when the stress distribution does not have a closed-form solution.

When computing K_I for the internal surface flaw, we must also take account of the pressure loading on the crack faces. Superimposing p on Eq. (9.9) and substituting the resulting coefficients (A_j) into Eq. (9.7) gives [7]:

$$K_I = \sqrt{\frac{\pi a}{Q}} \frac{p R_o^2}{R_o^2 - R_i^2} \left[2 G_o - 2\left(\frac{a}{R_i}\right) G_1 + 3\left(\frac{a}{R_i}\right)^2 G_2 - 4\left(\frac{a}{R_i}\right)^3 G_3 \right] \tag{9.10}$$

Applying a similar approach to an external surface flaw leads to [7]:

$$K_I = \sqrt{\frac{\pi a}{Q}} \frac{p R_i^2}{R_o^2 - R_i^2} \left[2 G_o + 2\left(\frac{a}{R_i}\right) G_1 + 3\left(\frac{a}{R_i}\right)^2 G_2 + 4\left(\frac{a}{R_i}\right)^3 G_3 \right] \tag{9.11}$$

The origin in this case was defined at the outer surface of the cylinder, and a series expansion was performed as before. Thus K_I for a surface flaw in a pressurized cylinder can be obtained by substituting the appropriate influence coefficients into Eq. (9.10) or Eq. (9.11).

Of course one can also infer the stress intensity factor by performing a full finite element analysis with the actual loading conditions. Figure 9.7 compares the K_I for an external crack estimated from Eq. (9.11) with a solution published by Atluri and Kathiresan [12]. The full solution and the estimate from influence coefficients differ by less than 10%. Raju and Newman [7] state that the discrepancies may be due to differences in numerical techniques, rather than inherent errors in the influence coefficient approach.

Influence coefficients are useful for inferring K_I values for cracks that emanate from stress concentrations. Figure 9.8 schematically illustrates a surface crack at the toe of a fillet weld. This geometry produces a local stress gradients that affect the K_I of the crack. Performing a three-dimensional finite element analysis of this structural detail with crack would be costly and time-consuming, and may be unnecessary. If the stress distribution in this detail is known for the uncracked

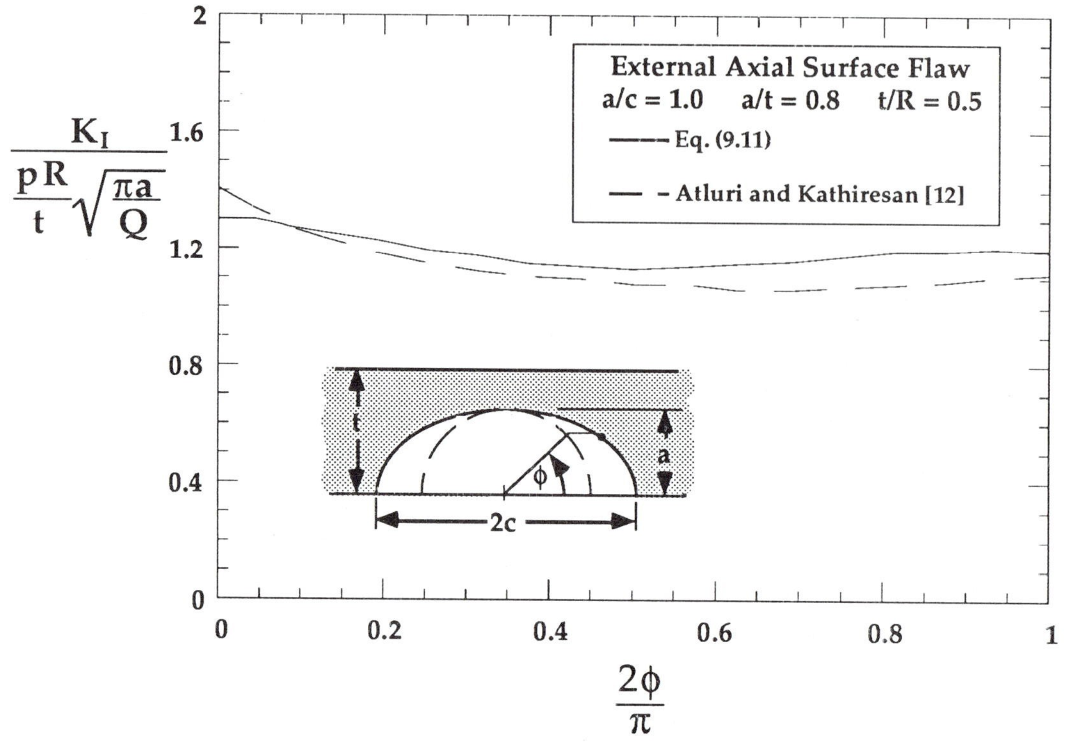

FIGURE 9.7 Comparison of stress intensity solutions from the influence coefficient approach [7] with finite element analysis of the actual geometry [12].

case, these stresses can be fit to a cubic polynomial (Eq. 9.3), and K_I can be estimated by substituting the influence coefficients into Eq. (9.7). The uncracked stress distribution can be inferred from a two-dimensional elastic finite element analysis with a relatively coarse mesh.

The previous example contains several sources of potential error. Since the influence coefficients in Chapter 12 were not derived from the fillet weld geometry, there may be slight errors if these Gj values are applied in this case. The influence coefficients for surface flaws depend on a/t and a/c, but are insensitive to the radius of curvature in the cross section. Thus as long as the crack shape and depth are taken into account, the G_j values in Chapter 12 should be reasonably accurate. One can minimize errors by applying the G_j values for an internal flaw in a cylinder, since the concave shape comes closest to matching the profile of the fillet weld. Another potential source of error is stress redistribution due to the presence of a flaw. In order to apply the influence coefficients to the stress field for the uncracked configuration, the cross sectional area of the flaw must be small compared to the cross section of the weld. Otherwise, the loads may redistribute, and the stress analysis of the unflawed configuration will no longer apply.

Since the flaw in Fig. 9.8 is near a weld, there is a possibility that weld residual stresses will be present. These stresses must be taken into account in order to obtain an accurate estimate of K_I. Weld residual stresses are an example of secondary stresses, as discussed below.

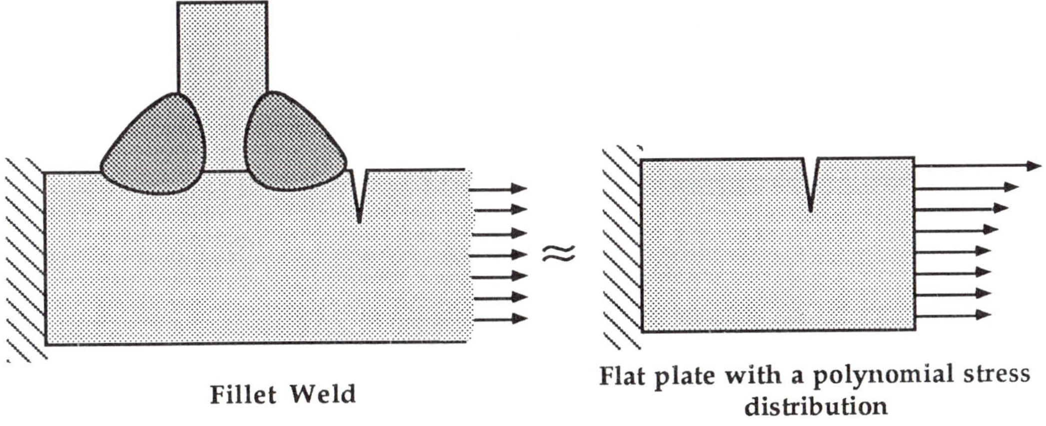

FIGURE 9.8 Application of the influence coefficient approach to a complex structural detail such as a fillet weld.

9.1.2 Primary and Secondary Stresses

The loading in a structure can be divided into primary and secondary stresses. Primary stresses generally arise from externally applied loads and moments, while secondary stresses are localized and are self-equilibrating through the cross section. Primary stresses, if sufficiently large, are capable of leading to plastic collapse, but secondary stresses cannot cause collapse of the structure. The latter can, however, contribute to fracture if large tensile secondary stress occur near a crack. Examples of secondary stresses include weld residual stresses and thermal stresses. In some cases, however, thermal loading can produce primary stresses. A stress should be classified as primary when it is not clear which category is appropriate.

In linear elastic analyses, primary and secondary stresses are treated in an identical fashion. The total stress intensity is simply the sum of the primary and secondary components:

$$K_I^{total} = K_I^p + K_I^s \tag{9.12}$$

where the superscripts p and s denote primary and secondary quantities, respectively.

The distinction between primary and secondary stresses is important only in elastic-plastic and fully plastic analyses. Sections 9.3, 9.4, and 9.7 describe the treatment of primary and secondary stresses in such cases.

9.1.3 Plasticity Corrections

Section 2.8 describes approaches for incorporating small amounts of crack tip plasticity into the estimation of the stress intensity. The Irwin approach [13] defines an effective crack length as the sum of the actual crack size, a, and a plastic zone correction, r_y. The effective stress intensity factor is given by

$$K_{eff} = C\{a + r_y\}\, \sigma \sqrt{\pi\,(a + r_y)} \tag{9.13}$$

where $C\{a + r_y\}$ denotes that the geometry correction factor is a *function* of the effective crack size (not C times $a + r_y$). The Irwin plastic zone corrections are as follows:

$$r_y = \frac{1}{2\pi}\left(\frac{K_{eff}}{\sigma_{YS}}\right)^2 \quad \text{for plane stress} \tag{9.14a}$$

and

$$r_y = \frac{1}{6\pi}\left(\frac{K_{eff}}{\sigma_{YS}}\right)^2 \quad \text{for plane strain} \tag{9.14b}$$

This correction becomes significant at applied stresses greater than approximately half the yield strength, and is inaccurate above ~ 0.7 σ_{YS} (see Fig. 2.30).

The strip yield correction for a through crack in an infinite plate in plane stress is given by

$$K_{eff} = \sigma_{YS} \sqrt{\pi a} \left[\frac{8}{\pi^2} \ln \sec \left(\frac{\pi \sigma}{2 \sigma_{YS}} \right) \right]^{1/2} \tag{9.15}$$

Equation (9.15) does not result from adding a plastic zone correction to the crack size, but is based on an analysis by Burdekin and Stone [14] (see Appendix 3.1).[1]

The strip yield correction for a through crack in an infinite plate does not apply to other configurations. The strip yield model can be applied to other geometries, but each configuration requires a separate analysis [15]. There is, however, an approximate method of generalizing the strip yield model to a single equation that describes all cracked geometries (see Section 9.4).

Both the Irwin and strip yield plastic zone corrections have the effect of increasing K_{eff} over the linear elastic value. Failure to apply an appropriate plasticity correction could, therefore, result in an underestimate of the crack driving force, which would lead to a nonconservative analysis.

9.1.4 K_{IC} from J_{crit}: Advantages and Pitfalls

For plane strain, small scale yielding conditions,
K_{IC} and a critical J value (defined at the same point on the load-displacement curve) are related as follows:

$$K_{IC} = \sqrt{\frac{J_{crit} E}{1 - \nu^2}} \tag{9.16}$$

where J_{crit} can either be a J_{IC} value, defined near the initiation of ductile crack crack growth, or a critical J for cleavage. When a test speci-

[1]Recall that the size of the strip yield zone was derived by requiring that the singularity vanish. Thus the effective K cannot be defined in terms of a singularity amplitude. Instead, Burdekin and Stone derived the CTOD from Westergaard functions, and converted CTOD to an effective K through Eq. (3.7). A strip yield equation for J can be derived by evaluating J contour integral along the strip yield zone (Eq. (3.43)).

men behaves in a predominantly linear elastic manner, either K_{IC} or J_{crit} can be measured, but K ceases to be valid when the plastic zone becomes too large.

Recall from Chapter 7 that the size requirements for valid J_{IC} tests and J_{crit} values for cleavage are much less strict than the requirements for a valid K_{IC} test. Thus size-independent fracture toughness values in terms of J can be obtained on much smaller specimens than are required for K_{IC} tests. A J_{crit} value that meets the necessary size requirements can be converted to an *equivalent* K_{IC} through Eq. (9.16). *This quantity can be viewed as the* K_{IC} *that would be measured, given a sufficiently large specimen.*

The equivalent K_{IC} value, which is usually given the designation K_{JC}, can be applied to a structure that behaves in a linear elastic fashion. Quantifying the toughness in terms of K_{JC} enables the designer to apply linear elastic relationships between stress, flaw size, and toughness. Linear elastic approaches are much simpler and more versatile than a fracture design methodology based on the J integral (Section 9.5).

A conversion to K_{JC} is only appropriate when the critical J value is a size-independent measure of fracture toughness for the material. A J_{IC} value for ductile crack growth must satisfy Eq. (7.14), while J_{crit} for cleavage must satisfy Eq. (7.19). Equation (7.14), however, may not be adequate if the temperature is near the ductile-brittle transition of the material. A situation could arise where a test specimen experiences ductile initiation at a J_{IC} near the size limit of Eq. (7.14), but would have failed by cleavage (at a lower J_{crit}) if the specimen were sufficiently large to satisfy Eq. (7.19).

9.1.5 A Warning About LEFM

Performing a purely linear elastic fracture analysis and *assuming* that LEFM is valid is potentially dangerous, because the analysis gives no warning when it becomes invalid. The user must rely on experience to know whether or not plasticity effects need to be considered. A general rule of thumb is that plasticity becomes important at around 50% of yield, but this is by no means a universal rule.

The safest approach is to adopt an analysis that spans the entire range from linear elastic to fully plastic behavior. Such an analysis accounts for the two extremes of brittle fracture and plastic collapse. At low stresses, the analysis reduces to LEFM, but predicts collapse if the stresses are sufficiently high. At intermediate stresses, the analysis automatically applies a plasticity correction when necessary; the user does not have to decide whether or not such a correction is needed.

Sections 9.4 to 9.7 give examples of fracture analyses that span the range of material behavior.

9.2 THE ASME REFERENCE CURVES

The American Society of Mechanical Engineers (ASME) Boiler and Pressure Vessel Code is a comprehensive guide for designers, fabricators, and operators of pressure vessels and related components. Section XI of the code, "Rules for Inservice Inspection of Nuclear Power Plant Components [16]," contains guidelines for computing allowable flaw sizes based on fracture mechanics principles.

Since fracture toughness data are not always available for a particular heat of steel, Section XI of the ASME code includes reference curves that give conservative estimates of toughness versus temperature. These curves were generated by compiling K_{IC}, K_{Id}, and K_{Ia} data for several heats of steel over a range of temperatures, and plotting these results relative to a reference temperature, RT_{NDT}.

As Fig. 9.9 schematically illustrates, different heats of pressure vessel steel typically display ductile-brittle transitions at different temperatures; the reference temperature is an attempt to collapse all data onto a single curve. The indexing temperature, RT_{NDT}, is assigned through a combination of the drop weight nil-ductility transition temperature (NDTT) and Charpy properties; RT_{NDT} is defined as the higher of the following cases:

(1) The drop weight NDTT.

(2) 33°C (60°F) below the minimum temperature at which the lowest of three Charpy results is at least 68 J (50 ft-lb).

The RT_{NDT} in a typical pressure vessel steel occurs near the lower "knee" of the fracture toughness transition curve.

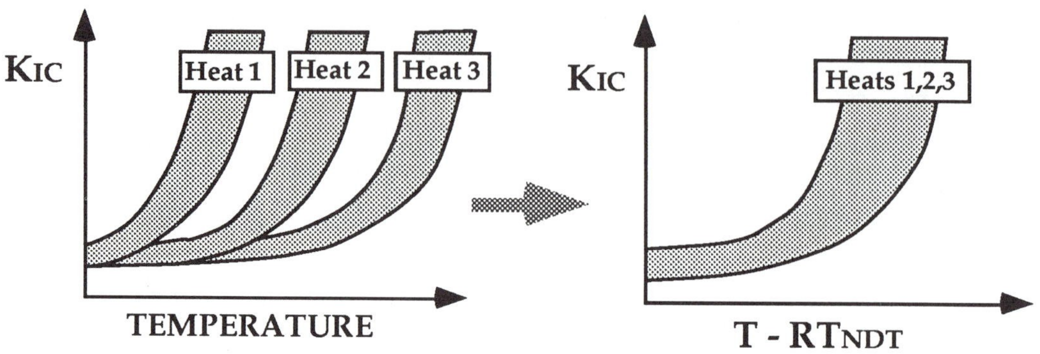

FIGURE 9.9 The ASME Section XI [16] approach for indexing multiple heats of steel.

Two reference toughness curves were originally developed: the K_{IC} curve and the K_{IR} curve. The former curve describes the lower envelope to a large set of K_{IC} data, while the latter is a lower envelop to a combined set of K_{IC}, K_{Id}, and K_{Ia} data. Since dynamic and crack arrest toughness values are generally lower than static initiation (K_{IC}) values, the K_{IR} curve is the more conservative of the two. The K_{IC} and K_{IR} curves, in SI units, are given by

$$K_{IC} = 36.5 + 3.084 \exp\left[0.036\left(T - RT_{NDT} + 56\right)\right] \tag{9.17a}$$

$$K_{IR} = 29.5 + 1.344 \exp\left[0.026\left(T - RT_{NDT} + 89\right)\right] \tag{9.17b}$$

where temperatures are in °C, and K_{IC} and K_{IR} are in MPa $\sqrt{m}$. Figure 9.10 shows a plot of Eqs. (9.17a) and (9,17b), together with the experimental data that defined the curves.

According to the exponential curve fits for K_{IC} and K_{IR}, the toughness increases without bound; these curves do not predict an upper shelf. In the late 1960s and early 1970s, when these curves were defined, there was no way to quantify upper shelf toughness with a fracture mechanics test. A number of very large specimens, up to 305 mm (12 in) thick, were required to quantify K_{IC} in the transition region[2]; upper shelf K_{IC} measurements were simply not possible. Because the

---

[2]The original K_{IC} curve is often referred to as the "Million Dollar Curve" which reflects the cost of testing such large specimens.

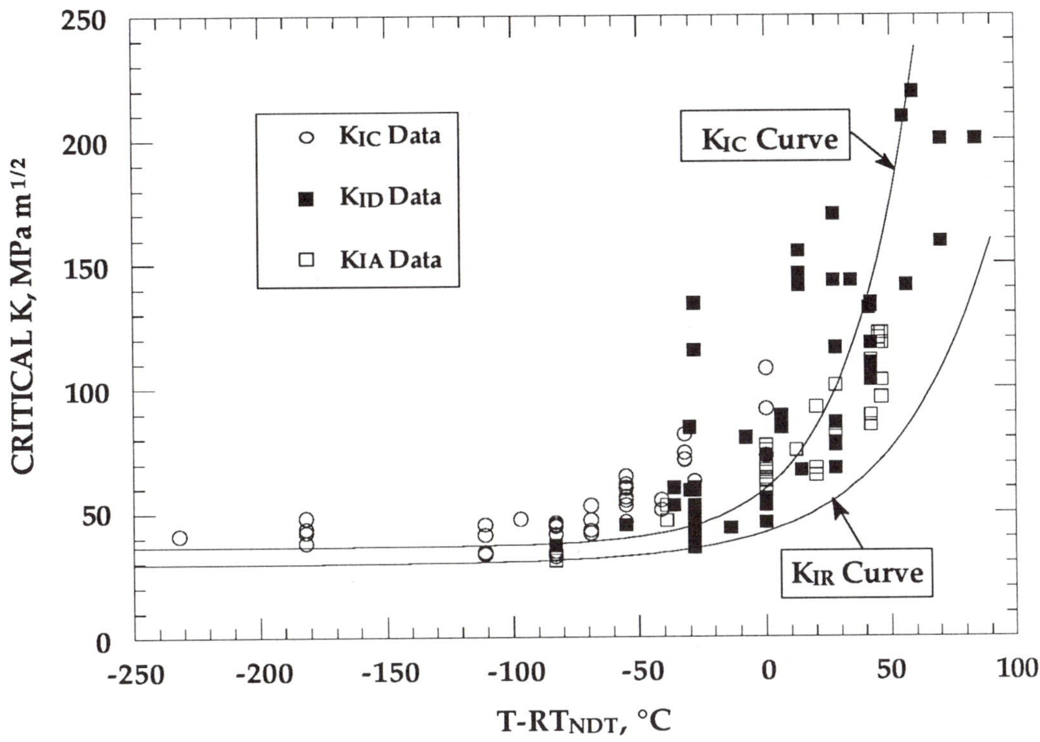

FIGURE 9.10 **The K_{IC} and K_{IR} curves, with the original data that was used to define the lower envelop curves [17].**

upper shelf could not be quantified, a cut-off was imposed at 220 MPa $\sqrt{m}$ (200 ksi $\sqrt{in}$).

Section XI of the ASME Code also gives guidelines for computing the *applied* K_I in a pressure vessel. For a surface flaw, the stresses are linearized and divided into bending and membrane components (Fig. 9.2), and stress intensity is estimated from Eq. (9.2).

9.3 THE CTOD DESIGN CURVE

The CTOD concept was applied to structural steels beginning in the late 1960s. The British Welding Research Association (now known as The Welding Institute) and other laboratories performed CTOD tests on structural steels and welds. At that time there was no way to apply these results to welded structures because CTOD driving force equations did not exist. Burdekin and Stone [14] developed the CTOD equivalent of the strip yield model in 1966. Although their model provides a basis for a CTOD driving force relationship, they were unable to modify the strip yield model to account for residual stresses and stress concentrations. (These difficulties were later overcome when a strip yield approach became the basis of the R6 design method, as discussed in the next section)

In 1971, Burdekin and Dawes [18] developed the CTOD design curve, a semi-empirical driving force relationship, that was based on an idea that Wells [19] originally proposed. For linear elastic conditions, fracture mechanics theory was reasonably well developed, but the theoretical framework required to estimate the driving force under elastic-plastic and fully plastic conditions did not exist until the late 1970s. Wells, however, suggested that global strain should scale linearly with CTOD under large scale yielding conditions. Burdekin and Dawes based their elastic-plastic driving force relationship on Wells' suggestion and an empirical correlation between small scale CTOD tests and wide double-edge notched tension panels made from the same material. The wide plate specimens were loaded to failure, and the failure strain and crack size of a given large scale specimen were correlated with the critical CTOD in the corresponding small scale test.

The correlation that resulted in the CTOD design curve is illustrated schematically in Fig. 9.11. The critical CTOD is nondimensionalized by the half crack length, a, of the wide plate and is shown on the ordinate of the graph. The nondimensional CTOD is plotted against the failure strain in the wide plate, normalized by the elastic yield strain, ε_y. Based on a plot similar to Fig. 9.11, Burdekin and Dawes [18,20] proposed the following two-part relationship:

$$\Phi = \frac{\delta_{crit}}{2\pi\,\varepsilon_y\,a} = \left(\frac{\varepsilon_f}{\varepsilon_y}\right)^2 \qquad \text{for } \frac{\varepsilon_f}{\varepsilon_y} \leq 0.5 \tag{9.18a}$$

and

$$\Phi = \frac{\delta_{crit}}{2\pi\,\varepsilon_y\,a} = \frac{\varepsilon_f}{\varepsilon_y} - 0.25 \qquad \text{for } \frac{\varepsilon_f}{\varepsilon_y} > 0.5 \tag{9.18b}$$

where Φ is the nondimensional CTOD. Equation (9.18a), which was derived from LEFM theory, includes a safety factor of 2.0 on crack size. Equation (9.18b) represents an upper envelope of the the experimental data.

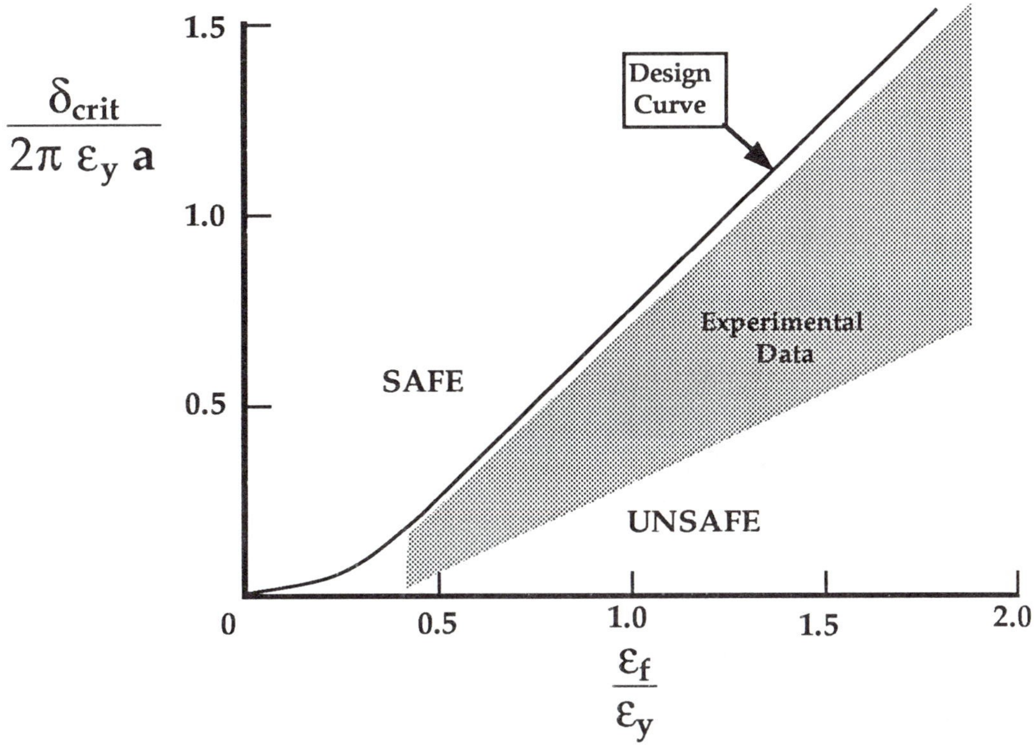

FIGURE 9.11 The CTOD design curve.

The applied strain and flaw size in a structure, along with the critical CTOD for the material, can be plotted on Fig. 9.11. If the point lies above the design curve, the structure is considered safe because all observed failures are below the design line. Equations (9.18a) and (9.18b) conform to the classical view of a fracture mechanics analysis, in relating stress (or strain in this case) to fracture toughness (δ_{crit}) and flaw size (a). The CTOD design curve is conservative, however, and does not relate *critical* combinations of these variables.

In 1980, the CTOD design curve approach was incorporated into the British Standards document PD 6493 [21]. This document addresses flaws of various shapes by relating them back to an equivalent through-thickness dimension, $\bar{a}$. For example, if a structure contains a surface flaw of length 2c and depth a, the equivalent through-thickness flaw produces the same stress intensity when loaded to the same stress as the structure with the surface flaw. Thus $\bar{a}$ is a generalized measure of a flaw's severity. The CTOD design curve can be applied to any flaw by replacing a with $\bar{a}$ in Eq. (9.18).

The original CTOD design curve was based on correlations with flat plates loaded in tension. Real structures, however, often include complex shapes that result in stress concentrations. In addition the structure may be subject to bending and residual stresses, as well as tensile (membrane) stresses. The PD 6493:1980 approach accounts for complex stress distributions simply and conservatively by estimating the maximum total strain in the cross section and assuming that this strain acts through the entire cross section. The maximum strain can be estimated from the following equation:

$$\varepsilon_1 = \frac{1}{E}\left[k_t(P_m + P_b) + S\right] \tag{9.19}$$

where k_t is the elastic stress concentration factor, P_m is the primary membrane stress, P_b is the primary bending stress, and S is the secondary stress, which may include thermal or residual stresses. Since the precise distribution of residual stresses is usually unknown, Q is often assumed to equal the yield strength in an as-welded weldment.

When Burdekin and Dawes developed the CTOD design curve, the CTOD and wide plate data were limited; the curve they constructed lay above all available data. In 1979, Kamath [22] reassessed the design curve approach with additional wide plate and CTOD data generated between 1971 and 1979. In most cases, there were three CTOD tests for a given condition. Kamath used the lowest measured CTOD value to predict failure in the corresponding wide plate specimen. When he plotted the results in the form of Fig. 9.11, a few data points fell above the design curve, indicating Eq. (9.18) was nonconservative in these instances. The CTOD design curve, however, was conservative in most cases. Kamath estimated the average safety factor on crack size to be 1.9, although individual safety factors ranged from less than 1 to greater than 10. With this much scatter, the concept of a safety factor is of little value. A much more meaningful quantity is the confidence level. Kamath estimated that the CTOD design curve method corresponds to a 97.5% confidence of survival. That is, the method in PD 6493:1980 is conservative approximately 97.5% of the time.

9.4 FAILURE ASSESSMENT DIAGRAMS

Structures made from materials with sufficient toughness may not be susceptible to brittle fracture, but they can fail by plastic collapse if they are overloaded. The CTOD design curve does not explicitly address collapse, and can be nonconservative if a separate collapse check is not applied.

Dowling and Townley [23] and Harrison, et al. [24] introduced the concept of a two-criteria failure assessment diagram (FAD) to describe the interaction between fracture and collapse. The first FAD was derived from a modified version of the strip yield model, as described below.

Equation (9.15) is the effective stress intensity factor for a through crack in an infinite plate, according to the strip yield model. As discussed earlier, this relationship is asymptotic to the yield strength. Equation (9.15) can be modified for real structures by replacing σ_{YS} with the collapse stress, σ_c, for the structure. This would ensure that the strip yield model predicts failure as the applied stress approaches the collapse stress. For a structure loaded in tension, collapse occurs when the stress on the net cross section reaches the flow stress of the material. Thus σ_c depends on the tensile properties of the material and the flaw size relative to the total cross section of the structure. The next step in deriving a failure assessment diagram from the strip yield model entails dividing the effective stress intensity by the linear elastic K:

$$\frac{K_{eff}}{K_I} = \frac{\sigma_c}{\sigma}\left[\frac{8}{\pi^2}\ln\sec\left(\frac{\pi}{2}\frac{\sigma}{\sigma_c}\right)\right]^{1/2} \tag{9.20}$$

This modification not only expresses the driving force in a dimensionless form but also eliminates the square root term that contains the the half length of the through crack. Thus Eq. (9.20) removes the geometry dependence of the strip yield model[3]. This is analogous to the PD 6493 approach, where the driving force relationship was generalized by

[3]This generalization of the strip yield model is not rigorously correct for all configurations, but it is a good approximation.

defining an equivalent through thickness flaw, ā. As a final step, we can define the stress ratio, S_r, and the K ratio, K_r, as follows:

$$K_r = \frac{K_I}{K_{eff}} \tag{9.21}$$

and

$$S_r = \frac{\sigma}{\sigma_c} \tag{9.22}$$

The failure assessment diagram is then obtained by inserting the above definitions into Eq. (9.20) and taking the reciprocal:

$$K_r = S_r\left[\frac{8}{\pi^2}\ln\sec\left(\frac{\pi}{2}S_r\right)\right]^{-1/2} \tag{9.23}$$

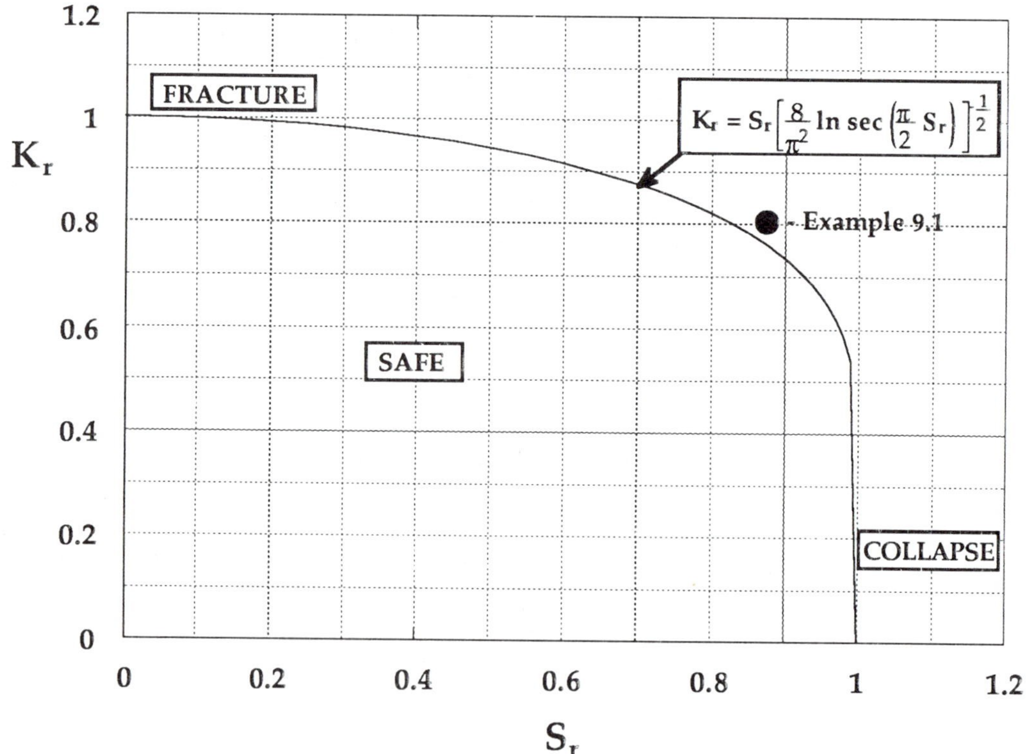

FIGURE 9.12 The strip yield failure assessment diagram [23,24].

Equation (9.23) is plotted in Fig. 9.12. The curve represents the locus of predicted failure points. Fracture is predicted when $K_{eff} = K_{IC}$. If the toughness is very large, the structure fails by collapse when $S_r = 1.0$. A brittle material will fail when $K_r = 1.0$. In intermediate cases, collapse and fracture interact, and both K_r and S_r are less than 1.0 at failure. All points inside of the FAD are considered safe; points outside of the diagram are unsafe.

In order to assess the significance of a particular flaw in a structure, one must determine the applied values of K_r and S_r, and plot the point on Fig. 9.12. The stress intensity ratio for the structure is given by

$$K_{r\,(\text{structure})} = \frac{K_I}{K_{IC}} \tag{9.24}$$

The applied stress ratio can be defined as the ratio of the applied stress to the collapse stress. Alternatively, the applied S_r can be defined in terms of axial forces or moments. If the applied conditions in the structure place it inside of the FAD, the structure is safe.

EXAMPLE 9.1

A middle tension (MT) panel (Fig. 7.1(e)) 1 m wide and 25 mm thick with a 200 mm crack must carry a 7.00 MN load. For the material, $K_{IC} = 200$ MPa $\sqrt{m}$, $\sigma_{YS} = 350$ MPa, and $\sigma_{TS} = 450$ MPa. Use the strip yield FAD to determine whether or not this panel will fail.

Solution: We can take account of work hardening by assuming a flow stress that is the average of yield and tensile strength. Thus $\sigma_{flow} = 400$ MPa. The collapse load is then defined when the stress on the remaining cross section reaches 400 MPa:

$$P_c = (400 \text{ MPa})(0.025 \text{ m})(1 \text{ m} - 0.200 \text{ m}) = 8.00 \text{ MN}$$

Therefore,

EXAMPLE 9.1 (Cont.)

$$Sr = \frac{7.00 \text{ MN}}{8.00 \text{ MN}} = 0.875$$

The applied stress intensity can be estimated from Eq. (2.46) (without the polynomial term):

$$K_I = \frac{7.00 \text{ MN}}{(0.025 \text{ m})(1.0 \text{ m})} \sqrt{\pi (0.100 \text{ m}) \sec\left(\frac{\pi (0.100 \text{ m})}{1.00 \text{ m}}\right)} = 161 \text{ MPa}\sqrt{\text{m}}$$

Thus

$$K_r = \frac{161}{200} = 0.805$$

The point (0.875, 0.805) is plotted in Fig. 9.12. Since this point falls outside of the failure assessment diagram, the panel will fail before reaching 7 MN. Note that a collapse analysis or brittle fracture analysis alone would have predicted a "safe" condition. Interaction of fracture and plastic collapse causes failure in this case.

In 1976, the Central Electricity Generating Board (CEGB) in Great Britain incorporated the strip yield failure assessment into a fracture analysis methodology, which became known as the R6 approach [24]. A revised version of the R6 document, which was published in 1980 [25], offers practical advise on how to apply the strip yield FAD to real structures. For example, it recommends that secondary stresses be taken into account through a secondary stress intensity. The total stress intensity is obtained by adding the primary and secondary components:

$$K_r = \frac{K_I}{K_{IC}} = \frac{K_I{}^P + K_I{}^S}{K_{IC}} \tag{9.25}$$

Only the primary stresses are used to compute S_r, because secondary stresses, by definition, do not contribute to collapse. Note that K_I is the LEFM stress intensity; it does not include a plastic zone correction. Plasticity effects are taken into account through the formulation of the failure assessment diagram (Eq. (9.23)).

The R6 procedure recommends that the fracture toughness input be obtained through testing the material according to ASTM E 399 or the equivalent British Standard (Chapter 7). When it is not possible to obtain a valid K_{IC} value experimentally, one can measure J_{IC} in the material and convert this toughness to an equivalent K_{IC} (or K_{JC}) by means of Eq. (9.16).

9.5 THE EPRI J ESTIMATION SCHEME

The R6 failure assessment diagram is based on a strip yield model. Since it assumes elastic-perfectly plastic material behavior, it is conservative when applied to strain hardening materials.

In 1976, Shih and Hutchinson [26] proposed a more advanced methodology for computing the fracture driving force that takes account of strain hardening. Their approach was developed further and validated at the General Electric Corporation in Schenectady, New York in the late 1970s and early 1980s, and was published as an engineering handbook by the Electric Power Research Institute (EPRI) in 1981 [27].

The EPRI procedure provides a means for computing the applied J integral under elastic-plastic and fully plastic conditions. The elastic and plastic components of J are computed separately and added to obtain the total J:

$$J_{tot} = J_{el} + J_{pl} \tag{9.26}$$

Figure 9.13 schematically illustrates a plot of J versus applied load. The plastic component of J is negligible at low loads, but dominates at high loads. The sum of elastic and plastic J values from the estimation scheme agrees well with an elastic-plastic finite element analysis.

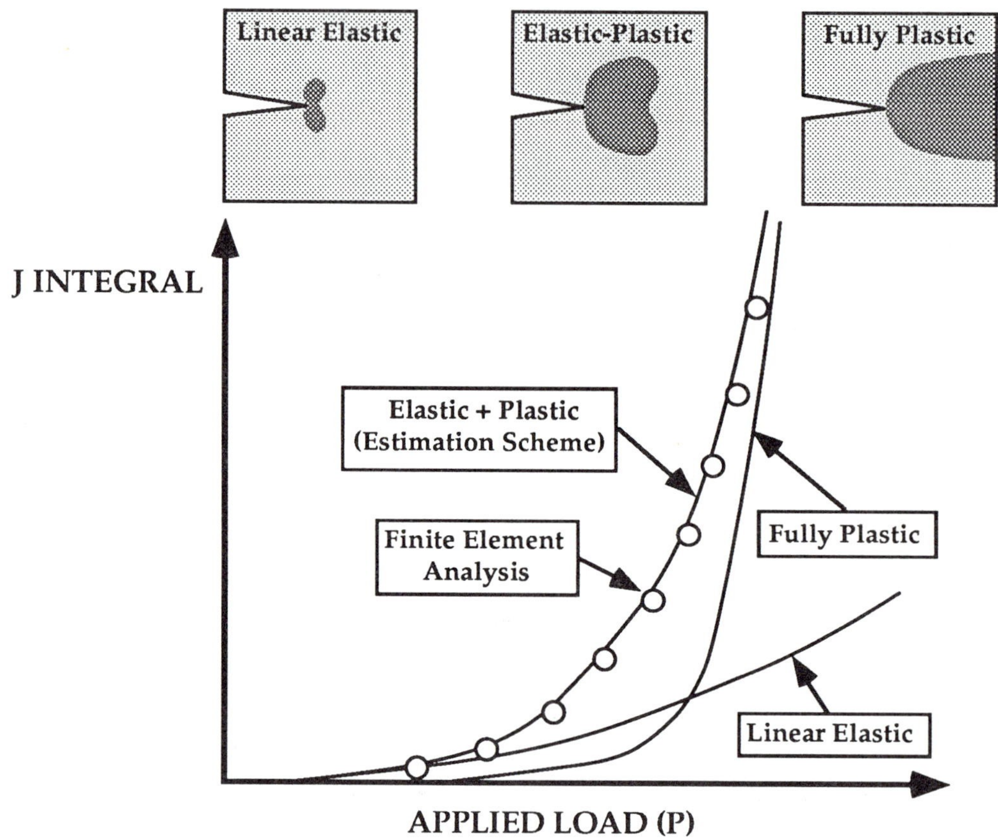

FIGURE 9.13 The EPRI J estimation scheme [27].

9.5.1 Theoretical Background

Consider a cracked structure with a fully plastic ligament, where elastic strains are negligible. Assume that the material follows a power-law stress-strain curve:

$$\frac{\varepsilon_{pl}}{\varepsilon_o} = \alpha \left(\frac{\sigma}{\sigma_o} \right)^n \tag{9.27}$$

which is the second term in the Ramberg-Osgood model (Eq. (3.22)). The parameters α, n, ε_o, and σ_o are defined in Section 3.2.3. Close to the crack tip, under J-controlled conditions, the stresses are given by the HRR singularity:

$$\sigma_{ij} = \sigma_o \left(\frac{J}{\alpha\, \varepsilon_o\, \sigma_o\, I_n\, r} \right)^{\frac{1}{n+1}} \tilde{\sigma}_{ij}(n,\theta) \tag{9.28}$$

which is a restatement of Eq. (3.24a). Solving for J in the HRR equation gives

$$J = \alpha\, \varepsilon_o\, \sigma_o\, I_n\, r \left(\frac{\sigma_{ij}}{\sigma_o} \right)^{n+1} \tilde{\sigma}_{ij}{}^{n+1} \tag{9.29}$$

For J controlled conditions, the loading must be proportional. That is, the local stresses must increase in proportion to the remote load, P. Therefore, Eq. (9.29) can be written in terms of P:

$$J = \alpha\, \varepsilon_o\, \sigma_o\, h\, L \left(\frac{P}{P_o} \right)^{n+1} \tag{9.30}$$

where h is a dimensionless function of geometry and n, L is a characteristic length for the structure, and P_o is a reference load. Both L and P_o can be defined arbitrarily, and h can be determined by numerical analysis of the configuration of interest.

It turns out that the assumptions of J dominance at the crack tip and proportional loading are not necessary to show that J scales with P^{n+1} for a power-law material, but these assumptions were useful for deriving the correct form of the J-P relationship.

Equation (9.30) is an estimate of the *fully plastic* J. Under linear elastic conditions, J must scale with P^2. The EPRI J estimation procedure assumes that the total J is equal to the sum of the elastic and plastic components (Eq. (9.26) and Fig. 9.13).

9.5.2 Estimation Equations

The fully plastic equations for J, crack mouth opening displacement (V_p), and load line displacement (Δ_p) have the following form for most geometries:

$$J_{pl} = \alpha\, \varepsilon_o\, \sigma_o\, b\, h_1(a/W, n) \left(\frac{P}{P_o} \right)^{n+1} \tag{9.31}$$

$$V_p = \alpha \varepsilon_o a h_2(a/W, n) \left(\frac{P}{P_o}\right)^n \tag{9.32}$$

$$\Delta_p = \alpha \varepsilon_o a h_3(a/W, n) \left(\frac{P}{P_o}\right)^n \tag{9.33}$$

where b is the uncracked ligament length, a is the crack length, and h_1, h_2, and h_3 are dimensionless parameters that depend on geometry and hardening exponent. The h factors for various geometries and n values, for both plane stress and plane strain, are tabulated in several EPRI reports [28-30], as well as Chapter 12.

The reference load, P_o, is usually defined by a limit load solution for the geometry of interest; P_o normally corresponds to the load at which the net cross section yields.

The plastic load line displacement, Δ_p, defined in Eq. 9.33 is only that component of plastic displacement that is due to the crack. Recall Section 3.2.5, where the displacement was divided into "crack" and "no crack" components; the latter is the displacement that would be measured if there were no crack, and the former is the *additional* displacement that results from the presence of the crack. The total displacement in a structure is the sum of the elastic and plastic "crack" and "no crack" components.

Several configurations have J expressions that are slightly different from Eq. (9.31). For example, the fully plastic J integral for a center cracked panel and a single edge notched tension panel is given by

$$J_{pl} = \alpha \varepsilon_o \sigma_o \frac{b\,a}{W} h_1(a/W, n) \left(\frac{P}{P_o}\right)^{n+1} \tag{9.34}$$

where, in the case of the center cracked panel, a is the half crack length and W is the half width. This modification was made in order to reduce the sensitivity of h_1 to the crack length/width ratio.

The elastic J is equal to $\mathcal{G}(a_{eff})$, the energy release rate for an effective crack length, which is based on a modified Irwin plastic zone correction:

$$a_{eff} = a + \frac{1}{1 + (P/P_o)^2} \frac{1}{\beta\pi} \left(\frac{n - 1}{n + 1}\right) \left(\frac{K_I}{\sigma_o}\right)^2 \qquad (9.35)$$

where $\beta = 2$ for plane stress and $\beta = 6$ for plane strain conditions. Equation (9.35) is a first-order correction, where a_{eff} is computed from the elastic K_I, rather than K_{eff}; thus iteration is not necessary.

The plastic zone correction that is applied to J_{el} does not have a theoretical basis, but it was incorporated to provide a smooth transition from linear elastic to fully plastic behavior. Estimated J values that include the plastic zone correction are closer to elastic-plastic finite element calculations than estimates of J without this correction. Equation (9.35) has a relatively small effect on the computed J value (Example 9.3); the effect is negligible at low loads, where the behavior is linear elastic, and at high loads, where the fully plastic term dominates.

The CTOD can be estimated from a computed J value as follows:

$$\delta = d_n \frac{J}{\sigma_o} \qquad (9.36)$$

where d_n is a dimensionless constant that depends on flow properties [31]. Figure 3.18 shows plots of d_n for both plane stress and plane strain. Equation (9.36) must be regarded as approximate in the elastic-plastic and fully plastic regimes, because the J-CTOD relationship is geometry dependent in large scale yielding [31].

EXAMPLE 9.2

Consider a single edge notched tensile panel with W = 1 m, B = 25 mm, and a = 125 mm. Calculate J versus applied load assuming plane stress conditions. Neglect the plastic zone correction.
Given: $\sigma_o = 414$ MPa; $n = 10$; $\alpha = 1.0$; $E = 207{,}000$ MPa $\varepsilon_o = \sigma_o/E = 0.002$

Solution: From Table 12.13, the reference load for this configuration is given by

EXAMPLE 9.2 (Cont.)

$$P_o = 1.072\,\eta\,\sigma_o\,b\,B$$

where

$$\eta = \sqrt{1 + (a/b)^2} - a/b = 0.867 \text{ for } a/b = 125/875 = 0.143$$

Solving for P_o gives

$$P_o = 8.42 \text{ MN}$$

For $a/W = 0.125$ and $n = 10$, $h_1 = 4.14$ (from Table 12.13). Thus the fully plastic J is given by

$$J_{pl} = (1.0)(0.002)(414{,}000 \text{ kPa})\frac{(0.875 \text{ m})(0.125 \text{ m})}{1.0 \text{ m}}(4.14)\left(\frac{P}{8.76 \text{ MN}}\right)^{11}$$

$$= 2.468 \times 10^{-8}\,P^{11}$$

where P is in MN and J_{pl} is in kJ/ m^2. The elastic J is given by

$$J_{el} = \frac{K_I^2}{E} = \frac{P^2 f^2(a/W)}{B^2 W E}$$

From the polynomial expression in Table 2.4, $f(a/W) = 0.770$ for $a/W = 0.125$. Thus

$$J_{el} = \frac{1000\,P^2\,(0.770)^2}{(0.025 \text{ m})^2\,(1.0 \text{ m})\,(207{,}000 \text{ MPa})} = 4.584\,P^2$$

where P is in MN and J_{el} is in kJ/m^2. The total J is the sum of J_{el} and J_{pl}:

$$J = 4.584\,P^2 + 2.468 \times 10^{-8}\,P^{11}$$

Figure 9.14 shows a plot of this equation. An analysis that includes the plastic zone correction (Eq. (9.35)) is also plotted for comparison.

EXAMPLE 9.3

For the panel in Example 9.2, determine the effect of the plastic zone correction at $P = P_o$.

Solution: From the previous problem, $P_o = 8.42$ MN. The elastic J without the Irwin correction is given by

$$J_{el}(a) = (4.584)\,(8.42\text{ MN})^2 = 325\text{ kJ/m}^2$$

The plastic J is not influenced by the Irwin correction:

$$J_{pl} = (2.468 \times 10\text{-}8)\,(8.42\text{ MN})^{11} = 372\text{ kJ/m}^2$$

The plastic zone correction is obtained by substituting the appropriate quantities into Eq. (9.35):

$$a_{eff} = 0.125\text{ m} + \frac{1}{1 + (1.0)^2}\frac{1}{2\pi}\left(\frac{9}{11}\right)\left(\frac{(8.42\text{ MN})\,(0.770)}{(0.025\text{ m})\sqrt{1\text{ m}}\,(414\text{ MPa})}\right)^2$$

$$= 0.151\text{ m}$$

For $a_{eff}/W = 0.151$, $f(a_{eff}/W) = 0.874$. Thus the corrected J_{el} is given by

$$J_{el}(a_{eff}) = 325\text{ kJ/m}^2\left(\frac{0.874}{0.770}\right)^2 = 418\text{ kJ/m}^2$$

The total J without the plastic zone correction is as follows:

$$J(a) = 325\text{ kJ/m}^2 + 372\text{ kJ/m}^2 = 697\text{ kJ/m}^2$$

and J with the correction is given by

$$J(a_{eff}) = 418 \text{ kJ/m}^2 + 372 \text{ kJ/m}^2 = 790 \text{ kJ/m}^2$$

EXAMPLE 9.3 (Cont.)

which is 13% higher than the estimate without the correction. This calculation represents a worst-case situation. The relative effect of the plastic zone correction is significantly less at both lower and higher loads. Also, the correction is smaller in plane strain than in plane stress.

9.5.3 Comparison with Experimental Estimates

Typical equations for estimating J from a laboratory specimen have the form

$$J = \frac{K^2}{E'} + \frac{\eta_p}{b} \int_0^{\Delta_p} P \, d\Delta_p \tag{9.37}$$

assuming unit thickness and a stationary crack. Equation (9.37) is convenient for experimental measurements because it relates J to the area under the load v. load line displacement curve, provided Δ_p does not contain a "no crack" component (see Section 3.2.5).

Since Eq. (9.33) gives an expression for the P-Δ_p curve for a stationary crack, it is possible to compare J_{pl} estimates from Eqs. (9.31) and (9.34) with Eq. (9.37). According to Eq. (9.33), the P-Δ_p curve follows a power law, where the exponent is the same as in a tensile test. The plastic energy absorbed by the specimen is as follows

$$\int_0^{\Delta_p} P \, d\Delta_p = \frac{n}{n+1} P \, \Delta_p$$

$$= \frac{n}{n+1} P_o \, \alpha \, \varepsilon_o \, a \, h_3 \left(\frac{P}{P_o}\right)^{n+1} \tag{9.38}$$

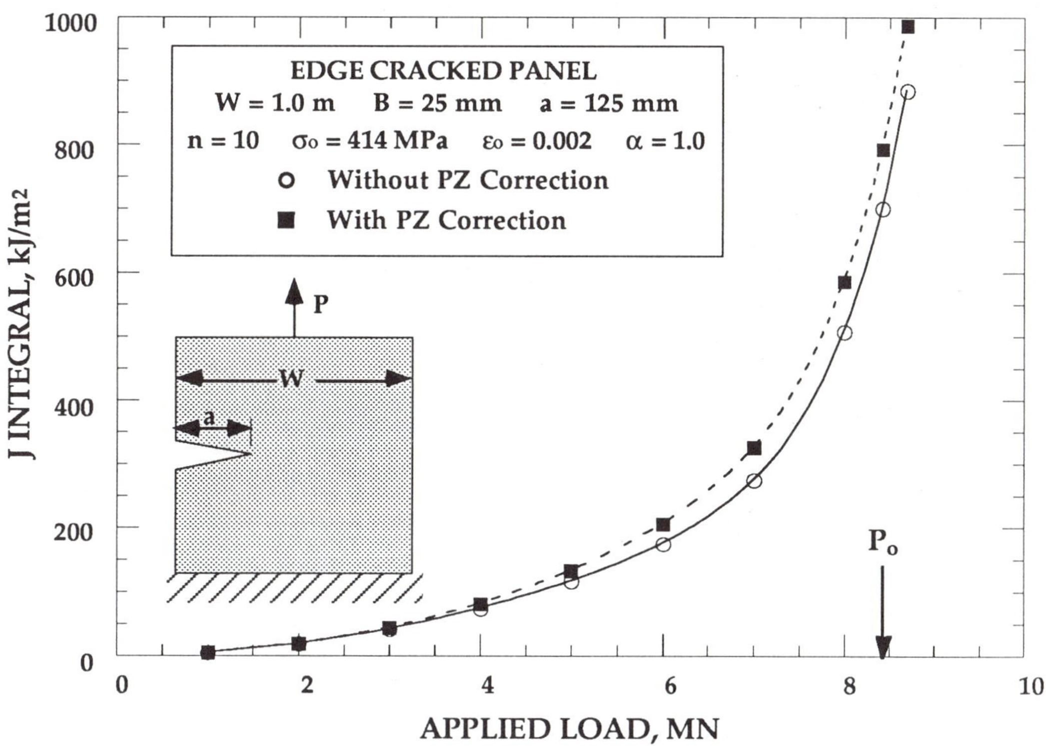

FIGURE 9.14 Applied J versus applied load in an edge cracked panel.

Thus the plastic J is given by

$$J_{pl} = \frac{n}{n+1}\eta_p P_o \alpha \varepsilon_o \frac{a}{b} h_3 \left(\frac{P}{P_o}\right)^{n+1} \tag{9.39}$$

Equating Eqs. (9.31) and (9.39) and solving for η_p gives

$$\eta_p = \frac{n+1}{n}\frac{\sigma_o b^2 h_1}{P_o a h_3} \tag{9.40a}$$

Alternatively, if J_{pl} is given by Eq. (9.34),

$$\eta_p = \frac{n+1}{n}\frac{\sigma_o b^2 h_1}{P_o W h_3} \tag{9.40b}$$

Consider an SENB specimen in plane strain. The reference load, assuming unit thickness and the standard span of 4W, is given by

$$P_o = \frac{0.364\,\sigma_o\,b^2}{W} \tag{9.41}$$

Substituting Eq. (9.41) into Eq. (9.40a) gives

$$\eta_p = \frac{n+1}{n}\,\frac{W}{a}\,\frac{h_1}{h_3} \tag{9.42}$$

Equation (9.42) is plotted in Fig. 9.15 for n = 5 and n = 10. According to the equation that was derived in Section 3.2.5, $\eta_p = 2$. This derivation, however, is only valid for deep cracks, since it assumes that the ligament length, b, is the only relevant length dimension. Figure 9.15 indicates that Eq. (9.31) approaches the deep crack limit with increasing a/W. For n = 10, the deep crack formula appears to be reasonably accurate beyond a/W ~ 0.3. Note that the η_p values computed from Eq. (9.42) for deep cracks fluctuate about an average of ~1.9, rather than the theoretical value of 2.0. These fluctuations may be indicative of numerical errors in the h_1 and h_3 values, while the average η_p slightly below 2.0 may indicate an a/W dependence that was not included in the dimensional analysis (Eq. (3.36)) in Section 3.2.5.

Equation (3.32) was derived for a double edge notched tension panel, but also applies to a deeply notched center cracked panel. A comparison of Eq. (9.34) with the second term of Eq. (3.32) leads to the following relationship for a center cracked panel in plane stress:

$$\frac{J_{EPRI}}{J_{DC}} = \frac{n+1}{n-1}\,\frac{b}{W}\,\frac{h_1}{h_3} \tag{9.43}$$

where J_{EPRI} is the plastic J computed from Eq. (9.34) and J_{DC} is the plastic J from the deep crack formula. Figure 9.16 is a plot of Eq. (9.43). The deep crack formula underestimates J at small a/W ratios, but coincides with J_{EPRI} when a/W is sufficiently large. Note that the deep crack formula applied to a wider range of a/W for n = 10. The deep crack formula assumes that all plasticity is confined to the ligament, a condition that is easier to achieve in low-hardening materials.

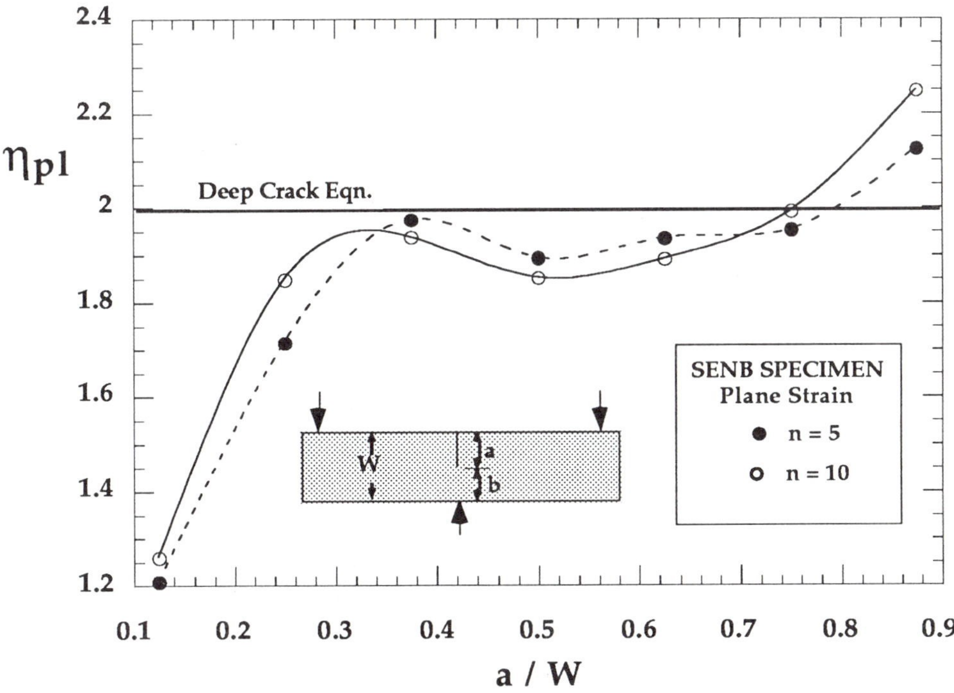

FIGURE 9.15 Comparison of the plastic η factor inferred from the EPRI Handbook with the deep crack value of 2.0 derived in Chapter 3

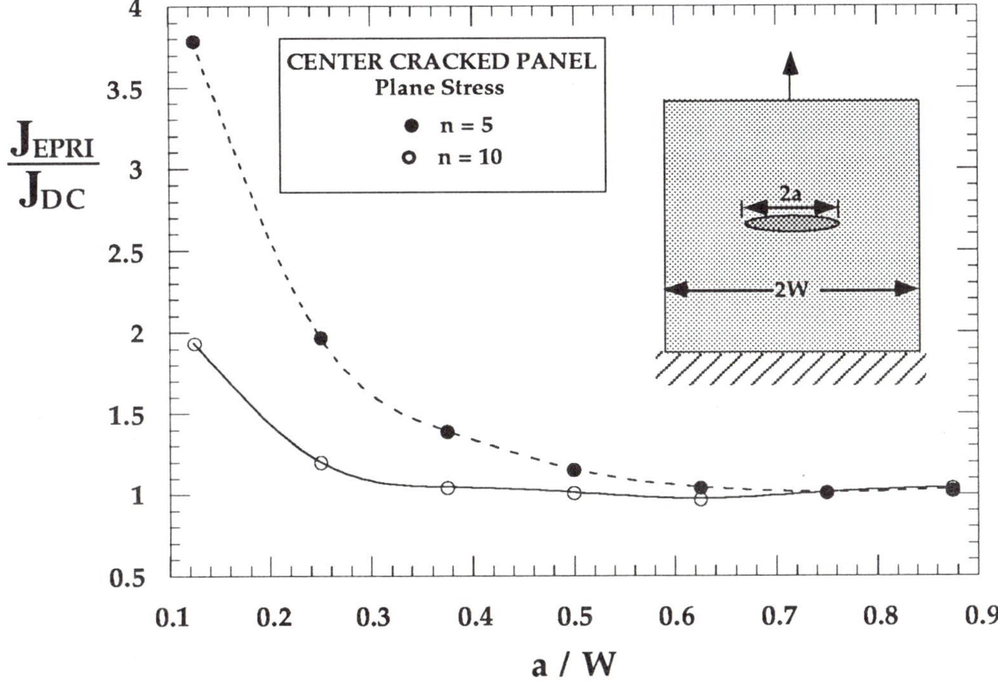

FIGURE 9.16 Comparison of J estimates from the EPRI handbook with the deep crack formula for a center cracked panel.

9.5.4 J-Based Failure Assessment Diagrams

The elastic-plastic driving force estimated from the EPRI procedure can also be expressed in terms of a failure assessment diagram, an idea first proposed by Bloom [32] and Shih, et al. [33]. The J ratio and stress ratio are defined as follows.

$$J_r = \frac{J_{el}(a)}{J_{el}(a_{eff}) + J_{pl}} \tag{9.44}$$

and

$$S_r = \frac{P}{P_o} \tag{9.45}$$

The equivalent K_r is equal to the square root of J_r.

Figure 9.17 shows the applied J for center cracked panels with a/W = 0.5 and n = 5, 10, and 20, plotted in terms of failure assessment diagrams. The strip yield diagram is included for comparison. Note that the shape of the FAD changes when strain hardening is taken into account. The strip yield diagram is conservative in this case because it assumes collapse will occur when the net section stresses equal σ_o; a panel made from a strain hardening material can withstand somewhat greater stresses. Figure 9.18 illustrates that failure assessment diagrams derived from the EPRI procedure are geometry dependent, while the strip yield diagram is geometry independent. The shape of the EPRI failure assessment diagram depends not only on the geometry of the cracked body, but also on the stress state (plane stress or plane strain). This makes fracture analyses with the EPRI approach more complicated, because a different FAD must be generated for each configuration analyzed.

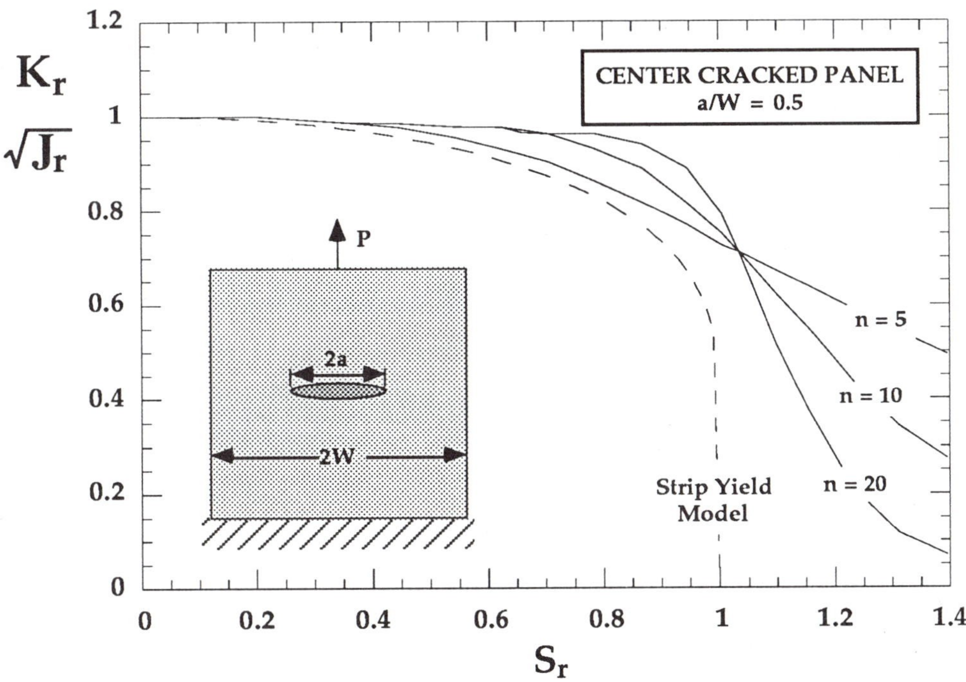

FIGURE 9.17 Failure assessment diagrams for a center cracked panel with n = 5, 10, and 20 computed from the EPRI Handbook.

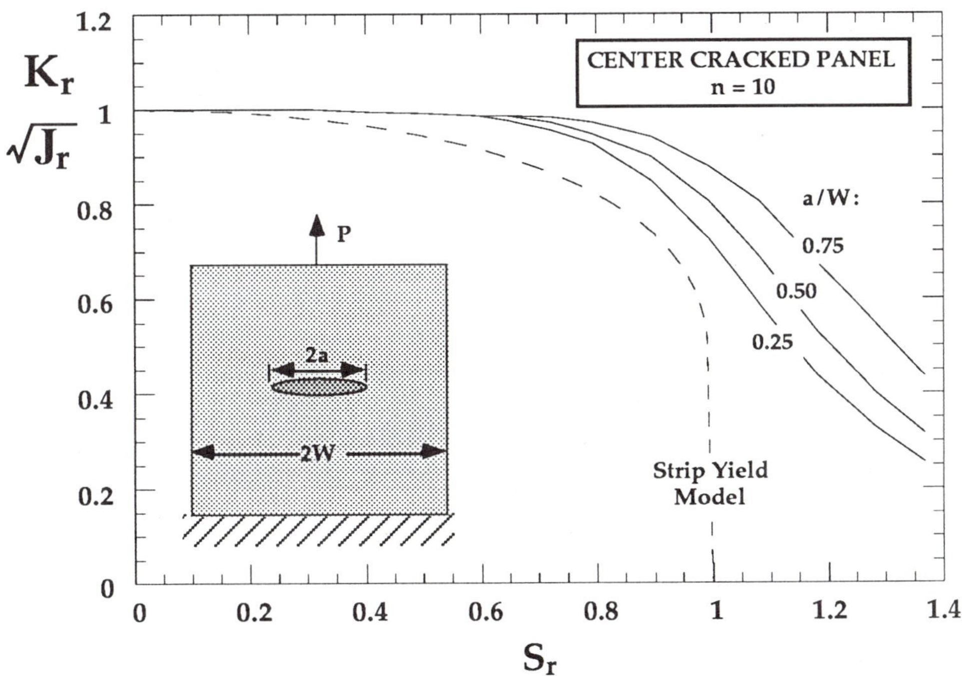

FIGURE 9.18 EPRI J-based failure assessment diagrams for a center cracked panel with various a/W ratios.

9.5.5 Ductile Instability Analysis

Section 3.4.1 outlined the theory of stability of J controlled crack growth. Crack growth is stable as long as the rate of change in the driving force (J) is less than or equal to the rate of change of the material resistance (J_R). Equations (3.49) and (3.50) defined the tearing modulus, which is a nondimensional representation of the derivatives of both the driving force and the resistance:

$$T_{app} = \frac{E}{\sigma_o^2}\left(\frac{dJ}{da}\right)_{\Delta_T} \quad \text{and} \quad T_R = \frac{E}{\sigma_o^2}\frac{dJ_R}{da} \tag{9.46}$$

where Δ_T is the remote displacement:

$$\Delta_T = \Delta + C_M P \tag{9.47}$$

where C_M is the system compliance (Fig. 2.12). Recall that the value of C_M influences the relative stability of the structure; $C_M = 0$ corresponds to dead loading, which tends to be unstable, while $C_M = \infty$ represents the other extreme of displacement control, which is more stable. Crack growth is unstable when

$$T_{app} > T_R \tag{9.48}$$

The rate of change in driving force at a fixed remote displacement is given by[4]

$$\left(\frac{dJ}{da}\right)_{\Delta_T} = \left(\frac{\partial J}{\partial a}\right)_P - \left(\frac{\partial J}{\partial P}\right)_a \left(\frac{\partial \Delta}{\partial a}\right)_P \left[C_m + \left(\frac{\partial \Delta}{\partial P}\right)_a\right]^{-1} \tag{9.49}$$

Since the EPRI J estimation approach provides expressions for J and Δ as a function of load and crack length, it is possible to evaluate the derivatives in Eq. (9.49) numerically at any P and a in the structure of

---

[4]The distinction between local and remote displacements (Δ and C_M P, respectively) is arbitrary, as long as all displacements due to the crack are included in Δ. The local displacement, Δ, can contain any portion of the "no crack" elastic displacements without affecting the term in square brackets in Eq. (9.49).

interest. The EPRI Handbook [27] recommends a forward difference approach for numerical differentiation:

$$\left(\frac{\partial J}{\partial a}\right)_P = \frac{J(a + \Delta a, P) - J(a, P)}{\Delta a} \tag{9.50a}$$

$$\left(\frac{\partial J}{\partial P}\right)_a = \frac{J(a, P + \Delta P) - J(a, P)}{\Delta P} \tag{9.50b}$$

$$\left(\frac{\partial \Delta}{\partial a}\right)_P = \frac{\Delta(a + \Delta a, P) - \Delta(a, P)}{\Delta a} \tag{9.50c}$$

$$\left(\frac{\partial \Delta}{\partial P}\right)_a = \frac{\Delta(a, P + \Delta P) - \Delta(a, P)}{\Delta P} \tag{9.50d}$$

One must exercise extreme caution in computing these derivatives: J and Δ are highly nonlinear functions of load and crack size, particularly in the fully plastic regime. The load and crack length increments, ΔP and Δa, should be chosen to minimize numerical errors. One possible approach is to chose progressively smaller increments until the numerical derivatives converge. Nonlinear differentiation is another alternative. For example, taking the logarithm of J, Δ, a, and P before differentiation may increase numerical accuracy.

The EPRI Handbook outlines two approaches for assessing structural stability: crack driving force diagrams and stability assessment diagrams. The former is a plot of J and J_R versus crack length, while a stability assessment diagram is a plot of tearing modulus versus J. These diagrams are merely alternative methods for plotting the same information.

Figure 3.20 shows a schematic driving force diagram for both load control and displacement control. In this example, the structure is unstable at P_3 and Δ_3 in load control, but the structure is stable in displacement control. Figure 9.19 illustrates driving force curves for this same structure, but with fixed remote displacement, Δ_T, and a finite system compliance, C_M. The structure is unstable at $\Delta_{T(4)}$ in this case.

Figure 9.20 illustrates the load-displacement curve for this hypothetical structure. A maximum load plateau occurs at P_3 and Δ_3, and the load decreases with further displacement. In load control, the structure is unstable at P_3, because the load cannot increase further. The structure is always stable in pure displacement control ($C_M = \infty$), but is unstable at Δ_4 (and $\Delta_{T(4)} = \Delta_4 + C_M P_4$) for the finite compliance case.

Figure 9.21 is a schematic stability assessment diagram. The applied and material tearing modulus are plotted against J and J_R, respectively. Instability occurs when the T_{app}-J curve crosses the T_R-J_R curve. The latter curve is relatively easy to obtain, since J_R depends only on the amount of crack growth:

$$J_R = J_R(a - a_o) \tag{9.51}$$

Thus there is a unique relationship between T_R and J_R, and the T_R-J_R curve can be defined unambiguously. Suppose, for example, that the J-R curve is fit to a power law:

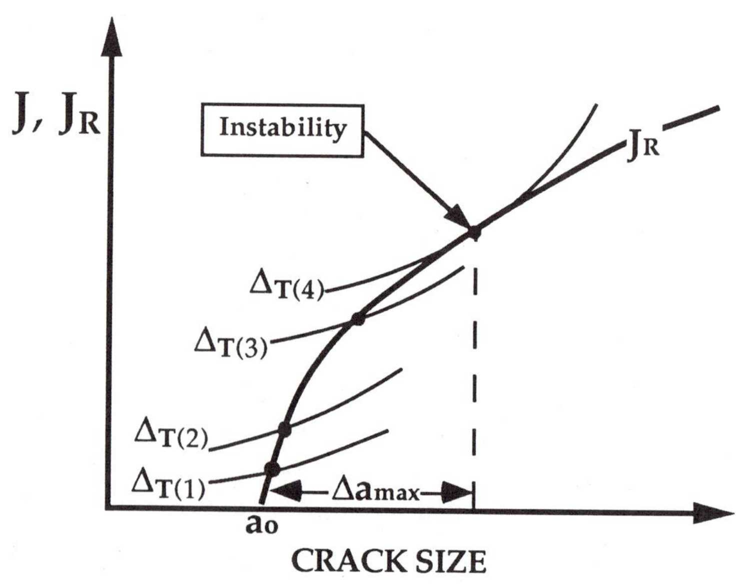

FIGURE 9.19 Schematic driving force diagram for a fixed remote displacement. Refer to Fig. 3.20 for the corresponding diagram for pure load control and displacement control.

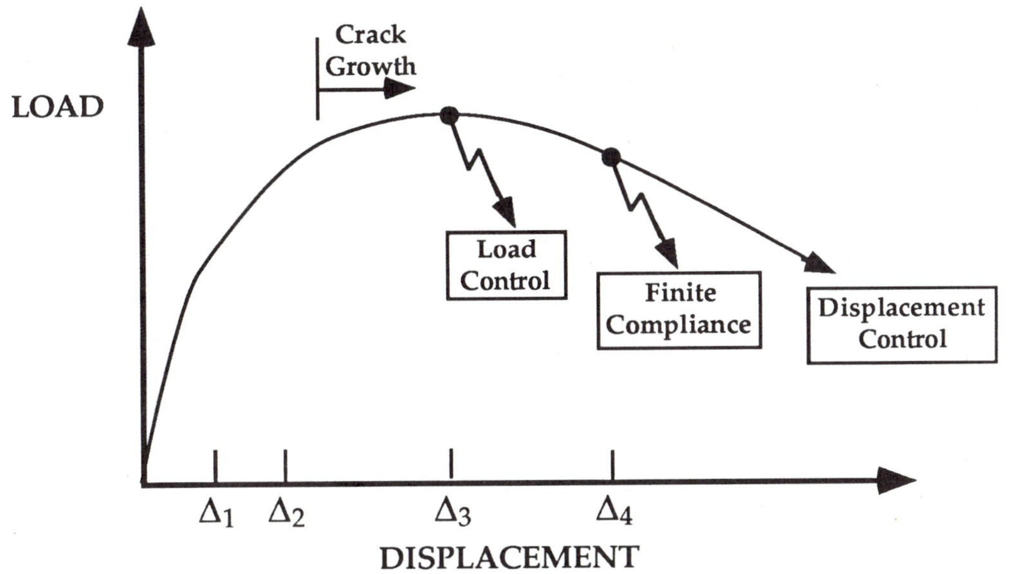

FIGURE 9.20 Schematic load-displacement curve for the material in Fig. 9.20.

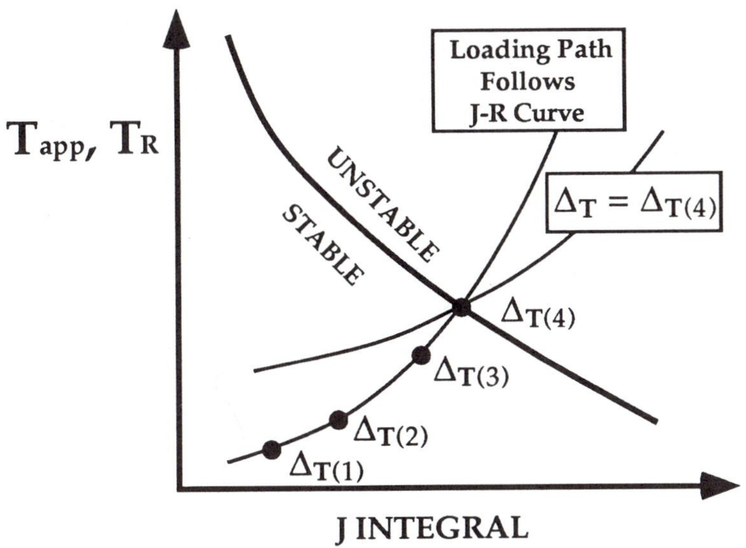

FIGURE 9.21 Schematic stability assessment diagram for the material in the two previous figures.

$$J_R = C_1(a - a_o)^{C_2} \tag{9.52}$$

The material tearing modulus is given by

$$T_R = \frac{E}{\sigma_o^2} \frac{C_2 J_R}{(a - a_o)} = \frac{E}{\sigma_o^2} C_2 C_1^{1/C_2} J_R^{\frac{C_2 - 1}{C_2}} \tag{9.53}$$

The applied tearing modulus curve is less clearly defined, however. There are a number of approaches for defining the T_{app}-J curve, depending on the application. Figure 9.21 illustrates two possible approaches, which are discussed below.

Suppose that the initial crack size, a_o, is known, and one wishes to determine the loading conditions (P, Δ, and Δ_T) at failure. In this case, the T_{app} should be computed at various points on the R curve. Since $J = J_R$ during stable crack growth, the applied J at a given crack size can be inferred from the J-R curve (Eq. (9.50). The remote displacement, Δ_T, increases as the loading progresses up the J-R curve (see Figs. 9.19 and 9.21); instability occurs at $\Delta_{T(4)}$. The final load, local displacement, crack size, and stable crack extension can be readily computed, once the critical point on the J-R curve has been identified.

The T_{app}-J curve can also be constructed by fixing one of the loading conditions (P, Δ, or Δ_T), and determining the critical crack size at failure, as well as a_o. For example, if we fix Δ_T at $\Delta_{T(4)}$ in the structure, we would predict the same failure point as the previous analysis but the T_{app}-J curve would follow a different path (Fig. 9.21). If, however, we fix the remote displacement at a different value, we would predict failure at another point on the T_R-J_R curve; the critical crack size, stable crack extension, and a_o would be different from the previous example.

9.5.6 Some Practical Considerations

If the material is sufficiently tough or if crack-like flaws in the structure are small, the structure will not fail unless it is loaded into the fully plastic regime. When performing fracture analyses in this regime, there are a number of important considerations that many practitioners overlook.

In the fully plastic regime, the J integral varies with P^{n+1}; a slight increase in load leads to a large increase in the applied J. The J versus crack length driving force curves are also very steep in this regime. Consequently, the failure stress and critical crack size are insensitive to toughness in the fully plastic regime; rather, failure is governed by the flow properties of the material. The problem is reduced to a limit load situation, where the main effect of the crack is to reduce the net cross section of the structure.

Predicting failure stress or critical crack size under fully plastic conditions need not be complicated. A detailed tearing instability analysis and a simple limit load analysis should lead to similar estimates of failure conditions.

Problems arise, however, when one tries to compute the applied J at a given load and crack size. Since J is very sensitive to load in the fully plastic regime, a slight error in P produces a significant error in the estimated J. For example, a 10% overestimate in the yield strength, σ_o, will produce a corresponding error in P_o, which will lead to an underestimate of J by a factor of 3.2 for n = 10. Since flow properties typically vary by several percent in different regions of a steel plate, and heat-to-heat variations can be much larger, accurate estimates of the applied J at a fixed load are virtually impossible.

If estimates of the applied J are required in the fully plastic regime, the *displacement*, not the load, should characterize conditions in the structure. While the plastic J is proportional to P^{n+1}, J_{pl} scales with $\Delta_p^{(n+1)/n}$, according to Eqs. (9.31) and (9.33). Thus a J-Δ plot is nearly linear in the fully plastic regime, and displacement is a much more sensitive indicator of the applied J in a structure. Figure 9.22 compares J-P and J-Δ plots for a center cracked panel with three strain hardening exponents.

Recall Section 3.3, where the empirical correlation of CTOD and wide plate data that resulted in the CTOD design curve was plotted in terms of strain (i.e., displacement over a fixed gage length) rather than stress [18,20]. A correlation based on stress would not have worked, because the failure stresses in the wide plate specimens were clustered around the flow stress of the material.

9.6 THE REFERENCE STRESS APPROACH

The EPRI equations for fully plastic J, Eqs. (9.31) and (9.34), assume that the material's stress-plastic strain curve follows a simple power law. Many materials, however, have flow behavior that deviates considerably from a power law. For example, most low carbon steels exhibit a plateau in the flow curve immediately after yielding. Applying Eq.

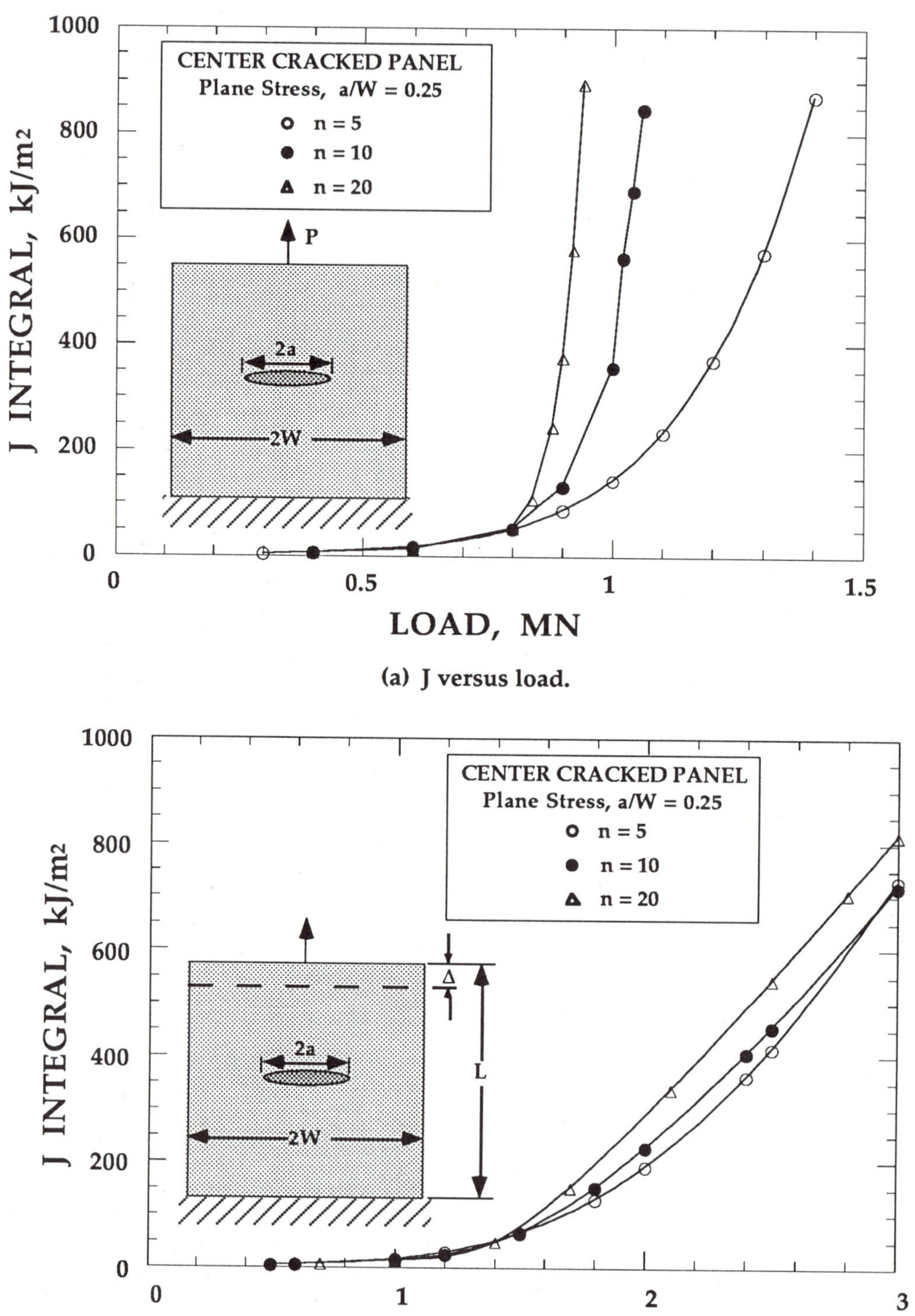

(a) J versus load.

(b) J versus load line displacement.

FIGURE 9.22 Comparison of J -load and J-displacement curves for a center cracked panel. (W = 50 mm, B = 25 mm, L = 400 mm, σ_o = 420 MPa, ε_o = 0.002, α = 1.0.)

(9.31) or (9.34) to such a material, results in significant errors. Ainsworth [34] modified the EPRI relationships to reflect more closely the flow behavior of real materials. He defined a reference stress as follows:

$$\sigma_{ref} = (P/P_o)\,\sigma_o \tag{9.54}$$

He further defined the reference strain as the total axial strain when the material is loaded to a uniaxial stress of σ_{ref}. Substituting these definitions into Eq. (9.31) gives

$$J_{pl} = \sigma_{ref}\, b\, h_1 \left(\varepsilon_{ref} - \frac{\sigma_{ref}\, \varepsilon_o}{\sigma_o} \right) \tag{9.55}$$

For materials that obey a power law, Eq. (9.55) agrees precisely with Eq. (9.31), but the former is more general, in that it is applicable to all types of stress-strain behavior.

Equation (9.55) still contains h_1, the geometry factor which depends on the power law hardening exponent n. Ainsworth proposed redefining P_o for a given configuration to produce another constant, h_1', that is insensitive to n. He noticed, however, that even without the modification of P_o, h_1 was relatively insensitive to n except at high n values (low hardening materials). Ainsworth was primarily interested in developing a driving force procedure for high hardening materials such as austenitic stainless steels. The strip yield failure assessment diagram was considered suitable for low hardening materials. He proposed the following approximation.

$$h_1(n) \cong h_1(1) \tag{9.56}$$

where $h_1(n)$ is the geometry constant for a material with a strain hardening exponent of n and $h_1(1)$ is the corresponding constant for a linear material. By substituting $h_1(1)$ into Eq. (9.31) (or (9.34)), Ainsworth was able to relate the plastic J to the linear elastic stress intensity factor:

$$J_{pl} = \frac{\mu\, K_I^2}{E} \left(\frac{E\, \varepsilon_{ref}}{\sigma_{ref}} - 1 \right) \tag{9.57}$$

where $\mu = 0.75$ for plane strain and $\mu = 1.0$ for plane stress.

Ainsworth's work has important ramifications. When applying the EPRI approach, one must obtain a stress intensity solution to compute the elastic J, and a separate solution for h_1 in order to compute the plastic term. The h_1 constant is a plastic geometry correction factor. However, Eq. (9.57) makes it possible to estimate J_{pl} from an elastic geometry correction factor. The original EPRI Handbook [27] and subsequent additions [28-30] contain h_1 solutions for a relatively small number of configurations, but there are hundreds of stress intensity solutions in handbooks and the literature. Thus Eq. (9.57) is not only simpler than Eq. (9.31), but also more widely applicable. The relative accuracy of Ainsworth's simplified equation is examined in Section 9.7.

Ainsworth made additional simplifications and modifications to the reference stress model in order to express it in terms of a failure assessment diagram. This FAD has been incorporated into a revision of the R6 procedure (see Section 9.4). The new document also contains more accurate procedures for analyzing secondary stresses. The revised R6 approach still permits application of the strip yield FAD to low hardening materials.

The reference stress FAD has also been included in the revised PD 6493 procedure, which was had yet to be published, as of this writing. Both the revised R6 and PD 6493 approaches are broadly similar, and are discussed in Section 9.8.

9.7 COMPARISON OF DRIVING FORCE EQUATIONS

The primary advantage of Ainsworth's reference stress approach is in accounting for the geometry of a cracked structure through a linear elastic stress intensity solution. The h_1 factor is replaced by an LEFM geometry factor. The other contribution of Ainsworth's analysis, the generalization to stress-strain laws other than power-law, is of secondary importance.

In most cases, the EPRI procedure and Ainsworth's simplified approach produce nearly identical estimates of critical flaw size and failure stress. This section presents the results of a parametric study of the accuracy of Ainsworth's approach relative to that of the EPRI procedure. The relative accuracy of the strip yield model was also evaluated.

A power law hardening material was assumed for all analyses, since the main purpose of this exercise was to evaluate the errors associated with the LEFM geometry correction factor in the elastic-plastic regime.

For a power-law material, the Ainsworth model gives the following expression for the total J:

$$J = \frac{K_I^2}{E}\left[(1-\nu^2) + \alpha\,\mu\left(\frac{P}{P_o}\right)^{n-1}\right] \tag{9.58}$$

if the Irwin plastic zone correction is neglected. Since stress intensity is proportional to load, this relationship has the form

$$J = C_\alpha P^2 + C_\beta P^{n+1}$$

The first term dominates under linear elastic conditions; the second term dominates under fully plastic conditions. The EPRI approach (Eqs. (9.31) and (9.34)) has the same form. The only difference between the EPRI equation and Eq. (9.58) is the value of the constant C_β; the equations agree precisely in the linear elastic range. Thus any discrepancies between the two approaches are observed only when the plastic term is significant.

Since load in the fully plastic range is insensitive to the applied J (Section (9.5.6), the predicted failure stress is insensitive to the differences between the EPRI approach and the Ainsworth model. The latter approach assumes that the geometry factor, $h_1(n)$, is equal to the linear elastic value, $h_1(1)$. Errors in J that result from applying Eq. (9.57) are proportional to the ratio $h_1(n)/h_1(1)$, which is plotted against n in Fig. 9.23 for a center cracked panel in plane strain with a/W = 0.75. Note that the h_1 ratio in this configuration is sensitive to the hardening exponent. Thus Eq. (9.57) leads to significant errors in J, particularly at high n values. However, when the h_1 ratio is raised to the power $^1/_{(n+1)}$, it is insensitive to n. This latter ratio is indicative of the differences in the predicted failure stress between the Ainsworth and EPRI approaches.

A design engineer often wishes to use a fracture mechanics analysis to estimate the critical flaw size at a given applied stress. To determine the sensitivity of critical flaw size estimates to the driving force equation, the author [35] performed a series of calculations with the EPRI,

Ainsworth, and strip yield models on center cracked panels and edge cracked bend specimens. The material was assumed to follow perfectly the power law expression (Eq. (9.27)) for stress versus plastic strain. The constants α and ε_o were fixed 1.0 and 0.002, respectively, for all n values; thus σ_o corresponds exactly to the 0.2% offset yield strength.

Two separate strip yield analyses were performed for each case: one assumed that the collapse stress was equal to the yield strength; the other based collapse on the flow stress, defined as the average of yield and tensile strengths. For a material whose true stress-true strain curve follows a power law, the flow stress can be estimated from the following expression:

$$\sigma_{flow} = \frac{\sigma_{YS}}{2}\left[1 + \frac{\left(\frac{N}{0.002}\right)^N}{\exp(N)}\right] \tag{9.59}$$

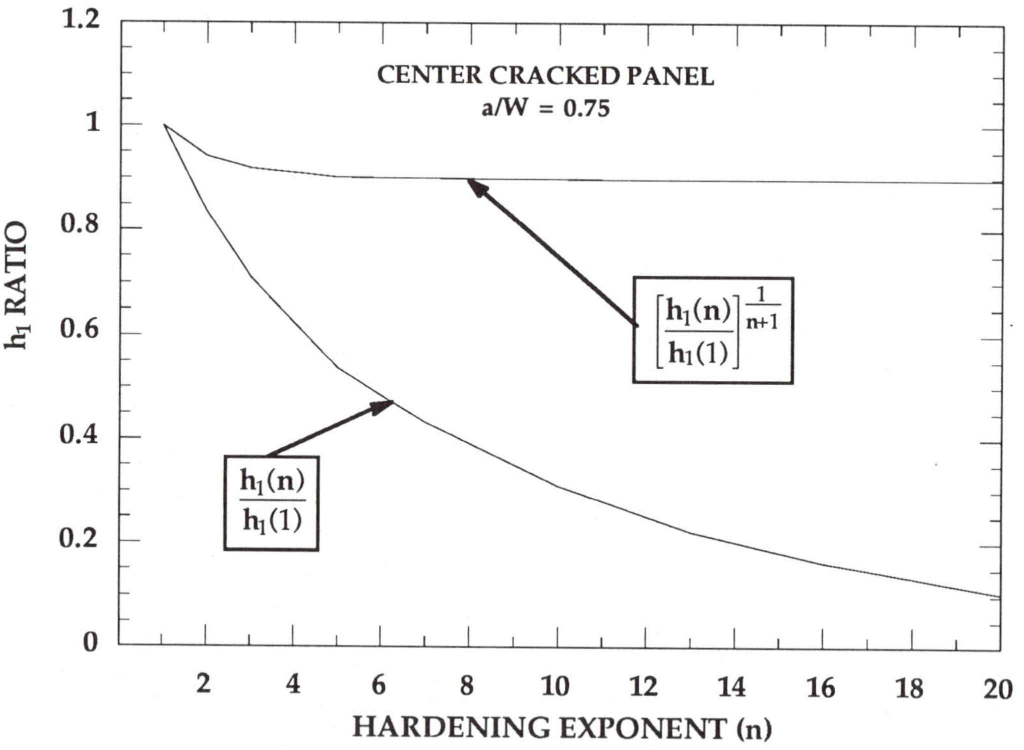

FIGURE 9.23 The effect of hardening exponent on the h_1 factor for a center cracked panel with a/W = 0.75. The h_1 ratio raised to the 1/n+1 power is indicative of the ratio of predicted failure stresses from the EPRI and reference stress approaches.

where N = 1/n. Equation (9.59) was derived by solving for the tensile instability point in Eq. (9.27) and converting true stress and strain to engineering values.

Figure 9.24 shows typical results from this analysis. The three driving force equations are applied to a center cracked panel with n = 5 and n = 10. Critical crack size, normalized by W, is plotted against critical J, normalized by width and yield strength. The nominal stress (P/BW) is fixed at $^2/_3$ yield. At low toughness levels, all predictions agree because linear elastic conditions prevail. At high toughness levels, the curves are relatively flat, indicating that critical crack size is insensitive to toughness. In this region, failure is controlled primarily by plastic collapse of the remaining cross section. The EPRI and Ainsworth equations agree well at all hardening rates.

The strip yield model is nonconservative for the high hardening material (Fig. 9.24(a)) when it is based on the flow stress. For n = 10, however, the strip yield model, with collapse defined at σ_{flow}, gives a good approximation of the other two curves; the agreement is even better at high n values. When the strip yield model is based on collapse at σ_{YS} it is always conservative.

This analysis indicates that the Ainsworth model can predict either critical crack size or failure stress in the elastic, elastic-plastic, and fully plastic regimes. The strip yield model gives reasonable results for low hardening materials.

9.8 THE REVISED PD 6493 METHOD

The CTOD design curve, which is the basis of the British Standards document PD 6493:1980, suffers from a number of shortcomings. For example, the driving force equation is mostly empirical and has a variable level of conservatism. In addition, this approach does not explicitly consider failure by plastic collapse. Improved driving force equations became available with the R6 and EPRI procedures, but the CTOD design curve had already been widely accepted by the welding fabrication industry in the United Kingdom and elsewhere. Many engineers are reluctant to discard PD 6493:1980. Structures designed according to the original PD 6493 method might have to be re-analyzed with the new procedure if the CTOD design curve were rendered obsolete. An additional problem with improving the procedure is the tendency for

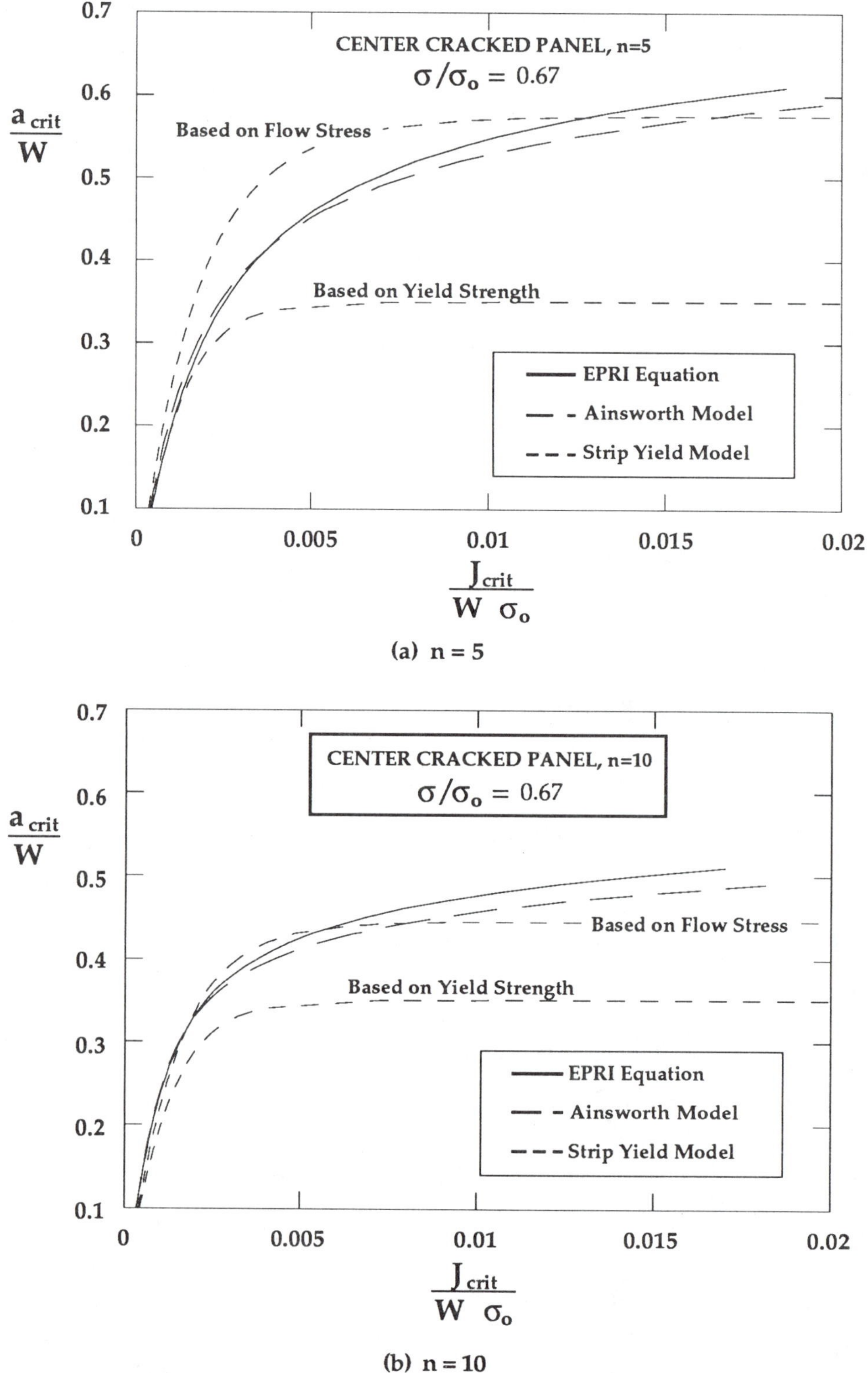

FIGURE 9.24 Comparison of predicted critical crack lengths from three driving force equations.

the more accurate elastic-plastic fracture analyses to be more complex and require more training.

The conflicting goals of improving PD 6493 and maintaining continuity with the past have been largely satisfied by a three-tier approach. The three tier philosophy assesses fracture problems at a level of complexity and accuracy appropriate for the situation. All three levels of PD 6493 are expressed as failure assessment diagrams. Level 1 is consistent with the CTOD design curve approach; Level 2 utilizes a strip yield model and Level 3 is based on the reference stress approach.

Although the new PD 6493 procedure has not been published (as of this writing), a number of recent articles [38,39] describe its salient features. Information in the following subsections is based on these articles.

9.8.1 Level 1

Level 1 is consistent with the CTOD design curve in the 1980 version of PD 6493. The main differences are that the equations are expressed in terms of a failure assessment diagram, and an explicit collapse analysis is included. Level 1, which is conservative, is intended as a screening tool.

If K_{IC} data are used (or equivalent K values from J data), the K ratio is defined by Eq. (9.21). For CTOD data, K_r is replaced by $\sqrt{\delta_r}$, defined as

$$\sqrt{\delta_r} = \sqrt{\frac{\delta_I}{\delta_{crit}}} \tag{9.60}$$

where δ_I is the applied CTOD obtained from a modified form of the CTOD design curve:

$$\delta_I = \frac{K_I^2}{\sigma_{YS} E} \quad \text{for } \sigma_1/\sigma_{YS} \leq 0.5 \tag{9.61a}$$

where $\sigma_1/E = \varepsilon_1$, the maximum membrane strain defined in Eq. (9.19). Recall that ε_1 (and thus σ_1) takes residual stresses, bending stresses, and stress concentrations into account by assuming that the maximum value of the total stress acts uniformly through the cross section.

Unlike Eq. (9.18a), the above expression does not include a safety factor of two on crack size. In the revised approach, this safety factor is included in the formulation of the FAD, which is a horizontal line at $\sqrt{\delta_r} = 1/\sqrt{2}$ for $\sigma_1/\sigma_{YS} \leq 0.5$. The Level 1 failure assessment diagram is illustrated in Fig. 9.25. For higher stress levels, the assessment line is defined from the empirical portion of the CTOD design curve:

$$\delta_I = \frac{K_I^2}{\sigma_{YS} E}\left(\frac{\sigma_{YS}}{\sigma_1}\right)^2\left(\frac{\sigma_1}{\sigma_{YS}} - 0.25\right) \quad \text{for } \sigma_1/\sigma_{YS} > 0.5 \tag{9.61b}$$

The influence of Eq. (9.61b) on the FAD is illustrated in Fig. 9.25. The revised CTOD design curve contains a conservative collapse check in the form of a maximum stress ratio, S_r. For Level 1, S_r is defined as

$$S_r = \frac{\sigma_n}{\sigma_{flow}} \tag{9.62}$$

where σ_n is the effective primary net section stress and σ_{flow} is the flow stress, defined as $(\sigma_{YS} + \sigma_{TS})/2$ or $1.2\ \sigma_{YS}$, whichever is less. As Fig. (9.25) indicates, the Level 1 approach is restricted to $0.8\ S_r$ because Eq. (9.61) can be nonconservative near limit load [39].

9.8.2 Level 2

Level 2 utilizes a strip yield failure assessment diagram. The assessment equation is identical to the original R6 relationship (Eq. (9.23)), except that it allows CTOD based analyses:

$$K_r\,,\sqrt{\delta_r} = S_r\left[\frac{8}{\pi^2}\ln\sec\left(\frac{\pi}{2}S_r\right)\right]^{-\frac{1}{2}} \tag{9.63}$$

Figure 9.25 compares the Level 1 and Level 2 failure assessment diagrams. Note that the Level 1 FAD is always conservative compared to the Level 2 method.

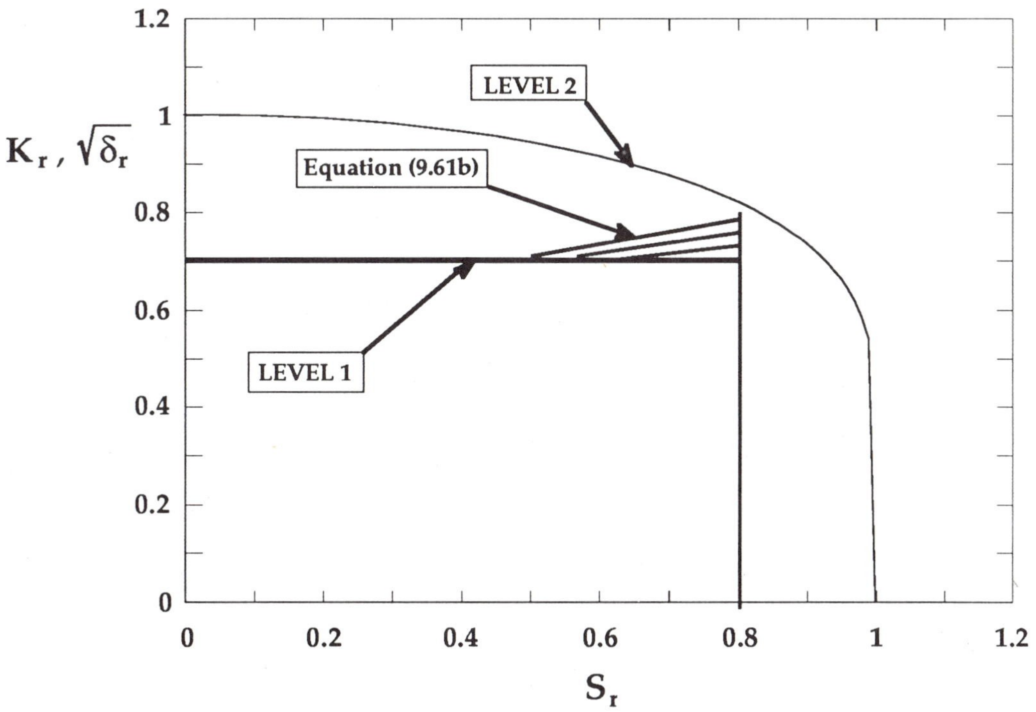

FIGURE 9.25 Failure assessment diagram for Levels 1 and 2 of the PD 6493 method.

The treatment of stress concentration effects and secondary stresses is more complex in the upper two levels. The procedure recommends that accurate stress intensity solutions be obtained for the actual primary and secondary stress distributions. If this is not feasible, an approximate solution can be obtained by linearizing the stress distribution and separating the stresses into bending and membrane components, as discussed in Section 9.1.1 For example, consider a surface crack of depth a. If the primary and secondary stresses are resolved into bending and membrane components, the approximate stress intensity factors are computed from the following expressions:

$$K_I = (P_m + H P_b) F \sqrt{\frac{\pi a}{Q}} \tag{9.64a}$$

$$K_I = (S_m + H S_b) F \sqrt{\frac{\pi a}{Q}} \tag{9.64b}$$

where Q is the flaw shape parameter, P_m and P_b are the primary membrane and bending stresses, S_m and S_b are the secondary stresses, and F and H are constants obtained from the Newman and Raju stress intensity solutions [10], which are given in Table 12.22. If, as in many cases, the actual distribution of secondary stresses is unknown, one should assume that S acts uniformly across the section. The British Standards document recommends that S be assumed to equal the material's yield strength in the case of as-welded components. For thoroughly stress relieved weldments, the estimate of S can be reduced to 30% of yield parallel to the weld and 15% of yield transverse to the weld.

The total K_I is the sum of the primary and secondary contributions. For assessments based on CTOD, δ_I is estimated from K_I by assuming plane stress conditions:

$$\delta_I = \frac{K_I^2}{\sigma_{YS} E} \tag{9.65}$$

There is a plastic interaction between primary and secondary interaction that must be taken into account in Level 2. This is achieved with the correction factor, ρ, based on the work of Ainsworth [40]. The applied toughness ratios for the structure are given by

$$K_r = \frac{K_I}{K_{crit}} + \rho \tag{9.66}$$

and

$$\sqrt{\delta_r} = \sqrt{\frac{\delta_I}{\delta_{crit}}} + \rho \tag{9.67}$$

The procedure for determining ρ is as follows:

$$\rho = \rho_1 \qquad \frac{\sigma_n}{\sigma_{YS}} < 0.8 \tag{9.68a}$$

$$\rho = 4\rho_1\left(1.05 - \frac{\sigma_n}{\sigma_{YS}}\right) \qquad 0.8 < \frac{\sigma_n}{\sigma_{YS}} < 1.05 \tag{9.68b}$$

$$\rho = 0 \qquad \frac{\sigma_n}{\sigma_{YS}} \geq 1.05 \tag{9.68c}$$

where

$$\rho_1 = 0 \qquad \chi < 0 \tag{9.68d}$$

$$\rho_1 = 0.1\,\chi^{0.714} - 0.007\,\chi^2 + 0.00003\,\chi^5 \qquad 0 < \chi \leq 5.2 \tag{9.68e}$$

$$\rho_1 = 0.25 \qquad \chi > 5.2 \tag{9.68f}$$

where $$\chi = \frac{K_I^s}{K_I^p}\frac{\sigma_n}{\sigma_{YS}} \tag{9.68g}$$

When a structure is loaded by primary stresses, a portion of the residual stresses are relieved by plastic strain. A simple way to model this mechanical stress relief is to assume that the sum of the primary and residual stresses cannot exceed the flow stress. For yield magnitude residual stresses in the *unloaded* state, the revised PD 6493 approach permits the user to incorporate the benefits of mechanical stress relief as follows:

$$\sigma_R = (1.4 - S_r)\sigma_{YS} \tag{9.69}$$

where σ_R is the residual stress which is used to compute K_I^s.

The stress ratio, S_r, is defined as the ratio of the effective net section stress to the flow stress, as in Level 1. Refer to Fig. 9.24, which shows that the strip yield model is nonconservative for high hardening materials; restricting the assumed flow stress in S_r to 1.2 σ_{YS} prevents the nonconservatism that is seen in Fig. 9.24(a).

When the point defined by Eqs (9.62), (9.66) and (9.67) falls inside of the Level 2 assessment line (Eq. (9.63)), the structure is considered safe.

9.8.3 Level 3

The Level 3 failure assessment diagram is based on Ainsworth's reference stress approach [34][5]. The FAD is related to the material's stress-strain behavior:

$$K_r, \sqrt{\delta_r} = \left(\frac{E \ln (1 + \varepsilon_{ref})}{\sigma_{ref} (1 + \varepsilon_{ref})} + \frac{\sigma_{ref}^3 (1 + \varepsilon_{ref})^3}{2\, \sigma_{YS}^2\, E \ln (1 + \varepsilon_{ref})} \right)^{-1/2} \tag{9.70}$$

The above quantity is plotted against the load ratio, L_r, defined as

$$L_r = \frac{\sigma_{ref}}{\sigma_{YS}} = \frac{\sigma_n}{\sigma_{YS}} \tag{9.71}$$

Note that the reference stress and the effective net section stress are equivalent. Since the load ratio is defined in terms of the yield strength rather than the flow stress, L_r can be greater than 1. The load ratio cannot exceed $\sigma_{flow}/\sigma_{YS}$, where σ_{flow} is defined as the average between yield and tensile strengths. For Level 3, the alternate definition of flow stress ($\sigma_{flow} = 1.2\, \sigma_{YS}$) does not apply. For $L_r > \sigma_{flow}/\sigma_{YS}$, $K_r = 0$.

If the stress-strain curve for the material is not available, such as would be the case when analyzing a flaw in a weld heat affected zone, the following FAD equation can be applied at Level 3:

$$K_r, \sqrt{\delta_r} = \left(1 - 0.14\, L_r^2\right)\left[0.3 + 0.7 \exp\left(-0.65\, L_r^6\right)\right] \tag{9.72}$$

This expression also has a cut-off at $L_r = \sigma_{flow}/\sigma_{YS}$. This alternate FAD requires a knowledge of only the yield and tensile strengths of the material, but this relationship can be excessively conservative [38]. For many materials, Eq. (9.72) is more conservative than the Level 2 FAD.

---

[5]When the original reference stress equation is expressed in terms of a failure assessment diagram, the resulting diagram is geometry dependent, much like the EPRI approach (Fig. 9.18). Equation (9.70), which is geometry *independent*, is apparently a semi-empirical approximation of a range of geometries.

Figure 9.26 is a plot of Eq. (9.72). Note that the upper cut-off on L_r depends on the hardening characteristics of the material.

The Level 3 analysis of K_r (or $\sqrt{\delta_r}$) for the structure is identical to the Level 2 procedures (Eqs. (9.65) to (9.68)), but Level 3 includes guidelines for ductile instability and tearing analysis.

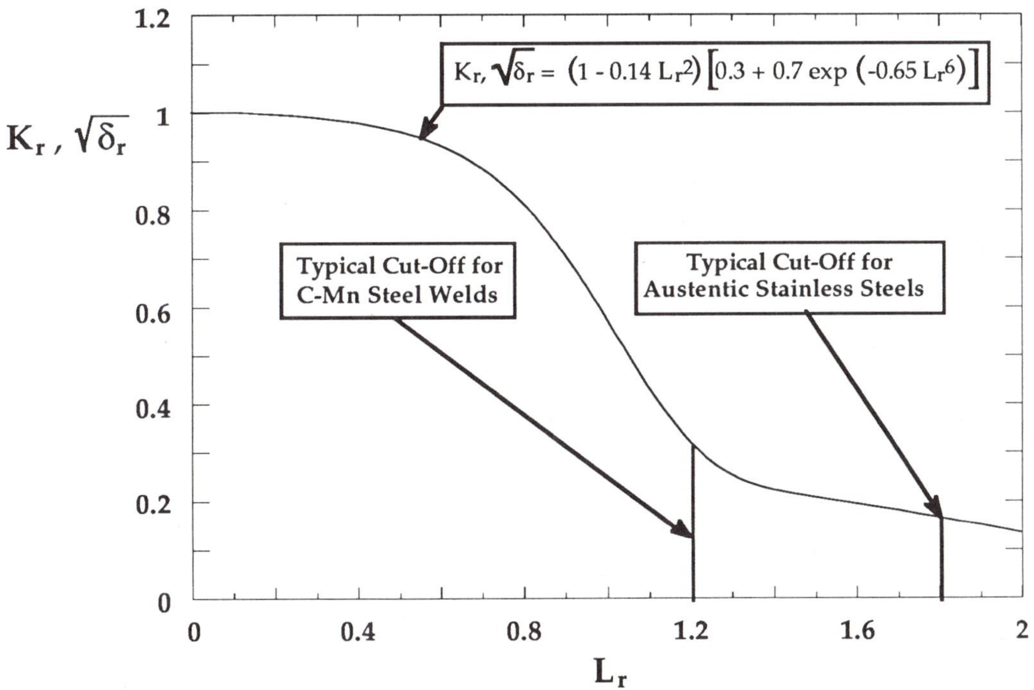

FIGURE 9.27 Optional Level 3 failure assessment diagram, for cases where a stress strain curve is not available.

9.9 THE REVISED R6 METHOD

A revised version of the R6 approach [36], the fracture analysis of nuclear power industry in the United Kingdom, bears a slight resemblance to the revised PD 6493 method. Both methods utilize failure assessment diagrams, and both have adopted the reference stress model as an option. Also, both approaches contain three levels of assessment, although the details at each level differ for the two documents. Space considerations preclude describing the minute details of R6 assessments; a brief overview is given below.

The R6 method contains three options, which are analogous to the three levels in PD 6493. The appropriate option depends on the available data and the desired accuracy.

Option 1 uses the lower-bound FAD defined by Eq. (9.72). This option is appropriate when the relevant stress-strain data are not available.

The *Option 2* FAD is based on the reference stress model. Thus it is necessary to have access to the stress-strain curve for the material in question. The failure assessment diagram for Option 2 is given by

$$K_r = \frac{E\,\varepsilon_{ref}}{L_r\,\sigma_{YS}} + \frac{L_r^3\,\sigma_{YS}}{2\,E\,\varepsilon_{ref}} \qquad \text{for } L_r \leq L_{r(max)} \tag{9.73}$$

Note that Eq. (9.73) differs from the reference stress FAD that is used in Level 3 of PD 6493 (Eq. (9.70)).

Option 3 provides the most accurate analysis. The Option 3 FAD is inferred from a J integral solution for the structure of interest. Normally, such a solution would require an elastic-plastic finite element analysis that incorporated the stress-strain response of the material of interest.

Within each option, there are three categories of analysis:

Category 1: Fracture initiation.

Category 2: Limited stable crack growth.

Category 3: Tearing instability.

The appropriate category depends on the intent of the analysis. For example, design would normally be based on avoiding fracture initiation (Category 1) under normal conditions, but analyses of potential accidents may consider stable and unstable crack growth (Categories 2 and 3).

9.10 PROBABILISTIC FRACTURE MECHANICS

Most fracture mechanics analyses are deterministic; i.e.,, a single value of fracture toughness is used to estimate failure stress or critical crack size. Much of what happens in the real world, however, is not predictable. Since fracture toughness data in the ductile-brittle transition region are widely scattered, it is not appropriate to view fracture toughness as a single-valued material constant. Other factors also introduce uncertainty into fracture analyses. A structure may contain a number of flaws of various sizes, orientations and locations. Extraordinary events such as hurricanes, tidal waves and accidents can result in stresses significantly above the intended design level. Because of these complexities, fracture should be viewed probabilistically rather than deterministically.

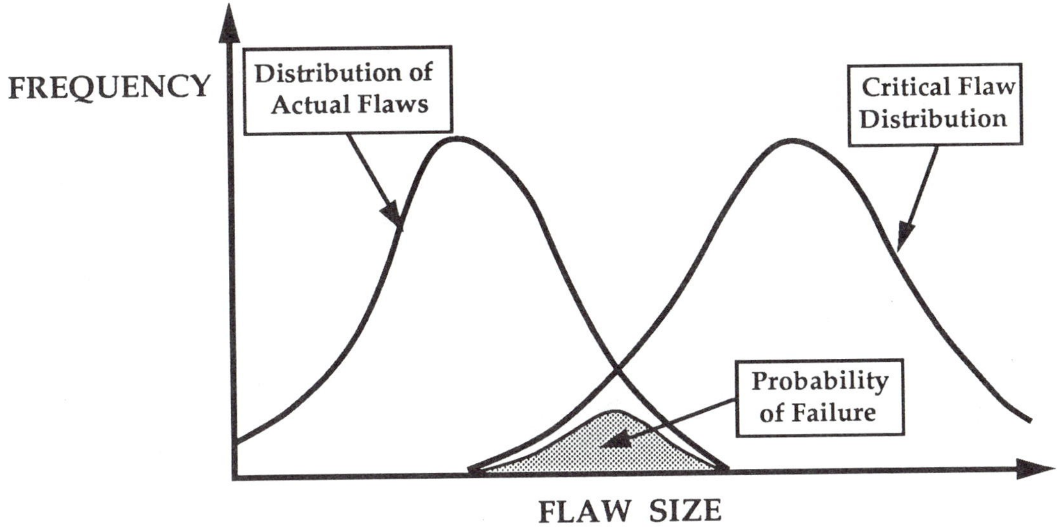

FIGURE 9.27 Schematic probabilistic fracture analysis.

Figure 9.27 is a schematic probabilistic fracture analysis. The curve on the left represents the distribution of flaws in the structure; the curve on the right is the critical flaw distribution. The latter distribution depends on the applied stress and the distribution of fracture toughness. The critical flaw size corresponding to a particular stress and toughness can be computed from one of the driving force equations described in previous sections. When the distributions of critical and actual flaws overlap, there is a finite probability of failure, indicated by the shaded area. Time-dependent crack growth, such as fatigue and stress corrosion cracking, can be taken into account by applying the appropriate growth law to the flaw distribution. This distribution moves to the right with time, increasing the failure probability.

The mathematics of probabilistic analysis is well established [41]. Reliability engineering is currently applied in a variety of circumstances, ranging from quality control in manufacturing to structural integrity. Probabilistic fracture analyses are rare, however, because the input data are usually not available. Scatter in fracture toughness data is one of the main culprits for uncertainty in critical flaw size. A probabilistic analysis traditionally requires performing a large number of fracture toughness tests to define the toughness distribution, but recent research results [42] may greatly reduce the amount of testing required.

9.11 LIMITATIONS OF EXISTING APPROACHES

Section 9.7 compared three driving force equations: the EPRI approach, the reference stress model, and the strip yield model. Each equation reduces to LEFM in the limit of small scale yielding, and each approaches a collapse limit under fully plastic conditions. Although the strip yield model and the reference stress model contain simplifying assumptions, they predict similar results to the more advanced EPRI approach that incorporates a fully plastic J analysis. Any errors in the simpler models, relative to the EPRI approach, are overshadowed by more serious shortcomings in *all* existing methods.

Although some of the fracture design analyses discussed in this chapter are complex, they still do not take into account all aspects of the problem. These analyses are two-dimensional and assume that the material is homogeneous. In addition, all fracture design analyses contain an inherent assumption that the computed driving force parameter (K, J, or CTOD) uniquely characterizes crack tip conditions.

The items discussed below do not constitute an exhaustive list of unresolved issues, but are key areas which need to be understood better before fracture analysis methods can be improved.

It should be noted that current methods of fracture analyses are generally safe. The effects discussed below tend to make current predictions conservative. A better understanding of these complexities would merely make analyses of critical conditions more accurate.

9.11.1 Driving Force in Weldments

A steel weld invariably has different flow properties than the parent metal. In most cases, the yield strength of the weld metal overmatches that of the parent metal, although undermatching sometimes occurs. Analyses such as the EPRI approach are unable to handle structures whose flow properties are heterogeneous. If a crack occurs in or near a weld, it is impossible to determine the driving force accurately without performing an elastic-plastic finite element analysis of the component.

9.11.2 Residual Stresses

The assumptions for secondary stresses have a significant effect on predictions with either the R6 method or PD 6493, but accurate information on the distribution of residual stresses is rarely available for the weld in question.

Conventional methods for measuring the through-thickness distribution of residual stress are destructive. Material from one side of the welded plate is typically removed by a milling machine while strain gage readings on the other side of the plate are recorded. Such an approach is obviously impractical for a structure in service. The center hole drilling technique does minimal damage to the structure, but it only provides information on the surface stresses. The only available nondestructive method for through-thickness residual stress measurement is neutron diffraction, a technique that is not portable.

A reliable, portable, nondestructive method for measuring residual stresses is desperately needed. Accurate finite element models that predict the residual stress distribution from the joint geometry and welding procedure are also desirable.

9.11.3 Three-Dimensional Effects

Existing elastic-plastic analyses do not account for the variation of the driving force along the crack front. Figures 3.36 to 3.38 demonstrate that both J and CTOD vary considerably along the tip of a crack. This effect is particularly pronounced in semielliptical surface flaws. Thus the crack tip conditions cannot be uniquely characterized with a single value of J or CTOD.

These three-dimensional effects influence both cleavage and ductile tearing. Since cleavage is statistical in nature (Chapter 5), good predictions will come only from summing the incremental failure probabilities along the crack front. Such a calculation must take account of the variation in the crack driving force with position. Since ductile crack growth occurs fastest where the driving force is highest, accurate tearing predictions are possible only with a three-dimensional analysis.

9.11.4 Crack Tip Constraint

Constraint is related to the three-dimensional issue. Plane strain fracture analyses assume that J or CTOD uniquely characterizes crack tip stresses and strains. If the entire crack front is not in plane strain, however, there are regions where a single parameter does not characterize crack tip conditions. Similarly, the single parameter assumption breaks down under large scale plasticity, which decrease the crack tip constraint. This constraint loss can occur at very low J (or CTOD) values in structures loaded predominantly in tension (see Section 3.6.1). In such cases, the structure has a higher apparent toughness than the small scale fracture toughness tests, which were loaded predominantly in bending (Fig. 3.34).

The analyses of surface cracks by Parks [43], the results of which are plotted in Figs. 3.38 and 3.39, demonstrate the complexities of constraint. The apparent driving force, quantified by either J or CTOD, was highest at the maximum crack depth; i.e., along an axis perpendicular to the plate surface. The J and CTOD values decreased smoothly along the crack front, reaching a minimum at the free surface, but the maximum normal stresses occur at approximately 30° from the free surface. The crack tip constraint had apparently relaxed at the point of maximum depth, which was close to the back surface of the plate. Because of constraint effects, the highest *true* driving force and the maximum *apparent* driving force were at two different locations.

9.11.5 Gross-Section Yielding

The presence of a crack reduces the cross section in the structure; the net section is defined as the portion of the cross section not occupied by the crack. Net-section yielding refers to the point when the plastic zone spreads throughout the net cross section. Gross-section yielding occurs when plasticity encompasses the entire cross section.

Net-section yielding tends to occur with deep cracks, while gross section yielding is more common with shallow flaws. Since deep cracks are usually avoided in structures, gross-section yielding is much more common. Unfortunately, most elastic-plastic fracture analyses are not equipped to handle this type of deformation.

Elastic-plastic fracture analyses such as the EPRI and R6 methods assume net-section yielding in the structure, and are conservative when gross-section yielding occurs. According to the EPRI approach, J in the fully plastic range should scale with P^{n+1}. However, experimental results of Read [44] indicate that gross-section yielding causes the applied J to increase much more gradually than expected. Thus an elastic-plastic fracture analysis can greatly overestimate the applied J in the case of gross-section yielding. Read's results indicate that gross-section yielding is most likely when the crack comprises less than 5% of the cross section.

REFERENCES

1. Rooke, D.P. and Cartwright, D.J., *Compendium of Stress Intensity Factors.* Her Majesty's Stationary Office, London, 1976.

2. Tada, H., Paris, P.C., and Irwin, G.R. *The Stress Analysis of Cracks Handbook.* Del Research Corporation, Hellertown, Pa, 1973.

3. Murakami, Y. *Stress Intensity Factors Handbook.* Pergamon Press, New York, 1987.

4. Sanford, R.J. and Dally, J.W., "A General Method for Determining Mixed-Mode Stress Intensity Factors from Isochromatic Fringe Patterns." *Engineering Fracture Mechanics,* Vol. 11, 1979, pp. 621-633.

5. Chona, R. , Irwin, G.R., and Shukla, A., "Two and Three Parameter Representation of Crack Tip Stress Fields." *Journal of Strain Analysis,* Vol. 17, 1982, pp. 79-86.

6. Kalthoff, J.F., Beinart, J., Winkler, S., and Klemm, W., "Experimental Analysis of Dynamic Effects in Different Crack Arrest Test Specimens." ASTM STP 711, American Society for Testing and Materials, Philadelphia, 1980, pp. 109-127.

7. Raju, I.S. and Newman J.C., Jr., "Stress-Intensity Factors for Internal and External Surface Cracks in Cylindrical Vessels." *Journal of Pressure Vessel Technology*, Vol. 104, 1972, pp. 293-298.

8. Rice, J.R., "Some Remarks on Elastic Crack-Tip Stress Fields." *International Journal of Solids and Structures*, Vol. 8, 1972, pp. 751-758.

9. Rice, J.R., "Weight Function Theory for Three-Dimensional Elastic Crack Analysis." ASTM STP 1020, American Society for Testing and Materials, Philadelphia, 1989, pp. 29-57.

10. Newman, J.C., Jr. and Raju, I.S., "An Empirical Stress Intensity Factor Equation for Surface Cracks." *Engineering Fracture Mechanics*, Vol 15, 1981, pp. 185-192.

11. Timoshenko, S., *Strength of Materials, Advanced Theory and Problems*. D. Van Nostrand Company, New York, 1956.

12. Atluri, S.N. and Kathiresan, K., "3-D Analyses of Surface Flaws in Thick-Walled Reactor Pressure Vessels Using Displacement-Hybrid Finite Element Method." *Nuclear Engineering and Design*, Vol. 51, 1979, pp. 163-176.

13. Irwin, G.R., "Plastic Zone Near a Crack and Fracture Toughness." *Sagamore Research Conference Proceedings*, Vol. 4, 1961.

14. Burdekin, F.M. and Stone, D.E.W., "The Crack Opening Displacement Approach to Fracture Mechanics in Yielding Materials." *Journal of Strain Analysis*, Vol. 1, 1966, pp. 144-153.

15. Hayes, D.J. and Williams, J.G., "A Practical Method for Determining Dugdale Model Solutions for Cracked Bodies of Arbitrary Shape." *International Journal of Fracture Mechanics*, Vol. 8, 1972, pp. 239-256.

16. *American Society of Mechanical Engineers (ASME) Boiler and Pressure Vessel Code. Section XI: Rules for Inservice Inspection of Nuclear Power Plant Components.* American Society of Mechanical Engineers, New York.

17. Marsdon, T.U., ed., "Flaw Evaluation Procedures: Background and Application of ASME Section XI, Appendix A." EPRI NP-719-SR, Electric Power Research Institute, Palo Alto, CA, 1978.

18. Burdekin, F.M. and Dawes, M.G., "Practical Use of Linear Elastic and Yielding Fracture Mechanics with Particular Reference to Pressure Vessels." *Proceedings of the Institute of Mechanical Engineers Conference*, London, May 1971, pp. 28-37.

19. Wells, A.A., "Application of Fracture Mechanics at and Beyond General Yielding." *British Welding Journal*, Vol 10, 1963, pp. 563-570.

20. Dawes, M.G., "Fracture Control in High Yield Strength Weldments." *Welding Journal*, Vol. 53, 1974, pp. 369-380.

21. PD 6493:1980, "Guidance on Some Methods for the Derivation of Acceptance Levels for Defects in Fusion Welded Joints." British Standards Institution, March 1980.

22. Kamath, M.S., "The COD Design Curve: An Assessment of Validity Using Wide Plate Tests." The Welding Institute Report 71/1978/E, September 1978.

23. Dowling, A.R. and Townley, C.H.A., "The Effects of Defects on Structural Failure: A Two-Criteria Approach." *International Journal of Pressure Vessels and Piping,* Vol 3, 1975, pp. 77-137.

24. Harrison, R.P., Loosemore, K., and Milne, I., "Assessment of the Integrity of Structures Containing Defects." CEGB Report R/H/R6, Central Electricity Generating Board, United Kingdom, 1976.

25. Harrison, R.P., Loosemore, K., Milne, I, and Dowling, A.R., "Assessment of the Integrity of Structures Containing Defects." CEGB Report R/H/R6-Rev 2, Central Electricity Generating Board, United Kingdom, 1980.

26. Shih, C.F. and Hutchinson, J.W., "Fully Plastic Solutions and Large-Scale Yielding Estimates for Plane Stress Crack Problems." *Journal of Engineering Materials and Technology,* Vol. 98, 1976, pp. 289-295.

27. Kumar, V., German, M.D., and Shih, C.F.,"An Engineering Approach for Elastic-Plastic Fracture Analysis." EPRI Report NP-1931, Electric Power Research Institute, Palo Alto, CA, 1981.

28. Kumar, V., German, M.D., Wilkening, W.W., Andrews, W.R., deLorenzi, H.G., and Mowbray, D.F., "Advances in Elastic-Plastic Fracture Analysis." EPRI Report NP-3607, Electric Power Research Institute, Palo Alto, CA, 1984.

29. Kumar, V. and German, M.D., "Elastic-Plastic Fracture Analysis of Through-Wall and Surface Flaws in Cylinders." EPRI Report NP-5596, Electric Power Research Institute, Palo Alto, CA, 1988.

30. Zahoor, A. "Ductile Fracture Handbook, Volume 1: Circumferential Throughwall Cracks." EPRI Report NP-6301-D, Electric Power Research Institute, Palo Alto, CA, 1989.

31. Shih, C.F. "Relationship between the J-Integral and the Crack Opening Displacement for Stationary and Extending Cracks." *Journal of the Mechanics and Physics of Solids,* Vol 29, 1981, pp. 305-326.

32. Bloom, J.M., "Prediction of Ductile Tearing Using a Proposed Strain Hardening Failure Assessment Diagram." *International Journal of Fracture,* Vol. 6., 1980, pp. R73-R77.

33. Shih, C.F., German, M.D., and Kumar, V., "An Engineering Approach for Examining Crack Growth and Stability in Flawed Structures." *International Journal of Pressure Vessels and Piping,* Vol. 9, 1981, pp. 159-196.

34. Ainsworth, R.A., "The Assessment of Defects in Structures of Strain Hardening Materials." *Engineering Fracture Mechanics,* Vol. 19, 1984, p. 633.

35. Anderson, T.L., "Application of Elastic-Plastic Fracture Mechanics to Welded Structures: A Critical Review." Mechanics and Materials Center Report MM 6165-89-1, Texas A&M University, College Station, TX, August 1989.

36. Milne, I., Ainsworth, R.A., Dowling, A.R., and Stewart, A.T., "Assessment of the Integrity of Structures Containing Defects." Central Electricity Generating Board Report R/H/R6-Rev 3, May 1986.

37. Burdekin, F.M., Garwood, S.J., and Milne, I, "The Basis for the Technical Revisions to the Fracture Clauses of PD 6493." Presented at the International Conference on Weld Failures, London, November 1988.

38. Garwood, S.J., Willoughby, A.A., Leggatt, R.H., and Jutla, T., "Crack Tip Opening Displacement (CTOD) Methods for Fracture Mechanics Assessments: Proposals for Revisions to PD 6493." Presented at ASFM 6, Ispra, Italy, October 1987.

39. Anderson, T.L., Leggatt, R.H., and Garwood, S.J., "The Use of CTOD Methods in Fitness for Purpose Analysis." *The Crack Tip Opening Displacement in Elastic-Plastic Fracture Mechanics,* Springer-Verlag, Berlin, 1986, pp. 281-313.

40. Ainsworth, R.A., "The Treatment of Thermal and Residual Stresses in Fracture Assessments." Central Electricity Generating Board Report TPRD/0479/N84, 1984.

41. Bain, L.J., *Statistical Analysis of Reliability and Life-Testing Models,* Marcel Dekker, Inc. New York, 1978.

42. Stienstra, D.I.A, "Stochastic Micromodeling of Cleavage Fracture in the Ductile-Brittle Transition Region." Ph.D. Dissertation, Texas A&M University, College Station, TX, August 1990.

43. Parks, D.M. and Wang, Y.-Y., "Elastic-Plastic Analysis of Part-Through Surface Cracks." Analytical, Numerical and Experimental Aspects of Three-Dimensional Fracture Processes, American Society of Mechanical Engineers, New York, 1988, pp. 19-32.

44. Read, D.T., "Applied J-Integral in HY130 Tensile Panels and Implications for Fitness for Service Assessment." Report NBSIR 82- 1670, National Bureau of Standards, Boulder, CO, 1982.

10. FATIGUE CRACK PROPAGATION

Most of the material in the preceding chapters has dealt with static or monotonic loading of cracked bodies. This chapter considers crack growth in the presence of cyclic stresses. The focus is on fatigue of metals, but many of the concepts presented in this chapter apply to other materials as well.

In the early 1960s, Paris, et al. [1,2] demonstrated that fracture mechanics is a useful tool for characterizing crack growth by fatigue. Since that time, the application of fracture mechanics to fatigue problems has become almost routine. There are, however, a number of controversial issues and unanswered questions in this field.

The procedures for analyzing constant amplitude fatigue[1] under small scale yielding conditions are fairly well established, although a number of uncertainties remain. Variable amplitude loading, large scale plasticity, and short cracks introduce additional complications that are not fully understood.

This chapter summarizes the fundamental concepts and practical applications of the fracture mechanics approach to fatigue crack propagation. Section 10.1 outlines the similitude concept, which provides the theoretical justification for applying fracture mechanics to fatigue problems. This is followed by a summary of the more common empirical and semiempirical equations for characterizing fatigue crack growth. Subsequent sections discuss crack closure, variable amplitude loading, retardation, and growth of short cracks in terms of the validity (or lack of validity) of the similitude assumption. The micromechanisms of fatigue are also discussed briefly. The final two sections are geared to practical applications; Section 10.7 outlines procedures for experimental measurements of fatigue crack growth, and Section 10.8 summarizes the damage tolerance approach to fatigue safe design. Appendix 10 at the end of this chapter gives the applicability of the J integral to cyclic loading.

---

[1]In this chapter, constant amplitude loading is defined as a constant *stress intensity* amplitude rather than constant stress amplitude.

10.1 SIMILITUDE IN FATIGUE

The concept of similitude, when it applies, provides the theoretical basis for fracture mechanics. Similitude implies that the crack tip conditions are uniquely defined by a single loading parameter such as the stress intensity factor. In the case of a stationary crack, two configurations will fail at the same critical K value, provided an elastic singularity zone exists at the crack tip (Section 2.10). Under certain conditions, fatigue crack growth can also be characterized by the stress intensity factor, as discussed below.

Consider a growing crack in the presence of a constant amplitude cyclic stress intensity (Fig. 10.1). A cyclic plastic zone forms at the crack tip, and the growing crack leaves behind a plastic wake. If the plastic zone is sufficiently small that it is embedded within an elastic singularity zone, the conditions at the crack tip are uniquely defined by the current K value[2], and the crack growth rate is characterized by K_{min} and K_{max}. It is convenient to express the functional relationship for crack growth in the following form:

$$\frac{da}{dN} = f_1(\Delta K, R) \tag{10.1}$$

where $\Delta K \equiv (K_{max} - K_{min})$, $R \equiv K_{min}/K_{max}$, and da/dN is the crack growth per cycle. The influence of the plastic zone and plastic wake on crack growth is implicit in Eq. (10.1), since the size of the plastic zone depends only on K_{min} and K_{max}.

A number of expressions for f_1 have been proposed, most of which are empirical. Section 10.2 outlines some of the more common fatigue crack growth relationships. Equation (10.1) can be integrated to estimate fatigue life. The number of cycles required to propagate a crack from an initial length, a_o, to a final length, a_f, is given by

---

[2]The justification for the similitude assumption in fatigue is essentially identical to the dimensional argument for steady state crack growth (Section 3.5.2 and Appendix 3.5.2). If the tip of the growing crack is sufficiently far from its initial position, and external boundaries are remote, the plastic zone size and width of the plastic wake will reach steady state values.

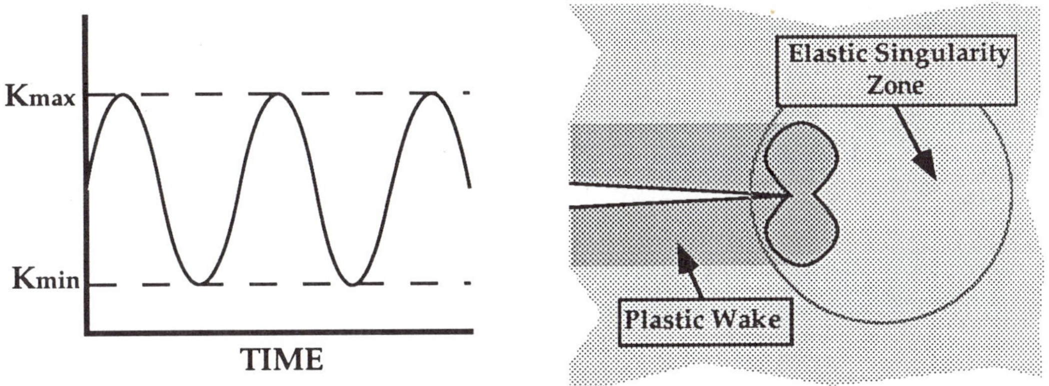

FIGURE 10.1 Constant amplitude fatigue crack growth under small scale yielding conditions.

$$N = \int_{a_o}^{a_f} \frac{da}{f_1(\Delta K, R)} \tag{10.2}$$

If K_{max} or K_{min} vary during cyclic loading, the crack growth in a given cycle may depend on the loading history as well as the current values of K_{min} and K_{max}:

$$\frac{da}{dN} = f_2(\Delta K, R, \mathcal{H}) \tag{10.3}$$

where $\mathcal{H}$ indicates the history dependence, which results from prior plastic deformation. Equation (10.3) violates the similitude assumption; two configurations cyclically loaded at the same ΔK and R will not exhibit the same crack growth rate unless both configurations are subject to the same prior history.

Figure 10.2 illustrates several examples where the similitude assumption is invalid. In each case, prior loading history influences the current conditions at the crack tip. Section 10.4 discusses the reasons for history-dependent fatigue, and gives an example of a model that accounts for loading history.

Fatigue crack growth analyses become considerably more complicated when prior loading history is taken into account. Consequently, equations of the form of Eq. (10.1) are applied whenever possible. It must be recognized, however, that such analyses are only approximate in the case of variable amplitude loading.

Excessive plasticity during fatigue can violate similitude, since K no longer characterizes the crack tip conditions in such cases. A number of researchers [3,4] have applied the J integral to fatigue accompanied by large scale yielding; they have assumed a growth law of the form

$$\frac{da}{dN} = f_3(\Delta J, R) \tag{10.4}$$

where ΔJ is a contour integral for cyclic loading, analogous to the J integral for monotonic loading (see Appendix 10). Equation (10.4) is valid in the case of constant amplitude fatigue in small scale yielding, because of the relationship between J and K under linear elastic conditions[3]. The validity of Eq. (10.4) in the presence of significant plasticity is less clear, however.

Recall from Chapter 3 that deformation plasticity (i.e., nonlinear elasticity) is an essential component of J integral theory. When unloading occurs in an elastic-plastic material, deformation plasticity theory no longer models the actual material response (see Fig. 3.7). Consequently, the ability of the J integral to characterize fatigue crack growth in the presence of large scale cyclic plasticity is questionable, to say the least.

There is, however, some theoretical and experimental evidence in favor of Eq. (10.4). If certain assumptions are made with respect to the loading and unloading branches of a cyclic stress-strain curve, it can be shown that ΔJ is path independent, and it uniquely characterizes the change in stresses and strains in a given cycle [5,6]. Appendix 10 summarizes this analysis. Experimental data [3,4] indicate that ΔJ correlates crack growth data reasonably well in certain cases. Several researchers have found that CTOD may also be a suitable parameter for fatigue under elastic-plastic conditions [7].

---

[3] $\Delta J = \Delta K^2/E'$ in the case of small-scale yielding. Thus ΔJ cannot be interpreted as the range of applied J values. That is, $\Delta J \neq J_{max} - J_{min}$ in general. See Appendix 10 for additional background on the definition of ΔJ.

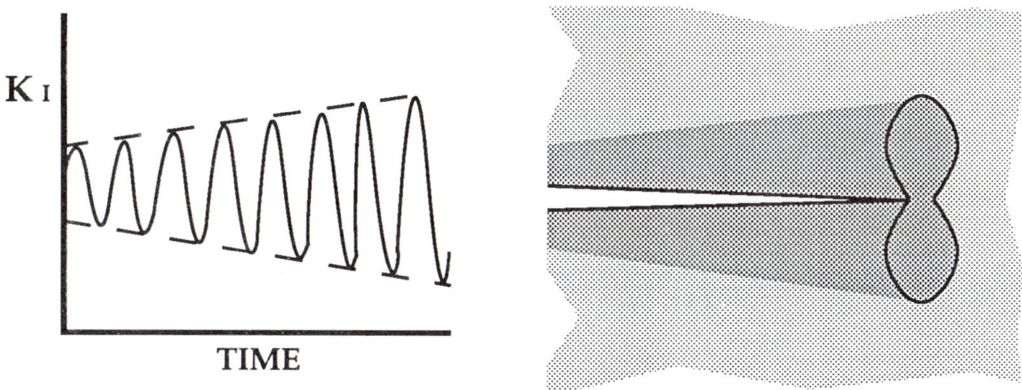

(a) K increasing.

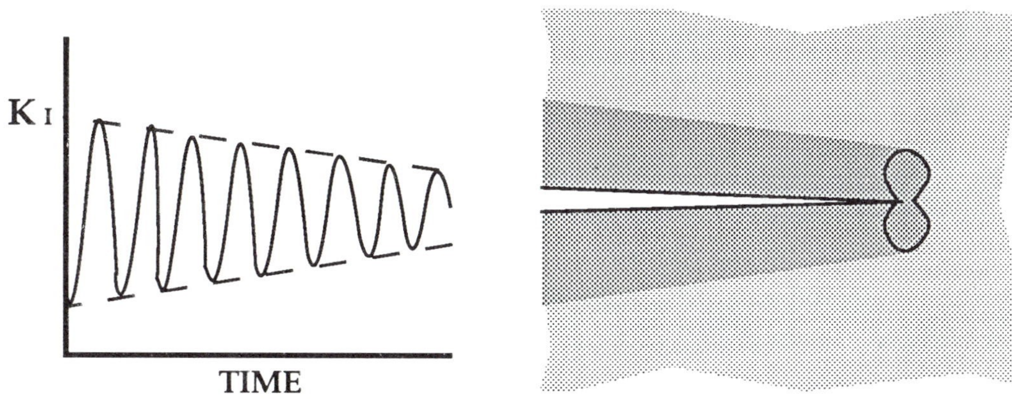

(b) K decreasing

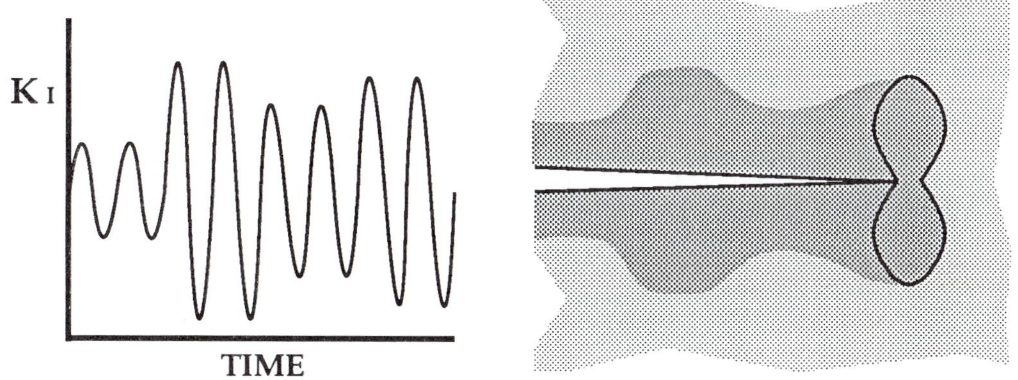

(c) Random loading.

FIGURE 10.2. Examples of cyclic loading that violate similitude.

The validity of Eq. (10.4) has not been proven conclusively, but this approach appears to be useful for many engineering problems. Of course, Eq. (10.4) is subject to the same restrictions on prior history as Eq. (10.1). The crack growth rate may exhibit a history effect if ΔJ or R vary during cyclic loading.

10.2 EMPIRICAL FATIGUE CRACK GROWTH EQUATIONS

Figure 10.3 is a schematic log-log plot of da/dN versus ΔK, which illustrates typical fatigue crack growth behavior in metals. The sigmoidal curve contains three distinct regions. At intermediate ΔK values, the curve is linear, but the crack growth rate deviates from the linear trend at high and low ΔK levels. In the former case, the crack growth rate accelerates as K_{max} approaches K_{crit}, the fracture toughness of the material. At the other extreme, da/dN approaches zero at a threshold ΔK; Section 10.3 explores the causes of this threshold.

The linear region of the log-log plot in Fig. 10.3 can be described by a power law:

$$\frac{da}{dN} = C\,\Delta K^m \tag{10.5}$$

where C and m are material constants that are determined experimentally. According to Eq. (10.5), the fatigue crack growth rate depends only on ΔK; da/dN is insensitive to the R ratio in Region II.

Paris and Erdogan [2] were apparently the first to discover the power law relationship for fatigue crack growth in Region II. They proposed an exponent of four, which was in line with their experimental data. Subsequent studies over the past three decades, however, have shown that m is not necessarily four, but ranges from two to seven for various materials. Equation (10.5) has become widely known as the Paris law.

A number of researchers have developed equations that model all or part of the sigmoidal da/dN - ΔK relationship. Many of these equations are empirical, although some are based on physical considerations. Forman [8] proposed the following relationship for Regions II and III:

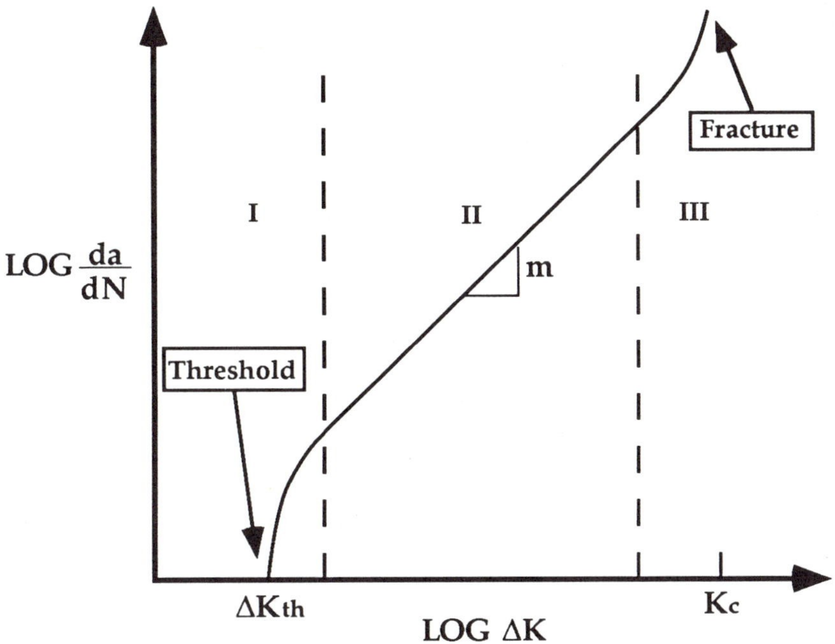

FIGURE 10.3 Typical fatigue crack growth behavior in metals.

$$\frac{da}{dN} = \frac{C\,\Delta K^m}{(1 - R)\,K_{crit} - \Delta K} \tag{10.6}$$

This equation can be rewritten in the following form:

$$\frac{da}{dN} = \frac{C\,\Delta K^{m-1}}{\dfrac{K_{crit}}{K_{max}} - 1} \tag{10.7}$$

Thus the crack growth rate becomes infinite as K_{max} approaches K_{crit}. Note that the above relationship accounts for R ratio effects, while Eq. (10.5) assumes that da/dN depends only on ΔK. Another important point is that the material constants C and m in the Forman equation do not have the same numerical values or units as in the Paris-Erdogan equation (Eq. (10.5)).

Weertman [9] proposed an alternative semiempirical equation for Regions II and III:

$$\frac{da}{dN} = \frac{C\,\Delta K^4}{K_{crit}^2 - K_{max}^2} \tag{10.8}$$

This equation can be made more general with a variable exponent, m, on ΔK. Again, the fitting parameters, C and m, do not necessarily have the same values or units in the various crack growth equations.

Both the Forman and Weertman equations are asymptotic to $K_{max} = K_{crit}$, but neither predicts a threshold. Klesnil and Lukas [10] modified Eq. (10.5) to account for the threshold:

$$\frac{da}{dN} = C\,(\Delta K^m - \Delta K_{th}^m) \tag{10.9}$$

Donahue [11] suggested a similar equation, but with the exponent, m, applied to the quantity $(\Delta K - \Delta K_{th})$. In both cases, the threshold is a fitting parameter to be determined experimentally. One problem with these equations is that ΔK_{th} often depends on the R ratio (see Section 10.3).

A number of equations attempt to describe the entire crack growth curve, taking account of both the threshold and K_{crit}. For example, Priddle proposed the following empirical relationship:

$$\frac{da}{dN} = C\left(\frac{\Delta K - \Delta K_{th}}{K_{crit} - K_{max}}\right)^m \tag{10.10}$$

McEvily [12] developed another equation that can be fit to the entire crack growth curve:

$$\frac{da}{dN} = C\,(\Delta K - \Delta K_{th})^2\left(1 + \frac{\Delta K}{K_{crit} - K_{max}}\right) \tag{10.11}$$

Equation (10.11) is based on a simple physical model rather than a purely empirical fit.

Equations (10.5) to (10.11) all have the form of Eq. (10.1). Each of these equations can be integrated to infer fatigue life (Eq. (10.2)). The most general of these expressions contain four material constants:[4] C, m, K_{crit} and ΔK_{th}. For a given material, the fatigue crack growth rate depends only on the loading parameters ΔK and R, at least according to the Eqs. (10.5) to (10.12). Thus all of the preceding expressions assume elastic similitude of the growing crack; none of these equations incorporate a history dependence, and thus are strictly valid only for constant (stress intensity) amplitude loading. Many of these formula, however, were developed with variable amplitude loading in mind. Although there are situations where similitude is approximately satisfied for variable amplitude loading, one must always bear in mind the potential for history effects. See Sections 10.3 and 10.4 for additional discussion of this issue.

Dowling and Begley [3] applied the J integral to fatigue crack growth under large scale yielding conditions where K is no longer valid. They fit the growth rate data to a power law expression in ΔJ:

$$\frac{da}{dN} = C\,\Delta J^m \tag{10.12}$$

Appendix 10 outlines the theoretical justification and limitations of J-based approaches.

EXAMPLE 10.1

Derive an expression for the number of stress cycles required to grow a semicircular surface crack from an initial radius a_o to a final size a_f, assuming the Paris-Erdogan equation describes the growth rate. Assume that a_f is small compared to plate dimensions, and that the stress amplitude, $\Delta\sigma$, is constant.

Solution: The stress intensity amplitude for a semicircular surface crack in an infinite plate (Eq. 2.44) is given by

$$\Delta K = \frac{2.24}{\pi}\,\Delta\sigma\,\sqrt{\pi\, a}$$

---

[4]The threshold stress intensity range, ΔK_{th}, is not a true material constant since it usually depends on the R ratio (Section 10.3).

EXAMPLE 10.1 (Cont.)

Substituting this expression into Eq. (10.5) gives

$$\frac{da}{dN} = C\left(\frac{2.24}{\pi}\Delta\sigma\right)^m (\pi\ a)^{m/2}$$

which can be integrated to determine fatigue life:

$$N = \frac{1}{C\left(\frac{2.24}{\sqrt{\pi}}\Delta\sigma\right)^m}\int_{a_o}^{a_f} a^{-m/2}\,da$$

$$= \frac{a_o^{1-m/2} - a_f^{1-m/2}}{C\left(\frac{m}{2} - 1\right)\left(\frac{2.24}{\sqrt{\pi}}\Delta\sigma\right)^m} \qquad \text{(for } m \neq 2\text{)}$$

Closed-form integration is possible in this case because the K expression is relatively simple. In most instances, numerical integration is required.

10.3 CRACK CLOSURE AND THE FATIGUE THRESHOLD

Soon after the so-called Paris law (Eq. (10.5)) gained wide acceptance as a predictor of fatigue crack growth, many researchers came to the realization that this simple expression was not universally applicable. As Fig 10.3 illustrates, a log-log plot of da/dN v. ΔK is sigmoidal rather than linear when crack growth data are obtained over a sufficiently wide range. Also, the fatigue crack growth rate exhibits a dependence on the R ratio, particularly at both extremes of the crack growth curve. As discussed in the previous section, the R ratio effects at the upper end of the curve can be explained in terms of the interaction between fatigue and ultimate failure at or near K_c. This section addresses the behavior at the lower end of the da/dN - ΔK curve.

A discovery by Elber [13] provided at least a partial explanation for both the fatigue threshold and R ratio effects. He noticed an anomaly in the elastic compliance of several fatigue specimens, which Fig. 10.4(a) schematically illustrates. At high loads, the compliance ($d\Delta/dP$) agreed with standard formulas for fracture mechanics specimens (see Chapters 7 and 12), but at low loads, the compliance was close to that of an uncracked specimen. Elber believed that this change in compliance was due to the contact between crack surfaces (i.e., crack closure) at loads that were low but greater than zero.

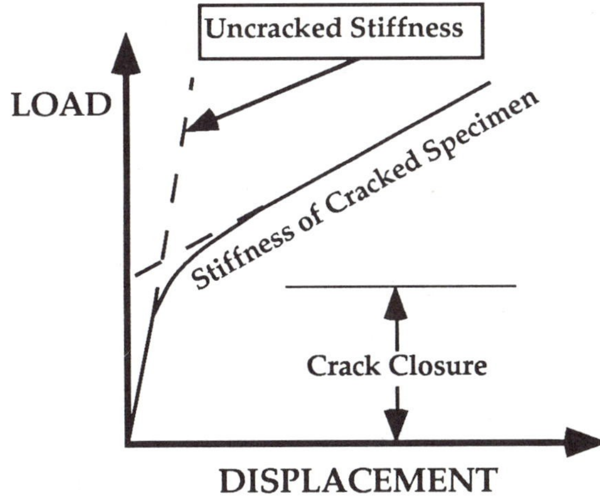

(a) Load-displacement behavior.

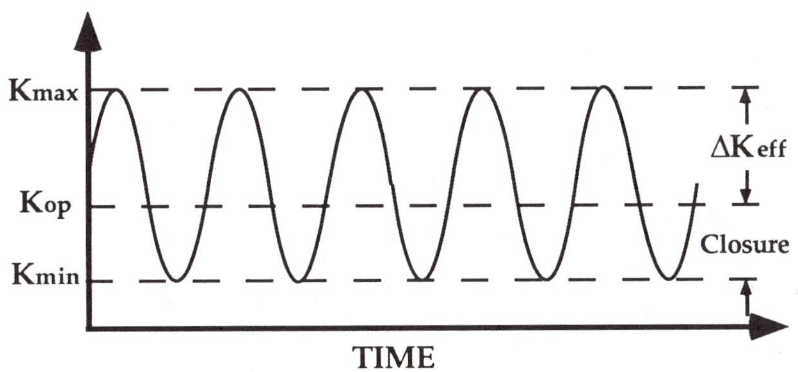

(b) Definition of effective stress intensity range.

FIGURE 10.4 Crack closure during fatigue crack growth. The crack faces contact at a positive load (a), resulting in a reduced driving force for fatigue, ΔK_{eff} (b).

Elber postulated that crack closure decreased the fatigue crack growth rate by reducing the effective stress intensity range. Figure 10.4(b) illustrates the closure concept. When a specimen is cyclically loaded at K_{max} and K_{min}, the crack faces are in contact below K_{op}, the stress intensity at which the crack opens. Elber assumed that the portion of the cycle that is below K_{op} does not contribute to fatigue crack growth. He defined an effective stress intensity range as follows:

$$\Delta K_{eff} \equiv K_{max} - K_{op} \tag{10.13}$$

He also introduced an effective stress intensity ratio:

$$U \equiv \frac{\Delta K_{eff}}{\Delta K}$$

$$= \frac{K_{max} - K_{op}}{K_{max} - K_{min}} \tag{10.14}$$

Elber then proposed a modified Paris-Erdogan equation:

$$\frac{da}{dN} = C\, \Delta K_{eff}^{m} \tag{10.15}$$

Equation (10.15) has been reasonably successful in correlating fatigue crack growth data at various R ratios.

Since Elber's original study, numerous researchers have confirmed that crack closure does in fact occur during fatigue crack propagation. Suresh and Ritchie [14] identified five mechanisms for fatigue crack closure, which are illustrated in Fig. 10.5

Plasticity-induced closure, Fig. 10.5(a), results from residual stresses in the plastic wake. Budiansky and Hutchinson [15] applied the Dugdale-Barenblatt strip yield model to this problem and showed that residual stretch in the plastic wake causes the crack faces to close at a positive remote stress. Although quantitative predictions from the Budiansky and Hutchinson model do not agree with experimental data [16], this model is useful for demonstrating qualitatively the effect of plasticity on crack closure. Recently, a several investigators [17,18] have studied plasticity-induced closure with finite element analysis.

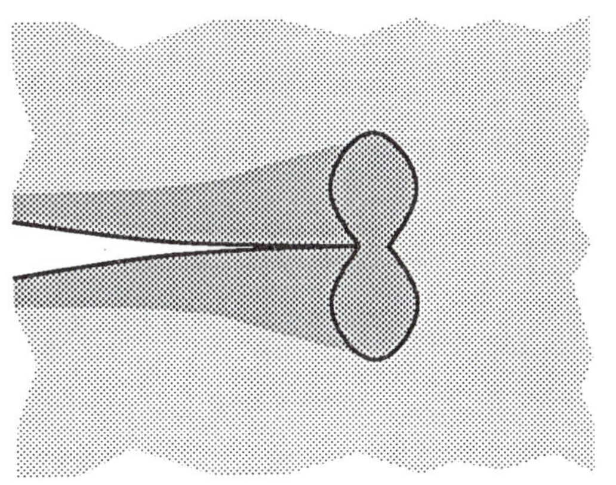

(a) Plasticity-induced closure.

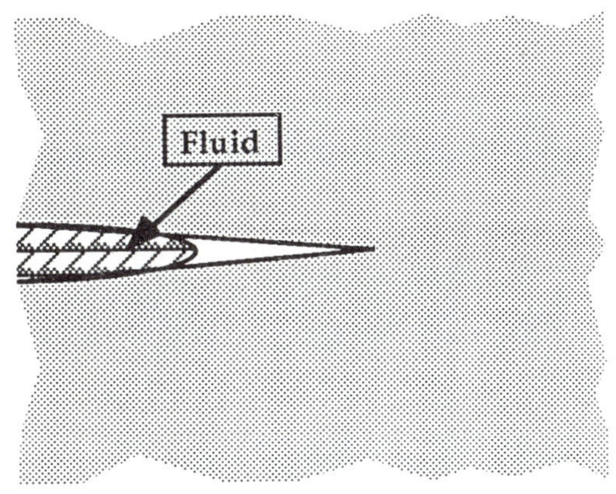

(d) Closure induced by a viscous fluid.

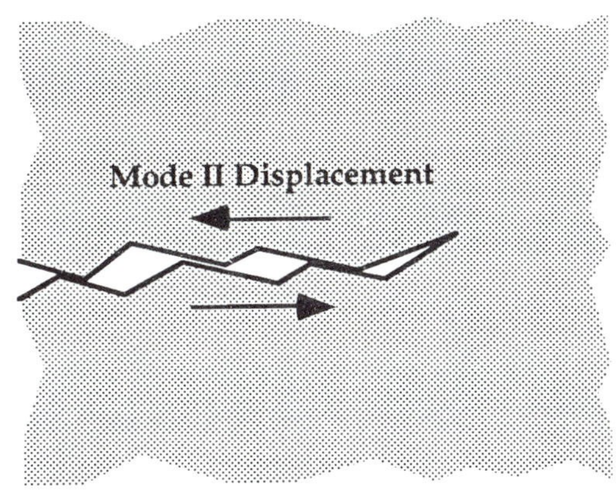

(b) Roughness-induced closure.

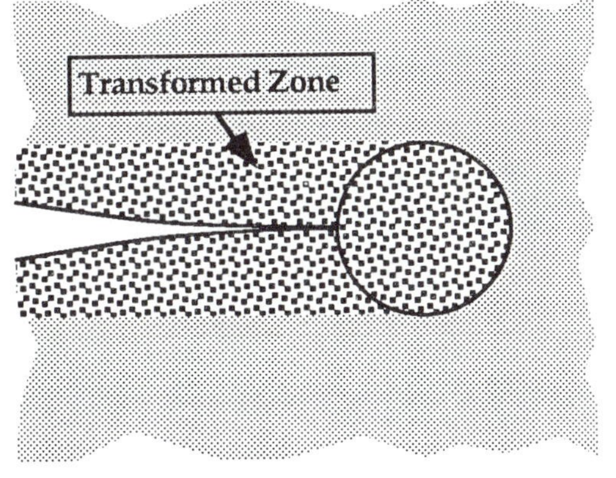

(e) Transformation-induced closure.

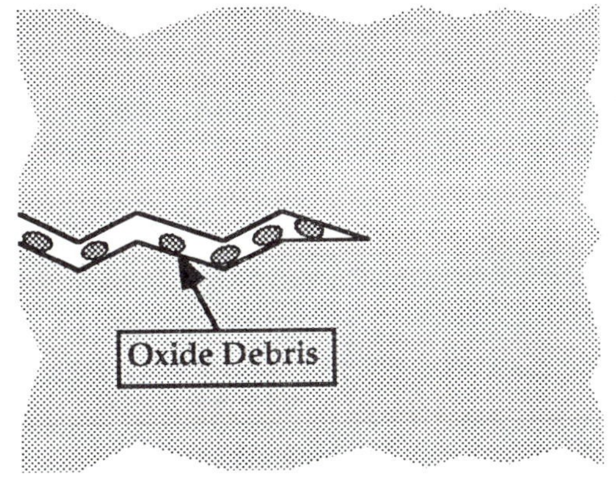

(c) Oxide-induced closure.

FIGURE 10.5 Fatigue crack closure mechanisms in metals [14].

Roughness-induced closure, which is illustrated in Fig. 10.5(b), is influenced by the microstructure. Although fatigue cracks propagate in pure Mode I conditions on a global scale, crack deflections due to microstructural heterogeneity can lead to mixed mode conditions on the microscopic level. When the crack path deviates from the Mode I symmetry plane, the crack is subject to Mode II displacements, as Fig. 10.5(b) illustrates. These displacements cause mismatch between upper and lower crack faces, which in turn results in a positive closure load. Coarse-grained materials usually produce a higher degree of surface roughness in fatigue, and correspondingly higher closure loads [19]. Figure 10.6 illustrates the effect of grain size on fatigue crack propagation in 1018 steel. At the lower R ratio, where closure effects are most pronounced (see below), the coarse grained material has a higher ΔK_{th}, due to a higher closure load that is caused by greater surface roughness (Fig. 10.6(b)). Note that grain size effects disappear when the data are characterized by ΔK_{eff} (Fig. (10.6(c)).

Oxide-induced closure, Fig. 10.5(c), is usually associated with an aggressive environment. Oxide debris or other corrosion products become wedged between crack faces.

Crack closure can also be introduced by a viscous fluid, as Fig. 10.5(d) illustrates. The fluid acts as a wedge between crack faces, somewhat like the oxide mechanism.

A stress induced martensitic transformation at the tip of the growing crack can result in a process zone wake[5], as Fig. 10.5(e) illustrates. Residual stresses in the transformed zone can lead to crack closure.

The relative importance of the various closure mechanisms depends on microstructure, yield strength, and environment.

10.3.1 A Simplistic View of Closure and ΔK_{th}

Crack closure reduces the fatigue crack growth rate and introduces a threshold. It is possible to estimate the effects the fatigue crack closure by making two simplifying assumptions:

---

[5]See Section 6.2.2 for a discussion of stress-induced martensitic transformations in ceramics.

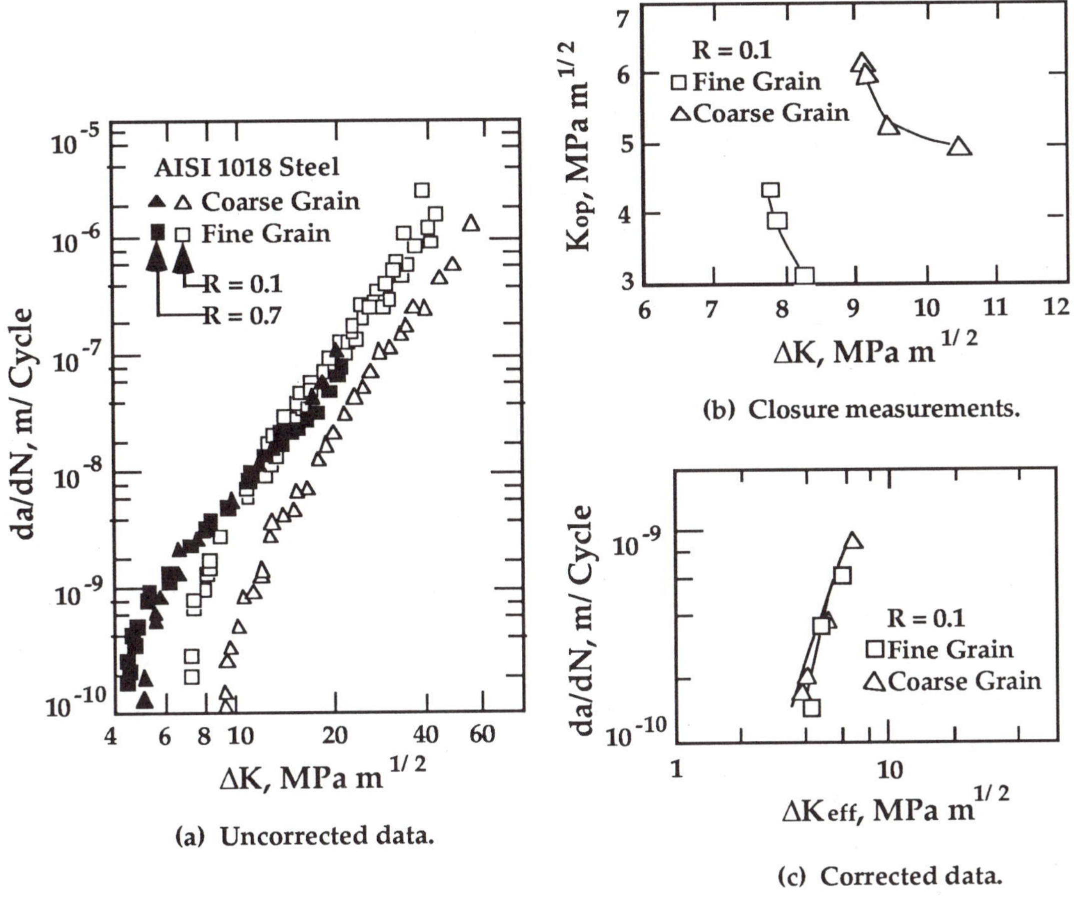

FIGURE 10.6 Effect of grain size on fatigue crack growth in mild steel [19].

(1) The opening stress intensity, K_{op}, is a material constant; i.e. it is independent of K_{min}, K_{max}, and prior history.

(2) There is no intrinsic ΔK_{th} for the material; Eq. (10.15) applies as long as $\Delta K_{eff} > 0$.

Although neither assumption is valid for real materials (Section 10.3.2 and 10.3.3), the intent of the present exercise is merely to illustrate the consequences of closure, independent of other factors.

The Elber crack growth expression (Eq. (10.15)) can be written in the following form:

$$\frac{da}{dN} = C\,(U\,\Delta K)^m \tag{10.16}$$

where U is defined by Eq. (10.14) for $K_{min} < K_{op}$. When $K_{min} \geq K_{op}$, U = 1, and closure does not influence the results; the crack growth law reduces to the Paris-Erdogan equation in this latter case.

Rewriting Eq. (10.14) in terms of ΔK and R gives

$$U = \frac{1}{1 - R} - \frac{K_{op}}{\Delta K} \tag{10.17}$$

The threshold can be inferred by setting U = 0:

$$\Delta K_{th} = K_{op}\,(1 - R) \tag{10.18}$$

Equation (10.17) applies to the following range of ΔK values:

$$K_{op}\,(1 - R) \leq \Delta K \leq K_{op}\left(\frac{1}{R} - 1\right)$$

Figure 10.7 is a nondimensional plot of Eq. (10.16) for various R values, with U given by Eq. (10.17). The curves in this figure exhibit typical Region I and Region II behavior. The threshold stress intensity range, ΔK_{th}, decreases with increasing R ratio, but the predicted fatigue crack growth rate in the Paris law regime is insensitive to R. Figure 10.8 shows experimental data for a mild steel [7], which exhibit the same trends as Fig. 10.7. Thus this simple analysis predicts behavior that is qualitatively consistent with experiment.

10.3.2 Effects of Loading Variables on Closure

The stress intensity for crack closure is not actually a material constant, but depends on a number of factors. Elber measured the closure stress intensity in 2023-T3 aluminum at various load levels and R ratios, and obtained the following empirical relationship:

$$U = 0.5 + 0.4\,R \quad (\text{for } -0.1 \leq R \leq 0.7) \tag{10.19}$$

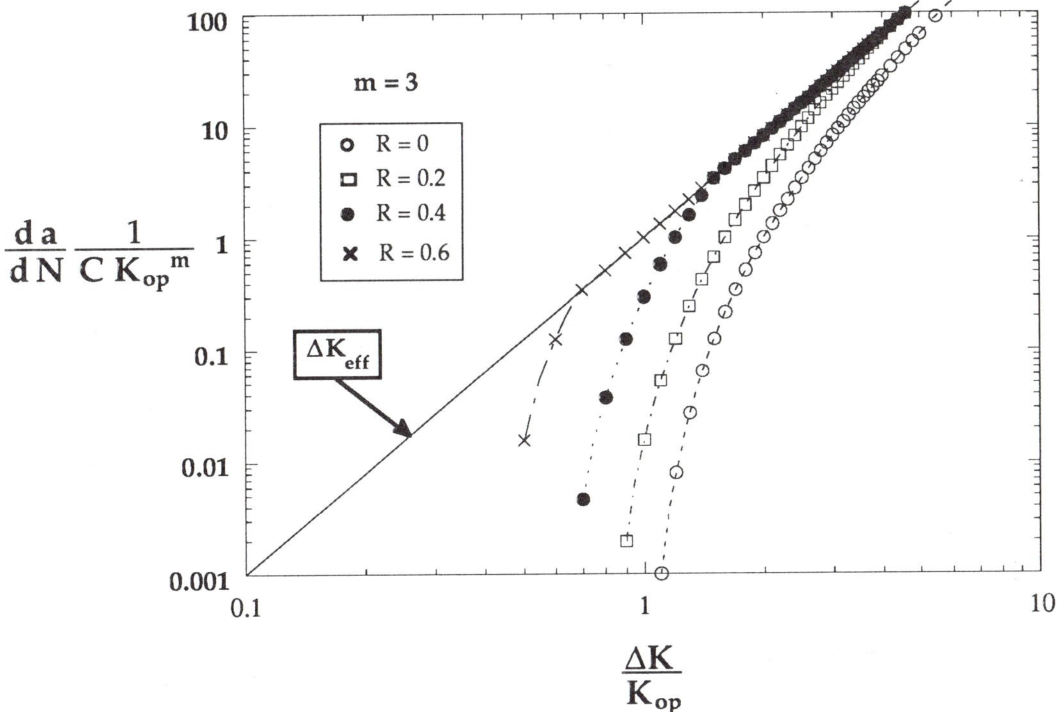

FIGURE 10.7 Nondimensional crack growth curves computed from Eqs. (10.16) and (10.17).

Subsequent researchers [20-22] inferred similar empirical expressions for other alloys.

According to Eq. (10.19), U depends only on R. Shih and Wei [23,24], however, argued that the Elber expression, as well as many of the subsequent equations, are over-simplified. Shih and Wei observed a dependence on K_{max} for crack closure in a Ti-6Al-4V titanium alloy. They also showed that experimental data of earlier researchers, when replotted, exhibits a definite K_{max} dependence.

There has been a great deal of confusion and controversy about the K_{max} dependence of U. Hudak and Davidson [16] cited contradictory examples from the literature where various researchers reported U to increase, decrease, or remain constant with increasing K_{max}. Hudak and Davidson performed closure measurements on a 7091 aluminum alloy and 304 stainless steel over a wide range of loading variables. For both materials, they inferred a closure relationship of the form:

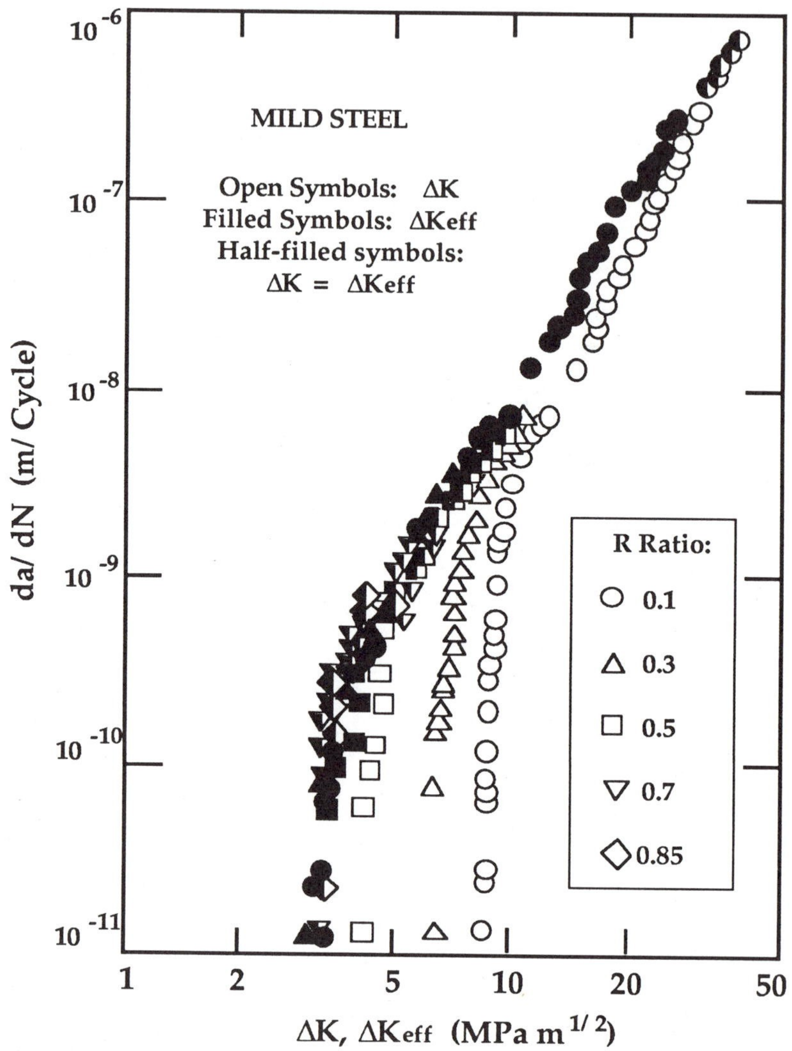

FIGURE 10.8. Fatigue crack growth data for mild steel at various R ratios [7].

$$U = 1 - \frac{K_o}{K_{max}}$$

$$= 1 - \frac{K_o(1 - R)}{\Delta K} \qquad (10.20)$$

where K_o is a material constant. Hudak and Davidson concluded that the inconsistent results from the literature could be attributed to a number of factors: (1) the range of ΔK values in the experiments was too narrow; (2) the ΔK values were not sufficiently close to the threshold; and (3) the measurement techniques lacked the required sensitivity.

The effect of loading variables on K_{op} can be inferred by substituting the relationships for U into Eq. (10.14). For example, the Elber equation implies the following relationship for K_{op}:

$$K_{op} = \Delta K \left(\frac{1}{1 - R} - 0.5 - 0.4\, R \right) \tag{10.21}$$

Note that the ratio $K_{op}/\Delta K$ depends only on the R ratio. Equation (10.21), however, leads to a different expression for K_{op}:

$$\begin{aligned} K_{op} &= K_o (1 - R) + K_{max} R \\ &= K_o (1 - R) + \frac{\Delta K\, R}{1 - R} \end{aligned} \tag{10.22}$$

Thus K_o is the opening stress intensity for R = 0.

Hudak and Davidson [16] attributed the confusion and controversy over the effect of loading variables on closure to experimental factors. More recently, McClung [25] conducted an extensive review of experimental and analytical closure results and concluded that there are three distinct regimes of crack closure. Near the threshold, closure levels decrease with increasing stress intensity, while U is independent of K_{max} at intermediate K levels. At high ΔK values, the specimen experiences a loss in constraint, and U decreases with increasing stress intensity. McClung found that no single equation could describe closure in all three regimes. Most of the seemingly contradictory data in the literature can be reconciled by considering the regimes in which the data were collected.

Microstructural effects can also lead to differences in observed closure behavior in various materials. Figure 10.6, for example, shows the effect of grain size on crack closure in 1018 steel. Equation (10.20) contains one fitting parameter (K_o) that must be obtained experimentally. It remains to be seen whether or not this empirical expression is sufficiently flexible to account for the full range of fatigue behavior in various alloys.

10.3.3 The Fatigue Threshold

Crack closure influences the fatigue threshold, but closure is not the only cause of ΔK_{th}. The threshold stress intensity range can be resolved into intrinsic and extrinsic components:

$$\Delta K_{th} = \Delta K_{th(eff)} + \Delta K_{th(cl)} \tag{10.23}$$

where $\Delta K_{th(eff)}$ is the intrinsic fatigue threshold and $\Delta K_{th(cl)}$ is the contribution from crack closure. The precise mechanism for the intrinsic threshold has not been established, but several researchers have developed models for $\Delta K_{th(eff)}$ based on dislocation emission from the crack tip [26] or blockage of slip bands by grain boundaries [27] (see Section 10.6.2).

Figure 10.8 shows fatigue crack growth data near the threshold for a mild steel [7]. The data exhibit the expected R ratio effects when characterized by ΔK. Expressing the data in terms of ΔK_{eff} removes the R dependence, but the threshold remains. Thus the Elber power law relationship (Eq. (10.15)) may not apply at low growth rates. The intrinsic threshold, however, is very small in this case (~ 3 MPa $\sqrt{m}$), and the crack growth rates must be less than 10^{-9} m/cycle for a noticeable deviation from Eq. (10.15). Therefore, the power law equation is acceptable in most practical circumstances.

Klesnil and Lukas [28] proposed the following empirical relationship between ΔK_{th} and R:

$$\Delta K = \Delta K_{tho} (1 - R)^{\gamma} \tag{10.24}$$

where ΔK_{tho} is the threshold for R = 0 and γ is a fitting parameter. When $\gamma = 1$, Eq. (10.24) is consistent with Eq. (10.18).

The effect of loading variables on the threshold can also be inferred from the crack closure relationships described in Section 10.3.2. Threshold conditions are obtained by setting $\Delta K_{eff} = \Delta K_{th(eff)}$ and $\Delta K = \Delta K_{th}$. For example, the Elber equation (Eq. (10.19)) leads to the following expression for ΔK_{th}:

$$\Delta K_{th} = \frac{\Delta K_{th(eff)}}{0.5 + 0.4\, R} \tag{10.25}$$

Solving for the threshold stress intensity from the Hudak and Davidson relationship (Eq. (10.20)) gives

$$\Delta K_{th} = \Delta K_{th(eff)} + K_o (1 - R) \tag{10.26}$$

If $\Delta K_{th(eff)}$ is small compared to ΔK_{th}, Eq. (10.26) coincides with Eq. (10.18), as well as Eq. (10.24) for $\gamma = 1$. Recall that Eq. (10.18) was derived by assuming K_{op} is a material constant. According to Eq. (10.26), however, such an assumption is not necessary to obtain a linear relationship between ΔK_{th} and R.

Figure 10.9 shows a compilation of threshold values for a variety of steels [7]. Aside from a high strength martensitic steel, where ΔK_{th} is apparently independent of R, the relationship between ΔK_{th} and R is reasonably linear between R = 0 and R = 0.8. Above R = 0.8, the data appear to reach a plateau, which may be indicative of the intrinsic threshold.

10.3.4 Pitfalls and Limitations of ΔK_{eff}

The effective stress intensity range, ΔK_{eff}, has been fairly successful in characterizing fatigue crack growth. Seemingly random trends in data can often be rationalized by taking account of crack closure. This approach contains a number of pitfalls, however, and there are instances where ΔK_{eff} fails to correlate fatigue data.

Since closure is influenced by microstructure, and distinct regimes of closure operate at various K levels [25], the empirical relationships for U and K_{op} described above are not reliable for estimating ΔK_{eff}. Empirical fits to a given set of data only apply to a particular loading regime (e.g. near-threshold behavior) and should not be extrapolated to other regimes or other materials.

Correlation of crack growth data at different R ratios is often cited as evidence of the validity of ΔK_{eff} estimates. Fatigue crack growth data alone, however, is not sufficient to define U and K_{op} uniquely [29]. Consider a set of da/dN - ΔK_{eff} data at various R ratios; assume that these data all follow a common trend. We can define a new quantity, ΔK_{eff}^*, as follows:

$$\Delta K_{eff}^* \equiv q\,\Delta K_{eff}$$

where q is an arbitrary constant. This new parameter would be equally successful at collapsing the data to a common curve; all data would be shifted an equal amount on the horizontal axis. Thus the ability to correlate fatigue data for various R ratios is not conclusive proof that the apparent ΔK_{eff} is the true driving force for fatigue. The distinction between the true ΔK_{eff} and an apparent value that differs by a factor q may be important when fatigue data are used to predict structural behavior.

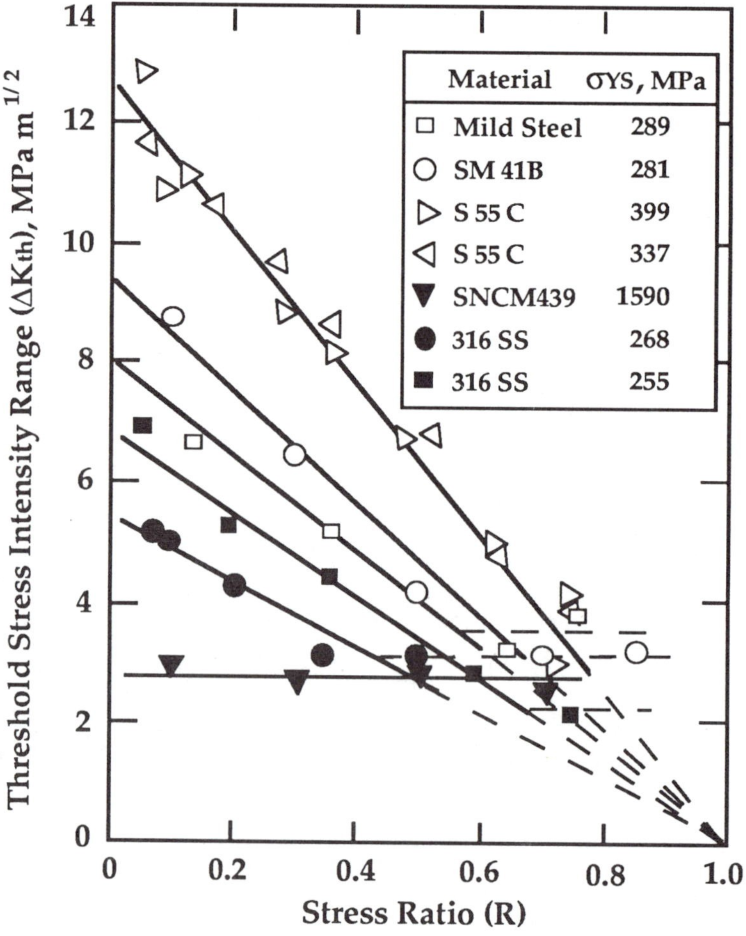

FIGURE 10.9 Effect of R ratio on the threshold stress intensity range [7].

The only reliable method for inferring closure loads is through direct measurement. Section 10.7 outlines several methods for measuring K_{op}. Such measurements, however, tend to be ambiguous. Crack opening and closure are progressive events that do not occur at a distinct load. Figure 10.3(a) schematically illustrates the typical load-displacement behavior, where the compliance gradually changes from the closed to fully open case. The definition of the closure load is somewhat arbitrary in such cases. The most appropriate definition of K_{op} is the subject of ongoing research.

Even without uncertainties in closure load measurements, the definition of ΔK_{eff} is complicated by history effects. All of the equations in Section 10.3 implicitly assume similitude. That is, the fatigue crack growth rate is assumed to be a function only of K_{min} and K_{max} (Eq. (10.1)). This assumption is strictly valid only for constant amplitude fatigue; i.e., $dK/da = 0$.

The fatigue crack growth rate depends on dK/da [29,30]. Consequently, a rising K field and a falling K field result in different fatigue crack growth rates for a given K_{min} and K_{max}. Some, but not all, of the history effects can be explained in terms of crack closure: K_{op} depends on prior history. Hertzberg, et al. [30] showed that fatigue tests at a constant R ratio but with a decreasing ΔK can result in overestimates of ΔK_{th}; the negative dK/da produces large closure loads, which reduce ΔK_{eff}. Applying such data to structures may be nonconservative, particularly if $dK/da > 0$ in the structure.

One method to account for history effects is to measure K_{op} throughout the test to ensure that ΔK_{eff} reflects the current closure behavior. Such an approach complicates fatigue experiments when K_{op} is determined intermittently, and a cycle-by-cycle evaluation of closure loads is totally impractical. Moreover, there is no guarantee that the instantaneous ΔK_{eff} uniquely characterizes crack growth rate when $dK/da \neq 0$. Even if similitude of ΔK_{eff} were valid in the general case, it would be of little practical use unless K_{cl} were known for the loading spectrum in the structure of interest. History effects and variable amplitude loading are explored further in Section 10.4.

10.4 VARIABLE AMPLITUDE LOADING AND RETARDATION

Similitude of crack tip conditions, which implies a unique relationship between da/dN, ΔK, and R, is strictly valid only for constant amplitude loading (i.e., dK/da = 0). Real structures, however, seldom conform to this ideal. A typical structure experiences a spectrum of stresses over its lifetime. In such cases, the crack growth rate at any moment in time depends not only on the current loading conditions, but also the prior history. Equation (10.3) is a general mathematical representation of the dependence on past and present conditions.

10.4.1 Reverse Plasticity at the Crack Tip

History effects in fatigue are a direct result of the history dependence of plastic deformation. Figure 10.10 schematically illustrates the cyclic stress-strain response of an elastic-plastic material which is loaded beyond yield in both tension and compression. If we desire to know the stress at a particular strain, ε^*, it is not sufficient merely to specify the strain. For the loading path in Fig. 10.10, there are three different stresses that correspond to ε^*; we must specify not only ε^*, but also the deformation history that preceded this strain.

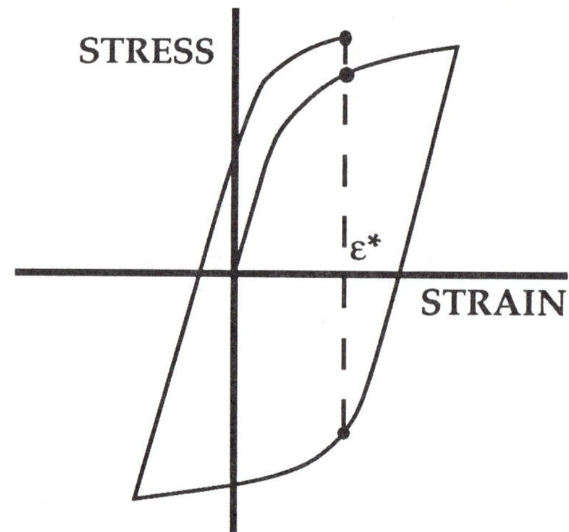

FIGURE 10.10. Schematic stress-strain response of a material that is yielded in both tension and compression. The stress at a given strain, ε^*, depends on prior loading history.

Figure 10.11 illustrates the crack tip plastic deformation that results from a single stress cycle. A plastic zone forms when the structure is loaded to K_{max}. Upon unloading, material near the crack tip exhibits reverse plasticity, which results in a compressive plastic zone. The compressive stress field at the crack tip influences subsequent deformation and crack growth. Retardation of crack growth after an overload (Section 10.4.2) is an example of this effect.

Following an approach proposed by Rice [31], we can analyze reverse plasticity by means of the Dugdale-Barenblatt strip yield model. The advantage of this model is that it permits superposition of loading and unloading stress fields. Refer to Fig. 10.11(a), where the structure is loaded to K_{max}. Assuming small scale yielding, the size of the plastic zone is given by

$$\rho = \frac{\pi}{8}\left(\frac{K_{max}}{\sigma_{YS}}\right)^2 \tag{10.27}$$

Let us now superimpose a compressive stress intensity, $-\Delta K$. The effective yield stress for reverse yielding is $-2\sigma_{YS}$, since the material in the compressive plastic zone must be stressed to $-\sigma_{YS}$ from an initial value of $+\sigma_{YS}$. Figure 10.11(b) illustrates the superimposed stress field, and Fig. 10.11(c) shows the net stress field after unloading. The estimated size of the compressive plastic zone is

$$\rho^* = \frac{\pi}{8}\left(\frac{\Delta K}{2\,\sigma_{YS}}\right)^2 \tag{10.28}$$

Because of the redistribution of stresses upon unloading, much of the material that was previously in the monotonic plastic zone is now in compression. According to Eq. (10.28), the compressive plastic zone is 1/4 the size of the monotonic zone. Finite element analysis, however, predicts a much smaller cyclic plastic zone [32].

A somewhat more complicated version of the above analysis forms the basis of the Budiansky and Hutchinson [15] crack closure model. They incorporated a plastic wake into the strip yield model, and showed that residual stretch in the wake results in positive closure loads.

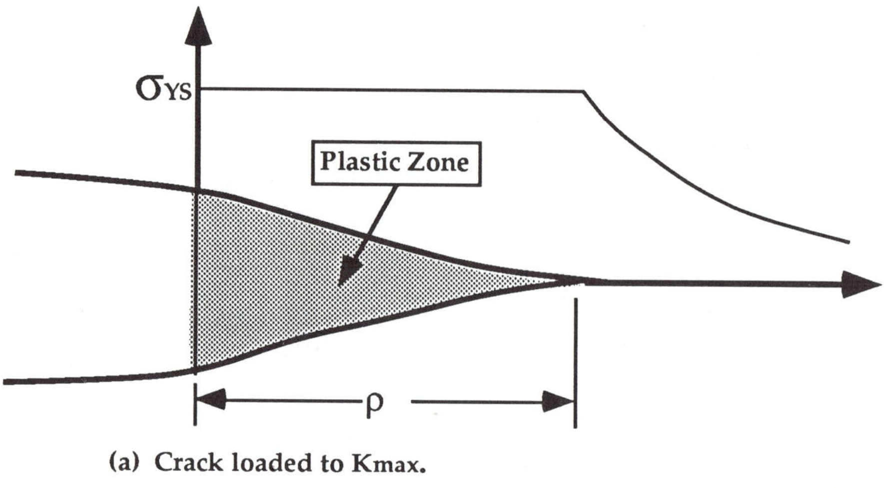

(a) Crack loaded to Kmax.

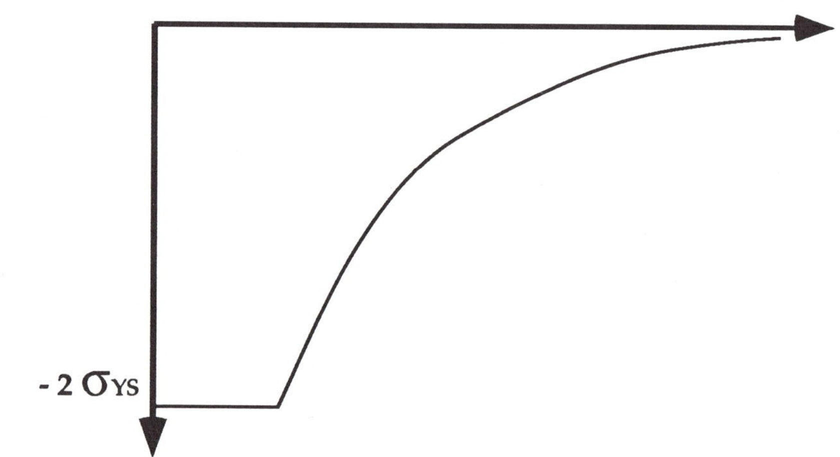

(b) Superimposed stress field.

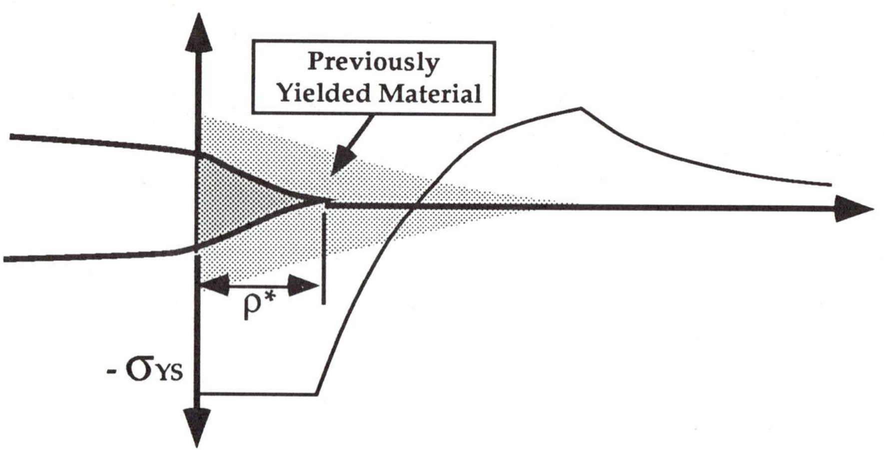

(c) Stress field after unloading.

FIGURE 10.11. Formation of a reverse plastic zone during cyclic loading.

10.4.2 The Effect of Overloads

Consider the fatigue loading history illustrated in Fig. 10.12. Constant amplitude loading is interrupted by a single overload, after which the K amplitude resumes its previous value. Prior to the overload, the plastic zone would have reached a steady state size, but the overload cycle produces a significantly larger plastic zone. When the load drops to the original K_{min} and K_{max}, the residual stresses that result from the the overload plastic zone (Fig. 10.11) are likely to influence subsequent fatigue behavior.

Figure 10.13 shows actual experimental data [33], where a single overload is imposed in an otherwise constant amplitude test. Immediately after application of the overload, da/dN drops dramatically. The overload results in compressive residual stresses at the crack tip, which retard fatigue crack growth. The growth rate resumes its earlier value once the crack has grown through the overload plastic zone. Figure 10.13 is an obvious example where similitude is violated, and the instantaneous values of ΔK and R are not sufficient to define da/dN.

Retardation following an overload is a complicated phenomenon that has so far eluded rigorous mathematical description. There are a number of empirical and semiempirical models for retardation, which contain one or more fitting parameters that must be obtained experimentally. Some models assume that crack closure effects are responsible for retardation, while others consider the plastic zone in front of the crack tip. The Wheeler model [34], which takes the latter approach, is one of the simplest and most widely used retardation analyses. This model relates the crack growth rate to the overload plastic zone size and the current plastic zone size (Fig. 10.14). The former quantity is given by

$$r_{y(o)} = \frac{1}{\beta\pi}\left(\frac{K_o}{\sigma_{YS}}\right)^2 \tag{10.29}$$

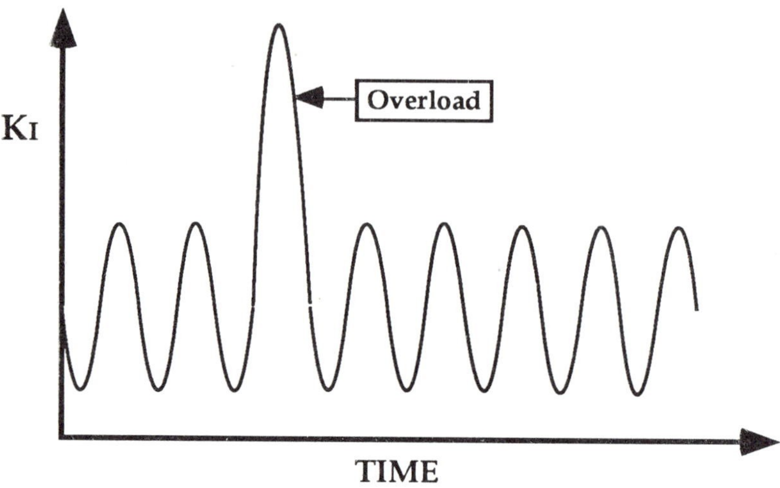

FIGURE 10.12 A single overload during cyclic loading.

where K_o is the stress intensity at the peak overload, and $\beta = 2$ for plane stress and $\beta = 6$ for plane strain. The plastic zone size that corresponds to the current K_{max} is given by

$$r_{y(c)} = \frac{1}{\beta\pi}\left(\frac{K_{max}}{\sigma_{YS}}\right)^2 \tag{10.30}$$

Wheeler assumed that retardation effects persist as long as $r_{y(c)}$ is contained within $r_{y(o)}$ (Fig. 10.14(a) and (b)), but the overload effects disappear when the current plastic zone touches the outer boundary of $r_{y(o)}$, as Fig. 10.14(c) illustrates. For a crack that has grown Δa since the the overload, Wheeler defined a retardation factor as follows:

$$\phi_R = \left(\frac{\Delta a + r_{y(c)}}{r_{y(o)}}\right)^\gamma \tag{10.31}$$

where γ is a fitting parameter. The crack growth rate is reduced from a baseline value by ϕ_R:

$$\left(\frac{da}{dN}\right)_R = \phi_R \frac{da}{dN} \tag{10.32}$$

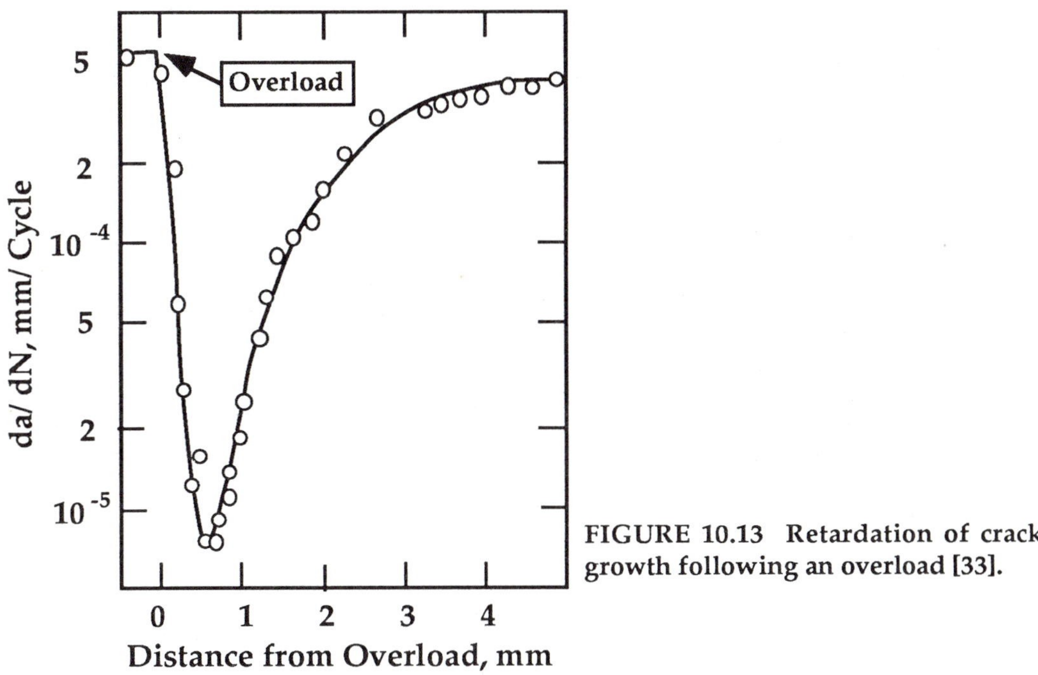

FIGURE 10.13 Retardation of crack growth following an overload [33].

The baseline crack growth rate is obtained from a growth law of the form of Eq. (10.1). Thus for a single overload, the number of cycles required to grow through the overload plastic zone can be integrated as follows:

$$N_R = \int_{a_o}^{a*} \frac{da}{\phi_R\{r_{y(o)}, r_{y(c)}, \Delta a\}\, f_1\{\Delta K, R\}} \qquad (10.33)$$

where a_o is the crack size at the application of the overload, f_1 is the baseline growth law (Eq. (10.1)), and $a^* = a_o + r_{y(o)} - r_{y(c)}$.

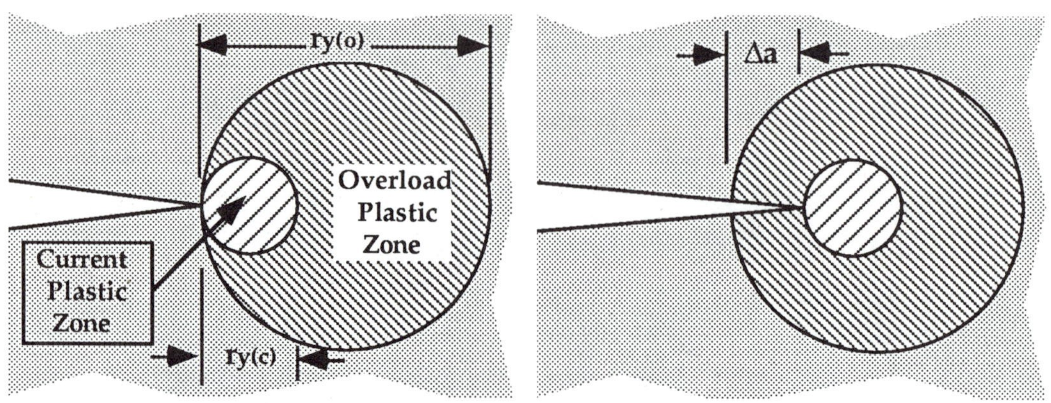

(a) Immediately following the overload. **(b) After the crack propagates Δa.**

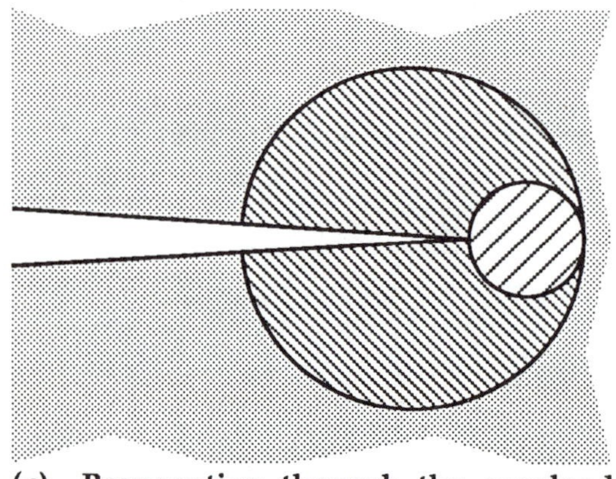

(c) Propagation through the overload plastic zone.

FIGURE 10.14 The Wheeler model for fatigue retardation. The crack growth rate depends on the size and position of the current plastic zone relative to the overload plastic zone.

10.4.3 Analysis of Variable Amplitude Fatigue

Variable amplitude fatigue can involve either a regular pattern of cyclic stresses or a random sequence of loads. Figure 10.2 illustrates several examples of variable amplitude loading. Similitude is not satisfied in such cases, and history effects can be quite pronounced.

In some instances, it may be appropriate to analyze variable amplitude fatigue with a growth law of the form of Eq. (10.1). In a rising or falling K field, for example, similitude may be approximately satisfied if dK/da is small. Similitude may also be valid when constant amplitude loading follows a single overload, provided the crack has propagated out of the overload plastic zone.

Simple fatigue growth laws that assume similitude are usually conservative when applied to variable amplitude loading. Retardation effects, which the simple equations do not consider, tend to extend the fatigue life of a structure.

Currently, the most accurate means of quantifying the fatigue life in variable amplitude loading involves a cycle-by-cycle integration of one of the retardation models. The procedure for applying the Wheeler model to variable amplitude problems is summarized below. Other retardation models can be applied in an analogous fashion.

The Wheeler retardation model is implemented in much the same manner as for overload cases, except that the overload plastic zone size, $r_{y(o)}$, and the current plastic size, $r_{y(c)}$, must be evaluated for each cycle in a variable amplitude problem. Figure 10.15 illustrates a typical scenario, where a very high stress cycle occurred earlier in the history, and a moderately high stress occurred in the previous cycle. The overload plastic zone is chosen such that the retardation factor, ϕ_R, is minimized. In this case, the more recent overload is considered, despite the fact that it produced a smaller plastic zone than the earlier peak stress.

Figure 10.16 shows a flow chart for computing fatigue crack growth with the Wheeler model. Although the algorithm is relatively simple, the analysis can be very time-consuming, since a cycle-by-cycle summation is required. The stress input consists of two components: the spectrum and the sequence. The former is a statistical distribution of stress amplitudes, which quantifies the relative frequency of low, medium and high stress cycles. The sequence, which defines the order of the various stress amplitudes, can be either random or a regular pattern.

An important point about the Wheeler model is that the exponent, γ, in Eq. (10.31) depends on material properties and stress spectrum. Therefore, this parameter must be obtained empirically from an experiment with a stress spectrum that has a similar character to that of the structure. A variable amplitude loading analysis must first be performed on the experiment to determine the γ value that gives the best prediction of crack growth. The model can then be applied to structural predictions. If a structure with a different stress spectrum is to be analyzed, the Wheeler model must be recalibrated with a new experiment. All empirical retardation models contain adjustable parameters and must be calibrated experimentally.

A detailed discussion of the development of computer codes for variable amplitude fatigue analysis is beyond the scope of this book. Broek [35] offers some practical advice on this subject. He describes procedures for obtaining stress spectra, and he gives examples of typical stress sequences for practical problems. He also mentions some of the common pitfalls that programmers encounter. For example, summation of crack growth often involves adding a very small number (the crack growth in a given cycle) to a large number (the current crack size). In such cases, double precision variables should be used in order to minimize round-off errors.

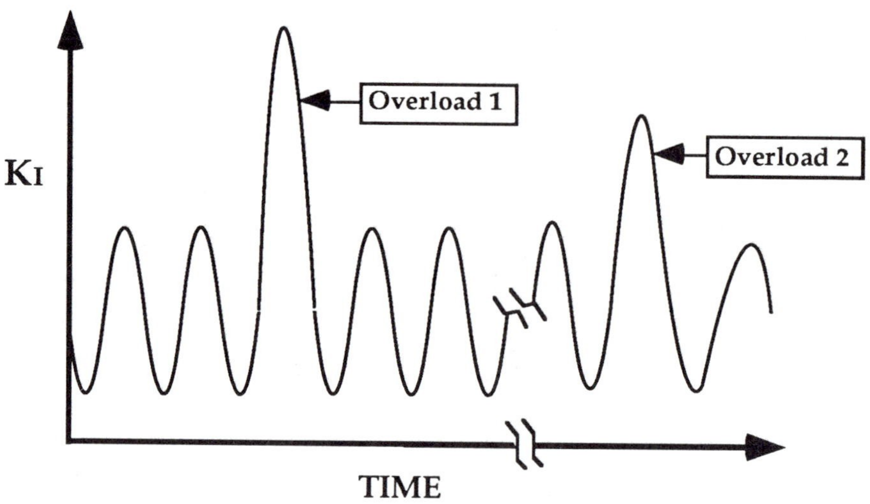

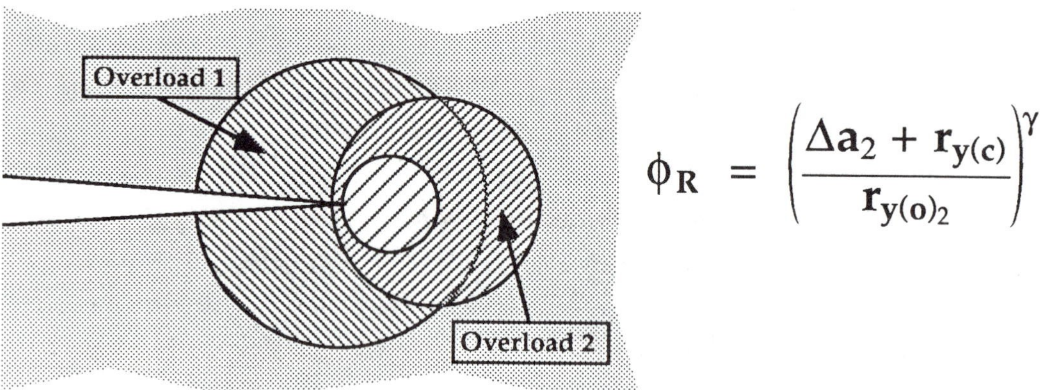

$$\phi_R = \left(\frac{\Delta a_2 + r_{y(c)}}{r_{y(o)_2}} \right)^{\gamma}$$

FIGURE 10.15 Analysis of variable amplitude fatigue with the Wheeler retardation model. The overload plastic zone size, $r_{y(o)}$ is chosen so as to minimize ϕ_R.

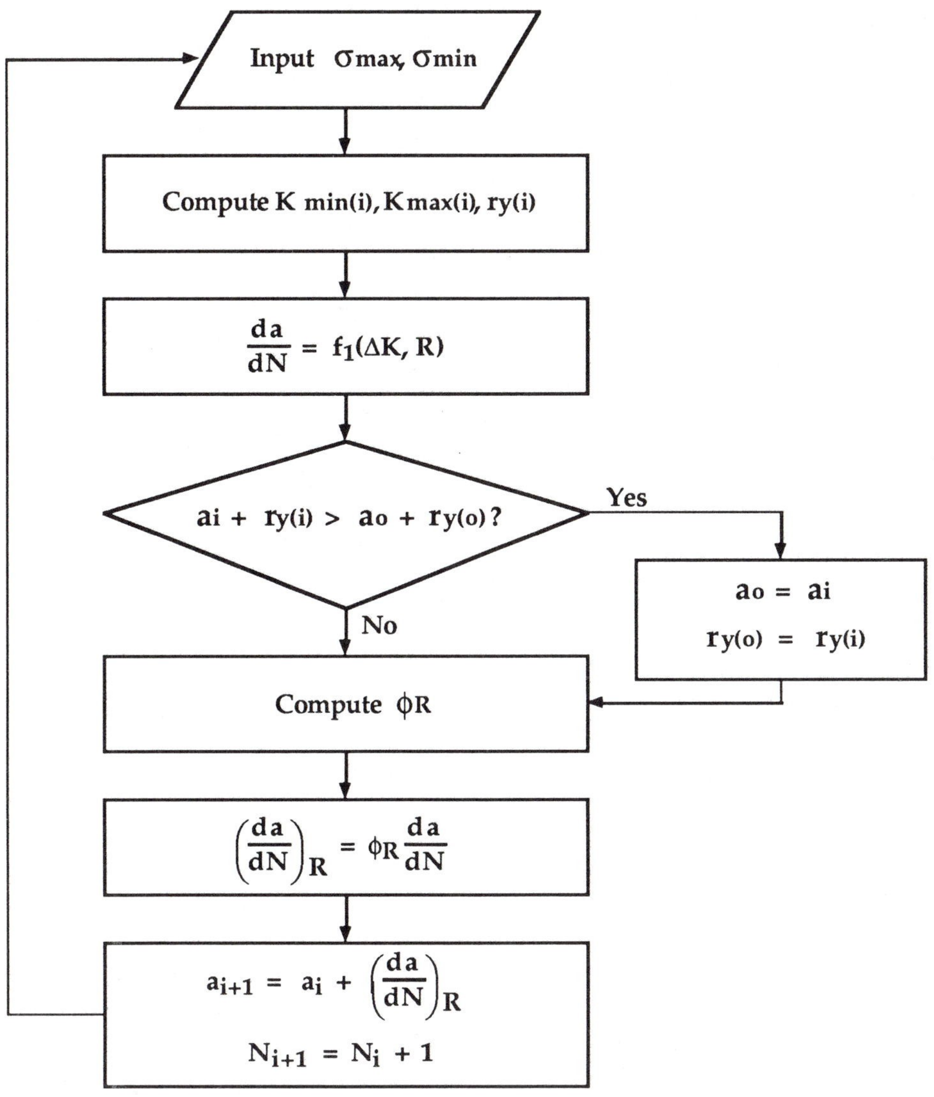

FIGURE 10.16 Flow chart for variable amplitude fatigue analysis with the Wheeler model.

10.5 GROWTH OF SHORT CRACKS

The fatigue behavior of short cracks is often very different from that of longer cracks. There is not a precise definition of what constitutes a "short" crack, but most experts consider cracks less than 1 mm long to be small.

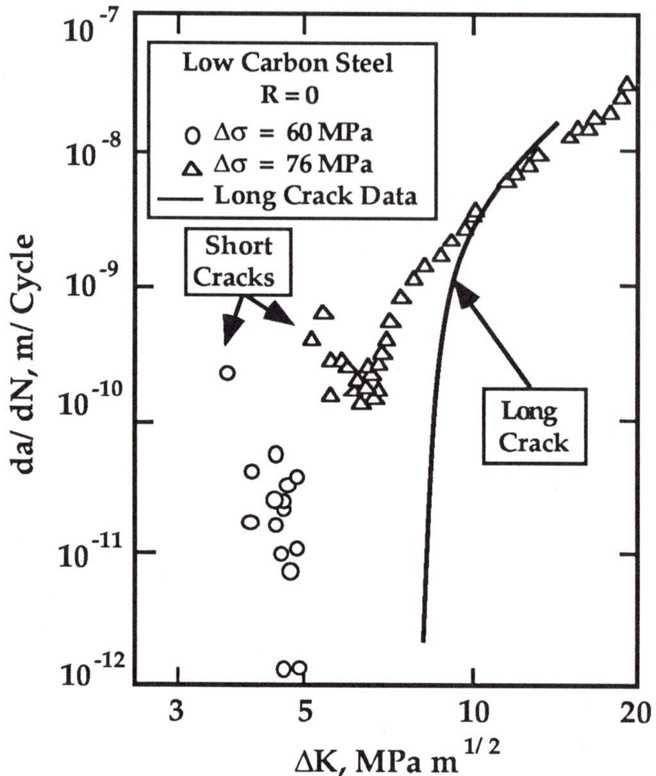

FIGURE 10.17 Growth of short cracks in a low carbon steel [36].

Figure 10.17 compares short crack data with long crack data near the threshold [36]. In this case, the short cracks were initiated from a blunt notch. Note that the short cracks exhibit finite growth rates well below ΔK_{th} for long cracks. Also, the trend in da/dN is inconsistent with expected behavior; the crack growth rate actually decreases with ΔK when the stress range is 60 MPa, and the da/dN - ΔK curve exhibits a minimum at the other stress level.

A number of factors can contribute to the anomalous behavior of small fatigue cracks. The fatigue mechanisms depend on whether the crack is microstructurally short or mechanically short, as described below.

10.5.1 Microstructurally Short Cracks

A microstructurally short crack has dimensions that are on the order of the grain size. Cracks less than 100 μm long are generally considered microstructurally short. The material no longer behaves as a homogeneous continuum at such length scales; the growth is strongly influenced by microstructural features in such cases. The growth of microstructurally short cracks is often very sporadic; the crack may grow rapidly at certain intervals, and then virtually arrest when it encounters barriers such as grain boundaries and second-phase particles [7].

10.5.2 Mechanically Short Cracks

A crack that is between 100 μm and 1 mm in length is mechanically short. The size is sufficient to apply continuum theory, but the mechanical behavior is not the same as in longer cracks. Mechanically short cracks typically grow much faster that long cracks at the same ΔK level, particularly near the threshold (Fig. 10.17).

Two factors have been identified as contributing to faster growth of short cracks: plastic zone size and crack closure.

When the plastic zone size is significant compared to the crack length, an elastic singularity does not exist at the crack tip, and K is invalid. The effective driving force can be inferred by adding an Irwin plastic zone correction. El Haddad, et al. [37] introduced an "intrinsic crack length" which, when added to the physical crack size, brings short crack data in line with the corresponding long crack results. The intrinsic crack length is merely a fitting parameter, however, and does not correspond to a physical length scale in the material. Tanaka [7], among others, proposed adjusting the data for crack tip plasticity by characterizing da/dN with ΔJ rather than ΔK.

According to the closure argument, short cracks exhibit different crack closure behavior than long cracks, and data for different crack sizes can be rationalized through ΔK_{eff}. Figure 10.18 [36] shows K_{op} measurements for the short and long crack data in Fig. 10.17. The closure loads are significantly higher in the long cracks, particularly at low ΔK levels. Figure 10.19 [36] shows the small and large crack data lie on a common curve when da/dN is plotted against ΔK_{eff}, thereby lending credibility to the closure theory of short crack behavior.

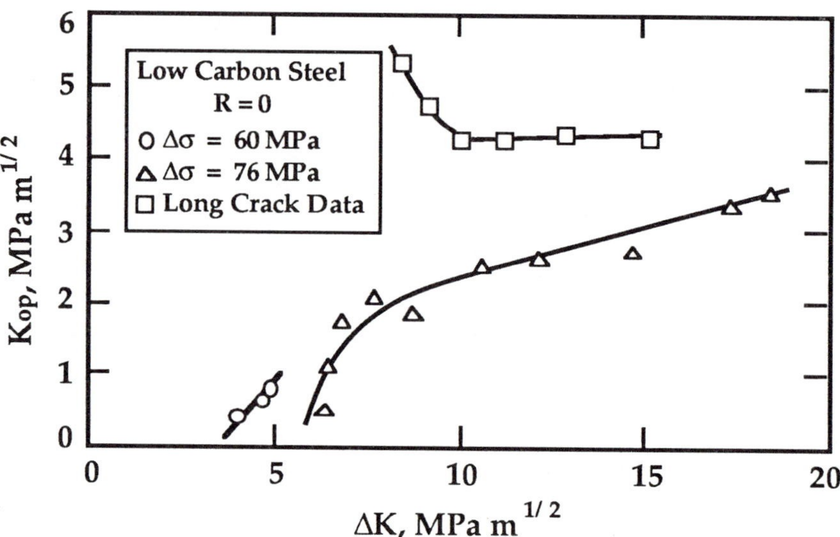

FIGURE 10.18 Crack closure data for short and long cracks in a low carbon steel [36].

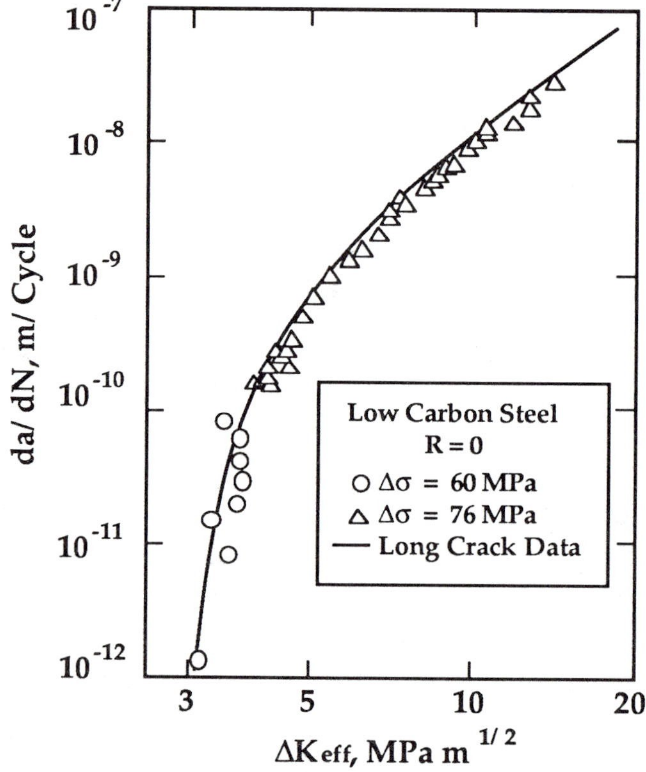

FIGURE 10.19 Short crack fatigue crack growth data from Fig. 10.17, corrected for closure [36].

10.6 MICROMECHANISMS OF FATIGUE

Figure 10.20 summarizes the failure mechanisms for metals in the three regions of the fatigue crack growth curve. In Region II, where da/dN follows a power law, the crack growth rate is relatively insensitive to microstructure and tensile properties, while da/dN at either extreme of the curve is highly sensitive to these variables. The fatigue mechanisms in each region are described in more detail below.

10.6.1 Fatigue in Region II

In Region II, the fatigue crack growth rate is not a strong function of microstructure or monotonic flow properties. Two aluminum alloys with vastly different mechanical properties, for instance, are likely to have very similar fatigue crack growth characteristics. Steel and aluminum, however, exhibit significantly different fatigue behavior. Thus da/dN is not sensitive to microstructure and tensile properties *within a given material system.*

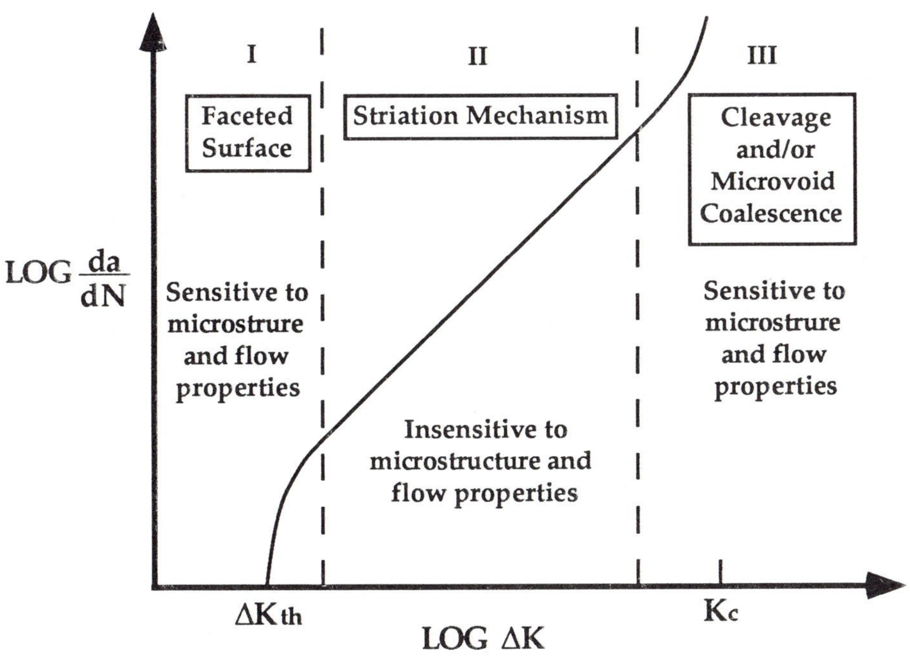

FIGURE 10.20 Micromechanisms of fatigue in metals.

One explanation for the lack of sensitivity to metallurgical variables is that *cyclic* flow properties, rather than monotonic tensile properties, control fatigue crack propagation. Figure 10.21 schematically compares monotonic and cyclic stress-strain behavior for two alloys of a given material. The low strength alloy tends to strain harden, while the strong alloy tends to strain *soften* with cyclic loading. In both cases, the cyclic stress-strain curve tends toward a steady-state hysteresis loop, which is relatively insensitive to the initial strength level.

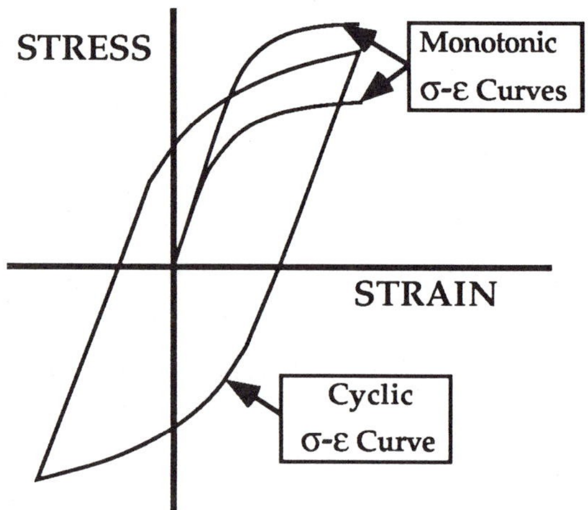

FIGURE 10.21 Schematic comparison of monotonic and cyclic stress-strain curves.

Propagating fatigue cracks often produce *striations* on the fracture surface. Striations are small ridges that are perpendicular to the direction of crack propagation. Figure 10.22 illustrates one proposed mechanism for striation formation during fatigue crack growth [38]. The crack tip blunts as the load increases, and an increment of growth occurs as a result of the formation of a stretch zone. Local slip is concentrated at ±45° from the crack plane. When the load decreases, the direction of slip reverses, and the crack tip folds inward. The process is repeated with subsequent cycles, and each cycle produces a striation on the upper and lower crack faces. The striation spacing, according to this mechanism, is equal to the crack growth per cycle (da/dN).

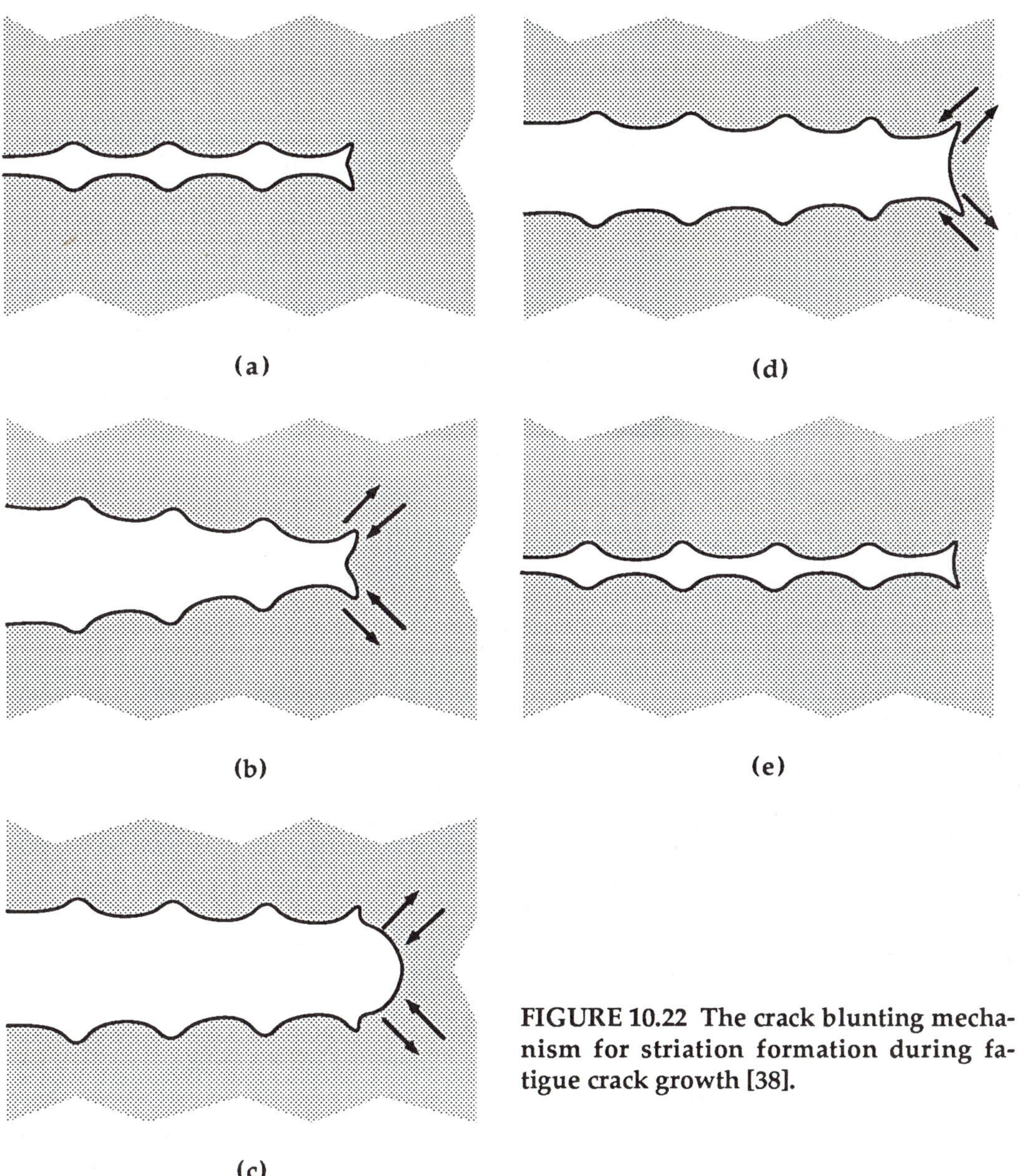

FIGURE 10.22 The crack blunting mechanism for striation formation during fatigue crack growth [38].

An alternative view of fatigue is the damage accumulation mechanism, which states that a number of cycles are required to produce a critical amount of damage, at which time the crack grows a small increment [39]. This mechanism was supported by Lankford and Davidson [40], who observed that the striation spacing did not necessarily correspond the crack growth after one cycle. Several cycles may be required to produce one striation, depending on the ΔK level; the number of cycles per striation apparently decreases with increasing ΔK, and striation spacing = da/dN at high ΔK values.

A number of researchers have attempted to relate the observed crack growth rate to the micromechanism of fatigue, with limited success. The blunting mechanism, where crack advance occurs through the formation of a stretch zone, implies that the crack growth per cycle is proportional to ΔCTOD. This, in turn, implies that da/dN should be proportional to ΔK^2. Actual Paris law exponents, however, are typically between three and four for metals. One possible explanation for this discrepancy is that the blunting mechanism is incorrect. An alternate explanation for exponents greater than two is that the shape of the blunted crack is not geometrically similar at high and low K values [7]. Figure 10.23 [7] shows the crack opening profile for copper at two load levels; note that the shapes of the blunted cracks are different.

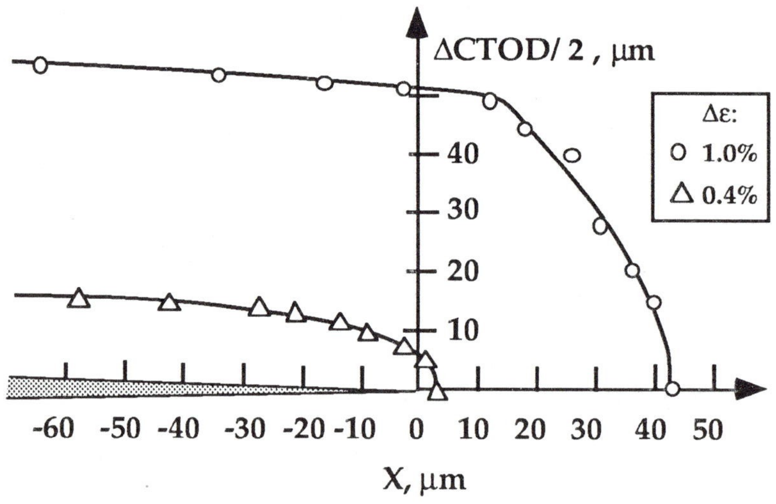

FIGURE 10.23 Crack opening profile in copper [7].

10.6.2 Micromechanisms Near the Threshold

The fracture surface that results from fatigue near the threshold has a flat, faceted appearance that resembles cleavage [41]. The crack apparently follows specific crystallographic planes, and changes directions when it encounters a barrier such as a grain boundary.

The fatigue crack growth rate in this region is sensitive to grain size, in part because coarse grained microstructures produce rough surfaces and roughness-induced closure (Fig. 10.6). Grain size can also affect the *intrinsic* threshold in certain cases. One model for $\Delta K_{th(eff)}$ [27] states that the threshold occurs when grain boundaries block slip bands and prevent them from propagating into the adjoining grain. This apparently happens when the plastic zone size is approximately equal to the average grain diameter, which suggests the following relationship between $\Delta K_{th(eff)}$ and grain size:

$$\sqrt{d} = A\frac{\Delta K_{th(eff)}}{\sigma_{YS}} \tag{10.34}$$

where d is average grain diameter and A is a constant. Thus the intrinsic threshold increases with grain size, assuming σ_{YS} is constant. The Hall-Petch relationship, however, predicts that yield strength decreases with grain coarsening:

$$\sigma_{YS} = \sigma_i + k_y d^{-1/2} \tag{10.35}$$

Consequently, the grain size dependence of yield strength offsets the tendency for the the intrinsic threshold to increase with grain coarsening.

10.6.3 Fatigue at High ΔK Values

In Region III, da/dN accelerates due to an interaction between fatigue and fracture mechanisms. Fracture surfaces in this region typically include a mixture of fatigue striations, microvoid coalescence, and (depending on the material and the temperature) cleavage facets. The overall growth rate can be estimated by summing the effects of the various mechanisms:

$$\left.\frac{da}{dN}\right|_{total} = \left.\frac{da}{dN}\right|_{fatigue} + \left.\frac{da}{dN}\right|_{MVC} + \left.\frac{da}{dN}\right|_{cleavage} \tag{10.36}$$

The relative contribution of fatigue decreases with increasing K_{max}. At K_c, crack growth is completely dominated by microvoid coalescence, cleavage, or both.

10.7 EXPERIMENTAL MEASUREMENT OF FATIGUE CRACK GROWTH

The American Society for Testing and Materials (ASTM) recently published Standard E 647-88a [42], which outlines a test method for fatigue crack growth measurements. The procedure, which is summarized below, does not take crack closure into account. A committee within ASTM, however, is currently studying crack closure measurement and analysis. Section 10.7.2 below describes some of the closure measurement techniques that are currently available.

10.7.1 ASTM Standard E 647-88a

The *Standard Test Method for Measurement of Fatigue Crack Growth Rates*, ASTM E 647-88a [42], describes how to determine da/dN as a function of ΔK from an experiment. The crack is grown by cyclic loading, and K_{min}, K_{max}, and crack length are monitored throughout the test.

The test fixtures and specimen design are essentially identical to those required for fracture toughness testing, which are described in Chapter 7. The E 647 document allows tests on compact specimens and middle tension panels (Fig. 7.1).

The ASTM standard for fatigue crack growth measurements requires that the behavior of the specimen be predominantly elastic during the tests. This standard specifies the following requirement for the uncracked ligament of a compact specimen:

$$W - a \geq \frac{4}{\pi}\left(\frac{K_{max}}{\sigma_{YS}}\right)^2 \tag{10.37}$$

There are no specific requirements on specimen thickness; this standard is often applied to thin sheet alloys for aerospace applications. The fatigue properties, however, can depend on thickness, much like fracture toughness is thickness dependent. Consequently, the thickness of the test specimen should match the section thickness of the structure of interest.

All specimens must be fatigue precracked prior to the actual test. The K_{max} at the end of fatigue precracking should not exceed the initial K_{max} in the fatigue test. Otherwise, retardation effects may influence the growth rate.

During the test, the crack length must be measured periodically. Crack length measurement techniques include optical, unloading compliance, and potential drop (see Chapter 7). Accurate optical crack length measurements require a traveling microscope. One disadvantage of this method is that it can only detect growth on the surface; in thick specimens, the crack length measurements must be corrected for tunneling, which cannot be done until the specimen is broken open after the test. Another disadvantage of the optical technique is that the crack length measurements are usually recorded manually[6], while the other techniques can be automated. The unloading compliance technique requires that the test be interrupted for each crack length measurement. If the specimen is statically loaded for a finite length of time, material in the plastic zone may creep. In an aggressive environment, long hold times may result in environmentally assisted cracking or the deposition of an oxide film on the crack faces. Consequently, the compliance measurements should be made as quickly as possible. The ASTM standard requires that hold times be limited to ten minutes; it should be possible to perform an unloading compliance measurement in less than one minute.

The ASTM standard E 647 outlines two types of fatigue tests: (1) constant load amplitude tests where K increases, and (2) K-decreasing tests. In the latter case, the load amplitude decreases during the test to achieve a negative K gradient. The K-increasing test is suitable for crack growth rates greater that 10^{-8} m/cycle, but is difficult to apply at lower rates because of fatigue precracking considerations (see above). The K-decreasing procedure is preferable when near-threshold data are required. Because of the potential for history effects when the K ampli-

---

[6]It may be possible to automate optical crack length measurements with image analysis hardware and software, but most mechanical testing laboratories do not have this capability.

tude varies, ASTM E 647 requires that the normalized K gradient be computed and reported:

$$G \equiv \frac{1}{K}\frac{dK}{da} = \frac{1}{\Delta K}\frac{d\Delta K}{da} = \frac{1}{K_{min}}\frac{dK_{min}}{da} = \frac{1}{K_{max}}\frac{dK_{max}}{da} \tag{10.38}$$

The K-decreasing test is more likely to produce history effects, because prior cycles produce larger plastic zones, which can retard crack growth. Retardation in a rising K test is not a significant problem, since the plastic zone produced by a given cycle is slightly larger than that in the previous cycle. A K-increasing test is not immune to history effects, however; the width of the plastic wake increases with crack growth, which may result in different closure behavior than in a constant K amplitude test.

The ASTM standard recommends that the algebraic value of G be greater than $-0.08\ mm^{-1}$ in a K-decreasing test. If the test is computer controlled, the load can be programmed to decrease continuously to give the desired K gradient. Otherwise, the load amplitude can be decreased in steps, provided the step size is less than 10% of the current ΔP. In either case, the load should be decreased until the desired crack growth rate is achieved. It is usually not practical to collect data below $da/dN = 10^{-10}$ m/cycle.

The E 647 standard outlines a procedure for assessing whether or not history effects have occurred in a K-decreasing test. First, the test is performed at a negative G value until the crack growth rate reaches the intended value. Then the K gradient is reversed, and the crack is grown until the growth rate is well out of the threshold region. The K-decreasing and K-increasing portions of the test should yield the same $da/dN - \Delta K$ curve. This two-step procedure is time consuming, but it need only be performed once for a given material and R ratio to ensure that the true threshold behavior is achieved by subsequent K-decreasing tests.

Figure 10.24 schematically illustrates typical crack length versus N curves. These curves must be differentiated to infer da/dN. The ASTM standard E 647 suggests two alternative numerical methods to compute the derivatives. A linear differentiation approach is the simplest, but it is subject to scatter. The derivative at a given point on the curve can also be obtained by fitting several neighboring points to a quadratic polynomial (i.e., a parabola).

The linear method computes the slope from two neighboring data points: (a_i, N_i) and (a_{i+1}, N_{i+1}). The crack growth rate for $a = \bar{a}$ is given by

$$\left(\frac{da}{dN}\right)_{\bar{a}} = \frac{a_{i+1} - a_i}{N_{i+1} - N_i} \tag{10.39}$$

where $\bar{a} = (a_{i+1} + a_i)/2$.

The incremental polynomial approach involves fitting a quadratic equation to a local region of the crack length versus N curve, and solving for the derivative mathematically. A group of (2n + 1) neighboring points are selected, where n is typically 1, 2, 3, or 4, and (a_i, N_i) is the middle value in the (2n + 1) points. The following equation is fitted to the range $a_{i-n} \leq a \leq a_{i+n}$:

$$\hat{a}_j = b_o + b_1\left(\frac{N_j - C_1}{C_2}\right) + b_2\left(\frac{N_j - C_1}{C_2}\right)^2 \quad (i-n \leq j \leq i+n) \tag{10.40}$$

where b_o, b_1, and b_2 are the curve fitting coefficients, and $\hat{a}_j$ is the fitted value of crack length at N_j. The coefficients $C_1 = (N_{i-n} + N_{i+n})/2$ and $C_2 = (N_{i-n} - N_{i+n})/2$ scale the data in order to avoid numerical difficulties. (N_j is often a large number.) The crack growth rate at $\hat{a}_i$ is determined by differentiating Eq. (10.40):

$$\left(\frac{da}{dN}\right)_{\hat{a}_i} = \frac{b_1}{C_2} + \frac{2\,B_2\,(N_i - C_1)}{C_2{}^2} \tag{10.41}$$

An appendix in ASTM E 647 lists a Fortran program which performs the curve fitting operation and solves for da/dN.

10.7.2 Closure Measurements

A number of experimental techniques for measurement of closure loads in fatigue are currently available. Allison [43] has reviewed the existing procedures. A brief summary of the more common techniques is given below.

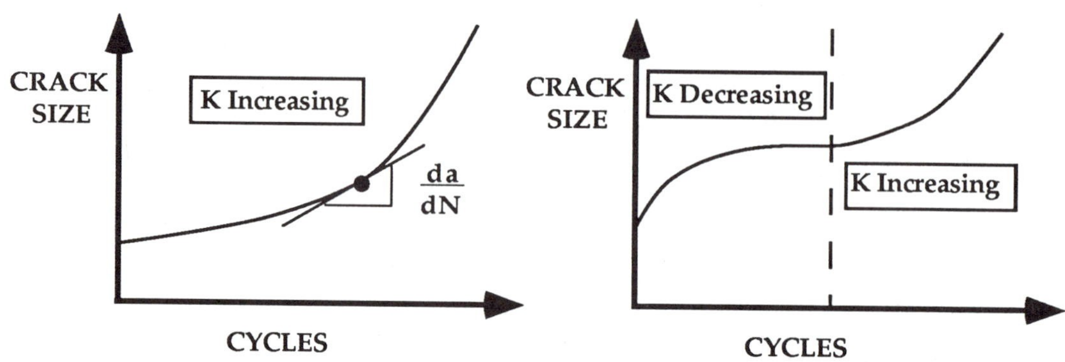

FIGURE 10.24 Schematic fatigue crack growth growth curves. da/dN is inferred from numerical differentiation of these curves.

Most measurements of closure conditions are inferred from compliance. Figure 10.25 schematically illustrates the load-displacement behavior of a specimen that exhibits crack closure. The precise closure load is ill-defined, because there is a often a significant range of loads where the crack is partiality closed. The closure load can be defined by a deviation in linearity in either the fully closed or fully open case (P_1 and P_3, respectively), or by extrapolating the fully closed and fully open load-displacement curve to the point of intersection (P_2).

Figure 10.26 illustrates the required instrumentation for the three most common compliance techniques for closure measurements. The closure load can be inferred from clip gage displacement at the crack mouth, back-face strain measurements, or laser interferometry applied to surface indentations. Specimen alignment is critical when inferring closure loads from compliance measurements.

Crack mouth opening displacement measurements with a clip gage are relatively simple, but extra care is necessary when in attaching the clip gage. Nonlinearity or hysteresis can result from improper gage attachment. Crack tip plasticity can also produce hysteresis effects in clip gage measurements. Displacement measurements remote from the crack tip often lack sensitivity. A signal processing technique called *differential compliance* can enhance sensitivity of global displacement measurements. A baseline compliance is inferred from the fully open portion of the load-displacement curve, as Fig. 10.25 illustrates. A differential clip gage displacement, ΔV, is defined as follows:

$$\Delta V = k(V - CP) \tag{10.42}$$

where C is the compliance when the crack is fully open and k is a gain factor. Figure 10.27 shows a schematic load-differential displacement curve. As the specimen is unloaded, the initially vertical line exhibits a finite slope when the crack closes. The sensitivity of this technique can be adjusted through the gain factor, k.

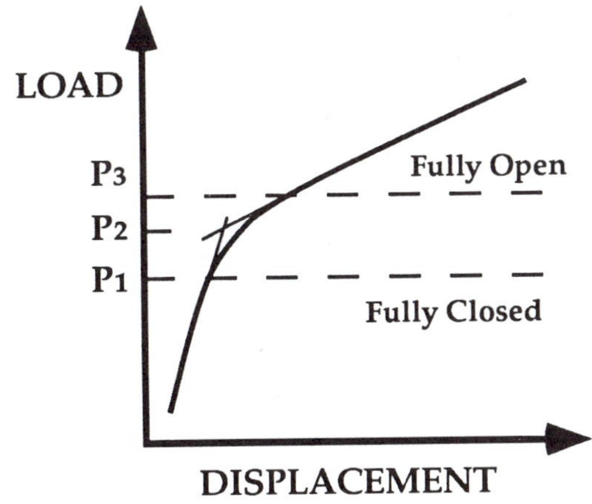

FIGURE 10.25 Alternative definitions of the closure load.

A back face strain gage has a relatively high degree of sensitivity. This measurement is not subject to hysteresis effects, provided the plastic zone is small.

Interferometric techniques provide a local measurement of crack closure [44]. Monochromatic light from a laser is scattered off of two indentations on either side of the crack. The two scattered beams interfere constructively and destructively, resulting in fringe patterns. The fringes change as the indentations move apart.

Crack closure is a three-dimensional phenomenon. The interior of a specimen exhibits different closure behavior than the surface. The clip cage and back-face strain gage methods provide a thickness-average measure of closure, while laser interferometry is strictly a surface measurement.

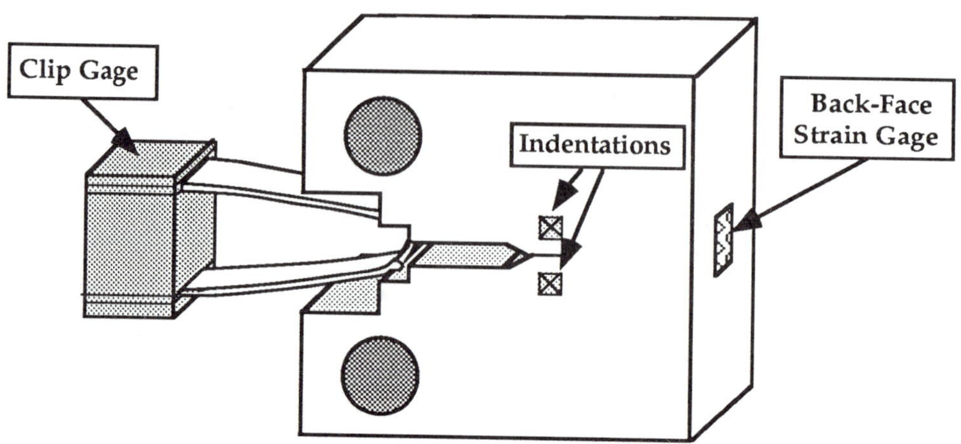

FIGURE 10.26 Instrumentation for the three most common closure measurement techniques.

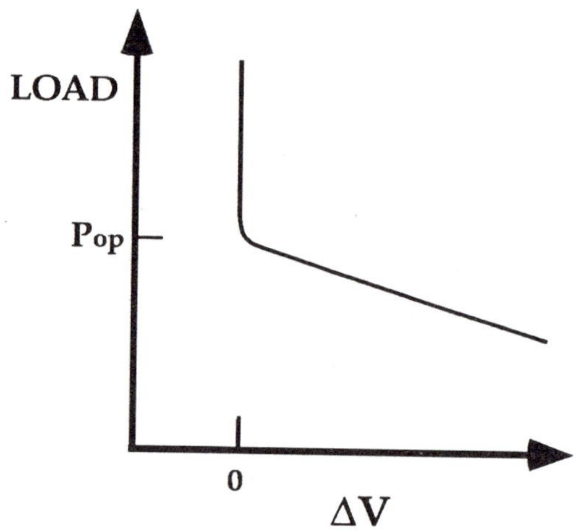

FIGURE 10.27 Schematic load versus differential displacement curve. This technique enhances the sensitivity of closure measurements from clip gage displacement.

More elaborate experimental techniques are available to study three-dimensional effects. For example optical interferometry [45] has been applied to transparent polymers to infer closure behavior through the thickness. Fleck [46] has developed a special gage to measure crack opening displacements at interior of a specimen.

10.8 DAMAGE TOLERANCE METHODOLOGY

The first six sections of this chapter described macroscopic and microscopic aspects of fatigue, and outlined various equations and analyses for characterizing crack growth. Section 10.7 addressed experimental measurements of fatigue behavior. This section describes how to apply fatigue data and growth models to structures, as part of a damage tolerance design scheme.

The term *damage tolerance* has a variety of meanings, but normally refers to a design methodology in which fracture mechanics analyses predict remaining life and quantify inspection intervals. This approach is usually applied to structures that are susceptible to time-dependent flaw growth (e.g.., fatigue, environmental-assisted cracking, creep crack growth). As its name suggests, the damage tolerance philosophy allows flaws to remain in the structure, provided they are well below the critical size.

Fracture control procedures vary considerably among various industries; a detailed description of each available approach is beyond the scope of this book. This section outlines a generic damage tolerance methodology and discusses some of the practical considerations. Although fatigue is the primary subject of this chapter, the approaches described below can, in principle, be applied to all types of time-dependent crack growth.

One of the first tasks of a damage tolerance analysis is the estimation of the critical flaw size, a_c. Chapter 9 describes approaches for computing critical crack size. Depending on material properties, ultimate failure may be governed by fracture or plastic collapse. Consequently, an elastic-plastic fracture mechanics analysis that includes the extremes of brittle fracture and collapse as special cases is preferable. (The possibility of geometric instabilities, such as buckling, should also be considered.)

Once the critical crack size has been estimated, a safety factor is normally applied to determine the *tolerable* flaw size, a_t. The safety factor is often chosen arbitrarily, but a more rational definition should be based on uncertainties in the input parameters (e.g.., stress and toughness) in the fracture analysis. Another consideration in specifying the tolerable flaw size is the crack growth rate; a_t should be chosen such that da/dt at this flaw size is relatively small, and a reasonable length of time is required to grow the flaw from a_t to a_c.

Fracture mechanics analysis is closely tied to nondestructive evaluation (NDE) in fracture control procedures. The NDE provides input to the fracture analysis, which in turn helps to define inspection intervals. A structure is inspected at the beginning of its life to determine the size of initial flaws. If no significant flaws are detected, the initial flaw size is set at an assumed value, a_o, which corresponds to the largest flaw *that might be missed* by NDE. This flaw size should not be confused with the NDE detectability limit, which is the smallest flaw *that can be detected* by the NDE technique (on a good day). In most cases, a_o is significantly larger than the detectability limit, due to the variability in operating conditions and the skill of the operator.

Figure 10.28(a) illustrates the procedure for determining the first inspection interval in the structure. The lower curve defines the "true" behavior of the worst flaw in the structure, while the predicted curve assumes the initial flaw size is a_o. The time required to grow the flaw from a_o to a_t (the tolerable flaw size) is computed. The first inspection interval, I_1, should be less than this time, in order to preclude flaw growth beyond a_t before the next inspection. If no flaws greater than a_o are detected, the second inspection interval, I_2, is equal to I_1, as Fig. 10.28(b) illustrates. Suppose that the next inspection reveals a flaw of length a_1, which is larger than a_o. In this instance, a flaw growth analysis must be performed to estimate the time required to grow from a_1 to a_t. The next inspection interval, I_3, might be shorter than I_2, as Fig. 10.28(c) illustrates. Inspection intervals would then become progressively shorter as the structure approaches the end of its life. The structure is repaired or taken out of service when the flaw size reaches the maximum tolerable size, or when required inspections become too frequent to justify continued operation.

In many applications, a variable inspection interval is not practical; inspections must be often carried out at regular times that can be scheduled well in advance. In such instances a variation of the above approach is required. The main purpose of any damage tolerance assessment is to ensure that flaws will not grow to failure between inspections. The precise methods for achieving this goal depend on practical circumstances.

The schematic in Fig. 10.28(c) illustrates a flaw growth analysis that is conservative. If retardation effects are not taken into account, the analysis will be considerably simpler and will tend to overestimate growth rates. If a more detailed analysis is applied, a comparison of actual and predicted flaw sizes after each inspection interval can be used to calibrate the analysis.

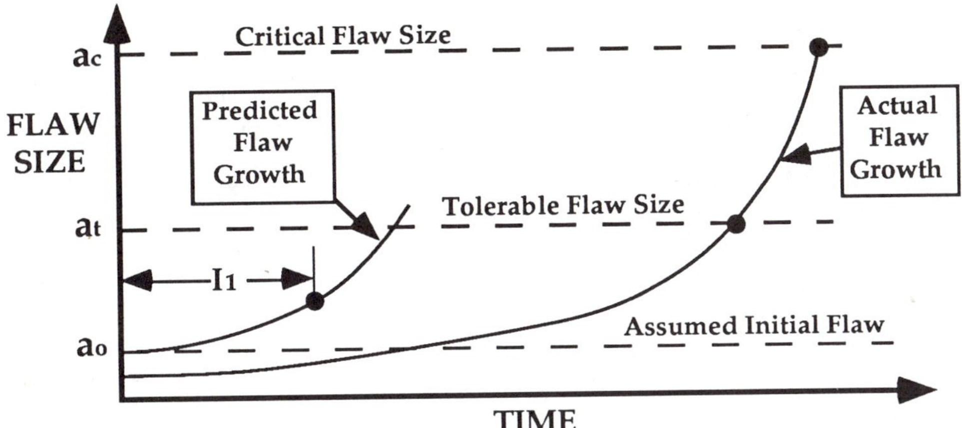

(a) Determination of first inspection interval, I_1.

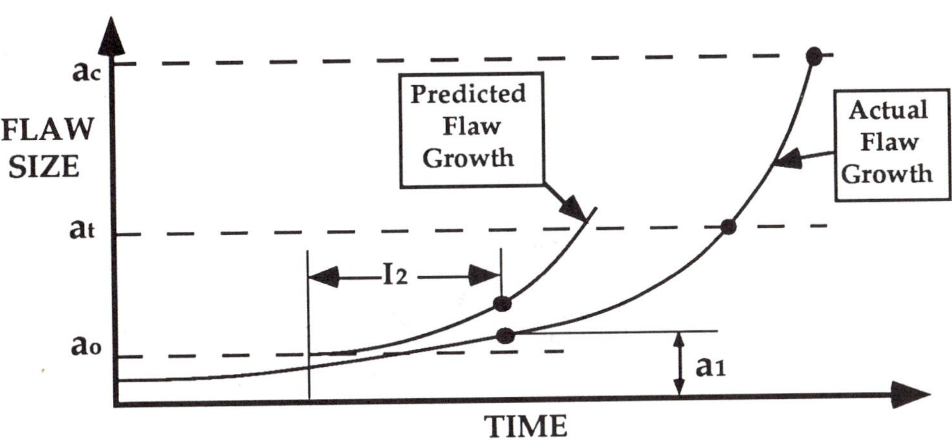

(b) Determination of second inspection interval, I_2.

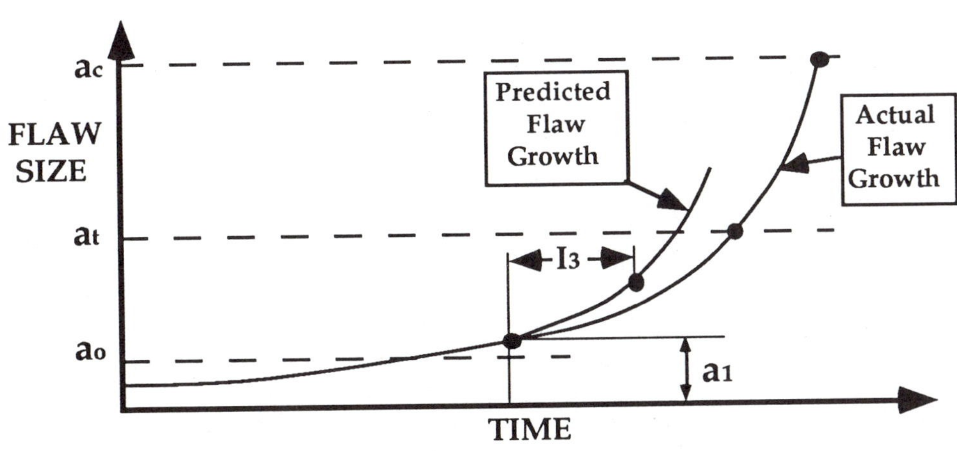

(c) Determination of third inspection interval, I_3.

FIGURE 10.28 Schematic damage tolerance analysis.

REFERENCES

1. Paris, P.C., Gomez, M.P., and Anderson, W.P., "A Rational Analytic Theory of Fatigue." *The Trend in Engineering*, Vol. 13, 1961, pp. 9-14.

2. Paris, P.C. and Erdogan, F., "A Critical Analysis of Crack Propagation Laws." *Journal of Basic Engineering*, Vol. 85, 1960, pp. 528-534.

3. Dowling, N.E. and Begley, J.A., "Fatigue Crack Growth During Gross Plasticity and the J Integral." ASTM STP 590, American Society for Testing and Materials, Philadelphia, 1976, pp. 82-103.

4. Lambert, Y., Saillard, P., and Bathias, C., "Application of the J Concept to Fatigue Crack Growth in Large-Scale Yielding." ASTM STP 969, American Society for Testing and Materials, Philadelphia, 1988, pp. 318-329.

5. Lamba, H.S., "The J-Integral Applied to Cyclic Loading." *Engineering Fracture Mechanics*, Vol. 7, 1975, pp. 693-703.

6. Wüthrich, C., "The Extension of the J-Integral Concept to Fatigue Cracks." *International Journal of Fracture*, Vol. 20, 1982, pp. R35-R37.

7. Tanaka, K., "Mechanics and Micromechanics of Fatigue Crack Propagation." ASTM STP 1020, American Society for Testing and Materials, Philadelphia, 1989, pp. 151-183.

8. Foreman, R.G., Keary, V.E., and Engle, R.M., "Numerical Analysis of Crack Propagation in Cyclic-Loaded Structures." *Journal of Basic Engineering*, Vol. 89, 1967, pp. 459-464.

9. Weertman, J., "Rate of Growth of Fatigue Cracks Calculated from the Theory of Infinitesimal Dislocations Distributed on a Plane." *International Journal of Fracture Mechanics*, Vol. 2, 1966, pp. 460-467.

10. Klesnil, M. and Lukas, P., "Influence of Strength and Stress History on Growth and Stabilisation of Fatigue Cracks." *Engineering Fracture Mechanics*, Vol. 4, 1972, pp. 77-92.

11. Donahue, R.J., Clark, H.M., Atanmo, P., Kumble, R., and McEvily, A.J., "Crack Opening Displacement and the Rate of Fatigue Crack Growth." *International Journal of Fracture Mechanics*, Vol 8, 1972, pp. 209-219.

12. McEvily, A.J., "On Closure in Fatigue Crack Growth." ASTM STP 982, American Society for Testing and Materials, Philadelphia, 1988, pp. 35-43.

13. Elber, W., "Fatigue Crack Closure Under Cyclic Tension." *Engineering Fracture Mechanics*, Vol. 2, 1970, pp. 37-45.

14. Suresh, S. and Ritchie, R.O., "Propagation of Short Fatigue Cracks." *International Metallurgical Reviews*, Vol. 29, pp. 445-476.

15. Budiansky, B. and Hutchinson, J.W., "Analysis of Closure in Fatigue Crack Growth." *Journal of Applied Mechanics,* Vol. 45, 1978, pp. 267-276.

16. Hudak, S.J., Jr. and Davidson, D.L., "The Dependence of Crack Closure on Fatigue Loading Variables." ASTM STP 982, American Society for Testing and Materials, Philadelphia, 1988, pp. 121-138.

17. Newman, J.C., "A Finite Element Analysis of Fatigue Crack Closure." ASTM STP 590, American Society for Testing and Materials, Philadelphia, 1976, pp. 281-301.

18. McClung, R.C. and Raveendra, S.T., "On the Finite Element Analysis of Fatigue Crack Closure - 1. Basic Modeling Issues." *Engineering Fracture Mechanics,* Vol. 33, 1989, pp. 342-360.

19. Gray, G.T., Williams, J.C., and Thompson, A.W., "Roughness Induced Crack Closure: An Explanation for Microstructurally Sensitive Fatigue Crack Growth." *Metallurgical Transactions,* Vol. 14A, 1983, pp. 421-433.

20. Schijve, J., "Some Formulas for the Crack Opening Stress Level." *Engineering Fracture Mechanics,* Vol. 14, 1981, pp. 461-465.

21. Gomez, M.P., Ernst, H., and Vazquez, J., "On the Validity of Elber's Results on Fatigue Crack Closure for 2024-T3 Aluminum." *International Journal of Fracture,* Vol. 12, 1976, pp. 178-180.

22. Clerivet, A. and Bathias, C., "Study of Crack Tip Opening under Cyclic Loading Taking into Account the Environment and R Ratio." *Engineering Fracture Mechanics,* Vol 12, 1979, pp. 599-611.

23. Shih, T.T. and Wei, R.P., "A Study of Crack Closure in Fatigue." *Engineering Fracture Mechanics,* Vol. 6, 1974, pp. 19-32.

24. Shih, T.T. and Wei, R.P., "Discussion." *International Journal of Fracture,* Vol. 13, 1977, pp. 105-106.

25. McClung, R.C., "The Influence of Applied Stress, Crack Length, and Stress Intensity Factor on Crack Closure." Submitted to *Metallurgical Transactions,* 1990.

26. Yokobori, T., Yokobori, A.T., Jr., and Kamei, A., "Dislocation Dynamic Theory for Fatigue Crack Growth." *International Journal of Fracture,* Vol. 11, 1975, pp. 781-788.

27. Tanaka, K., Akiniwa, Y., and Yamashita, M., "Fatigue Growth Threshold of Small Cracks." *International Journal of Fracture,* Vol. 17, 1981, pp. 519-533.

28. Klesnil, M. and Lucas. P., "Effect of Stress Cycle Asymmetry on Fatigue Crack Growth." *Materials Science and Engineering,* Vol. 9, 1972, pp. 231-240.

29. Schijve, J., "Fatigue Crack Closure: Observations and Technical Significance." ASTM STP 982, American Society for Testing and Materials, Philadelphia, 1988, pp. 5-34.

30. Hertzberg, R.W., Newton, C.H., and Jaccard, R., "Crack Closure: Correlation and Confusion." ASTM STP 982, American Society for Testing and Materials, Philadelphia, 1988, pp. 139-148.

31. Rice, J.R., "Mechanics of Crack-Tip Deformation and Extension by Fatigue." ASTM STP 415, American Society for Testing and Materials, Philadelphia, 1967, pp. 247-309.

32. McClung, R.C., "Crack Closure and Plastic Zone sizes in Fatigue." to appear in *Fatigue and Fracture of Engineering Materials and Structures.*

33. von Euw, E.F.J., Hertzberg, R.W., and Roberts, R., "Delay Effects in Fatigue-Crack Propagation." ASTM STP 513, American Society for Testing and Materials, Philadelphia, 1972, pp. 230-259.

34. Wheeler, O.E., "Spectrum Loading and Crack Growth." *Journal of Basic Engineering,* Vol 94, 1972, pp. 181-186.

35. Broek, D. *The Practical Use of Fracture Mechanics,* Kluwer Academic Publishers, Dordrect, Netherlands, 1988.

36. Tanaka, K. and Nakai, Y., "Propagation and Non-Propagation of Short Fatigue Cracks at a Sharp Notch." *Fatigue of Engineering Materials and Structures,* Vol. 6., 1983, pp. 315-327.

37. El Haddad, M.H., Topper, T.H., Smith, K.N., "Prediction of Non-Propagating Cracks." *Engineering Fracture Mechanics,* Vol. 11, 1979, pp. 573-584.

38. Laird, C., "Mechanisms and Theories of Fatigue." *Fatigue and Microstructure,* American Society for Metals, Metals Park, OH, 1979, pp. 149-203.

39. Starke, E.A. and Williams, J.C., "Microstructure and the Fracture Mechanics of Fatigue Crack Propagation." ASTM STP 1020, American Society for Testing and Materials, Philadelphia, 1989, pp. 184-205.

40. Lankford, J. and Davidson, D.L., "Fatigue Crack Micromechanisms in Ingot and Powder Metallurgy 7XXX Aluminum Alloys in Air and Vacuum." *Acta Metallurgica,* Vol 31, 1983, pp. 1273-1284.

41. Hertzberg, R.W., *Deformation and Fracture of Engineering Materials,* John Wiley and Sons, New York, 1989.

42. E 647-88a "Standard Method for Measurement of Fatigue Crack Growth Rates." American Society for Testing and Materials, Philadelphia, 1988.

43. Allison, J.E., "The Measurement of Crack Closure During Fatigue Crack Growth." ASTM STP 945, American Society for Testing and Materials, Philadelphia, 1988, pp. 913-933.

44. Sharpe, W.N., and Grandt, A.F., ASTM STP 590, American Society for Testing and Materials, Philadelphia, 1976, pp. 302-320.

45. Pitoniak, F.J., Grandt, A.F., Jr., Montulli, L.T., and Packman, P.F., "Fatigue Crack Retardation and Closure in Polymethylmethacrylate." *Engineering Fracture Mechanics*, Vol. 6, 1974, pp. 663-670.

46. Fleck, N.A. and Smith, R.A., "Crack Closure - Is it Just a Surface Phenomenon?" *International Journal of Fatigue*, Vol. 4, 1982, pp. 157-160.

APPENDIX 10: APPLICATION OF THE J CONTOUR INTEGRAL TO CYCLIC LOADING

A10.1 DEFINITION OF ΔJ

Material ahead of a growing fatigue crack experiences cyclic elastic-plastic loading, as Fig. A10.1 illustrates. The material deformation can be characterized by the stress range, $\Delta\sigma ij$, and the strain range, $\Delta\varepsilon ij$, in a given cycle.

Consider the *loading* branch of the stress-strain curve, where the stresses and strains have initial values $\sigma_{ij}^{(1)}$ and $\varepsilon_{ij}^{(1)}$, and increase to $\sigma_{ij}^{(2)}$ and $\varepsilon_{ij}^{(2)}$. It is possible to define a J-like integral as follows [3-6]:

$$\Delta J = \int_{\Gamma} \left(\psi(\Delta\varepsilon ij)\, dy - \Delta T_i \frac{\partial \Delta u_i}{\partial x} ds \right) \tag{A10.1}$$

where Γ defines the integration path around the crack tip, and ΔT_i and Δu_i are the changes in traction and displacement between points (1) and (2). The quantity ψ is analogous to the strain energy density:

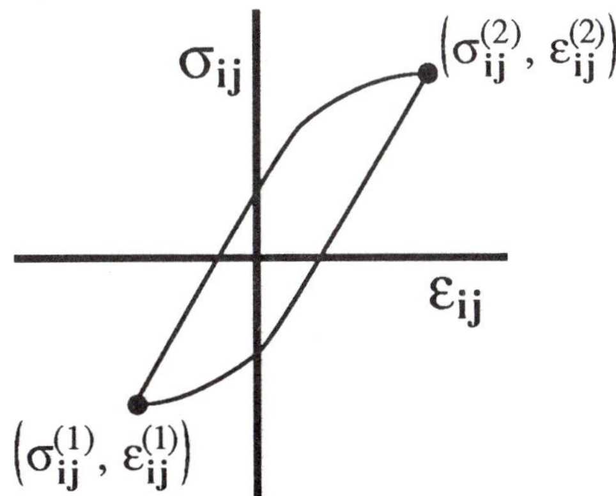

FIGURE A10.1 Schematic cyclic stress-strain behavior ahead of a growing fatigue crack.

$$\psi(\Delta\varepsilon_{kl}) = \int_0^{\Delta\varepsilon_{kl}} \Delta\sigma_{ij}\, d(\Delta\varepsilon_{ij})$$

$$= \int_{\varepsilon_{kl}^{(1)}}^{\varepsilon_{kl}^{(2)}} (\sigma_{ij} - \sigma_{ij}^{(1)})\, d\varepsilon_{ij} \qquad \text{(A10.2)}$$

Note that ψ represents the stress work per unit volume performed during *loading*, rather than the stress work in a complete cycle. The latter corresponds to the area inside the hysteresis loop (Fig. A10.1). For the special case where $\sigma_{ij}^{(1)} = \varepsilon_{ij}^{(1)} = 0$, $\Delta J = J$. Thus ΔJ is merely a generalization of the J integral, in which the origin is not necessarily at zero stress and strain.

Although ΔJ is normally defined from the loading branch of the cyclic stress-strain curve, it is also possible to define a ΔJ from the unloading branch. The two definitions coincide if the cyclic stress-strain curve forms a closed loop, and the loading and unloading branches are symmetric.

Just as it is possible to estimate J experimentally from a load-displacement curve (Chapters 3 and 7), ΔJ can be inferred from the cyclic load-displacement behavior. Consider a specimen with thickness B and uncracked ligament length b, that is cyclically loaded between the loads P_{min} and P_{max} and the load line displacements V_{min} and V_{max}, as Fig. A10.2(a) illustrates.[7] The ΔJ can be computed from an equation of the form

$$\Delta J = \frac{\eta}{B\,b} \int_0^{\Delta V} \Delta P\, d(\Delta V)$$

[7]The convention of previous chapters, where Δ represents the load line displacement, is suspended here to avoid confusion with the present use of this symbol.

$$= \frac{\eta}{B\,b} \int_{V_{min}}^{V_{max}} (P - P_{min})\, dV \qquad \text{(A10.3)}$$

where the dimensionless constant η has the same value as for monotonic loading. For example, $\eta = 2.0$ for a deeply notched bend specimen.

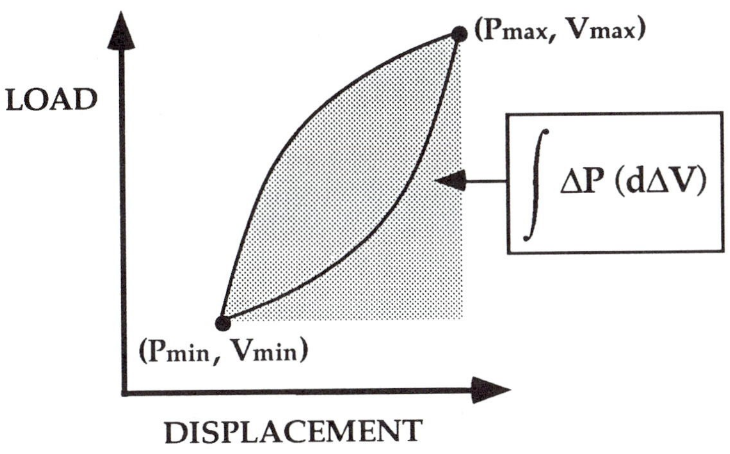

(a) No closure.

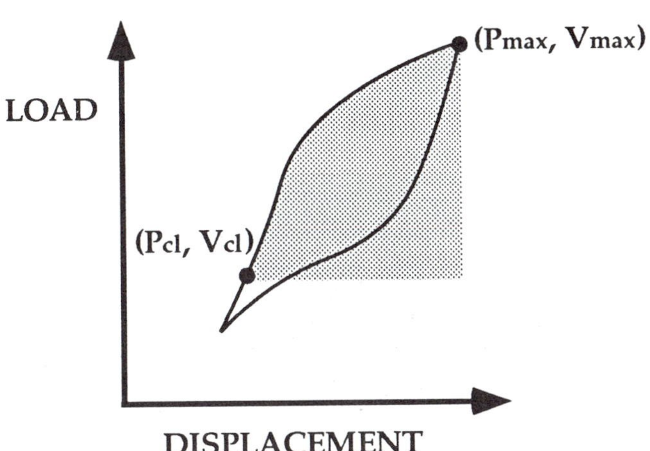

(b) With crack closure.

FIGURE A10.2 Cyclic load displacement behavior for fatigue under large scale yielding conditions.

Because the ΔJ parameter is often applied to crack growth under large scale yielding conditions, plasticity induced closure often has a significant effect on the results. If the crack is closed below P_{cl} and V_{cl} (Fig. 10.2(b)), Eq. (A10.3) can be modified as follows:[8]

$$\Delta J_{eff} = \frac{\eta}{B\,b} \int_{V_{cl}}^{V_{max}} (P - P_{cl})\, dV \qquad (A10.4)$$

A10.2 PATH INDEPENDENCE OF ΔJ

If ψ exhibits the properties of a potential, the stresses can be derived by differentiating ψ with respect to the strains:

$$\Delta\sigma_{ij} = \frac{\partial \psi}{\partial(\Delta\varepsilon_{ij})} \qquad (A10.5)$$

The validity of Eq. (A10.5) is both necessary and sufficient for path independence of ΔJ. The proof of path independence is essentially identical to the analysis in Appendix 3.2, except that stresses, strains and displacements are replaced by the *changes* in these quantities from states (1) to (2). Evaluating ΔJ along a closed contour, Γ^*, (Fig. A3.2) and invoking Green's theorem gives

$$\Delta J^* = \int_{A^*} \left[\frac{\partial \psi}{\partial x} - \frac{\partial}{\partial x_j}\left(\Delta\sigma_{ij} \frac{\partial \Delta u_j}{\partial x} \right) \right] dx\, dy \qquad (A10.6)$$

where A^* is the area enclosed by Γ^*. By assuming ψ displays the properties of a potential (Eq. A10.5), the first term in the integrand can be written as

[8]The global displacement at closure, V_{cl}, is not necessarily zero. The crack tip region may be closed while the crack mouth is open. Thus, V_{cl} is often positive.

$$\frac{\partial \psi}{\partial x} = \frac{\partial \psi}{\partial(\Delta\varepsilon_{ij})}\frac{\partial(\Delta\varepsilon_{ij})}{\partial x} = \Delta\sigma_{ij}\frac{\partial(\Delta\varepsilon_{ij})}{\partial x} \tag{A10.7}$$

By invoking the strain-displacement relationships for small strains, it can be shown that Eq. (A10.7) is equal to the absolute value of the second term in the integrand in Eq. (A10.6). (See Eqs. (A3.16) to (A3.18) for the mathematical details.) Thus $\Delta J^* = 0$ for any closed contour. Path independence of ΔJ evaluated along a crack tip integral can thus be readily demonstrated by considering the contour illustrated in Fig. A3.3, and noting that $J_1 = -J_2$.

The validity of Eq. (A10.5) is crucial in demonstrating path independence of ΔJ. This relationship is automatically satisfied when there is proportional loading on each branch of the cyclic stress-strain curves. That is, $\Delta\sigma_{ij}$ must increase (or decrease) in proportion to $\Delta\sigma_{kl}$, and the shapes of the $\Delta\sigma_{ij}$-$\Delta\varepsilon_{ij}$ hysteresis loops must be similar to one another.

Proportional loading also implies a single parameter characterization of crack tip conditions. Consequently, ΔJ uniquely defines the changes in stress and strain near the crack tip when there is proportional loading in this region.

In the case of monotonic loading, the J integral ceases to provide a single parameter description of crack tip conditions when there is excessive plastic flow or crack growth (Section 3.6). Similarly, one would not expect ΔJ to characterize fatigue crack growth beyond a certain level of plastic deformation. The limitations of ΔJ have yet to be established.

A10.3 SMALL SCALE YIELDING LIMIT

When the cyclic plastic zone is small compared to specimen dimensions, ΔJ should characterize fatigue crack growth, since it is related to ΔK. The precise relationship between ΔK and ΔJ under small scale yielding conditions can be inferred by evaluating Eq. (A10.1) along a contour in the elastic singularity dominated zone. For a given ΔK_I, the changes in the stresses, strains and displacements are given by

$$\Delta\sigma_{ij} = \frac{\Delta K_I}{\sqrt{2\pi r}} f_{ij}(\theta) \tag{A10.8a}$$

$$\Delta\varepsilon_{ij} = \frac{\Delta K_I}{\sqrt{2\pi r}} g_{ij}(\theta) \tag{A10.8b}$$

$$\Delta u_i = \frac{\Delta K_I}{2\mu} \sqrt{\frac{r}{2\pi}} h_{ij}(\theta) \tag{A10.8c}$$

where f_{ij} and h_{ij} are given in Tables 2.1 and 2.2, and g_{ij} can be inferred from Hooke's law or the strain-displacement relationships.

Inserting Eqs (A10.8a) to (A10.8c) into Eq. (A10.1) and evaluating J along a circular contour of radius r leads to

$$\Delta J = \frac{\Delta K_I^2}{E'} \tag{A10.9}$$

where E' = E for plane stress conditions and $E' = E/(1 - \nu^2)$ for plane strain. Note that although $\Delta K = (K_{max} - K_{min})$, $\Delta J \neq (J_{max} - J_{min})$ since

$$\Delta K^2 = (K_{max})^2 - K_{max}K_{min} + (K_{min})^2$$

11. COMPUTATIONAL FRACTURE MECHANICS

Computers have had an enormous influence in virtually all branches of engineering, and fracture mechanics is no exception. Numerical modeling has become an indispensable tool in fracture analysis, since relatively few practical problems have closed-form analytical solutions.

Stress intensity solutions for literally hundreds of configurations have been published, the majority of which were inferred from numerical models. Elastic-plastic analyses to compute the J integral and crack tip opening displacement (CTOD) are also becoming relatively common. In addition, researchers are applying advanced numerical techniques to special problems, such as fracture at interfaces, dynamic fracture, and ductile crack growth.

Rapid advances in computer technology are primarily responsible for the exponential growth in applications of computational fracture mechanics. The personal computers that most engineers have on their desks are more powerful than mainframe computers of 20 years ago. The latest supercomputers require only a few minutes to solve problems that would take months or even years on older machines.

Hardware does not deserve all of the credit for the success of computational fracture mechanics, however. More efficient numerical algorithms have greatly reduced solution times in fracture problems. For example, the domain integral approach (Section 11.4) enables one to generate K and J solutions from finite element models with surprisingly coarse meshes. Commercial numerical analysis codes have become relatively user friendly, and many codes have incorporated fracture mechanics routines.

This chapter will not turn the reader into an expert on computational fracture mechanics, but it should serve as an introduction to the subject. The sections that follow describe some of the traditional approaches in numerical analysis of fracture problems, as well as some recent innovations.

The format of this chapter differs from earlier chapters, in that the main body of this chapter contains several relatively complicated mathematical derivations; previous chapters confined such material to appendices. This information is unavoidable when explaining the basis of the common numerical techniques. Readers who are intimidated by the mathematical details should at least skim this material and attempt to understand its significance.

11.1 OVERVIEW OF NUMERICAL METHODS

It is often necessary to determine the distribution of stresses and strains in a body that is subject to external loads or displacements. In limited cases, it is possible to obtain a closed-form analytical solution for the stresses and strains. If, for example, the body is subject to either plane stress or plane strain loading and it is composed of an isotropic linear elastic material, it may be possible to find a stress function that leads to the desired solution. Westergaard [1] and Williams [2] used such an approach to derive solutions for the stresses and strains near the tip of a sharp crack in an elastic material (see Appendix 2). In most instances, however, closed-form solutions are not possible, and the stresses in the body must be estimated numerically.[1]

A variety of numerical techniques have been applied to problems in solid mechanics, including finite difference [3], finite element [4] and boundary integral equation methods [5]. In recent years, the latter two numerical methods have been applied almost exclusively. The vast majority of analyses of cracked bodies utilize finite elements, although the boundary integral method may be useful in limited circumstances.

11.1.1 The Finite Element Method

In the finite element method, the structure of interest is subdivided into discrete shapes called *elements*. Element types include one-dimensional beams, two-dimensional plane stress or plane strain elements, and three-dimensional bricks. The elements are connected at *node* points where continuity of the displacement fields is enforced. The dimensionality of the structure need not correspond to the ele-

[1]Experimental stress analysis methods, such as photoelasticity, Moiréinterferometry, and caustics are available, but even these techniques often require numerical analysis to interpret experimental observations.

ment dimension. For example, a three-dimensional truss can be constructed from beam elements.

The *stiffness finite element method* [4] is usually applied to stress analysis problems. This approach is outlined below for the two-dimensional case.

Figure 11.1 shows an *isoparametric* plate element for two-dimensional plane stress or plane strain problems, together with local and global coordinate axes. The local coordinates, which are also called *parametric* coordinates, vary from -1 to +1 over the element area; the node at the lower left-hand corner has parametric coordinates (-1, -1) while upper right-hand corner is at (+1, +1) in the local system. Note that the parametric coordinate system is not necessarily orthogonal. Consider a point on the element at (ξ, η). The global coordinates of this point are given by

$$x = \sum_{i=1}^{n} N_i(\xi, \eta)\, x_i$$

$$y = \sum_{i=1}^{n} N_i(\xi, \eta)\, y_i \tag{11.1}$$

where n is the number of nodes in the element and N_i are the shape functions corresponding to the node i, whose coordinates are (x_i, y_i) in the global system and (ξ_i, η_i) in the parametric system.

The shape functions are polynomials that interpolate field quantities within the element. The degree of the polynomial depends on the number of nodes in the element. If, for example, the element contains nodes only at the corners, N_i are linear. Figure 11.1 illustrates a four-sided, eight-node element, which requires a quadratic interpolation. Appendix 11 gives the shape functions for the latter case.

The displacements within an element are interpolated as follows:

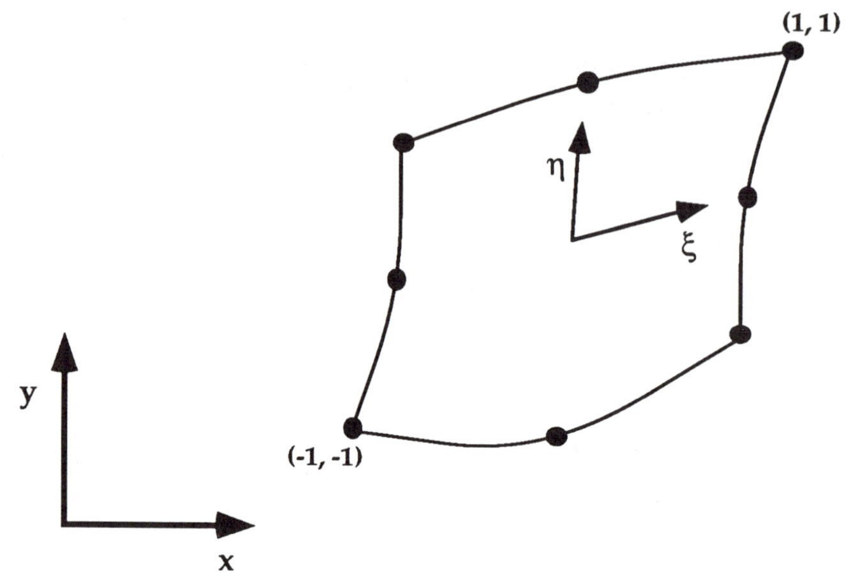

FIGURE 11.1 Local and global coordinates for a two-dimensional element.

$$u = \sum_{i=1}^{n} N_i(\xi, \eta)\, u_i$$
$$v = \sum_{i=1}^{n} N_i(\xi, \eta)\, v_i \tag{11.2}$$

where (u_i, v_i) are the nodal displacements in the x and y directions, respectively. The strain matrix at (x, y) is given by

$$\{\varepsilon\} = [\mathbf{B}] \begin{Bmatrix} u_i \\ v_i \end{Bmatrix} \tag{11.3}$$

where

$$[\mathbf{B}] = \begin{bmatrix} \dfrac{\partial N_i}{\partial x} & 0 \\ 0 & \dfrac{\partial N_i}{\partial y} \\ \dfrac{\partial N_i}{\partial y} & \dfrac{\partial N_i}{\partial x} \end{bmatrix} \tag{11.4}$$

and

$$\begin{Bmatrix} \dfrac{\partial N_i}{\partial x} \\ \dfrac{\partial N_i}{\partial y} \end{Bmatrix} = [\mathbf{J}]^{-1} \begin{Bmatrix} \dfrac{\partial N_i}{\partial \xi} \\ \dfrac{\partial N_i}{\partial \eta} \end{Bmatrix} \tag{11.5}$$

where [**J**] is the Jacobian matrix, which is given by

$$[\mathbf{J}] = \begin{bmatrix} \dfrac{\partial x}{\partial \xi} & \dfrac{\partial y}{\partial \xi} \\ \dfrac{\partial x}{\partial \eta} & \dfrac{\partial y}{\partial \eta} \end{bmatrix} = \begin{bmatrix} \cdots & \dfrac{\partial N_i}{\partial x} & \cdots \\ \cdots & \dfrac{\partial N_i}{\partial y} & \cdots \end{bmatrix} \begin{bmatrix} \vdots & \vdots \\ x_i & y_i \\ \vdots & \vdots \end{bmatrix} \tag{11.6}$$

The stress matrix is computed as follows:

$$\{\sigma\} = [\mathbf{D}]\,\{\varepsilon\} \tag{11.7a}$$

where [**D**] is the stress-strain constitutive matrix. For problems that incorporate incremental plasticity, stress and strain are computed incrementally and [**D**] is updated at each load step:

$$\{\Delta\sigma\} = [\mathbf{D}(\varepsilon, \sigma)]\,\{\Delta\varepsilon\} \tag{11.7b}$$

Thus the stress and strain distribution throughout the body can be inferred from the nodal displacements and the constitutive law. The stresses and strains are usually evaluated at several Gauss points or integration points within each elements. For 2-D elements, 2x2 Gaussian integration is typical, where there are four integration points on each element.

The displacements at the nodes depend on the element stiffness and the nodal forces. The elemental stiffness matrix is given by:

$$[k] = \int_{-1}^{1}\int_{-1}^{1} [\mathbf{B}]^T [\mathbf{D}] [\mathbf{B}] \det|\mathbf{J}| \, d\xi \, d\eta \tag{11.8}$$

where the subscript T denotes the transpose of the matrix. Equation (11.8) can be derived from the principle of minimum potential energy [4].

The elemental stiffness matrices are assembled to give the global stiffness matrix, $[\mathcal{K}]$. The global force, displacement, and stiffness matrices are related as follows:

$$[\mathcal{K}] [\mathbf{u}] = [\mathbf{F}] \tag{11.9}$$

11.1.2 The Boundary Integral Equation Method

Most problems in nature cannot be solved mathematically without specifying appropriate boundary conditions. In solid mechanics, for example, a well posed problem is one in which either the tractions or the displacements (but not both) are specified over the entire surface. In the general case, the surface of a body can be divided into two regions: S_u, where displacements are specified, and S_T, where tractions are specified. (One cannot specify both traction and displacement on the same area, since one quantity depends on the other.) Given these boundary conditions, it is theoretically possible to solve for the tractions on S_u and the displacements on S_T, as well as the stresses, strains, and displacements within the body.

The boundary integral equation (BIE) method [5-9] is a very powerful technique for solving for unknown tractions and displacements on the surface. This approach can also provide solutions for internal field quantities, but finite element analysis is more efficient for this purpose.

The BIE method stems from Betti's reciprocal theorem, which relates work done by two different loadings on the same body. In the absence of body forces, Betti's theorem can be stated as follows:

$$\int_S T_i^{(1)} u_i^{(2)} dS = \int_S T_i^{(2)} u_i^{(1)} dS \tag{11.10}$$

where T_i and u_i are components of the traction and displacement vectors, respectively, and the superscripts denote loadings (1) and (2). The standard convention is followed in this chapter, where repeated indices imply summation. Equation (11.10) can be derived from the principle of virtual work, together with the fact that $\sigma_{ij}^{(1)}\varepsilon_{ij}^{(2)} = \sigma_{ij}^{(2)}\varepsilon_{ij}^{(1)}$ for a linear elastic material.

Let us assume that (1) is the loading of interest and (2) is a reference loading with a known solution. Figure 11.2 illustrates the conventional reference boundary conditions for BIE problems. A unit force is applied at an interior point p in each of the three coordinate directions, x_i, resulting in displacements and tractions at surface point Q in the x_j direction[2]. For example, a unit force in the x_1 direction may produce displacements and tractions at Q in all three coordinate directions. Consequently, the resulting displacements and tractions at Q, $\mathbf{u}_{ij}$ and $\mathbf{T}_{ij}$, are second-order tensors. The quantities $\mathbf{u}_{ij}(p,Q)$ and $\mathbf{T}_{ij}(p,Q)$ have closed-form solutions for several cases, including a point force on the surface of a semi-infinite elastic body [5].

Applying the Betti reciprocal theorem to the boundary conditions described above leads to [5]:

$$u_i(p) = -\int_S \mathbf{T}_{ij}(p,Q)\, u_j(Q)\, dS + \int_S \mathbf{u}_{ij}(p,Q)\, T_j(Q)\, dS \tag{11.11}$$

where $u_i(p)$ is the displacement vector at the interior point p; $u_j(Q)$ and $T_j(Q)$ are the reference displacement and traction vectors at the boundary point Q. Note that $u_i(p)$, $u_j(Q)$, and $T_j(Q)$ correspond to the loading of interest; i.e., loading (1). At a given point Q on the boundary, either traction or displacement is known a priori, and it is necessary to solve

---

[2]For the remainder of this chapter, we will adopt the x_1-x_2-x_3 coordinate system, rather than x-y-z. The former notation is more convenient when manipulating tensor quantities.

for the other quantity. If we let $p \rightarrow P$, where P is a point on the surface, Eq. (11.11) becomes [5]:

$$\frac{1}{2} u_i(P) + \int_S T_{ij}(P,Q)\, u_j(Q)\, dS = \int_S u_{ij}(P,Q)\, T_j(Q)\, dS \tag{11.12}$$

assuming the surface is smooth. (This relationship is modified slightly when P is near a corner or other discontinuity.) Equation (11.12) represents a set of integral constraint equations that relate surface displacements to surface tractions. In order to solve for the unknown boundary data, the surface must be subdivided into segments (i.e., elements), and Eq. (11.12) approximated by a system of algebraic equations. If it is assumed that u_i and T_i vary linearly between discrete nodal points on the surface, Eq. (11.12) can be written as

$$\left(\left[\frac{1}{2}\delta_{ij}\right] + \left[\Delta T_{ij}\right]\right)\left\{u_j\right\} = \left[\Delta U_{ij}\right]\left\{T_j\right\} \tag{11.13}$$

where δ_{ij} is the Kronecker delta. Equation (11.13) represents a set of 3n equations for a three-dimensional problem, where n is the number of nodes. Once all of the boundary quantities are known, displacements at internal points can be inferred from Eq. (11.11).

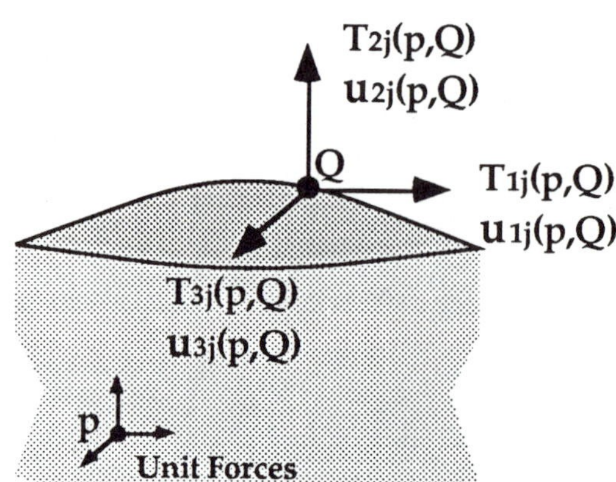

FIGURE 11.2 Reference boundary conditions for a boundary integral element problem. Unit forces are applied in each of the coordinate directions at point p, resulting in tractions and displacements on the surface point Q.

The boundary elements have one less dimension than the body being analyzed. That is, the boundary of a two-dimensional problem is surrounded by one-dimensional elements, while the surface a three-dimensional solid is paved with two-dimensional elements. Consequently, boundary element analysis can be very efficient, particularly when the boundary quantities are of primary interest. This method tends to be inefficient, however, when solving for internal field quantities.

The boundary integral equation method is usually applied to linear elastic problems, but this technique can also be utilized for elastic-plastic analysis [6,9]. As with the finite element technique, nonlinear BIE analyses are typically performed incrementally, and the stress-strain relationship is assumed to be linear within each increment.

11.2 TRADITIONAL COMPUTATIONAL METHODS FOR LEFM

This section describes several of the earlier approaches for inferring fracture mechanics parameters from numerical analysis. Most of these methods have been made obsolete by more recent techniques that are significantly more accurate and efficient (Sections 11.3 and 11.4).

The approaches outlined below can be divided into two categories: point matching and energy methods. The former technique entails inferring the stress intensity factor from the stress or displacement fields in the body, while energy methods compute the energy release rate in the body and relate $\mathcal{G}$ to stress intensity. One advantage of energy methods is that they can be applied to nonlinear material behavior; a disadvantage is that it is often very difficult to separate energy release rate into mixed-mode K components.

Most of the techniques described below can be implemented with either finite element or boundary element methods. The stiffness derivative approach (Section 11.2.4), however, was formulated in terms of the finite element stiffness matrix, and thus is not compatible with boundary element analysis.

11.2.1 Stress and Displacement Matching

Consider a cracked body subject to pure Mode I loading. On the crack plane ($\theta = 0$), K_I is related to the stress in the x_2 direction as follows:

$$K_I = \lim_{r \to 0} \left[\sigma_{22} \sqrt{2\pi r} \right] \qquad (\theta = 0) \tag{11.14}$$

The stress intensity factor can be inferred by plotting the quantity in square brackets against distance from the crack tip, and extrapolating to $r = 0$. Alternatively, K_I can be estimated by from a similar extrapolation of crack opening displacement:

$$K_I = \frac{2\mu}{\kappa + 1} \lim_{r \to 0} \left[u_2 \sqrt{\frac{2\pi}{r}} \right] \qquad (\theta = \pi) \tag{11.15}$$

Equation (11.15) tends to give more accurate estimates of K than Eq. (11.14) because nodal displacements can be inferred with a higher degree of precision than stresses. Both methods, however, are vastly inferior to current approaches. These extrapolation approaches require a high degree of mesh refinement for reasonable accuracy. For example, with a two-dimensional finite element mesh with 2000 degrees of freedom, the extrapolation methods typically give errors in K_I of around 5% [10]; present-day energy methods (Sections 11.3 and 11.4) provide much better accuracy and do not require such fine meshes.

The boundary collocation method [11,12] is an alternative point matching technique for computing stress intensity factors. This approach entails finding stress functions that satisfy the boundary conditions at various nodes, and inferring the stress intensity factor from these functions. For plane stress or plane strain problems, the Airy stress function (Appendix 2) can be expressed in terms of two complex analytic functions, which can be represented as polynomials in the complex variable z ($= x_1 + ix_2$). In a boundary collocation analysis, the coefficients of the polynomials are inferred from nodal quantities. The minimum number of nodes utilized in the analysis corresponds to the number of unknown coefficients in the polynomials. The results can be improved by analyzing more than the minimum number of nodes and solving for the unknowns by least squares. This approach can be highly cumbersome; energy methods are preferable in most instances.

Early researchers in computational fracture mechanics attempted to reduce the mesh size requirements for point matching analyses by introducing special elements at the crack tip that exhibit the $1/\sqrt{r}$ singularity [13]. Barsoum [14] later showed that this same effect could be achieved by a slight modification to conventional isoparametric elements (see Section 11.5 and Appendix 11).

11.2.2 Elemental Crack Advance

Recall from Chapter 2 that the energy release rate can be inferred from the rate of change in global potential energy with crack growth. If two separate numerical analyses of a given geometry are performed, one with crack length a, and the other with crack length $a + \Delta a$, the energy release rate is given by

$$\mathcal{G} = -\left(\frac{\Delta\Pi}{\Delta a}\right)_{\text{fixed boundary conditions}} \tag{11.16}$$

assuming a two-dimensional body with unit thickness.

This technique requires minimal post-processing, since total strain energy is output by many commercial analysis codes. This technique is also more efficient than the point matching methods, since global energy estimates do not require refined meshes.

One disadvantage of the elemental crack advance method is that multiple solutions are required in this case, while other methods infer the desired crack tip parameter from a single analysis. This may not be a serious shortcoming if the intention is to compute G (or K) as a function of crack size. The numerical differentiation in Eq. (11.16), however, can result in significant errors unless the crack length intervals (Δa) are small.

11.2.3 Contour Integration

The J integral can be evaluated numerically along a contour surrounding the crack tip. The advantages of this method are that it can be applied both to linear and nonlinear problems, and path independence (in elastic materials) enables the user to evaluate J at a remote contour, where numerical accuracy is greater. For problems that include path-dependent plastic deformation or thermal strains, it is still possible to compute J at a remote contour, provided an appropriate correction term (i.e., an area integral) is applied [15,16].

For three-dimensional problems, however, the contour integral becomes a surface integral, which is extremely difficult to evaluate numerically.

More recent formulations of J apply an area integration for two-dimensional problems and a volume integration for three-dimensional problems. Area and volume integrals provide much better accuracy that contour and surface integrals, and are much easier to implement numerically. The first such approach was the stiffness derivative formulation of the virtual crack extension method, which is described below. This approach has since been improved and made more general, as Sections 11.3 and 11.4 discuss.

11.2.4 Virtual Crack Extension: Stiffness Derivative Formulation

In 1974, Parks [10] and Hellen [17] independently proposed the following finite element method for inferring energy release rate in elastic bodies. Several years later Parks [18] extended this method to nonlinear behavior and large deformation at the crack tip. Although the stiffness derivative method is now outdated, it was the precursor to the modern approaches described in Sections 11.3 and 11.4.

Consider a two-dimensional cracked body with unit thickness, subject to Mode I loading. The potential energy of the body, in terms of the finite element solution, is given by

$$\Pi = \frac{1}{2}[\mathbf{u}]^T[\mathcal{K}][\mathbf{u}] - [\mathbf{u}]^T[\mathbf{F}] \tag{11.17}$$

where Π is the potential energy, and the other quantities are as defined in Section 11.1.1. Recall from Chapter 2 that the energy release rate is the derivative of Π with respect to crack area, for both fixed load and fixed displacement conditions. It is convenient in this instance to evaluate $\mathcal{G}$ under fixed load conditions:

$$\mathcal{G} = -\left(\frac{\partial \Pi}{\partial a}\right)_{\text{load}}$$

$$= -\frac{\partial[\mathbf{u}]^T}{\partial a}\{[\mathcal{K}][\mathbf{u}] - [\mathbf{F}]\} - \frac{1}{2}[\mathbf{u}]^T\frac{\partial[\mathcal{K}]}{\partial a}[\mathbf{u}] + [\mathbf{u}]^T\frac{\partial[\mathbf{F}]}{\partial a} \tag{11.18}$$

Comparing Eq. (11.9) to the above result, we see that the first term in Eq. (11.18) must be zero. In the absence of tractions on the crack face, the third term must also vanish, since loads are held constant. Thus the energy release rate is given by

$$\mathcal{G} = \frac{K_I^2}{E'} = -\frac{1}{2}[\mathbf{u}]^T\frac{\partial[\mathcal{K}]}{\partial a}[\mathbf{u}] \tag{11.19}$$

Thus the energy release rate is proportional to the derivative of the stiffness matrix with respect to crack length.

Suppose that we have generated a finite element mesh for a body with crack length *a* and we wish to extend the crack by Δa. It would not be necessary to change all of the elements in the mesh; we could accommodate the crack growth by moving elements near the crack tip and leaving the rest of the mesh intact. Figure 11.3 illustrates such a process, where elements inside the contour Γ_o are shifted by Δa, and elements outside of the contour Γ_1 are unaffected. Each of the elements between Γ_o and Γ_1 is distorted, such that its stiffness changes. The energy release rate is related to this change in element stiffness:

$$\mathcal{G} = -\frac{1}{2}[\mathbf{u}]^T\left(\sum_{i=1}^{N_c}\frac{\partial[k_i]}{\partial a}\right)[\mathbf{u}] \tag{11.20}$$

where $[k_i]$ are the elemental stiffness matrices and N_c is the number of elements between the contours Γ_o and Γ_1. Parks [10] demonstrated that Eq. (11.20) is equivalent to the J integral. The value of G (or J) is independent of the choice of the inner and outer contours.

It is important to note that in a virtual crack extension analysis, it is not necessary to generate a second mesh with a slightly longer crack. It is sufficient merely to calculate the change in elemental stiffness matrices corresponding to shifts in the nodal coordinates.

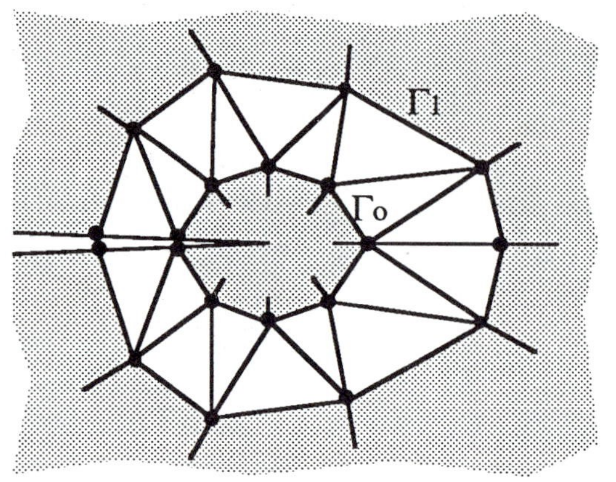

(a) Initial conditions.

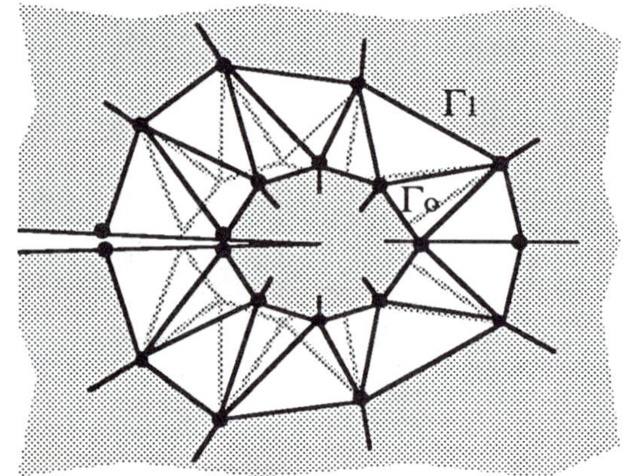

(b) After virtual crack advance.

FIGURE 11.3 Virtual crack extension in a finite element model [10,17]. Elements between Γ_1 and Γ_0 are distorted to accommodate a crack advance.

One problem with the stiffness derivative approach is that it involves cumbersome numerical differencing. Also, this formulation is poorly suited to problems that include thermal strain. A more recent formulation of the virtual crack extension method overcomes these difficulties, as discussed below.

11.3 VIRTUAL CRACK EXTENSION: MODERN APPROACH

Parks [10] and Hellen [17] formulated the virtual crack extension approach in terms of finite element stiffness and displacement matrices. deLorenzi [19,20] improved the virtual crack extension method by considering the energy release rate of a continuum. The main advantages of the continuum approach are two-fold: first, the methodology is not restricted to the finite element method; and second, deLorenzi's approach does not require numerical differencing.

Figure 11.4 illustrates a virtual crack advance in a two-dimensional continuum. Material points inside Γ_0 experience rigid body translation a distance Δa in the x_1 direction, while points outside of Γ_1 remain fixed. In the region between contours, virtual crack extension causes material points to translate by Δx_1. For an elastic material, or one that obeys deformation plasticity theory, deLorenzi showed that energy release rate is given by

$$\mathcal{G} = \frac{1}{\Delta a} \int_A \left(\sigma_{ij} \frac{\partial u_j}{\partial x_1} - w\, \delta_{i1} \right) \frac{\partial \Delta x_1}{\partial x_i}\, dA \tag{11.21}$$

where w is the strain energy density. Equation (11.21) assumes unit thickness, crack growth in the x_1 direction, no body forces within Γ_1, and no tractions on the crack faces. Note that $\partial \Delta x_1 / \partial x_i = 0$ outside of Γ_1 and within Γ_o; thus the integration need only be performed over the annular region between Γ_o and Γ_1.

deLorenzi actually derived a more general expression that considers a three-dimensional body, tractions on the crack surface, and body forces:

$$\mathcal{G} = \frac{1}{\Delta A_c} \int_V \left[\left(\sigma_{ij} \frac{\partial u_j}{\partial x_k} - w\, \delta_{ik} \right) \frac{\partial \Delta x_k}{\partial x_i} - F_i \frac{\partial u_i}{\partial x_j} \Delta x_j \right] dV$$

$$- \frac{1}{\Delta A_c} \int_S T_i \frac{\partial u_i}{\partial x_k} \Delta x_j\, dS \tag{11.22}$$

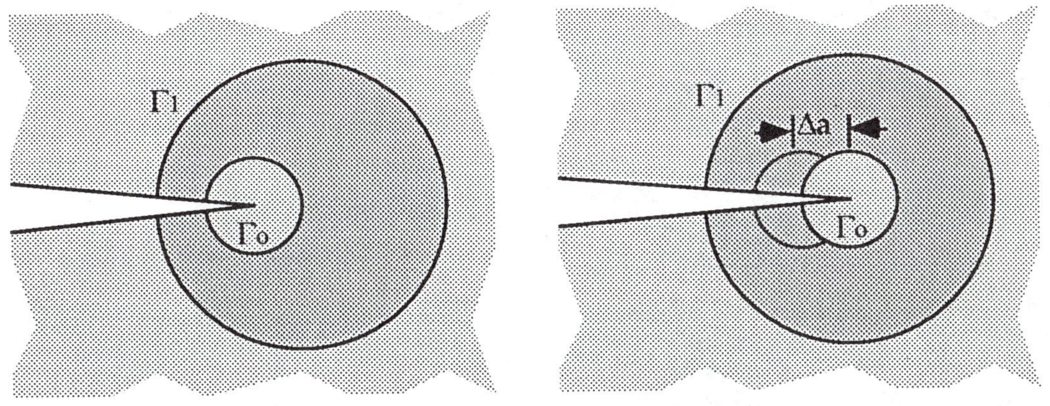

(a) Initial state **(b) After virtual crack advance.**

FIGURE 11.4 Virtual crack extension in a two-dimensional elastic continuum.

where ΔA_c is the increase in crack area generated by the virtual crack advance, V is the volume of the body, and F_i are the body forces. In this instance, two surfaces enclose the crack front. Material points within the inner surface, S_o, are displaced by Δa_i, while the material outside of the outer surface, S_1, remains fixed. The displacement vector between S_o and S_1 is Δx_i, which ranges from 0 to Δa_i. Equation (11.22) assumes a fixed coordinate system; consequently, the virtual crack advance, Δa_i, is not necessarily in the x_1 direction when the crack front is curved. The above expression, however, only applies to virtual crack advance normal to the crack front, in the plane of the crack.

In a three-dimensional problem, $\mathcal{G}$ may vary along the crack front. In computing $\mathcal{G}$, one can consider a uniform virtual crack advance over the entire crack front or a crack advance over a small increment, as Fig. 11.5 illustrates. In the former case, $\Delta A_c = \Delta a\ L$, and the computed energy release rate would be a weighted average. Defining A_c incrementally along the crack front would result in a local measure of $\mathcal{G}$.

For two-dimensional problems, the virtual crack extension formulation of $\mathcal{G}$ requires an area integration, while three-dimensional problems require a volume integration. Such an approach is easier to implement numerically and is more accurate than contour and surface integrations for two- and three-dimensional problems, respectively.

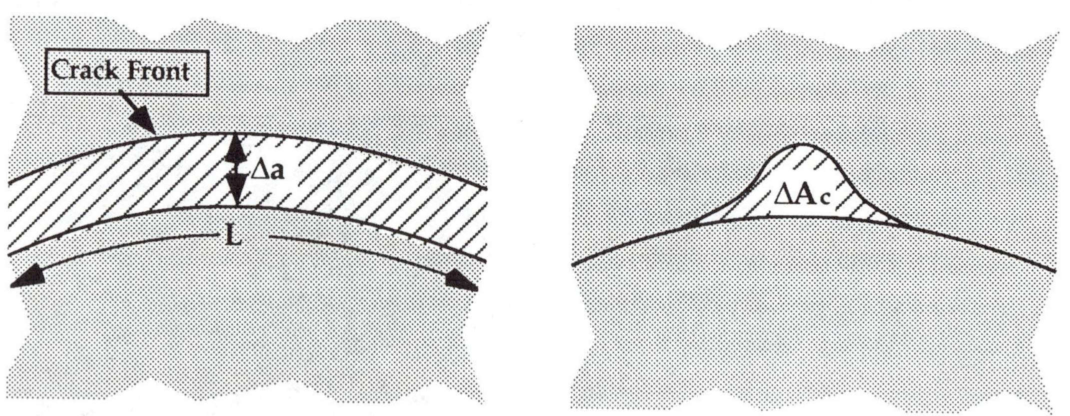

(a) Uniform crack advance.

(b) Advance over an increment of crack front.

FIGURE 11.5 Virtual crack extension along a three-dimensional crack front.

Numerical implementation of the virtual crack extension method entails applying a virtual displacement to nodes within a specified contour. Since the domain integral formulation is very similar to the above method, further discussion on numerical implementation is deferred to Section 11.4.3.

11.4 THE ENERGY DOMAIN INTEGRAL

Shih, et. al. [21,22] have recently formulated the energy domain integral methodology, which is a general framework for numerical analysis of the J integral. This approach is extremely versatile, as it can be applied to both quasistatic and dynamic problems with elastic, plastic, or viscoplastic material response, as well as thermal loading. Moreover, the domain integral formulation is relatively simple to implement numerically, and it is very efficient. This approach is very similar to the virtual crack extension method.

11.4.1 Theoretical Background

Appendix 4.2 presents a derivation of a general expression for the J integral that includes the effects of inertia as well as inelastic material behavior. The generalized definition of J requires that the contour surrounding the crack tip be vanishingly small:

$$J = \lim_{\Gamma_o \to 0} \int_{\Gamma_o} \left[(w + T)\,\delta_{1i} - \sigma_{ij} \frac{\partial u_j}{\partial x_1} \right] n_i \, d\Gamma \tag{11.23}$$

where T is the kinetic energy density. Various material behavior can be taken into account through the definition of w, the stress work.

Consider an elastic-plastic material loaded under quasistatic conditions (T = 0). If thermal strains are present, the total strain is given by

$$\varepsilon_{ij}^{total} = \varepsilon_{ij}^{e} + \varepsilon_{ij}^{p} + \alpha\,\Theta\,\delta_{ij} = \varepsilon_{ij}^{m} + \varepsilon_{kk}^{t} \tag{11.24}$$

where α is the coefficient of thermal expansion and Θ is the temperature, relative to ambient. The superscripts e, p, m, and t denote elastic, plastic, mechanical, and thermal strains, respectively. The mechanical strain is equal to the sum of elastic and plastic components. The stress work is given by

$$w = \int_0^{\varepsilon_{kl}^m} \sigma_{ij} \, d\varepsilon_{ij}^m \tag{11.25}$$

The form of Eq. (11.23) is not suitable for numerical analysis, since it is not feasible to evaluate stresses and strains along a vanishingly small contour. Let us construct a closed contour by connecting inner and outer contours, as Fig. 11.6 illustrates. The outer contour, Γ_1, is finite, while Γ_o is vanishingly small. For a linear or nonlinear elastic material under quasistatic conditions, J could be evaluated along either Γ_1 or Γ_o, but only the inner contour gives the correct value of J in the general case. For quasistatic conditions, where T = 0, Eq. (11.23) can be written in terms of the following integral around the closed contour, $\Gamma^* = \Gamma_1 + \Gamma_+ + \Gamma_- - \Gamma_o$ [21,22]:

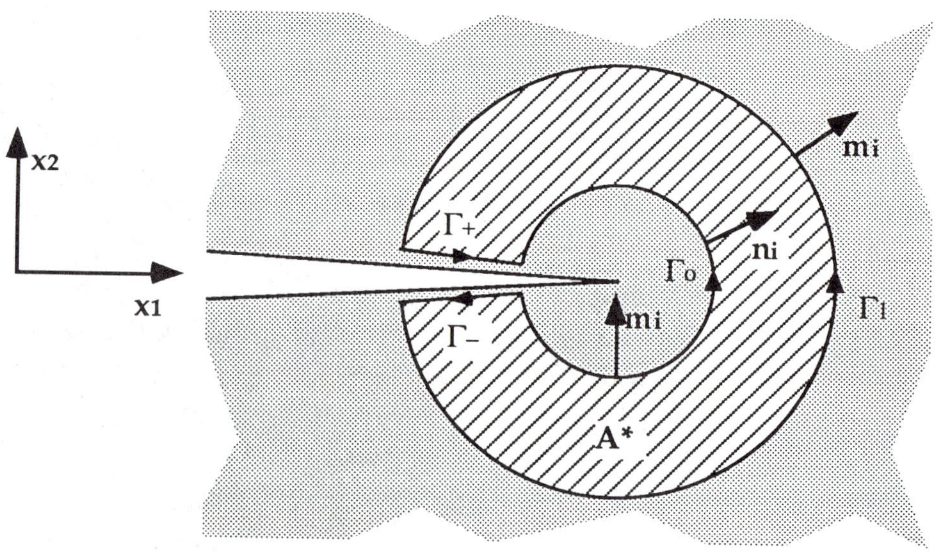

FIGURE 11.6 Inner and outer contours, which form a closed contour around the crack tip when connected by Γ_+ and Γ_-.

$$J = \int_{\Gamma^*} \left[\sigma_{ij} \frac{\partial u_j}{\partial x_1} - w\,\delta_{1i} \right] q\, m_i\, d\Gamma - \int_{\Gamma_+ + \Gamma_-} \sigma_{2j} \frac{\partial u_j}{\partial x_1} q\, d\Gamma \tag{11.26}$$

where Γ^+ and Γ^- are the upper and lower crack faces, respectively, m_i is the outward normal on Γ^*, and q is an arbitrary but smooth function that is equal to unity on Γ_o and zero on Γ_1. Note that $m_i = -n_i$ on Γ_o; also, $m_1 = 0$ and $m_2 = \pm 1$ on Γ^+ and Γ^-. In the absence of crack face tractions, the second integral in Eq. (11.26) vanishes.

For the moment, assume that the crack faces are traction free. Applying the divergence theorem to Eq. (11.26) gives

$$J = \int_{A^*} \frac{\partial}{\partial x_i} \left\{ \left[\sigma_{ij} \frac{\partial u_j}{\partial x_1} - w\,\delta_{1i} \right] q \right\} dA$$

$$= \int_{A^*} \left[\sigma_{ij} \frac{\partial u_j}{\partial x_1} - w\,\delta_{1i} \right] \frac{\partial q}{\partial x_i} dA + \int_{A^*} \left[\frac{\partial}{\partial x_i} \left(\sigma_{ij} \frac{\partial u_j}{\partial x_1} \right) - \frac{\partial w}{\partial x_1} \right] q\, dA \tag{11.27}$$

where A^* is the area enclosed by Γ^*. Referring to Appendix 3.2, we see that

$$\frac{\partial}{\partial x_i} \left(\sigma_{ij} \frac{\partial u_j}{\partial x_1} \right) - \frac{\partial w}{\partial x_1} = 0 \tag{11.28}$$

when there are no body forces and w exhibits the properties of an elastic potential:

$$\sigma_{ij} = \frac{\partial w}{\partial \varepsilon_{ij}} \tag{11.29}$$

It is convenient at this point to divide w into elastic and plastic components:

$$w = w^e + w^p = \int_0^{\varepsilon_{kl}^e} \sigma_{ij}\, d\varepsilon_{ij}^e + \int_0^{\varepsilon_{kl}^p} S_{ij}\, d\varepsilon_{ij}^p \tag{11.30}$$

where S_{ij} is the deviatoric stress, defined in Eq. (A3.62). While the elastic components of w and ε_{ij} satisfy Eq. (11.29), plastic deformation does not, in general, exhibit the properties of a potential. (Equation (11.29) may be approximately valid for plastic deformation when there is no unloading.) Moreover, thermal strains would cause the left side of Eq. (11.28) to be nonzero. Thus the second integrand in Eq. (11.27) vanishes in limited circumstances, but not in general. Taking account of plastic strain, thermal strain, body forces, and crack face tractions leads to the following general expression for J in two dimensions:

$$J = \int_{A^*} \left\{ \left[\sigma_{ij} \frac{\partial u_j}{\partial x_1} - w\,\delta_{1i} \right] \frac{\partial q}{\partial x_i} + \left[\sigma_{ij} \frac{\partial \varepsilon_{ij}^p}{\partial x_1} - \frac{\partial w^p}{\partial x_1} + \alpha \sigma_{ii} \frac{\partial \Theta}{\partial x_1} - F_i \frac{\partial u_j}{\partial x_1} \right] q \right\} dA$$

$$- \int_{\Gamma_+ + \Gamma_-} \sigma_{2j} \frac{\partial u_j}{\partial x_1}\, q\, d\Gamma \tag{11.31}$$

where the body force contribution is inferred from the equilibrium equations, and the contribution from thermal loading is obtained by substituting Eqs. (11.24) and (11.30) into Eq. (11.27). Inertia can be taken into account by incorporating T into the group of terms that are multiplied by q. For a linear or nonlinear elastic material under quasistatic conditions, in the absence of body forces, thermal strains, and crack face tractions, Eq. (11.31) reduces to

$$J = \int_{A^*} \left[\sigma_{ij} \frac{\partial u_j}{\partial x_1} - w\,\delta_{1i} \right] \frac{\partial q}{\partial x_i}\, dA \tag{11.32}$$

Equation (11.32) is equivalent to Rice's path-independent J integral (Chapter 3). When sum of the additional terms in the more general expression (Eq. (11.31)) is nonzero, J is path dependent.

Comparing Eqs. (11.21) and (11.32) we see that the two expressions are identical if $q = \Delta x_1 / \Delta a$. Thus q can be interpreted as a normalized virtual displacement, although the above derivation does not require such an interpretation. The q function is merely a mathematical device that enables the generation of an area integral, which is better suited to numerical calculations. Section 11.4.3 provides guidelines for defining q.

11.4.2 Generalization to Three Dimensions

Equation (11.23) defines the J integral in both two and three dimensions, but the form of this equation is poorly suited to numerical analysis. In the previous section, J was expressed in terms of an area integral in order to facilitate numerical evaluation. For three-dimensional problems, it is necessary to convert Eq. (11.23) into a volume integral.

Figure 11.7 illustrates a planar crack in a three-dimensional body; η corresponds to the position along the crack front. Suppose that we wish to evaluate J at a particular η on the crack front. It is convenient to define a local coordinate system at η, with x_1 normal to the crack front, x_2 normal to the crack plane, and x_3 tangent to the crack front. The J integral at η is defined by Eq. (11.23), where the contour Γ_o lies in the x_1-x_2 plane.

Let us now construct a tube of length ΔL and radius r_o that surrounds a segment of the crack front, as Fig. 11.7 illustrates. Assuming quasistatic conditions, we can define a *weighted average* J over the crack front segment ΔL as follows:

$$\bar{J}\,\Delta L = \int_{\Delta L} J(\eta)\, q\, d\eta$$

$$= \lim_{r_o \to 0} \int_{S_o} \left[w\,\delta_{1i} - \sigma_{ij} \frac{\partial u_j}{\partial x_1} \right] q\, n_i\, dS \qquad (11.33)$$

where $J(\eta)$ is the point-wise value of J, S_o is the surface area of the tube in Fig. 11.7, and q is a weighting function that was introduced in the previous section. Note that the integrand in Eq. (11.33) is evaluated in terms of the local coordinate system, where x_3 is tangent to the crack front at each point along ΔL.

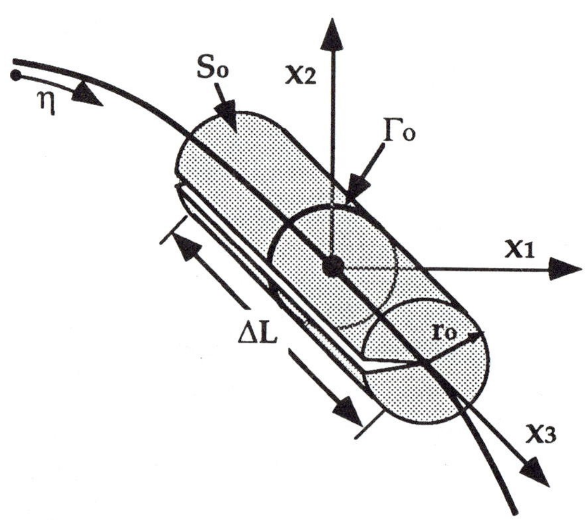

FIGURE 11.7 Surface enclosing an increment of a three-dimensional crack front.

Recall from the previous section that q can be interpreted as a virtual crack advance. For example, Fig. 11.8 illustrates an incremental crack advance over ΔL, where q is defined by

$$\Delta a(\eta) = q(\eta)\, \Delta a_{max} \tag{11.34}$$

and the incremental area of the virtual crack advance is given by

$$\Delta A_c = \Delta a_{max} \int_{\Delta L} q(\eta)\, d\eta \tag{11.35}$$

The q function need not be defined in terms of a virtual crack extension, but attaching a physical significance to this parameter may aid in understanding.

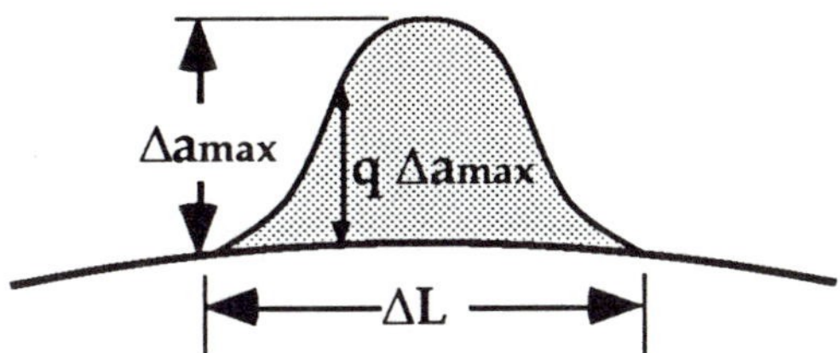

FIGURE 11.8 Interpretation of q in terms of a virtual crack advance along ΔL.

If we construct a second tube of radius r_1 around the crack front (Fig. 11.9), it is possible to define the weighted average J in terms of a closed surface, analogous to the two-dimensional case (Fig. 11.6 and Eq. (11.26)):

$$\bar{J}\,\Delta L = \int_{S^*}\left[\sigma_{ij}\frac{\partial u_j}{\partial x_1} - w\,\delta_{1i}\right] q\, m_i\, dS - \int_{S_+ + S_-}\sigma_{2j}\frac{\partial u_j}{\partial x_1} q\, dS \tag{11.36}$$

where the closed surface $S^* = S_1 + S_+ + S_- - S_o$, and S_+ and S_- are the upper and lower crack faces, respectively, that are enclosed by S_1. From this point, the derivation of the domain integral formulation is essentially identical to the two-dimensional case, except that Eq. (11.34) becomes a volume integral:

$$\bar{J}\,\Delta L = \int_{V^*}\left\{\left[\sigma_{ij}\frac{\partial u_j}{\partial x_1} - w\,\delta_{1i}\right]\frac{\partial q}{\partial x_i} + \left[\sigma_{ij}\frac{\partial \varepsilon_{ij}{}^p}{\partial x_1} - \frac{\partial w^p}{\partial x_1} + \alpha\sigma_{ii}\frac{\partial \Theta}{\partial x_1} - F_i\frac{\partial u_j}{\partial x_1}\right] q\right\} dV$$

$$- \int_{S_+ + S_-}\sigma_{2j}\frac{\partial u_j}{\partial x_1} q\, d\Gamma \tag{11.37}$$

Equation (11.37) requires that q = 0 at either end of ΔL; otherwise, there may be a contribution to J from the end surfaces of the cylinder. The virtual crack advance interpretation of q (Fig. 11.8) fulfills this requirement.

If the point-wise value of the J integral does not vary appreciably over ΔL, to a first approximation, J(η) is given by

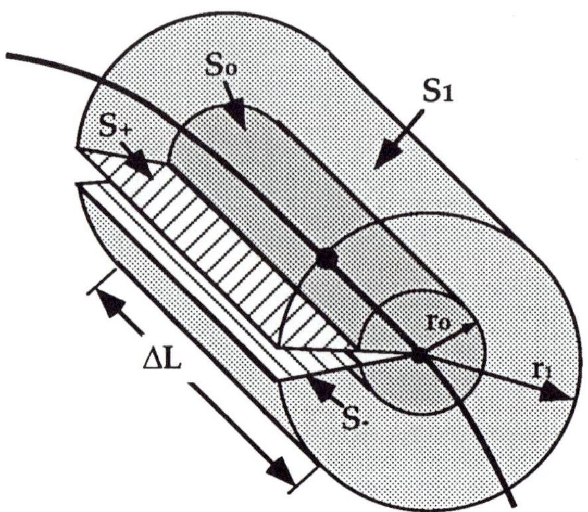

FIGURE 11.9 Inner and outer surfaces, S_o and S_1, which enclose V*.

$$J(\eta) \approx \frac{\bar{J}\,\Delta L}{\int_{\Delta L} q(\eta, r_o)\, d\eta} \tag{11.38}$$

Equation (11.38) is a reasonable approximation if the q gradient along the crack front is steep relative to the variation in $J(\eta)$.

Recall that Eq. (11.22) was defined in terms of a fixed coordinate system, while Eq. (11.37) assumes a fixed coordinate system. The domain integral formulation can be expressed in terms of a fixed coordinate system by replacing q with a vector quantity, q_i, and evaluating the partial derivatives in the integrand with respect to x_i rather than x_1, where the vectors q_i and x_i are parallel to the direction of crack growth. Several commercial codes that incorporate the domain integral definition of J require that the q function be defined with respect to a fixed origin.

11.4.3 Finite Element Implementation

Shih, et al. [21] and Dodds and Vargas [23] give detailed instructions for implementing the domain integral approach. Their recommendations are summarized briefly below.

In two-dimensional problems, one must define the area over which the integration is to be performed. The inner contour, Γ_o is often taken as the crack tip, in which case A* corresponds to the area inside of Γ_1. The boundary of Γ_1 should coincide with element boundaries. An analogous situation applies in three dimensions, where it is necessary to define the volume of integration. The latter situation is somewhat more complicated, however, since $J(\eta)$ is usually evaluated at a number of locations along the crack front.

The q function must be specified at all nodes within the area or volume of integration. The shape of the q function is arbitrary, as long has q has the correct values on the domain boundaries. In a plane stress or plane strain problem, for example, $q = 1$ at Γ_o (which is usually the crack tip) and $q = 0$ at the outer boundary. Figure 11.10 illustrates two common examples of q functions for two-dimensional problems, with the corresponding virtual nodal displacements. This example shows 4-node square elements and rectangular domains for the sake of simplicity. The pyramid function (Fig. 11.10(a)) is equal to 1 at the crack tip but varies linearly to zero in all directions, while the plateau function (Fig. 11.10(b)) equals 1 in all regions except the outer ring of elements. Shih, et al. [21] have shown that the computed value of J is insensitive to the assumed shape of the q function.

Figure 11.11 illustrates the pyramid function along a three-dimensional crack front, where the crack tip node of interest is displaced a unit amount, and all other nodes are fixed. If desired, $J(\eta)$ can be evaluated at each node along the crack front.

The value of q within an element can be interpolated as follows:

$$q(x_i) = \sum_{I=1}^{n} N_I q_I \tag{11.39}$$

where n is the number of nodes per element, q_I are the nodal values of q, and N_I are the element shape functions, which were introduced in Section 1.1.1.

The spatial derivatives of q are given by

$$\frac{\partial q}{\partial x_i} = \sum_{I=1}^{n} \sum_{k=1}^{2 \text{ or } 3} \frac{\partial N_I}{\partial \xi_k} \frac{\partial \xi_k}{\partial x_j} \tag{11.40}$$

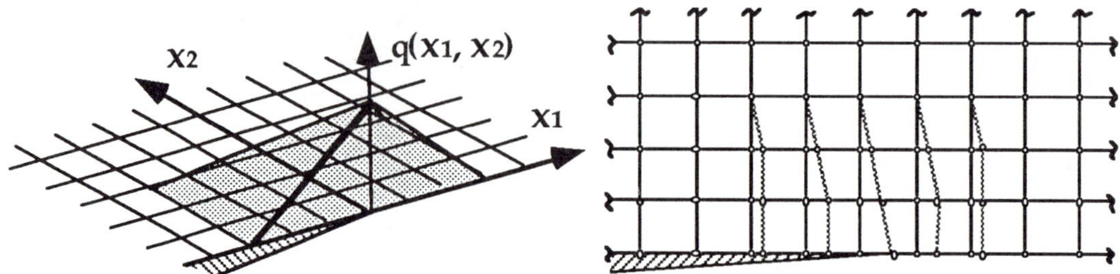

(a) The pyramid function.

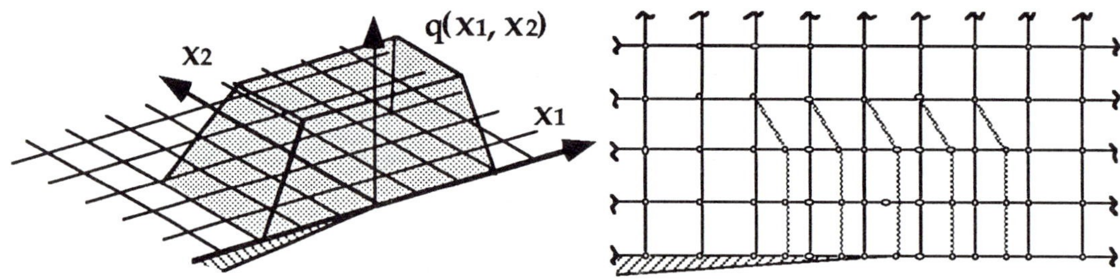

(b) The plateau function.

FIGURE 11.10 Examples of q functions in two dimensions, with the corresponding virtual nodal displacements [21].

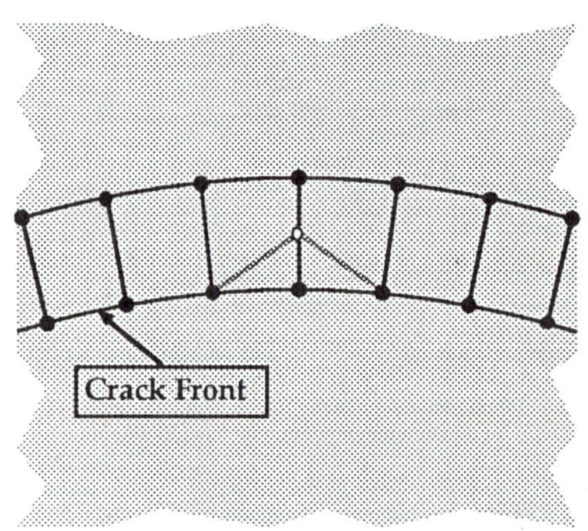

FIGURE 11.11 Definition of q in terms of a virtual nodal displacement along a three-dimensional crack front.

where ξ_i are the parametric coordinates for the element.

In the absence of thermal strains, path-dependent plastic strains, and body forces within the integration volume or area, the discretized form of the domain integral is as follows:

$$J = \sum_{A^* \text{ or } V^*} \sum_{p=1}^{m} \left\{ \left[\left(\sigma_{ij} \frac{\partial u_j}{\partial x_1} - w\,\delta_{1i} \right) \frac{\partial q}{\partial x_i} \right] \det \left(\frac{\partial x_j}{\partial \xi_k} \right) \right\}_p w_p$$

$$- \sum_{\text{crack faces}} \left(\sigma_{2j} \frac{\partial u_j}{\partial x_1} q \right) w \qquad (11.41)$$

where m is the number of Gaussian points per element, and w_p and w are weighting factors. The quantities within $\{\ \}_p$ are evaluated at the Gaussian points. Note that the integration over crack faces is only necessary when there are nonzero tractions.

11.5 MESH DESIGN

The design of a finite element mesh is as much an art form as it is a science. Although many commercial codes have automatic mesh generation capabilities, construction of a properly designed finite element model invariably requires some human intervention. Crack problems, in particular, require a certain amount of skill on the part of the user.

This section gives a brief overview of some of the considerations that should govern the construction of a mesh for analysis of crack problems. It is not possible to address in a few pages all of the situations that may arise. Readers with limited experience in this area should consult the published literature, which contains numerous examples of finite element meshes for crack problems.

Figure 11.12 illustrates several common element types for crack problems. Shih, et al. [21] recommend 9-node biquadradic Lagrangian elements for two-dimensional problems and 27-node triquadradic Lagrangian elements in three dimensions. The 8- and 20-node 2-D and brick elements are also common in crack problems.

At the crack tip, four-sided elements (in 2-D problems) are often degenerated down to triangles, as Fig. 11.13 illustrates. Note that three nodes occupy the same point in space. Figure 11.14 shows the analogous situation for three dimensions, where a brick element is degenerated to a wedge.

In elastic problems, the nodes at the crack tip are normally tied, and the mid-side nodes moved to the $^1/_4$ points (Fig. 11.15(a)). Such a modification results in a $1/\sqrt{r}$ strain singularity in the element, which enhances numerical accuracy.[3] A similar result can be achieved by moving the mid-side nodes to $^1/_4$ points in 4-sided elements, but the singularity would only exist on the element edges [14,24]; triangular elements are preferable in this case because the singularity exists within the element as well as on the edges. Appendix 11 presents a mathematical derivation that explains why moving the mid-side nodes results in the desired singularity for elastic problems.

When a plastic zone forms, the $1/\sqrt{r}$ singularity no longer exists at the crack tip. Consequently, elastic singular elements are not appropriate for elastic-plastic analyses. Figure 11.15(b) shows an element that exhibits the desired strain singularity under fully plastic conditions. The element is degenerated to a triangle as before, but the crack tip nodes are untied and the location of the mid-side nodes is unchanged. This element geometry produces a $1/r$ strain singularity, which corresponds to the actual crack tip strain field for fully plastic, nonhardening materials.

One side benefit of the plastic singular element design is that it allows the crack tip opening displacement (CTOD) to be computed from the deformed mesh, as Fig. 11.16 illustrates. The untied nodes initially occupy the same point in space, but move apart as the elements deform. The CTOD can be inferred from the deformed crack profile by means of the 90° intercept method (See Fig. 3.4).

For most problems, the most efficient mesh design for the crack tip region has proven to be the "spider web" configuration, which consists of concentric rings of four sided elements that are focused toward the crack tip. The inner-most ring of elements are degenerated to triangles, as described above. Since the crack tip region contains steep stress and strain gradients, the mesh refinement should be greatest at the crack tip. The spider web design facilitates a smooth transition from a fine mesh at the tip to a coarser mesh remote to the tip. Figure 11.17 shows a half-symmetric model of a simple cracked body, in which a spider-web mesh transitions to coarse rectangular elements.

---

[3]In principle, the desired singularity can be achieved with ordinary elements, but a great deal of mesh refinement would be required to capture the crack tip fields. Moving the mid-side nodes forces the element to exhibit the intended behavior.

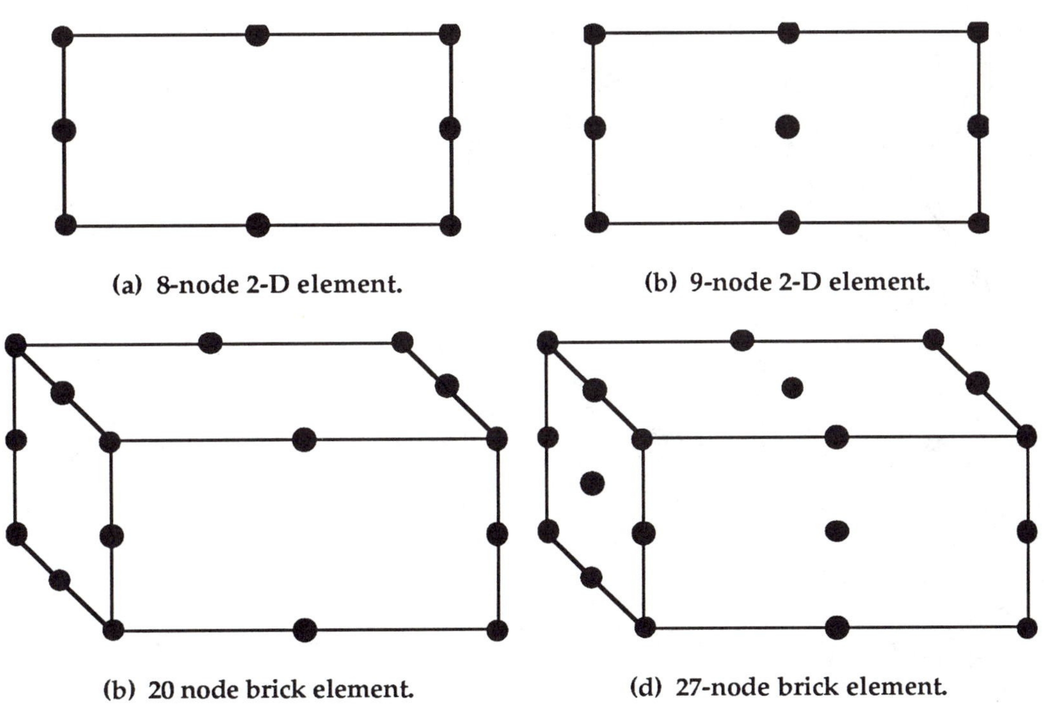

(a) 8-node 2-D element.

(b) 9-node 2-D element.

(b) 20 node brick element.

(d) 27-node brick element.

FIGURE 11.12 Isoparametric elements that are commonly used in two- and three-dimensional crack problems.

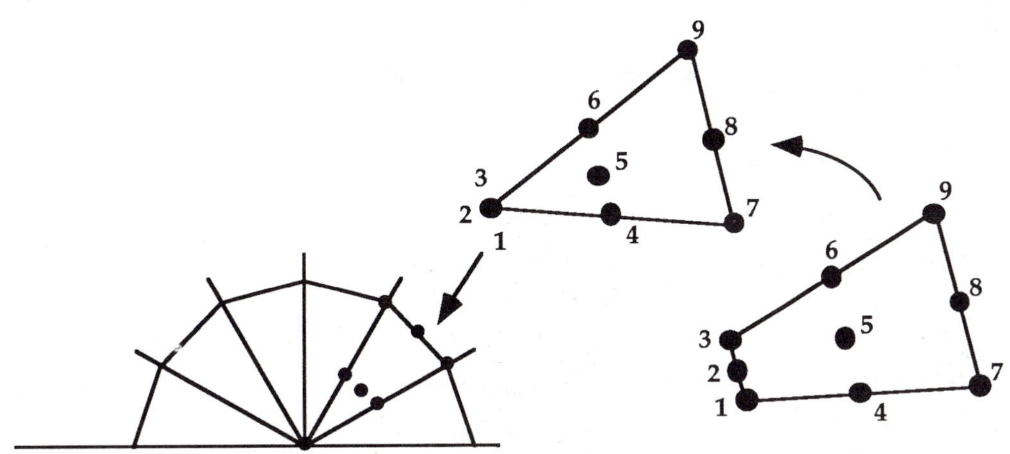

FIGURE 11.13 Degeneration of a quadrilateral element into a triangle at the crack tip.

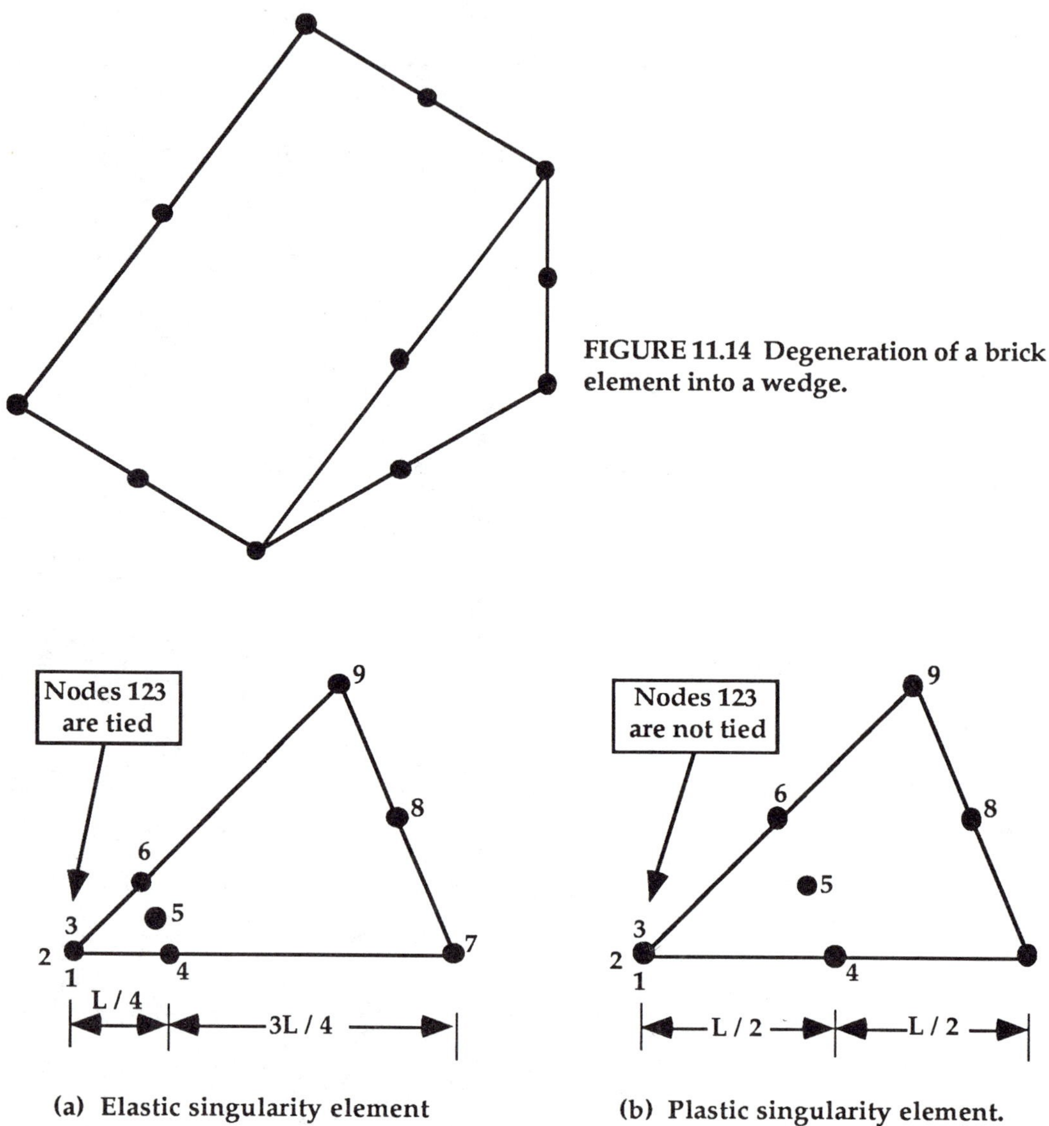

FIGURE 11.14 Degeneration of a brick element into a wedge.

(a) Elastic singularity element

(b) Plastic singularity element.

FIGURE 11.15 Crack tip elements for elastic and elastic-plastic analyses. Element (a) produces a $1/\sqrt{r}$ strain singularity, while (b) exhibits a 1/r strain singularity.

The appropriate level of mesh refinement depends on the purpose of analysis. Elastic analyses of stress intensity or energy release can be accomplished with relatively coarse meshes since modern methods, such as the domain integral approach, eliminate the need to resolve local crack tip fields accurately. The area and volume integrations in the newer approaches are relatively insensitive to mesh size for elastic problems. The mesh should include singularity elements at the crack

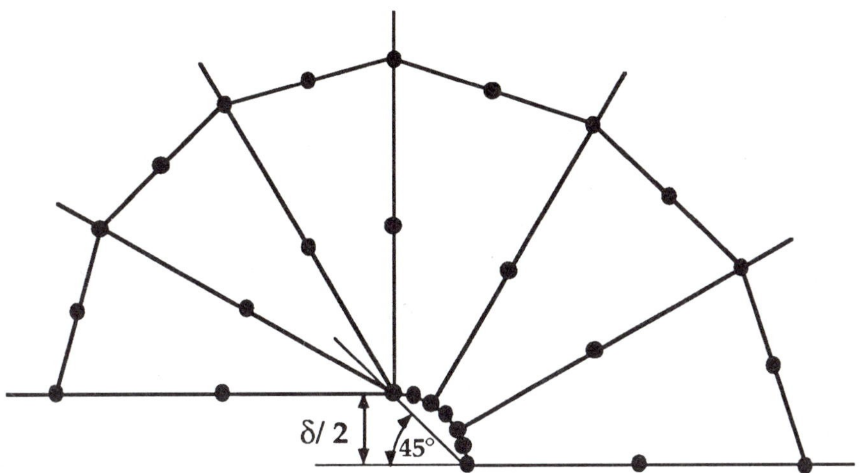

FIGURE 11.16 Deformed shape of plastic singularity elements (Fig. 11.15(b)). The crack tip elements model blunting, and it is possible to measure CTOD.

tip, however, when the domain is defined as a small region near the crack tip. If the domain is defined over a large portion of the mesh, singularity elements are unnecessary, because the crack tip elements contribute little to J. The relative contribution of the crack tip elements can be adjusted through the definition of the q function. For example, in elastic problems, the crack tip elements do not contribute to J when the plateau function (Fig. 11.10(b)) is adopted, since $dq/dx_1 = 0$ at the crack tip.

Elastic-plastic problems require more mesh refinement in regions of the body where yielding occurs. When a body experiences net section yielding, narrow deformation bands often propagate across the specimen (Fig. 3.26). The high level of plastic strain in these bands will make a significant contribution to the J integral; the finite element mesh must be sufficiently refined in these regions to capture this deformation accurately.

When the purpose of the analysis is to analyze crack tip stresses and strains, a very high level of mesh refinement is required [25,26]. As a general rule, it is desirable to have at least 10 elements on a radial line in the region of interest. In addition, if it is necessary to infer crack tip fields at distances less that twice the CTOD from the crack tip, the analysis code must incorporate large strain theory. McMeeking and

Parks [26] were among the first to apply a large strain analysis to the crack tip region. Figure 3.13 is a plot of some of their results.

In a large strain analysis, it is advisable to begin with a finite radius at the crack tip, as Fig. 11.18 illustrates. Note that the crack tip elements are not collapsed to triangles in this case. Provided the CTOD after deformation is at least 5 times the initial value, the results should be not be affected by the initial blunt notch [26].

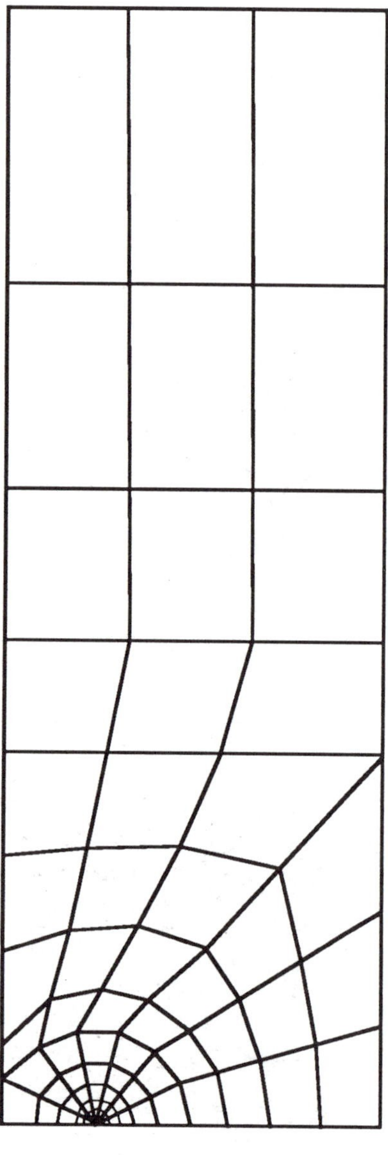

FIGURE 11.17 **Half-symmetric model of a cracked panel.**

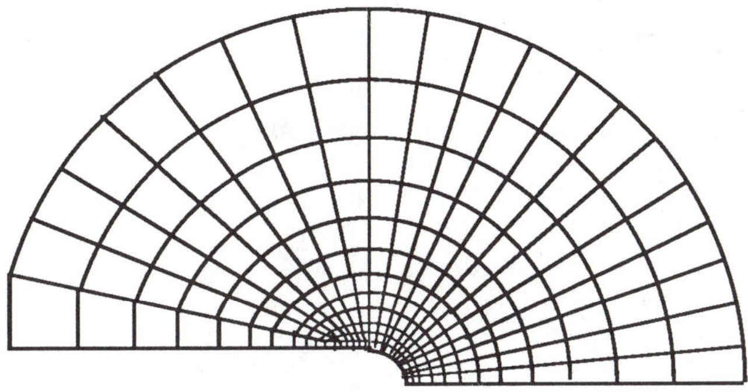

FIGURE 11.18 Crack tip region of a mesh for large strain analysis. Note that the initial crack tip radius is finite and the crack tip elements are not degenerated.

We will not address boundary conditions in detail, but it is worth mentioning a common pitfall. Many problems require forces to be applied at the boundaries of the body. For example, a single edge notched bend specimen is loaded in three point bending, with a load applied at mid-span, and appropriate restraints at each end. In elastic-plastic problems, the manner in which the load is applied can be very important. Figure 11.19 shows both acceptable and unacceptable ways of applying this boundary condition. If the load is applied to a single node (Fig. 11.19(a)), a local stress and strain concentration will occur, and the element connected to this node will yield almost immediately. The analysis code will spend an inordinate amount of time solving a punch indentation problem at this node, while the events of interest may be remote from the boundary. A better way of applying this boundary condition might be to distribute the load over several nodes, and specifying that the elements on which the load acts remain elastic (Fig. 11.19(b)); the load will then be transferred to the body without wasting computer time solving a local indentation problem. If, however, the local indentation is of interest, (e.g., if one wants to simulate the effects of the loading fixture) the load can be applied by a rigid or elastic indenter with a finite radius, as Fig. 11.19(c) illustrates. Note, however, that greater mesh refinement is required to resolve the plastic deformation at the indenter.

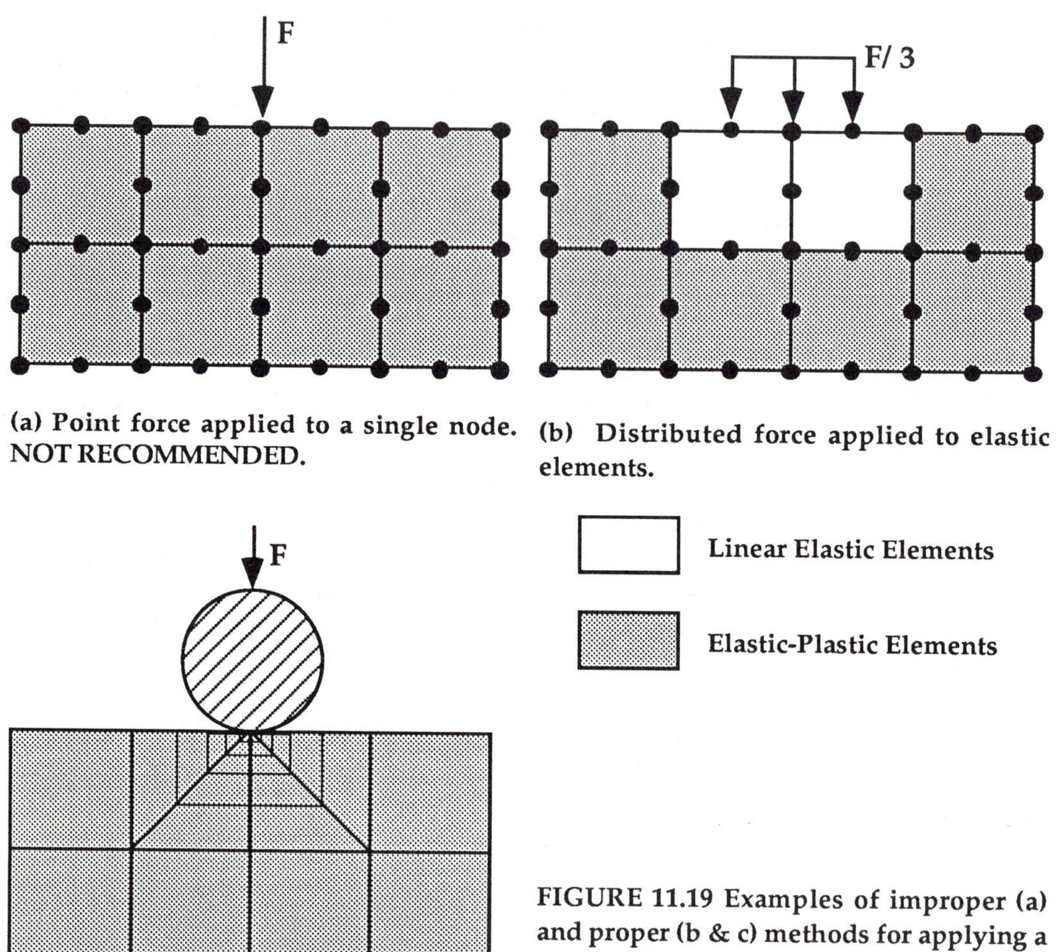

FIGURE 11.19 Examples of improper (a) and proper (b & c) methods for applying a force to a boundary.

11.6 INTRODUCTION TO WEIGHT FUNCTIONS

When one performs an analysis to infer a stress intensity factor for a cracked body, the K value that is computed applies only to one particular set of boundary conditions; different loading conditions would result in a different stress intensity factor for that geometry. It turns out, however, that the solution to one set of boundary conditions contains sufficient information to infer K for *any other* boundary conditions on that same geometry.

Consider two arbitrary loading conditions on an isotropic elastic cracked body in plane stress or plane strain. For now, we assume that both loadings are symmetric with respect to the crack plane, such that pure Mode I loading is achieved in each case. Suppose that we know the stress intensity factor for loading (1) and we wish to solve for $K_I^{(2)}$, the stress intensity factor for the second set of boundary conditions. Rice [27] showed that $K_I^{(1)}$ and $K_I^{(2)}$ are related as follows:

$$K_I^{(2)} = \frac{E'}{2\,K_I^{(1)}}\left[\int_{\Gamma} T_i \frac{\partial u_i^{(1)}}{\partial a}\, d\Gamma + \int_{A} F_i \frac{\partial u_i^{(1)}}{\partial a}\, dA\right] \tag{11.42}$$

where Γ and A are the perimeter and area of the body, respectively. Since loading systems (1) and (2) are arbitrary, it follows that $K_I^{(2)}$ cannot depend on $K_I^{(1)}$ and $u_i^{(1)}$. Therefore, the function

$$h(x_i) = \frac{E'}{2\,K_I^{(1)}} \frac{\partial u_i^{(1)}}{\partial a} \tag{11.43}$$

must be independent of the nature of loading system (1). Bueckner [28] derived a similar result to Eq. (11.42) two years before Rice, and referred to h as a *weight function*.

Weight functions are first order tensors that depend only on the geometry of the cracked body. Given the weight function for a particular configuration, it is possible to compute K_I from Eq. (11.42) for any boundary conditions. Example 11.1, which was taken from Rice's paper [27], illustrates this concept for a simple configuration.

EXAMPLE 11.1

Derive an expression for K_I for an arbitrary traction on the face of a through crack in an infinite plate.

Solution: We already know K_I for this configuration when a uniform remote tensile stress is applied:

EXAMPLE 11.1 (Cont.)

$$K_I = \sigma\sqrt{\pi a}$$

where a is the half crack length. From Eq. (A2.43), the opening displacement of the crack faces in this case is given by

$$u_y = \pm\frac{2\sigma}{E'}\sqrt{a^2 - x^2}$$

where the x-y coordinate axis is defined in Fig 11.20(a). The since the crack length is 2a we must differentiate u_y with respect to 2a rather than a:

$$\frac{\partial u_y}{\partial(2a)} = \pm\frac{2\sigma}{E'}\frac{a}{\sqrt{a^2 - x^2}}$$

Thus the weight function for this crack geometry is given by

$$h(x,0) = \pm\sqrt{\frac{a}{\pi}}\frac{1}{\sqrt{a^2 - x^2}}$$

If we apply a surface traction of ± p(x) on the crack faces, the Mode I stress intensity factor for the two crack tips is as follows:

$$K_I(x = +a) = \sqrt{\frac{a}{\pi}}\int_{-a}^{+a}\frac{p(x)\,dx}{\sqrt{a^2 - x^2}}$$

$$K_I(x = -a) = \sqrt{\frac{a}{\pi}}\int_{-a}^{+a}\frac{p(-x)\,dx}{\sqrt{a^2 - x^2}}$$

where the second expression is obtained from the first by changing the direction of integration and substituting x = -x. The above results can also be derived by integrating Eq. (2.68) over the crack face.

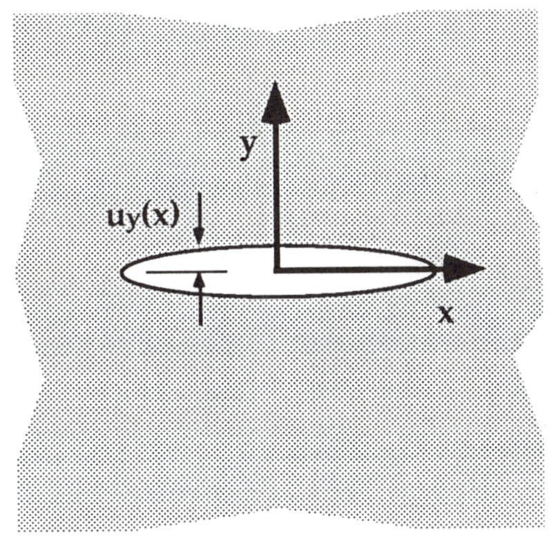

(a) Definition of coordinate axes

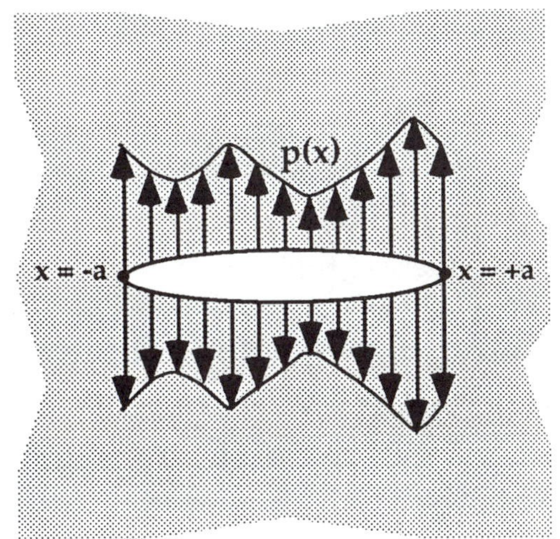

(b) Arbitrary traction applied to crack faces.

FIGURE 11.20 Through crack configuration analyzed in Example 11.1.

The weight function concept is not restricted to two-dimensional bodies, Mode I loading, or isotropic elastic materials. In their early work on weight functions, Rice [27] extended the theory to three dimensions, Bueckner [28] considered combined Mode I/II loading, and both allowed for anisotropy in the elastic properties. Subsequent researchers [29-34] have shown that the theory applies to all linear elastic bodies that contain an arbitrary number of cracks.

For mixed-mode problems, separate weight functions are required for each mode: h_I, h_{II}, and h_{III}. Since the stress intensity factors can vary along a three-dimensional crack front, the weight functions also vary along the crack front. That is,

$$h_\alpha = h_\alpha(x_i, \eta)$$

where α (= 1,2,3) indicates the mode of loading and η is the crack front position, as defined in Fig. 11.7.

The general mixed-mode three-dimensional formulation of the weight function approach can be expressed in the following form:

$$K_\alpha(\eta) = \int_S T_i\, h_\alpha(x_i, \eta)\, dS + \int_V F_i\, h_\alpha(x_i, \eta)\, dV \tag{11.44}$$

For Mode I loading, one simple (though inefficient) numerical method for determining h_I is the elemental crack advance procedure described in Section 11.2.2. For a given stress distribution, K_I and displacement fields can be computed as a function of crack length, and the Mode I weight function can be inferred from Eq. (11.43).

Parks and Kamentzky [30] applied the stiffness derivative virtual crack extension technique (Section 11.2.4) to the computation of Mode I weight functions from Eq. (11.43). Subsequent researchers [31,32] adapted this approach to combined Mode I/II loading. As Section 11.2.4 discussed, however, the stiffness derivative technique is relatively cumbersome and inefficient since it requires numerical differencing.

Sham [34] has developed a generalized finite element technique for determining weight functions in mixed-mode and three-dimensional problems. His approach does not utilize Eq. (11.43), where the weight functions are related to the stress intensity factor and displacement derivative for a particular set of boundary conditions. Rather, Sham solves for h_α by taking advantage of its unusual properties, which are discussed briefly below.

Weight functions can be interpreted as displacement fields which obey all equations of elasticity but one: h exhibits a $1/\sqrt{r}$ singularity near the crack tip, which gives rise to unbounded energy in any finite area surrounding the crack tip. Thus weight functions do not fall into the category of ordinary displacement fields, for which proof of uniqueness requires that the strain energy be finite. The stresses produced by the weight functions are self-equilibrating, and the h fields do not produce any body forces or surface tractions. Bueckner [28] referred to the stresses, strains that result from the weight functions as *fundamental fields*.

According to Eq. (11.44), stress intensity is related to work-like products of tractions and body forces with the weight functions. Using variational methods, Sham [34] derived a methodology that is analogous to the minimum potential energy principle. Since the stiffness finite element technique is based on the concept of minimum potential energy, Sham's method is well suited to finite element analysis. One problem, however, is that the minimum energy approach does not work when the energy is unbounded. Sham circumvented this problem by considering the singular and nonsingular portions of h separately. The full derivation of Sham's approach is rather lengthy and complicated, and will not be given here.

11.7 LIMITATIONS OF NUMERICAL FRACTURE ANALYSIS

Although computational methods are very useful in fracture mechanics, they cannot replace experiments. A numerical fracture simulation of a cracked body can compute crack tip parameters, but such an analysis alone cannot predict when fracture will occur. Techniques such as finite element analysis and boundary integral elements rely on continuum theory. A continuum does not contain voids, microcracks, second-phase particles, grain boundaries, dislocations, atoms, or any the other microscopic or submicroscopic features that control fracture behavior in engineering materials (see Chapters 5 and 6).

A numerical analysis of a cracked body can provide information on local stresses and strains at the crack tip, as well as global fracture parameters. Existing analyses, however, model only the deformation of the material. Fracture can be modeled, but a separate failure criterion is required. For example, one might model cleavage fracture by imposing a stressed-based failure criterion, in which the analysis would predict failure when a user-specified stress is reached at a particular point ahead of the crack tip. Predictions of fracture could not be made a priori in such cases, but would require one or more experiments to infer material-dependent parameters in the local fracture model.

Several researchers have attempted to combine flow and fracture behavior into a single constitutive model, and have incorporated such approaches into finite element analyses. The Gurson model, for example (Chapter 5), was intended to model both plastic flow and ductile fracture in metals. Because this approach is a continuum model and does not include voids, however, it does not capture the important microscopic events that lead to fracture, and it is unable to predict failure in real materials. A number of adjustable parameters have recently been added to this model in order to bring predictions in line with experimental data, but such parameters are based on curve fitting rather than sound physics.

Numerical analysis will undoubtedly play a major role in developing micromechanical models for fracture. Computer simulation of processes such as microcrack nucleation, void growth, and interface fracture should lead to new insights into fracture and damage mechanisms. Such research may then lead to rational failure criteria that can be incorporated into global continuum models of cracked bodies.

To reiterate the statement at the beginning of this section: computer modeling cannot replace experimentation. Any mathematical model, regardless of how sophisticated it is, must omit much of the real world in its formulation. Models often leave out the very feature that controls the physical process. Unlike a mathematical model, an experiment is obliged to obey all laws of nature, down to the quantum level. Thus an experiment often conveys important information that a simulation overlooks.

REFERENCES

1. Westergaard, H.M., "Bearing Pressures and Cracks." *Journal of Applied Mechanics,* Vol. 6, 1939, pp. 49-53.

2. Williams, M.L., "On the Stress Distribution at the Base of a Stationary Crack." *Journal of Applied Mechanics,* Vol. 24, 1957, pp. 109-114.

3. Lapidus, L. and Pinder, G.F., *Numerical Solution of Partial Differential Equations in Science and Engineering.* John Wiley and Sons, New York, 1982.

4. Zienkiewicz, O.C. and Taylor, R.L., *The Finite Element Method.* (Fourth Edition) McGraw-Hill, New York, 1989.

5. Rizzo, F.J., "An Integral Equation Approach to Boundary Value Problems of Classical Elastostatics." *Quarterly of Applied Mathematics,* Vol. 25, 1967, pp. 83-95.

6. Cruse, T.A., *Boundary Element Analysis in Computational Fracture Mechanics.* Kluwer Academic Publishers, Dordrect, Netherlands, 1988.

7. Blandford, G.E. and Ingraffea, A.R., "Two-Dimensional Stress Intensity Factor Computations Using the Boundary Element Method." *International Journal for Numerical Methods in Engineering,* Vol 17, 1981, pp. 387-404.

8. Cruse, T.A., "An Improved Boundary-Integral Equation for Three Dimensional Elastic Stress Analysis." *Computers and Structures,* Vol. 4, 1974, pp. 741-754.

9. Mendelson, A. and Albers, L.U., "Application of Boundary Integral Equations to Elastoplastic Problems." *Boundary Integral Equation Method: Computational*

Applications in Applied Mechanics, AMD-Vol 11, American Society of Mechanical Engineers, New York, 1975, pp. 47-84.

10. Parks, D.M., "A Stiffness Derivative Finite Element Technique for Determination of Crack Tip Stress Intensity Factors." *International Journal of Fracture,* Vol. 10, 1974, pp. 487-502.

11. Kobayashi, A.S., Cherepy, R.B., and Kinsel, W.C., "A Numerical Procedure for Estimating the Stress Intensity Factor of a Crack in a Finite Plate." *Journal of Basic Engineering,* Vol. 86, 1964, pp. 681-684.

12. Gross, B. and Srawley, J.E., "Stress Intensity Factors of Three Point Bend Specimens by Boundary Collocation." NASA Technical Note D-2603, 1965.

13. Tracey, D.M., "Finite Element Methods for Determination of Crack Tip Elastic Stress Intensity Factors." *Engineering Fracture Mechanics,* Vol. 3, 1971, pp. 255-266.

14. Barsoum, R.S, "On the Use of Isoparametric Finite Elements in Linear Fracture Mechanics." *International Journal for Numerical Methods in Engineering,* Vol. 10, 1976, pp. 25-37.

15. Budiansky, B. and Rice, J.R., "Conservation Laws and Energy Release Rates." *Journal of Applied Mechanics,* Vol. 40, 1973, pp. 201-203.

16. Carpenter, W.C., Read, D.T., and Dodds, R.H. Jr., "Comparison of Several Path Independent Integrals Including Plasticity Effects." *International Journal of Fracture,* Vol. 31, 1986, pp. 303-323.

17. Hellen, T.K., "On the Method of Virtual Crack Extensions." *International Journal for Numerical Methods in Engineering,* Vol 9, 1975, pp. 187-207.

18. Parks, D.M.. "The Virtual Crack Extension Method for Nonlinear Material Behavior." *Computer Methods in Applied Mechanics and Engineering,* Vol. 12, 1977, pp. 353-364.

19. deLorenzi, H.G., "On the Energy Release Rate and the J-Integral of 3-D Crack Configurations." *International Journal of Fracture,* Vol. 19, 1982, pp. 183-193.

20. deLorenzi, H.G., "Energy Release Rate Calculations by the Finite Element Method," *Engineering Fracture Mechanics,* Vol. 21, 1985, pp. 129-143.

21. Shih, C.F., Moran, B., and Nakamura, T., "Energy Release Rate Along a Three-Dimensional Crack Front in a Thermally Stressed Body." *International Journal of Fracture,* Vol. 30, 1986, pp. 79-102.

22. Moran, B. and Shih, C.F., "A General Treatment of Crack Tip Contour Integrals." *International Journal of Fracture,* Vol. 35, 1987, pp. 295-310.

23. Dodds, R.H., Jr. and Vargas, P.M., "Numerical Evaluation of Domain and Contour Integrals for Nonlinear Fracture Mechanics." Report UILU-ENG-88-2006, University of Illinois, Urbana, IL, August 1988.

24. Henshell, R.D. and Shaw, K.G., "Crack Tip Finite Elements are Unnecessary." *International Journal for Numerical Methods in Engineering*, Vol. 9, 1975, pp. 495-507.

25. Dodds, R.H. Jr., Anderson T.L., and Kirk, M.T. "A Framework to Correlate a/W Effects on Elastic-Plastic Fracture Toughness (J_c)." to be published in *International Journal of Fracture.*

26. McMeeking, R.M. and Parks, D.M., "On Criteria for J-Dominance of Crack Tip Fields in Large-Scale Yielding." ASTM STP 668, American Society for Testing and Materials, Philadelphia, 1979, pp. 175-194.

27. Rice, J.R., "Some Remarks on Elastic Crack-Tip Stress Fields." *International Journal of Solids and Structures*, Vol. 8, 1972, pp. 751-758.

28. Bueckner, H.F., "A Novel Principle for the Computation of Stress Intensity Factors." *Zeitschrift für Angewandte Mathematik und Mechanik*, Vol. 50, 1970, pp. 529-545.

29. Rice, J.R., "Weight Function Theory for Three-Dimensional Elastic Crack Analysis." ASTM STP 1020, American Society for Testing and Materials, Philadelphia, 1989, pp. 29-57.

30. Parks, D.M. and Kamentzky, E.M., "Weight Functions from Virtual Crack Extension." *International Journal for Numerical Methods in Engineering*, Vol. 14, 1979, pp. 1693-1706.

31. Vainshtok, V.A., "A Modified Virtual Crack Extension Method of the Weight Functions Calculation for Mixed Mode Fracture Problems." *International Journal of Fracture*, Vol. 19, 1982, pp. R9-R15.

32. Sha, G.T. and Yang, C.-T., "Weight Function Calculations for Mixed-Mode Fracture Problems with the Virtual Crack Extension Technique." *Engineering Fracture Mechanics*. Vol. 21, 1985, pp. 1119-1149.

33. Atluri, S.N. and Nishoika, T., "On Some Recent Advances in Computational Methods in the Mechanics of Fracture." *Advances in Fracture Research: Seventh International Conference on Fracture*, Pergamon Press, Oxford, 1989 pp. 1923-1969.

34. Sham, T.-L., "A Unified Finite Element Method for Determining Weight Functions in Two and Three Dimensions." *International Journal of Solids and Structures*, Vol. 23, 1987, pp. 1357-1372.

APPENDIX 11: PROPERTIES OF SINGULARITY ELEMENTS

Certain element/node configurations produce strain singularities. While such behavior is undesirable for most analyses, it is ideal for elastic crack problems. Forcing the elements at the crack tip to exhibit a $1/\sqrt{r}$ strain singularity greatly improves accuracy and reduces the need for a high degree of mesh refinement at the crack tip.

The derivations that follow show that the desired singularity can be produced in quadratic isoparametric elements by moving the mid-side nodes to the $^1/_4$ points. This behavior was first noted by Barsoum [14] and Henshell and Shaw [24].

From Eqs. (11.3) and (11.4), the strain matrix for a two-dimensional element can be written in the following form:

$$\{\varepsilon\} = [\mathbf{J}]^{-1}\,[\mathbf{B}^*]\begin{Bmatrix} u_i \\ v_i \end{Bmatrix} \tag{A11.1}$$

where

$$[\mathbf{B}^*] = \begin{bmatrix} \dfrac{\partial N_i}{\partial \xi} & 0 \\ 0 & \dfrac{\partial N_i}{\partial \eta} \\ \dfrac{\partial N_i}{\partial \eta} & \dfrac{\partial N_i}{\partial \xi} \end{bmatrix} \tag{A11.2}$$

where (ξ, η) are the parametric coordinates of a point on the element. Since the nodal displacements, $\{u_i, v_i\}$, are bounded, the strain matrix can only be singular if either $[\mathbf{B}^*]$ or $[\mathbf{J}]^{-1}$ is singular.

Consider an 8-noded quadratic isoparametric 2-D element (Fig 11.12(a)). The shape functions for this element are as follows [4]:

$$N_i = [(1 + \xi\,\xi_i)(1 + \eta\,\eta_i) - (1 - \xi^2)(1 + \eta\,\eta_i) - (1 - \eta^2)(1 + \xi\,\xi_i)]\,\frac{\xi_i^2\,\eta_i^2}{4}$$

$$+ (1 - \xi^2)(1 + \eta\,\eta_i)(1 - \xi_i^2)\frac{\eta_i^2}{2} + (1 - \eta^2)(1 + \xi\,\xi_i)(1 - \eta_i^2)\frac{\xi_i^2}{2} \qquad \text{(A11.3)}$$

where (ξ, η) are the parametric coordinates of a point in the element and (ξ_i, η_i) are the coordinates of the ith node.

In general, the shape functions are polynomials. Equation (A11.3), for example, is a quadratic equation. Thus N_i, $\partial N_i/\partial\xi$, or $\partial N_i/\partial\eta$ are all nonsingular, and [J] must be the cause of the singularity.

A strain singularity can arise if the determinant of the Jacobian matrix vanishes at the crack tip:

$$\det|\mathbf{J}| = \frac{\partial(x,y)}{\partial(\xi,\eta)} = 0 \qquad (11.4)$$

A11.1 QUADRILATERAL ELEMENT

Consider an 8-noded quadrilateral element with the mid-side nodes at the $^1/_4$ point, as Fig. A11.1 illustrates. For convenience, the origin of the x-y global coordinate system is placed at node 1. Let us evaluate the element boundary between nodes 1 and 2. From Eq. (A11.3), the shape functions along this line at nodes 1, 2, and 5 are given by

$$N_1 = -\frac{1}{2}\xi(1 - \xi)$$

$$N_2 = \frac{1}{2}\xi(1 - \xi) \qquad \text{(A11.5)}$$

$$N_5 = (1 - \xi^2)$$

Inserting these results into Eq. (A11.1) gives

$$x = -\frac{1}{2}\xi(1 - \xi)\,x_1 + \frac{1}{2}\xi(1 - \xi)\,x_2 + (1 - \xi^2)\,x_5 \qquad \text{(A11.6)}$$

Setting $x_1 = 0$, $x_2 = L$, and $x_5 = {}^L/_4$ results in

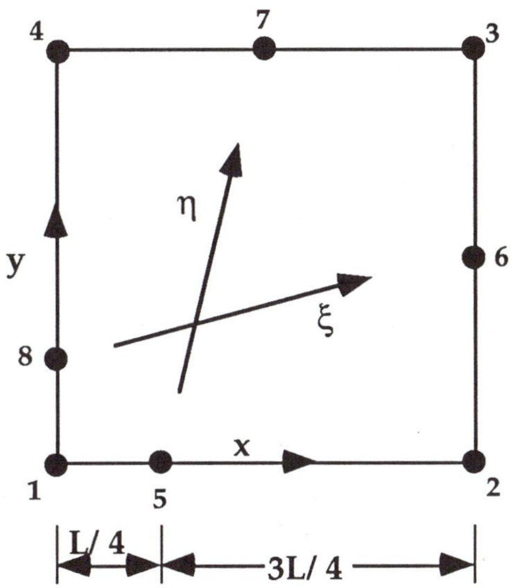

FIGURE A11.1 Quadrilateral isoparametric element with mid-side nodes moved to the quarter points.

$$x = \frac{1}{2}\xi(1-\xi)L + (1-\xi^2)\frac{L}{4} \tag{A11.7}$$

where L is the length of the element between nodes 1 and 2. Solving for ξ gives

$$\xi = -1 + 2\sqrt{\frac{x}{L}} \tag{A11.8}$$

The relevant term of the Jacobian is given by

$$\frac{\partial x}{\partial \xi} = \frac{L}{2}(1+\xi) = \sqrt{\frac{x}{L}} \tag{A11.9}$$

which vanishes at $x = 0$; thus the strain must be singular at this point. Considering only the displacements of points 1, 2, and 5, the displacements along the element edge are as follows:

$$x = -\frac{1}{2}\xi(1-\xi)x_1 + \frac{1}{2}\xi(1-\xi)x_2 + (1-\xi^2)x_5 \tag{A11.10}$$

Substituting Eq. (A11.8) into Eq. (A11.10) gives

$$u = -\frac{1}{2}\left(-1 + 2\sqrt{\frac{x}{L}}\right)\left(2 - 2\sqrt{\frac{x}{L}}\right)u_1$$

$$+ \left(-1 + 2\sqrt{\frac{x}{L}}\right)\left(2\sqrt{\frac{x}{L}}\right)u_2 + 4\left(\sqrt{\frac{x}{L}} - \frac{x}{L}\right)u_5 \qquad \text{(A11.11)}$$

Solving for the strain in the x direction leads to

$$\varepsilon_x = \frac{\partial u}{\partial x} = \frac{\partial \xi}{\partial x}\frac{\partial u}{\partial x}$$

$$= -\frac{1}{2}\left(\frac{3}{\sqrt{xL}} - \frac{4}{L}\right)u_1 + \frac{1}{2}\left(-\frac{1}{\sqrt{xL}} + \frac{4}{L}\right)u_2 + \left(\frac{2}{\sqrt{xL}} - \frac{4}{L}\right)u_5 \qquad \text{(A11.12)}$$

Therefore, the strain exhibits a $1/\sqrt{r}$ singularity along the element boundary.

A11.2 TRIANGULAR ELEMENT

Let us now construct a triangular element by collapsing nodes 1, 4, and 8 (Fig. A11.2). Nodes 5 and 7 are moved to the quarter points in this case. The $1/\sqrt{r}$ strain singularity exists along the 1-5-2 and 4-7-3 edges, as with the quadrilateral element. In this instance, however, the singularity also exists within the element.

Consider the x axis, where $\eta = 0$. The relationship between x and ξ is given by

$$x = (\xi^2 + 2\xi + 1)\frac{L_1}{4} \qquad \text{(A11.13)}$$

where L_1 is the length of the element in the x direction. Solving for ξ gives

$$\xi = -1 + 2\sqrt{\frac{x}{L_1}} \qquad \text{(A11.14)}$$

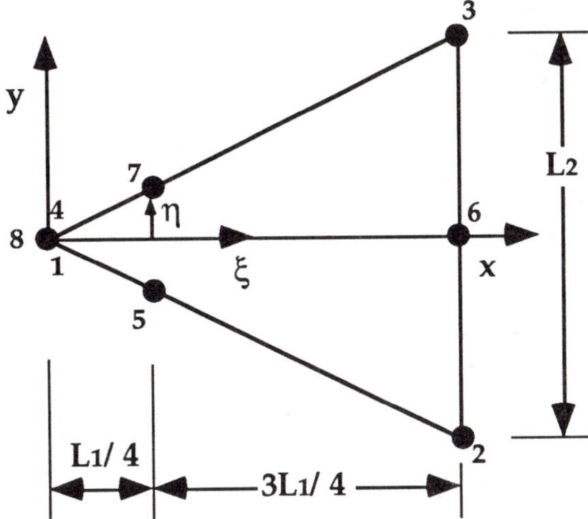

FIGURE A11.2 Degenerated isoparametric element, with mid-side nodes at the quarter points.

which is identical to Eq. (A11.8). Therefore, the strain is singular along the x axis in this element. By solving for strain as before (Eqs. (A11.10) to (A11.12)) it can easily be shown that the singularity is the desired $1/\sqrt{r}$ type.

PART V: REFERENCE MATERIAL

12. COMPILATION OF K, J, COMPLIANCE AND LIMIT LOAD SOLUTIONS

This chapter lists stress intensity, compliance, limit load, plastic J and plastic displacement solutions for selected configurations. Table 12.1 summarizes the geometries and solutions that are provided. The geometries are divided into three categories: (1) through-thickness cracks in flat plates, (2) part-through cracks in flat plates, and (3) flawed cylinders. Note that K_I solutions are given in all cases but that fully plastic J solutions are available for only a limited number of configurations.

12.1 THROUGH-THICKNESS CRACKS–FLAT PLATES

Figure 12.1 illustrates eight flat plate/through thickness crack configurations. Note that this geometry category includes most of the common test specimens.

12.1.1 Stress Intensity and Elastic Compliance

Table 12.2 gives polynomial K_I expressions for the eight configurations shown in Fig. 12.1. The load line compliance solutions for some of these geometries are listed in Table 12.3. It is important to note that the compliance values in Table 12.3 correspond to the *total* load line displacement. Recall from Chapter 3 that the load line displacement can be divided into *crack* and *no crack* components:

$$\Delta_{tot} = \Delta_{nc} + \Delta_c \qquad (12.1)$$

where Δ_{nc} is the displacement (at a given load) that would be measured in the absence of a crack, and Δ_c is the additional displacement that is due to the crack. The *no crack compliance* for each of the configurations in Table 12.3 can be readily inferred by setting $a/W = 0$.

TABLE 12.1
Summary of solutions in Chapter 12.

Geometry	Loading	K_I	Compl.	P_L	J_{pl}	Δ_p, V_p
(a) Through Cracks, Flat Plates.						
Edge crack	Tension	√	√	√	√	√
"	3 Pt. Bend	√	√	√	√	√
"	Pure Bend	√				
"	T + B	√		√	√	√
Center crack	Tension	√	√	√	√	√
Double edge notch	Tension	√	√	√	√	√
Compact spec.	Tension	√	√	√	√	√
Disk compact	Tension	√				
Arc specimen	Tension	√				
(b) Part-Through Cracks, Flat Plates.						
Surface flaw	T + B	√		√		
"	Polynom.	√				
3. Flawed Cylinders.						
Thr-wall, circum.	Tension	√		√	√	
"	Bending	√		√	√	
"	T + B	√		√	√	
Long, circum.	Tension	√	√	√	√	√
Part-thr., circum.	Tension	√				
Thr-wall, axial	Pressure	√				
Long, axial	Pressure	√	√	√	√	√
Part-thr., axial	Pressure	√				
"	Polynom.	√				

K_I - Mode I stress intensity.
Comp. - Elastic compliance.
P_L - Limit load.
J_{pl} - Plastic J integral.
Δ_p, V_p - Plastic displacements.

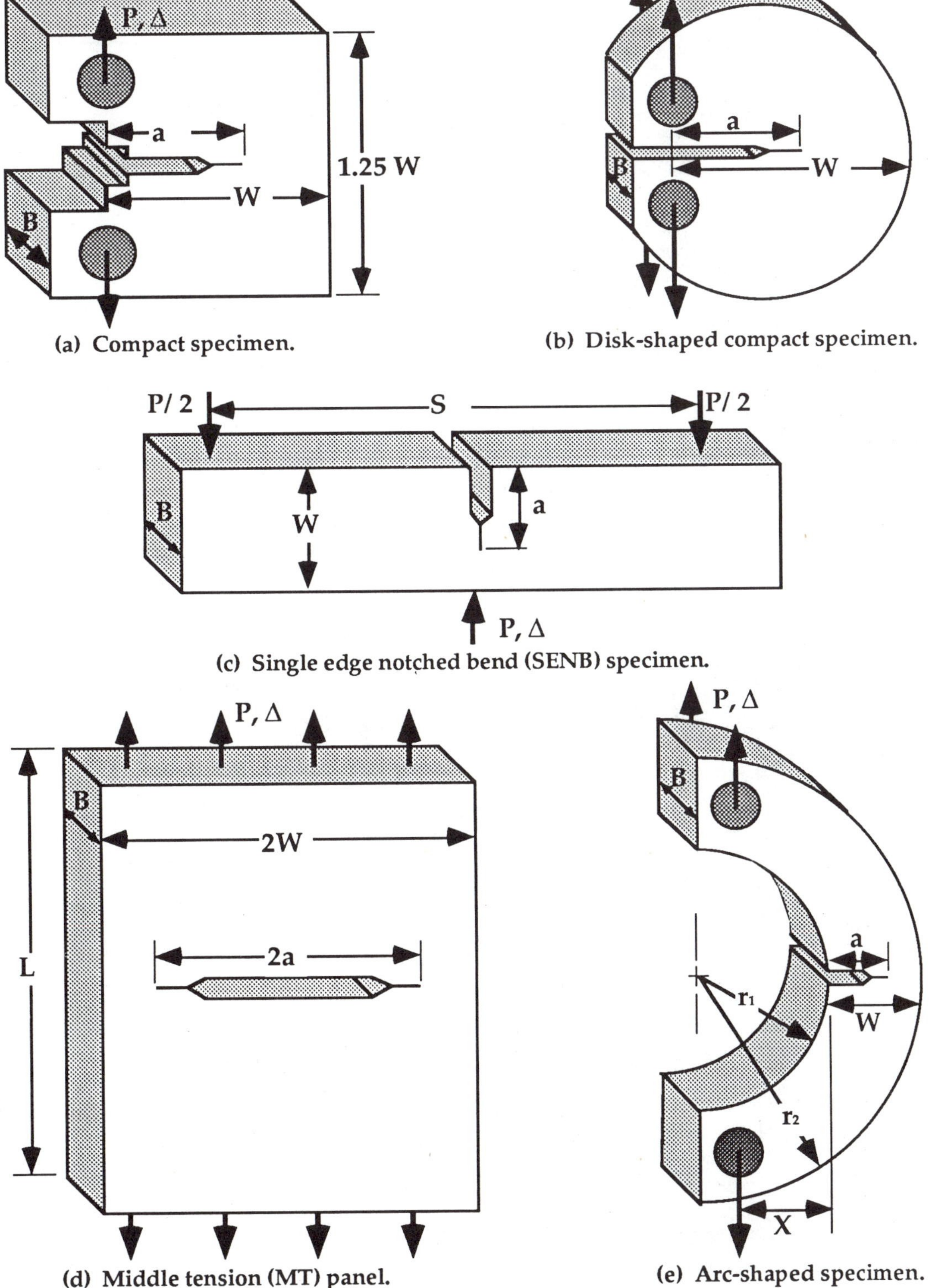

(a) Compact specimen.

(b) Disk-shaped compact specimen.

(c) Single edge notched bend (SENB) specimen.

(d) Middle tension (MT) panel.

(e) Arc-shaped specimen.

FIGURE 12.1 Flat plates with through-thickness cracks.

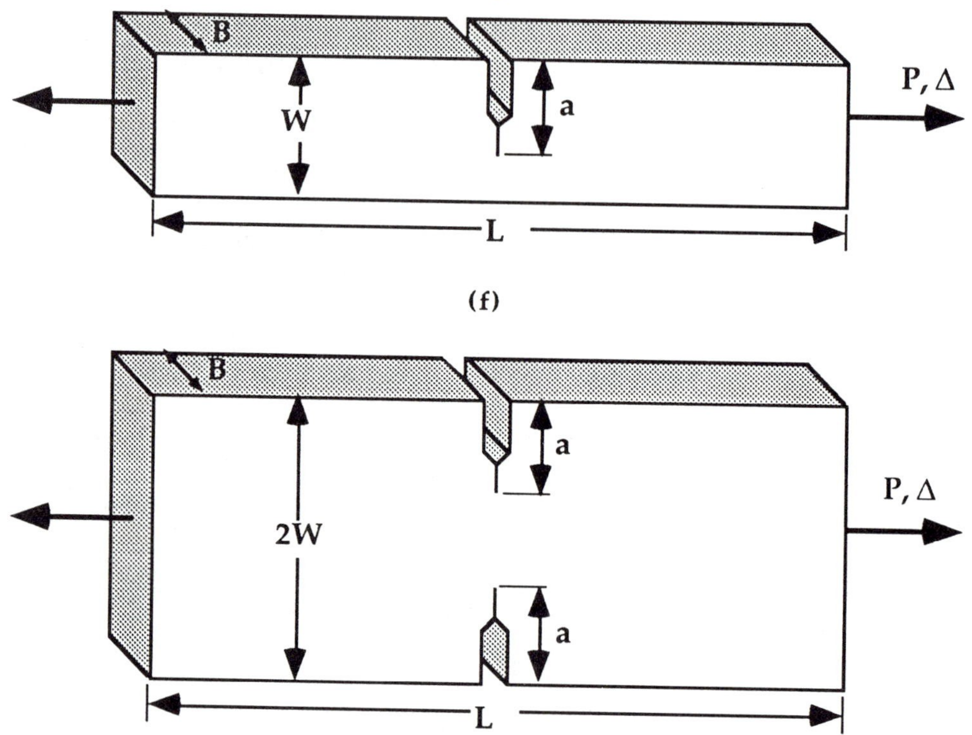

(f)

(g) Double edge notched tension (DENT) panel.

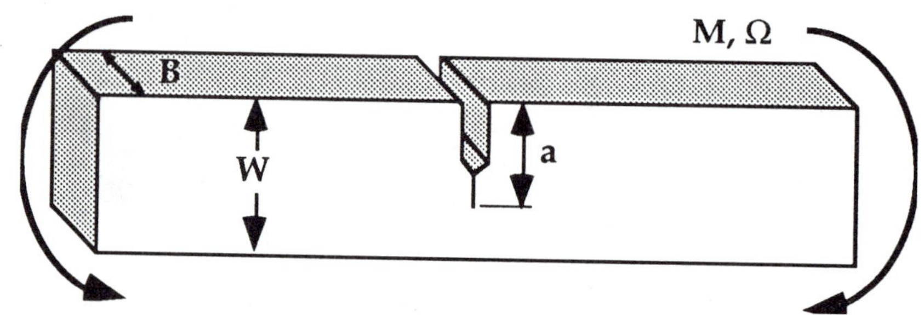

(h) Edge cracked plate in pure bending.

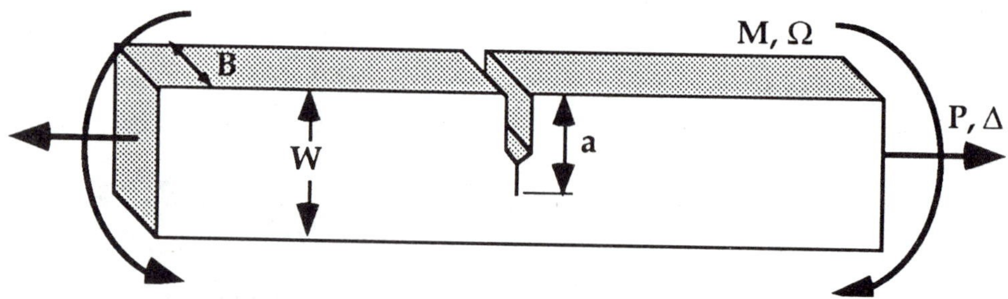

(i) Edge cracked plate in combined bending and tension.

FIGURE 12.1 (Cont.)

TABLE 12.2

Nondimensional K_I solutions for through-thickness cracks in flat plates [1,2]. See Fig. 12.1 for a definition of dimensions for each configuration.

$$f\left(\frac{a}{W}\right) = \frac{K_I B \sqrt{W}}{P}$$

(a) Compact specimen.

$$\frac{2 + \frac{a}{W}}{\left(1 - \frac{a}{W}\right)^{3/2}} \left[0.886 + 4.64\left(\frac{a}{W}\right) - 13.32\left(\frac{a}{W}\right)^2 + 14.72\left(\frac{a}{W}\right)^3 - 5.60\left(\frac{a}{W}\right)^4\right]$$

(b) Disk shaped compact specimen.

$$\frac{2 + \frac{a}{W}}{\left(1 - \frac{a}{W}\right)^{3/2}} \left[0.76 + 4.8\left(\frac{a}{W}\right) - 11.58\left(\frac{a}{W}\right)^2 + 11.43\left(\frac{a}{W}\right)^3 - 4.08\left(\frac{a}{W}\right)^4\right]$$

(c) Single edge notched bend specimen loaded in three-point bending.

$$\frac{3\frac{S}{W}\sqrt{\frac{a}{W}}}{2\left(1 + 2\frac{a}{W}\right)\left(1 - \frac{a}{W}\right)^{3/2}} \left[1.99 - \frac{a}{W}\left(1 - \frac{a}{W}\right)\left\{2.15 - 3.93\left(\frac{a}{W}\right) + 2.7\left(\frac{a}{W}\right)^2\right\}\right]$$

(d) Middle tension (MT) panel.

$$\sqrt{\frac{\pi a}{2W} \sec \frac{\pi a}{2W}} \left[1 - 0.025\left(\frac{a}{W}\right)^2 + 0.06\left(\frac{a}{W}\right)^4\right]$$

(e) Arc shaped specimen.

$$\left(\frac{3X}{W} + 1.9 + 1.1\frac{a}{W}\right)\left[1 + 0.25\left(1 - \frac{a}{W}\right)^2\right] g\left(\frac{a}{W}\right)$$

where

$$g\left(\frac{a}{W}\right) = \frac{\sqrt{\frac{a}{W}}}{\left(1 - \frac{a}{W}\right)^{3/2}} \left[3.74 - 6.30\left(\frac{a}{W}\right) + 6.32\left(\frac{a}{W}\right)^2 - 2.43\left(\frac{a}{W}\right)^3\right]$$

TABLE 12.2 (Cont.)

$$f\left(\frac{a}{W}\right) = \frac{K_I B \sqrt{W}}{P}$$

(f) Single edge notched tension (SENT) panel.

$$\frac{\sqrt{2 \tan \frac{\pi a}{2W}}}{\cos \frac{\pi a}{2W}} \left[0.752 + 2.02 \left(\frac{a}{W}\right) + 0.37 \left(1 - \sin \frac{\pi a}{2W}\right)^3 \right]$$

(g) Double edge notched tension (DENT) panel.

$$\frac{\sqrt{\frac{\pi a}{2W}}}{\sqrt{1 - \frac{a}{W}}} \left[1.122 - 0.561 \left(\frac{a}{W}\right) - 0.205 \left(\frac{a}{W}\right)^2 + 0.471 \left(\frac{a}{W}\right)^3 + 0.190 \left(\frac{a}{W}\right)^4 \right]$$

(h) Edge cracked plate subject to pure bending.

$$f\left(\frac{a}{W}\right) = \frac{K_I B W^{3/2}}{M} = \frac{6\sqrt{2\tan\left(\frac{\pi a}{2W}\right)}}{\cos\left(\frac{\pi a}{2W}\right)} \left[0.923 + 0.199 \left\{1 - \sin\left(\frac{\pi a}{2W}\right)\right\}^4 \right]$$

(i) Edge cracked plate subject to combined bending and tension.

$$K_I = \frac{1}{B\sqrt{W}} \left[P f_t\left(\frac{a}{W}\right) + \frac{M}{W} f_b\left(\frac{a}{W}\right) \right]$$

where f_t and f_b are given above in (f) and (h), respectively.

TABLE 12.3

Nondimensional load line compliance solutions for through-thickness cracks in flat plates [3]*. See Fig. 12.1 for a definition of dimensions for each configuration.

Nondimensional Compliance: $Z_{LL} = \dfrac{\Delta B E'}{P}$ where $\Delta = \Delta_c + \Delta_{nc}$

(a) Compact specimen.

$$\left(\frac{1+\frac{a}{W}}{1-\frac{a}{W}}\right)^2\left[2.163+12.219\left(\frac{a}{W}\right)-20.065\left(\frac{a}{W}\right)^2-0.9925\left(\frac{a}{W}\right)^3+20.609\left(\frac{a}{W}\right)^4+9.9314\left(\frac{a}{W}\right)^5\right]$$

(b) Single edge notched bend (SENB) specimen loaded in three-point bending.

$$\frac{S^3}{W^3(1-\nu^2)}\left[0.25+0.6\left(\frac{W}{S}\right)^2(1+\nu)\right]+1.5\left(\frac{S}{W}\right)^2\left(\frac{\frac{a}{W}}{1-\frac{a}{W}}\right)^2\left[5.58-19.57\left(\frac{a}{W}\right)+36.82\left(\frac{a}{W}\right)^2-34.94\left(\frac{a}{W}\right)^3+12.77\left(\frac{a}{W}\right)^4\right]$$

(c) Middle tension (MT) panel.

$$\frac{L}{2W(1-\nu^2)}+\left(\frac{a}{W}\right)\left[-1.071+0.250\left(\frac{a}{W}\right)-0.357\left(\frac{a}{W}\right)^2+0.121\left(\frac{a}{W}\right)^3-0.047\left(\frac{a}{W}\right)^4+0.008\left(\frac{a}{W}\right)^5-\frac{1.071}{\left(\frac{a}{W}\right)}\ln\left(1-\frac{a}{W}\right)\right]$$

(d) Single edge notched tension (SENT) specimen.

$$\frac{L}{2W(1-\nu^2)}+\frac{4\left(\frac{a}{W}\right)^2}{\left(1-\frac{a}{W}\right)^2}\left\{0.99-\frac{a}{W}\left(1-\frac{a}{W}\right)\left[1.3-1.2\left(\frac{a}{W}\right)+0.7\left(\frac{a}{W}\right)^2\right]\right\}$$

(e) Double edge notched tension (DENT) panel.

$$\frac{L}{2W(1-\nu^2)}+\frac{4}{\pi}\left[0.0629-0.0610\left(\cos\frac{\pi a}{2W}\right)^4-0.0019\left(\cos\frac{\pi a}{2W}\right)^8+\ln\left(\sec\frac{\pi a}{2W}\right)\right]$$

*For side-grooved specimens, B should be replaced by an effective thickness:

$$B_e = B - \frac{(B-B_N)^2}{B}$$ **where B_N is the net thickness.**

TABLE 12.4

Crack length-compliance relationships for compact and three-point bend specimens [4].*

Compact Specimen.
$$\frac{a}{W} = 1.00196 - 4.06319\,U_{LL} + 11.242\,U_{LL}^2 - 1.06043\,U_{LL}^3 + 464.335\,U_{LL}^4 - 650.677\,U_{LL}^5$$ where $$U_{LL} = \frac{1}{\sqrt{Z_{LL}} + 1} \quad \text{and} \quad Z_{LL} = \frac{B\,E\,\Delta}{P}$$
Single Edge Notched Bend Specimen Loaded in Three-Point Bending.
$$\frac{a}{W} = 0.999748 - 3.9504U_V + 2.9821U_V^2 - 3.21408U_V^3 + 51.51564U_V^4 - 113.031U_V^5$$ where $$U_V = \frac{1}{\sqrt{\frac{4Z_V\,W}{S}} + 1} \quad \text{and} \quad Z_V = \frac{BEV}{P}$$

***For side-grooved specimens, B should be replaced by an effective thickness:**

$$B_e = B - \frac{(B - B_N)^2}{B} \qquad \text{where } B_N \text{ is the net thickness.}$$

For large rotations, a compliance correction is required for the compact specimen. (See ASTM E 813-87 [4].)

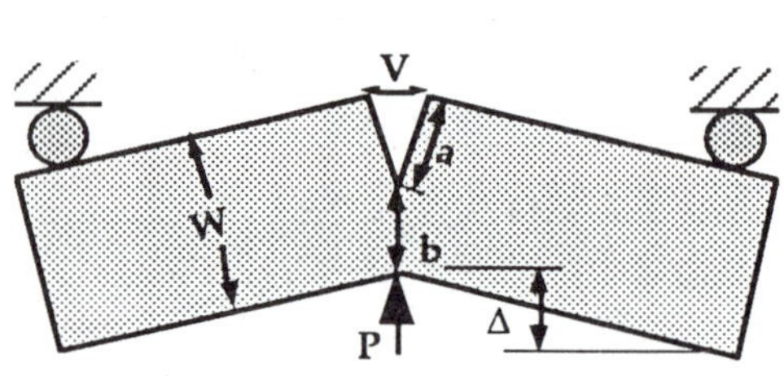

SENB Specimen.

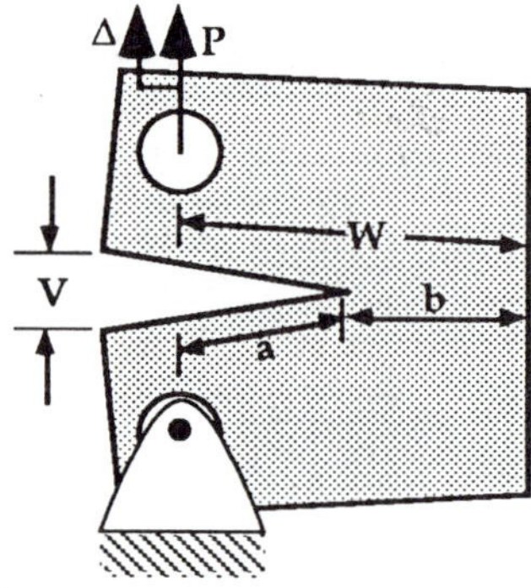

Compact specimen.

12.1.2 Limit Load

TABLE 12.5

Limit load solutions for through-thickness cracks in flat plates [5,6].* See Fig. 12.1 for a definition of dimensions for each configuration.

(a) Compact specimen.

$$P_L = 1.455\,\eta\,B\,b\,\sigma_Y \quad \text{(Plane Strain)}$$

$$P_L = 1.072\,\eta\,B\,b\,\sigma_Y \quad \text{(Plane Stress)}$$

where

$$\eta = \sqrt{\left(\frac{2a}{b}\right)^2 + \frac{4a}{b} + 2} - \left(\frac{2a}{b} + 1\right)$$

(b) Single edge notched bend (SENB) specimen loaded in three-point bending.

$$P_L = \frac{1.455\,B\,b^2\,\sigma_Y}{S} \quad \text{(Plane Strain)}$$

$$P_L = \frac{1.072\,B\,b^2\,\sigma_Y}{S} \quad \text{(Plane Stress)}$$

(c) Middle tension (MT) panel.

$$P_L = \frac{4}{\sqrt{3}}\,B\,b\,\sigma_Y \quad \text{(Plane Strain)}$$

$$P_L = 2\,B\,b\,\sigma_Y \quad \text{(Plane Stress)}$$

(d) Single edge notched tension (SENT) specimen.

$$P_L = 1.455\,\eta\,B\,b\,\sigma_Y \quad \text{(Plane Strain)}$$

$$P_L = 1.072\,\eta\,B\,b\,\sigma_Y \quad \text{(Plane Stress)}$$

where

$$\eta = \sqrt{1 + \left(\frac{a}{b}\right)^2} - \frac{a}{b}$$

(e) Double edge notched tension (DENT) panel.

$$P_L = \left(0.72 + 1.82\,\frac{b}{W}\right) B\,b\,\sigma_Y \quad \text{(Plane Strain)}$$

$$P_L = \frac{4}{\sqrt{3}}\,B\,b\,\sigma_Y \quad \text{(Plane Stress)}$$

***The flow stress, σ_Y, is normally taken as the average of σ_{YS} and σ_{TS}.**

TABLE 12.5 (Cont.)*

Edge Crack Subject to Combined Bending and Tension.

$$P_L = \frac{2\,B\,b\,\sigma_Y}{\sqrt{3}}\left[-\left(2\lambda + \frac{a}{W}\right) + \sqrt{\left(2\lambda + \frac{a}{W}\right)^2 + \left(\frac{a}{b}\right)^2}\;\right] \quad \text{(Plane Strain)}$$

$$P_L = B\,b\,\sigma_Y\left[-\left(2\lambda + \frac{a}{W}\right) + \sqrt{\left(2\lambda + \frac{a}{W}\right)^2 + \left(\frac{a}{b}\right)^2}\;\right] \quad \text{(Plane Stress)}$$

where

$$\lambda = \frac{M}{P\,W}$$

*The flow stress, σ_Y, is normally taken as the average of σ_{YS} and σ_{TS}.

12.1.3 Fully Plastic J and Displacement

The Electric Power Research Institute (EPRI) J estimation scheme [5] provides a means for computing the J integral in a variety of configurations and materials. A fully plastic solution is combined with the stress intensity solution to obtain an estimate of the elastic-plastic J. Section 9.5 describes the theoretical background and applications of this approach.

The fully plastic J integral in the EPRI J estimation scheme is normally expressed in the following form:

$$J_{pl} = \alpha\,\varepsilon_o\,\sigma_o\,b\,h_1\left(\frac{P}{P_o}\right)^{n+1} \tag{12.2}$$

where b is a characteristic length dimension, h_1 is a geometry factor, P is a characteristic load, and P_o is a reference load, which is usually inferred from the limit load solution for the geometry of interest. The other parameters in Eq. (12.2) are flow properties that are defined by a Ramberg-Osgood fit to the material stress-strain curve:

$$\frac{\varepsilon}{\varepsilon_o} = \frac{\sigma}{\sigma_o} + \alpha\left(\frac{\sigma}{\sigma_o}\right)^n \tag{12.3}$$

where σ and ε are uniaxial stress and strain, respectively, σ_o is a reference stress (usually defined at yield), $\varepsilon_o = \sigma_o/E$, α is a fitting constant, and n is a hardening exponent.

The geometry factor, h_1, has been tabulated for a variety of configurations and hardening exponents (Tables 12.6-12.21, 12.34-12.51) [5-7].

The total J is given by the sum of the fully plastic value and an effective elastic J:

$$J_{tot} = J_{el} + J_{pl} \tag{12.4}$$

where J_{el} is computed from the elastic stress intensity factor of an effective crack size:

$$J_{el} = \frac{K_I^2(a_{eff})}{E'} \tag{12.5}$$

where E' = E for plane stress and $E = E/(1 - \nu^2)$ for plane strain conditions. The parentheses in Eq. (12.5) indicates that K_I is a function of a_{eff}, rather than a multiplication product. The effective crack size is inferred from a first order Irwin correction:

$$a_{eff} = a + \frac{1}{1 + (P/P_o)^2} \frac{1}{\beta\pi} \left(\frac{n - 1}{n + 1}\right) \left(\frac{K_I(a)}{\sigma_o}\right)^2 \tag{12.6}$$

where $\beta = 2$ for plane stress and $\beta = 6$ for plane strain conditions.

Elastic-plastic displacements can also be computed from the EPRI approach by adding the elastic displacement, inferred from the compliance solution with crack size = a_{eff}, to the plastic displacement. The fully plastic crack mouth opening displacement, V_p, and plastic load line displacement, Δ_p are normally expressed in the following form:

$$V_p = \alpha\, \varepsilon_o\, a\, h_2(a/W, n) \left(\frac{P}{P_o}\right)^n \tag{12.7}$$

$$\Delta_p = \alpha\, \varepsilon_o\, a\, h_3(a/W, n) \left(\frac{P}{P_o}\right)^n \tag{12.8}$$

where h_2 and h_3 are geometry factors.

TABLE 12.6
Fully plastic J and displacement for a compact specimen in plane strain [5].

a/W:		n=1	n=2	n=3	n=5	n=7	n=10	n=13	n=16	n=20
0.250	h_1	2.23	2.05	1.78	1.48	1.33	1.26	1.25	1.32	1.57
	h_2	17.9	12.5	11.7	10.8	10.5	10.7	11.5	12.6	14.6
	h_3	9.85	8.51	8.17	7.77	7.71	7.92	8.52	9.31	10.9
0.375	h_1	2.15	1.72	1.39	0.970	0.693	0.443	0.276	0.176	0.098
	h_2	12.6	8.18	6.52	4.32	2.97	1.79	1.10	0.686	0.370
	h_3	7.94	5.76	4.64	3.10	2.14	1.29	0.793	0.494	0.266
0.500	h_1	1.94	1.51	1.24	0.919	0.685	0.461	0.314	0.216	0.132
	h_2	9.33	5.85	4.30	2.75	1.91	1.20	0.788	0.530	0.317
	h_3	6.41	4.27	3.16	2.02	1.41	0.888	0.585	0.393	0.236
0.625	h_1	1.76	1.45	1.24	0.974	0.752	0.602	0.459	0.347	0.248
	h_2	7.61	4.57	3.42	2.36	1.81	1.32	0.983	0.749	0.485
	h_3	5.52	3.43	2.58	1.79	1.37	1.00	0.746	0.568	0.368
0.750	h_1	1.71	1.42	1.26	1.033	0.864	0.717	0.575	0.448	0.345
	h_2	6.37	3.95	3.18	2.34	1.88	1.44	1.12	0.887	0.665
	h_3	4.86	3.05	2.46	1.81	1.45	1.11	0.869	0.686	0.514
→1	h_1	1.57	1.45	1.35	1.18	1.08	0.950	0.850	0.730	0.630
	h_2	5.39	3.74	3.09	2.43	2.12	1.80	1.57	1.33	1.14
	h_3	4.31	2.99	2.47	1.95	1.79	1.44	1.26	1.07	0.909

$$J_{pl} = \alpha \varepsilon_o \sigma_o b \, h_1(a/W, n) \left(\frac{P}{P_o}\right)^{n+1}$$

$$V_P = \alpha \varepsilon_o a \, h_2(a/W, n) \left(\frac{P}{P_o}\right)^{n}$$

$$\Delta_P = \alpha \varepsilon_o a \, h_3(a/W, n) \left(\frac{P}{P_o}\right)^{n}$$

$$P_o = 1.455 \, \eta \, B b \sigma_o$$

where

$$\eta = \sqrt{\left(\frac{2a}{b}\right)^2 + \frac{4a}{b} + 2} - \left(\frac{2a}{b} + 1\right)$$

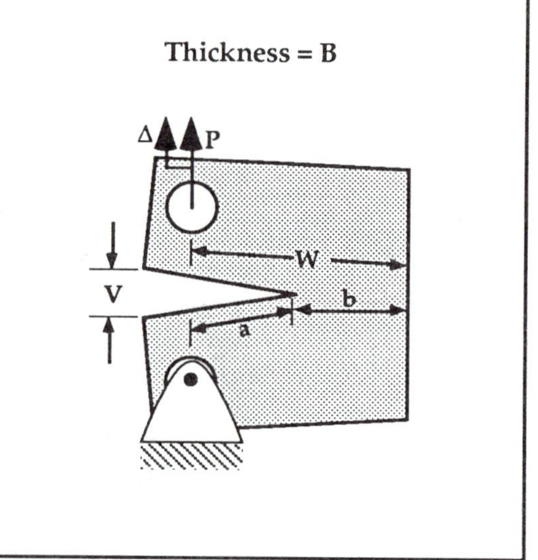

TABLE 12.7
Fully plastic J and displacement for a compact specimen in plane stress [5].

a/W:		n=1	n=2	n=3	n=5	n=7	n=10	n=13	n=16	n=20
0.250	h_1	1.61	1.46	1.28	1.06	0.903	0.729	0.601	0.511	0.395
	h_2	17.6	12.0	10.7	8.74	7.32	5.74	4.63	3.75	2.92
	h_3	9.67	8.00	7.21	5.94	5.00	3.95	3.19	2.59	2.023
0.375	h_1	1.55	1.25	1.05	0.801	0.647	0.484	0.377	0.284	0.220
	h_2	12.4	8.20	6.54	4.56	3.45	2.44	1.83	1.36	1.02
	h_3	7.80	5.73	4.62	3.25	2.48	1.77	1.33	0.990	0.746
0.500	h_1	1.40	1.08	0.901	0.686	0.558	0.436	0.356	0.298	0.238
	h_2	9.16	5.67	4.21	2.80	2.12	1.57	1.25	1.03	0.814
	h_3	6.29	4.15	3.11	2.09	1.59	1.18	0.938	0.774	0.614
0.625	h_1	1.27	1.03	0.875	0.695	0.593	0.494	0.423	0.370	0.310
	h_2	7.47	4.48	3.35	2.37	1.92	1.54	1.29	1.12	0.928
	h_3	5.42	3.38	2.54	1.80	1.47	1.18	0.988	0.853	0.710
0.750	h_1	1.23	0.977	0.833	0.683	0.598	0.506	0.431	0.373	0.314
	h_2	6.25	3.78	2.89	2.14	1.78	1.44	1.20	1.03	0.857
	h_3	4.77	2.92	2.24	1.66	1.38	1.12	0.936	0.800	0.666
→1	h_1	1.13	1.01	0.775	0.680	0.650	0.620	0.490	0.470	0.420
	h_2	5.29	3.54	2.41	1.91	1.73	1.59	1.23	1.17	1.03
	h_3	4.23	2.83	1.93	1.52	1.39	1.27	0.985	0.933	0.824

$$J_{pl} = \alpha \varepsilon_o \sigma_o b\, h_1(a/W, n) \left(\frac{P}{P_o}\right)^{n+1}$$

$$V_p = \alpha \varepsilon_o a\, h_2(a/W, n) \left(\frac{P}{P_o}\right)^{n}$$

$$\Delta_p = \alpha \varepsilon_o a\, h_3(a/W, n) \left(\frac{P}{P_o}\right)^{n}$$

$$P_o = 1.072\, \eta B b \sigma_o$$

where

$$\eta = \sqrt{\left(\frac{2a}{b}\right)^2 + \frac{4a}{b} + 2} - \left(\frac{2a}{b} + 1\right)$$

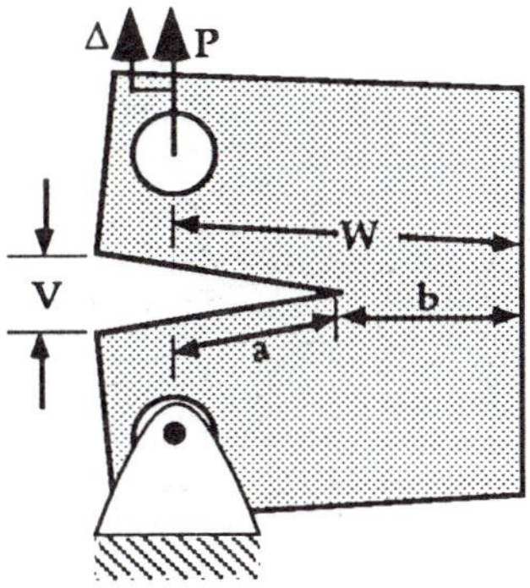

TABLE 12.8
Fully plastic J and displacement for a single edge notched bend (SENB) specimen in plane strain subject to three-point bending [5].

a/W:		n=1	n=2	n=3	n=5	n=7	n=10	n=13	n=16	n=20
0.125	h_1	0.936	0.869	0.805	0.687	0.580	0.437	0.329	0.245	0.165
	h_2	6.97	6.77	6.29	5.29	4.38	3.24	2.40	1.78	1.19
	h_3	3.00	22.1	20.0	15.0	11.7	8.39	6.14	4.54	3.01
0.250	h_1	1.20	1.034	0.930	0.762	0.633	0.523	0.396	0.303	0.215
	h_2	5.80	4.67	4.01	3.08	2.45	1.93	1.45	1.09	0.758
	h_3	4.08	9.72	8.36	5.86	4.47	3.42	2.54	1.90	1.32
0.375	h_1	1.33	1.15	1.02	0.084	0.695	0.556	0.442	0.360	0.265
	h_2	5.18	3.93	3.20	2.38	1.93	1.47	1.15	0.928	0.684
	h_3	4.51	6.01	5.03	3.74	3.02	2.30	1.80	1.45	1.07
0.500	h_1	1.41	1.09	0.922	0.675	0.495	0.331	0.211	0.135	0.0741
	h_2	4.87	3.28	2.53	1.69	1.19	0.773	0.480	0.304	0.165
	h_3	4.69	4.33	3.49	2.35	1.66	1.08	0.669	0.424	0.230
0.625	h_1	1.46	1.07	0.896	0.631	0.436	0.255	0.142	0.084	0.0411
	h_2	4.64	2.86	2.16	1.37	0.907	0.518	0.287	0.166	0.0806
	h_3	4.71	3.49	2.70	1.72	1.14	0.652	0.361	0.209	0.102
0.750	h_1	1.48	1.15	0.974	0.693	0.500	0.348	0.223	0.140	0.0745
	h_2	4.47	2.75	2.10	1.36	0.936	0.618	0.388	0.239	0.127
	h_3	4.49	3.14	2.40	1.56	1.07	0.704	0.441	0.272	0.144
0.875	h_1	1.50	1.35	1.20	1.02	0.855	0.690	0.551	0.440	0.321
	h_2	4.36	2.90	2.31	1.70	1.33	1.00	0.782	0.613	0.459
	h_3	4.15	3.08	2.45	1.81	1.41	1.06	0.828	0.646	0.486

$$J_{pl} = \alpha \varepsilon_o \sigma_o b\, h_1(a/W, n) \left(\frac{P}{P_o}\right)^{n+1}$$

$$V_p = \alpha \varepsilon_o a\, h_2(a/W, n) \left(\frac{P}{P_o}\right)^{n}$$

$$\Delta_p = \alpha \varepsilon_o a\, h_3(a/W, n) \left(\frac{P}{P_o}\right)^{n}$$

$$P_o = \frac{1.455\, B\, b^2 \sigma_o}{S}$$

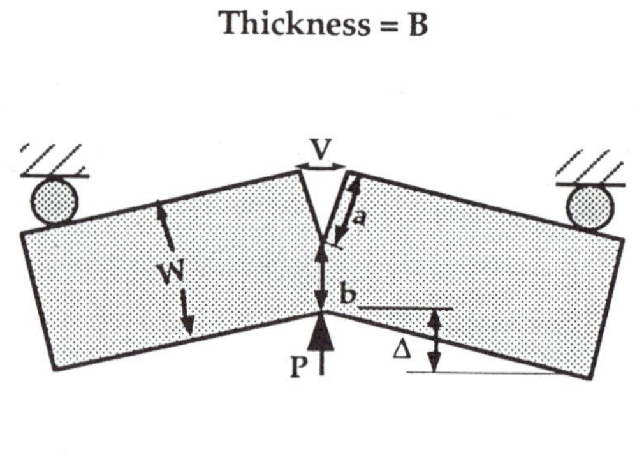

Thickness = B

V
a
W
b
Δ
P

TABLE 12.9
Fully plastic J and displacement for a single edge notched bend (SENB) specimen in plane stress subject to three-point bending [5].

a/W:		n=1	n=2	n=3	n=5	n=7	n=10	n=13	n=16	n=20
0.125	h_1	0.676	0.600	0.548	0.459	0.383	0.297	0.238	0.192	0.148
	h_2	6.84	6.30	5.66	4.53	3.64	2.72	2.12	1.67	1.26
	h_3	2.95	20.1	14.6	12.2	9.12	6.75	5.20	4.09	3.07
0.250	h_1	0.869	0.731	0.629	0.479	0.370	0.246	0.174	0.117	0.0593
	h_2	5.69	4.50	3.68	2.61	1.95	1.29	0.897	0.603	0.307
	h_3	4.01	8.81	7.19	4.73	3.39	2.20	1.52	1.01	0.508
0.375	h_1	0.963	0.797	0.680	0.527	0.418	0.307	0.232	0.174	0.105
	h_2	5.09	3.73	2.93	2.07	1.58	1.13	0.841	0.626	0.381
	h_3	4.42	5.53	4.48	3.17	2.41	1.73	1.28	0.948	0.575
0.500	h_1	1.02	0.767	0.621	0.453	0.324	0.202	0.128	0.0813	0.0298
	h_2	4.77	3.12	2.32	1.55	1.08	0.655	0.410	0.259	0.0974
	h_3	4.60	4.09	3.09	2.08	1.44	0.874	0.545	0.344	0.129
0.625	h_1	1.05	0.786	0.649	0.494	0.357	0.235	0.173	0.105	0.0471
	h_2	4.55	2.83	2.12	1.46	1.02	0.656	0.472	0.286	0.130
	h_3	4.62	3.43	2.60	1.79	1.26	0.803	0.577	0.349	0.158
0.750	h_1	1.07	0.786	0.643	0.474	0.343	0.230	0.167	0.110	0.0442
	h_2	4.39	2.66	1.97	1.33	0.928	0.601	0.427	0.280	0.114
	h_3	4.39	3.01	2.24	1.51	1.05	0.680	0.483	0.316	0.129
0.875	h_1	1.086	0.928	0.810	6.46	0.538	0.423	0.332	0.242	0.205
	h_2	4.28	2.76	2.16	1.56	1.23	0.922	0.702	0.561	0.428
	h_3	4.07	2.93	2.29	1.65	1.30	0.975	0.742	0.592	0.452

$$J_{pl} = \alpha \varepsilon_o \sigma_o b h_1(a/W, n) \left(\frac{P}{P_o}\right)^{n+1}$$

$$V_p = \alpha \varepsilon_o a h_2(a/W, n) \left(\frac{P}{P_o}\right)^{n}$$

$$\Delta_p = \alpha \varepsilon_o a h_3(a/W, n) \left(\frac{P}{P_o}\right)^{n}$$

$$P_o = \frac{1.072 B b^2 \sigma_o}{S}$$

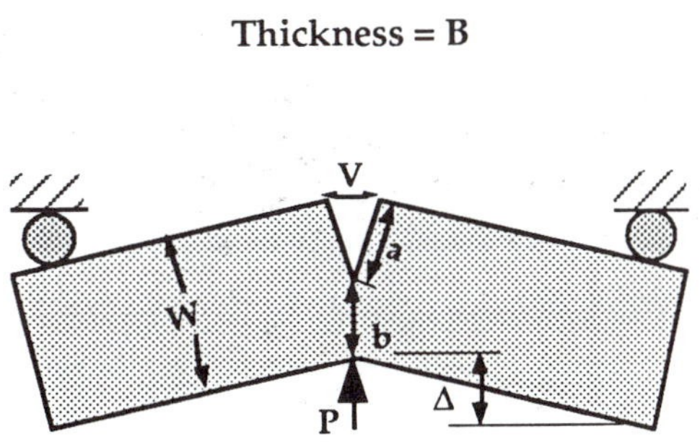

Thickness = B

TABLE 12.10
Fully plastic J and displacement for a middle tension (MT) specimen in plane strain [5].

a/W:		n=1	n = 2	n=3	n=5	n=7	n=10	n=13	n=16	n=20
0.125	h_1	2.80	3.61	4.06	4.35	4.33	4.02	3.56	3.06	2.46
	h_2	3.05	3.62	3.91	4.06	3.93	3.54	3.07	2.60	2.06
	h_3	0.303	0.574	0.840	1.30	1.63	1.95	2.03	1.96	1.77
0.250	h_1	2.54	3.01	3.21	3.29	3.18	2.92	2.63	2.34	2.03
	h_2	2.68	2.99	3.01	2.85	2.61	2.30	1.97	1.71	1.45
	h_3	0.536	0.911	1.22	1.64	1.84	1.85	1.80	1.64	1.43
0.375	h_1	2.34	2.62	2.65	2.51	2.28	1.97	1.71	1.46	1.19
	h_2	2.35	2.39	2.23	1.88	1.58	1.28	1.07	0.890	0.715
	h_3	0.699	1.06	1.28	1.44	1.40	1.23	1.05	0.888	0.719
0.500	h_1	2.21	2.29	2.20	1.97	1.76	1.52	1.32	1.16	0.978
	h_2	2.03	1.86	1.60	1.23	1.00	0.799	0.664	0.564	0.466
	h_3	0.803	1.07	1.16	1.10	0.968	0.796	0.665	0.565	0.469
0.625	h_1	2.12	1.96	1.76	1.43	1.17	0.863	0.628	0.458	0.300
	h_2	1.71	1.32	1.04	0.707	0.524	0.358	0.250	0.178	0.114
	h_3	0.844	0.937	0.879	0.701	0.522	0.361	0.251	0.178	0.115
0.750	h_1	2.07	1.73	1.47	1.11	0.895	0.642	0.461	0.337	0.216
	h_2	1.35	0.857	0.596	0.361	0.254	0.167	0.114	0.0810	0.0511
	h_3	0.805	0.700	0.555	0.359	0.254	0.168	0.114	0.0813	0.0516
0.875	h_1	2.08	1.64	1.40	1.14	0.987	0.814	0.688	0.573	0.461
	h_2	0.889	0.428	0.287	0.181	0.139	0.105	0.0837	0.0682	0.0533
	h_3	0.632	0.400	0.291	0.182	0.140	0.106	0.0839	0.0683	0.0535

$$J_{pl} = \alpha \varepsilon_o \sigma_o \frac{b\,a}{W} h_1(a/W, n) \left(\frac{P}{P_o}\right)^{n+1}$$

$$V_p = \alpha \varepsilon_o a\, h_2(a/W, n) \left(\frac{P}{P_o}\right)^{n}$$

$$\Delta_{p(c)} = \alpha \varepsilon_o a\, h_3(a/W, n) \left(\frac{P}{P_o}\right)^{n}$$

$$P_o = \frac{4}{\sqrt{3}} B\, b\, \sigma_o$$

$$\Delta_{p(nc)} = \frac{\sqrt{3}}{2} \alpha \varepsilon_o L \left(\frac{\sqrt{3}\,P}{4\,W\,\sigma_o}\right)^{n}$$

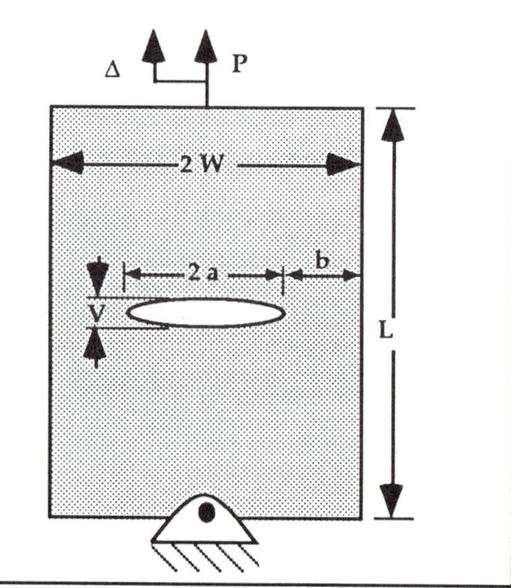

TABLE 12.11

Fully plastic J and displacement for a middle tension (MT) specimen in plane stress [5].

a/W:		n=1	n=2	n=3	n=5	n=5	n=7	n=10	n=16	n=20
0.125	h_1	2.80	3.57	4.01	4.47	4.65	4.62	4.41	4.13	3.72
	h_2	3.53	4.09	4.43	4.74	4.79	4.63	4.33	4.00	3.55
	h_3	0.350	0.661	0.997	1.55	2.05	2.56	2.83	2.95	2.92
0.250	h_1	2.54	2.97	3.14	3.20	3.11	2.86	2.65	2.47	2.20
	h_2	3.10	3.29	3.30	3.15	2.93	2.56	2.29	2.08	1.81
	h_3	0.619	1.01	1.35	1.83	2.08	2.19	2.12	2.01	1.79
0.375	h_1	2.34	2.53	2.52	2.35	2.17	1.95	1.77	1.61	1.43
	h_2	2.71	2.62	2.41	2.03	1.75	1.47	1.28	1.13	0.988
	h_3	0.807	1.20	1.43	1.59	1.57	1.43	1.27	1.13	0.994
0.750	h_1	2.21	2.20	2.06	1.81	1.63	1.43	1.30	1.17	1.00
	h_2	2.34	2.01	1.70	1.30	1.07	0.871	0.757	0.666	0.557
	h_3	0.927	1.19	1.26	1.18	1.04	0.867	0.758	0.668	0.560
0.625	h_1	2.12	1.91	1.69	1.41	1.22	1.01	0.853	0.712	0.573
	h_2	1.97	1.46	1.13	0.785	0.617	0.474	0.383	0.313	0.256
	h_3	0.975	1.05	0.970	0.763	0.620	0.478	0.386	0.318	0.273
0.750	h_1	2.07	1.71	1.46	1.21	1.08	0.867	0.745	0.646	0.532
	h_2	1.55	0.970	0.685	0.452	0.361	0.262	0.216	0.183	0.148
	h_3	0.929	0.802	0.642	0.450	0.361	0.263	0.216	0.183	0.149
0.875	h_1	2.08	1.57	1.31	1.08	0.972	0.862	0.778	0.715	0.630
	h_2	1.03	0.485	0.310	0.196	0.157	0.127	0.109	0.0971	0.0842
	h_3	0.730	0.452	0.313	0.198	0.157	0.127	0.109	0.0973	0.0842

$$J_{pl} = \alpha \varepsilon_o \sigma_o \frac{b\,a}{W} h_1(a/W, n) \left(\frac{P}{P_o}\right)^{n+1}$$

$$V_p = \alpha \varepsilon_o a\, h_2(a/W, n) \left(\frac{P}{P_o}\right)^{n}$$

$$\Delta_{p(c)} = \alpha \varepsilon_o a\, h_3(a/W, n) \left(\frac{P}{P_o}\right)^{n}$$

$$P_o = 2 B b \sigma_o$$

$$\Delta_{p(nc)} = \alpha \varepsilon_o L \left(\frac{P}{2 W \sigma_o}\right)^{n}$$

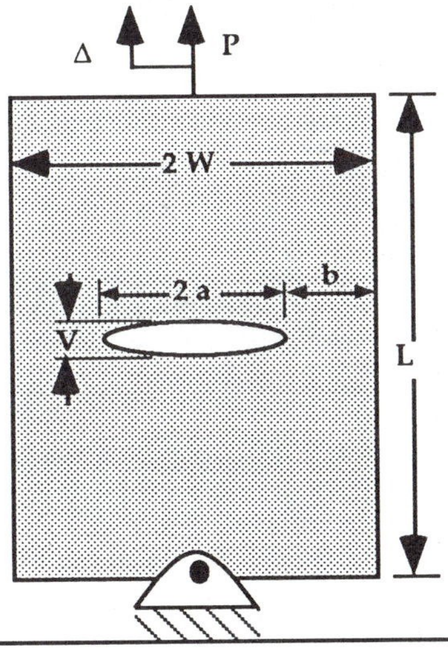

TABLE 12.12
Fully plastic J and displacement for a single edge notched tension (SENT) specimen in plane strain [5].

a/W:		n=1	n=2	n=3	n=5	n=7	n=10	n=13	n=16	n=20
0.125	h_1	4.95	6.93	8.57	11.5	13.5	16.1	18.1	19.9	21.2
	h_2	5.25	6.47	7.56	9.46	11.1	12.9	14.4	15.7	16.8
	h_3	26.6	25.8	25.2	24.2	23.6	23.2	23.2	23.5	23.7
0.250	h_1	4.34	4.77	4.64	3.82	3.06	2.17	1.55	1.11	0.712
	h_2	4.76	4.56	4.28	3.39	2.64	1.81	1.25	0.875	0.552
	h_3	10.3	7.64	5.87	3.70	2.48	1.50	0.970	0.654	0.404
0.375	h_1	3.88	3.25	2.63	1.68	1.06	0.539	0.276	0.142	0.0595
	h_2	4.54	3.49	2.67	1.57	0.946	0.458	0.229	0.116	0.048
	h_3	5.14	2.99	1.90	0.923	0.515	0.240	0.119	0.060	0.0246
0.500	h_1	3.40	2.30	1.69	0.928	0.514	0.213	0.0902	0.0385	0.0119
	h_2	4.45	2.77	1.89	0.954	0.507	0.204	0.0854	0.0356	0.0110
	h_3	3.15	1.54	0.912	0.417	0.215	0.085	0.0358	0.0147	0.0045
0.625	h_1	2.86	1.80	1.30	0.697	0.378	0.153	0.0625	0.0256	0.0078
	h_2	4.37	2.44	1.62	0.0806	0.423	0.167	0.0671	0.0272	0.0082
	h_3	2.31	1.08	0.681	0.329	0.171	0.067	0.0268	0.0108	0.0033
0.750	h_1	2.34	1.61	1.25	0.769	0.477	0.233	0.116	0.059	0.0215
	h_2	4.32	2.52	1.79	1.03	0.619	0.296	0.146	0.0735	0.0267
	h_3	2.02	1.10	0.765	0.435	0.262	0.125	0.0617	0.0312	0.0113
0.875	h_1	1.91	1.57	1.37	1.10	0.925	0.702			
	h_2	4.29	2.75	2.14	1.55	1.23	0.921			
	h_3	2.01	1.27	0.988	0.713	0.564	0.424			

$$J_{pl} = \alpha \varepsilon_o \sigma_o \frac{b\,a}{W} h_1(a/W, n) \left(\frac{P}{P_o}\right)^{n+1}$$

$$V_p = \alpha \varepsilon_o a\, h_2(a/W, n) \left(\frac{P}{P_o}\right)^{n}$$

$$\Delta_{p(c)} = \alpha \varepsilon_o a\, h_3(a/W, n) \left(\frac{P}{P_o}\right)^{n}$$

$$P_o = 1.455\,\eta\, B\, b\, \sigma_o$$

where

$$\eta = \sqrt{1 + \left(\frac{a}{b}\right)^2} - \frac{a}{b}$$

$$\Delta_{p(nc)} = \frac{\sqrt{3}}{2} \alpha \varepsilon_o L \left(\frac{\sqrt{3}\,P}{4\,W\,\sigma_o}\right)^{n}$$

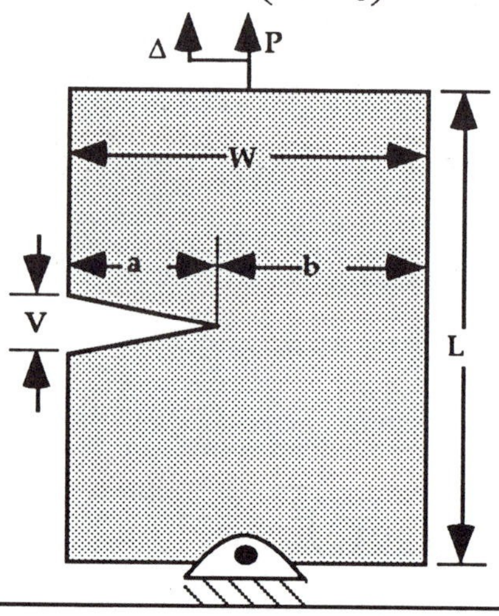

Δ P W a b V L

TABLE 12.13
Fully plastic J and displacement for a single edge notched tension (SENT) specimen in plane stress [5].

a/W:		n=1	n=2	n=3	n=5	n=7	n=10	n=13	n=16	n=20
0.125	h_1	3.58	4.55	5.06	5.30	4.96	4.14	3.29	2.60	1.92
	h_2	5.15	5.43	6.05	6.01	5.47	4.46	3.48	2.74	2.02
	h_3	26.1	21.6	18.0	12.7	9.24	5.98	3.94	2.72	2.0
0.250	h_1	3.14	3.26	2.92	2.12	1.53	0.960	0.615	0.400	0.230
	h_2	4.67	4.30	3.70	2.53	1.76	1.05	0.656	0.419	0.237
	h_3	10.1	6.49	4.36	2.19	1.24	0.630	0.362	0.224	0.123
0.375	h_1	2.88	2.37	1.94	1.37	1.01	0.677	0.474	0.342	0.226
	h_2	4.47	3.43	2.63	1.69	1.18	0.762	0.524	0.372	0.244
	h_3	5.05	2.65	1.60	0.812	0.525	0.328	0.223	0.157	0.102
0.500	h_1	2.46	1.67	1.25	0.776	0.510	0.286	0.164	0.0956	0.0469
	h_2	4.37	2.73	1.91	1.09	0.694	0.380	0.216	0.124	0.0607
	h_3	3.10	1.43	0.871	0.461	0.286	0.155	0.088	0.0506	0.0247
0.625	h_1	2.07	1.41	1.105	0.755	0.551	0.363	0.248	0.172	0.107
	h_2	4.30	2.55	1.84	1.16	0.816	0.523	0.353	2.42	0.150
	h_3	2.27	1.13	0.771	0.478	0.336	0.215	0.146	0.100	0.0616
0.750	h_1	1.70	1.14	0.910	0.624	0.447	0.280	0.181	0.118	0.0670
	h_2	4.24	2.47	1.81	1.15	0.798	0.490	0.314	0.203	0.115
	h_3	1.98	1.09	0.784	0.494	0.344	0.211	0.136	0.0581	0.0496
0.875	h_1	1.38	1.11	0.962	0.792	0.677	0.574			
	h_2	4.22	2.68	2.08	1.54	1.27	1.04			
	h_3	1.97	1.25	0.969	0.716	0.591	0.483			

$$J_{pl} = \alpha \varepsilon_o \sigma_o \frac{b\,a}{W} h_1(a/W, n) \left(\frac{P}{P_o}\right)^{n+1}$$

$$V_p = \alpha \varepsilon_o a\, h_2(a/W, n) \left(\frac{P}{P_o}\right)^{n}$$

$$\Delta_{p(c)} = \alpha \varepsilon_o a\, h_3(a/W, n) \left(\frac{P}{P_o}\right)^{n}$$

$$P_o = 1.072\, \eta\, B\, b\, \sigma_o$$

where

$$\eta = \sqrt{1 + \left(\frac{a}{b}\right)^2} - \frac{a}{b}$$

$$\Delta_{p(nc)} = \alpha \varepsilon_o L \left(\frac{P}{2\, W \sigma_o}\right)^{n}$$

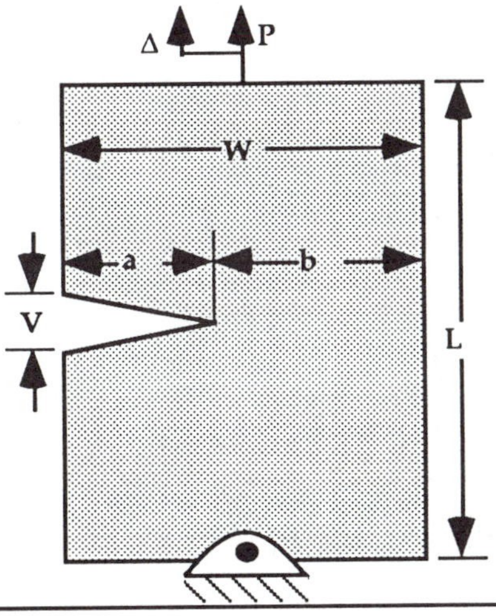

TABLE 12.14
Fully plastic J and displacement for a double edge notched tension (DENT) specimen in plane strain [5].

a/W:		n=1	n=2	n=3	n=5	n=7	n=10	n=13	n=16	n=20
0.125	h_1	0.572	0.772	0.922	1.13	1.35	1.61	1.86	2.08	2.44
	h_2	0.732	0.852	0.961	1.14	1.29	1.50	1.70	1.94	2.17
	h_3	0.063	0.126	0.200	0.372	0.571	0.911	1.30	1.74	2.29
0.250	h_1	1.10	1.32	1.38	1.65	1.75	1.82	1.86	1.89	1.92
	h_2	1.56	1.63	1.70	1.78	1.80	1.81	1.79	1.78	1.76
	h_3	0.267	0.479	0.698	1.11	1.47	1.92	2.25	2.49	2.73
0.375	h_1	1.61	1.83	1.92	1.92	1.84	1.68	1.49	1.32	1.12
	h_2	2.51	2.41	2.35	2.15	1.94	1.68	1.44	1.25	1.05
	h_3	0.637	1.05	1.40	1.87	2.11	2.20	2.09	1.92	1.67
0.500	h_1	2.22	2.43	2.48	2.43	2.32	2.12	1.91	1.60	1.51
	h_2	3.73	3.40	3.15	2.71	2.37	2.01	1.72	1.40	1.38
	h_3	1.26	1.92	2.37	2.79	2.85	2.68	2.40	1.99	1.94
0.625	h_1	3.16	3.38	3.45	3.42	3.28	3.00	2.54	2.36	2.27
	h_2	5.57	4.76	4.23	3.46	2.97	2.48	2.02	1.82	1.66
	h_3	2.36	3.29	3.74	3.90	3.68	3.23	2.66	2.40	2.19
0.750	h_1	5.24	6.29	7.17	8.44	9.46	10.9	11.9	11.3	17.4
	h_2	9.10	7.76	7.14	6.64	6.83	7.48	7.79	7.14	11.1
	h_3	4.73	6.26	7.03	7.63	8.14	9.04	9.40	8.58	13.5
0.875	h_1	14.2	24.8	39.0	78.4	140.0	341.0	777.0	1570.0	3820.0
	h_2	20.1	19.4	22.7	36.1	58.9	133.0	294.0	585.0	1400.0
	h_3	12.7	18.2	24.1	40.4	65.8	149.0	327.0	650.0	1560.0

$$J_{pl} = \alpha \varepsilon_o \sigma_o \frac{b\,a}{W} h_1(a/W, n) \left(\frac{P}{P_o}\right)^{n+1}$$

$$V_p = \alpha \varepsilon_o a\, h_2(a/W, n) \left(\frac{P}{P_o}\right)^{n}$$

$$\Delta_{p(c)} = \alpha \varepsilon_o a\, h_3(a/W, n) \left(\frac{P}{P_o}\right)^{n}$$

$$P_o = \left(0.72 + 1.82\frac{b}{W}\right) B\, b\, \sigma_o$$

$$\Delta_{p(nc)} = \frac{\sqrt{3}}{2} \alpha \varepsilon_o L \left(\frac{\sqrt{3}\,P}{4\,W\,\sigma_o}\right)^{n}$$

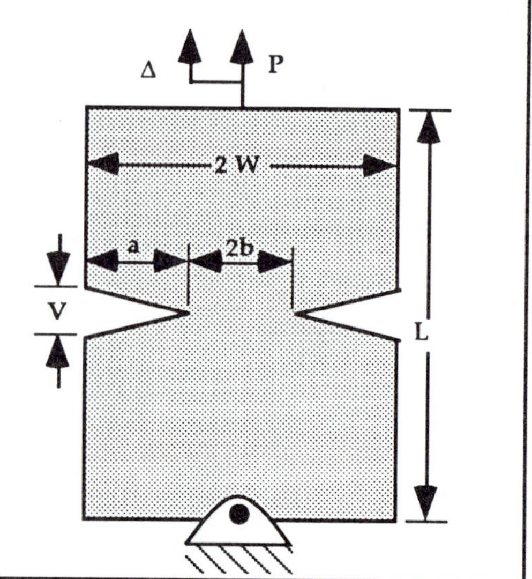

TABLE 12.15
Fully plastic J and displacement for a double edge notched tension (DENT) specimen in plane stress [5].

a/W:		n=1	n=2	n=3	n=5	n=7	n=10	n=13	n=16	n=20
0.125	h_1	0.583	0.825	1.02	1.37	1.71	2.24	2.84	3.54	4.62
	h_2	0.853	1.05	1.23	1.55	1.87	2.38	2.96	3.65	4.70
	h_3	0.0729	0.159	0.26	0.504	0.821	1.41	2.18	3.16	4.73
0.250	h_1	1.01	1.23	1.36	1.48	1.54	1.58	1.59	1.59	1.59
	h_2	1.73	1.82	1.89	1.92	1.91	1.85	1.80	1.75	1.70
	h_3	0.296	0.537	0.770	1.17	1.49	1.82	2.02	2.12	2.20
0.375	h_1	1.29	1.42	1.43	1.34	1.24	1.09	0.970	0.873	0.674
	h_2	2.59	2.39	2.22	1.86	1.59	1.28	1.07	0.922	0.709
	h_3	0.658	1.04	1.30	1.52	1.55	1.41	1.23	1.07	0.830
0.500	h_1	1.48	1.47	1.38	1.17	1.01	0.845	0.732	0.625	0.208
	h_2	3.51	2.82	2.34	1.67	1.28	0.944	0.762	0.630	0.232
	h_3	1.18	1.58	1.69	1.56	1.32	1.01	0.809	0.662	0.266
0.625	h_1	1.59	1.45	1.29	1.04	0.882	0.737	0.649	0.466	0.0202
	h_2	4.56	3.15	2.32	1.45	1.06	0.790	0.657	0.473	0.0277
	h_3	1.93	2.14	1.95	1.44	1.09	0.809	0.665	0.487	0.0317
0.750	h_1	1.65	1.43	1.22	0.979	0.834	0.701	0.630	0.297	
	h_2	5.90	3.37	2.22	1.30	0.966	0.741	0.636	0.312	
	h_3	3.06	2.67	2.06	1.31	0.978	0.747	0.638	0.318	
0.875	h_1	1.69	1.43	1.22	0.979	0.845	0.738	0.664	0.614	0.562
	h_2	8.02	3.51	2.14	1.27	0.971	0.775	0.663	0.596	0.535
	h_3	5.07	3.18	2.16	1.30	0.980	0.779	0.665	0.597	0.538

$$J_{pl} = \alpha \varepsilon_o \sigma_o \frac{b\,a}{W} h_1(a/W, n) \left(\frac{P}{P_o}\right)^{n+1}$$

$$V_p = \alpha \varepsilon_o a\, h_2(a/W, n) \left(\frac{P}{P_o}\right)^{n}$$

$$\Delta_{p(c)} = \alpha \varepsilon_o a\, h_3(a/W, n) \left(\frac{P}{P_o}\right)^{n}$$

$$P_o = \frac{4}{\sqrt{3}} B\, b\, \sigma_o$$

$$\Delta_{p(nc)} = \alpha \varepsilon_o L \left(\frac{P}{2\,W\,\sigma_o}\right)^{n}$$

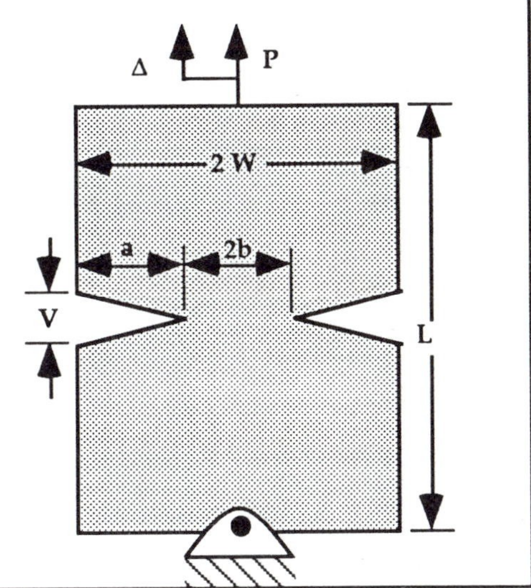

TABLE 12.16

Fully plastic J, displacement, and rotation for an edge cracked plate in plane strain subject to combined tension and bending [6].

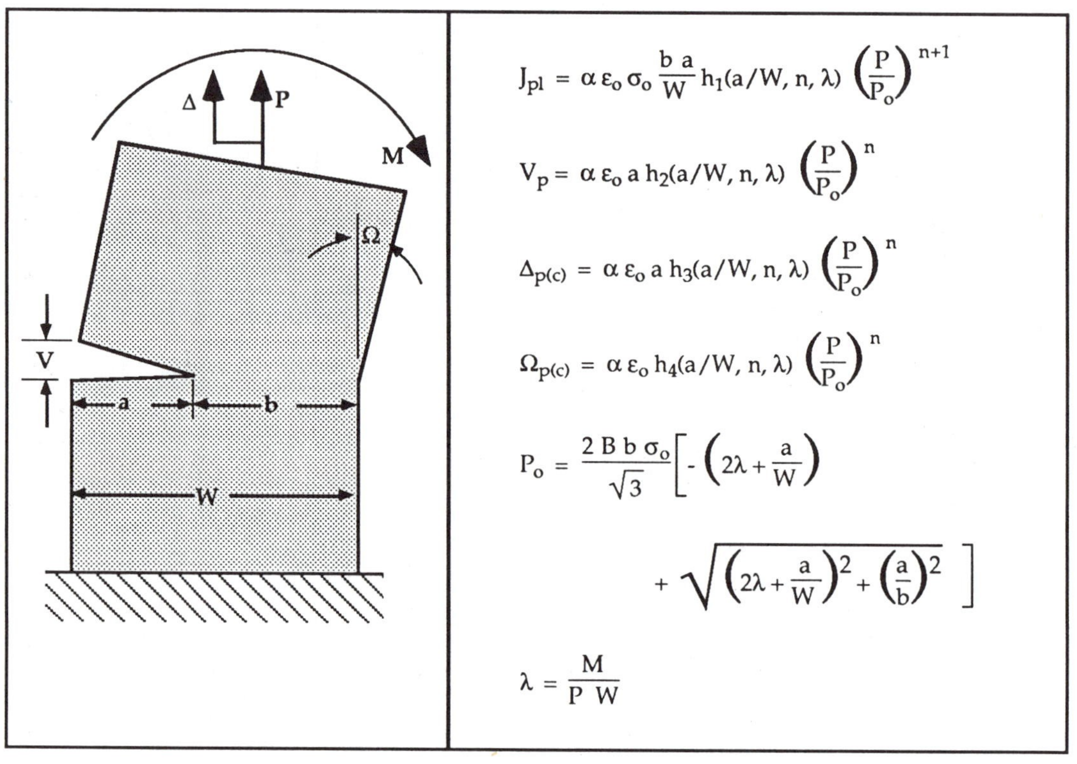

$$J_{pl} = \alpha \varepsilon_o \sigma_o \frac{b\,a}{W} h_1(a/W, n, \lambda) \left(\frac{P}{P_o}\right)^{n+1}$$

$$V_p = \alpha \varepsilon_o a\, h_2(a/W, n, \lambda) \left(\frac{P}{P_o}\right)^{n}$$

$$\Delta_{p(c)} = \alpha \varepsilon_o a\, h_3(a/W, n, \lambda) \left(\frac{P}{P_o}\right)^{n}$$

$$\Omega_{p(c)} = \alpha \varepsilon_o h_4(a/W, n, \lambda) \left(\frac{P}{P_o}\right)^{n}$$

$$P_o = \frac{2\,B\,b\,\sigma_o}{\sqrt{3}} \left[- \left(2\lambda + \frac{a}{W}\right) + \sqrt{\left(2\lambda + \frac{a}{W}\right)^2 + \left(\frac{a}{b}\right)^2}\; \right]$$

$$\lambda = \frac{M}{P\,W}$$

Tables 12.17 to 12.21 list h_1, h_2, h_3, and h_4 values for various λ.

TABLE 12.17

h factors for an edge cracked plate in plane strain subject to combined bending and tension with λ = 0.125 [6].

a/W:		n=1	n=2	n=3	n=5	n=7	n=10
0.125	h_1	4.761	4.544	3.881	2.632	1.734	0.905
	h_2	5.275	4.988	9.314	2.872	1.873	0.962
	h_3	0.394	0.779	0.925	0.898	0.748	0.488
	h_4	0.328	0.309	0.324	0.289	0.240	0.158
0.250	h_1	3.568	2.536	1.773	0.843	0.392	0.119
	h_2	4.510	3.193	2.195	1.006	0.445	0.138
	h_3	0.745	0.868	0.765	0.443	0.227	0.075
	h_4	0.889	0.682	0.548	0.321	0.165	0.054
0.375	h_1	2.742	1.657	1.016	0.373	0.136	0.029
	h_2	4.027	2.316	1.357	0.468	0.163	0.034
	h_3	0.961	0.787	0.539	0.204	0.073	0.015
	h_4	1.536	0.952	0.612	0.237	0.085	0.018
0.500	h_1	2.200	1.305	0.804	0.310	0.120	0.030
	h_2	3.767	1.977	1.131	0.399	0.149	0.036
	h_3	1.147	0.724	0.465	0.157	0.057	0.013
	h_4	2.116	1.192	0.714	0.259	0.097	0.023
0.625	h_1	1.738	1.056	0.678	0.290	0.129	0.039
	h_2	3.518	1.781	1.055	0.411	0.174	0.051
	h_3	1.275	0.707	0.441	0.162	0.067	0.019
	h_4	2.560	1.341	0.830	0.320	0.135	0.040
0.750	h_1	1.401	0.901	0.609	0.293	0.142	0.050
	h_2	3.373	1.727	1.073	0.459	0.214	0.073
	h_3	1.390	0.741	0.462	0.193	0.090	0.030
	h_4	2.881	1.513	0.946	0.402	0.188	0.064
0.875	h_1	1.168	0.785	0.551	0.283	0.153	0.060
	h_2	3.348	1.751	1.113	0.512	0.260	0.100
	h_3	1.532	0.808	0.514	0.235	0.119	0.046
	h_4	3.171	1.664	1.059	0.486	0.247	0.095

TABLE 12.18
h factors for an edge cracked plate in plane strain subject to combined bending and tension with λ = 0.0625 [6].

a/W:		n=1	n=2	n=3	n=5	n=7	n=10
0.125	h_1	1.910	1.781	1.494	1.101	0.865	0.692
	h_2	3.111	2.437	1.823	1.098	0.683	0.344
	h_3	0.279	0.353	0.364	0.294	0.310	0.428
	h_4	0.181	0.253	0.252	0.139	0.007	-0.063
0.250	h_1	2.014	1.943	1.714	1.253	0.782	0.471
	h_2	3.017	2.424	1.952	1.299	0.864	0.448
	h_3	0.516	0.618	0.601	0.503	0.398	0.251
	h_4	0.621	0.712	2.474	0.385	0.235	0.101
0.375	h_1	2.053	1.635	1.145	0.530	0.242	0.074
	h_2	3.069	2.119	1.387	0.592	0.257	0.076
	h_3	0.766	0.738	0.581	0.305	0.147	0.046
	h_4	1.200	0.946	0.604	0.215	0.083	0.022
0.500	h_1	1.985	1.232	0.751	0.277	0.156	0.023
	h_2	3.184	1.742	0.983	0.331	0.170	0.025
	h_3	1.002	0.707	0.440	0.157	0.079	0.012
	h_4	1.828	1.006	0.549	0.175	0.091	0.013
0.625	h_1	1.737	1.012	0.613	0.232	0.089	0.022
	h_2	3.209	1.589	0.880	0.305	0.113	0.027
	h_3	1.188	0.673	0.379	0.128	0.047	0.011
	h_4	2.354	1.176	0.643	0.222	0.083	0.020
0.750	h_1	1.443	0.886	0.572	0.250	0.113	0.035
	h_2	3.217	1.607	0.940	0.376	0.163	0.049
	h_3	1.341	0.710	0.408	0.161	0.069	0.021
	h_4	2.754	1.370	0.804	0.324	0.141	0.043
0.875	h_1	1.192	0.784	0.538	0.268	0.136	
	h_2	3.284	1.697	1.044	0.470	0.229	
	h_3	1.509	0.789	0.482	0.217	0.105	
	h_4	3.107	1.602	0.987	0.445	0.217	

TABLE 12.19
h factors for an edge cracked plate in plane strain subject to combined bending and tension with λ = - 0.125 [6].

a/W:		n=1	n=2	n=3	n=5	n=7	n=10
0.375	h_1	1.490	1.364	1.180	0.804	0.509	0.241
	h_2	2.296	1.646	1.276	0.786	0.472	0.213
	h_3	0.598	0.565	0.507	0.374	0.260	0.137
	h_4	0.928	0.798	0.628	0.327	0.166	0.057
0.500	h_1	1.723	1.235	0.816	0.346	0.147	0.042
	h_2	2.685	1.646	1.001	0.386	0.155	0.043
	h_3	0.869	0.689	0.469	0.210	0.091	0.029
	h_4	1.574	0.985	0.547	0.179	0.066	0.016
0.625	h_1	1.682	0.988	0.594	0.217	0.080	0.019
	h_2	2.949	1.491	0.825	0.279	0.100	0.024
	h_3	1.111	0.671	0.395	0.136	0.048	0.013
	h_4	2.178	1.093	0.571	0.185	0.067	0.015
0.750	h_1	1.456	0.874	0.553	0.231	0.099	0.029
	h_2	3.101	1.556	0.899	0.345	0.143	0.042
	h_3	1.303	0.724	0.412	0.154	0.063	0.019
	h_4	2.659	1.279	0.740	0.289	0.120	0.035
0.875	h_1	1.217	0.784	0.532	0.259	0.130	0.065
	h_2	3.259	1.694	1.032	0.454	0.219	0.104
	h_3	1.500	0.794	0.481	0.210	0.101	0.048
	h_4	3.081	1.585	0.971	0.429	0.207	0.099

TABLE 12.20

h factors for an edge cracked plate in plane strain subject to combined bending and tension with λ = - 0.1875 [6].

a/W:		n=1	n=2	n=3	n=5	n=7	n=10
0.500	h_1	1.246	1.148	0.973	0.598	0.350	0.157
	h_2	1.812	1.289	0.988	0.547	0.300	0.127
	h_3	0.636	0.521	0.438	0.285	0.179	0.084
	h_4	-1.203	-5.153	-2.895	-0.862	-0.238	-0.022
0.625	h_1	1.526	0.976	0.609	0.234	0.090	0.044
	h_2	2.471	1.352	0.777	0.272	0.100	0.023
	h_3	0.962	0.612	0.379	0.136	0.050	0.011
	h_4	-2.744	-0.601	0.045	0.132	0.059	0.014
0.750	h_1	1.456	0.855	0.525	0.204	0.081	0.021
	h_2	2.899	1.440	0.814	0.287	0.110	0.028
	h_3	1.234	0.685	0.387	0.129	0.049	0.012
	h_4	0.758	0.929	0.617	0.238	0.092	0.023
0.875	h_1	1.223	0.781	0.522	0.245	0.119	0.042
	h_2	3.181	1.645	0.991	0.420	0.196	0.066
	h_3	1.469	0.776	0.464	0.194	0.090	0.031
	h_4	2.671	1.516	0.926	0.396	0.185	0.063

TABLE 12.21

h factors for an edge cracked plate in plane strain subject to combined bending and tension with λ = - 0.250 [6].

a/W:		n=1	n=2	n=3	n=5	n=7	n=10
0.625	h_1	1.148	0.989	0.737	0.382	0.194	0.071
	h_2	1.521	1.074	0.736	0.348	0.168	0.059
	h_3	0.659	0.516	0.385	0.201	0.102	0.036
	h_4	-7.060	-3.118	-1.246	-0.133	0.024	0.023
0.750	h_1	1.405	0.828	0.497	0.182	0.067	0.015
	h_2	2.491	1.264	0.705	0.236	0.084	0.019
	h_3	1.091	0.612	0.352	0.112	0.039	0.009
	h_4	-1.008	0.419	0.426	0.182	0.067	0.015
0.875	h_1	1.445	0.977	0.703	0.375	0.207	0.087
	h_2	3.384	1.895	1.228	0.595	0.316	0.130
	h_3	1.572	0.902	0.580	0.277	0.147	0.060
	h_4	2.533	1.693	1.126	0.555	0.296	0.122

12.2 PART-THROUGH CRACKS–FLAT PLATES

TABLE 12.22
Stress intensity solution for a semielliptical surface flaw in a flat plate [8].

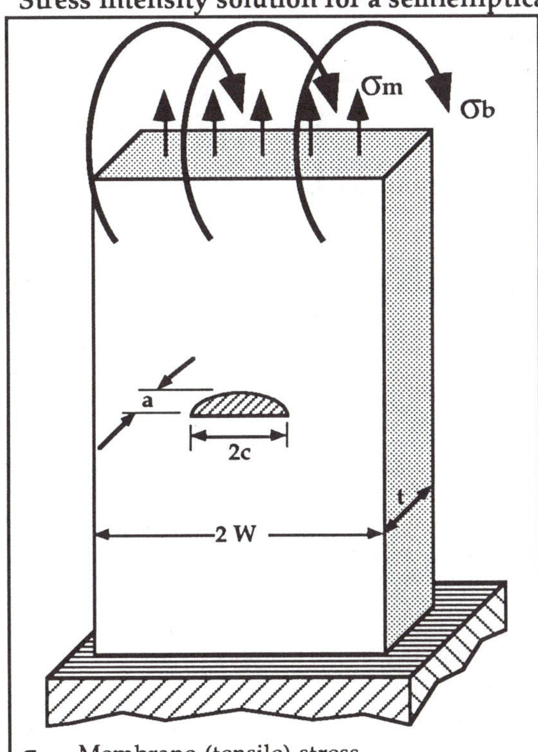

σ_m - Membrane (tensile) stress.

σ_b - Bending stress

$$= \frac{M\,t}{2\,I}$$

where

$$I = \frac{W\,t^3}{6}$$

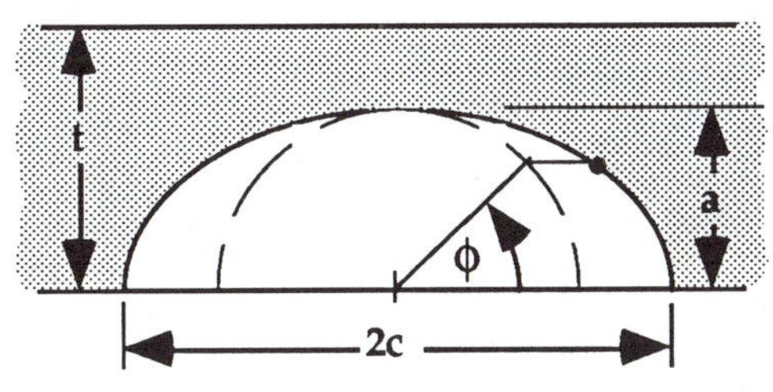

$$K_I = (\sigma_m + H\,\sigma_b)\sqrt{\frac{\pi a}{Q}}\;F\left(\frac{a}{t}, \frac{a}{c}, \frac{c}{W}, \phi\right)$$

where

$$Q = 1 + 1.464\left(\frac{a}{c}\right)^{1.65}$$

$$F = \left[M_1 + M_2\left(\frac{a}{t}\right)^2 + M_3\left(\frac{a}{t}\right)^4\right] f_\phi\, f_w\, g$$

$$M_1 = 1.13 - 0.09\left(\frac{a}{c}\right)$$

$$M_2 = -0.54 + \frac{0.89}{0.2 + \frac{a}{c}}$$

$$M_3 = 0.5 - \frac{1.0}{0.65 + \frac{a}{c}} + 14\left(1.0 - \frac{a}{c}\right)^{24}$$

$$f_\phi = \left[\left(\frac{a}{c}\right)^2 \cos^2\phi + \sin^2\phi\right]^{1/4}$$

$$f_w = \left[\sec\left(\frac{\pi c}{2W}\right)\sqrt{\frac{a}{t}}\right]^{1/2}$$

$$g = 1 + \left[0.1 + 0.35\left(\frac{a}{t}\right)^2\right](1 - \sin\phi)^2$$

$$H = H_1 + (H_2 - H_1)(\sin\phi)^p$$

$$p = 0.2 + \frac{a}{c} + 0.6\left(\frac{a}{t}\right)$$

$$H_1 = 1 - 0.34\frac{a}{t} - 0.11\frac{a}{c}\left(\frac{a}{t}\right)$$

$$H_2 = 1 + G_1\left(\frac{a}{t}\right) + G_2\left(\frac{a}{t}\right)^2$$

$$G_1 = -1.22 - 0.12\left(\frac{a}{c}\right)$$

$$G_2 = 0.55 - 1.05\left(\frac{a}{c}\right)^{0.75} + 0.47\left(\frac{a}{c}\right)^{1.5}$$

TABLE 12.23

Influence coefficients for a semi-elliptical surface crack in a flat plate, where the remote opening mode stress is fit to a cubic polynomial over the range $0 \le x \le a$ [9]. See Section 9.1.1 for an example of this approach.

		$\frac{a}{c} = 0.2$			$\frac{a}{c} = 0.4$			$\frac{a}{c} = 1.0$		
G_i:	$\frac{a}{t}$: / $\frac{2\phi}{\pi}$:	0.2	0.5	0.8	0.2	0.5	0.8	0.2	0.5	0.8
	0	0.611	0.816	1.262	0.784	0.965	1.283	1.150	1.247	1.400
G_o	0.25	0.748	0.967	1.382	0.818	0.979	1.222	1.076	1.148	1.233
	0.5	0.958	1.240	1.670	0.951	1.112	1.287	1.039	1.090	1.106
	0.75	1.090	1.432	1.840	1.051	1.220	1.372	1.025	1.068	1.090
	1	1.134	1.498	1.861	1.086	1.258	1.388	1.021	1.062	1.086
	0	0.080	0.145	0.275	0.127	0.185	0.275	0.200	0.229	0.268
G_1	0.25	0.208	0.278	0.400	0.248	0.301	0.371	0.362	0.384	0.406
	0.5	0.426	0.519	0.646	0.445	0.498	0.549	0.543	0.559	0.558
	0.75	0.609	0.726	0.866	0.612	0.670	0.728	0.671	0.686	0.701
	1	0.680	0.807	0.948	0.676	0.736	0.800	0.717	0.733	0.756
	0	0.023	0.055	0.113	0.044	0.073	0.112	0.075	0.089	0.104
G_2	0.25	0.076	0.110	0.165	0.098	0.124	0.155	0.154	0.165	0.174
	0.5	0.239	0.285	0.342	0.258	0.284	0.306	0.334	0.341	0.338
	0.75	0.432	0.491	0.561	0.443	0.472	0.504	0.514	0.522	0.531
	1	0.518	0.583	0.662	0.526	0.556	0.596	0.589	0.597	0.615
	0	0.010	0.029	0.060	0.022	0.038	0.059	0.038	0.046	0.054
G_3	0.25	0.032	0.052	0.082	0.045	0.060	0.077	0.076	0.082	0.086
	0.5	0.147	0.173	0.205	0.162	0.177	0.188	0.219	0.223	0.219
	0.75	0.334	0.369	0.411	0.348	0.364	0.384	0.417	0.421	0.427
	1	0.431	0.470	0.522	0.442	0.460	0.488	0.513	0.516	0.530

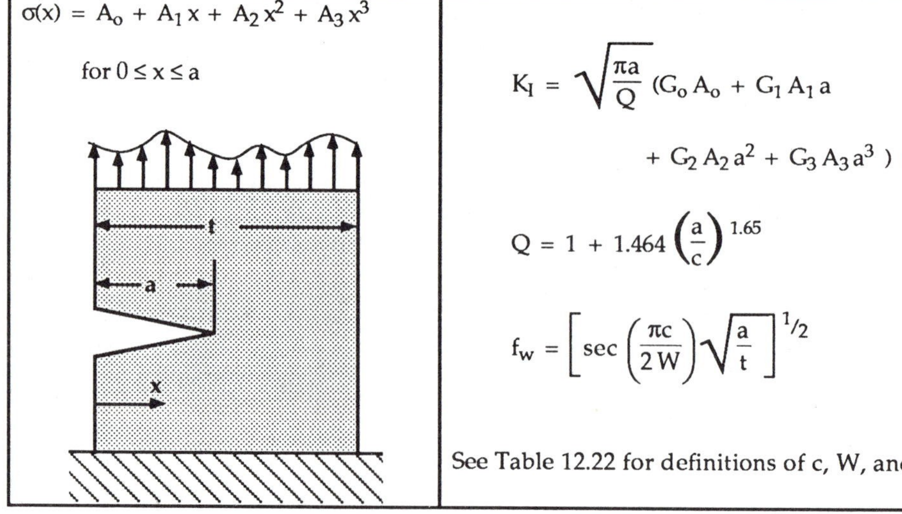

$$\sigma(x) = A_o + A_1 x + A_2 x^2 + A_3 x^3$$

for $0 \le x \le a$

$$K_I = \sqrt{\frac{\pi a}{Q}} \left(G_o A_o + G_1 A_1 a + G_2 A_2 a^2 + G_3 A_3 a^3 \right) f_w$$

$$Q = 1 + 1.464 \left(\frac{a}{c}\right)^{1.65}$$

$$f_w = \left[\sec\left(\frac{\pi c}{2W}\right) \sqrt{\frac{a}{t}} \right]^{1/2}$$

See Table 12.22 for definitions of c, W, and ϕ.

TABLE 12.24
Limit load solution for a semielliptical surface crack in a flat plate subject to combined bending and tension [10].

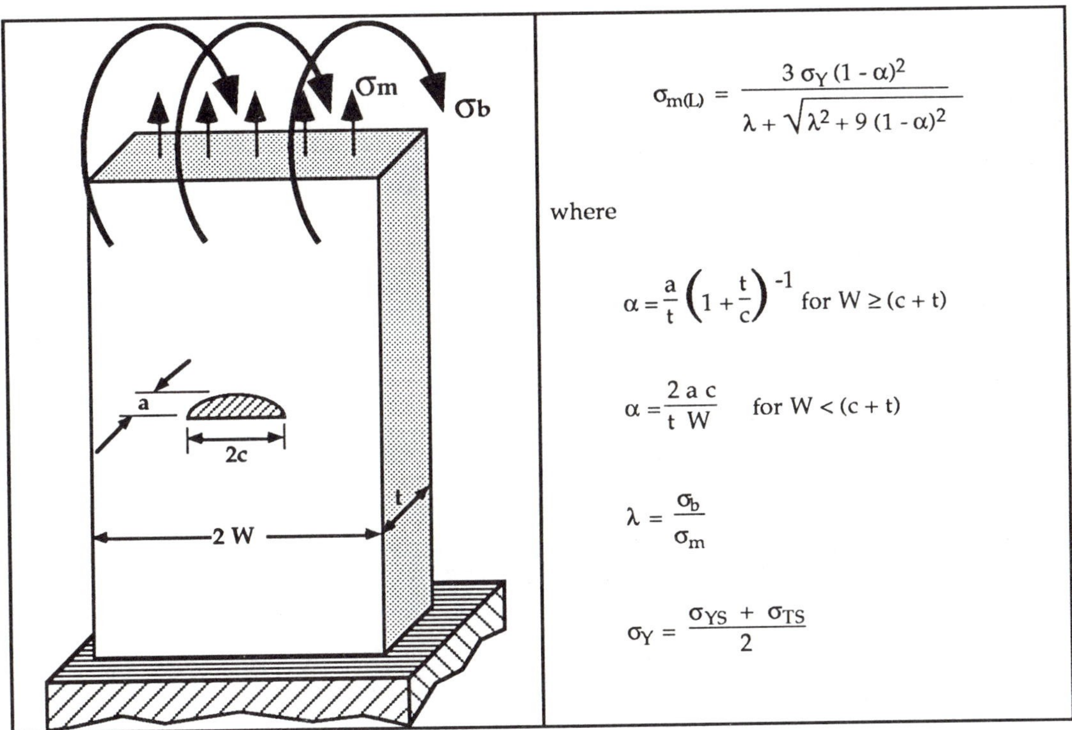

$$\sigma_{m(L)} = \frac{3\,\sigma_Y (1-\alpha)^2}{\lambda + \sqrt{\lambda^2 + 9(1-\alpha)^2}}$$

where

$$\alpha = \frac{a}{t}\left(1 + \frac{t}{c}\right)^{-1} \quad \text{for } W \geq (c+t)$$

$$\alpha = \frac{2\,a\,c}{t\,W} \quad \text{for } W < (c+t)$$

$$\lambda = \frac{\sigma_b}{\sigma_m}$$

$$\sigma_Y = \frac{\sigma_{YS} + \sigma_{TS}}{2}$$

12.3 FLAWED CYLINDERS

This section contains K_I, J, and displacement solutions for flawed cylinders. Part-through and through-wall flaw geometries are included, in both axial and circumferential orientations. Loading cases include uniform tension, bending, and internal pressure. Influence coefficients for part-through axial cracks are included, which enable the user to determine K_I for a wide range of loading.

12.3.1 Stress Intensity Factor

TABLE 12.25
Stress intensity solutions for circumferential through-wall flaws in cylinders [7].

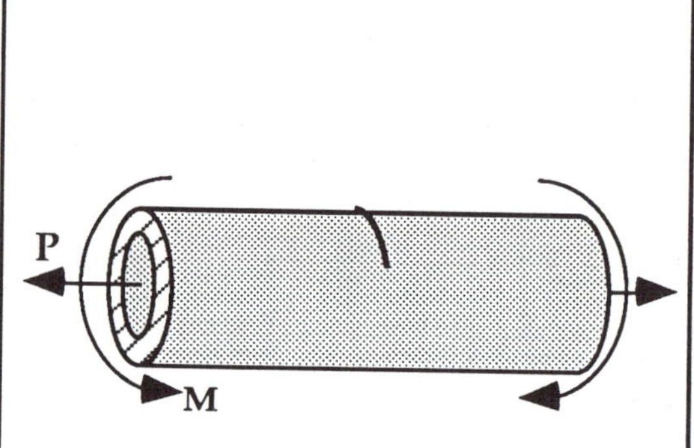

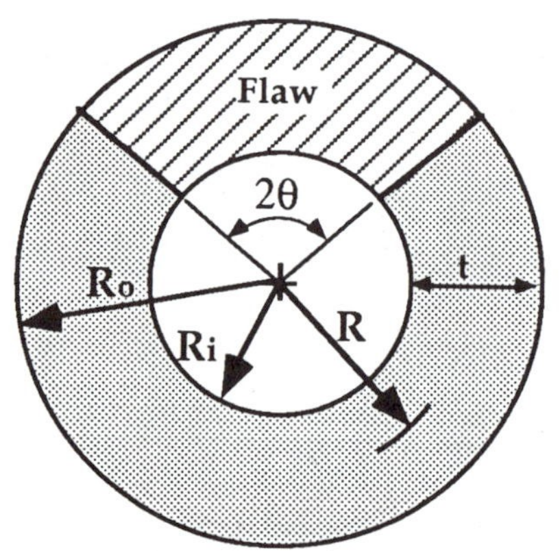

R - mean radius:

$$R = \frac{R_o + R_i}{2}$$

θ -crack half angle in radians

Axial Tension.

$$K_I = \sigma_t \sqrt{\pi R \theta}\ F_t$$

where $\sigma_t = \dfrac{P}{2\pi R t}$

$$F_t = 1 + A\left[5.3303\left(\frac{\theta}{\pi}\right)^{1.5} + 18.773\left(\frac{\theta}{\pi}\right)^{4.24}\right]$$

$$A = \left(0.125\frac{R}{t} - 0.25\right)^{0.25} \qquad \text{for } 5 \le \frac{R}{t} \le 10$$

$$A = \left(0.4\frac{R}{t} - 3.0\right)^{0.25} \qquad \text{for } 10 \le \frac{R}{t} \le 20$$

Bending Moment.

$$K_I = \sigma_b \sqrt{\pi R \theta}\ F_b$$

where $\sigma_b = \dfrac{M}{\pi R^2 t}$

$$F_b = 1 + A\left[4.5967\left(\frac{\theta}{\pi}\right)^{1.5} + 2.6422\left(\frac{\theta}{\pi}\right)^{4.24}\right]$$

where A is as defined above for the pure tension case.

Internal Pressure.

$$K_I = \sigma_m \sqrt{\pi R \theta}\ F_m$$

where $\sigma_m = \dfrac{pR}{2t}$

$$F_m = 1 + 0.1501\,\gamma^{1.5} \qquad \text{for } \gamma \le 2$$

$$F_m = 0.8875 + 0.2625\,\gamma \qquad \text{for } 2 \le \gamma \le 5$$

$$\gamma = \theta\sqrt{\frac{R}{t}}$$

TABLE 12.26
Stress intensity solutions for part-through internal circumferential flaws in cylinders subject to uniform tension [11]

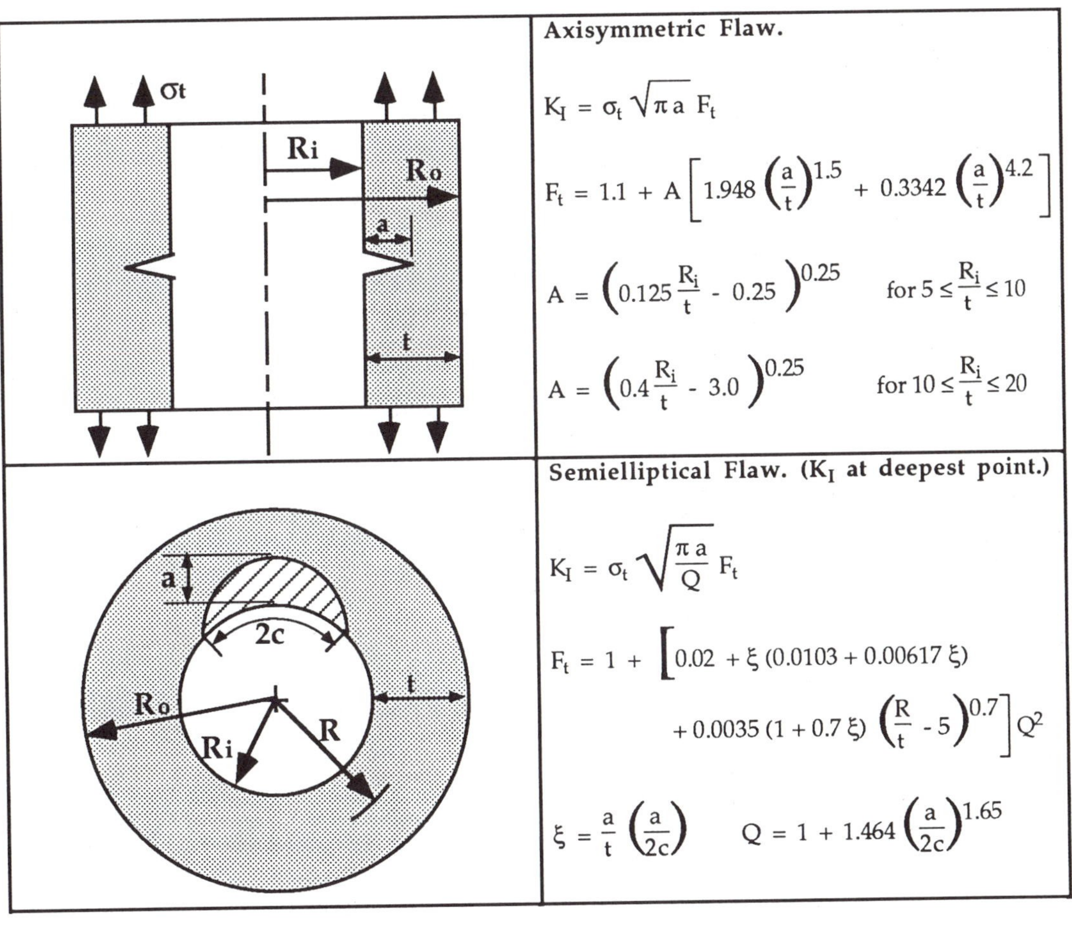

Axisymmetric Flaw.

$$K_I = \sigma_t \sqrt{\pi a}\, F_t$$

$$F_t = 1.1 + A\left[1.948\left(\frac{a}{t}\right)^{1.5} + 0.3342\left(\frac{a}{t}\right)^{4.2}\right]$$

$$A = \left(0.125\frac{R_i}{t} - 0.25\right)^{0.25} \qquad \text{for } 5 \le \frac{R_i}{t} \le 10$$

$$A = \left(0.4\frac{R_i}{t} - 3.0\right)^{0.25} \qquad \text{for } 10 \le \frac{R_i}{t} \le 20$$

Semielliptical Flaw. (K_I at deepest point.)

$$K_I = \sigma_t \sqrt{\frac{\pi a}{Q}}\, F_t$$

$$F_t = 1 + \left[0.02 + \xi\,(0.0103 + 0.00617\,\xi) + 0.0035\,(1 + 0.7\,\xi)\left(\frac{R}{t} - 5\right)^{0.7}\right]Q^2$$

$$\xi = \frac{a}{t}\left(\frac{a}{2c}\right) \qquad Q = 1 + 1.464\left(\frac{a}{2c}\right)^{1.65}$$

TABLE 12.27

Stress intensity factors for axial flaws in cylinders subject to internal pressure [1,11].

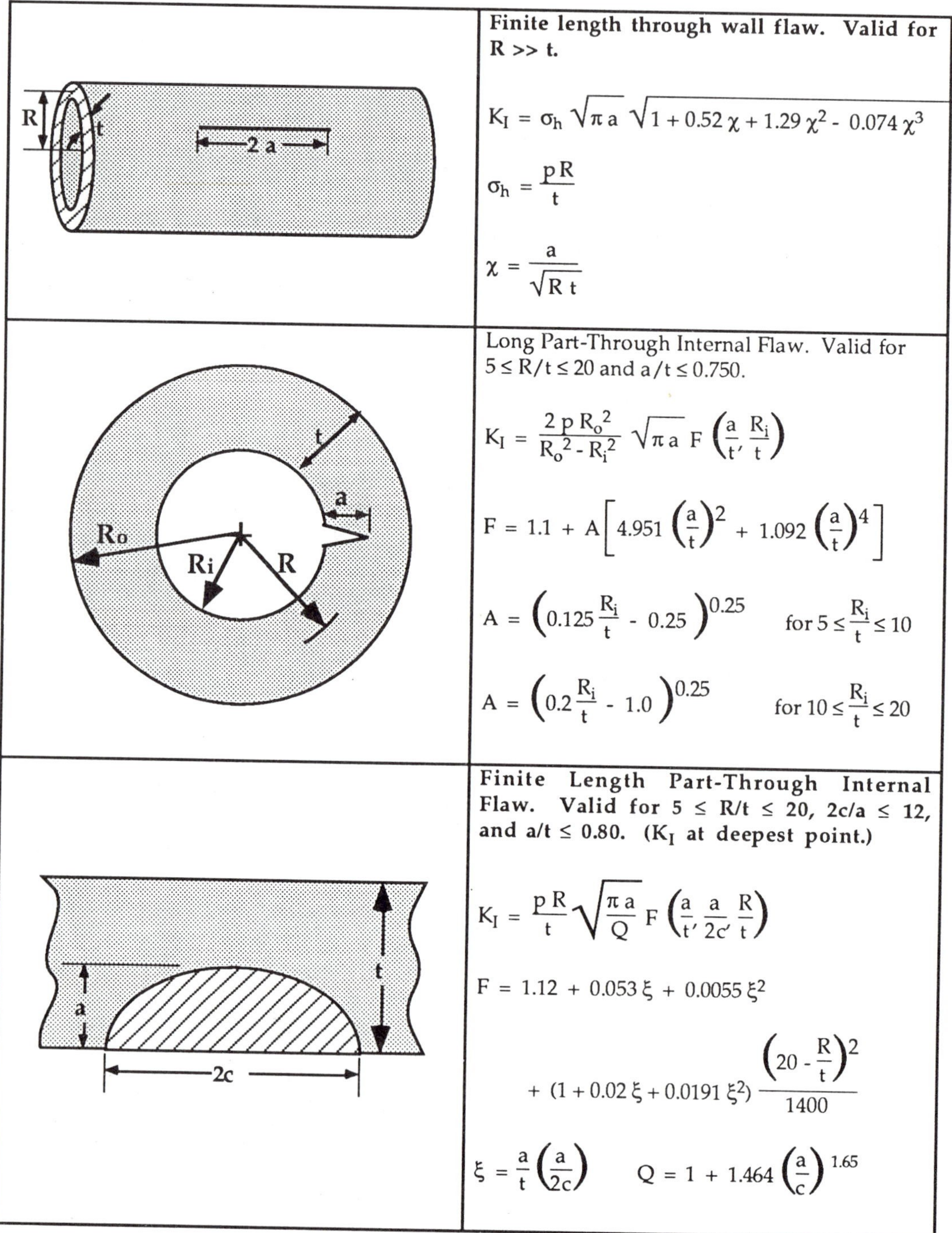

Geometry	Solution
	Finite length through wall flaw. Valid for R >> t. $K_I = \sigma_h \sqrt{\pi a} \sqrt{1 + 0.52\,\chi + 1.29\,\chi^2 - 0.074\,\chi^3}$ $\sigma_h = \frac{p R}{t}$ $\chi = \frac{a}{\sqrt{R t}}$
	Long Part-Through Internal Flaw. Valid for $5 \le R/t \le 20$ and $a/t \le 0.750$. $K_I = \frac{2 p R_o^2}{R_o^2 - R_i^2} \sqrt{\pi a}\, F\left(\frac{a}{t}, \frac{R_i}{t}\right)$ $F = 1.1 + A\left[4.951\left(\frac{a}{t}\right)^2 + 1.092\left(\frac{a}{t}\right)^4\right]$ $A = \left(0.125\frac{R_i}{t} - 0.25\right)^{0.25}$ for $5 \le \frac{R_i}{t} \le 10$ $A = \left(0.2\frac{R_i}{t} - 1.0\right)^{0.25}$ for $10 \le \frac{R_i}{t} \le 20$
	Finite Length Part-Through Internal Flaw. Valid for $5 \le R/t \le 20$, $2c/a \le 12$, and $a/t \le 0.80$. (K_I at deepest point.) $K_I = \frac{p R}{t}\sqrt{\frac{\pi a}{Q}}\, F\left(\frac{a}{t}, \frac{a}{2c}, \frac{R}{t}\right)$ $F = 1.12 + 0.053\,\xi + 0.0055\,\xi^2 + (1 + 0.02\,\xi + 0.0191\,\xi^2)\frac{\left(20 - \frac{R}{t}\right)^2}{1400}$ $\xi = \frac{a}{t}\left(\frac{a}{2c}\right)$ $\quad Q = 1 + 1.464\left(\frac{a}{c}\right)^{1.65}$

TABLE 12.28

Influence coefficients for a semi-elliptical surface crack on the inside of a cylinder with $t/R_i = 0.10$ [9]. See Section 9.1.1 for an example of this approach.

		$\frac{a}{c} = 0.2$			$\frac{a}{c} = 0.4$			$\frac{a}{c} = 1.0$		
	$\frac{a}{t}$:	0.2	0.5	0.8	0.2	0.5	0.8	0.2	0.5	0.8
G_i:	$\frac{2\phi}{\pi}$:									
G_o	0	0.607	0.791	1.179	0.777	0.936	1.219	1.140	1.219	1.348
	0.25	0.740	0.932	1.284	0.810	0.948	1.164	1.068	1.126	1.200
	0.5	0.945	1.188	1.568	0.940	1.076	1.243	1.033	1.074	1.091
	0.75	1.073	1.366	1.798	1.038	1.180	1.357	1.019	1.055	1.090
	1	1.115	1.427	1.872	1.072	1.217	1.393	1.015	1.050	1.090
G_1	0	0.079	0.138	0.253	0.125	0.176	0.259	0.197	0.221	0.255
	0.25	0.206	0.268	0.374	0.246	0.291	0.356	0.359	0.377	0.397
	0.5	0.422	0.503	0.619	0.442	0.487	0.538	0.541	0.554	0.555
	0.75	0.603	0.705	0.859	0.608	0.657	0.727	0.669	0.683	0.703
	1	0.673	0.783	0.960	0.672	0.723	0.806	0.715	0.729	0.760
G_2	0	0.023	0.052	0.104	0.043	0.069	0.106	0.074	0.085	0.099
	0.25	0.075	0.105	0.154	0.097	0.119	0.149	0.153	0.162	0.170
	0.5	0.237	0.277	0.331	0.256	0.279	0.302	0.333	0.339	0.337
	0.75	0.429	0.480	0.560	0.441	0.466	0.505	0.514	0.520	0.533
	1	0.514	0.571	0.671	0.523	0.549	0.601	0.588	0.596	0.618
G_3	0	0.010	0.027	0.056	0.021	0.036	0.056	0.038	0.044	0.051
	0.25	0.032	0.049	0.077	0.044	0.058	0.074	0.075	0.080	0.085
	0.5	0.146	0.169	0.199	0.161	0.174	0.187	0.218	0.222	0.219
	0.75	0.332	0.363	0.412	0.346	0.360	0.385	0.417	0.420	0.429
	1	0.438	0.462	0.529	0.441	0.456	0.493	0.512	0.515	0.532

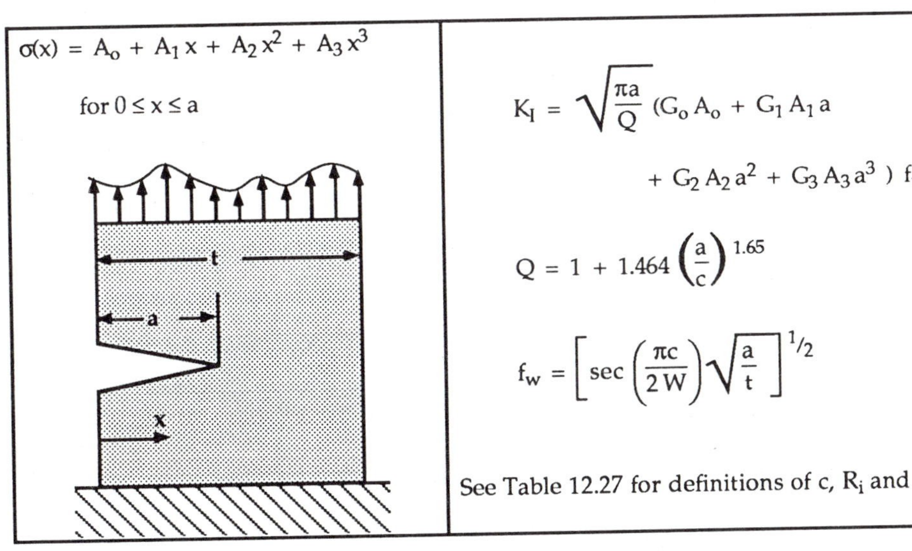

$$\sigma(x) = A_o + A_1 x + A_2 x^2 + A_3 x^3$$

for $0 \le x \le a$

$$K_I = \sqrt{\frac{\pi a}{Q}} \left(G_o A_o + G_1 A_1 a + G_2 A_2 a^2 + G_3 A_3 a^3 \right) f_w$$

$$Q = 1 + 1.464 \left(\frac{a}{c} \right)^{1.65}$$

$$f_w = \left[\sec \left(\frac{\pi c}{2W} \right) \sqrt{\frac{a}{t}} \right]^{1/2}$$

See Table 12.27 for definitions of c, R_i and ϕ.

TABLE 12.29

Influence coefficients for a semi-elliptical surface crack on the inside of a cylinder with $t/R_i = 0.25$ [9]. See Section 9.1.1 for an example of this approach.

		$\frac{a}{c} = 0.2$			$\frac{a}{c} = 0.4$			$\frac{a}{c} = 1.0$		
G_i:	$\frac{a}{t}$: / $\frac{2\phi}{\pi}$:	0.2	0.5	0.8	0.2	0.5	0.8	0.2	0.5	0.8
G_o	0	0.606	0.797	1.201	0.770	0.924	1.219	1.128	1.191	1.316
	0.25	0.736	0.925	1.270	0.801	0.932	1.154	1.058	1.105	1.180
	0.5	0.935	1.170	1.549	0.928	1.056	1.241	1.025	1.060	1.088
	0.75	1.057	1.343	1.838	1.024	1.157	1.385	1.013	1.045	1.099
	1	1.097	1.405	1.959	1.057	1.193	1.443	1.009	1.041	1.105
G_1	0	0.079	0.141	0.262	0.123	0.174	0.263	0.194	0.214	0.248
	0.25	0.205	0.268	0.372	0.243	0.287	0.356	0.356	0.371	0.393
	0.5	0.419	0.498	0.615	0.438	0.481	0.540	0.538	0.550	0.556
	0.75	0.598	0.698	0.876	0.603	0.650	0.740	0.667	0.680	0.708
	1	0.666	0.776	0.996	0.666	0.715	0.828	0.713	0.726	0.768
G_2	0	0.023	0.054	0.108	0.042	0.068	0.109	0.072	0.082	0.097
	0.25	0.075	0.106	0.154	0.096	0.118	0.150	0.152	0.159	0.169
	0.5	0.236	0.275	0.330	0.254	0.276	0.304	0.332	0.338	0.339
	0.75	0.426	0.477	0.571	0.439	0.462	0.513	0.512	0.519	0.537
	1	0.511	0.567	0.692	0.520	0.545	0.614	0.588	0.594	0.623
G_3	0	0.010	0.028	0.059	0.021	0.036	0.059	0.037	0.043	0.050
	0.25	0.032	0.050	0.077	0.044	0.057	0.075	0.075	0.079	0.085
	0.5	0.145	0.168	0.199	0.160	0.173	0.188	0.217	0.221	0.220
	0.75	0.330	0.361	0.419	0.345	0.358	0.391	0.416	0.419	0.431
	1	0.426	0.460	0.542	0.439	0.454	0.509	0.511	0.515	0.536

$$\sigma(x) = A_o + A_1 x + A_2 x^2 + A_3 x^3$$

for $0 \le x \le a$

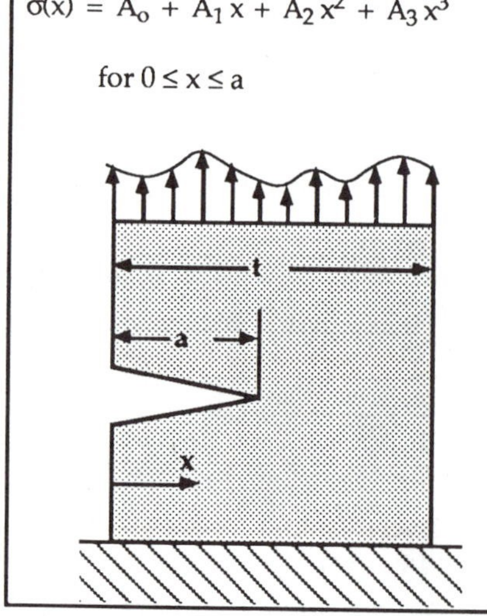

$$K_I = \sqrt{\frac{\pi a}{Q}}\,(G_o A_o + G_1 A_1 a + G_2 A_2 a^2 + G_3 A_3 a^3)\, f_w$$

$$Q = 1 + 1.464 \left(\frac{a}{c}\right)^{1.65}$$

$$f_w = \left[\sec\left(\frac{\pi c}{2W}\right)\sqrt{\frac{a}{t}}\right]^{1/2}$$

See Table 12.27 for definitions of c, R_i and ϕ.

TABLE 12.30

Influence coefficients for a semi-elliptical surface crack on the outside of a cylinder with $t/R_i = 0.10$ [9]. See Section 9.1.1 for an example of this approach.

		$\frac{a}{c} = 0.2$			$\frac{a}{c} = 0.4$			$\frac{a}{c} = 1.0$		
G_i:	$\frac{a}{t}$: / $\frac{2\phi}{\pi}$:	0.2	0.5	0.8	0.2	0.5	0.8	0.2	0.5	0.8
G_o	0	0.612	0.806	1.262	0.788	0.984	1.378	1.156	1.266	1.453
	0.25	0.750	0.968	1.432	0.823	1.002	1.325	1.082	1.165	1.278
	0.5	0.965	1.272	1.867	0.958	1.147	1.425	1.044	1.106	1.144
	0.75	1.102	1.502	2.208	1.061	1.267	1.541	1.029	1.083	1.125
	1	1.147	1.584	2.298	1.096	1.310	1.565	1.025	1.078	1.118
G_1	0	0.080	0.142	0.277	0.128	0.192	0.309	0.202	0.236	0.286
	0.25	0.208	0.279	0.419	0.250	0.309	0.406	0.363	0.390	0.421
	0.5	0.428	0.530	0.715	0.448	0.511	0.595	0.544	0.565	0.570
	0.75	0.614	0.752	0.993	0.616	0.687	0.784	0.673	0.692	0.712
	1	0.685	0.839	1.099	0.680	0.755	0.858	0.718	0.738	0.765
G_2	0	0.023	0.053	0.114	0.045	0.076	0.129	0.076	0.092	0.113
	0.25	0.076	0.110	0.175	0.099	0.128	0.173	0.155	0.168	0.181
	0.5	0.240	0.290	0.377	0.259	0.290	0.329	0.335	0.344	0.344
	0.75	0.434	0.504	0.626	0.445	0.481	0.531	0.515	0.524	0.536
	1	0.521	0.600	0.739	0.528	0.565	0.625	0.590	0.600	0.619
G_3	0	0.010	0.028	0.062	0.022	0.040	0.070	0.039	0.048	0.059
	0.25	0.032	0.052	0.088	0.046	0.063	0.088	0.077	0.084	0.091
	0.5	0.147	0.177	0.226	0.163	0.181	0.202	0.219	0.224	0.222
	0.75	0.335	0.378	0.450	0.349	0.370	0.400	0.418	0.422	0.430
	1	0.432	0.480	0.568	0.444	0.466	0.505	0.513	0.518	0.533

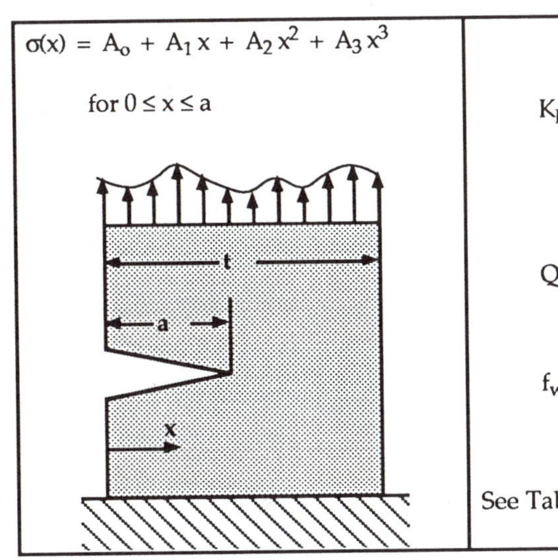

$$K_I = \sqrt{\frac{\pi a}{Q}}\,(G_o A_o + G_1 A_1 a + G_2 A_2 a^2 + G_3 A_3 a^3)\, f_w$$

$$Q = 1 + 1.464 \left(\frac{a}{c}\right)^{1.65}$$

$$f_w = \left[\sec\left(\frac{\pi c}{2W}\right)\sqrt{\frac{a}{t}}\right]^{1/2}$$

See Table 12.27 for definitions of c, R_i and ϕ.

TABLE 12.31

Influence coefficients for a semi-elliptical surface crack on the outside of a cylinder with $t/R_i = 0.25$ [9]. See Section 9.1.1 for an example of this approach.

		$\frac{a}{c} = 0.2$			$\frac{a}{c} = 0.4$			$\frac{a}{c} = 1.0$		
G_i:	$\frac{a}{t}$: / $\frac{2\phi}{\pi}$:	0.2	0.5	0.8	0.2	0.5	0.8	0.2	0.5	0.8
G_o	0	0.612	0.786	1.160	0.793	0.994	1.400	1.163	1.286	1.498
	0.25	0.752	0.952	1.346	0.828	1.016	1.365	1.088	1.184	1.320
	0.5	0.972	1.278	1.860	0.967	1.175	1.513	1.049	1.123	1.183
	0.75	1.114	1.541	2.344	1.072	1.311	1.682	1.034	1.100	1.163
	1	1.162	1.640	2.510	1.109	1.360	1.727	1.030	1.094	1.156
G_1	0	0.080	0.134	0.242	0.130	0.195	0.318	0.204	0.243	0.302
	0.25	0.209	0.272	0.389	0.252	0.315	0.421	0.365	0.396	0.435
	0.5	0.430	0.532	0.713	0.451	0.521	0.626	0.546	0.570	0.583
	0.75	0.618	0.767	1.044	0.620	0.702	0.833	0.674	0.698	0.724
	1	0.691	0.861	1.178	0.685	0.773	0.914	0.720	0.743	0.777
G_2	0	0.023	0.049	0.097	0.045	0.078	0.134	0.077	0.096	0.122
	0.25	0.076	0.106	0.159	0.100	0.130	0.180	0.156	0.171	0.188
	0.5	0.241	0.291	0.376	0.261	0.295	0.345	0.336	0.347	0.350
	0.75	0.437	0.513	0.654	0.447	0.489	0.556	0.516	0.527	0.542
	1	0.524	0.613	0.782	0.530	0.575	0.653	0.591	0.603	0.625
G_3	0	0.010	0.025	0.051	0.022	0.041	0.073	0.040	0.051	0.064
	0.25	0.032	0.050	0.079	0.046	0.064	0.093	0.077	0.086	0.095
	0.5	0.148	0.177	0.225	0.164	0.184	0.212	0.220	0.226	0.226
	0.75	0.337	0.383	0.468	0.350	0.375	0.416	0.418	0.424	0.433
	1	0.434	0.488	0.596	0.445	0.472	0.523	0.513	0.520	0.536

$\sigma(x) = A_o + A_1 x + A_2 x^2 + A_3 x^3$

for $0 \le x \le a$

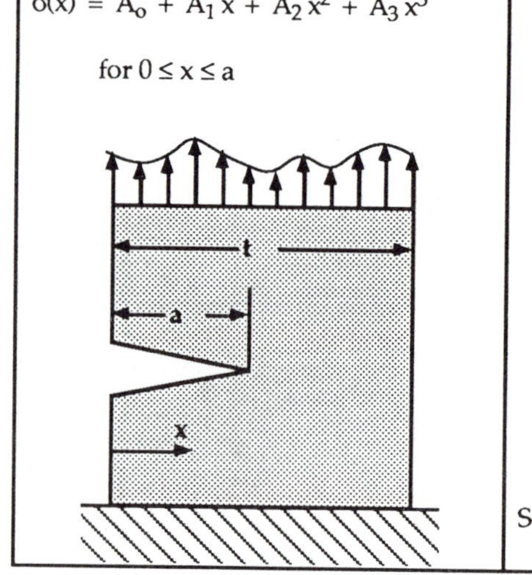

$$K_I = \sqrt{\frac{\pi a}{Q}}\,(G_o A_o + G_1 A_1 a + G_2 A_2 a^2 + G_3 A_3 a^3)\, f_w$$

$$Q = 1 + 1.464 \left(\frac{a}{c}\right)^{1.65}$$

$$f_w = \left[\sec\left(\frac{\pi c}{2W}\right)\sqrt{\frac{a}{t}}\right]^{1/2}$$

See Table 12.27 for definitions of c, R_i and ϕ.

12.3.2 Limit Load

TABLE 12.32
Limit load solutions for circumferential through-wall flaws in cylinders [7].

Axial Tension.

$$P_L = \sigma_Y (2 R_o t - t^2) [2 \cos^{-1}(0.5 \cos \theta) - \theta]$$

$$\sigma_Y = \frac{\sigma_{YS} + \sigma_{TS}}{2}$$

Bending Moment.

$$M_L = 4 \sigma_Y R_o^2 t \left(1 - \xi + \frac{\xi^2}{3}\right)(\cos \omega - 0.5 \sin \theta)$$

$$\xi = \frac{t}{R_o} \qquad \omega = \frac{0.5\, \theta (1 - \xi)(1 + 0.5\, \xi)}{(1 - \xi)(1 - 0.5\, \xi)}$$

Combined Tension and Bending.

$$M_L = 4 \sigma_Y R_o^2 t \left(1 - \xi + \frac{\xi^2}{3}\right)(\cos \omega - 0.5 \sin \theta)$$

$$\xi = \frac{t}{R_o} \qquad \omega = \frac{0.5\, \theta (1 - \xi)(1 + 0.5\, \xi)}{(1 - \xi)(1 - 0.5\, \xi)} + \frac{P}{4 \sigma_Y R_o t (1 - \xi)}$$

12.3.3 Fully Plastic J and Displacement

The pages that follow give fully plastic J and displacement solutions for several configurations of flawed cylinders. See Sections 9.5 and 12.1.3 for the appropriate elastic-plastic estimation formulae.

TABLE 12.32

Fully plastic J and crack mouth opening displacement for an axially cracked cylinder under internal pressure [5].

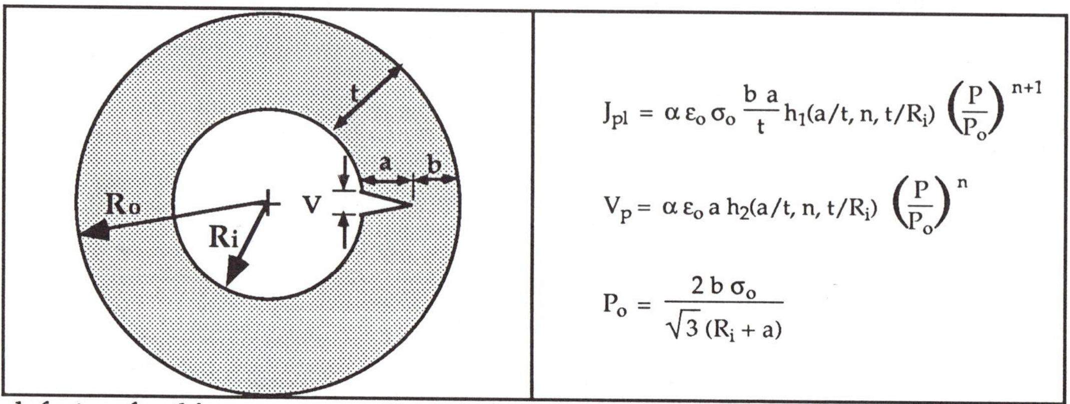

h factors for this geometry are listed in Tables 12.34 to 12.36.

TABLE 12.33

Fully plastic J and displacement for a circumferentially cracked cylinder in tension [5].

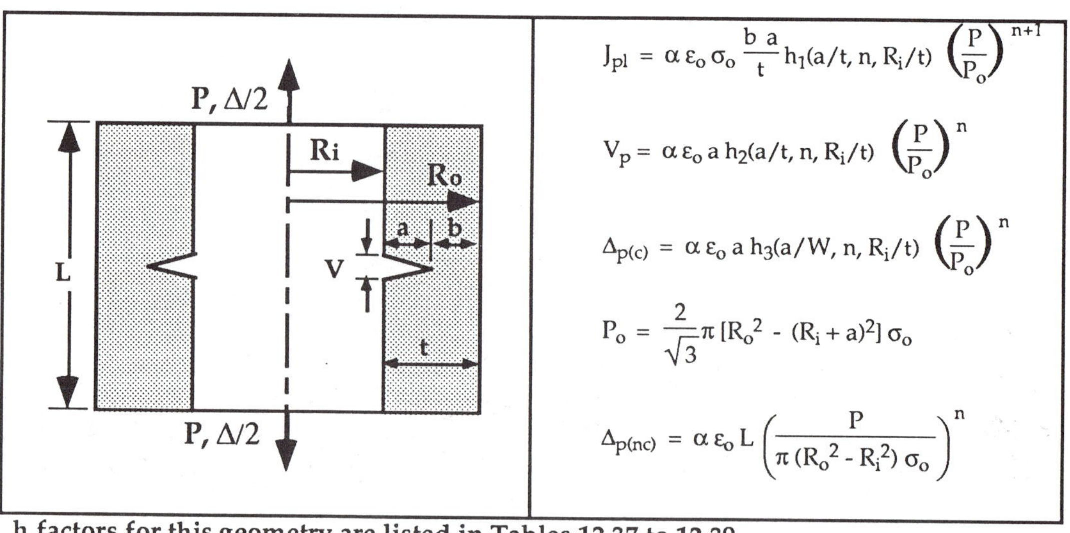

h factors for this geometry are listed in Tables 12.37 to 12.39.

TABLE 12.34
h factors for an axially cracked cylinder under internal pressure with $t/R_i = 0.20$ [5].

a/t:		n=1	n=2	n=3	n=5	n=7	n=10
0.125	h_1	6.32	7.93	9.32	11.5	13.12	14.94
	h_2	5.83	7.01	7.96	9.49	10.67	11.96
0.250	h_1	7.00	8.34	9.03	9.59	9.71	9.45
	h_2	5.92	8.72	7.07	7.26	7.14	6.71
0.500	h_1	9.79	10.37	9.07	5.61	3.52	2.11
	h_2	7.05	6.97	6.01	3.70	2.28	1.25
0.750	h_1	11.00	5.54	2.84	1.24	0.83	0.493
	h_2	7.35	3.86	0.186	0.556	0.261	0.129

TABLE 12.35
h factors for an axially cracked cylinder under internal pressure with $t/R_i = 0.10$. [5]

a/t:		n=1	n=2	n=3	n=5	n=7	n=10
0.125	h_1	5.22	6.64	7.59	8.76	9.34	9.55
	h_2	5.31	6.25	6.88	7.65	8.02	8..09
0.250	h_1	6.16	7.49	7.96	8.08	7.78	6.98
	h_2	5.56	6.31	6.52	6.40	6.01	5.27
0.500	h_1	10.5	11.6	10.7	6.47	3.95	2.27
	h_2	7.48	7.72	7.01	4.29	2.58	1.37
0.750	h_1	16.1	8.19	3.87	1.46	1.05	0.787
	h_2	9.57	5.40	2.57	0.706	0.370	0.232

TABLE 12.36
h factors for an axially cracked cylinder under internal pressure with $t/R_i = 0.10$ [5].

a/t:		n=1	n=2	n=3	n=5	n=7	n = 10
0.125	h_1	4.50	5.79	6.62	7.65	8.07	7.75
	h_2	4.96	5.71	6.20	6.82	7.02	6.66
0.250	h_1	5.57	6.91	7.37	7.47	7.21	6.53
	h_2	5.29	5.98	6.16	6.01	5.63	4.93
0.500	h_1	10.8	12.8	12.8	8.16	4.88	2.62
	h_2	7.66	8.33	8.13	5.33	3.20	1.65
0.750	h_1	23.1	13.1	5.87	1.90	1.23	0.883
	h_2	12.1	7.88	3.84	1.01	0.454	0.240

TABLE 12.37
h factors for a circumferentially cracked cylinder in tension with $t/R_i = 0.20$ [5].

a/t:		n=1	n=2	n=3	n=5	n=7	n = 10
	h_1	3.78	5.00	5.94	7.54	8.99	11.1
0.125	h_2	4.56	5.55	6.37	7.79	9.10	11.0
	h_3	0.369	0.700	1.07	1.96	3.04	4.94
	h_1	3.88	4.95	5.64	6.49	6.94	7.22
0.250	h_2	4.40	5.12	5.57	6.07	6.28	6.30
	h_3	0.673	1.25	1.79	2.79	3.61	4.52
	h_1	4.40	4.78	4.59	3.79	3.07	2.34
0.500	h_2	4.36	4.30	3.91	3.00	2.26	1.55
	h_3	1.33	1.93	2.21	2.23	1.94	1.46
	h_1	4.12	3.03	2.23	1.546	1.30	1.11
0.750	h_2	3.46	2.19	1.36	0.638	0.436	0.325
	h_3	1.54	1.39	1.04	0.686	0.508	0.366

TABLE 12.38
h factors for a circumferentially cracked cylinder in tension with $t/R_i = 0.10$ [5].

a/t:		n=1	n=2	n=3	n=5	n=7	n=10
0.125	h_1	4.00	5.13	6.09	7.69	9.09	11.1
	h_2	4.71	5.63	6.45	7.85	9.09	10.9
	h_3	0.548	0.733	1.13	2.07	3.16	5.07
0.250	h_1	4.17	5.35	6.09	6.93	7.30	7.41
	h_2	4.58	5.36	5.84	6.31	6.44	6.31
	h_3	0.757	1.35	1.93	2.96	3.78	4.60
0.500	h_1	5.40	5.90	5.63	4.51	3.49	2.47
	h_2	4.99	5.01	4.59	3.48	2.56	1.67
	h_3	1.555	2.26	2.59	2.57	2.18	1.56
0.750	h_1	5.18	3.78	2.57	1.59	1.31	1.10
	h_2	4.22	2.79	1.67	0.725	0.48	0.300
	h_3	1.86	1.73	1.26	0.775	0.561	0.360

TABLE 12.39
h factors for a circumferentially cracked cylinder in tension with $t/R_i = 0.05$ [5].

a/t:		n=1	n=2	n=3	n=5	n=7	n=10
0.125	h_1	4.04	5.23	6.22	7.82	9.19	11.1
	h_2	4.82	5.69	6.52	7.90	9.11	10.8
	h_3	0.680	0.759	1.17	2.13	3.23	5.12
0.250	h_1	4.38	5.68	6.45	7.29	7.62	7.65
	h_2	4.71	5.56	6.05	6.51	6.59	6.39
	h_3	0.818	1.43	2.03	3.10	3.91	4.69
0.500	h_1	6.55	7.17	6.89	5.46	4.13	2.77
	h_2	5.67	5.77	5.36	4.08	2.97	1.88
	h_3	1.80	2.59	2.99	2.98	2.50	1.74
0.750	h_1	6.64	4.87	3.08	1.68	1.30	1.07
	h_2	5.18	3.57	2.07	0.808	0.472	0.316
	h_3	2.36	2.18	1.53	0.772	0.494	0.330

TABLE 12.40
Fully plastic J integral for circumferential through-wall flaws in cylinders [7].

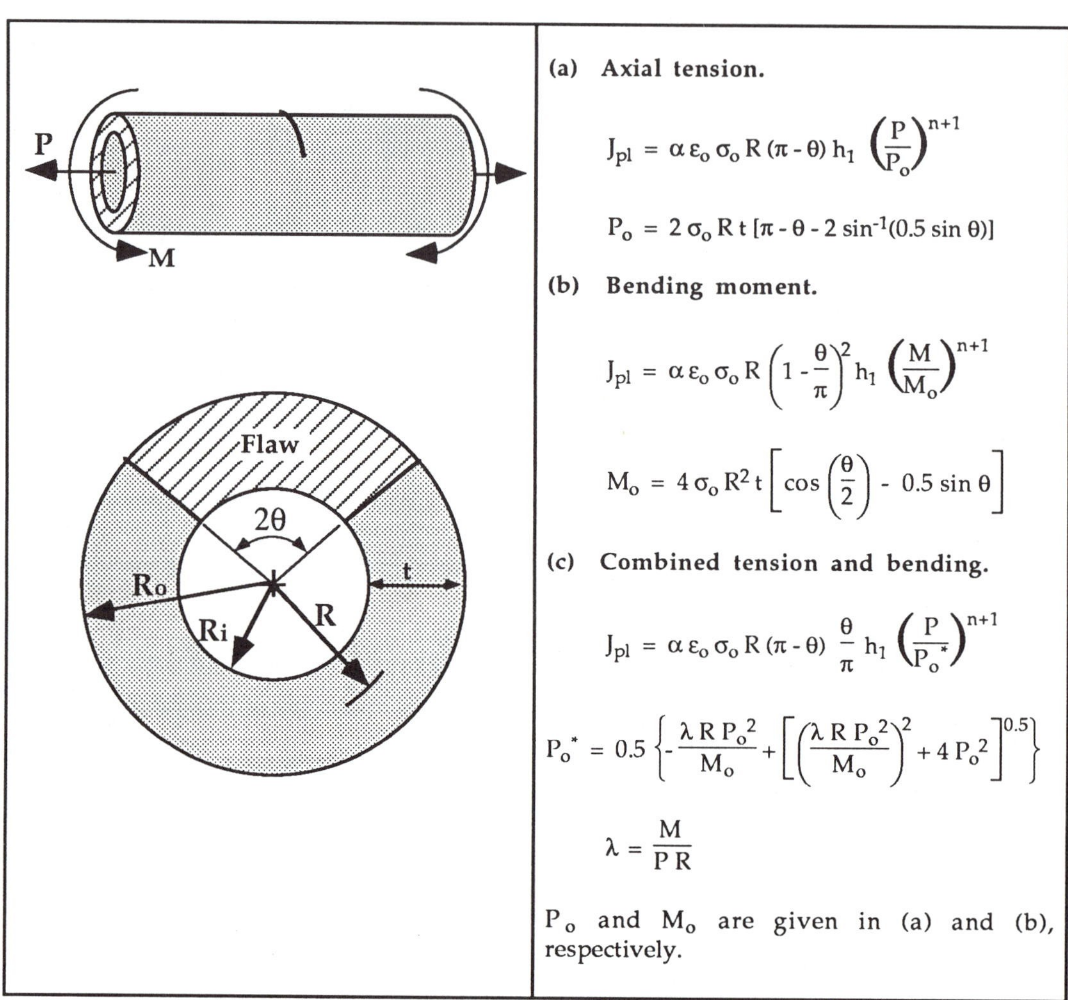

(a) **Axial tension.**

$$J_{pl} = \alpha \varepsilon_o \sigma_o R (\pi - \theta) h_1 \left(\frac{P}{P_o}\right)^{n+1}$$

$$P_o = 2 \sigma_o R t \left[\pi - \theta - 2 \sin^{-1}(0.5 \sin \theta)\right]$$

(b) **Bending moment.**

$$J_{pl} = \alpha \varepsilon_o \sigma_o R \left(1 - \frac{\theta}{\pi}\right)^2 h_1 \left(\frac{M}{M_o}\right)^{n+1}$$

$$M_o = 4 \sigma_o R^2 t \left[\cos\left(\frac{\theta}{2}\right) - 0.5 \sin \theta\right]$$

(c) **Combined tension and bending.**

$$J_{pl} = \alpha \varepsilon_o \sigma_o R (\pi - \theta) \frac{\theta}{\pi} h_1 \left(\frac{P}{P_o^*}\right)^{n+1}$$

$$P_o^* = 0.5 \left\{ -\frac{\lambda R P_o^2}{M_o} + \left[\left(\frac{\lambda R P_o^2}{M_o}\right)^2 + 4 P_o^2\right]^{0.5} \right\}$$

$$\lambda = \frac{M}{P R}$$

P_o and M_o are given in (a) and (b), respectively.

h_1 factors for this geometry are listed in Tables 12.41 to 12.51.

TABLE 12.41
h_1 factors for through-wall flaws in cylinders in uniform tension, R/t = 5 [7].

θ/π	n= 1	n=2	n= 3	n= 5	n= 7
0.000	0.000	0.000	0.000	0.000	0.000
0.063	0.177	0.230	0.265	0.307	0.326
0.100	0.280	0.355	0.400	0.433	0.425
0.125	0.352	0.437	0.479	0.500	0.478
0.150	0.427	0.520	0.560	0.554	0.494
0.175	0.504	0.600	0.630	0.580	0.492
0.200	0.578	0.674	0.695	0.595	0.485
0.225	0.652	0.740	0.742	0.596	0.477
0.250	0.731	0.804	0.765	0.596	0.464
0.275	0.793	0.840	0.786	0.595	0.455
0.300	0.858	0.850	0.795	0.590	0.440
0.325	0.918	0.853	0.793	0.580	0.421
0.350	0.976	0.856	0.786	0.562	0.404
0.375	1.025	0.859	0.778	0.545	0.387
0.400	1.075	0.862	0.760	0.525	0.370
0.425	1.117	0.865	0.743	0.506	0.348
0.450	1.154	0.868	0.720	0.480	0.330
0.475	1.184	0.873	0.696	0.454	0.308
0.500	1.224	0.879	0.669	0.426	0.288

TABLE 12.42
h_1 factors for through-wall flaws in cylinders in uniform tension, R/t = 10 [7].

θ/π	n= 1	n= 2	n= 3	n= 5	n= 7
0.000	0.000	0.000	0.000	0.000	0.000
0.063	0.186	0.248	0.291	0.323	0.382
0.100	0.330	0.415	0.472	0.518	0.595
0.125	0.403	0.520	0.589	0.645	0.638
0.150	0.530	0.620	0.683	0.730	0.660
0.175	0.626	0.720	0.790	0.777	0.666
0.200	0.723	0.822	0.884	0.800	0.670
0.225	0.824	0.930	0.970	0.808	0.661
0.250	0.919	1.040	1.008	0.810	0.652
0.275	1.005	1.073	1.030	0.804	0.635
0.300	1.084	1.094	1.037	0.794	0.615
0.325	1.163	1.100	1.035	0.778	0.597
0.350	1.235	1.105	1.021	0.759	0.573
0.375	1.300	1.107	1.007	0.734	0.550
0.400	1.360	1.110	0.982	0.704	0.524
0.425	1.420	1.110	0.959	0.674	0.496
0.450	1.460	1.110	0.925	0.640	0.465
0.475	1.505	1.110	0.890	0.603	0.439
0.500	1.546	1.110	0.857	0.568	0.408

TABLE 12.43
h_1 factors for through-wall flaws in cylinders in uniform tension, R/t = 20 [7].

θ/π	n= 1	n= 2	n= 3	n= 5	n= 7
0.000	0.000	0.000	0.000	0.000	0.000
0.063	0.204	0.281	0.339	0.426	0.480
0.100	0.368	0.504	0.584	0.684	0.710
0.125	0.489	0.657	0.767	0.873	0.801
0.150	0.630	0.800	0.928	1.000	0.880
0.175	0.770	0.945	1.100	1.050	0.900
0.200	0.920	1.085	1.241	1.080	0.907
0.225	1.065	1.240	1.310	1.099	0.902
0.275	1.340	1.460	1.370	1.088	0.875
0.300	1.448	1.498	1.375	1.075	0.855
0.325	1.548	1.496	1.364	1.049	0.824
0.350	1.640	1.488	1.350	1.019	0.800
0.375	1.728	1.479	1.328	0.984	0.760
0.400	1.800	1.470	1.294	0.945	0.720
0.425	1.868	1.464	1.260	0.904	0.684
0.450	1.930	1.458	1.220	0.855	0.650
0.475	1.970	1.452	1.176	0.804	0.606
0.500	2.020	1.445	1.121	0.758	0.564

TABLE 12.44
h_1 factors for through-wall flaws in cylinders in bending, R/t = 5 [7].

θ/π	n= 1	n= 2	n= 3	n= 5	n= 7	n= 10	n= 20
0.000	0.000	0.000	0.000	0.000	0.000	0.000	0.000
0.063	0.313	0.369	0.403	0.433	0.432	0.431	0.431
0.100	0.520	0.560	0.559	0.558	0.557	0.536	0.496
0.125	0.664	0.708	0.713	0.675	0.611	0.577	0.529
0.150	0.823	0.836	0.848	0.760	0.638	0.582	0.510
0.175	0.983	0.960	0.960	0.802	0.664	0.586	0.490
0.200	1.142	1.080	1.056	0.880	0.691	0.591	0.471
0.250	1.461	1.333	1.198	0.941	0.744	0.600	0.432
0.300	1.760	1.526	1.328	0.990	0.754	0.604	0.428
0.350	2.048	1.720	1.420	1.016	0.760	0.608	0.425
0.400	2.320	1.872	1.488	1.036	0.774	0.612	0.421
0.450	2.576	1.990	1.556	1.040	0.780	0.616	0.418
0.500	2.795	2.059	1.598	1.048	0.794	0.620	0.414

TABLE 12.45
h_1 factors for through-wall flaws in cylinders in bending, R/t = 10 [7].

θ/π	n = 1	n = 2	n = 3	n = 5	n = 7	n = 10	n = 20
0.000	0.000	0.000	0.000	0.000	0.000	0.000	0.000
0.063	0.333	0.401	0.450	0.508	0.531	0.549	0.570
0.100	0.600	0.760	0.770	0.788	0.750	0.710	0.715
0.125	0.766	0.855	0.897	0.902	0.857	0.816	0.777
0.150	0.987	1.040	1.080	1.072	0.930	0.834	0.759
0.175	1.209	1.202	1.240	1.160	0.988	0.852	0.742
0.200	1.430	1.400	1.400	1.244	1.020	0.876	0.724
0.250	1.873	1.771	1.629	1.323	1.080	0.907	0.688
0.300	2.280	2.003	1.760	1.343	1.100	0.889	0.640
0.350	2.680	2.260	1.970	1.364	1.100	0.880	0.592
0.400	3.040	2.440	1.880	1.383	1.080	0.840	0.524
0.450	3.360	2.572	1.970	1.400	1.052	0.800	0.472
0.500	3.646	2.682	2.105	1.424	1.035	0.760	0.392

TABLE 12.46
h_1 factors for through-wall flaws in cylinders in bending, R/t = 20 [7].

θ/π	n = 1	n = 2	n = 3	n = 5	n = 7	n = 10	n = 20
0.000	0.000	0.000	0.000	0.000	0.000	0.000	0.000
0.063	0365	0.456	0.527	0.631	0.696	0.757	0.864
0.100	0.650	0.850	0.938	0.985	1.025	1.010	1.010
0.125	0.933	1.091	1.187	1.241	1.208	1.177	1.143
0.150	1.242	1.350	1.415	1.463	1.271	1.195	1.114
0.175	1.552	1.600	1.650	1.610	1.334	1.213	1.085
0.200	1.861	1.863	1.865	1.708	1.397	1.231	1.056
0.250	2.480	2.391	2.224	1.834	1.524	1.267	1.010
0.300	2.960	2.700	2.435	1.857	1.539	1.277	1.005
0.350	3.450	3.020	2.575	1.886	1.560	1.285	1.003
0.400	3.950	3.260	2.670	1.904	1.570	1.299	1.002
0.450	4.400	3.445	2.750	1.925	1.595	1.303	1.001
0.500	4.859	3.571	2.821	1.950	1.600	1.320	1.000

TABLE 12.47

h_1 factors for through-wall flaws in cylinders in combined tension and bending, R/t = 10, θ/π = 0.0625 [7].

$\lambda/(1+\lambda)$	n = 2	n = 5	n = 7	n = 10
0.00	3.967	5.567	6.104	6.510
0.05	4.313	6.500	6.500	7.969
0.10	4.736	7.375	7.080	9.721
0.15	5.125	8.250	7.875	11.250
0.20	5.614	9.080	8.787	12.937
0.25	6.000	9.750	9.875	14.250
0.30	6.438	10.501	11.078	15.463
0.35	6.789	11.000	12.125	16.375
0.40	7.140	11.457	13.188	17.063
0.45	7.500	11.875	14.000	17.500
0.50	7.901	12.150	14.610	17.839
0.55	8.094	12.313	15.000	17.550
0.60	8.287	12.236	15.130	17.241
0.65	8.344	11.938	14.875	16.375
0.70	8.257	11.642	14.408	15.500
0.75	8.125	11.125	13.625	14.500
0.80	7.811	10.617	12.729	13.366
0.85	7.500	9.875	11.688	12.125
0.90	7.063	9.190	10.447	10.738
0.95	6.563	8.500	9.250	9.313
1.00	6.018	7.620	8.160	7.928

TABLE 12.47
h_1 factors for through-wall flaws in cylinders in combined tension and bending, R/t = 10, θ/π = 0.125 [7].

$\lambda/(1+\lambda)$	n = 2	n = 5	n = 7	n = 10
0.00	4.157	5.163	5.102	4.750
0.05	4.625	5.750	5.719	5.250
0.10	5.085	6.331	6.331	5.982
0.15	5.531	6.875	6.906	6.813
0.20	5.956	7.323	7.518	7.688
0.25	6.375	7.875	8.000	8.563
0.30	6.750	8.309	8.500	9.416
0.35	7.125	8.688	9.000	10.000
0.40	7.500	9.008	9.381	10.543
0.45	7.719	9.313	9.630	10.813
0.50	7.930	9.500	9.875	10.935
0.55	8.000	9.625	10.000	10.844
0.60	8.064	9.566	10.011	10.567
0.65	8.063	9.406	9.813	10.219
0.70	7.926	9.165	9.495	9.702
0.75	7.750	8.813	9.000	9.125
0.80	7.435	8.390	8.535	8.578
0.85	7.125	7.938	7.875	7.844
0.90	6.772	7.391	7.313	7.158
0.95	6.375	6.875	6.625	6.375
1.00	5.987	6.311	5.996	5.688

TABLE 12.49

h_1 factors for through-wall flaws in cylinders in combined tension and bending, R/t = 10, θ/π = 0.250 [7].

λ/(1+λ)	n= 2	n= 5	n= 7	n= 10
0.00	4.159	3.238	2.605	3.000
0.05	4.522	3.550	3.000	3.225
0.10	4.885	3.988	3.490	3.589
0.15	5.250	4.400	4.050	4.000
0.20	5.614	4.882	4.491	4.540
0.25	6.000	5.400	5.200	5.050
0.30	6.347	5.908	5.816	5.700
0.35	6.650	6.350	6.350	6.400
0.40	6.900	6.800	6.800	7.264
0.45	7.150	7.100	7.150	7.707
0.50	7.363	7.333	7.375	8.150
0.55	7.500	7.450	7.500	8.150
0.60	7.508	7.397	7.397	7.900
0.65	7.450	7.250	7.075	7.500
0.70	7.307	7.018	6.688	7.018
0.75	7.100	6.619	6.150	6.300
0.80	6.885	6.220	5.641	5.670
0.85	6.519	5.700	5.000	4.950
0.90	6.152	5.108	4.372	4.353
0.95	5.700	4.550	3.800	3.700
1.00	5.312	3.969	3.240	3.125

TABLE 12.50

h_1 factors for through-wall flaws in cylinders in combined tension and bending, R/t = 10, θ/π = 0.375 [7].

$\lambda/(1+\lambda)$	n = 2	n = 5	n = 7	n = 10
0.00	2.892	1.992	1.496	2.000
0.05	3.315	2.000	1.700	2.100
0.10	3.739	2.044	1.894	2.243
0.15	4.164	2.213	2.094	2.342
0.20	4.589	2.441	2.294	2.441
0.25	4.957	2.651	2.484	2.558
0.30	5.325	2.862	2.675	2.675
0.35	5.651	3.091	2.791	2.708
0.40	5.977	3.233	2.906	2.740
0.45	6.237	3.285	2.945	2.750
0.50	6.497	3.336	2.985	2.739
0.55	6.586	3.300	2.883	2.650
0.60	6.674	3.200	2.781	2.447
0.65	6.536	3.045	2.629	2.260
0.70	6.399	2.890	2.477	2.064
0.75	6.093	2.646	2.251	1.856
0.80	5.786	2.401	2.025	1.649
0.85	5.360	2.120	1.767	1.444
0.90	4.934	1.838	1.509	1.238
0.95	4.459	1.603	1.301	1.049
1.00	3.984	1.368	1.092	0.860

TABLE 12.51

h_1 factors for through-wall flaws in cylinders in combined tension and bending, R/t = 10, θ/π = 0.500 [7].

$\lambda/(1+\lambda)$	n = 2	n = 5	n = 7	n = 10
0.00	2.220	1.137	0.816	1.500
0.05	2.300	1.300	0.950	1.525
0.10	2.525	1.496	1.122	1.570
0.15	2.850	1.700	1.350	1.650
0.20	3.320	1.953	1.611	1.757
0.25	3.784	2.200	1.800	1.800
0.30	4.247	2.475	2.050	1.900
0.35	4.800	2.700	2.300	2.000
0.40	5.230	2.900	2.438	2.050
0.45	5.600	3.100	2.575	2.100
0.50	5.830	3.160	2.550	2.107
0.55	5.900	3.200	2.525	2.100
0.60	5.850	3.170	2.447	2.000
0.65	5.700	3.050	2.297	1.900
0.70	5.367	2.890	2.147	1.734
0.75	5.000	2.700	2.000	1.600
0.80	4.542	2.430	1.823	1.446
0.85	4.100	2.183	1.589	1.200
0.90	3.600	1.935	1.354	1.083
0.95	3.100	1.700	1.200	0.900
1.00	2.682	1.424	1.008	0.760

REFERENCES

1. Tada, H., Paris, P.C., and Irwin, G.R. *The Stress Analysis of Cracks Handbook.* (2nd Ed.) Paris Productions, Inc., St. Louis, 1985.

2. E 399-83, "Standard Test Method for Plane-Strain Fracture Toughness of Metallic Materials." American Society for Testing and Materials, Philadelphia, 1983.

3. Towers, O.L., "Stress Intensity Factors, Compliances, and Elastic η Factors for Six Test Geometries." Report 136/1981, The Welding Institute, Abington, UK, 1981.

4. E 813-87, "Standard Test Method for J_{Ic}, a Measure of Fracture Toughness." American Society for Testing and Materials, Philadelphia, 1987.

5. Kumar, V., German, M.D., and Shih, C.F.,"An Engineering Approach for Elastic-Plastic Fracture Analysis." EPRI Report NP-1931, Electric Power Research Institute, Palo Alto, CA, 1981.

6. Kumar, V., German, M.D., Wilkening, W.W., Andrews, W.R., deLorenzi, H.G., and Mowbray, D.F., "Advances in Elastic-Plastic Fracture Analysis." EPRI Report NP-3607, Electric Power Research Institute, Palo Alto, CA, 1984.

7. Zahoor, A. "Ductile Fracture Handbook, Volume 1: Circumferential Throughwall Cracks." EPRI Report NP-6301-D, Electric Power Research Institute, Palo Alto, CA, 1989.

8. Newman, J.C., Jr. and Raju, I.S., "An Empirical Stress Intensity Factor Equation for Surface Cracks." *Engineering Fracture Mechanics,* Vol 15, 1981, pp. 185-192.

9. Raju, I.S. and Newman J.C., Jr., "Stress-Intensity Factors for Internal and External Surface Cracks in Cylindrical Vessels." *Journal of Pressure Vessel Technology,* Vol. 104, 1972, pp. 293-298.

10. Miller, A.G., "Review of Limit Loads of Structures Containing Defects." *International Journal of Pressure Vessels and Piping,* Vol. 32, 1988, pp. 197-327

11. Zahoor, A., "Closed Form Expressions for Fracture Mechanics Analysis of Cracked Pipes." *Journal of Pressure Vessel Technology,* Vol. 107, 1985, pp. 203-205.

13. PRACTICE PROBLEMS

This chapter contains practice problems that correspond to material in Chapters 1 to 11. Some of the problems for Chapters 7 to 10 require a computer program or spreadsheet macro. This level of complexity was necessary in order to make the application-oriented problems realistic.

All quantitative data are given in SI units, although the corresponding values in English units are also provided in many cases.

13.1 CHAPTER 1

1.1 Compile a list of five mechanical or structural failures that have occurred within the last 20 years. Describe the factors that led to each failure and identify the failures that resulted from misapplication of existing knowledge (Type 1) and those that involved new technology or a significant design modification (Type 2).

1.2 A flat plate with a through-thickness crack (Fig. 1.8) is subject to a 100 MPa (14.5 ksi) tensile stress and has a fracture toughness (K_{IC}) of 50.0 MPa $\sqrt{m}$ (45.5 ksi $\sqrt{in}$). Determine the critical crack length for this plate, assuming the material is linear elastic.

1.3 Compute the critical energy release rate ($\mathcal{G}_c$) of the material in the previous problem for E = 207,000 MPa (30,000 ksi).

1.4 Suppose that you plan to drop a bomb out of an airplane and that you are interested in the time of flight before it hits the ground, but you cannot remember the appropriate equation from your undergraduate physics course. Your decide to infer a relationship for time of flight of a falling object by experimentation. You reason that the time of flight, t, must depend on the height above the ground, h, and the weight of the object, mg, where m is the mass and g is the gravitational acceleration. Therefore, neglecting aerodynamic drag, the time of flight is given by the following function:

$$t = f(h, m, g)$$

Apply dimensional analysis to this equation and determine how many experiments would be required to determine f to a reasonable approximation, assuming you know the numerical value of g. Does the time of flight depend on the mass of the object?

13.2 CHAPTER 2

2.1 According to Eq. (2.25), the energy required to increase the crack area a unit amount is equal to *twice* the fracture work per unit surface area, w_f. Why is the factor of 2 in this equation necessary?

2.2 Derive Eq. (2.30) for both load control and displacement control by substituting Eq. (2.29) into Eqs. (2.27) and (2.28), respectively.

2.3 Figure 2.10 illustrates that the driving force is linear for a through-thickness crack in an infinite plate when the stress is fixed. Suppose that a remote displacement (rather than load) were fixed in this configuration. Would the driving force curves be altered? Explain. (Hint: see Section 2.5.3).

2.4 A plate 2W wide contains a centrally located crack 2a long and is subject to a tensile load, P. Beginning with Eq. (2.24), derive an expression for the elastic compliance, C $(= \Delta/P)$ in terms of the plate dimensions and elastic modulus, E. The stress in Eq. (2.24) is the nominal value; i.e., $\sigma = P/2BW$ in this problem. (Note: Eq. (2.24) only applies when $a \ll W$; the expression you derive is only approximate for a finite width plate.)

2.5 A material exhibits the following crack growth resistance behavior:

$$R = 6.95\,(a - a_o)^{0.5}$$

where a_o is the initial crack size. R has units of kJ/m^2 and crack size is in millimeters. Alternatively,

$$R = 200\,(a - a_o)^{0.5}$$

where R has units of $in\text{-}lb/in^2$ and crack size is in inches. The elastic modulus of this material = 207,000 MPa (30,000 ksi). Consider a wide plate with a through crack $(a \ll W)$ that is made from this material.

(a) If this plate fractures at 138 MPa (20.0 ksi), compute the following:

(i) The half crack size at failure (a_c).

(ii) The amount of stable crack growth (at each crack tip) that precedes failure (a_c - a_o).

(b) If this plate has an initial crack length ($2a_o$) of 50.8 mm (2.0 in) and the plate is loaded to failure, compute the following:

(i) The stress at failure.

(ii) The half crack size at failure.

(iii) The stable crack growth at each crack tip.

2.6 Example 2.3 showed that the energy release rate, $\mathcal{G}$, of the double cantilever beam (DCB) specimen increases with crack growth when the specimen is held at a constant load. Describe (qualitatively) how you could alter the design of the DCB specimen such that a growing crack in load control would experience a constant $\mathcal{G}$.

2.7 Beginning with Eq. (2.20), derive an expression for the potential energy of a plate subject to a tensile stress σ with a penny-shaped flaw of radius a. Assume that a << plate dimensions.

2.8 Beginning with Eq. (2.20), derive expressions for the energy release rate and Mode I stress intensity factor of a penny-shaped flaw subject to a remote tensile stress. (Your K_I expression should be identical to Eq. (2.44).)

2.9 Calculate K_I for a rectangular bar containing an edge crack loaded in three point bending.

P = 35.0 kN (7870 lb); W = 50.8 mm (2.0 in); B = 25 mm (1.0 in); a/W = 0.2; S = 203 mm (8.0 in).

2.10 A large block of steel is loaded to a stress of 345 MPa (50 ksi). If the fracture toughness is 44 MPa $\sqrt{m}$ (40 ksi $\sqrt{in}$), determine the critical radius of a penny-shaped crack.

2.11 A semicircular surface crack in a pressure vessel is 10 mm (0.394 in) deep. The crack is on the inner wall of the pressure vessel and is oriented such that the hoop stress is perpendicular to the crack plane. Calculate K_I if the local hoop stress = 200 MPa (29.0 ksi) and the internal pressure = 20 MPa (2900 psi). Assume that the wall thickness >> 10 mm.

2.12 Calculate K_I for a semielliptical surface flaw at $\phi = 0°, 30°, 60°, 90°$.

$\sigma = 150$ MPa (21.8 ksi); $a = 8.00$ mm (0.315 in); $2c = 40$ mm (1.57 in).

2.13 Consider a plate subject to biaxial tension with a through crack of length 2a, oriented at an angle β from the σ_2 axis (Fig. 13.1). Derive expressions for K_I and K_{II} for this configuration. What happens to each K expression when $\sigma_1 = \sigma_2$?

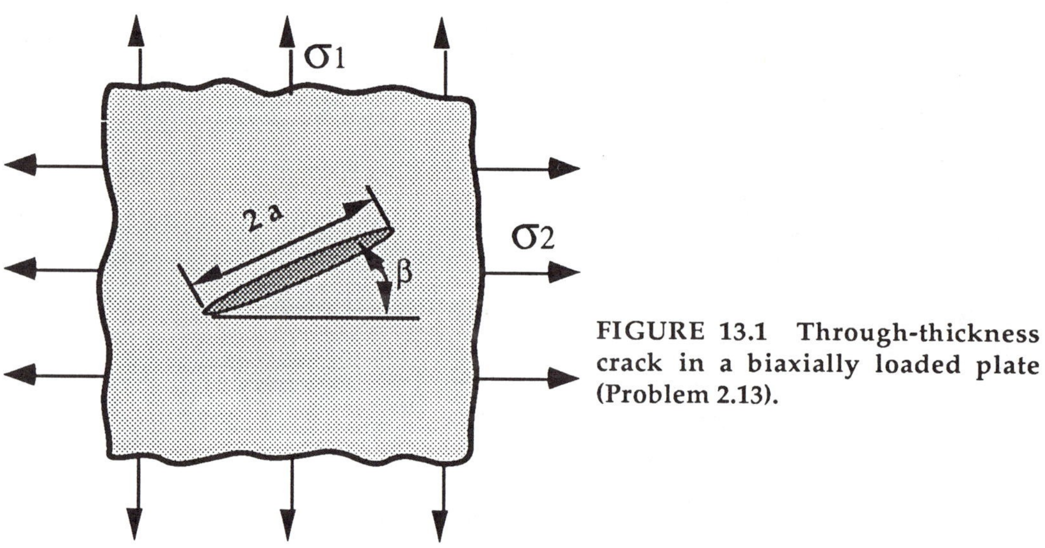

FIGURE 13.1 Through-thickness crack in a biaxially loaded plate (Problem 2.13).

2.14 Calculate K_{eff} (Irwin correction) for a through crack in a plate of width 2W (Fig. 2.20(b)). Assume plane stress conditions and the following stress intensity relationship:

$$K_{eff} = \sigma\sqrt{\pi a_{eff}}\left[\sec\left(\frac{\pi a_{eff}}{2W}\right)\right]^{1/2}$$

$\sigma = 250$ MPa (36.3 ksi); $\sigma_{YS} = 350$ MPa (50.8 ksi); $2W = 203$ mm (8.0 in); $2a = 50.8$ mm (2.0 in).

2.15 For an infinite plate with a through crack 50.8 mm (2.0 in) long, compute and tabulate K_{eff} v. stress using the three methods indicated below. Assume σ_{YS} = 250 MPa (36.3 ksi).

Stress, MPa (ksi)	K_{eff}, MPa $\sqrt{m}$ or ksi $\sqrt{in}$		
	LEFM	Irwin Correction	Strip Yield Model
25 (3.63)			
50 (7.25)			
100 (14.5)			
150 (21.8)			
200 (29.0)			
225 (32.6)			
249 (36.1)			
250 (36.3)			

2.16 A material has a yield strength of 345 MPa (50 ksi) and a plane strain linear elastic fracture toughness of 110 MPa $\sqrt{m}$ (100 ksi $\sqrt{in}$). Determine the minimum specimen dimensions (B, a, W) required to perform a valid K_{IC} test on this material. Comment on the feasibility of testing a specimen of this size.

2.17 You have been given a set of fracture mechanics test specimens, all of the same size and geometry. These specimens have been fatigue precracked to various crack lengths. The stress intensity of this specimen configuration can be expressed as follows:

$$K_I = \frac{P}{B\sqrt{W}} f(a/W)$$

where P is load, B is thickness, W is width, a is crack length, and f(a/W) is a dimensionless geometry correction factor.

Describe a set of experiments you could perform to determine f(a/W) for this specimen configuration. Hint: you may want to take advantage of the relationship between K_I and energy release rate for linear elastic materials.

2.18 Beginning with Eq. (2.68), derive an expression for K_I for a through crack subject to an arbitrary stress distribution, $\sigma(x)$, as illustrated in Fig. 13.2. Note: you will need to obtain a separate expression for each crack tip.

2.19 Derive the Griffith-Inglis result for the potential energy of a through crack in an infinite plate subject to a remote tensile stress (Eq. (2.16)). Hint: solve for the work required to close the crack faces; Eq. (A2.43) gives the crack opening displacement for this configuration.

2.20 Using the Westergaard stress function approach, derive the stress intensity factor relationship for an infinite array of collinear cracks in a plate subject to biaxial tension (Fig. 2.21).

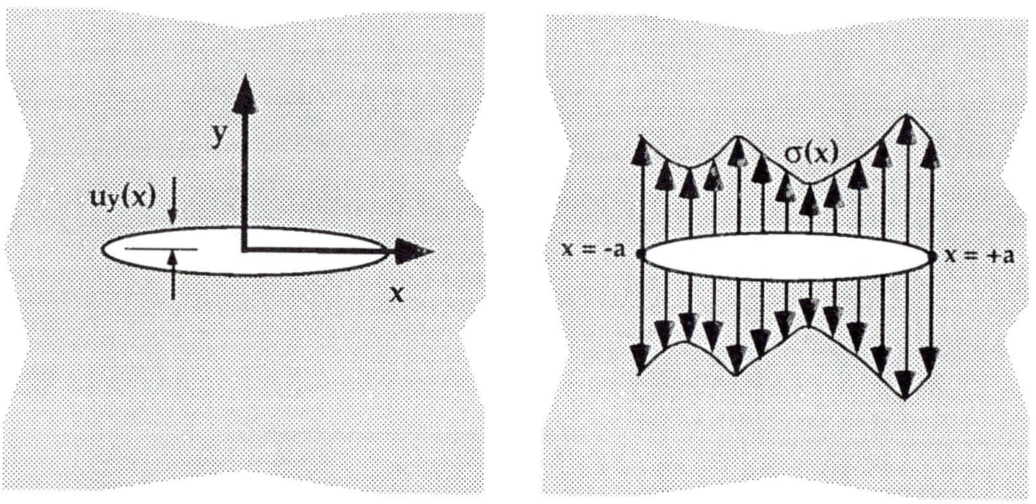

FIGURE 13.2 Through-thickness crack subject to an arbitrary crack face traction (Problem 2.18).

13.3 CHAPTER 3

3.1 Repeat the derivation of Eqs. (3.1) to (3.3) for the plane strain case.

3.2 A CTOD test is performed on a three point bend specimen. Figure 13.3 shows the deformed specimen after it has been unloaded. That is, the displacements shown are the *plastic components*.

(a) Derive an expression for plastic CTOD (δ_p) in terms of Δ_p and specimen dimensions.

(b) Suppose that V_p and Δ_p are measured on the same specimen, but that the plastic rotational factor, r_p, is unknown. Derive an expression for r_p in terms of Δ_p, V_p and specimen dimensions, assuming the angle of rotation is small.

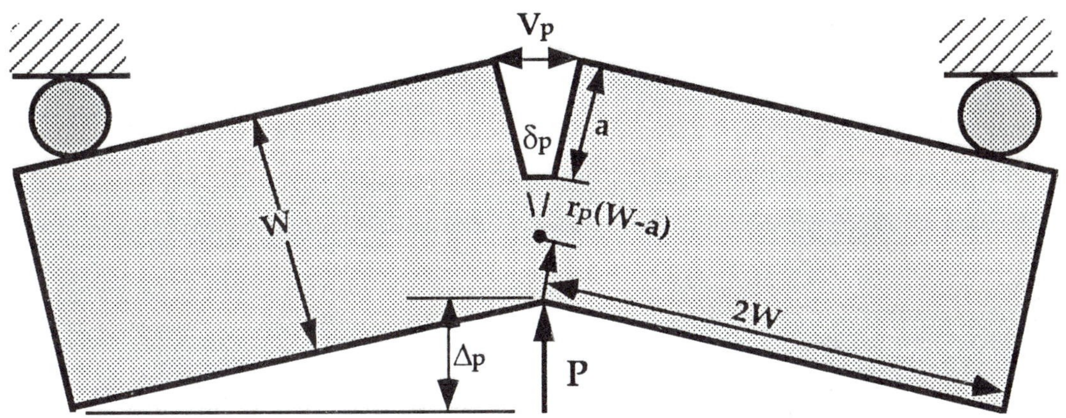

FIGURE 13.3 Three-point bend specimen rotating about a plastic hinge (Problem 3.2).

3.3 Fill in the missing steps between Eqs. (3.36) and (3.37)

3.4 Derive an expression for the J integral for a deeply notched three-point bend specimen, loaded over a span S, in terms of the area under the load-displacement curve and ligament length, b. Figure 13.3 illustrates two displacement measurements on a bend specimen: the load line displacement (Δ) and the crack mouth opening displacement (V). Which of these two displacement measurements is more appropriate for inferring the J integral? Explain.

3.5 Derive an expression for the J integral for an axisymmetrically notched bar in tension (Fig. 13.4), where the notch depth is sufficient to confine plastic deformation to the ligament.

3.6 Derive an expression for the J integral in a deeply notched three-point bend specimen in terms of the area under the load-crack mouth opening displacement curve. Begin with the corresponding formula for the P-Δ curve (given below), and assume rotation about a plastic hinge (Fig. 13.3).

$$J = \frac{K^2}{E'} + \frac{2}{b} \int_0^{\Delta_p} P \, d\Delta_p$$

for a specimen with unit thickness.

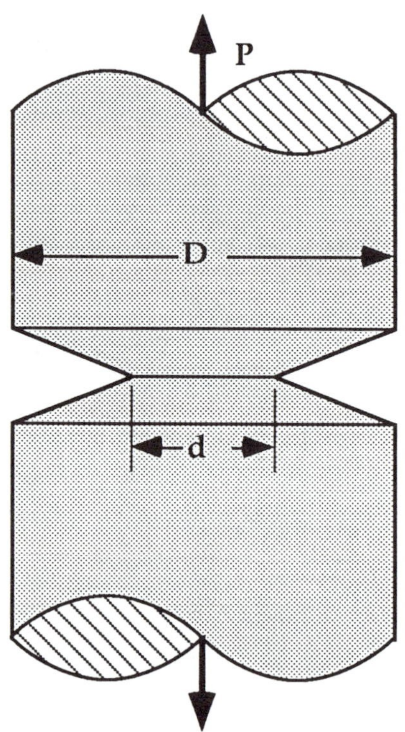

FIGURE 13.4 Axisymmetrically notched bar loaded in tension (Problem 3.5).

3.7 A bend specimen fails by cleavage fracture at J_c = 180 kJ/m^2. If a/W = 0.50 and W = 50.8 mm (2.0 in), estimate the toughness of the material (according to Fig. 3.29) if the specimen were infinite. Assume n = 10.

3.8 A pressure vessel steel has a cleavage fracture toughness (K_{IC}) of 220 MPa $\sqrt{m}$ (200 ksi $\sqrt{in}$) at the design temperature. Estimate the minimum specimen dimensions required to meet the size requirement of Eq. (3.74) with M = 200. Compare this specimen size to the K_{IC} size requirements (Chapter 2).

σ_{YS} = 350 MPa (51 ksi); σ_{TS} = 450 MPa (65 ksi); E = 207,000 MPa (30,000 ksi).

3.9 Derive an expression for the dimensionless constant γ in Eq. (3.75). Begin with Eq. (3.40) and solve for the partial derivative of J_{pl} with respect to crack size.

13.4 CHAPTER 4

4.1 A high rate fracture toughness test is to be performed on a high strength steel with K_{Id} = 110 MPa $\sqrt{m}$ (100 ksi $\sqrt{in}$). A three-point bend specimen will be used, with W = 50.8 mm (2.0 in), a/W = 0.5, B = W/2, and span = 4W. Also, c_1 = 5940 m/sec (19,500 ft/sec) for steel. Estimate the maximum loading rate at which the quasistatic formula for estimating K_{Id} is approximately valid.

4.2 Unstable fracture initiates in a steel specimen and arrests after the crack propagates 8.0 mm (0.32 in). The total propagation time was 7.52 x 10^{-6} sec. The initial ligament length in the specimen was 30.0 mm (1.18 in) and c_1 for steel = 5940 m/sec (19,500 ft/sec). Determine whether or not reflected stress waves influenced the propagating crack.

4.3 Fracture initiates at an edge crack in a 2.0 m (78.7 in) wide steel plate and rapidly propagates through the material. The stress in the plate is fixed at 300 MPa (43.5 ksi). Plot the crack speed versus crack size for crack lengths ranging from 10 to 60 mm (or 0.4 to 2.4 in). The dynamic fracture toughness of the material is given by

$$K_{ID} = \frac{K_{IA}}{1 - \left(\frac{V}{V_\ell}\right)^2}$$

where K_{IA} = 55 MPa $\sqrt{m}$ (50 ksi $\sqrt{in}$) and V_ℓ = 1500 m/sec (4920 ft/sec) Use the Rose approximation (Eqs. (4.17) and (4.18)) for the driving force. The elastic wave speeds for steel are given below.

c_1	5940 m/sec	19,500 ft/sec
c_2	3220 m/sec	10,600 ft/sec
c_r	2980 m/sec	9780 ft/sec

4.4 Derive an expression for C* in a double edge notched tension panel in terms of specimen dimensions, creep exponent, load, and displacement rate. See Section 3.2.5 for the corresponding J expression.

4.5 A three-point bend specimen is tested in displacement control at an elevated temperature. The displacement rate is increased in steps as the test progresses. The load, load line displacement rate, a/W, and crack velocity are tabulated below. Compute C*, and construct a log-log plot of crack velocity versus C*. The specimen thickness and width are 25 mm and 50 mm, respectively. The creep exponent = 5.0 for the material.

$\dot{\Delta}$, m/s	Load, kN	a/W	$\dot{a}$, m/s
1.0×10^{-7}	10.8	0.52	3.67×10^{-9}
5.0×10^{-7}	13.8	0.54	1.79×10^{-8}
1.0×10^{-6}	14.9	0.56	3.49×10^{-8}
5.0×10^{-6}	19.0	0.58	1.71×10^{-7}
1.0×10^{-5}	20.4	0.60	3.37×10^{-7}
5.0×10^{-5}	24.0	0.65	1.65×10^{-6}

4.6 In a linear viscoelastic material, the pseudo elastic displacement and the physical displacement are related through a hereditary integral:

$$\Delta^e = \{E\, d\Delta\}$$

Simplify this expression for the case of a constant displacement rate.

4.7 Consider a fracture toughness test on a nonlinear viscoelastic material at a constant displacement rate. Assume that the load is related to the pseudo elastic displacement by a power law:

$$P = M\,(\Delta^e)^N$$

where M and N are constants that do not vary with time. Show that the viscoelastic J integral and the conventional J integral are related as follows:

$$J_v = J\,\phi(t)$$

where ϕ is a function of time. Derive an expression for $\phi(t)$. Hint: begin with Eqs. (3.17) and (4.75). Also, the result from the previous problem may be useful.

4.8 A fracture toughness test on a linear viscoelastic material results in a nonlinear load-displacement curve in a constant rate test. Yielding is restricted to a very small region near the crack tip. Why is the curve nonlinear? Does the stress intensity factor characterize the crack tip conditions in this case? Explain. What is the relationship between J and K_I for a linear viscoelastic material? Hint: refer to the second equation in the previous problem.

13.5 CHAPTER 5

5.1 A body-centered cubic (BCC) material contains second phase particles. The size of these particles can be controlled through thermal treatment. Discuss the anticipated effect of particle size on the material's resistance to both cleavage fracture and microvoid coalescence, assuming the volume fraction of the second phase remains constant.

5.2 An aluminum alloy fails by microvoid coalescence when the average void size reaches ten times the initial value. If the voids grow according to Eq. (5.11), with σ_{YS} replaced by σ_e, plot the equivalent plastic strain (ε_{eq}) at failure versus σ_m/σ_e for σ_m/σ_e ranging from 0 to 2.5. Assume the triaxiality ratio remains constant during deformation of a given sample; i.e.,

$$\ln\left(\frac{\bar{R}}{R_o}\right) = 0.283\exp\left(\frac{1.5\,\sigma_m}{\sigma_e}\right)\int_0^{\varepsilon_{eq}} d\varepsilon_{eq}$$

5.3 The critical microstructural feature for cleavage initiation in a steel sample is a 2.12 μm diameter spherical carbide; failure occurs when this particle forms a microcrack that satisfies the Griffith criterion (Eq. (5.18)), where γ_p = 14 J/m^2, E = 207,000 MPa, and ν = 0.30 for the material. Assuming Fig. 5.14 describes the stress distribution ahead of the macroscopic crack, where σ_o = 350 MPa, estimate the critical J value of the sample if the particle is located 0.1 mm ahead of the crack tip, on the crack plane. Repeat this calculation for the case where the critical particle is 0.4 mm ahead of the crack tip.

5.4 Cleavage initiates in a ferritic steel at 3.0 μm diameter spherical particles. The fracture energy on a single grain, γ_p, is 14 J/m^2 and the fracture energy required for propagation across grain boundaries, γ_{gb}, is 50 J/m^2. At what grain size does propagation across grain boundaries become the controlling step for cleavage fracture?

5.5 Compute the relative size of the 90% confidence band of K_{IC} data (as in Example 5.1), assuming Eq. (5.24) describes the toughness distribution. Compute the confidence band width for K_o/Θ_K = 0, 0.5, 1.0, 2.0, and 5.0. What is the effect of the threshold toughness, K_o, on the relative scatter? What is the physical significance of Θ_K in this case?

5.6 Compute the relative size of the 90% confidence band of K_{IC} data (as in Example 5.1), assuming Eq. (5.26) describes the toughness distribution. Compute the confidence band width for K_o/Θ_K = 0, 0.5, 1.0, 2.0, and 5.0. What is the effect of the threshold toughness, K_o, on the relative scatter?

13.6 CHAPTER 6

6.1 For the Maxwell spring and dashpot model (Fig. 6.6) derive an expression for the relaxation modulus.

6.2 Fill in the missing steps in the derivation of Eq. (6.14).

6.3 At room temperature, tensile specimens of polycarbonate show 60% elongation and no stress whitening, while thick compact specimens used in fracture toughness testing show stress whitening at the crack tip. Explain these observations. Polycarbonate is an amorphous glassy polymer at room temperature.

6.4 A wide and thin specimen of PMMA has a 15 mm (0.59 in) long through crack with a 1.5 mm (0.059 in) long craze at each crack tip. If the applied stress is 3.5 MPa (508 psi), calculate the crazing stress in this material.

6.5 When a macroscopic crack grows in a ceramic specimen, a process zone 0.2 mm wide forms. This process zone contains 10,000 penny-shaped microcracks/mm^3 with an average radius of 10 μm. Estimate the increase in toughness due to the release of strain energy by these microcracks. The surface energy of the material = 25 J/m^2.

13.7 CHAPTER 7

7.1 A fracture toughness test is performed on a compact specimen. Calculate K_Q and determine whether or not $K_Q = K_{IC}$.

B = 25.4 mm (1.0 in); W = 50.8 mm (2.0 in); a = 27.7 mm (1.09 in) P_Q = 42.3 kN (9.52 kip); P_{max} = 46.3 kN (10.4 kip); σ_{YS} = 759 kN (110 ksi)

7.2 You have been asked to perform a K_{IC} test on a material with σ_{YS} = 690 MPa (100 ksi). The toughness of this material is expected to lie between 40 MPa $\sqrt{m}$ and 60 MPa $\sqrt{m}$ (1 ksi $\sqrt{in}$ = 1.10 MPa $\sqrt{m}$). Design an experiment to measure K_{IC} in this material using a compact specimen. Specify the following quantities: (a) specimen dimensions, (b) precracking loads, and (c) required load capacity of the test machine.

7.3 A titanium alloy is supplied in 15.9 mm (0.625 in) thick plate. If σ_{YS} = 807 MPa (117 ksi), calculate the maximum valid K_{IC} that can be measured in this material.

7.4 Recall Problem 2.16, where a material with K_{IC} = 110 MPa $\sqrt{m}$ (100 ksi $\sqrt{in}$) required a 254 mm (10.0 in) thick specimen for a valid K_{IC} test. Suppose that a compact specimen of the appropriate dimensions has been fabricated. Estimate the required load capacity of the test machine for such a test.

7.5 A fracture toughness test is performed on a compact specimen fabricated from a 5 mm thick sheet aluminum alloy. The specimen width (W) = 50.0 mm and B = 5 mm (the sheet thickness). The initial crack length is 26.0 mm. Young's modulus = 70,000 MPa. Compute the K-R curve from the load-displacement data tabulated below. Assume that all nonlinearity in the P-Δ curve is due to crack growth. (See Chapter 12 for the appropriate compliance and stress intensity relationships.)

Load, kN	Load Line Displacement, mm	Load, kN	Load Line Displacement, mm
0	0	2.851	0.3698
0.5433	0.0635	2.913	0.3860
1.087	0.1270	2.903	0.3971
1.630	0.1906	2.850	0.4113
2.161	0.2552	2.749	0.4191
2.361	0.2817	2.652	0.4274
2.541	0.3096	2.553	0.4355
2.699	0.3392	2.457	0.4443

1 kN = 224.8 lb   25.4 mm = 1 in   1 MPa = 0.145 ksi

7.6 A number of fracture toughness specimens have been loaded to various points and then unloaded. Values of J and crack growth were measured in each specimen and are tabulated below. Plot the R curve for this material and determine J_Q and, if possible, J_{IC}.

σ_{YS} = 350 MPa; σ_{TS} = 450 MPa; B = 25 mm; b_o = 22 mm;

Specimen	J, kJ/m2	Crack Extension, mm
1	100	0.30
2	175	0.40
3	185	0.80
4	225	1.20
5	250	1.60
6	300	1.70

25.4 mm = 1 in   1 MPa = 0.145 ksi   1 kJ/m^2 = 5.71 in-lb/in^2

7.7 Recall Problem 2.16, where a material with K_{IC} = 110 MPa $\sqrt{m}$ (100 ksi $\sqrt{in}$) and σ_{YS} = 345 MPa (50 ksi) required a specimen 254 mm (10 in) thick for a valid K_{IC} test. Estimate the thickness required for a valid J_{IC} test on this material.

σ_{TS} = 483 MPa (70 ksi); E = 207,000 MPa (30,000 ksi); ν = 0.3.

7.8 An unloading compliance test has been performed on a 3-point bend specimen. The data obtained at each unloading point is tabulated below.

(a) Compute and plot the J resistance curve according to ASTM E 1152-87.

(b) Determine J_{IC} according to ASTM E 813-87.

B = 25.0 mm; W = 50.0 mm; a_o = 26.1 mm; E = 210,000 MPa, ν = 0.3

LOAD , kN	Plastic Displacement, mm	Crack Extension, mm
20.8	0	0.013
31.2	0.0032	0.020
35.4	0.011	0.023
37.4	0.020	0.025
41.6	0.056	0.031
43.7	0.092	0.036
45.7	0.146	0.044
47.6	0.228	0.055
49.9	0.349	0.071
51.6	0.525	0.091
53.5	0.777	0.128
55.3	1.13	0.183
56.6	1.63	0.321
56.7	2.32	0.723
56.5	2.66	0.928
55.8	3.25	1.29
54.7	3.96	1.74
53.7	4.51	2.08
52.5	5.13	2.48
50.1	6.20	3.17
44.4	8.43	4.67
40.0	10.09	5.81
36.6	11.37	6.70
30.9	13.54	8.23
26.8	15.19	9.41

1 kN = 224.8 lb   25.4 mm = 1 in   1 MPa = 0.145 ksi

7.9 A CTOD test was performed on a three point bend specimen with B = W = 25.4 mm (1.0 in). The crack depth, a, was 12.3 mm (0.484 in). Examination of the fracture surface revealed that the specimen failed by cleavage with no prior stable crack growth. Compute the critical CTOD in this test. Be sure to use the appropriate notation (i.e., δ_c, δ_u, δ_i, or δ_m).

V_p = 1.05 mm (0.0413 in); $P_{critical}$ = 24.6 kN (5.53 kip); E = 207,000 MPa (30,000 ksi); σ_{YS} = 400 MPa (58.0 ksi); ν = 0.3.

7.10 A crack arrest test has been performed in accordance with ASTM E 1221. The side-grooved compact crack arrest specimen has the following dimensions: W = 100 mm (3.94 in), B = 25.4 mm (1.0 in), and B_N = 19.1 mm (0.75 in). The initial crack length = 46.0

mm (1.81 in) and the crack length at arrest = 63.0 mm (2.48 in). The corrected clip gage displacements at initiation and arrest are V_o = 0.582 mm (0.0229 in) and V_a = 0.547 (0.0215 in), respectively. E = 207,000 MPa (30,000 ksi) and $\sigma_{YS(static)}$ = 483 MPa (70 ksi). Calculate the stress intensity at initiation, K_o, and the arrest toughness, K_a. Determine whether or not this test satisfies the validity criteria in Eq. (7.25). The stress intensity solution for the compact crack arrest specimen is given below.

$$K_I = \frac{E\,V\,f(x)\,\sqrt{B/B_N}}{\sqrt{W}}$$

where

$$x = a/W$$

$$f(x) = \frac{2.24\,(1.72 - 0.9\,x + x^2)\,\sqrt{1 - x}}{9.85 - 0.17\,x + 11\,x^2}$$

13.8 CHAPTER 8

8.1 A 25.4 mm (1.0 in) thick plate of PVC has a yield strength of 60 MPa (8.70 ksi). The anticipated fracture toughness (K_{IC}) of this material is 5 MPa $\sqrt{m}$ (4.5 ksi $\sqrt{in}$) Design an experiment to measure K_{IC} of a compact specimen machined from this material. Determine the appropriate specimen dimensions (B, W, a) and estimate the required load capacity of the test machine.

8.2 A 15.9 mm (0.625 in) thick plastic plate has a yield strength of 50 MPa (7.25 ksi). Determine the largest valid K_{IC} value that can be measured on this material.

8.3 A K_{IC} test is to be performed on a polymer with a time-dependent relaxation modulus which has been fit to the following equation:

$$E(t) = [0.417 + 0.037\,t^{0.65}]^{-1}$$

where E is in GPa and t is in seconds. Assuming P_Q is determined from a 5% secant construction, estimate the test duration (i.e. the time to reach P_Q) at which 90% of the nonlinearity in the load-displacement curve at P_Q is due to viscoelastic effects. Does the 5% secant load give an appropriate indication of material toughness in this case? Explain.

8.4 Derive a relationship between the conventional J integral and the isochronous J integral, J_t, in a constant displacement rate test on a viscoelastic material for which Eqs. (8.10) and (8.15) describe the load-displacement behavior.

8.5 A 500 mm wide plastic plate contains a through-thickness center crack that is initially 50 mm long. The crack velocity in this material is given by

$$\dot{a} = 10^{-40} K^{10}$$

where K is in kPa $\sqrt{m}$ and $\dot{a}$ is in mm/sec (1 psi $\sqrt{in}$ = 1.1 kPa $\sqrt{m}$, 1 in = 24.5 mm). Calculate the time to failure in this plate assuming remote tensile stresses of 5 MPa and 10 MPa (1 ksi = 6.897 MPa). Comment on the sensitivity of the time to failure on the applied stress. (As a first approximation, neglect the finite width correction on K. For an optional exercise, repeat the calculations with this correction to assess its effect on the computed failure times.)

8.6 A composite double cantilever beam (DCB) specimen is loaded to 445 N (100 lb) at which time crack growth begins. Calculate $\mathcal{G}_{IC}$ for this material assuming linear beam theory.

E = 124,000 MPa (18,000 ksi); a = 76.2 mm (3.0 in); h = 0.254 mm (0.10 in); B = 25.4 mm (1.0 in).

8.7 One of the problems with testing brittle materials is that crack growth tends to be unstable in conventional test specimens and test machines. Consider, for example, a single edge notched bend (SENB) specimen loaded in three point bending. The influence of the test machine can be represented by a spring in series, as Fig. 13.5 illustrates. Show that the stress intensity factor for this specimen can be expressed as a function of crosshead displacement and compliance as follows:

$$K_I = \frac{\Delta_t \, f(a/W)}{(C + C_m) B \sqrt{W}}$$

where Δ_t is the crosshead displacement, C is the specimen compliance, C_m is the machine compliance, and f(a/W) is defined in

Table 12.2. Construct a nondimensional plot of K_I versus crack size for a fixed crosshead displacement and a/W ranging from 0.25 to 0.75. Develop a family of these curves for a range of machine compliance. (You will have to express C_m in an appropriate nondimensional form.) What is the effect of machine compliance on the relative stability of the specimen? At what machine compliance would a growing crack experience a relatively constant K_I between a/W = 0.5 and a/W = 0.6 ?

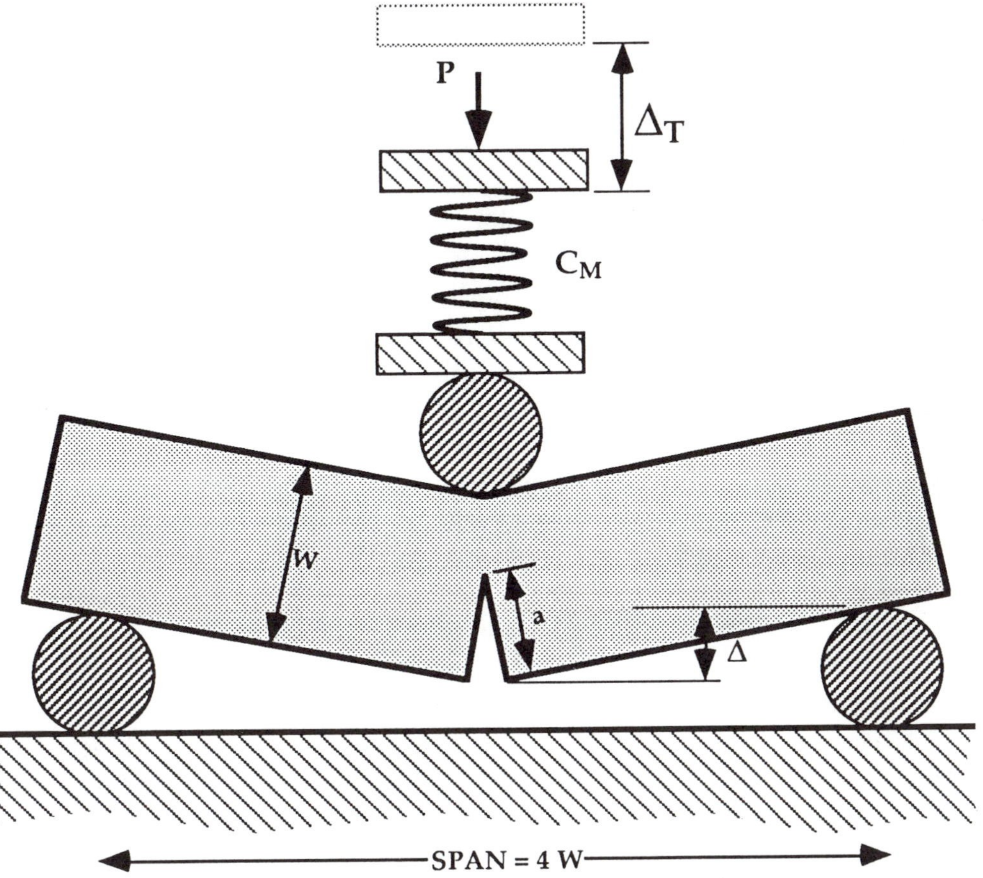

FIGURE 13.5 Single edge notch bend specimen loaded in crosshead control (Problem 8.7). The effect of machine compliance is schematically represented by a spring in series.

13.9 CHAPTER 9

9.1 Develop a computer program or spreadsheet macro to calculate stress intensity factors for semielliptical surface flaws in flat plates subject to linear stress distributions (Table 12.22). Plot a family of curves for F and H as a function of a/t for a/c = 0, 0.2, 0.4, 0.6, 0.8, and 1.0, where $\phi = 90°$ and c << W.

9.2 Calculate the stress intensity factor at the deepest point ($\phi = 90°$) of an axial flaw in a pressure vessel using both Eq. 9.10 and Table 12.27.

p = 17.2 MPa (2500 psi); R_i = 1.00 m; t/R_i = 0.10; a/t = 0.20; a/c = 0.40.

9.3 A nuclear reactor pressure vessel operates at an internal pressure of 17.2 MPa (2500 psi) and a temperature of 200°C (392°F). The steel in this pressure vessel has an RT_{NDT} of 100°C (212°F), and thus is relatively brittle at room temperature. Consequently, the full design pressure is not applied when the reactor is cold. Upon start-up, the temperature and pressure must be increased in tandem in order to avoid brittle fracture.

(a) Determine the maximum allowable pressure-temperature curve, ranging from ambient to the design temperature. As a worst case, assume the vessel contains an internal axial surface flaw with a/t = 0.25 and a/c = 0.50, and that the fracture toughness is given by the K_{IR} curve (Eq. (9.17b)). Assume linear elastic conditions. The vessel dimensions are given below.

R_i = 2.16 m (85.0 in); t = 21.6 mm (8.50 in)

(b) As the reactor operates over a period of several years, the steel becomes embrittled due to radiation damage, and the RT_{NDT} increases with time. Estimate the RT_{NDT} at which it is no longer safe to start up the reactor.

(c) The pressure vessel is made from A 533 Grade B steel, which has a yield strength of 460 MPa (66.7 ksi). Was the assumption of linear elastic conditions acceptable in this case?

9.4 A structure contains a through-thickness crack 20 mm long. Strain gages indicate an applied normal strain of 0.0042 when the structure is loaded to its design limit. The structure is made of a steel with ε_y = 0.0020 and δ_{crit} = 0.15 mm. Is this structure safe, according to the CTOD design curve?

9.5 A welded structure is loaded in combined bending and tension, with P_m = 200 MPa and P_b = 150 MPa. The structure is in the as-welded condition; the precise residual stress distribution in the weldment is unknown. Determine the maximum allowable flaw size, $\bar{a}$, according to the 1980 version of the PD 6493 approach (Eqs. (9.18) and (9.19)).

σ_{YS} = 400 MPa; E = 207,000 MPa; δ_{crit} = 0.23 mm

9.6 A flat plate 1.0 m (39.4 in) wide which contains a semi-elliptical surface flaw is loaded in uniaxial tension to 0.75 σ_{YS}. Assuming the ratio a/2c = constant = 0.3, plot K_r and S_r values on a strip yield failure assessment diagram for various flaw sizes. Estimate the critical flaw size for failure. (See Tables 12.22 and 12.24 for K_I and limit load solutions.)

σ_{YS} = 345 MPa (50 ksi); σ_{TS} = 448 MPa (65 ksi); E = 207,000 MPa (30,000 ksi); K_{IC} = 110 MPa $\sqrt{m}$ (100 ksi $\sqrt{in}$)

9.7 A pipe with 1.10 m (43.3 in) outside diameter and 50 mm (1.97 in) thick wall contains a long internal axial flaw 10 mm (0.394 in) deep. The material flow properties have been fit to a Ramberg-Osgood equation:

σ_o = 450 MPa (65.3 ksi); $\varepsilon_o = \sigma_o/E$; α = 1.25; n = 9.72
E = 207,000 MPa (30,000 ksi)

(a) Plot the applied J versus internal pressure.

(b) If J_{IC} for this material is 300 kJ/m^2 (1.71 in-kip/in^2), determine the pressure required to initiate ductile crack growth.

(Disregard the Irwin plastic zone correction for all calculations.)

9.8 Suppose the edge cracked plate in Examples 9.2 and 9.3 is subject to a 5 MN tensile load.

(a) Calculate the applied J integral, both with and without the Irwin plastic zone correction.

(b) Calculate the load line displacement over a 5 m gage length.

(c) Calculate the load line displacement over a 50 mm gage length.

9.9 For the plate in the previous problem, estimate the following:

(a) dJ/da for fixed load (5 MN)

(b) dJ/da for fixed displacement at 5 m gage length. (P = 5 MN when a = 225 mm.)

(c) dJ/da for fixed displacement at 50 mm gage length. (P = 5 MN when a = 225 mm.)

9.10 When a single edge notched bend specimen is loaded in the fully plastic range, the deformation can be described by a simple hinge model (Fig. 13.3). The plastic rotational factor can be estimated from load line displacement and crack mouth opening displacement as follows:

$$r_p = \frac{1}{W - a}\left(\frac{W\,V_p}{\Delta_p} - a\right)$$

assuming a small angle of rotation. Beginning with Eqs (9.32) and (9.33) solve for r_p in terms of h_2, h_3, and specimen dimensions. Use the resulting expression to compute r_p for n = 10 and a/W = 0.250, 0.375, 0.500, 0.625, and 0.750. Repeat for n = 3 and the same a/W values. Assume plane strain for all calculations. How do the r_p values estimated from the EPRI Handbook compare with the assumed value of 0.44 in ASTM E 1290-89?

9.11 A welded panel 5 m wide and 50 mm thick contains a semielliptical surface crack (in the weld metal) with a = 10 mm and 2c = 54 mm. The primary membrane stress is 260 MPa and the primary bending stress is 60 MPa. The panel is in the as-welded condition; the residual stress distribution is unknown. Perform a PD 6493 Level 2 assessment on the weld flaw to determine whether or not it is acceptable. The material properties are as follows:

$\sigma_{YS} = 480$ MPa; $\sigma_{TS} = 610$ MPa; $E = 207{,}000$ MPa; $\delta_{crit} = 0.15$ mm

9.12 Suppose that the panel in Problem 9.11 is thermally stress relieved. Repeat the Level 2 assessment, assuming the residual stresses after the heat treatment are equal to one third the yield strength.

13.10 CHAPTER 10

10.1 Using the Paris-Erdogan equation for fatigue crack propagation, calculate the number of fatigue cycles corresponding to the combinations of initial and final crack radius for a semicircular surface flaw tabulated below. Assume that the crack radius is small compared to the cross section of the structure.

$\frac{da}{dN} = 6.87 \times 10^{-9} (\Delta K)^3$, where da/dN is in mm/cycle and ΔK is in MPa $\sqrt{m}$. Also, $\Delta\sigma$ = 200 MPa.

Initial Crack Radius	Final Crack Radius
1 mm	10 mm
1 mm	20 mm
2 mm	10 mm
2mm	20 mm

1 .1 MPa $\sqrt{in}$ = 1 ksi $\sqrt{in}$ 25.4 mm = 1 in 1 MPa = 0.145 ksi

Discuss the relative sensitivity of N_{tot} to:

- initial crack size.
- final crack size.

10.2 A structural component made from a high strength steel is subject to cyclic loading, with σ_{max} = 210 MPa and σ_{min} = 70 MPa. This component experiences 100 stress cycles per day. Prior to going into service, the component was inspected by nondestructive evaluation (NDE), and no flaws were found. The material has the following properties: σ_{YS} = 1000 MPa, K_{IC} = 25 MPa $\sqrt{m}$. The fatigue crack growth rate in this material is the same as in Problem 10.1.

(a) The NDE technique can find flaws ≥ 2 mm deep. Estimate the maximum safe design life of this component, assuming that subsequent in-service inspections will not be performed. Assume that any flaws that may be present are semicircular surface cracks and that they are small relative to the cross section of the component.

(b) Repeat part (a), assuming an NDE detectability limit of 10 mm.

10.3 Fatigue tests are performed on two samples of an alloy for aerospace applications. In the first experiment, R = 0, while R = 0.8 in the second experiment. Sketch the expected trends in the data for the two experiments on a schematic log(da/dN) v. log(ΔK) plot. Assume that the experiments cover a wide range of ΔK values. Briefly explain the trends in the curves.

10.4 Write a program or spreadsheet macro to compute fatigue crack growth behavior in a compact specimen, assuming the fatigue crack growth is governed by the Paris-Erdogan equation.

Consider a 1T compact specimen (see Section 7.1.1) that is loaded cyclically at a constant load amplitude with P_{max} = 18 kN and P_{min} = 5 kN. Using the fatigue crack growth data in Problem 10.1, calculate the number of cycles required to grow the crack from a/W = 0.35 to a/W = 0.60. Plot crack size versus cumulative cycles for this range of a/W.

10.5 Write a program or spreadsheet macro to compute the fatigue crack growth behavior in a flat plate that contains a semielliptical surface flaw and is subject to a cyclic membrane (tensile) stress. Assume that the flaw remains semielliptical, but take account of the variation in K around the circumference of the flaw. Also, assume that c << W, but that a/t is finite. Use the Paris-Erdogan equation to compute the crack growth rate.

Consider a 25.4 mm (1.0 in) thick plate that is loaded cyclically at a constant stress amplitude of 200 MPa (29 ksi). Given an initial flaw with a/t = 0.1 and a/2c = 0.1, calculate the number of cycles required to grow the crack to a/t = 0.8, using the fatigue crack growth data in Problem 10.1. Construct a contour plot that shows the crack size and shape at a/t = 0.1, 0.2, 0.4, 0.6, and 0.8. What happens to the a/2c ratio as the crack grows?

10.6 Estimate U and K_{op} as a function of R and ΔK for the data in Fig. 10.8. Does Eq. (10.19) fit the data adequately or does U depend on

K_{max}? Does Eq. (10.20) adequately describe the data? If so, determine the parameter K_o.

10.7 Suppose that the 1T compact specimen in Problem 10.4 experiences a single overload of 36 kN when a/W = 0.45. During all other cycles the load amplitude is constant, with P_{max} = 18 kN and P_{min} = 5 kN. Using the Wheeler retardation model with γ = 1.5, estimate the number of cycles required to grow the crack from a/W = 0.35 to a/W = 0.60. Plot crack size versus cumulative cycles, comparing the present case to the constant load amplitude case of Problem 10.4. Assume plane strain conditions at the crack tip.

10.8 You have been asked to perform K-decreasing tests on a material to determine the near-threshold behavior at R = 0.1. Your laboratory has a computer-controlled test machine that can be programmed to vary P_{max} and P_{min} on a cycle-by-cycle basis.

(a) Compute and plot P_{max} and P_{min} versus crack length for the range $0.5 \leq a/W \leq 0.75$ corresponding to a normalized K gradient of - 0.07 mm in a 1T compact specimen.

(b) Suppose that the material exhibits the following crack growth behavior near the threshold:

$$\frac{da}{dN} = 4.63 \times 10^{-12} \left(\Delta K^3 - \Delta K_{th}^3\right)$$

where da/dN is in m/cycle and ΔK is in MPa $\sqrt{m}$. For R = 0.1, ΔK_{th} = 8.50 MPa $\sqrt{m}$. When the test begins, a/W = 0.520 and da/dN = 5.53×10^{-7} m/cycle. As the test continues in accordance with the loading history determined in part (a), the crack growth rate decreases. You stop the test when da/dN reaches 10^{-10} m/cycle. Calculate the following:

(i) The number of cycles required to complete the test.

(ii) The final crack length.

(iii) The final ΔK.

13.11 CHAPTER 11

11.1 A series of finite element meshes have been generated that model compact specimens with various crack lengths. Plane stress linear elastic analyses have been performed on these models. Nondimensional compliance values as a function of a/W are tabulated below. Estimate the nondimensional stress intensity for the compact specimen from these data and compare your estimates to the polynomial solution in Table 12.2.

$\frac{a}{W}$	$\frac{\Delta B E}{P}$	$\frac{a}{W}$	$\frac{\Delta B E}{P}$	$\frac{a}{W}$	$\frac{\Delta B E}{P}$
0.20	8.61	0.45	29.0	0.70	123
0.25	11.2	0.50	37.0	0.75	186
0.30	14.3	0.55	47.9	0.80	306
0.35	18.1	0.60	63.3	0.85	577
0.40	22.9	0.65	86.3	0.90	1390

11.2 A finite element analysis is performed on a through crack in a wide plate (Fig. 2.3). The remote stress is 100 MPa, and the half crack length = 25 mm. The stress normal to the crack plane (σ_{22}) at $\theta = 0$ is determined at node points near the crack tip and is tabulated below. Estimate K_I by means of the stress matching approach (Eq. 11.14) and compare your estimate to the exact solution for this geometry. Is the mesh refinement sufficient to obtain an accurate solution in this case?

$\frac{r}{a}$ $(\theta = 0)$	$\frac{\sigma_{22}}{\sigma^{\infty}}$	$\frac{r}{a}$ $(\theta = 0)$	$\frac{\sigma_{22}}{\sigma^{\infty}}$
0.005	11.0	0.080	3.50
0.010	8.07	0.100	3.24
0.020	6.00	0.150	2.83
0.040	4.54	0.200	2.58
0.060	3.89	0.250	2.41

11.3 Displacements at nodes along the upper crack face (u_2 at $\theta = \pi$) in the previous problem are tabulated below. The elastic constants are as follows: $\mu = 80{,}000$ MPa and $\kappa = 1.80$. Estimate K_I by means of the displacement matching approach (Eq. 11.15) and compare your estimate to the exact solution for this geometry. Is the mesh refinement sufficient to obtain an accurate solution in this case?

$\frac{r}{a}$ $(\theta = \pi)$	$\frac{u_2}{a}$
0.005	9.99×10^{-5}
0.010	1.41×10^{-4}
0.020	1.99×10^{-4}
0.040	2.80×10^{-4}
0.060	3.41×10^{-4}

$\frac{r}{a}$ $(\theta = \pi)$	$\frac{u_2}{a}$
0.080	3.92×10^{-4}
0.100	4.36×10^{-4}
0.150	5.27×10^{-4}
0.200	6.00×10^{-4}
0.250	6.61×10^{-4}

11.4 Figure 13.6 illustrates a one-dimensional element with three nodes. Consider two cases: (1) Node 2 at $x = 0.50L$ and (2) Node 2 at at $x = 0.25L$.

(a) Determine the relationship between the global and parametric coordinates, $x(\xi)$, in each case.

(b) Compute the axial strain, $\varepsilon(\xi)$ for each case in terms of the nodal displacements and parametric coordinate.

(c) Show that $x_2 = 0.25L$ leads to a $1/\sqrt{x}$ singularity in the axial strain.

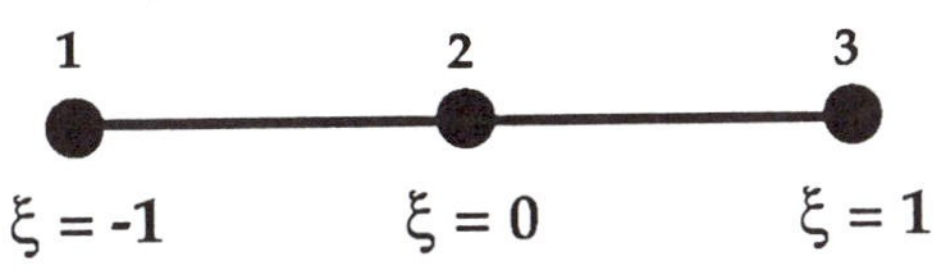

(a) Parametric coordinates.

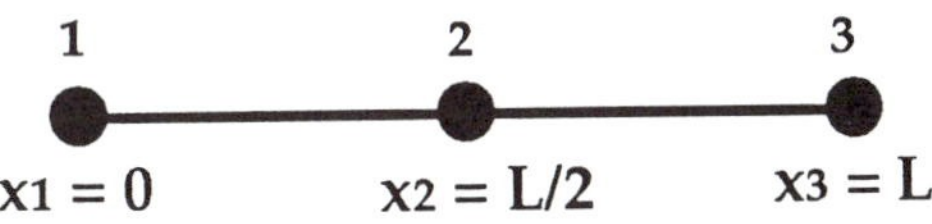

(b) Global coordinates, Case (1).

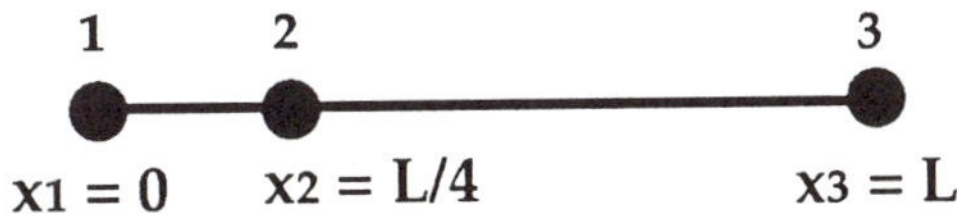

(c) Global coordinates, Case (2).

FIGURE 13.6 One-dimensional element with 3 nodes (Problem 11.4).

INDEX